BIRKHÄUSER

Operator Theory: Advances and Applications

Vol. 186

Methods of Spectral Analysis in Mathematical Physics

Conference on Operator Theory, Analysis and Mathematical Physics (OTAMP) 2006, Lund, Sweden

Jan Janas
Pavel Kurasov
Ari Laptev
Sergei Naboko
Günter Stolz
Editors

Birkhäuser
Basel · Boston · Berlin

Editors:

Jan Janas
Institut of Mathematics
Polish Academy of Sciences
ul. Sw. Tomasza 30/7
31-027 Krakow, Poland
e-mail: najanas@cyf-kr.edu.pl

Pavel Kurasov
Department of Mathematics
Lund Institute of Technology
221 00 Lund, Sweden
e-mail: kurasov@maths.lth.se

Sergei Naboko
Department of Mathematical Physics
St. Petersburg State University
Ulyanovskaya 1
198904 St. Petersburg, Russia
e-mail: naboko@snoopy.phys.spbu.ru

Ari Laptev
Department of Mathematics
South Kensington Campus
Imperial College London
SW7 2AZ London, UK
e-mail:a.laptev@imperial.ac.uk
and
Department of Mathematics
Royal Institute of Technology
100 44 Stockholm, Sweden
e-mail: laptev@math.kth.se

Günter Stolz
Department of Mathematics
University of Alabama
452 Campbell Hall
Birmingham, AL 35294-1170, USA
e-mail: stolz@math.uab.edu

2000 Mathematical Subject Classification: 05C25, 34A55, 34B, 34L40, 35J10, 35K, 35P, 37K, 42C05, 47, 49N45, 81Q, 81T99, 82B43

Library of Congress Control Number: 2008935517

Bibliographic information published by Die Deutsche Bibliothek.
Die Deutsche Bibliothek lists this publication in the Deutsche Nationalbibliografie; detailed bibliographic data is available in the Internet at http://dnb.ddb.de

ISBN 978-3-7643-8754-9 Birkhäuser Verlag AG, Basel - Boston - Berlin

Basel · Boston · Berlin
P.O. Box 133, CH-4010 Basel, Switzerland
Part of Springer Science+Business Media
Printed on acid-free paper produced from chlorine-free pulp. TCF∞
Printed in Germany
ISBN 978-3-7643-8754-9
9 8 7 6 5 4 3 2 1

e-ISBN 978-3-7643-8755-6
www.birkhauser.ch

Contents

Introduction

This volume contains mainly the lectures delivered by the participants of the International Conference: Operator Theory, Analysis and Mathematical Physics – OTAMP2006, held in Lund. As in the previous conferences of the OTAMP series, the main lectures presented an overview of current research which uses operator methods in analysis and mathematical physics.

The topics of the Proceedings belong to various different fields of mathematical physics. Among others (but there is much more in this volume) the following subjects are presented: inverse spectral and scattering problems on graphs, review articles on quadratic Hamiltonians and Born-Oppenheimer approximations, recursive construction of the Ablowitz-Ladik Hierarchy, spectral properties of finite difference and one (or two) dimensional Schrödinger operators, eigenvalues and their estimates for Jacobi matrices and Aharonov-Bohm operator.

Most papers of the volume contain original material and were refereed by acknowledged experts. The Editors thank all the referees whose job helped to improve the clarity of the collected material. The Organizing Committee of the conference also thanks all session organizers for the interesting choice of the scientific programme and to all participants who delivered fine lectures. We greatly appreciate financial support of the **ESF** programme **SPECT**, without whose financial support the OTAMP2006 would never been so well organized. Special thanks go to other agencies supported the conference financially: Vetenskapsrådet, Wenner-Gren Foundation, as well as to Lund University and to the Institute of Mathematics of the Polish Academy of Sciences.

Last but not least, we thank the Editorial Board and especially Professor I. Gohberg for including this volume (as the previous ones of OTAMP meetings) into the series Operator Theory: Advances and Applications and to Birkhäuser-Verlag for help in preparation of the volume.

Birmingham-Krakow-Lund | April 2008
St. Petersburg-Stockholm | The Editors

Operator Theory:
Advances and Applications, Vol. 186, 1–29

Spectral Asymptotics of Percolation Hamiltonians on Amenable Cayley Graphs

Tonći Antunović and Ivan Veselić

Abstract. In this paper we study spectral properties of adjacency and Laplace operators on percolation subgraphs of Cayley graphs of amenable, finitely generated groups. In particular we describe the asymptotic behaviour of the integrated density of states (spectral distribution function) of these random Hamiltonians near the spectral minimum.

The first part of the note discusses various aspects of the quantum percolation model, subsequently we formulate a series of new results, and finally we outline the strategy used to prove our main theorem.

Mathematics Subject Classification (2000). 05C25, 82B43, 05C80, 37A30, 35P15.

Keywords. Amenable groups, Cayley graphs, random graphs, percolation, random operators, spectral graph theory, phase transition.

1. Introduction

The results on which we report rely on tools and ideas from various mathematical fields. In the introduction we sketch the role they play in our study of percolation Hamiltonians.

1.1. Integrated density of states

For a large class of random operators which are ergodic with respect to a group of translations on the configuration space an *integrated density of states* (or *spectral density function*) can be defined using an exhaustion of the whole space by subsets of finite volume. Since this fact relies on an ergodic theorem, it is not surprising that the underlying group needs to be amenable. Note that periodic operators are a special class of ergodic ones and thus also posses a well-defined integrated density of states (in the following abbreviated as IDS). For ergodic random operators the spectrum is deterministic, i.e., for any two realizations of the operator the spectrum (as a set) coincides almost surely. The same statement holds for the measure theoretic components of the spectrum. However, there are other spectral quantities, like eigenvalues or eigenvectors, which are highly dependent on the randomness. For random Schrödinger type operators on $L^2(\mathbb{R}^d)$ and $\ell^2(\mathbb{Z}^d)$ all these fundamental results can be found, e.g., in the monographs [17, 94]. For the underlying original results of Pastur and Shubin for random, respectively almost-periodic operators see, e.g., [91, 93] and [102, 103].

Once the existence of the IDS is established, it is natural to ask whether, for specific models, one can describe some of its characteristic features in more detail. Among them the most prominent are: the continuity and discontinuity properties of the IDS, and the asymptotic behaviour near the spectral boundaries. The interest in these questions has a motivation stemming from the quantum theory of solids. A good understanding of the features of the IDS is an important step towards the determination of the spectral types of the considered Hamilton operators. Of all the measure theoretic spectral types of random operators the pure point part is so far understood best. In particular, for certain random Hamiltonians it is established that there exists an energy interval with dense eigenvalues and no continuous spectral component, a phenomenon called *localisation*. For detailed expositions of this subject see, for instance, [3, 5, 37, 101]. All the proofs of localisation known so far use as an essential ingredient estimates on the IDS or some closely related quantity. Expository accounts of the IDS of various types of random Hamiltonians can be found among others in [67], [53] or [113].

1.2. Lifshitz asymptotics

The IDS of lower bounded periodic operators in Euclidean space exhibits typically a polynomial behaviour at the minimum of the spectrum. This is for instance the case for the discrete Laplace operator on $\ell^2(\mathbb{Z}^d)$ and for Schrödinger operators on $L^2(\mathbb{R}^d)$ with a periodic potential. Analogous results can be proven for uniformly elliptic divergence type operators with periodic coefficients using [51] and for quantum waveguides using, e.g., [15] or [60]. In the mentioned cases the low energy behaviour of the IDS is characterised by the so-called *van Hove singularity*. This means that the IDS $N(\cdot)$ behaves for $0 < E \ll 1$ asymptotically like $N(E) \sim E^{d/2}$, i.e., polynomially with the *van Hove exponent* equal to $d/2$.

A random perturbation of the Hamilton operator changes drastically the low energy asymptotics. For many random models in d-dimensional Euclidean space it has been proven that the asymptotic behaviour of the IDS is exponential in the sense that $N(E) \sim \exp(-\operatorname{const}(E - E_0)^{-d/2})$. Here E_0 denotes the minimum of the spectrum of the random operator, which is by ergodicity independent of the realization almost surely. The exponential behaviour of the IDS has been first deduced on physical grounds by I.M. Lifšic in [69–71] and is accordingly called *Lifshitz asymptotics* or *Lifshitz tail*. The most precise bounds of this type have been obtained for random Schrödinger operators with a potential generated by impurities which are distributed randomly in space according to a Poisson process, see, e.g., [35, 59, 86, 92, 105]. Similar results hold for a discrete relative of this operator, namely the *Anderson model* on $\ell^2(\mathbb{Z}^d)$, see, e.g., [7, 16, 36, 57, 79, 109]. The reason why these models are amenable to a very precise analysis is the applicability of Brownian motion, respectively random walk techniques and Feynman-Kac functionals. For several other types of random operators the weaker form

$$\lim_{E \searrow E_0} \frac{\log|\log(N(E) - N(E_0))|}{|\log(E - E_0)|} = \frac{d}{2} \tag{1}$$

of the exponential law has been established using rather simple estimates based on inequalities by Thirring [107] and Temple [106]. In particular, this method can be applied to show that the asymptotics (1) holds for the Anderson model on $\ell^2(\mathbb{Z}^d)$ and its continuum counterpart, the *alloy type model* on $L^2(\mathbb{R}^d)$. This strategy of proof was pursued in the 1980s in several papers [52, 55, 80–82, 98, 99] by Kirsch, Martinelli, Simon and Mezincescu. These two models will serve as a point of reference in the present note, since they are closely related to percolation Hamiltonians. Moreover, the methods used to study the low-energy asymptotics of percolation models are, at least in part, parallel to those of the last mentioned series of papers. The main difference is that one needs to replace the Temple or Thirring bound by some inequality from spectral geometry. We will call the right-hand side of (1) the *Lifshitz exponent.*

Let us relate the asymptotic behaviour (1) to two spectral features mentioned already in §1.1, namely spectral localisation and continuity of the IDS. If (1) holds then the spectral edge E_0 is called a *fluctuation boundary.* This term stems from the fact, that if the system is restricted to a large, finite volume, eigenvalues close to E_0 correspond to very particular and rare realizations of the randomness. Thus the spectral edge on finite volumes is highly sensitive to fluctuations of the random configurations. This feature is closely related to the phenomenon of localisation. Indeed, for many models existence of pure point spectrum has been proven precisely in the energy regimes where the density of states is very sparse, see, e.g., the characterisation given in [38]. The Lifshitz asymptotics implies that the IDS is extremely thin near the bottom of the spectrum, thus this energy region is a typical candidate for pure point spectrum.

One expects that the IDS is continuous at the fluctuation boundaries. Indeed, the Lifshitz asymptotics implies continuity at the minimum E_0 of the spectrum and moreover that the limes superior and inferior of difference quotients of any order vanish at E_0. This is in remarkable contrast to the dense set of discontinuities of the IDS exhibited by many percolation Hamiltonians, see, e.g., [18, 27, 112].

1.3. Percolation Hamiltonians

Hamiltonians on site percolation subgraphs of the lattice were introduced by P.-G. de Gennes, P. Lafore and J. Millot in [26, 27] as quantum mechanical models for binary alloys. The resulting random operator bears a strong similarity to the Anderson model considered by P.W. Anderson in [6]. The difference is that in the site percolation Hamiltonian of [27] the random potential may assume only two values, namely zero and plus infinity. The lattice sites where the potential equals infinity are deleted from the graph, and thus the Hamiltonian is restricted to the space of vertices where the potential vanishes. The resulting Hamiltonian may be understood as a single band approximation of an Anderson model whose potential is a family of Bernoulli distributed random variables. More precisely, if one equips the Bernoulli-Anderson potential with a coupling constant and lets this one tend to infinity, the corresponding quadratic form converges to the one of the quantum percolation model, cf. Remarks 11 and 12. This limit is in the physical literature, cf., e.g., [50], sometimes understood as the strong scattering limit of the tight binding model. It may be possible to estimate efficiently the convergence of the associated IDS using the techniques and estimates introduced in [64].

A series of papers in theoretical [18, 50] and computational physics, e.g., [44–47, 97], analysed the spectral features of the quantum percolation model. A particular point of interest in the numerical studies was to compare the spectral localisation of percolation Hamiltonians with the one of the Anderson model.

More recently there has been interest for such models in the mathematics community. The paper [16], which studies the intermittency behaviour of the parabolic Anderson model, establishes also Lifshitz tails for the quantum site percolation model. Given the above-mentioned relation between the Anderson and the quantum percolation Hamiltonian, it is not surprising that a detailed analysis of the former gives also results about the latter. Some related properties of random walks on percolation graphs have been analysed rigorously already in [7].

The existence of the IDS for a rather general class of site percolation Hamiltonians on graphs with a quasi transitive, free[1] amenable group action was established in [111]. This result relies on Lindenstrauss' pointwise ergodic theorem [72] for locally compact second countable amenable groups. Likewise, the non-randomness of the spectrum and its components is valid also in this general setting, see [65, 112]. These results hold in particular for the Anderson model and periodic Laplacians on such graphs. In fact, these features of the percolation Hamiltonians and their proofs are quite analogous to those of the Anderson model on $\ell^2(\mathbb{Z}^d)$. However, there are some distinct features of the quantum percolation model which distinguish it sharply from random Hamiltonians on the full lattice. The latter have a continuous IDS [24, 28], while percolation Hamiltonians have a dense set of discontinuities, which survives even if one restricts the operator to infinite percolation clusters [18, 50, 112]. In [54] Kirsch & Müller analysed basic spectral features of bond percolation Laplacians on the lattice and moreover carried out a thorough study of the low-lying spectrum of these operators in the non-percolating regime. These results were complemented by Müller & Stollmann in a paper [83] where the percolating regime is studied. The present note continues the analysis [112] of site percolation Hamiltonians on general graphs and at the same time extends the results of [54] to bond percolation models on amenable Cayley graphs. In particular, part of our proofs relies on ideas introduced in [54] and extends them to a more general geometric setting.

[1]In the references [111, 112] is was forgotten to spell out the assumption that the group acts freely on the graph.

It is not surprising that the analysis of (certain) spectral properties of percolation Hamiltonians relies on the proper understanding of the underlying 'classical' percolation problem. An exposition of this theory applied to independent bond percolation on the lattice can be found in Grimmett's book [41]. Some other aspects are covered in the monograph [49] by Kesten. In the more recent literature quite a body of work is devoted to percolation processes on more general graphs than $\mathbb{Z}^d$, see for instance [13, 14, 76, 96]. Still, much more is known for percolation on lattices than on general graphs. For the purposes of the results presented in this paper, we extended the theorems in [2, 4, 77] on the sharpness of the phase transition and the exponential decay of the cluster size in the subcritical regime to quasi transitive graphs, see [8].

Since in the subcritical regime no infinite percolation cluster exists almost surely, the entire spectrum of the corresponding Hamiltonian consists of eigenvalues. In the supercritical regime an infinite cluster exists and thus the Hamiltonian may have continuous spectrum. It is however conjectured [97] that for values of the percolation parameter just above the critical point the spectrum will still have no continuous component, although an infinite cluster exists. Thus there may be a second 'quantum' critical value of the percolation parameter, strictly larger than the classical critical value. Let us note that there is no rigorous proof of Anderson localisation for percolation Hamiltonians (in the percolating regime). The reason is that this model has Bernoulli distributed randomness, as it is the case for the Bernoulli-Anderson Hamiltonian. Due to the singular nature of the distribution of the randomness the known proofs of localisation do not apply.

1.4. Spectral graph theory and geometric L^2-invariants

To make sense of the term 'low energy asymptotics' one has to know where the minimum of the spectrum lies. In the case of Cayley graphs of amenable groups it is known from Kesten's theorem [48] that the bottom of the spectrum of the Laplacian is equal to zero. If the graph under consideration happens to be bipartite the spectrum of the adjacency operator is symmetric with respect to the origin, see, e.g., [23, 98]. This allows one to translate results about the lower spectral boundary of adjacency and Laplace operators to statements on the upper boundary.

The low energy behaviour of the IDS is determined by the first non-zero eigenvalue of the Hamiltonian on percolation clusters with 'optimal shape'. The minimizing configurations are determined by inequalities from spectral graph theory like the isoperimetric, the Cheeger and the Faber & Krahn inequality, cf. [19–22]. They relate the lowest eigenvalue on a finite subgraph to its volume and the volume growth behaviour of the Cayley graph. The growth of balls on Cayley graphs, in turn, can be classified using a result of Bass [12] and Gromov's theorem [42]. We use an improved version of the latter due to van den Dries & Wilkie [108]. Let us mention that even if one is interested in purely geometric properties of the automorphism group of a graph, like amenability or unimodularity, percolation may be a tool of choice, see for instance [13]. There several equivalences are established, each connecting a geometric property of a graph with a probabilistic property of an associated percolation process.

For several types of Hamiltonians on the lattice the Lifshitz exponent of the random operator equals the exponent of the van Hove singularity of its periodic counterpart. It turns out that the same holds for certain percolation models on Cayley graphs. To formulate this more properly one needs to characterise the high energy behaviour of the adjacency operator (resp. low energy behaviour of the Laplacian) on the full Cayley graph. This can be done by relating it to known results on random walks on groups, cf. [115], or on geometric L^2-invariants, cf., e.g., [75]. In fact, it turns out that the van Hove exponent of a Cayley graph equals the first *Novikov-Shubin invariant* of the Laplacian on the graph. These invariants have been introduced in [87, 88] and studied, e.g., in [43]. For various analogies between geometric L^2-invariants and properties of the IDS see the workshop report [31]. The above-mentioned equality of the Lifshitz and the van Hove exponent is encountered also in other contexts, see for instance Klopp's analysis [56] of Lifshitz tails at internal spectral gap edges.

1.5. The main result

Let us loosely state the main result of this note. Consider a Cayley graph of a finitely generated amenable group. Assume that the volume of balls of radius n in the graph behaves like n^d with the convention that $d = \infty$ corresponds to superpolynomial growth. Each deleted site (respectively bond) induces a new boundary in the graph, at which we may impose a certain type of boundary condition giving rise to different Laplace operators. More precisely, since we are dealing with bounded operators, the boundary condition is not a restriction on the domain of the operator, but rather an additional boundary term. As in [54] we consider Dirichlet, Neumann, and adjacency (or pseudo-Dirichlet) percolation Laplacians.

For the adjacency and Dirichlet Laplacian the low energy asymptotics of the IDS is given by

$$\lim_{E\searrow 0} \frac{\log|\log N(E)|}{|\log E|} = \frac{d}{2}. \tag{2}$$

Thus we have a Lifshitz type behaviour and zero is a fluctuation boundary of the spectrum. The Lifshitz exponent coincides with the van Hove exponent of the underlying full Cayley graph. For the Neumann Laplacian the low energy asymptotics of the IDS is given by

$$\lim_{E\searrow 0} \frac{\log|\log(N(E) - N(0))|}{|\log E|} = \frac{1}{2} \tag{3}$$

where $N(0)$ is a non-zero value corresponding to the number of open clusters per vertex in the percolation graph.

2. Model and results

We present several results which we have obtained for the IDS of Hamiltonians on full Cayley graphs and on graphs diluted by a percolation process. For the analysis of the latter operators it was necessary to establish certain properties of the 'classical percolation model' on Cayley graphs, which are detailed below.

2.1. The Laplace Hamiltonian on a full Cayley graph

We consider Cayley graphs of finitely generated, amenable groups and the corresponding Laplace operator. Its IDS exhibits a van Hove asymptotics at the lower spectral edge whose exponent is determined by the volume growth behaviour of the group.

To formulate this more precisely we explain the geometric setting in detail. Let Γ be a finitely generated group. Each choice of a finite, symmetric set of generators S not containing the unit element ι of Γ gives rise to Cayley graph. Note that this graph is regular (i.e., all vertices have the same degree) with degree equal to the number of elements in S and that the vertex set of G can be identified with Γ. Using the distance function on the Cayley graph we define the ball $B(n)$ of radius n around the unit element ι in Γ and set $V(n) := |B(n)|$. For a positive integer d we use the notation $V(n) \sim n^d$ to signify that there exist constants $0 < a, b < \infty$ such that $a\,n^d \leq V(n) \leq b\,n^d$.

From now on we tacitly assume that all considered groups are finitely generated. The growth of all such groups can be classified using deep results of Bass [12], Gromov [42] and van den Dries & Wilkie [108].

Theorem 1. *Let G be a Cayley graph of a finitely generated group. Then exactly one of the following is true:*

(a) *G has polynomial growth, i.e., $V(n) \sim n^d$ holds for some $d \in \mathbb{N}$,*
(b) *G has superpolynomial growth, i.e., for every $d \in \mathbb{N}$ and every $b \in \mathbb{R}$ there exists only finitely many integers n such that $V(n) \leq b\,n^d$.*

The growth behaviour (in particular the exponent d) is independent of the chosen set of generators S.

More precisely, the theorem follows from the following results. Bass has shown that every nilpotent group satisfies $V(n) \sim n^d$ for some $d \in \mathbb{N}$, sharpening earlier upper and lower bounds of Wolf [116]. Gromov's result in [42] is that every group of polynomially bounded volume growth is virtually nilpotent. Finally, van den Dries & Wilkie have shown that the conclusion of Gromov's theorem still holds, if one requires the polynomial bound $V(n_k) \leq b\,n_k^d$ merely along a strictly increasing sequence $(n_k)_{k\in\mathbb{N}}$ of integers $n_k \in \mathbb{N}$. We note that Pansu [90] has shown that $c := \lim_{n\to\infty} V(n)/n^d$ exists.

Let us now define the Laplacian on G. For future reference we consider a more general situation than needed at this stage. Let $G = (V, E)$ be a connected regular graph of degree k with vertex set V and edge set E, and $G' = (V', E')$ an arbitrary subgraph of G. Note that G' is in general not regular. We denote the degree of the vertex $x \in V'$ in G' by $d_{G'}(x)$. If two vertices $x, y \in V'$ are adjacent in the subgraph G' we write $y \sim_{G'} x$.

Definition 2. For G and G' as above we define the following operators on $\ell^2(V')$.

(a) The identity operator on G' is denoted by $\mathrm{Id}_{V'}$. If there is no danger of confusion we drop the subscript V'.

(b) The *degree operator* of G' acts on $\varphi \in \ell^2(V')$ according to

$$[D(G')\varphi](x) := d_{G'}(x)\varphi(x).$$

(c) The *adjacency operator* on G' is defined as

$$[A(G')\varphi](x) := \sum_{\substack{y \in V' \\ y \sim_{G'} x}} \varphi(y).$$

(d) The *adjacency Laplacian* on G' is defined as

$$\Delta^A(G') := k\,\mathrm{Id}_{V'} - A(G').$$

(e) In the special case $G' = G$, the *Laplacian* or *free Hamiltonian* on G is defined as

$$\Delta(G) := k\,\mathrm{Id}_V - A(G).$$

(Thus it coincides with the adjacency Laplacian on G.)

There are several different names used for the adjacency Laplacian $\Delta^A(G')$ in the literature. Our terminology is motivated by the fact that up to a multiplicative and an additive constant it is equal to the adjacency operator on the subgraph G'. This will be different for operators with an additional Dirichlet or Neumann boundary term introduced in Subsection 2.3. For induced subgraphs G' of G one can consider the *restriction of the Laplacian* which is defined as $\Delta^P(G') := P_{V'}\Delta(G)P_{V'}^*$. Here $P_{V'}\colon \ell^2(V) \to \ell^2(V')$, $P_{V'}\varphi(x) := \chi_{V'}(x)\varphi(x)$ denotes the orthogonal projection on V'. It turns out that $\Delta^P(G') = k\,\mathrm{Id}_{V'} - A(G') = \Delta^A(G')$.

The IDS of the free Hamiltonian on a Cayley graph G of an amenable group Γ can be defined as $N_0(E) := \langle \chi_{]-\infty,E]}(\Delta(G))\delta_\iota, \delta_\iota\rangle$. Here the function δ_ι has value 1 at ι and 0 everywhere else. It is possible to construct N_0 via an exhaustion procedure, see, e.g., [32, 33, 73].

The next Theorem characterises the asymptotic behaviour of the IDS at the spectral bottom. For groups of polynomial growth it exhibits a van Hove singularity. For groups of superpolynomial growth one encounters a type of generalised van Hove asymptotics with the van Hove exponent equal to infinity.

Theorem 3. *Let Γ be a finitely generated, amenable group, $\Delta(G)$ the Laplace operator on a Cayley graph of Γ and N_0 the associated IDS. If Γ has polynomial growth of order d then*

$$\lim_{E \searrow 0} \frac{\log N_0(E)}{\log E} = \frac{d}{2}. \tag{4}$$

and if Γ *has superpolynomial growth then*

$$\lim_{E\searrow 0} \frac{\log N_0(E)}{\log E} = \infty. \tag{5}$$

2.2. Classical percolation

Next we briefly introduce percolation on graphs. More precisely, we will consider independent site percolation, as well as independent bond percolation on *quasi transitive* graphs. These are graphs whose vertex set decomposes into finitely many equivalence classes under the action of the automorphism group. Such graphs have uniformly bounded vertex degree. If there is only one orbit, the graph is called *transitive*. Cayley graphs are particular examples of transitive graphs.

Let $G = (V, E)$ be an infinite, connected, quasi transitive graph and p some fixed real number between 0 and 1. For every site $x \in V$ we say it is *open* with probability p and *closed* with probability $1 - p$ independently of all other sites. This is the most simple site percolation model. More formally, we consider for every vertex x the probability space $\Omega_x := \{0, 1\}$, with the sigma algebra consisting of the power set $\mathcal{P}(\Omega_x)$ of Ω_x and the probability measure $\mathbb{P}_x$ on $(\Omega_x, \mathcal{P}(\Omega_x))$ defined by $\mathbb{P}_x(0) = 1 - p$ and $\mathbb{P}_x(1) = p$. The probability space $(\Omega_V, \mathcal{F}_V, \mathbb{P}_V)$ associated to site percolation is defined as the product $\prod_{x\in V}(\Omega_x, \mathcal{P}(\Omega_x), \mathbb{P}_x)$. The coordinate map $\Omega_V \ni \omega \mapsto \omega_x$ defines the *site percolation process* $\Omega_V \times V \to \{0, 1\}$. We call $V(\omega) := \{x \in V \mid \omega_x = 1\}$ the set of open or active sites. The induced subgraph of G with vertex set $V(\omega)$ is denoted by $G(\omega)$ and called the *percolation subgraph* in the configuration ω. The connected components of $G(\omega)$ are called *clusters.*

The bond percolation process is defined completely analogously. The bond percolation probability space is $(\Omega_E, \mathcal{F}_E, \mathbb{P}_E) := \prod_{e\in E}(\Omega_e, \mathcal{P}(\Omega_e), \mathbb{P}_e)$, where for every edge $e \in E$ we have $\Omega_e := \{0, 1\}$, with power set $\mathcal{P}(\Omega_e)$ and probability measure $\mathbb{P}_e$ defined by $\mathbb{P}_e(0) = 1 - p$ and $\mathbb{P}_e(1) = p$. For a given configuration $\omega \in \Omega_E$ we define the *percolation subgraph* $G(\omega)$ as the graph whose edge set $E(\omega)$ is the set of all $e \in E$ with $\omega_e = 1$ and whose vertex set $V(\omega)$ consist of all vertices in V which are incident to an element of $E(\omega)$. Connected components of $G(\omega)$ are called clusters. We will denote the expectation with respect to either $\mathbb{P}_V$ or $\mathbb{P}_E$ by $\mathbb{E}\{\dots\}$.

The most basic result in percolation theory is that for both the site and the bond model there exists a critical parameter $0 < p_c \le 1$ such that the following statement holds:

- if $p < p_c$ there is no infinite cluster almost surely (*subcritical phase*),
- if $p > p_c$ there is an infinite cluster almost surely (*supercritical phase*).

Of course, the value of p_c depends on the graph and on the type of percolation process considered.

It turns out that more can be said about the size of clusters in the subcritical phase. Let $o \in V$ be an arbitrary, but fixed vertex in a quasi transitive graph G. Consider site or bond percolation on G and denote by $C_o(\omega)$ the cluster containing o and by $|C_o(\omega)|$ the number of vertices in $C_o(\omega)$.

Theorem 4. *For any $p < p_c$ there exist a constant $\tau_p > 0$ such that $\mathbb{P}(|C_o(\omega)| \geq n) \leq e^{-\tau_p n}$ for all $n \in \mathbb{N}$. In particular, the expected number of vertices in $C_o(\omega)$ is finite for all $p < p_c$.*

This extends earlier results of Aizenman & Newman [4], Menshikov [77] (see also [78]), and Aizenman & Barsky [2]. The generalisation to bond percolation on quasi transitive graphs is given in [8] and relies on the differential inequalities for order parameters established in [2]. The proofs of [8] apply to the site percolation process, too. In our later applications, we will need this result only for Cayley graphs of amenable finitely generated groups.

2.3. Percolation Hamiltonians on Cayley graphs

We introduce percolation Hamiltonians with and without boundary terms and establish the existence of a selfaveraging IDS. This enables us to state our main result about the Lifshitz asymptotics and to compare it with the van Hove singularities of the Hamiltonian on the full Cayley graph.

Consider an arbitrary subgraph $G' = (V', E')$ of an infinite, k-regular Cayley graph $G = (V, E)$. The multiplication operator

$$W^{b.c.}(G') = k \operatorname{Id}_{V'} - D(G')$$

is non negative and has support in the interior vertex boundary of G'. Thus $\pm W^{b.c.}(G')$ can be understood as a potential due to the repulsion/attraction of the boundary. When added to the adjacency Laplacian Δ^A it gives rise to Dirichlet and Neumann boundary condition. Here we follow the nomenclature of [54, 81, 98].

Definition 5. We define the following operators on $\ell^2(V')$

(a) The *Dirichlet Laplacian* is defined as

$$\Delta^D(G') := \Delta^A + W^{b.c.}(G') = 2k \operatorname{Id}_{V'} - D(G') - A(G').$$

(b) The *Neumann Laplacian* is defined as

$$\Delta^N(G') := \Delta^A - W^{b.c.}(G') = D(G') - A(G').$$

The Laplacian $\Delta(G)$ on the full graph will be abbreviated by Δ. Note that in this case all three versions of the Laplacian coincide since there is no boundary. Let us collect certain basic properties of the adjacency, Dirichlet and Neumann Laplacian.

Remark 6.

a) All operators introduced in Definitions 2 and 5 are bounded and selfadjoint.
b) Since $W^{b.c.}(G') \geq 0$ we have the following inequalities in the sense of quadratic forms:

$$\Delta^N(G') \leq \Delta^A(G') \leq \Delta^D(G') \tag{6}$$

This is complemented by $0 \leq \Delta^N(G')$ and $\Delta^D(G') \leq 2k \operatorname{Id}$, which can be show by direct calculation in the same manner as in [54]. Later we will see, that there is a more complete chain of inequalities between five operators.

c) The operators $\Delta^A(G')$ and $\Delta^D(G')$ are injective. The Neumann Laplacian $\Delta^N(G')$ is injective if and only if G' has no finite components. In particular, the dimension of the kernel of $\Delta^N(G')$ is the number of the finite components of G'. Let us note that this relation between the number of zero Neumann eigenvalues and the number of clusters makes it possible to refine the analysis of the large time behaviour of random walks on finite percolation clusters, see, e.g., [100].

d) As we said in §1.4, it is a classical result [48] that zero is the lower spectral edge of Δ if and only if G is an amenable graph. Finding lower bounds in the nonamenable case is a more difficult task, see for instance [10, 84, 117].

e) A graph G is called *bipartite* if there is a partition V_1, V_2 of the vertex set such that there is no edge in G which joins two elements of V_i for $i = 1, 2$. For bipartite graphs it is useful to consider conjugation with the operator $\mathcal{U}$ on $\ell^2(V)$ which is given by the multiplication with the function $\chi_{V_1} - \chi_{V_2}$. This operator is unitary, selfadjoint and an involution. We will denote the restriction of $\mathcal{U}$ to some subset $\ell^2(V') \subset \ell^2(V)$ again by the same symbol. The conjugation with $\mathcal{U}$ relates the upper spectral edge to the lower one, see, e.g., [54, 81, 98]. For any subgraph G' we have $D(G') = \mathcal{U}D(G')\mathcal{U}$ and $A(G') = -\mathcal{U}A(G')\mathcal{U}$. This implies the relations

$$\begin{aligned} \Delta^A(G') &= 2k\,\mathrm{Id} - \mathcal{U}\Delta^A(G')\mathcal{U} \\ \text{and} \quad \Delta^{N(D)}(G') &= 2k\,\mathrm{Id} - \mathcal{U}\Delta^{D(N)}(G')\mathcal{U}. \end{aligned} \tag{7}$$

Consequently the spectrum of $\Delta^A(G')$ is symmetric with respect to k, while the spectrum of $\Delta^N(G')$ is a set obtained from the spectrum of $\Delta^D(G')$ by the symmetry with respect to k. It follows that in the special case of amenable bipartite graphs the upper spectral edge of Δ is equal to $2k$. Bipartiteness is not only sufficient, but also necessary for the mentioned symmetry. Claim 4.5 from [89] shows that on amenable non-bipartite graphs $2k$ is not in the spectrum of Δ, thus the symmetry fails.

Now we introduce percolation Laplacians associated to the configuration ω as bounded, selfadjoint operators on $\ell^2(V(\omega))$.

Definition 7. Let $(\Omega, \mathcal{F}, \mathbb{P})$ be a probability space corresponding either to site or bond percolation on G. For $\omega \in \Omega_V$ (resp. $\omega \in \Omega_E$) the operator

$$\Delta_\omega^\# := \Delta^\#(G(\omega)) \colon \ell^2(V(\omega)) \to \ell^2(V(\omega)), \quad \# \in \{A, D, N\}$$

is called *adjacency, resp. Dirichlet, resp. Neumann percolation Laplacian.*

For brevity we introduce the notation

$$A_\omega := A(G(\omega)) \quad \text{and} \quad W_\omega^{b.c.} := W^{b.c.}(G(\omega)).$$

Note that in the site percolation model we have

$$d_{G(\omega)}(x) := \omega_x \sum_{y \sim x} \omega_y,$$

while for bond percolation

$$d_{G(\omega)}(x) := \chi_{V(\omega)}(x) \sum_{e \sim x} \omega_e$$

where $e \sim x$ signifies that in the graph G the edge e is incident to x. Note that if the distance of $x, y \in V$ is greater than two, the random variables $d_{G(\omega)}(x)$ and $d_{G(\omega)}(y)$ are independent. These facts imply that in both models the stochastic field $(\omega, x) \mapsto W_\omega^{b.c.}(x)$ gives rise to a random potential which is stationary and ergodic with respect to the action of the group Γ.

Due to the structure of the underlying percolation process the considered random operators satisfy an equivariance relation which we explain next. For a group Γ and an associated Cayley graph G there is a natural action of Γ on G by multiplication from the left. This gives rise to a Γ-action both on the corresponding percolation probability space and on the bounded operators on the ℓ^2 space over the graph. The operation on the site percolation probability space $(\Omega_V, \mathbb{P}_V)$ by measure preserving transformations is given by $(\tau_\gamma(\omega))_x := \omega_{\gamma^{-1}x}$, for $\gamma \in \Gamma$ and $x \in V$. Note that γ and x are from the same set, but the first one is considered as a group element, while the second as a vertex. The family of transformations $(\tau_\gamma)_{\gamma \in \Gamma}$ acts ergodically on $(\Omega_V, \mathbb{P}_V)$. In the same way there is an ergodic action of measure preserving transformations indexed by $\gamma \in \Gamma$, which we again denote by τ_γ, on the bond percolation probability space $(\Omega_E, \mathbb{P}_E)$. Here the transformations are given by $(\tau_\gamma(\omega))_e := \omega_{\gamma^{-1}e}$, where $\gamma^{-1}[y, z] = [\gamma^{-1}y, \gamma^{-1}z]$.

The group Γ acts by unitary translation operators $(U_\gamma \varphi)(y) := \varphi(\gamma^{-1}y)$ on $\ell^2(V)$. If we restrict these operators to the ℓ^2 spaces over the active sites they act consistently with the shift on the probability space, more precisely $U\colon \ell^2(V(\omega)) \to \ell^2(\gamma V(\omega)) = \ell^2(V(\tau_\gamma \omega))$. Due to the transformation behaviour of the adjacency operator A_ω and the boundary term potential $W_\omega^{b.c.}$ one has the equivariance relation $\Delta^\#(\tau_\gamma \omega) = U_\gamma \Delta^\#(\omega) U_\gamma^{-1}$ for $\# \in \{A, D, N\}$. Since $(\tau_\gamma)_{\gamma \in \Gamma}$ acts ergodically on Ω, $(\Delta_\omega^\#)_\omega$ falls into the class of random operators studied in [65].

This yields the non-randomness of the spectrum and spectral components. In the following let us denote by σ the spectrum, and by σ_{disc}, σ_{ess}, σ_{pp}, σ_{sc}, σ_c and σ_{ac}, the discrete, the essential, the pure point, the singular continuous, the continuous, and the absolutely continuous component of the spectrum respectively. Let us recall that the pure point spectrum of an operator is the *closure* of the set of its eigenvalues. We use the notation $\sigma_{fin}(H)$ for the set of eigenvalues of H which posses an eigenfunction with compact, i.e., finite, support.

The following theorem holds for site and bond percolation Hamiltonians and for all values of $p \in [0, 1]$.

Theorem 8. *Let $\Delta_\omega^\#$ be one of the percolation Laplacians defined above. Then there exists for each $\bullet \in \{\text{disc}, \text{ess}, \text{pp}, \text{sc}, \text{c}, \text{ac}, \text{fin}\}$ a subset $\Sigma_\bullet^\# \subset \mathbb{R}$ and a subset $\Omega' \subset \Omega_V$ (resp. $\Omega' \subset \Omega_E$) of full measure, such that $\sigma_\bullet(\Delta_\omega^\#) = \Sigma_\bullet^\#$ holds for every $\omega \in \Omega'$ and for all $\bullet \in \{\text{disc}, \text{ess}, \text{pp}, \text{sc}, \text{c}, \text{ac}, \text{fin}\}$. In particular $\sigma(\Delta_\omega^\#) = \Sigma^\#$ for all $\omega \in \Omega'$.*

This has been proven for site percolation models in [65, 111, 112], and holds with the same proofs for bond percolation. For similar results in the case of bond percolation on the lattice see also [54].

Remark 9. (a) All the results stated so far hold for arbitrary quasi transitive graphs. In particular it is not necessary to assume that the graph is amenable or infinite. For infinite graphs the discrete spectrum Σ_{disc} is empty.

(b) The percolation Laplacians $\Delta^\#_\omega$ are defined on $\ell^2(V(\omega))$. Thus, technically $(\Delta^\#_\omega)_{\omega\in\Omega}$ is associated to a direct integral operator with non-constant fibre. One can modify the operator in such a way that the fibres become constant and that the spectrum changes in a controlled way. First note that for a fixed configuration ω and a fixed cluster $C(\omega)$ any percolation Hamiltonian leaves $\ell^2(C(\omega))$ invariant. Consequently, the full operator decomposes according to the clusters:

$$\Delta^\#_\omega = \bigoplus_{C(\omega) \text{ cluster of } G(\omega)} \Delta^\#(C(\omega))) \tag{8}$$

For any constant K we may add to (8) the operator

$$\bigoplus_{x\in V\setminus V(\omega)} K \,\mathrm{Id}_x = K\,\mathrm{Id}_{V\setminus V(\omega)}\,.$$

Denote the resulting sum by $\widetilde{\Delta^\#_\omega}$. It acts on $\ell^2(V)$ and leaves the subspaces $\ell^2(V(\omega))$ and $\ell^2(V\setminus V(\omega))$ invariant. In particular, $\widetilde{\Delta^\#_\omega}$ can be written in the same way as (8) with the only difference that the direct sum extends over all $\tilde{C}(\omega)$ which are either a cluster in $G(\omega)$ or a vertex in $V\setminus V(\omega)$. If the configuration ω is such that $V(\omega)=V$ then $\widetilde{\Delta^\#_\omega}$ coincides with $\Delta^\#_\omega$. Note however that these configurations form a set of measure zero, both in the site and the bond percolation model. In all other cases we have

$$\begin{aligned}\sigma(\widetilde{\Delta^\#_\omega}) &= \sigma(\Delta^\#_\omega)\cup\{K\},\\ \sigma_{pp}(\widetilde{\Delta^\#_\omega}) &= \sigma_{pp}(\Delta^\#_\omega)\cup\{K\} \quad \text{and}\\ \sigma_c(\widetilde{\Delta^\#_\omega}) &= \sigma_c(\Delta^\#_\omega).\end{aligned}$$

(c) In the subcritical regime there are no infinite clusters. The decomposition (8) implies that the operator is a direct sum of finite dimensional operators, thus the spectrum consists entirely of eigenvalues in $\sigma_{\text{fin}}(\Delta^\#_\omega)$. In particular, $\Sigma_{\text{sc}} = \Sigma_{\text{ac}} = \emptyset$ and $\Sigma_{\text{pp}} = \Sigma_{\text{fin}}$.

(d) We are not able to calculate the deterministic spectrum $\Sigma^\#$ but only to give partial information about it. Since we can find arbitrarily large clusters in $G(\omega)$ almost surely, strong resolvent convergence gives that $\Sigma^\# \supset \sigma(\Delta)$, see [54] for details. Together with $\sigma(\Delta^\#_\omega)\subset[0,2k]$ for all ω this implies that in the case of amenable Cayley graphs zero is the lower spectral edge of the operators $\Delta^\#_\omega$ almost surely for any $\# \in \{A,D,N\}$. Similar inclusions for the deterministic spectrum of Anderson models have been given in [63]. If the Cayley graph is bipartite and the spectrum of the free Laplacian Δ has no gaps we have $\Sigma^\# = [0,2k]$. If $\sigma(\Delta)$

has several components, it might be that $\Sigma^{\#}$ contains values in a spectral gap of Δ. This is related to the phenomenon called *spectral pollution* encountered when using strong convergence to approximate spectral values of the limiting operator, see for instance [25, 68] and the references therein.

Our next aim is to introduce the IDS and state its main properties. An abstract IDS may be defined without an exhaustion procedure by setting

$$N^{\#}(E) = \mathbb{E}\{\langle \chi_{]-\infty,E]}(\Delta^{\#}_{\omega})\delta_\iota, \delta_\iota\rangle\}$$

see [65]. However, the physically interesting situation is when the IDS is *selfaveraging*, i.e., when spatial averages of the normalised eigenvalue counting functions converge to the expectation with respect to the randomness. To be able to show that the IDS has indeed this property we have to require that the group Γ is amenable. Each such group has a sequence of finite, non-empty subsets $(I_j)_j$ which satisfies for every finite set $F \subset \Gamma$ the property

$$\lim_{j\to\infty} \frac{|I_j \triangle F \cdot I_j|}{|I_j|} = 0.$$

Such an $(I_j)_j$ is called a *Følner sequence.* Here $\triangle$ denotes the symmetric difference of two sets. Various properties of such sequences are discussed in [1], and their role in the construction of the IDS of random operators in [95] and [110, §2.3]. Each Følner sequence has a tempered subsequence $(I_{j_k})_k$, meaning that there exists a constant C such that $|I_{j_k} I_{j_{k-1}}^{-1}| \le C|I_{j_k}|$. We may consider I_{j_k} as a subset of the vertices in the Cayley graph. Denote by P_k the projection on the vertex set $I_{j_k} \cap V(\omega)$ and by $\Delta^{\#,k}_{\omega}$ the restricted operator $P_k \Delta^{\#}_{\omega} P_k^*$ for any $\# \in \{A, D, N\}$. Thus we obtain a selfadjoint operator on a finite dimensional space which has a finite number of real eigenvalues. Hence the trace of the spectral projection $\operatorname{Tr}\left[\chi_{]-\infty,E]}(\Delta^{\#,k}_{\omega})\right]$ is finite. We define the eigenvalue counting distribution function as

$$N^{\#,k}_{\omega}(E) := \frac{1}{|I_k|} \operatorname{Tr}\left[\chi_{]-\infty,E]}(\Delta^{\#,k}_{\omega})\right].$$

The next theorem holds for site and bond percolation Hamiltonians and for all values of $p \in [0,1]$.

Theorem 10. *Let G be a Cayley graph of an amenable group Γ and $(I_k)_k$ a tempered Følner sequence. There exists a set $\Omega' \subset \Omega_V$ (resp. Ω_E) of full measure such that for every $\omega \in \Omega'$ and every $E \in \mathbb{R}$ which is a continuity point of $N^{\#}(\cdot)$ we have*

$$\lim_{k\to\infty} N^{\#,k}_{\omega}(E) = N^{\#}(E).$$

The support of the measure associated to the distribution function $N^{\#}$ equals $\Sigma^{\#}$.

For site percolation, the theorem is proven in [111], the modification to bond percolation is not hard. See also [54] for the case of bond-percolation on the lattice.

Thus we have obtained an selfaveraging IDS. The theorem implies that zero is the lower edge of the support of $N^{\#}$ for $\# \in \{A, D, N\}$ and that $N^A(0) =$

$N^D(0) = 0$ while $N^N(0) > 0$. The distribution function $N^\#$ has a discontinuity at $E \in \mathbb{R}$ if and only if $E \in \Sigma^\#_{\mathrm{fin}}$. This will be discussed in more detail in §3.1.

Let us denote the density of vertices in $V \setminus V(\omega)$ by $g(\omega)$. By ergodicity there is a value $g \in [0,1]$ such that $g(\omega) = g$ almost surely. In the site percolation model $g = 1 - p$ and in the bond percolation model $g = (1-p)^k$. Then the IDS $\widetilde{N^\#}$ of $(\widetilde{\Delta^\#_\omega})_\omega$ is related to $N^\#$ by

$$\widetilde{N^\#} = N^\# + g\,\chi_{[K,\infty[}.$$

Remark 11. Let μ be a non-trivial probability measure on a compact interval $[0,s]$ and $(w_\omega(x))_{x\in V}$ an i.i.d. family of random variables with distribution μ. The Anderson model on the graph G is given by $H_\omega := \Delta + W_\omega \colon \ell^2(V) \to \ell^2(V)$. Here W_ω is the multiplication operator with $w_\omega(x)$. This is again an ergodic operator with non-random spectrum and selfaveraging IDS. It is very convenient to compare it with the site percolation Laplacian. For this purpose we specialise to the case $\mu = p\delta_0 + (1-p)\delta_1$ and introduce an additional, global coupling constant $\lambda \geq 0$. We denote the quadratic form of the operator $H^{BA}_{\omega,\lambda} := \Delta + \lambda W_\omega$ by $q^{BA}_{\omega,\lambda}$. Here BA stands for the *Bernoulli-Anderson* or *binary-alloy* type of the Hamiltonian. Denote its IDS by N^{BA}_λ.

Recall that one may compare two (lower-bounded, closed) quadratic forms q_1 and q_2 on a Hilbert space by setting $q_1 \leq q_2$ if and only if the domains satisfy the inclusion $\mathcal{D}(q_1) \supset \mathcal{D}(q_2)$ and for all $\varphi \in \mathcal{D}(q_2)$ we have $q_1[\varphi] \leq q_2[\varphi]$. This notion of inequality of quadratic forms is consistent with extending each form to the whole Hilbert space by the rule $q_i[\varphi] = +\infty$ for $\varphi \notin \mathcal{D}(q_i)$.

Denote the quadratic form associated to any of the Hamiltonians $\Delta^\#_\omega$ by $q^\#_\omega$. Comparing the Anderson and the adjacency site percolation Hamiltonian we see that $q^{BA}_{\omega,\lambda} \leq q^A_\omega$ for all $\lambda \geq 0$. This relation holds pointwise for all ω if we introduce the obvious coupling between the random potential $(w_\omega(x))_{x\in V}$ and the percolation process. Moreover for $\lambda \to \infty$ the quadratic form $q^{BA}_{\omega,\lambda}$ converges monotonously to q^A_ω. On the other hand, since the potential $(w_\omega(x))_{x\in V}$ is non-negative we have $q_G \leq q^{BA}_{\omega,\lambda}$ where q_G denotes the quadratic form of Δ on the full graph. Using the argument in [98, §2] one sees that $\widetilde{q^N_\omega} \leq q_G$, where $\widetilde{q^N_\omega}$ is the form corresponding to the operator $\widetilde{\Delta^N_\omega} = \Delta^N_\omega \oplus 0 \cdot \mathrm{Id}_{V\setminus V(\omega)}$.

Summarising all quadratic form inequalities obtained so far for the site percolation case we obtain the chain

$$\widetilde{q^N_\omega} \leq q_G \leq q^{BA}_{\omega,\lambda} \leq q^A_\omega \leq q^D_\omega$$

which implies for the corresponding IDS'

$$\widetilde{N^N}(E) \geq N_G(E) \geq N^{BA}_\lambda(E) \geq N^A(E) \geq N^D(E) \quad \text{for all } E \in \mathbb{R}.$$

Now it is clear why the study of the Anderson Hamiltonian in [16] gave also results on the (adjacency) percolation Laplacian. Upper bounds on the IDS N^{BA}_λ imply upper estimates for the IDS of the adjacency and Dirichlet site percolation model.

There is also a relation between bond percolation Hamiltonians on the one hand and random hopping models [58] and discrete Schrödinger operators with random magnetic field [85] on the other hand. All three models share the feature that the randomness enters the Hamiltonian in the off-diagonal matrix elements. In fact, in [58, 85] certain spectral properties of such models are analysed by relating them to Hamiltonians with diagonal disorder.

Remark 12. The Anderson model on $\ell^2(\mathbb{Z}^d)$ is interpreted by physicists as a single spectral band approximation of a continuum random Schrödinger operator on $L^2(\mathbb{R}^d)$. Due to the i.i.d. assumption on the random variables in the Anderson model its spectrum as a set equals almost surely

$$\sigma(\Delta) + \operatorname{supp}\mu := \{t + s \in \mathbb{R} \mid t \in \sigma(\Delta), s \in \operatorname{supp}\mu\}$$

see Theorem III.2 in [63]. Here as before μ denotes the distribution measure of the potential values of the Anderson model and $\operatorname{supp}\mu$ its (topological) support. Now, if the support of μ is split into several connected components and if the gaps between them are large enough, the a.s. spectrum of the Anderson model also contains gaps and consequently has a (sub)band structure. Internal Lifshitz tails for such models have been studied in [81, 99].

In particular, if we consider the Bernoulli-Anderson model $H^{BA}_{\omega,\lambda}$ introduced above, we see that for large enough λ the almost sure spectrum of $H^{BA}_{\omega,\lambda}$ splits into subbands. As we let $\lambda \to \infty$, one of the bands diverges to infinity. On the other hand, we know that in the sense of quadratic forms $q^{BA}_{\omega,\lambda} \nearrow q^{A}_{\omega}$ for $\lambda \to \infty$. Thus the resulting site percolation Hamiltonian Δ^A_ω may be understood as an approximation of $H^{BA}_{\omega,\lambda}$ (with large values of λ) associated to a spectral (sub)band.

Now we are in the position to state our main results in the next two theorems. In both cases we have the same setting as before: we consider a Cayley graph G of an amenable finitely generated group Γ, site or bond percolation on G and the percolation Laplacians $\Delta^{\#}_\omega$ with associated IDS $N^{\#}$. We restrict ourselves now to the subcritical phase $p < p_c$. The asymptotic behaviour of the IDS of the adjacency and the Dirichlet percolation Laplacian at low energies is as follows:

Theorem 13. *Assume that G has polynomial growth and $V(n) \sim n^d$. Then there are positive constants $\alpha^+_D(p)$ and $\alpha^-_D(p)$ such that for all positive E small enough*

$$e^{-\alpha^-_D(p)E^{-d/2}} \le N^D(E) \le N^A(E) \le e^{-\alpha^+_D(p)E^{-d/2}}. \tag{9}$$

Assume that G has superpolynomial growth. Then

$$\lim_{E\searrow 0} \frac{\ln|\ln N^D(E)|}{|\ln E|} = \lim_{E\searrow 0} \frac{\ln|\ln N^A(E)|}{|\ln E|} = \infty. \tag{10}$$

In particular, the IDS is very sparse near $E = 0$ and consequently the bottom of the spectrum is a fluctuation boundary. Relations (4) and (9) imply that in the case of polynomial growth the Lifshitz exponent coincides with the van Hove exponent of the Laplacian on the full Cayley graph. In the case of superpolynomial growth we have that both exponents are infinite. One may ask whether the limits

(5) and (10), which define the two exponents, diverge at the same rate. To express this in a quantitative way let us note that in the case of polynomial growth we have

$$\lim_{E\searrow 0} \frac{\ln|\ln N^A(E)|}{|\ln N_0(E)|} = 1$$

and the analogous relation for the Dirichlet Laplacian. In the case of superpolynomial growth one may hope to prove at least

$$\lim_{E\searrow 0} \frac{\ln\ln|\ln N^A(E)|}{\ln|\ln N_0(E)|} = 1.$$

We are not able prove that in general, but at least for the case of the Lamplighter group, see §3.3.

In the case of Neumann boundary conditions we know that $\widetilde{\Delta_\omega^N} \le \Delta$. Consequently, the IDS $\widetilde{N^N}$ is at the bottom of the spectrum at least as 'thick' as the one of the free Laplacian, which exhibits a van Hove singularity. In particular, the energy zero is not a fluctuation boundary. This remains true if we pass from $\widetilde{N^N}$ to N^N by removing the point mass $g\delta_0$ from the density of states measure.

Theorem 14. *There exist positive constants $\alpha_N^+(p)$ and $\alpha_N^-(p)$ such that for all positive E small enough*

$$e^{-\alpha_N^-(p)E^{-1/2}} \le N^N(E) - N^N(0) \le e^{-\alpha_N^+(p)E^{-1/2}}. \tag{11}$$

The value $N^N(0)$ coincides with the average number of clusters per vertex in the random graph $G(\omega)$. After subtracting this value we can speak of (11) as a kind of 'renormalised' Lifshitz asymptotics with exponent $1/2$.

Remark 15. (a) In the next section we give a sketch of the proof of the theorems, while the full version will appear elsewhere.
(b) The lower bounds in (9) and (11) are actually true for all values of p.
(c) In the special case of bipartite Cayley graphs there is a relation between the behaviour of the IDS near the upper and lower spectral edges of the spectrum, see Remark 6. In that case we are able to characterise the asymptotic behaviour of the IDS at the upper spectral boundary in the same way as done in [54].

3. Discussion, additional results and sketch of proofs

In this final section we conclude the paper with a discussion of further spectral properties of percolation Hamiltonians, an outline of the proof of the Theorems 13 and 14, and an important example. The example concerns the Lamplighter group, which is amenable, but of exponential growth. The spectral properties we mentioned are related to jumps of the IDS, to finitely supported eigenfunctions and to the *unique continuation principle*, see [112, 114] and [66].

3.1. Discontinuities of the IDS

Let us recall a result on the location of the set of discontinuities of the IDS, established in [112] for site percolation Laplacians and generalisations thereof. The result and its proof apply verbatim for the bond percolation model. Furthermore for this result it is not necessary to assume that the percolation process is independent at different sites, but merely that it is ergodic with respect to the transformations $(\tau_\gamma)_{\gamma\in\Gamma}$ introduced in §2.2.

Proposition 16. *Let G be a Cayley graph of an amenable group Γ and $(\Delta_\omega^\#)_\omega$, $\# \in \{A, D, N\}$ a percolation Laplacian associated either to site or to bond percolation. Then the following two properties are equivalent:*

(i) *the IDS of $(\Delta_\omega^\#)_\omega$ is discontinuous at $E \in \mathbb{R}$,*
(ii) $E \in \Sigma_{\mathrm{fin}}^\#$.

Remark 17. This result can be strengthened to describe the size of the jumps also. In a forthcoming paper [66] of Lenz and the second named author ergodic Hamiltonians on discrete structures are analysed. The abstract setting considered there covers in particular Anderson Hamiltonians and site and bond percolation Laplacians $(\Delta_\omega^\#)_\omega$ on a Cayley graph G of an amenable group Γ. Again one has to assume that the stochastic process which determines the random potential, respectively the percolation process entering the percolation Laplacian, is ergodic with respect to the transformation group $(\tau_\gamma)_{\gamma\in\Gamma}$. Then the eigenspace corresponding to the eigenvalue E is spanned by finitely supported eigenfunctions almost surely. See also [61, 62] for a similar result for periodic operators on graphs with an Abelian group structure.

The jumps of the IDS of invariant operators on Cayley graphs have been studied in the literature on L^2-invariants. The kth L^2-Betti number is the trace per unit volume of the kernel of the Laplacian on k-forms. It has been proven in great generality that it does not matter whether one defines it via the continuum or the combinatorial Laplacian, see, e.g., [30]. Likewise it is known that the zeroth L^2-Betti number vanishes on amenable Cayley graphs [74, Thm. 1.7], which is the same as saying that the IDS is continuous at the bottom of the spectrum. Using the same terminology for random ergodic operators, we may say that the Neumann percolation Laplacian has non vanishing (zero order) L^2-Betti number. With this regard it would be interesting to find an interpretation of higher order Betti numbers in terms of quantities of mathematical physics. The sizes of jumps of the IDS at discontinuity points have be discussed for Abelian periodic models in [34, 104]. The task of characterising the sizes of jumps of the IDS is related to the Atiyah conjecture, see, e.g., [39, 75] or [74, Conjecture 2.1], and the references therein.

3.2. Outline of the proof of Theorems 13 and 14

While the full proofs of our results will be given in [9] we present here certain key estimates. In particular, we state upper and lower bounds for the IDS of percolation Laplacians in a neighbourhood of the lower spectral edge. The bounds are given in terms of $\lambda^{\#}(G')$, the lowest non-zero eigenvalue of $\Delta^{\#}(G')$. These estimates are generalisations of Lemmata 2.7 and 2.9 in [54]. The proof of Theorem 18 uses the exponential decay from Theorem 4.

Theorem 18. *Let G be an amenable Cayley graph and $\# \in \{A, D, N\}$. Assume that there is a continuous strictly decreasing function $f\colon [1,\infty[\to \mathbb{R}^+$ such that $\lim_{s\to\infty} f(s) = 0$ and $\lambda^{\#}(G') \geq f(|G'|)$ for any finite subgraph G'. Then, for every $0 < p < p_c$ the IDS satisfies the following inequality*

$$N^{\#}(E) - N^{\#}(0) \leq e^{-a_p f^{-1}(E)}, \tag{12}$$

for some positive constant a_p, when E is small enough. Here f^{-1} denotes the inverse function of f.

Theorem 19. *Let G be an amenable Cayley graph and $\# \in \{A, D, N\}$. Suppose that there is a sequence of connected subgraphs $(G'_n)_n$ and a sequence $(c_n)_n$ in $\mathbb{R}^+$ such that*

(i) $\lim_{n\to\infty} |G'_n| = \infty$
(ii) $\lim_{n\to\infty} c_n = 0$
(iii) $\lambda^{\#}(G'_n) \leq c_n$

For every $E > 0$ small enough define $n(E) := \min\{n; c_n \leq E\}$. Then for every $0 < p < 1$ there is a positive constant b_p such that the following inequality holds for all $E > 0$ small enough

$$N^{\#}(E) - N^{\#}(0) \geq e^{-b_p |G'_{n(E)}|}. \tag{13}$$

Hence our problem is reduced to finding efficient bounds for $\lambda^{\#}(G')$ in terms of the geometric properties of G'. For this we will use, following [54], the Cheeger and Faber & Krahn inequalities. Since we are considering general Cayley graphs of amenable groups with polynomial growth, these two inequalities are not sufficient, but we will need additionally an appropriate version of the isoperimetric inequality.

The function V was defined in Theorem 1. We also define

$$\phi(t) := \min\{n \geq 0; V(n) > t\}.$$

Moreover we will denote the linear subgraph with n vertices by L_n.

Proposition 20. *For a Cayley graph $G = (V, E)$ there are positive constants α_D, β_D, γ_D, α_N and γ_N such that the following are true*

(i) *For every finite connected subgraph G'*

$$\lambda^{A}(G') \geq \frac{\alpha_D}{\phi(\beta_D |G'|)^2} \quad \text{and} \quad \lambda^{N}(G') \geq \frac{\alpha_N}{|G'|^2}. \tag{14}$$

(ii) *For every positive integer n*

$$\lambda^{D}(B(n)) \leq \frac{\gamma_D V(n)}{n^2 V(\lfloor n/2 \rfloor)} \quad \text{and} \quad \lambda^{N}(L_n) \leq \frac{\gamma_N}{n^2}. \tag{15}$$

Sketch of the proof. (i) The inequality for adjacency Laplacian can be proved following the arguments in the proofs of Lemma 2.4 from [54] and Proposition 7.1 from [19] and using the isoperimetric inequality from Théorème 1 in [22].

The inequality for Neumann Laplacian is a simple consequence of the Cheeger inequality (see Théorème 3.1 in [21]).

(ii) Both inequalities can be obtained by inserting an appropriate test function to the sesquilinear form and using the mini-max principle. For the Dirichlet Laplacian we choose a test function which has value 0 outside the ball $B(n)$, value i on the sphere of radius $n-i$, for $i = 0, \dots, \lceil n/2 \rceil$ and value $\lceil n/2 \rceil$ inside the ball $B(\lfloor n/2 \rfloor)$.

The operator $\Delta^N(L_n)$ is not injective and thus the test function must be orthogonal to the kernel. This means that the test function φ must have support inside L_n and $\sum_{x \in L_n} \varphi(x) = 0$. The appropriate test function is the one which grows linearly from $\frac{-n+1}{2}$ to $\frac{n-1}{2}$ along the vertices of L_n and is 0 outside of L_n. □

Remark 21. (a) In the case of polynomial growth the first bound in (14) and the first bound in (15) are of the same order in n if $G' = B(n)$. Likewise the second bounds in (14) and (15) are of the same order of magnitude for $G' = L_n$. Thus the proposition shows that the optimal subgraph configuration for adjacency and Dirichlet Laplacians is a ball, and for the Neumann Laplacian is a line graph. For Dirichlet and Neumann boundary conditions this can be motivated in terms of the potential $\pm W^{b.c.}$. Since $W^{b.c.}$ is non-negative (repulsive) at the interior vertex boundary, a test function φ with low expectation value $\langle \varphi, \Delta^D(G')\varphi \rangle$ has to make both $\langle \varphi, \Delta^A(G')\varphi \rangle$ and $\langle \varphi, W^{b.c.}\varphi \rangle$ small. The second condition pushes the mass of φ away form the boundary, while the first one minimizes the variation of φ. This is best realised when G' is a ball. On the contrary, in the case of Neumann boundary conditions $-W^{b.c.} \leq 0$ is attracting φ to the boundary. For fixed volume the graph with most boundary is a linear graph.

(b) Note that if Γ is a group of polynomial growth, then the balls $B(n), n \in \mathbb{N}$ form a Følner sequence: Since every finite F satisfies $F \subset B(k)$ for some $k \in \mathbb{N}$ we have

$$|F \cdot B(n) \setminus B(n)| \leq V(n+k) - V(n) \leq (c + \mathfrak{o}(1))\,(n+k)^d - (c - \mathfrak{o}(1))\,n^d = \mathfrak{o}(V(n))$$

and similarly $|B(n) \setminus F \cdot B(n)| \leq V(n) - V(n-k) = \mathfrak{o}(V(n))$. Since $V(2n) \leq 2^d \frac{b}{a} V(n)$ the sequence of balls satisfies the *doubling property* and is in particular tempered. Note that although every amenable group contains a Følner sequence, balls may not form one.

3.3. The Lamplighter group

In this section we will explain how the ideas and methods we used to study the behaviour of the IDS in the case of groups of polynomial growth, can be used to give sharp bounds on the IDS asymptotics in the case of a particular group of superpolynomial growth. Namely, we consider certain Lamplighter groups, which are examples of amenable groups with exponential growth (i.e., there exists a constant $c > 1$ such that $V(n) \geq c^n$).

We define the Lamplighter group as the wreath product $\mathbb{Z}_m \wr \mathbb{Z}$, where m is an arbitrary positive integer. Elements of this group are ordered pairs (φ, x), where φ is a function $\varphi\colon \mathbb{Z} \to \mathbb{Z}_m$ with finite support and $x \in \mathbb{Z}$. The multiplication is given by $(\varphi_1, x_1) * (\varphi_2, x_2) := (\varphi_1 + \varphi_2(\cdot - x_1), x_1 + x_2)$. More generally, a lamplighter group can be defined as a wreath product of two Abelian groups, one of which is finite. However, here we shall only consider lamplighter groups of the form $\mathbb{Z}_m \wr \mathbb{Z}$.

Since Lamplighter groups are amenable it makes sense to consider the integrated density of states in both the deterministic and random setting, as defined in Section 2. The next theorem concerns the Laplace operator on the full Cayley graph. It establishes a Lifshitz-type asymptotics in the sense that

$$\lim_{E \searrow 0} \frac{\ln |\ln N_0(E)|}{|\ln E|} = \frac{1}{2}. \tag{16}$$

Of course the underlying operator is not random, but the IDS is exponentially thin at the bottom of the spectrum as it is the case for spectral fluctuation boundaries of random operators. The quantity (16) may be called a Lifshitz exponent or maybe more appropriately secondary Novikov-Shubin invariant of the Laplacian, using the terminology of [89].

Theorem 22. *Let G be a Cayley graph of the Lamplighter group. There are positive constants a_1^+ and a_2^+ such that*

$$N_0(E) \leq a_1^+ e^{-a_2^+ E^{-1/2}}, \quad \textit{for all } E \textit{ small enough.}$$

Moreover for every $r > 1/2$ there are positive constants $a_{r,1}^-$ and $a_{r,2}^-$ such that

$$N_0(E) \geq a_{r,1}^- e^{-a_{r,2}^- E^{-r}}, \quad \textit{for all } E \textit{ small enough.}$$

The proof of the preceding theorem follows from the proof of Theorem 4.4 (parts (ii) and (iii)) in [89]. As an input we need the following inequalities for $\mu^{(2n)}(\iota)$, the return probability of the simple random walk after $2n$ steps:

$$a_1 e^{-a_2 n^{1/3}} \leq \mu^{(2n)}(\iota) \leq A_1 e^{-A_2 n^{1/3}},$$

for some positive constants a_i, A_i, $i = 1, 2$, and all positive integers n. For the reference see Theorem 15.15 in [115].

For the percolation case we shall once again use Theorems 18 and 19. Exponential growth and Proposition 20 i) give lower bounds for the lowest eigenvalues $\lambda^A(G')$ which are of the form $\mathrm{const}/(\ln |G'|)^2$. Now Theorem 18 implies the following result.

Theorem 23. *Let G be an arbitrary Cayley graph of the lamplighter group $\mathbb{Z}_m \wr \mathbb{Z}$. For every $p < p_c$ there are positive constants b_1 and b_2 such that the IDS of the adjacency and Dirichlet percolation Laplacian satisfies the following inequality*

$$N^A(E) \leq N^D(E) \leq e^{-b_1 e^{b_2 E^{-1/2}}}, \quad \textit{for all } E \textit{ small enough.} \tag{17}$$

The upper bounds from Proposition 20 are not applicable in the Theorem 19 any more. Instead, we shall use results from [11], where Bartholdi & Woess proved that the Laplacian Δ on Diestel-Leader graphs has only pure point spectrum, expanding on earlier results of Grigorchuk & Żuk [40] and Dicks and Schick [29] concerning certain Cayley graphs of Lamplighter groups. Diestel-Leader graphs are a generalisation of Cayley graphs of Lamplighter groups with a particular set of generators. This 'natural' set of generators is given by

$$S_0 := \{(l \cdot \delta_1, 1), l \in \mathbb{Z}_m\} \cup \{(l \cdot \delta_0, -1), l \in \mathbb{Z}_m\}\,. \tag{18}$$

Here $l \cdot \delta_z$ denotes the function which has value l in z and 0 everywhere else. In the following we formulate certain facts about the spectrum of adjacency Laplacians on certain finite, connected subgraphs of the Lamplighter graph, called *tetrahedrons*, which were established in [11], see also [29, 40]. We will not give a definition of these subgraphs, since it would require a quite comprehensive description of horocyclic products of homogeneous trees. Rather, we refer to [11] for precise definitions and background information.

We will need three facts concerning tetrahedrons and corresponding eigenvalues of adjacency Laplacians:

(a) the tetrahedron of depth n (denoted by T_n) has $(n+1)m^n$ vertices,
(b) $2m(1-\cos\frac{\pi}{n})$ is an eigenvalue of the operator $\Delta^A(T_n)$,
(c) there is an eigenvector of $\Delta^A(T_n)$, corresponding to the eigenvalue $2m(1-\cos\frac{\pi}{n})$, that has value 0 on the inner vertex boundary of T_n.

Claim a) follows directly from the definition of tetrahedron. For the proofs of claims b) and c) see Lemma 2 and Corollary 1 in [11].

Now Theorem 19, with $G'_n = T_n$ and $c_n = 2m(1-\cos\frac{\pi}{n})$, implies the following bound.

Theorem 24. *Let G_{S_0} be the Cayley graph of the lamplighter group $\mathbb{Z}_m \wr \mathbb{Z}$ defined with respect to the set of generators S_0. For every $0 < p < 1$ there are positive constants c_1 and c_2 such that the IDS of percolation Laplacians satisfies the following inequality*

$$e^{-c_1 e^{c_2 E^{-1/2}}} \le N^D(E) \le N^A(E), \text{ for all } E \text{ small enough.} \tag{19}$$

Theorem 24 is in fact valid for all Cayley graphs of the lamplighter group $\mathbb{Z}_m \wr \mathbb{Z}$ (with constants c_1 and c_2 possibly depending on the choice of the generator set). This can be seen as follows:

Assume we are given a Cayley graph G_S defined with respect to a generator set S. A natural candidate for the sequence of subgraphs G'_n in Theorem 19 would be $G_S(V_n)$, the subgraphs in G_S induced by V_n (V_n being the vertex set of a tetrahedron with depth n). Since this subgraph is not necessarily connected, to each vertex x in V_n we add a ball (in G_S) of some large, but fixed radius R, centered at x. In this way we get the vertex set V_n^R, which is connected in G_S. Because of fact (c) above, the adjacency Laplacian on the subgraph of G_{S_0} induced by the vertex set V_n^R is again bounded above by $2m(1-\cos\frac{\pi}{n})$. Now calculations similar to those in the proof of Theorem 3.2 in [115] show that the lowest eigenvalue of the adjacency Laplacian on the $G_S(V_n^R)$ has an upper bound of the form const$(1 - \cos\frac{\pi}{n})$. Thus Theorem 19 with $G'_n = G_S(V_n^R)$ proves the claim in Theorem 24 for the Cayley graph G_S.

Acknowledgements

It is a pleasure to thank D. Lenz, I. Nakić and F. Sobieczky for stimulating discussions related to the topics of this paper. Visits of the authors to the TU Graz were supported by the ESF through a Short Visit Grant and the Östereichischer Fonds zur Förderung der wissenschaftlichen Forschung.

References

[1] T. Adachi. A note on the Følner condition for amenability. *Nagoya Math. J.*, 131:67–74, 1993.

[2] M. Aizenman and D.J. Barsky. Sharpness of the phase transition in percolation models. *Comm. Math. Phys.*, 108(3):489–526, 1987.

[3] M. Aizenman, A. Elgart, S. Naboko, J.H. Schenker, and G. Stolz. Moment analysis for localization in random Schrödinger operators. *Invent. Math.*, 163(2):343–413, 2006. http://www.ma.utexas.edu/mp_arc/03-377.

[4] M. Aizenman and C.M. Newman. Tree graph inequalities and critical behavior in percolation models. *J. Statist. Phys.*, 36(1-2):107–143, 1984.

[5] M. Aizenman, J.H. Schenker, R.M. Friedrich, and D. Hundertmark. Finite-volume fractional-moment criteria for Anderson localization. *Comm. Math. Phys.*, 224(1):219–253, 2001.

[6] P. Anderson. Absence of diffusion in certain random lattices. *Phys. Rev.*, 109:1492, 1958.

[7] P. Antal. Enlargement of obstacles for the simple random walk. *Ann. Probab.*, 23(3):1061–1101, 1995.

[8] T. Antunović and I. Veselić. Sharpness of the phase transition and exponential decay of the subcritical cluster size for percolation on quasi-transitive graphs. *J. Statist. Phys.* 130(5):983–1009, http://www.arxiv.org/abs/0707.1089.

[9] T. Antunović and I. Veselić. Equality of Lifshitz and van Hove exponents on amenable Cayley graphs. http://www.arxiv.org/abs/0706.2844.

[10] L. Bartholdi, S. Cantat, T. Ceccherini-Silberstein, and P. de la Harpe. Estimates for simple random walks on fundamental groups of surfaces. *Colloq. Math.*, 72(1):173–193, 1997.

[11] L. Bartholdi and W. Woess. Spectral computations on lamplighter groups and Diestel-Leader graphs. *J. Fourier Anal. Appl.*, 11(2):175–202, 2005.

[12] H. Bass. The degree of polynomial growth of finitely generated nilpotent groups. *Proc. London Math. Soc.*, 25:603–614, 1972.

[13] I. Benjamini, R. Lyons, Y. Peres, and O. Schramm. Group-invariant percolation on graphs. *Geom. Funct. Anal.*, 9(1):29–66, 1999.

[14] I. Benjamini and O. Schramm. Percolation beyond $\mathbf{Z}^d$, many questions and a few answers. *Electron. Comm. Probab.*, 1(8):71–82, 1996. revised 1999 version: http://research.microsoft.com/~schramm/pyondrep/.

[15] M.S. Birman and M. Solomyak. On the negative discrete spectrum of a periodic elliptic operator in a waveguide-type domain, perturbed by a decaying potential. *J. Anal. Math.*, 83:337–391, 2001.

[16] M. Biskup and W. König. Long-time tails in the parabolic Anderson model with bounded potential. *Ann. Probab.*, 29(2):636–682, 2001.

[17] R. Carmona and J. Lacroix. *Spectral Theory of Random Schrödinger Operators.* Birkhäuser, Boston, 1990.

[18] J.T. Chayes, L. Chayes, J.R. Franz, J.P. Sethna, and S.A. Trugman. On the density of states for the quantum percolation problem. *J. Phys. A*, 19(18):L1173–L1177, 1986.

[19] F. Chung, A. Grigor′yan, and S.-T. Yau. Higher eigenvalues and isoperimetric inequalities on Riemannian manifolds and graphs. *Comm. Anal. Geom.*, 8(5):969–1026, 2000.

[20] F.R.K. Chung. *Spectral graph theory*, volume 92 of *CBMS Regional Conference Series in Mathematics.* Conference Board of the Mathematical Sciences, Washington, DC, 1997.

[21] Y. Colin de Verdière. *Spectres de graphes*, volume 4 of *Cours Spécialisés.* Société Mathématique de France, Paris, 1998.

[22] T. Coulhon and L. Saloff-Coste. Isopérimétrie pour les groupes et les variétés. *Rev. Mat. Iberoamericana*, 9(2):293–314, 1993.

[23] C. Coulson and G. Rushbrooke. Note on the method of molecular orbitals. *Proc. Cambridge Philos. Soc,*, 36:193–200, 1940.

[24] W. Craig and B. Simon. Log Hölder continuity of the integrated density of states for stochastic Jacobi matrices. *Commun. Math. Phys.*, 90:207–218, 1983.

[25] E.B. Davies and M. Plum. Spectral pollution. *IMA J. Numer. Anal.*, 24(3):417–438, 2004.

[26] P.-G. de Gennes, P. Lafore, and J. Millot. Amas accidentels dans les solutions solides désordonnées. *J. of Phys. and Chem. of Solids*, 11(1–2):105–110, 1959.

[27] P.-G. de Gennes, P. Lafore, and J. Millot. Sur un phénomène de propagation dans un milieu désordonné. *J. Phys. Rad.*, 20:624, 1959.

[28] F. Delyon and B. Souillard. Remark on the continuity of the density of states of ergodic finite-difference operators. *Commun. Math. Phys.*, 94:289–291, 1984.

[29] W. Dicks and T. Schick. The spectral measure of certain elements of the complex group ring of a wreath product. *Geom. Dedicata*, 93:121–134, 2002. www.arxiv.org/math/0107145.

[30] J. Dodziuk. de Rham-Hodge theory for L^2-cohomology of infinite coverings. *Topology*, 16(2):157–165, 1977.

[31] J. Dodziuk, N. Lenz, D. Peyerimhoff, T. Schick, and I. Veselić, editors. *L^2-Spectral Invariants and the Integrated Density of States*, volume 3 of *Oberwolfach Rep.*, 2006. http://www.mfo.de/programme/schedule/2006/08b/OWR_2006_09.pdf.

[32] J. Dodziuk, P. Linnell, V. Mathai, T. Schick, and S. Yates. Approximating L^2-invariants, and the Atiyah conjecture. *Comm. Pure Appl. Math.*, 56(7):839–873, 2003.

[33] J. Dodziuk and V. Mathai. Approximating L^2 invariants of amenable covering spaces: a combinatorial approach. *J. Funct. Anal.*, 154(2):359–378, 1998.

[34] H. Donnelly. On L^2-Betti numbers for abelian groups. *Canad. Math. Bull.*, 24(1):91–95, 1981.

[35] M.D. Donsker and S.R.S. Varadhan. Asymptotics for the Wiener sausage. *Commun. Pure Appl. Math.*, 28:525–565, 1975.

[36] M.D. Donsker and S.R.S. Varadhan. On the number of distinct sites visited by a random walk. *Comm. Pure Appl. Math.*, 32(6):721–747, 1979.

[37] F. Germinet and A. Klein. Explicit finite volume criteria for localization in continuous random media and applications. *Geom. Funct. Anal.*, 13(6):1201–1238, 2003. http://www.ma.utexas.edu/mp_arc/c/02/02-375.ps.gz.

[38] F. Germinet and A. Klein. A characterization of the Anderson metal-insulator transport transition. *Duke Math. J.*, 124(2):309–350, 2004. http://www.ma.utexas.edu/mp_arc/c/01/01-486.pdf.

[39] R.I. Grigorchuk, P. Linnell, T. Schick, and A. Żuk. On a question of Atiyah. *C.R. Acad. Sci. Paris Sér. I Math.*, 331(9):663–668, 2000.

[40] R.I. Grigorchuk and A. Żuk. The lamplighter group as a group generated by a 2-state automaton, and its spectrum. *Geom. Dedicata*, 87:209–244, 2001.

[41] G. Grimmett. *Percolation*, volume 321 of *Grundlehren der Mathematischen Wissenschaften*. Springer, Berlin, 1999.

[42] M. Gromov. Groups of polynomial growth and expanding maps. *Inst. Hautes Études Sci. Publ. Math.*, 53:53–73, 1981.

[43] M. Gromov and M.A. Shubin. von Neumann spectra near zero. *Geom. Funct. Anal.*, 1(4):375–404, 1991.

[44] J.W. Kantelhardt and A. Bunde. Electrons and fractons on percolation structures at criticality: Sublocalization and superlocalization. *Phys. Rev. E*, 56:6693–6701, 1997.

[45] J.W. Kantelhardt and A. Bunde. Extended fractons and localized phonons on percolation clusters. *Phys. Rev. Lett.*, 81:4907–4910, 1998.

[46] J.W. Kantelhardt and A. Bunde. Wave functions in the Anderson model and in the quantum percolation model: a comparison. *Ann. Phys. (8)*, 7(5-6):400–405, 1998.

[47] J.W. Kantelhardt and A. Bunde. Sublocalization, superlocalization, and violation of standard single-parameter scaling in the Anderson model. *Phys. Rev. B*, 66, 2002.

[48] H. Kesten. Symmetric random walks on groups. *Trans. Amer. Math. Soc.*, 92:336–354, 1959.

[49] H. Kesten. *Percolation theory for mathematicians*, volume 2 of *Progress in Probability and Statistics*. Birkhäuser, Boston, 1982.

[50] S. Kirkpatrick and T.P. Eggarter. Localized states of a binary alloy. *Phys. Rev. B*, 6:3598, 1972.

[51] W. Kirsch. Estimates on the difference between succeeding eigenvalues and Lifshitz tails for random Schrödinger operators. In *Stochastic processes – mathematics and*

physics, II (Bielefeld, 1985), volume 1250 of *Lecture Notes in Math.*, pages 138–151. Springer, Berlin, 1987.

[52] W. Kirsch and F. Martinelli. Large deviations and Lifshitz singularity of the integrated density of states of random Hamitonians. *Commun. Math. Phys.*, 89:27–40, 1983.

[53] W. Kirsch and B. Metzger. The integrated density of states for random Schrödinger operators. In *Spectral Theory and Mathematical Physics: A Festschrift in Honor of Barry Simon's* 60*th Birthday,* volume 76 of *Proceedings of Symposia in Pure Mathematics,* pages 649–698. AMS, 2007. http://www.arXiv.org/abs/math-ph/0608066.

[54] W. Kirsch and P. Müller. Spectral properties of the Laplacian on bond-percolation graphs. *Math. Zeit.*, 252(4):899–916, 2006. http://www.arXiv.org/abs/math-ph/0407047.

[55] W. Kirsch and B. Simon. Lifshitz tails for periodic plus random potentials. *J. Statist. Phys.*, 42:799–808, 1986.

[56] F. Klopp. Internal Lifshits tails for random perturbations of periodic Schrödinger operators. *Duke Math. J.*, 98(2):335–396, 1999.

[57] F. Klopp. Precise high energy asymptotics for the integrated density of states of an unbounded random Jacobi matrix. *Rev. Math. Phys.*, 12(4):575–620, 2000.

[58] F. Klopp and S. Nakamura. A note on Anderson localization for the random hopping model. *J. Math. Phys.*, 44(11):4975–4980, 2003.

[59] F. Klopp and L. Pastur. Lifshitz tails for random Schrödinger operators with negative singular Poisson potential. *Commun. Math. Phys.*, 206(1):57–103, 1999.

[60] S. Kondej and I. Veselić. Spectral gap of segments of periodic waveguides. *Lett. Math. Phys.*, 79(1):95–98, 2007. http://arxiv.org/abs//math-ph/0608005.

[61] P. Kuchment. Quantum graphs. II. Some spectral properties of quantum and combinatorial graphs. *J. Phys. A*, 38(22):4887–4900, 2005.

[62] P.A. Kuchment. On the Floquet theory of periodic difference equations. In *Geometrical and algebraical aspects in several complex variables* (*Cetraro,* 1989), volume 8 of *Sem. Conf.*, pages 201–209. EditEl, Rende, 1991.

[63] H. Kunz and B. Souillard. Sur le spectre des opérateur aux différence finies aléatoires. *Commun. Math. Phys.*, 78:201–246, 1980.

[64] D. Lenz, P. Müller, and I. Veselić. Uniform existence of the integrated density of states for models on $\mathbb{Z}^d$. *Positivity,* in press. http://arxiv.org/abs/math-ph/0607063.

[65] D. Lenz, N. Peyerimhoff, and I. Veselić. Von Neumann algebras, groupoids and the integrated density of states. arXiv.org/abs/math-ph/0203026, *Math. Phys. Anal. Geom.* http://dx.doi.org/10.1007/s11040-007-9019-2.

[66] D. Lenz and I. Veselić. Hamiltonians on discrete structures: jumps of the integrated density of states and uniform convergence. http://arxiv.org/abs/0709.2836.

[67] H. Leschke, P. Müller, and S. Warzel. A survey of rigorous results on random Schrödinger operators for amorphous solids. *Markov Process. Related Fields,* 9(4):729–760, 2003. http://www.arXiv.org/cond-mat/0210708).

[68] M. Levitin and E. Shargorodsky. Spectral pollution and second-order relative spectra for self-adjoint operators. *IMA J. Numer. Anal.*, 24(3):393–416, 2004.

[69] I.M. Lifshitz. Structure of the energy spectrum of the impurity bands in disordered solid solutions. *Sov. Phys. JETP*, 17:1159–1170, 1963. [Russian original: Zh. Eksp. Ter. Fiz. 44: 1723–1741].

[70] I.M. Lifshitz. The energy spectrum of disordered systems. *Adv. Phys.*, 13:483–536, 1964.

[71] I.M. Lifshitz. Energy spectrum and the quantum states of disordered condensed systems. *Sov. Phys. Usp.*, 7:549–573, 1965. [Russian original: Uspehi Fiz. Nauk 83: 617–663 (1964).]

[72] E. Lindenstrauss. Pointwise theorems for amenable groups. *Invent. Math.*, 146(2): 259–295, 2001.

[73] W. Lück. Approximating L^2-invariants by their finite-dimensional analogues. *Geom. Funct. Anal.*, 4(4):455–481, 1994.

[74] W. Lück. L^2-invariants of regular coverings of compact manifolds and CW-complexes. In *Handbook of geometric topology*, pages 735–817. North-Holland, Amsterdam, 2002. http://www.math.uni-muenster.de/u/lueck/publ/lueck/hand.pdf.

[75] W. Lück. *L^2-invariants: theory and applications to geometry and K-theory*, volume 44 of *Ergebnisse der Mathematik und ihrer Grenzgebiete. 3rd Series.* Springer-Verlag, Berlin, 2002.

[76] R. Lyons. Probability on trees and networks. (Cambridge University Press, Cambridge, in preparation). Written with assistance from Y. Peres. Current version available at http://php.indiana.edu/~rdlyons/.

[77] M. Men'shikov. Coincidence of critical points in percolation problems. *Sov. Math., Dokl.*, 33:856–859, 1986.

[78] M.V. Men′shikov, S.A. Molchanov, and A.F. Sidorenko. Percolation theory and some applications. In *Probability theory. Mathematical statistics. Theoretical cybernetics, Vol.* 24 (*Russian*), Itogi Nauki i Tekhniki, pages 53–110, i. Akad. Nauk SSSR Vsesoyuz. Inst. Nauchn. i Tekhn. Inform., Moscow, 1986. Translated in J. Soviet Math. **42** (1988), no. 4, 1766–1810, http://dx.doi.org/10.1007/BF01095508.

[79] B. Metzger. *Assymptotische Eigenschaften im Wechselspiel von Diffusion und Wellenausbreitung in zufälligen Medien.* PhD thesis, TU Chemnitz, 2005.

[80] G.A. Mezincescu. Bounds on the integrated density of electronic states for disordered Hamiltonians. *Phys. Rev. B*, 32:6272–6277, 1985.

[81] G.A. Mezincescu. Internal Lifshitz singularities of disordered finite-difference Schrödinger operators. *Commun. Math. Phys.*, 103:167–176, 1986.

[82] G.A. Mezincescu. Lifschitz singularities for periodic operators plus random potentials. *J. Statist. Phys.*, 49(5-6):1181–1190, 1987.

[83] P. Müller and P. Stollmann. Spectral asymptotics of the Laplacian on supercritical bond-percolation graphs. *J. Funct. Anal.*, 252(1):233–246, 2007. http://www.arxiv.org/math-ph/0506053.

[84] T. Nagnibeda. An upper bound for the spectral radius of a random walk on surface groups. *Zap. Nauchn. Sem. S.-Peterburg. Otdel. Mat. Inst. Steklov.* (*POMI*), 240 (Teor. Predst. Din. Sist. Komb. i Algoritm. Metody. 2):154–165, 293–294, 1997.

[85] S. Nakamura. Lifshitz tail for 2D discrete Schrödinger operator with random magnetic field. *Ann. Henri Poincaré*, 1(5):823–835, 2000.

[86] S. Nakao. On the spectral distribution of the Schrödinger operator with random potential. *Japan. J. Math. (N.S.)*, 3(1):111–139, 1977.

[87] S.P. Novikov and M.A. Shubin. Morse inequalities and von Neumann invariants of non-simply connected manifolds. *Uspekhi Matem. Nauk*, 41(5):222, 1986. (In Russian.)

[88] S.P. Novikov and M.A. Shubin. Morse inequalities and von Neumann II_1-factors. *Dokl. Akad. Nauk SSSR*, 289(2):289–292, 1986.

[89] S.-I. Oguni. The secondary Novikov-Shubin invariants of groups and quasi-isometry. *J. Math. Soc. Japan*, 59(1):223–237, 2007. http://www.math.kyoto-u.ac.jp/preprint/preprint2005.html.

[90] P. Pansu. Croissance des boules et des géodésiques fermées dans les nilvariétés. *Ergodic Theory Dynam. Systems*, 3(3):415–445, 1983.

[91] L.A. Pastur. Selfaverageability of the number of states of the Schrödinger equation with a random potential. *Mat. Fiz. i Funkcional. Anal.*, (Vyp. 2):111–116, 238, 1971.

[92] L.A. Pastur. Behaviour of some Wiener integrals as $t \to \infty$ and the density of states of the Schrödinger equation with a random potential. *Teor. Mat. Fiz.*, 32:88–95, 1977.

[93] L.A. Pastur. Spectral properties of disordered systems in the one-body approximation. *Commun. Math. Phys.*, 75:179–196, 1980.

[94] L.A. Pastur and A.L. Figotin. *Spectra of Random and Almost-Periodic Operators.* Springer Verlag, Berlin, 1992.

[95] N. Peyerimhoff and I. Veselić. Integrated density of states for ergodic random Schrödinger operators on manifolds. *Geom. Dedicata*, 91(1):117–135, 2002.

[96] A. Procacci and B. Scoppola. Infinite graphs with a nontrivial bond percolation threshold: some sufficient conditions. *J. Statist. Phys.*, 115(3-4):1113–1127, 2004.

[97] Y. Shapir, A. Aharony, and A.B. Harris. Localization and quantum percolation. *Phys. Rev. Lett.*, 49(7):486–489, 1982.

[98] B. Simon. Lifschitz tails for the Anderson model. *J. Statist. Phys.*, 38:65–76, 1985.

[99] B. Simon. Internal Lifschitz tails. *J. Statist. Phys.*, 46(5-6):911–918, 1987.

[100] F. Sobieczky. An interlacing technique for spectra of random walks and its application to finite percolation clusters. www.arXiv.org/math/0504518.

[101] P. Stollmann. *Caught by disorder: A Course on Bound States in Random Media*, volume 20 of *Progress in Mathematical Physics*. Birkhäuser, 2001.

[102] M.A. Shubin. Almost periodic functions and partial differential operators. *Uspehi Mat. Nauk*, 33(2(200)):3–47, 247, 1978.

[103] M.A. Shubin. Spectral theory and the index of elliptic operators with almost-periodic coefficients. *Uspekhi Mat. Nauk*, 34(2(206)):95–135, 1979. [English translation: Russ. Math. Surveys, **34**:109-157, 1979.]

[104] T. Sunada. Fundamental groups and Laplacians. In *Geometry and analysis on manifolds (Katata/Kyoto*, 1987), pages 248–277. Springer, Berlin, 1988.

[105] A.-S. Sznitman. *Brownian motion, obstacles and random media.* Springer-Verlag, Berlin, 1998.

[106] G. Temple. The theory of Rayleigh's principle as applied to continuous systems. *Proc. Roy. Soc. London A*, 119:276–293, 1028.

[107] W. Thirring. *Lehrbuch der mathematischen Physik. Band* 3. Springer-Verlag, Vienna, second edition, 1994. Quantenmechanik von Atomen und Molekülen.

[108] L. van den Dries and A. Wilkie. Gromov's theorem on groups of polynomial growth and elementary logic. *J. Algebra*, 89:349–374, 1984.

[109] R. van der Hofstad, W. König, and P. Mörters. The universality classes in the parabolic Anderson model. *Comm. Math. Phys.*, 267(2):307–353, 2006.

[110] I. Veselić. Integrated density of states and Wegner estimates for random Schrödinger operators. *Contemp. Math.*, 340:98–184, 2004. http://arXiv.org/math-ph/0307062.

[111] I. Veselić. Quantum site percolation on amenable graphs. In *Proceedings of the Conference on Applied Mathematics and Scientific Computing*, pages 317–328, Dordrecht, 2005. Springer. http://arXiv.org/math-ph/0308041.

[112] I. Veselić. Spectral analysis of percolation Hamiltonians. *Math. Ann.*, 331(4):841–865, 2005. http://arXiv.org/math-ph/0405006.

[113] I. Veselić. *Existence and regularity properties of the integrated density of states of random Schrödinger Operators*, Habilitation thesis, TU Chemnitz, 2006. Lecture Notes in Mathematics 1917, Springer, 2007.

[114] I. Veselić. Spectral properties of Anderson-percolation Hamiltonians. In *Oberwolfach Rep.* [31], pages 545–547.
http://www.mfo.de/programme/schedule/2006/08b/OWR_2006_09.pdf.

[115] W. Woess. *Random walks on infinite graphs and groups*, volume 138 of *Cambridge Tracts in Mathematics*. Cambridge University Press, Cambridge, 2000.

[116] J.A. Wolf. Growth of finitely generated solvable groups and curvature of Riemaniann manifolds. *J. Differential Geometry*, 2:421–446, 1968.

[117] A. Żuk. A remark on the norm of a random walk on surface groups. *Colloq. Math.*, 72(1):195–206, 1997.

Tonći Antunović and Ivan Veselić
Emmy-Noether-Programme of the Deutsche Forschungsgemeinschaft
& Fakultät für Mathematik
09107 TU Chemnitz, Germany
URL: www.tu-chemnitz.de/mathematik/schroedinger/members.php

Operator Theory:
Advances and Applications, Vol. 186, 31–48

On Inverse Problems for Finite Trees

B.M. Brown and R. Weikard

Abstract. In this paper two classical theorems by Levinson and Marchenko for the inverse problem of the Schrödinger equation on a compact interval are extended to finite trees. Specifically, (1) the Dirichlet eigenvalues and the Neumann data of the eigenfunctions determine the potential uniquely (a Levinson-type result) and (2) the Dirichlet eigenvalues and a set of generalized norming constants determine the potential uniquely (a Marchenko-type result).

Mathematics Subject Classification (2000). Primary 34A55, Secondary: 05C05.

Keywords. Dirichlet-to-Neumann map; inverse spectral problems; trees.

1. Introduction

Some 60 years ago Borg, Levinson, and Marchenko established the now famous inverse spectral theory for the Schrödinger equation on a compact interval. Denote the solution of the initial value problem $-y'' + qy = \lambda y$, $y(0) = 0$ and $y'(0) = 1$ on the interval $[0, 1]$ by $s(\lambda, \cdot)$. Then the real integrable potential q is uniquely determined by any of the following sets of data:

- Borg [5] (1946): The Dirichlet-Dirichlet eigenvalues and the Dirichlet-Neumann eigenvalues, i.e., the zeros of $s(\cdot, 1)$ and of $s'(\cdot, 1)$.
- Levinson [17] (1949): The Dirichlet-Dirichlet eigenvalues (denoted by λ_n) and the Neumann data of the associated eigenfunctions, i.e., the numbers $s'(\lambda_n, 1)$ (recall that $s'(\lambda_n, 0) = 1$).
- Marchenko [18] (1950): The Dirichlet-Dirichlet eigenvalues and the norming constants $\int_0^1 s(\lambda_n, t)^2 dt$ of the associated eigenfunctions.

Since the 1980s spectral problems on graphs and trees have also been investigated (we are interested here solely in so-called metric trees where edges are homeomorphic to real intervals and may support potentials). We refer the reader

This material is based upon work supported by the National Science Foundation under Grant No. DMS-0304280.

to the excellent surveys [13] and [14] by Kuchment for an overview of these developments and their applications. Investigation of inverse problems on such graphs and trees are not quite so numerous. Without a claim to completeness we list here the works of Gerasimenko [11]; Curtis and Morrow [9], [10]; Carlson [7]; Pivovarchik [21], [20]; Kurasov and Stenberg [15]; Belishev [2]; Brown and Weikard [6]; Harmer [12]; Kurasov and Nowaczyk [16]; Yurko [22]; Belishev and Vakulenko [3]; and Avdonin and Kurasov [1]. We emphasize that the results in [2], [22], and [1] are particularly close to ours.

In this paper we address generalizations of the above mentioned results of Levinson and Marchenko to the case of finite trees with r edges and n_0 boundary vertices. The generalization of Borg's result has been established recently by Yurko [22]: spectra for n_0 (specific) boundary conditions are sufficient to determine the potential on the tree.

In this paper we focus for simplicity on real and bounded potentials on trees whose edge lengths are one. Our methods generalize to the case of integrable complex potentials on trees with varying edge lengths. We plan to address such generalizations in a subsequent paper.

Under the given circumstances the Dirichlet eigenvalues, which we denote by $\lambda_1, \lambda_2, \ldots$, are real and discrete since they are zeros of an entire function which can not vanish away from the real axis. We use the letter Σ to signify the set of all Dirichlet eigenvalues. Geometric multiplicities of Dirichlet eigenvalues may be larger than one. The multiplicity of the eigenvalue E is denoted $\mu(E)$.

The generalization of Levinson's result reads as follows.

Theorem 1.1. *Let q be a real-valued bounded potential supported on a finite metric tree whose edge lengths are one. Then the Dirichlet eigenvalues, their multiplicities, and the Neumann data of an orthonormal set of Dirichlet eigenfunctions uniquely determine the potential almost everywhere on the tree.*

We remark here that eigenfunctions are not automatically orthonormal since eigenspaces may have dimensions larger than one. Theorem 1.1 will be proved in Section 6.

A particularly important role is played in this paper by the Weyl solutions. They are uniquely defined away from the Dirichlet eigenvalues by requiring that they satisfy homogeneous Dirichlet boundary conditions at all but one boundary vertex where they assume the value 1. At the Dirichlet eigenvalues the Weyl solutions cease to exist but multiplying them with the "minimal function" $\chi(\lambda) = \prod_{E\in\Sigma}^{\infty}(1 - \lambda/E)$ (as opposed to the characteristic function which takes the multiplicities of the eigenvalues into account) gives globally defined functions $\omega(k, \lambda, \cdot)$ where $k \in \{1, \ldots, n_0\}$ signifies the boundary vertex at which the function value is prescribed to be $\chi(\lambda)$. If E is a Dirichlet eigenvalue of multiplicity $\mu(E)$ then there will be $\mu(E)$ of the functions $\omega(k, \lambda, \cdot)$ which are linearly independent. Let $K(E) \subset \{1, \ldots, n_0\}$ be a maximal set such that $\{\omega(k, \lambda, \cdot) : k \in K(E)\}$ is linearly independent. We also introduce the quantities $N_{j,k}(E)$ which are the inner products of the functions $\omega(j, E, \cdot)$ and $\omega(k, E, \cdot)$. For $j = k$ these are norming constants of eigenfunctions.

We can now formulate the generalization of Marchenko's result.

Theorem 1.2. *Let q be a real-valued bounded potential supported on a finite metric tree whose edge lengths are one. Then the Dirichlet eigenvalues, their multiplicities, and the quantities $N_{j,k}(E)$, $j \in \{1, \dots, n_0\}$, $k \in K(E)$, $E \in \Sigma$ uniquely determine the potential almost everywhere on the tree.*

This theorem will be proved in Section 7.

The paper is organized as follows. Section 2 gives formal definitions of trees, interface conditions, and operators under consideration. It also provides basic results on initial value and Dirichlet boundary value problems. Weyl solutions and Green's function play (as is to be expected) an important role. They are studied in Section 3 and Section 4, respectively. The Dirichlet-to-Neumann map is introduced in Section 5. Sections 6 and 7 are devoted to the proofs of Theorems 1.1 and 1.2, respectively.

2. Preliminaries

2.1. Trees

A finite tree is given by a Hausdorff space T and a set of r homeomorphisms $\epsilon_j : [0, 1] \to T$, $j = 1, \dots, r$, such that the following conditions are met:

1. $T = \bigcup_{j=1}^{r} \{\epsilon_j(t) : t \in [0, 1]\}$.
2. If $\epsilon_j(t) = \epsilon_k(s)$ and $j \neq k$ then $t, s \in \{0, 1\}$.
3. T is simply connected.

In this paper we will be concerned only with finite trees and we will always mean a finite tree when we speak of a tree.

The set $V = \{\epsilon_1(0), \epsilon_1(1), \dots, \epsilon_r(0), \epsilon_r(1)\}$ has precisely $r+1$ elements called vertices of the tree. The homeomorphisms ϵ_j are called edges of the tree. We say a vertex v belongs to an edge ϵ_j or that edge ϵ_j is attached to v if $v = \epsilon_j(0)$, the initial vertex of ϵ_j, or $v = \epsilon_j(1)$, the terminal vertex of ϵ_j. A vertex is called a boundary vertex if it belongs to only one edge. Such an edge will be called a boundary edge. A vertex which belongs to several edges is called an internal vertex. An edge both of whose endpoints are internal vertices is called an internal edge.

The fact that edges are homeomorphic to the metric space $[0, 1]$ turns the space T in a natural way into a metric space.

Since the interval $[0, 1]$ has an orientation associated with it, so does each of the edges. However, we are not interested in the orientations of the edges, but only in the metric structure provided by the homeomorphisms ϵ_j. Thus, given a finite tree we obtain another tree by replacing the homeomorphism $t \mapsto \epsilon_j(t)$ by $t \mapsto \epsilon_j(1-t)$ for some (or none or all) of the indices j. The new tree is, of course, still associated with the Hausdorff space T and the trees, so related, are called equivalent. This is an equivalence relation and we will henceforth choose the labels and orientations for the edges as convenience suggests. Often times we will call the Hausdorff space T a tree, assuming a tacit understanding of its metric structure.

We take here (for the most part) the point of view that all boundary vertices play identical roles. Assuming there are n_0 boundary vertices we will then assign labels $1, \dots, n_0$ to the boundary vertices and the corresponding boundary edges. We assume also that the boundary edges are oriented so that $v_j = \epsilon_j(0)$ for $j = 1, \dots, n_0$.

However, in some intermediate results one boundary vertex of the tree is singled out. In this case it is convenient to assign labels and orientations in the following way. The special boundary vertex is called the root and is designated as v_0. All other boundary vertices are called branch tips. The edge attached to the root, denoted by ϵ_0, will be called the trunk. All edges are oriented so that their initial vertex is closer to the root than their terminal vertex. In particular, $\epsilon_0(0) = v_0$, the root of the tree. The edges other than the trunk attached to the terminal vertex of the trunk $v_1 = \epsilon_0(1)$ are called limbs. Each one of them is the trunk of a subtree with root v_1. A tree labeled and oriented this way will be called a rooted tree.

We emphasize that all results obtained are independent of the particular way in which edges and vertices are labeled and in which edges are oriented; these designations only serve to communicate proofs more easily.

2.2. The interface conditions

A function y defined on T may be represented as $\vec{y} = (y_1, \dots, y_r)^\top$ where $y_j(t) = y(\epsilon_j(t))$. We say that y is integrable on T (or square integrable on T or $y \in L^p(T)$) if the y_j are integrable on $[0,1]$ (or square integrable on $[0,1]$ or $y_j \in L^p([0,1])$) for $j = 1, \dots, r$.

We define $\mathcal{C}$ to be the set of all functions y defined on T which satisfy the following conditions:

1. For each j the functions y_j and y_j' are absolutely continuous on $[0,1]$.
2. y is continuous on T.
3. For each internal vertex v the Kirchhoff condition

$$\sum_{\epsilon_j(1)=v} y_j'(1) - \sum_{\epsilon_j(0)=v} y_j'(0) = 0$$

holds.

Conditions 2 and 3 are called interface conditions. Note that $\mathcal{C}$ is independent of the orientation or the labeling of the edges.

In order to deal conveniently with the interface conditions we introduce[1] the operator

$$\mathfrak{I} = E_0\mathfrak{E}_0 + E_1\mathfrak{E}_1 + D_0\mathfrak{D}_0 + D_1\mathfrak{D}_1,$$

[1]This notation is different from the one we used in [6]

where E_0, E_1, D_0 and D_1 are certain $(2r - n_0) \times r$ matrices whose entries are 0 or ± 1 and $\mathfrak{E}_0$, etc. are evaluation operators defined by $\mathfrak{E}_0 \vec{y} = \vec{y}(0)$, $\mathfrak{E}_1 \vec{y} = \vec{y}(1)$, $\mathfrak{D}_0 \vec{y} = \vec{y}\,'(0)$, and $\mathfrak{D}_1 \vec{y} = \vec{y}\,'(1)$. Thus y satisfies the interface conditions if and only if $\mathfrak{I} \vec{y} = 0$.

2.3. The differential equations

If q is an integrable function on T define $Q = \operatorname{diag}(q_1, \dots, q_r)$. We will consider the differential expression $\vec{y} \mapsto -\vec{y}\,'' + Q\vec{y}$ and the differential equation $-\vec{y}\,'' + Q\vec{y} = \lambda \vec{y}$. Define

$$\mathcal{D} = \{y \in \mathcal{C} : -y_j'' + q_j y_j \in L^2([0,1])\}.$$

We now define the operator $L : \mathcal{D} \to L^2(T)$ by $(Ly)(\epsilon_j(t)) = -y_j''(t) + q_j(t) y_j(t)$. Again, this definition is independent of the orientation or the labeling of the edges.

If $y \in \mathcal{D}$ satisfies $Ly = \lambda y$ we will call it a solution of $Ly = \lambda y$ signifying that both the differential equations and the interface conditions are satisfied. Functions solving the differential equations form a $2r$-dimensional vector space. Since there are $2r - n_0$ interface conditions it is reasonable to expect that the space of solutions of $Ly = \lambda y$ is n_0-dimensional. This fact will be proved below.

Denote by $c_j(\lambda, \cdot)$ and $s_j(\lambda, \cdot)$ the basis of solutions of $-y_j'' + q_j y_j = \lambda y_j$ defined by initial conditions $c_j(\lambda, 0) = s_j'(\lambda, 0) = 1$ and $c_j'(\lambda, 0) = s_j(\lambda, 0) = 0$. We collect these functions in the $r \times r$-matrices $C(\lambda, t) = \operatorname{diag}(c_1(\lambda, t), \dots, c_r(\lambda, t))$ and $S(\lambda, t) = \operatorname{diag}(s_1(\lambda, t), \dots, s_r(\lambda, t))$. A function y satisfying the differential equations may now be expressed as $\vec{y} = (C(\lambda, \cdot), S(\lambda, \cdot))\xi$ for an appropriate $\xi \in \mathbb{C}^{2r}$.

In particular, the function $\vec{y} = (C(\lambda, \cdot), S(\lambda, \cdot))\xi$ satisfies the interface conditions (and hence is a solution of $Ly = \lambda y$) precisely if ξ is in the kernel of the $(2r - n_0) \times 2r$-matrix

$$J(\lambda) = \mathfrak{I}(C(\lambda, \cdot), S(\lambda, \cdot)).$$

2.4. Initial value problems

Initial value problems do, in general, not have unique solutions on trees. This causes the main differences in the treatment of inverse problems on trees when compared to intervals. However, it is still useful to investigate the set of all solutions for the initial value problem.

One of the boundary vertices is being singled out as the "initial point". Thus, it is now convenient to treat the tree as a rooted tree.

Lemma 2.1. *The initial value problem $Ly = \lambda y$, $y_0(0) = a$, and $y_0'(0) = b$ has a solution for any choice of $a, b \in \mathbb{C}$.*

Proof. This is obvious when the tree has only one internal vertex and follows by induction over the number of internal vertices for general trees. In fact, if there are k limbs attached to v_1 (i.e., $k+1$ edges), then there are k subtrees for which we know the existence of solutions of the initial value problem by induction hypothesis. For subtree j, where $1 \le j \le k$, we use initial conditions $a_j = ac_0(\lambda, 1) + bs_0(\lambda, 1)$ and some value for b_j. These provide then for a solution on the full tree provided that $b_1 + \dots + b_k = ac_0'(\lambda, 1) + bs_0'(\lambda, 1)$. □

Corollary 2.2. *Suppose T has n_0 boundary vertices. The vector space of solutions of $Ly = \lambda y$ satisfying homogeneous initial conditions $y_0(0) = 0$ and $y_0'(0) = 0$ has dimension $n_0 - 2$.*

Proof. Again, this holds obviously for a tree with one internal vertex. Suppose it is true for trees with less than ℓ internal vertices and consider a tree with exactly ℓ internal vertices. Suppose the tree has k limbs. On each of the k subtrees which start at the end of the trunk there is a solution of the initial value problem for function value 0 and derivative 1. These give rise to $k-1$ linearly independent solutions of the homogeneous initial value problem on the full tree. Now, given any solution of the homogeneous initial value problem, one can subtract a suitable linear combination of the $k-1$ basis functions just constructed to obtain a solution on the full tree which is identically zero on the trunk and on the k limbs. It is thus a linear combination of all possible basis functions for each of the homogeneous initial value problems on the subtrees. If subtree j has $n_j + 1$ boundary vertices then we have, by the induction hypothesis, $n_j - 1$ such basis functions. Since $n_0 = 1 + \sum_{j=1}^k n_j$ we get for the total number of basis functions on the full tree $k - 1 + \sum_{j=1}^k (n_j - 1) = n_0 - 2$. □

These observations give us now immediately the following theorem.

Theorem 2.3. *Suppose T has r edges and n_0 boundary vertices. Then the vector space of solutions of $Ly = \lambda y$ has dimension n_0. Moreover, the matrix $J(\lambda)$ has full rank $2r - n_0$ for every complex λ.*

Proof. There are linearly independent solutions which are equal to $c_0(\lambda, \cdot)$ and $s_0(\lambda, \cdot)$, respectively, when restricted to the trunk of the tree. Let y be any solution and subtract a suitable combination of the functions just mentioned so that the resulting function satisfies homogeneous initial conditions. That function may expressed as a linear combination of the $n_0 - 2$ basis functions constructed in the previous corollary.

The last statement follows now from the fundamental theorem of linear algebra. □

2.5. The Dirichlet boundary value problem

Given a vector $f = (f_1, \dots, f_{n_0})^\top$ in $\mathbb{C}^{n_0}$ we are looking for a solution of $Ly = \lambda y$ satisfying the nonhomogeneous Dirichlet boundary conditions $y(v_j) = f_j$ when $v_1, \dots, v_{n_0}$ denote the boundary vertices. These conditions are expressed as $\mathfrak{A}\vec{y} = f$ where $\mathfrak{A} = A_0 \mathfrak{E}_0$ and $A_0 = (I_{n_0 \times n_0}, 0_{n_0 \times (r-n_0)})$ assuming that the boundary edges are oriented such that $v_j = \epsilon_j(0)$ for $j = 1, \dots, n_0$. Here $I_{n\times n}$ and $0_{n\times m}$ denote the identity matrix and the zero matrix of the given dimensions, respectively. We might subsequently drop the subscripts if the dimensions are clear from the context. We also note that the $n_0 \times 2r$-matrix $\mathfrak{A}(C(\lambda, \cdot), S(\lambda, \cdot)) = (I_{n_0\times n_0}, 0_{n_0\times(2r-n_0)})$.

Finally, we introduce the $2r \times 2r$ matrix

$$M(\lambda) = \begin{pmatrix} \mathfrak{A} \\ \mathfrak{J} \end{pmatrix} (C(\lambda, \cdot), S(\lambda, \cdot)) = \begin{pmatrix} I & 0 \\ R(\lambda) & P(\lambda) \end{pmatrix}$$

where R represents the first n_0 columns and P the last $2r - n_0$ columns of the matrix J introduced in Section 2.3. Thus solutions of the equation $M(\lambda)\xi = (f, 0)^\top$ provide solutions of $Ly = \lambda y$ satisfying the given boundary conditions.

The entries of M are entire functions. Therefore the zeros of the determinant of M are isolated unless it is identically equal to zero.

This latter possibility can be ruled out most easily under our assumption of a real potential since the zeros of the determinant are eigenvalues of the Dirichlet operator L_D, the restriction of L to the set of all $y \in \mathcal{D}$ which satisfy homogeneous Dirichlet conditions at the boundary vertices. An integration by parts, using the interface and boundary conditions, shows, that the operator L_D is self-adjoint so that its eigenvalues are real.

Thus, the equation $Ly = \lambda y$ has a unique solution satisfying given (possibly nonhomogeneous) Dirichlet boundary conditions unless λ is one of the countably many real eigenvalues of L_D.

The geometric multiplicity of a Dirichlet eigenvalue is strictly less than n_0 for there is at least one solution of $Ly = \lambda y$ which does not satisfy homogeneous Dirichlet conditions. Since the problem is self-adjoint the geometric multiplicity of a Dirichlet eigenvalue equals its algebraic multiplicity. The algebraic multiplicity in turn equals the order of the eigenvalue as a zero of $\det M$. This may be seen by a generalization of Naimark's argument for a differential equation with scalar coefficients to one with matrix coefficients.

3. Weyl solutions

A solution ψ of $Ly = \lambda y$ is called a Weyl solution if ψ assumes the value one at precisely one of the boundary vertices and the value zero at each of the others. With one of the boundary vertices singled out it is convenient, in this section, to treat the tree as a rooted tree (see Section 2.1). If x and x' are two points in T we denote their distance by $d(x, x')$ (recall that T is a metric space). The number $h = \max\{d(x, v_0) : x \in T\}$ is called the height of the tree with respect to the root v_0. (The height of a tree depends on which boundary vertex is designated as root).

Lemma 3.1. *Let $\psi(\lambda, \cdot)$ be the Weyl solution for a tree T which assumes the value one at the root and the value zero at each of the branch tips. Then*

$$\psi_0'(\lambda, 0) = -\sqrt{-\lambda} + o(1)$$

as λ tends to infinity on the negative real axis. Furthermore, if s is the label of a boundary edge leading to branch tip $\epsilon_s(1)$, then

$$\psi_s'(\lambda, 1) = -2\sqrt{-\lambda}\exp(-d(v_0, \epsilon_s(1))\sqrt{-\lambda})(1 + o(1))$$

again as λ tends to infinity on the negative real axis.

Proof. The proof is by induction on the height of the tree. If the tree has height one, then $\psi_0'(\lambda, 0) = m(\lambda)$ is the Titchmarsh-Weyl m-function for the problem (recall that $\psi_0(\lambda, 0) = 1$). It is well known that $m(\lambda) = iz + o(1)$ as $z = i\sqrt{-\lambda}$

tends to infinity on the positive imaginary axis. Also for the branch tip $\epsilon_0(1)$ we have $\psi_0'(\lambda,1) = -1/s_0(\lambda,1) = 2ize^{iz} + O(e^{iz} + ze^{3iz})$.

Next assume the validity of our claim for all trees of height at most n and let T be a tree of height $n+1$. Let $v_1 = \epsilon_0(1)$. Then v_1 is the root of several, say k, subtrees, whose height is at most n. These will be labeled $T_1, \dots, T_k$ and their trunks, viewed as limbs of the full tree, are correspondingly labeled $\epsilon_1, \dots, \epsilon_k$. The Weyl solution for the tree T_j is denoted by $\psi(j,\lambda,\cdot)$. Note that the Weyl solution $\psi(\lambda,\cdot)$ for the full tree, restricted to T_j, is a multiple of $\psi(j,\lambda,\cdot)$, in fact

$$\psi_\sigma(\lambda,\cdot) = \psi(\lambda,v_1)\psi_s(j,\lambda,\cdot) \tag{3.1}$$

when s denotes the label of an edge of T_j and σ denotes the label of the same edge when viewed as an edge of T. In particular,

$$\psi_j(\lambda,\cdot) = \psi(\lambda,v_1)\psi_0(j,\lambda,\cdot).$$

Now we will determine $\psi_0(\lambda,\cdot)$, the Weyl solution for T on its trunk, by solving a Riccati equation. To this end we define

$$\mu(\lambda,\cdot) = \frac{\psi_0'(\lambda,\cdot)}{\psi_0(\lambda,\cdot)}.$$

Then $\mu(\lambda,\cdot)$ satisfies the differential equation

$$\mu'(\lambda,x) + \mu(\lambda,x)^2 + \lambda - q_0(x) = 0$$

and the initial condition

$$\mu(\lambda,1) = \frac{\sum_{j=1}^k \psi_j'(\lambda,0)}{\psi(\lambda,v_1)} = \sum_{j=1}^k \psi_0'(j,\lambda,0) = ikz + o(1).$$

This problem will be solved by emulating Bennewitz's approach in [4]. The first step is to find the solution $\mu_0(\lambda,\cdot)$ for vanishing q_0 but with the correct initial condition. Thus

$$\mu_0(\lambda,x) = iz + \frac{2iz(\mu(\lambda,1)-iz)}{(iz+\mu(\lambda,1))\exp(2iz(x-1)) + iz - \mu(\lambda,1)}.$$

Now recall that $q_0 \in L^1([0,1])$ and define

$$\mu_1(\lambda,x) = -\int_x^1 e^{2\int_x^t \mu_0(\lambda,u)du} q_0(t)dt$$

and

$$\mu_n(\lambda,x) = \int_x^1 e^{2\int_x^t \sum_{j=0}^{n-1}\mu_j(\lambda,u)du} \mu_{n-1}(\lambda,t)^2 dt, \quad n = 2,3,\dots.$$

Then

$$\begin{aligned} \mu_1'(\lambda,x) &= q_0(x) - 2\mu_0(\lambda,x)\mu_1(\lambda,x), \\ \mu_n'(\lambda,x) &= -\mu_{n-1}(\lambda,x)^2 - 2\mu_n(\lambda,x)\sum_{j=0}^{n-1}\mu_j(\lambda,x), \quad n = 2,3,\dots, \end{aligned}$$

and $\mu_n(\lambda,1) = 0$ for all $n \in \mathbb{N}$.

Thus, assuming uniform convergence of $\sum_{n=0}^{\infty}\mu_n$ and $\sum_{n=0}^{\infty}\mu_n'$, we have (after some algebra) that the series $\sum_{n=0}^{\infty}\mu_n(\lambda,\cdot)$ satisfies the same initial value problem (the Riccati equation and the initial condition at $x=1$) as $\mu(\lambda,\cdot)$ and hence is equal to it, i.e.,

$$\mu(\lambda,x)=\sum_{n=0}^{\infty}\mu_n(\lambda,x). \tag{3.2}$$

In order to investigate convergence we first realize that

$$\frac{\mu_0(\lambda,x)}{iz}=1+o(1/z)$$

for $k=1$ and

$$\frac{\mu_0(\lambda,x)}{iz}=1+\frac{2\mathrm{e}^{2iz(1-x)}(k-1)}{k+1-(k-1)\mathrm{e}^{2iz(1-x)}}(1+o(1/z))$$

for $k>1$. Since iz is negative we may estimate in either case that

$$\operatorname{Re}(\mu_0(\lambda,x))\le 3iz/4=-3\operatorname{Im}(z)/4$$

for sufficiently large z. Thus

$$|\mu_1(\lambda,x)|\le\int_x^1\mathrm{e}^{-3\operatorname{Im}(z)(t-x)/2}|q_0(t)|dt.$$

Given $\varepsilon>0$, there are complex numbers $\alpha_1,\dots,\alpha_N$ and intervals $I_1,\dots,I_N$ such that $\tilde q_0=\sum_{j=1}^{N}\alpha_j\chi_{I_j}$ is a step function on $[0,1]$ with $\|q_0-\tilde q_0\|\le\varepsilon$. Then

$$|\mu_1(\lambda,x)|\le\varepsilon+\sum_{j=1}^{N}|\alpha_j|\int_x^1\mathrm{e}^{-3\operatorname{Im}(z)(t-x)/2}\chi_{I_j}(t)dt\le\varepsilon+\sum_{j=1}^{N}\frac{2|\alpha_j|}{3\operatorname{Im}(z)}.$$

This estimate holds regardless of x and proves that $|\mu_1(\lambda,x)|$ tends to zero uniformly in x as $\operatorname{Im}(z)$ tends to infinity.

Now let

$$a(\lambda,x)=\sup\{|\mu_1(\lambda,t)|:x\le t\le 1\}$$

and assume that z is large enough so that $\operatorname{Im}(z)\ge 1$ and $a(\lambda,x)/\operatorname{Im}z\le 1/2$.

One then shows by induction that

$$|\mu_n(\lambda,x)|\le\left(\frac{a(\lambda,x)}{\operatorname{Im}(z)}\right)^{2^{n-1}}\operatorname{Im}(z)$$

using that $a(\lambda,\cdot)$ is non-increasing and that

$$\sum_{j=1}^{n-1}|\mu_j(\lambda,x)|\le\operatorname{Im}(z)\sum_{j=1}^{n-1}\left(\frac{a(\lambda,x)}{\operatorname{Im}(z)}\right)^{2^{j-1}}\le\operatorname{Im}(z)\sum_{j=1}^{\infty}\left(\frac{a(\lambda,x)}{\operatorname{Im}(z)}\right)^{j}\le 2a(\lambda,0).$$

These estimates show also that $\sum_{n=0}^{\infty}\mu_n(\lambda,\cdot)$ may be differentiated term by term, thus proving the validity of equation (3.2).

The Weyl solution on the trunk of T is now given as

$$\psi_0(\lambda, x) = \exp\left(\int_0^x \mu(\lambda, t)dt\right).$$

In particular,

$$\psi_0'(\lambda, 0) = \mu(\lambda, 0) = iz + o(1) \tag{3.3}$$

and

$$\psi(\lambda, v_1) = \psi_0(\lambda, 1) = \exp\left(\int_0^1 \mu(\lambda, t)dt\right) = \mathrm{e}^{iz}(1 + o(1)).$$

Finally, consider a boundary edge s on tree T_j, different from the trunk of T_j, and denote the distance of the corresponding boundary vertex from the root of T_j by N. Then, according to our induction hypothesis,

$$\psi_s'(j, \lambda, 1) = 2iz\mathrm{e}^{izN}(1 + o(1)).$$

This edge is also a boundary edge of T with label σ, say. Using (3.1) we find

$$\psi_\sigma'(\lambda, 1) = \psi(\lambda, v_1)\psi_s'(j, \lambda, 1) = 2iz\mathrm{e}^{iz(N+1)}(1 + o(1)).$$

This and equation (3.3) complete our proof. □

4. Green's function

In Section 2.5 we introduced the matrix $M(\lambda)$ to describe both boundary and interface conditions. Now we introduce also the operators $\mathfrak{I}_0 = E_0\mathfrak{C}_0 + D_0\mathfrak{D}_0$ and $\mathfrak{A}_0 = A_0\mathfrak{C}_0$ as well as the matrix

$$M_0(\lambda) = \begin{pmatrix} \mathfrak{A}_0 \\ \mathfrak{I}_0 \end{pmatrix} (C(\lambda, \cdot), S(\lambda, \cdot)).$$

The solution of the nonhomogeneous system of equations $(L - \lambda)u = h$ where $h \in L^2(T)$ and where u is subject to the homogeneous boundary conditions $\mathfrak{A}\vec{u} = 0$ as well as the interface conditions $\mathfrak{I}\vec{u} = 0$ is given by (see [6])

$$\vec{u}(t) = \int_0^1 \Gamma(\lambda, t, t')\vec{h}(t')dt'$$

where

$$\Gamma(\lambda, t, t') = (C(\lambda, t), S(\lambda, t)) \left(M(\lambda)^{-1}M_0(\lambda) - H(t' - t)\right) \begin{pmatrix} S(\lambda, t') \\ -C(\lambda, t') \end{pmatrix}$$

with H being the Heaviside function, i.e., $H(t)$ equals zero or one depending on whether t is negative or positive.

We remark that $\Gamma(\lambda, \cdot, \cdot)$ represents a scalar function $G(\lambda, \cdot, \cdot)$ on $T \times T$ via $G(\lambda, x, y) = \Gamma_{j,k}(t, t')$ when $x = \epsilon_j(t)$ and $y = \epsilon_k(t')$.

Lemma 4.1. *The function Γ has the following properties:*

1. *For any $t, t' \in [0, 1]$ the function $\Gamma(\cdot, t, t')$ is analytic away from the eigenvalues associated with the Dirichlet problem for T.*

2. *If λ is not an eigenvalue associated with the Dirichlet problem for T then the function $\Gamma(\lambda,\cdot,\cdot)$ has the following properties:*
   (a) *$\Gamma(\lambda,\cdot,\cdot)$ is continuous on $[0,1]^2$.*
   (b) *$\Gamma^{(0,0,1)}(\lambda,\cdot,\cdot)$ is continuous in either of the triangles $\{(t,t')\in[0,1]^2 : t<t'\}$ and $\{(t,t')\in[0,1]^2 : t>t'\}$. On the line $t=t'$ there is a jump discontinuity:*
$$\lim_{\tau\downarrow 0}(\Gamma^{(0,0,1)}(\lambda,t,t+\tau)-\Gamma^{(0,0,1)}(\lambda,t,t-\tau))=-I.$$
   (c) *For fixed $t'\in[0,1]$ the function $\Gamma(\lambda,\cdot,t')$ satisfies the equation*
$$-\Gamma(\lambda,\cdot,t')''+(Q-\lambda)\Gamma(\lambda,\cdot,t')=0$$
   on $(0,t')$ as well as $(t',1)$.
   (d) *For fixed $t\in[0,1]$ the function $\Gamma(\lambda,x,\cdot)$ satisfies the equation*
$$-\Gamma(\lambda,t,\cdot)''+\Gamma(\lambda,t,\cdot)(Q-\lambda)=0$$
   on $(0,t)$ as well as $(t,1)$.
   (e) *If $k\le n_0$ then $\Gamma_{k,j}(\lambda,0,t')=0$.*
   (f) *If $j\le n_0$ then $\Gamma_{k,j}(\lambda,t,0)=0$.*

Proof. Statements (a) through (d) follow easily from the properties of the Heaviside function and the functions c_j and s_j.

To see the validity of statement (e) note first that M and M_0 are both of the form
$$\begin{pmatrix} I_{n_0\times n_0} & 0_{n_0\times(2r-n_0)}\\ * & ** \end{pmatrix}$$
where $*$ indicates an appropriate $(2r-n_0)\times n_0$-matrix and $**$ indicates an appropriate $(2r-n_0)\times(2r-n_0)$-matrix. This implies that $M^{-1}M_0$ is also of that same form. Hence, when $t'>0$, the first n_0 rows of $M^{-1}M_0-H(t')$ are all zero proving (e) in that case. By continuity this is also true when $t'=0$.

To prove (f) note that, if $t>0$, we have $H(-t)=0$. For this case we need to know more about the structure of M_0. Since the bottom $2r-n_0$ rows contain only information about interior vertices we know that the entries in the first n_0 columns and the last $2r-n_0$ rows of $(\mathfrak{A}_0,\mathfrak{I}_0)^\top$ are zero. Hence M_0 has the form
$$\begin{pmatrix} I_{n_0\times n_0} & 0_{n_0\times(r-n_0)} & 0_{n_0\times n_0} & 0_{n_0\times(r-n_0)}\\ 0_{(2r-n_0)\times n_0} & * & 0_{(2r-n_0)\times n_0} & ** \end{pmatrix}$$
where $*$ and $**$ indicate appropriate $(2r-n_0)\times(r-n_0)$-matrices. Thus
$$M_0\begin{pmatrix} S(\lambda,t')\\ -C(\lambda,t')\end{pmatrix}=\begin{pmatrix} S_e(\lambda,t') & 0_{n_0\times(r-n_0)}\\ 0_{(2r-n_0)\times n_0} & *** \end{pmatrix}$$
where $***$ indicates a $(2r-n_0)\times(r-n_0)$-matrix and where $S_e(\lambda,t')=\operatorname{diag}(s_1(\lambda,t'),\dots,s_{n_0}(\lambda,t'))$. The first n_0 columns, corresponding to the requirement $j\le n_0$ are zero when $t'=0$. □

We now express the Weyl solutions in terms of Green's function. Let h be a function in $\mathcal{D}$, the domain of L satisfying the boundary conditions $\mathfrak{A}h = e_\ell$. Next let y be the unique solution of the following inhomogeneous problem

$$(L-\lambda)y = (\lambda - L)h, \quad y \in \mathcal{D}, \quad \mathfrak{A}y = 0.$$

Then $\psi(\ell,\lambda,t) = (y+h)(t)$. We proceed to compute y. By partial integration

$$\begin{aligned}\vec{y}(t) &= \int_0^1 \Gamma(\lambda,t,t')(\lambda - Q(t'))\vec{h}(t')dt' + \int_0^1 \Gamma(\lambda,t,t')\vec{h}''(t')dt' \\ &= -\int_0^t \Gamma^{(0,0,2)}(\lambda,t,t')\vec{h}(t')dt' - \int_t^1 \Gamma^{(0,0,2)}(\lambda,t,t')\vec{h}(t')dt' \\ &\quad + \int_0^1 \Gamma(\lambda,t,t')\vec{h}''(t')dt' \\ &= -\Gamma^{(0,0,1)}(\lambda,t,\cdot)\vec{h}|_0^t - \Gamma^{(0,0,1)}(\lambda,t,\cdot)\vec{h}|_t^1 + \Gamma(\lambda,t,\cdot)\vec{h}'|_0^t + \Gamma(\lambda,t,\cdot)\vec{h}'|_t^1.\end{aligned}$$

In particular, for $h_\ell(t') = (1-t')^2$ and $h_j(t') = 0$ for $j \neq \ell$ we get

$$\begin{aligned}\vec{y}(t) = &\lim_{\tau\to 0}[\Gamma^{(0,0,1)}(\lambda,t,t+\tau)\vec{h}(t+\tau) - \Gamma^{(0,0,1)}(\lambda,t,t-\tau)\vec{h}(t-\tau)] \\ &+ \Gamma^{(0,0,1)}(\lambda,t,0)\vec{h}(0) - \Gamma(\lambda,t,0)\vec{h}'(0).\end{aligned}$$

Since $\vec{h}(0) = e_\ell$, since column ℓ of $\Gamma(\lambda,t,0)$ is a zero column, and since the term in brackets tends to $-\vec{h}(t)$ we obtain

$$\psi_j(k,\lambda,t) = \Gamma^{(0,0,1)}_{j,k}(\lambda,t,0). \tag{4.1}$$

4.1. Eigenfunction expansion of Green's function

Let L_D be the Dirichlet operator introduced in Section 2.5. The eigenvalues λ_k of L_D may have geometric multiplicity larger than one. We will label them in such a way that they are repeated according to their multiplicity and such that $\lambda_1 \leq \lambda_2 \leq \cdots$. The corresponding orthonormalized eigenfunctions are denoted by φ_1, φ_2 etc. Then (see Coddington and Levinson [8], pp. 298/299 for a similar situation)

$$G(\lambda,x,y) = \sum_{n=1}^\infty \frac{\varphi_n(x)\overline{\varphi_n(y)}}{\lambda_n - \lambda}$$

or, equivalently,

$$\Gamma_{j,\ell}(\lambda,t,s) = \sum_{n=1}^\infty \frac{\varphi_{n;j}(t)\overline{\varphi_{n;\ell}(s)}}{\lambda_n - \lambda}.$$

Integration by parts shows that

$$\lambda_n = \lambda_n \int_T |\varphi_n|^2 = \sum_{j=1}^r \int_0^1 (-\varphi''_{n;j} + q_j\varphi_{n;j})\overline{\varphi_{n;j}} = \sum_{j=1}^r \int_0^1 (|\varphi'_{n;j}|^2 + q_j|\varphi_{n;j}|^2)$$

since $\sum_{j=1}^{r}(\varphi'_{n;j}\overline{\varphi_{n;j}})|_0^1 = 0$ due to the boundary and interface conditions satisfied by φ_n. Because q is real and bounded by a constant C this gives

$$\sum_{j=1}^{r}\int_0^1 |\varphi'_{n;j}|^2 \le C + \lambda_n.$$

Also

$$\sum_{j=1}^{r}\int_0^1 |\varphi''_{n;j}|^2 \le \sum_{j=1}^{r}\int_0^1 |(q_j - \lambda_n)\varphi_{n;j}|^2 \le (C + |\lambda_n|)^2$$

again using the boundedness of q. The elementary estimate

$$|f(x)|^2 \le 4\int_0^1 (|f|^2 + |f'|^2),$$

which holds for an absolutely continuous function defined on $[0,1]$, shows then the existence of a constant C' such that

$$|\varphi'_{n;j}(t)| \le C'|\lambda_n|$$

for all sufficiently large n. Thus the series $\sum_{n=1}^{\infty}\varphi'_{n;j}(t)\overline{\varphi'_{n;\ell}(s)}(\lambda_n - \lambda)^{-m-1}$ is absolutely and uniformly convergent for $m \ge 2$, since the λ_n, as zeros of an entire function of growth order 1/2 satisfy $\sum_{n=1}^{\infty}|\lambda_n|^{-1} < \infty$. Therefore

$$\Gamma_{j,\ell}^{(m,1,1)}(\lambda,t,s) = m!\sum_{n=1}^{\infty}\frac{\varphi'_{n;j}(t)\overline{\varphi'_{n;\ell}(s)}}{(\lambda_n - \lambda)^{m+1}} \tag{4.2}$$

if $m \ge 2$.

5. The Dirichlet to Neumann Map

Recall that there is a unique solution of $Ly = \lambda y$ satisfying the nonhomogeneous Dirichlet boundary conditions $y_j(0) = f_j$ for $j = 1,\dots,n_0$ unless λ is one of the isolated Dirichlet eigenvalues. One may then compute the values $g_j = -y'_j(0)$, $j = 1,\dots,n_0$, (Neumann data). The relationship between the f_j and the g_j is linear and is called the Dirichlet-to-Neumann map. We denote this map by Λ. Recall that $\psi(k,\lambda,\cdot)$ denotes the Weyl solution which assumes the value one at the boundary vertex v_k (and zero at every other boundary vertex). Thus we have

$$\Lambda_{j,k} = -\psi'_j(k,\lambda,0)$$

and, using equation (4.1),

$$\Lambda_{j,k} = -\Gamma_{j,k}^{(0,1,1)}(\lambda,0,0). \tag{5.1}$$

In [6] the following theorem was proved.

Theorem 5.1. *Let q be a complex-valued integrable potential supported on a finite metric tree whose edge lengths are one. Then the associated Dirichlet-to-Neumann map uniquely determines the potential almost everywhere on the tree.*

This theorem is the analogue for trees of a result of Nachman, Sylvester, and Uhlmann [19] who consider a Schrödinger equation on a bounded domain in $\mathbb{R}^n$.

Our proof, given in [6], proceeds in two steps. First, we show that the diagonal elements of the Dirichlet-to-Neumann map determine the potential on the boundary edges. Then we show that the Dirichlet-to-Neumann map of the tree determines the Dirichlet-to-Neumann map for a smaller tree with some of the boundary edges pruned off.

As the case of a finite interval can be considered a tree with two boundary vertices, it is clear that not all the information provided by the Dirichlet-to-Neumann map is necessarily needed. In that case it is sufficient to know one of the diagonal entries. Indeed, as Yurko proves in [22], this generalizes to the present case: knowing $n_0 - 1$ of the diagonal elements of the Dirichlet-to-Neumann map is already sufficient to determine the potential almost everywhere on the tree. Yurko shows this by a more careful investigation of the tree pruning procedure than we had performed.

6. A Levinson-type theorem

In this section we prove Theorem 1.1. The proof is divided in two lemmas following the idea of Nachman, Sylvester, and Uhlmann [19]. Assume two potentials q and $\tilde{q}$ are given both satisfying the hypotheses of the theorem, that is to be real-valued and bounded. To each is associated a Dirichlet-to-Neumann map, denoted by Λ and $\tilde{\Lambda}$, respectively. Both problems have the same Dirichlet eigenvalues and the same Neumann data for an orthonormal set of Dirichlet eigenfunctions. We prove in Lemma 6.1 that, under these circumstances, $\Lambda - \tilde{\Lambda}$ is a polynomial and in Lemma 6.2 that this polynomial must be zero. The conclusion of the theorem follows then by applying Theorem 5.1, i.e., that the Dirichlet-to-Neumann map determines the potential almost everywhere on the tree.

Lemma 6.1. *$\Lambda - \tilde{\Lambda}$ is a polynomial of degree at most one.*

Proof. By equation (5.1) we have $\Lambda_{j,k}(\lambda) = -\Gamma_{j,k}^{(0,1,1)}(\lambda, 0, 0)$. Hence, employing equation (4.2),

$$\Lambda_{j,k}^{(m)}(\lambda) = -\Gamma_{j,k}^{(m,1,1)}(\lambda, 0, 0) = -m! \sum_{n=1}^{\infty} \frac{\varphi'_{n;j}(0)\overline{\varphi'_{n;k}(0)}}{(\lambda_n - \lambda)^{m+1}}$$

provided $m \geq 2$. The right-hand side is determined by the information provided, i.e., the Dirichlet eigenvalues and the Neumann data of the Dirichlet eigenfunctions. Therefore the exact same expression is obtained for $\tilde{\Lambda}_{j,k}^{(m)}$. □

Lemma 6.2. *As λ tends to negative infinity $\Lambda - \tilde{\Lambda}$ tends to zero.*

Proof. By Lemma 3.1 the quantities $\Lambda_{j,k} = -\psi'_j(k, \lambda, 0)$ are exponentially small except for $j = k$ in which case we have $\Lambda_{k,k} = \psi'_k(k, \lambda, 0) = -\sqrt{-\lambda} + o(1)$. But since also $\tilde{\Lambda}_{k,k} = -\sqrt{-\lambda} + o(1)$, we have that $(\Lambda - \tilde{\Lambda})_{k,k} = o(1)$. □

7. A Marchenko-type theorem

In this section we prove Theorem 1.2. The proof relies on establishing a relationship between the eigenfunctions associated with a Dirichlet eigenvalue E and the Weyl solutions for λ near E.

We denote the set of eigenvalues by Σ, i.e., $\Sigma = \{\lambda_n : n \in \mathbb{N}\}$. If $E \in \Sigma$ we denote its geometric multiplicity by $\mu(E)$. Recall that the solution of a Dirichlet boundary value problem is determined by an equation of the form $M\xi = f$. The characteristic function associated with the operator L_D is (a multiple of) the determinant of M, an entire function of growth order 1/2. Hence, by Hadamard's factorization theorem[2]

$$\det(M(\lambda)) = C\prod_{n=1}^{\infty}(1-\lambda/\lambda_n) = C\prod_{E\in\Sigma}(1-\lambda/E)^{\mu(E)}$$

where C is an appropriate constant. Since our problem is self-adjoint geometric and algebraic multiplicities coincide. In particular, every eigenvalue has index one[3]. Therefore we define

$$\chi(\lambda) = \prod_{E\in\Sigma}(1-\lambda/E)$$

the "minimal function" associated with the operator L_D.

Lemma 7.1. *Suppose that E is a Dirichlet eigenvalue. Let*

$$\xi(j,\lambda) = (a_1(j,\lambda),\dots,a_r(j,\lambda),b_1(j,\lambda),\dots,b_r(j,\lambda))^\top$$

be the unique solution of $M(\lambda)\xi = \chi(\lambda)e_j$ for λ near E and $1 \le j \le n_0$. Then the $\xi(j,\lambda)$ have a nonzero limit as λ tends to E and the functions

$$(C(E,\cdot),S(E,\cdot))\xi(j,E), \quad j = 1,\dots,n_0$$

span the space of Dirichlet eigenfunctions for E.

Proof. We know that $\mu(E)$, the multiplicity of the Dirichlet eigenvalue E, is also the multiplicity of E as a zero of $\det M$. Because of the structure of M this means that the lower right $(2r-n_0)\times(2r-n_0)$ block $P(E)$ of $M(E)$ has rank $2r-n_0-\mu(E)$. Without loss of generality, employing elementary row operations, we may assume that the top $2r-n_0-\mu(E)$ rows of $P(E)$ are independent and, consequently, that all entries in the bottom $\mu(E)$ rows of $P(\lambda)$ tend to zero as λ tends to E. This structure is exhibited if we write M in the following way:

$$M = \begin{pmatrix} I & 0 & 0 \\ R_1 & P_1 & P_2 \\ R_2 & P_3 & P_4 \end{pmatrix}$$

where the blocks in the diagonal are of size $n_0 \times n_0$, $(2r-n_0-\mu(E))\times(2r-n_0-\mu(E))$, and $\mu(E)\times\mu(E)$, respectively. As we just pointed out we know that $P_3(\lambda)$

[2]We are assuming here that zero is not an eigenvalue. This is without loss of generality since a zero eigenvalue would only change the notation but not the essence of the argument.

[3]The index of an eigenvalue is the length of the longest possible Jordan chain.

and $P_4(\lambda)$ tend to zero as λ tends to E. Moreover, $R_2(E)$ has full rank (equal to $\mu(E)$) since, by Theorem 2.3, this is true for

$$J = \begin{pmatrix} R_1 & P_1 & P_2 \\ R_2 & P_3 & P_4 \end{pmatrix}$$

regardless of λ.

We now multiply the equation $M(\lambda)\xi = \chi(\lambda)e_j$ from the left with the invertible matrix

$$T = \begin{pmatrix} I & 0 & 0 \\ -P_1^{-1}R_1 & P_1^{-1} & 0 \\ P_3P_1^{-1}R_1 - R_2 & P_3P_1^{-1} & I \end{pmatrix}$$

(suppressing the dependence on λ wherever it is convenient). Thereby we obtain

$$\begin{pmatrix} I & 0 & 0 \\ 0 & I & P_1^{-1}P_2 \\ 0 & 0 & P_4 - P_3P_1^{-1}P_2 \end{pmatrix} \xi = \chi(\lambda)Te_j.$$

Writing $\xi = (x, y, z)^\top$ with appropriately sized columns x and y, and z, we will first consider the equation $(P_4 - P_3P_1^{-1}P_2)z = \chi g_j$ where g_j is the jth column of $P_3P_1^{-1}R_1 - R_2$. Since all entries of the matrix $P_4 - P_3P_1^{-1}P_2$ tend to zero as λ tends to E we may write $(P_4 - P_3P_1^{-1}P_2)(\lambda) = (\lambda - E)F(\lambda)$ and $Fz = \tilde{\chi}g_j$ where $\tilde{\chi}(\lambda) = \chi(\lambda)/(\lambda - E)$. Since $\det(T) = 1/\det(P_1)$ assumes a finite nonzero value at E we obtain that $\det(TM) = \det(P_4 - P_3P_1^{-1}P_2)$ has a zero at E of the same order as $\det M$, i.e., $\mu(E)$. This implies that $\det F(E) \neq 0$. Also, since $\tilde{\chi}(E) \neq 0$ and $g_j(E) \neq 0$ we obtain that $z(j, \lambda)$ tends to a nontrivial vector $z(j, E)$. We now may determine also $x(j, E)$ (which will always be zero) and $y(j, E)$. This proves the first statement of the lemma.

The second statement follows from the observation that $(P_3P_1^{-1}R_1 - R_2)(E) = -R_2(E)$ has rank $\mu(E)$ and hence $\mu(E)$ linearly independent columns $j_1, \ldots, j_{\mu(E)}$, giving rise to linearly independent vectors $\xi(j_1, E), \ldots, \xi(j_{\mu(E)}, E)$ which in turn provide $\mu(E)$ linearly independent Dirichlet eigenfunctions. □

Corollary 7.2. *The functions* $\omega(k, \lambda, \cdot) = \chi(\lambda)\psi(k, \lambda, \cdot)$, $k = 1, \ldots, n_0$ *are defined for all values of* $\lambda \in \mathbb{C}$. *If* λ *is not a Dirichlet eigenvalue these functions are linearly independent and span the space of all solutions of* $Ly = \lambda y$. *If* $\lambda = E$ *is a Dirichlet eigenvalue they span the space of all solutions of* $L_D y = Ey$, *i.e., the eigenspace associated with the Dirichlet eigenvalue* E.

In the following we denote derivatives with respect to the spectral parameter by a dot and (as before) derivatives with respect to the spatial variable with a prime. It is easy to check that $uv = (u'\dot{v} - u\dot{v}')'$ if $u(\lambda, \cdot)$ and $v(\lambda, \cdot)$ both satisfy the equation $-y'' + qy = \lambda y$ for all λ. Hence we get

$$\int_T \omega(k, \lambda, \cdot)\omega(\ell, \lambda, \cdot) = \sum_{j=1}^{r} (\omega_j'(k, \lambda, \cdot)\dot{\omega}_j(\ell, \lambda, \cdot) - \omega_j(k, \lambda, \cdot)\dot{\omega}_j'(\ell, \lambda, \cdot))\Big|_0^1.$$

Note that both $\omega(k,\lambda,\cdot)$ and $\dot\omega(\ell,\lambda,\cdot)$ satisfy the interface conditions. Moreover, the boundary conditions, $\omega_j(k,\lambda,0)=\chi(\lambda)\delta_{j,k}$ and $\dot\omega_j(\ell,\lambda,0)=\dot\chi(\lambda)\delta_{j,\ell}$ also hold. Hence we get

$$\int_T \omega(k,\lambda,\cdot)\omega(\ell,\lambda,\cdot)=\chi(\lambda)\dot{\omega_k}'(\ell,\lambda,0)-\dot\chi(\lambda)\omega_\ell'(k,\lambda,0)$$

and, in particular,

$$\int_T \omega(k,E,\cdot)\omega(\ell,E,\cdot)=-\dot\chi(E)\omega_\ell'(k,E,0). \tag{7.1}$$

We introduce the symmetric matrix $N(E)$ by setting

$$N_{k,\ell}(E)=\int_T \omega(k,E,\cdot)\omega(\ell,E,\cdot).$$

Proof of Theorem 1.2. We will prove this theorem by computing the Neumann data of an orthonormal basis of Dirichlet eigenfunctions from the given data. Let E be a Dirichlet eigenvalue of multiplicity $\mu(E)$. Since the naming of the boundary vertices is unimportant we assume now that $K(E)=\{1,\dots,\mu(E)\}$. We may introduce an orthonormal basis $\varphi(k,E,\cdot)$, $k=1,\dots,\mu(E)$ of the eigenspace of E by writing

$$\begin{pmatrix}\omega(1,E,\cdot)\\ \vdots\\ \omega(\mu(E),E,\cdot)\end{pmatrix}=C\begin{pmatrix}\varphi(1,E,\cdot)\\ \vdots\\ \varphi(\mu(E),E,\cdot)\end{pmatrix}$$

with a lower triangular matrix whose diagonal elements are non-zero. In fact, $CC^\top$ is the LU-factorization of the upper left $\mu(E)\times\mu(E)$ block of $N(E)$. Thus, this block determines the matrices C and C^{-1} uniquely.

Now, the Neumann data of the orthonormal Dirichlet eigenfunctions $\varphi(k,E,\cdot)$ are given as linear combinations (in terms of the matrix C) of the Neumann data of the $\omega(k,E,\cdot)$, $k=1,\dots,\mu(E)$ which in turn are uniquely determined, according to equation (7.1) by the first $\mu(E)$ columns of $N(E)$ and the known function χ. □

References

[1] S. Avdonin and P. Kurasov. Inverse problems for quantum trees. Preprint, 2006.

[2] M.I. Belishev. Boundary spectral inverse problem on a class of graphs (trees) by the BC method. *Inverse Problems*, 20(3):647–672, 2004.

[3] M.I. Belishev and A.F. Vakulenko. Inverse problems on graphs: recovering the tree of strings by the BC-method. *J. Inverse Ill-Posed Probl.*, 14(1):29–46, 2006.

[4] C. Bennewitz. A note on the Titchmarsh-Weyl m-function. Preprint ANL-87-26, Vol. 2, 105–111. Argonne National Laboratory, Argonne, Illinois, USA, 1988.

[5] G. Borg. Eine Umkehrung der Sturm-Liouvilleschen Eigenwertaufgabe. Bestimmung der Differentialgleichung durch die Eigenwerte. *Acta Math.*, 78:1–96, 1946.

[6] B.M. Brown and R. Weikard. A Borg-Levinson theorem for trees. *Proc. R. Soc. Lond. Ser. A Math. Phys. Eng. Sci.*, 461(2062):3231–3243, 2005.

[7] R. Carlson. Inverse eigenvalue problems on directed graphs. *Trans. Amer. Math. Soc.*, 351(10):4069–4088, 1999.

[8] E.A. Coddington and N. Levinson. *Theory of Ordinary Differential Equations.* Robert E. Krieger, Malabar, Florida, 1984.

[9] E.B. Curtis and J.A. Morrow. Determining the resistors in a network. *SIAM J. Appl. Math.*, 50(3):918–930, 1990.

[10] E.B. Curtis and J.A. Morrow. The Dirichlet to Neumann map for a resistor network. *SIAM J. Appl. Math.*, 51(4):1011–1029, 1991.

[11] N.I. Gerasimenko. The inverse scattering problem on a noncompact graph. *Teoret. Mat. Fiz.*, 75(2):187–200, 1988.

[12] M. Harmer. Inverse scattering on matrices with boundary conditions. *J. Phys. A*, 38(22):4875–4885, 2005.

[13] P. Kuchment. Graph models for waves in thin structures. *Waves Random Media*, 12(4):R1–R24, 2002.

[14] P. Kuchment. Quantum graphs. I. Some basic structures. *Waves Random Media*, 14(1):S107–S128, 2004. Special section on quantum graphs.

[15] P. Kurasov and F. Stenberg. On the inverse scattering problem on branching graphs. *J. Phys. A*, 35(1):101–121, 2002.

[16] P. Kurasov and M. Nowaczyk. Inverse spectral problem for quantum graphs. *J. Phys. A*, 38(22):4901–4915, 2005.

[17] N. Levinson. The inverse Sturm-Liouville problem. *Mat. Tidsskr. B.*, 1949:25–30, 1949.

[18] V.A. Marčenko. Concerning the theory of a differential operator of the second order. *Doklady Akad. Nauk SSSR. (N.S.)*, 72:457–460, 1950.

[19] A. Nachman, J. Sylvester, and G. Uhlmann. An n-dimensional Borg-Levinson theorem. *Comm. Math. Phys.*, 115(4):595–605, 1988.

[20] V.N. Pivovarchik. Ambartsumyan's theorem for the Sturm-Liouville boundary value problem on a star-shaped graph. *Funktsional. Anal. i Prilozhen.*, 39(2):78–81, 2005.

[21] V.N. Pivovarchik. Inverse problem for the Sturm-Liouville equation on a simple graph. *SIAM J. Math. Anal.*, 32(4):801–819 (electronic), 2000.

[22] V. Yurko. Inverse spectral problems for Sturm-Liouville operators on graphs. *Inverse Problems*, 21(3):1075–1086, 2005.

B.M. Brown
School of Computer Science, University of Wales
Cardiff, PO Box 916
Cardiff CF2 3XF, UK
e-mail: `Malcolm.Brown@cs.cf.ac.uk`

R. Weikard
Department of Mathematics
University of Alabama at Birmingham
Birmingham
Alabama 35294-1170, USA
e-mail: `rudi@math.uab.edu`

Operator Theory:
Advances and Applications, Vol. 186, 49–68

Wave Operators for Nonlocal Sturm-Liouville Operators with Trivial Potential

E.V. Cheremnikh

Abstract. The wave operators for the pair of Sturm-Liouville operators corresponding to different nonlocal boundary conditions are considered. By analogy with an absolutely continuous subspace in the selfadjoint case a so-called essential subspace is defined. This definition is given for a pair of operators (not one operator only). In the case of the same finite sets of spectral singularities of two operators the similarity of corresponding semigroups is proved. The domain of definition of the semigroup has finite codimension.

Mathematics Subject Classification (2000). 47B38, 34L25.

Keywords. Wave operator, spectral singularity, absolutely continuous subspace.

1. Introduction

In [1] the perturbation of a boundary condition within the scattering theory was considered for selfadjoint operators. Among the works concerning dissipative operators we note [2] where Sturm-Liouville operators on a half-line with the usual boundary condition were considered. Here the space which plays the role of an absolutely continuous subspace was defined (in terms of the Hardy space H^2) like the construction within the well-known functional model (see, e.g., [3–4]).

In the present article we consider similar questions for the Sturm-Liouville operator with trivial potential which is non-selfadjoint because of the nonlocal boundary condition. Apropos, in [5] in the space $L^2(0,\infty)$ the operator $-id/dx$ (i.e., the operator with trivial potential too) with nonlocal condition $\int_0^\infty u(x)d\sigma(x) = 0$ was studied, here in the first place the point spectrum was investigated.

It is necessary to note that local wave operators $W_\pm(\Delta)$ (Δ an arbitrary interval without singularities) are investigated for much more general non-selfadjoint operators (see, e.g., [6–7]).

In [8] where the non-selfadjoint Friedrichs' model was considered the function of the operator just in presence of spectral singularities was studied. Nonlocal perturbation of a boundary condition was introduced within Friedrichs' model in [9]. So, one can understand the present article as some development of [9] too.

We use the method of calculus which belongs to the direction defined by [6]. Our version is more closed to [10] and in the present article itself we emphasize the use of the Laplace transformation. We note too that the present article is a direct prolongation of [11] where an exponential function is given for the operator indicated in our title.

2. Certain notations and auxiliary statements

Let $\mathcal{L}y = -y''$ and

$$\begin{cases} D(B) = \{y \in L^2(0,\infty) : \mathcal{L}y \in L^2(0,\infty),\, y(0) = 0\} \\ By = \mathcal{L}y,\, y \in D(B). \end{cases} \tag{2.1}$$

Let $\eta \in L^2(0,\infty)$ be an arbitrary function, $\Phi(y) = (y,\eta)_{L^2(0,\infty)}$ and

$$\begin{cases} D(A) = \{z \in L^2(0,\infty) : \mathcal{L}z \in L^2(0,\infty),\, z(0) + \Phi(z) = 0\} \\ Az = \mathcal{L}z,\, z \in D(A). \end{cases} \tag{2.2}$$

We call A a perturbed operator with respect to the operator B.

We denote by $e \in L^2(0,\infty)$ an auxiliary function such that

$$\mathcal{L}e, \mathcal{L}^2 e \in L^2(0,\infty), \quad e(0) = 1, \quad \Phi(e) = -2 \tag{2.3}$$

About the element η we suppose that

$$\int_0^\infty |B^3\eta(x)| e^{\varepsilon x}\, dx < \infty, \quad \varepsilon > 0. \tag{2.4}$$

The element η satisfies the conditions of Lemma 2.4 if, for example

$$\eta(0) = \eta''(0) = \eta^{IV}(0) = 0, \quad \eta, \eta'', \eta^{IV} \in L^2(0,\infty) \tag{2.5}$$

All statements of Section 2 were proved in [11].

Lemma 2.1. *Let the operator $I_A : L^2(0,\infty) \to L^2(0,\infty)$ be defined by the relation*

$$D(I_A) = D(B); \quad I_A y \equiv y + \Phi(y)e,\, y \in D(B)$$

Then $I_A D(B) = D(A)$ and $I_A^{-1} z = z + \Phi(z)e$, $z \in D(A)$.

We need the Fourier transformation $\mathcal{F} : L^2(0,\infty) \to H \equiv L^2_\rho(0,\infty)$, $\rho(\tau) = \frac{1}{\pi}\sqrt{\tau}$ where

$$\mathcal{F}y(\tau) = \int_0^\infty y(x) \frac{\sin x\sqrt{\tau}}{\sqrt{\tau}}\, dx, \quad \tau > 0$$

and we need the element $e_\zeta \in L^2(0,\infty)$, $\zeta \notin [0,\infty)$ such that

$$\mathcal{F}e_\zeta(\tau) = \frac{1}{\tau - \zeta}, \quad \tau > 0, \quad \zeta \notin [0,\infty) \tag{2.6}$$

Let

$$\delta(\zeta) = 1 + \Phi(e_\zeta) = 1 + \int_0^\infty \frac{\overline{\mathcal{F}\eta(s)\rho(s)}}{s-\zeta} ds, \quad \rho(s) = \frac{1}{\pi}\sqrt{s}, \quad \zeta \notin [0,\infty). \tag{2.7}$$

Let $\delta_\pm(\sigma) = \lim\limits_{\tau \to \pm 0} \delta(\sigma + i\tau)$, $\sigma > 0$.

Lemma 2.2. *If the conditions* (2.3), (2.4) *hold then the function* $\delta(\zeta), \zeta \notin [0,\infty)$ (*see* (2.7)) *admits an analytic prolongation* $\delta_\pm(\zeta)$ *in the domain* $|\Im\sqrt{\zeta}| < \varepsilon$ *and* $\lim\limits_{|\zeta|\to\infty} (\delta_\pm(\zeta) - 1) = 0$ *uniformly in the corresponding domain* $\Im\zeta > -\varepsilon_1$ *or* $\Im\zeta < \varepsilon_1$ *for every* $\varepsilon_1 > 0$.

We denote the resolvent of A by $R_\zeta(A) = (A - \zeta)^{-1}$ and the resolvent set by $\rho(A)$. There exists a linear subspace L dense in $L^2(0,\infty)$ such that the limit values

$$(R_\sigma(A)f, g)_\pm = \lim_{\tau \to \pm 0} (R_{\sigma+i\tau}(A)f, g)_{L^2(0,\infty)}, \quad f, g \in L, \quad \sigma > 0 \tag{2.8}$$

are smooth functions.

A generalized pole of at least one of the function (2.8) is called spectral singularity of the operator A.

Theorem 2.3.
1) *Let* $\zeta \notin [0,\infty)$, *then* $\zeta \in \rho(A)$ *iff* $\delta(\zeta) \neq 0$, *in this case*

$$R_\zeta(A)f = R_\zeta(B)f - \frac{1}{\delta(\zeta)} (R_\zeta(B)f, \eta)_{L^2(0,\infty)} e_\zeta \tag{2.9}$$

where the element $e_\zeta \in L^2(0,\infty)$ *is defined by the relation* (2.6). *A value* $\zeta \notin [0,\infty)$ *is an eigenvalue of the operator* A *iff* $\delta(\zeta) = 0$.
2) *Let* $\sigma > 0$ *and* $\delta_\pm(\sigma) = 0$. *Then* σ *belongs to the continuous spectrum of* A. *A value* $\sigma > 0$ *is a spectral singularity of the operator* A *iff* $\delta_+(\sigma) = 0$ *or* $\delta_-(\sigma) = 0$.
3) *The set of eigenvalues and spectral singularities of the operator* A *is finite.*

We denote

$$m(s) = i\left(\delta_0', e^{isB}\eta\right), \quad n(s) = -\left(y_0 + \Phi(y_0)e, e^{isB}\eta\right)_{L^2(0,\infty)} \tag{2.10}$$

where y_0 is an element from $L^2(0,\infty)$ and $(\delta_0', f) \equiv \overline{f'(0)}$.

Lemma 2.4. *Let* $\eta \in D(B^3) \cap D(A^*)$. *Let us consider the problem*

$$\dot{z} = -iAz, \quad z|_{t=0} = z_0 \in D(A), \quad t > 0 \tag{2.11}$$

where $z(t)$ is an unknown function and the problem (see (2.10))

$$\dot{y} = -iBy + f(t), \quad y|_{t=0} = y_0 \in D(B), \quad t > 0 \tag{2.12}$$

$$k(t) + \int_0^t m(t-s)k(s)ds = n(t) \tag{2.13}$$

where

$$f(t) \equiv -\dot{k}(t)e - ik(t)\mathcal{L}e$$

and $(y(t), k(t))$ are two unknown functions.

If $z(t)$ is a solution of problem (2.11) then the functions

$$y(t) = z(t) + \Phi(z(t))e, \quad k(t) = -\Phi(z(t))$$

define the solution of the problem (2.12)–(2.13) where $y_0 = z_0 + \Phi(z_0)e$. Inversely, if $(y(t), k(t))$ is a solution of problems (2.12)–(2.13) then $z(t) = y(t) + \Phi(y(t))e$ is a solution of problem (2.11) where $z_0 = y_0 + \Phi(y_0)e$.

Let $k(\cdot)$ be some locally integrable functions on $(0,\infty)$ and let

$$E_t^{\pm} k(\tau) \equiv \int_0^t e^{\pm is\tau} k(s)ds, \quad t, \tau > 0. \tag{2.14}$$

Obviously, if $k \in C^1[0,\infty)$ then $E_t^{\pm} k \in L^2_{\rho}(0,\infty)$, $t > 0$.

We denote by

$$\mathcal{I}(k,p) \equiv \int_0^{\infty} e^{-sp} k(s)ds$$

the Laplace transformation of the function k. The following theorem represents the exponential function of the non-selfadjoint operator A.

Theorem 2.5. *Suppose that the element $\eta \in L^2(0,\infty)$ satisfies conditions (2.4), (2.5). Then the problem (2.11) is uniformly correct and its solution for $t > 0$ takes the form (see (2.14))*

$$z(t) = e^{-itB} z_0 + ie^{-itB}\mathcal{F}^{-1} E_t^{+} k, \; t > 0 \tag{2.15}$$

where the function $k(s)$, $s > 0$ is defined by the relation

$$\mathcal{I}(k,p) = i\left(R_{ip}(A)z_0, \eta\right)_{L^2(0,\infty)}, \; \Re p > p_0 \tag{2.16}$$

and for $t < 0$ takes the form

$$z(t) = e^{-itB} z_0 - ie^{-itB}\mathcal{F}^{-1} E_t^{-} k_1, \quad t < 0 \tag{2.17}$$

where the function $k^{-}(s) = k_1(-s)$, $s > 0$ is defined by the relation

$$\mathcal{I}(k^{-},p) = -i\left(R_{-ip}(A)z_0, \eta\right)_{L^2(0,\infty)}. \tag{2.18}$$

Expression (2.15) represents the exponential function of the operator $\exp(-itA)$, $t > 0$ (see (2.17) in the case $t < 0$). This explicit form permits later to study corresponding wave operators.

3. Wave operators

Formally relations (2.15)–(2.17) give immediately the wave operators for the pair A, B or the pair A, A_1 where the operator A_1 is defined by analogy with operator A. But such wave operators in the general case are unbounded and our aim is to indicate the sufficient conditions of boundedness of these operators.

Function k (see (2.15)) is given by its Laplace's transformation (2.16). In view of equality (2.16) we introduce the following definition.

Recall that according to the relation (2.9) we have for $\lambda \in \rho(A)$

$$(R_\lambda(A)z, u)_{L^2(0,\infty)} = (R_\lambda(B)z, u)_{L^2(0,\infty)} - \frac{(R_\lambda(B)z, \eta)_{L^2(0,\infty)}(e_\lambda, u)_{L^2(0,\infty)}}{\delta(\lambda)}. \tag{3.1}$$

We denote

$$H_0 = L^2(0,\infty),\ H = L^2_\rho(0,\infty),\ \rho(\tau) = \frac{1}{\pi}\sqrt{\tau}.$$

Definition 3.1. Let $\mathcal{H}_{\text{ess}}(A) \subset H_0$ be a linear subspace of all elements $z \in H_0$ such that for every element $w \in H_0$ the function $\zeta \to (R_\zeta(A)z, w)_{H_0}$, $\zeta \in \rho(A)$ belongs to the Hardy space in both half-plains $\Im\zeta > 0$ and $\Im\zeta < 0$, i.e.,

$$\mathcal{H}_{\text{ess}}(A) = \{z \in H_0 : (R_\zeta(A)z, w)_{H_0} \in H^2(\Im\zeta > 0) \cap H^2(\Im\zeta < 0)\} \tag{3.2}$$

Further we write "belongs to H^2" instead "belongs to $H^2(\Im\zeta > 0)$ and belongs to $H^2(\Im\zeta < 0)$".

Let

$$U = \{u \in H_0 : \exists M(u) > 0 : |(e_\zeta, u)_{H_0}| \leqslant M(u),\ \zeta \notin [0,\infty)\}. \tag{3.3}$$

Note that if $u \in U$ then

$$\pi|(\mathcal{F}u)(\tau)|\rho(\tau) \leqslant M(u), \quad \tau > 0. \tag{3.4}$$

Therefore

$$(R_\zeta(B)u, w)_{H_0} \in H^2, \quad w \in H_0, \quad u \in U, \tag{3.5}$$

i.e., for the non-perturbed operator B we have $U \subset H_{\text{ess}}(B)$. We can rewrite (3.5) as

$$(R_\zeta(B)z, u)_{H_0} \in H^2, \quad z \in H_0, \quad u \in U. \tag{3.6}$$

The analogue relation for the perturbed operator A holds for z from some subspace in H_0 only. But this subspace does not depend from an element $u \in U$.

Proposition 3.2. *Let $z \in H_0$, then*

$$(R_\zeta(A)z, u)_{H_0} \in H^2,\ \forall u \in U \Longleftrightarrow (R_\zeta(A)z, \eta)_{H_0} \in H^2. \tag{3.7}$$

Proof. As the elements η satisfy (2.4) then $\eta \in U$. So, it remains to consider the direction ($\Longleftarrow$). Let $(R_\zeta(A)z, \eta)_{H_0} \in H^2$. Due to the relation

$$(R_\zeta(A)z, u)_{H_0} = (R_\zeta(B)z, u)_{H_0} - (R_\zeta(A)z, \eta)_{H_0}(e_\zeta, u)_{H_0}$$

and (3.3), (3.6) too we obtain $(R_\zeta(A)z, u)_{H_0} \in H^2$. The proposition is proved.

Note that the subspace $\mathcal{H}_{\text{ess}}(A) \subset H_0$ has a finite codimension, $\operatorname{codim}\overline{H_{\text{ess}}(A)} < \infty$.

Taking into account Proposition 3.2 we introduce

Definition 3.3. Let $D(\Gamma_\pm) = \mathcal{H}_{\text{ess}}(A)$ and

$$(\Gamma_\pm z)(\tau)(R_\tau(A)z,\eta)_\pm \frac{1}{\delta_\pm(\tau)}(R_\tau(B)z,\eta)_\pm, \quad z \in \mathcal{H}_{\text{ess}}(B). \tag{3.8}$$

Obviously the operator $\Gamma_\pm$ acts from $\mathcal{H}_{\text{ess}}$ in H_0.

Proposition 3.4. *Let $z \in D(A) \cap \mathcal{H}_{\text{ess}}(A)$. Then the limit value*

$$\lim_{t\to\pm\infty} e^{itB}e^{-itA}z \in L^2(0,\infty) = H_0$$

exists iff $\Gamma_\pm z \in L^2(0,\infty) \cap L^2_\rho(0,\infty) = H_0 \cap H$.

Proof. As $z \in D(A)\cap\mathcal{H}_{\text{ess}}(A)$ then we have $p_0 = 0$ in relation (2.16) and according to (3.8)

$$\int_0^\infty e^{i\lambda s}k(s)ds = \frac{i}{\delta_+(\lambda)}(R_\lambda(B)z,\eta)_+ = i(\Gamma_+ z)(\lambda), \quad \Im\lambda = 0. \tag{3.9}$$

Consequently $k \in L^2(0,\infty)$.

As $z \in D(A)$ in view of (2.14)–(2.15) we have

$$e^{itB}e^{-itA}z = z + i\mathcal{F}^{-1}\left(\int_0^t e^{is\tau}k(s)ds\right), \quad t > 0. \tag{3.10}$$

Therefore the limit value in (3.10) in the case $t \to +\infty$ exists iff the improper integral

$$\int_0^\infty e^{is\tau}k(s)ds = \lim_{t\to+\infty}\int_0^t e^{is\tau}k(s)ds \tag{3.11}$$

converges in the space $h = L^2_\rho(0,\infty)$. The case $t < 0$ is considered by analogy.

The proposition is proved.

Definition 3.5. Let

$$D(W^0_\pm(B,A)) = \{z \in D(A)\cap\mathcal{H}_{\text{ess}}(A) : \Gamma_\pm z \in H_0 \cap H\} \tag{3.12}$$

and

$$W^0_\pm(B,A)z = s - \lim_{t\to\pm\infty} e^{itB}e^{-itA}z, \quad z \in D(W^0_\pm(B,A)), \tag{3.13}$$

then the operators $W^0_\pm(B,A) : H_0 \to H_0$ are called **initial wave operators**. The closure of the operators $W^0_\pm(B,A)$ in the space H_0,

$$W_\pm(B,A) = \overline{W^0_\pm(B,A)} \tag{3.14}$$

are called **wave operators**.

We need too the notation

$$L^2_\varepsilon(0,\infty) = \Big\{ z \in H_0 : \mathcal{F}z(\tau) = O(\tau^{-2}),\ \tau \to \infty, \\ \text{uniformaly for } |\Im \sqrt{\tau}| \leqslant \frac{\varepsilon}{2} \Big\}. \tag{3.15}$$

Theorem 3.6. *If the element* $\eta \in H_0$ *satisfies conditions* (2.4), (2.5) *then* (*see* (3.13))

$$W^0_\pm(B,A)z = z - \mathcal{F}^{-1}\Gamma_\pm z, \quad z \in D(W^0_\pm(B,A)) \tag{3.16}$$

and for every $\varepsilon > 0$

$$D(A) \cap D(\Gamma_\pm) \cap L^2_\varepsilon(0,\infty) \subset D(W^0_\pm(B,A)). \tag{3.17}$$

Further we need the following additional condition on the element $\eta \in L^2(0,\infty)$: the transformation $\mathcal{F}\eta(\tau)$ is a rational function which is bounded on the half-line $(0,\infty)$.

As

$$\mathcal{F}(e^{i\sqrt{\zeta}x})(\tau) = \frac{1}{\tau - \zeta}, \quad \tau > 0, \quad \Im\sqrt{\zeta} > 0$$

then such an element $\eta \in H_0$ is a finite sum

$$\eta(x) = \sum_j p_j(x) e^{\alpha_j x} \tag{3.18}$$

where $p_j(x)$ is an arbitrary polynomial and α_j an arbitrary complex number, $\Re\,\alpha_j < 0$. In view of (2.7) and the relation

$$\int_0^\infty \frac{\sqrt{s}}{(s-a_j)(s-\zeta)} ds = \frac{\pi i}{\sqrt{a_j} + \sqrt{\zeta}}, \quad \Im\sqrt{a_j},\ \Im\sqrt{\zeta} > 0$$

we obtain that $\delta_\pm(\tau)$ are rational functions on $\sqrt{\tau}$, which are bounded on $[0,\infty)$.

We will suppose too that the value $\delta_\pm(0)$ is finite and

$$\delta_\pm(0) \neq 0. \tag{3.19}$$

Lemma 3.7. *Suppose conditions* (3.18)–(3.19) *hold.*

If $W_\pm(B,A) = \overline{W^0_\pm(B,A)}$ (*see* (3.12), (3.16)) *then*

$$D(W_\pm(B,A)) = \{z \in \mathcal{H}_{\text{ess}}(A) : \Gamma_\pm z \in H_0 \cap H\} \tag{3.20}$$

and

$$W_\pm(B,A)z = z - \mathcal{F}^{-1}\Gamma_\pm z, \quad z \in D(W_\pm(B,A)). \tag{3.21}$$

Corollary 3.8. $\mathcal{H}_{\text{ess}}(A) \cap L^2_\varepsilon(0,\infty) \subset D(W_\pm(B,A))$.

Proof. Condition $z \in D(A)$ in Theorem 3.6 serves for the existence itself of the representation (3.16) only. So, we can repeat the proof of Theorem 3.6 for $z \in \mathcal{H}_{\text{ess}}(A) \cap L^2_\varepsilon(0,\infty)$.

4. Properties of wave operators

We will make more precise the form of $W_\pm(B,A)$, give the connection with singular integral operators and prove the invertibility of the wave operator.

Theorem 4.1. *If $z \in D(W_\pm(B,A))$ and $\psi = \mathcal{F}z \in C^1(0,\infty)$ then*

$$\mathcal{F}W_\pm(B,A)\mathcal{F}^{-1}\psi(\tau) = (L_{p,q}\psi(\tau) + R\psi(\tau))/\delta_\pm(\tau) \tag{4.1}$$

where

$$\begin{cases} L_{p,q}\psi(\tau) = p(\tau)\psi(\tau) + q(\tau)\mathcal{H}[\psi](\tau),\ \mathcal{H}[\psi](\tau) = \dfrac{1}{\pi}V.p.\displaystyle\int_0^\infty \frac{\psi(s)}{s-\tau}ds \\ p(\tau) = \frac{1}{2}(\delta_+(\tau)+\delta_-(\tau)), \quad q(\tau) = -\frac{1}{2i}(\delta_+(\tau)-\delta_-(\tau)) \end{cases} \tag{4.2}$$

and

$$R\psi(\tau) = \int_0^\infty r(s,\tau)\psi(s)ds, \quad r(s,\tau) = -\mathcal{R}_\tau(\overline{\gamma}\rho)(s) \tag{4.3}$$

where $\gamma = \mathcal{F}\eta$, $\mathcal{R}_\tau h(s) \equiv (h(s)-h(\tau))/(s-\tau)$.

Proof. In view of relations (3.8), (3.21) we have

$$\mathcal{F}W_\pm(B,A)\mathcal{F}^{-1}\psi(\tau) = \psi(\tau) - \Gamma_\pm\mathcal{F}^{-1}\psi(\tau) = \psi(\tau) - \frac{1}{\delta_\pm(\tau)}(R_\tau(B)\mathcal{F}^{-1}\psi, \mathcal{F}^{-1}\gamma)_\pm$$

$$= \psi(\tau) - \frac{1}{\delta_\pm(\tau)}\left(\int_0^\infty \frac{\psi(s)\overline{\gamma(s)}}{s-\tau}\rho(s)ds\right)_\pm$$

$$= \frac{1}{\delta_\pm(\tau)}\left[(\delta_\pm(\tau) \mp \pi i\overline{\gamma(\tau)}\rho(\tau))\psi(\tau) - V.p.\int_0^\infty \frac{\psi(s)\overline{\gamma(s)}}{s-\tau}\rho(s)ds\right].$$

Using the equality (see (2.9))

$$\delta_\pm(\tau) = 1 + \left(\int_0^\infty \frac{\overline{\gamma(s)}}{s-\tau}\rho(s)ds\right)_\pm = 1 \pm \pi i\overline{\gamma(\tau)}\rho(\tau) + V.p.\int_0^\infty \frac{\overline{\gamma(s)}}{s-\tau}\rho(s)ds$$

and the elementary relation

$$V.p.\int_0^\infty \frac{\psi(s)\overline{\gamma(s)}}{s-\tau}\rho(s)ds\pi\overline{\gamma(\tau)}\rho(\tau)\mathcal{H}[\psi](\tau) + \int_0^\infty \mathcal{R}_\tau(\overline{\gamma}\rho)(s)\psi(s)ds$$

we obtain equalities (4.1)–(4.3). The corollary is proved.

Lemma 4.2. *Let p, q, a, b be arbitrary functions from the space $C^1[0, \infty)$. Then the operators $L_{p,q}$, $L_{a,b}$ (see (4.2)) are bounded in the space H and*

$$L_{p,q}L_{a,b}\theta = L_{pa-qb,pb+qa}\theta + qK_{a,b}\theta, \; K_{a,b}\theta(\tau) = \int_0^\infty k_{a,b}(\tau, s)\theta(s)ds \tag{4.4}$$

where

$$k_{a,b}(\tau, s) = \frac{1}{\pi}\mathcal{R}_\tau a(s) - \frac{1}{\pi}\mathcal{H}[\mathcal{R}_\tau b](s) + b(\tau)\mathcal{R}_\tau \ln s$$

Note that the proof uses the relation

$$\mathcal{H}^2[\theta](\tau) = -\theta(\tau) + \int_0^\infty \mathcal{R}_\tau \ln s\, \theta(s)ds.$$

Lemma 4.3. *The operator $\delta_\pm(\cdot)\mathcal{F}W_\pm(B, A)\mathcal{F}^{-1}$ admits a continuous extension as bounded singular integral operator*

$$\delta_+(\tau)\mathcal{F}W_\pm(B, A)\mathcal{F}^{-1}\psi(\tau) = Lp, q\psi(\tau) + R\psi(\tau), \quad \psi \in \mathcal{F}\overline{D(W_+(B, A))} \tag{4.5}$$

on the subspace $\mathcal{F}\overline{D(W_+(B, A))}$, where $\overline{D(W_+(B, A))}$ is a closure in H_0 of the domain of definition $D(W_+(B, A))$.

Proof. As it was mentioned the function $\delta_\pm(\tau)$ is a rational function on $\sqrt{\tau}$ bounded on $[0, \infty)$ (see (2.9), (3.18)). So, the functions $p(\tau)$, $q(\tau)$ are bounded on $[0, \infty)$ too (see (4.2)). The Hilbert transformation $\mathcal{H}$ is bounded in the space $L^2_\rho(0, \infty)$ where $\rho(s) = \dfrac{1}{\pi}\sqrt{s}$. Therefore the operator $L_{p,q}$ (see (4.2)) is bounded in $L^2_\rho(0, \infty)$.

It remains to prove that the integral operator

$$R\psi(\tau) = -\int_0^\infty \mathcal{R}_\tau(\overline{\gamma}\rho)(s)\psi(s)ds, \quad \gamma = \mathcal{F}\eta$$

is compact in the space $L^2_\rho(0, \infty)$. Recall that an operator

$$Lf(\tau) = \int_0^\infty l(\tau, s)f(s)\rho(s)ds \tag{4.6}$$

is compact in the space $L^2_\rho(0, \infty)$ if

$$\int_0^\infty\int_0^\infty |l(\tau, s)|^2\rho(\tau)\rho(s)d\tau ds < \infty \tag{4.7}$$

(see [13], Ch. 10.2).

So, we must prove that

$$\int_0^\infty\int_0^\infty |\mathcal{R}_\tau(\overline{\gamma}\rho)(s)|^2\frac{\rho(\tau)}{\rho(s)}d\tau ds < \infty. \tag{4.8}$$

But the function $\mathcal{R}_\tau(\overline{\gamma}\rho)(s)$, $\rho(s) = \frac{1}{\pi}\sqrt{s}$, $s > 0$ is a linear combination of terms under the form

$$\mathcal{R}_\tau\left(\frac{1}{\sqrt{s}+a}\right) = \frac{1}{(\sqrt{s}+a)(\sqrt{s}+\sqrt{\tau})(\sqrt{\tau}+a)} \tag{4.9}$$

and their derivatives with respect to a, what proves (4.8). The lemma is proved.

Proposition 4.4. *The "jump" of the resolvent of the operator A on the continuous spectrum has the form*

$$(R_\sigma(A)u, v)_+ - (R_\sigma(A)u, v)_- = (u, b_\sigma)(a_\sigma, v), \quad \delta_\pm \neq 0 \tag{4.10}$$

where (see (2.7))

$$\begin{cases} (a_\sigma, v) = \dfrac{1}{\delta_+(\sigma)}(e_\sigma, v)_+ - \dfrac{1}{\delta_-(\sigma)}(e_\sigma, v)_- \\ (u, b_\sigma) = \delta_-(\sigma)\mathcal{F}u(\sigma) - (R_\sigma(B)u, \eta)_- \end{cases} \tag{4.11}$$

for $\mathcal{F}u, \mathcal{F}v \in C^1(0,\infty)$. If in addition $u \in D(W_+(B,A)) \cap D(W_-(B,A))$, then

$$(u, b_\sigma)\delta_+(\sigma)W_+(B,A)u(\sigma)\delta_-(\sigma)W_-(B,A)u(\sigma). \tag{4.12}$$

Corollary 4.5. *The operators $W_\pm(B,A)$ are invertible.*

Really, if $W_\pm(B,A)u = 0$ then "jump" is zero, therefore $u = 0$.

5. Similarity of semigroups

In this section we will discuss the similarity of two semigroups $\exp(itA)$ and $\exp(itA_1)$ for the operators A and A_1 with the same set of spectral singularities.

We will use the notation

$$\mathfrak{S}(A) = \{\sigma_j, k_j^\pm\} \tag{5.1}$$

where $\{\sigma_j\}$ denotes the set of all spectral singularities of the operator A and $k_j^\pm$ denotes the multiplicity σ_j as root of the function $\delta_\pm(\sigma)$, $\sigma > 0$.

Let A_1 be an other operator defined like the operator A (see (2.3)) by an element $\eta_1 \in L^2(0,\infty)$ under the form (3.18) and let $\delta_1(\zeta) = 1 + (e_\zeta, \eta_1)_{L^2(0,\infty)}$, $\zeta \notin [0,\infty)$ (compare with (2.81)). We introduce the following notation:

$$S(\tau) = \frac{\delta_-(\tau)}{\delta_+(\tau)}, \quad S_1(\tau) = \frac{\delta_{1,-}(\tau)}{\delta_{1,+}(\tau)}, \quad \Delta_+(\tau) = \frac{\delta_+(\tau)}{\delta_{1,+}(\tau)} \tag{5.2}$$

and

$$m(\tau) = \frac{1}{2}\left(1 + \frac{S_1(\tau)}{S(\tau)}\right), \quad n(\tau) = -\frac{1}{2i}\left(1 - \frac{S_1(\tau)}{S(\tau)}\right), \quad \tau > 0. \tag{5.3}$$

We need some additional notation too. Let $P(s) = \prod(s-\sigma_j)^{k_j}$, $n = \sum k_j$ be some polynomials and $Q(s) = P(s)/(1+s)^n$. Using the coefficients of the decomposition

$$\frac{1}{Q(s)} = 1 + \sum_{j,k} \frac{p_{jk}}{(s-\sigma_j)^k}, \; p_{jk} = \text{const}$$

we introduce the functional

$$D[F(\sigma)] = \sum_{j,k} \frac{p_{jk}}{(k-1)!} F^{(k-1)}(\sigma_j), \quad F \in C^n(0,\infty),$$

the functions

$$Q_{lj}(s) = \frac{1}{l!} Q(s) \sum_{k=l+1}^{k_j} p_{jk}(s-\sigma_j)^{l-k}, \quad l = 0,1,\ldots,k_j - 1$$

and the operator $\pi_{\mathfrak{S}} : C^n(0,\infty) \to C^n(0,\infty)$

$$\pi_{\mathfrak{S}} h(s) = \sum_j \sum_{l=0}^{k_j-1} h^{(l)}(\sigma_j) Q_{lj}(s).$$

An elementary calculation results in the following proposition.

Proposition 5.1. *The operator $\pi_{\mathfrak{S}}$ is the projection, $\pi_{\mathfrak{S}}^2 = \pi_{\mathfrak{S}}$, and*

$$h(s) = \pi_{\mathfrak{S}} h(s) + Q(s)\mathcal{R}h(s), \quad h \in C^n(0,\infty) \tag{5.4}$$

where

$$\mathcal{R}h(s) = h(s) + D[\mathcal{R}_\sigma h](s), \quad \mathcal{R}_\sigma h(s) = (h(s) - h(\sigma))/(s-\sigma).$$

Note that the decomposition (5.4) is unique in the following sense: the equalities

$$h(s) = h_1(s) + Q(s)h_2(s), \quad h_1 \in R(\pi_{\mathfrak{S}}) \tag{5.5}$$

and

$$h_1 = \pi_{\mathfrak{S}} h, \quad h_2 = \mathcal{R}h \tag{5.6}$$

are equivalent.

Definition 5.2. Let $X_\pm(A, A_1) : \mathcal{H}_{\text{ess}}(A_1) \to \mathcal{H}_{\text{ess}}(A)$ be the operator with the domain of definition

$$\begin{aligned} D(X_\pm(A, A_1)) = \{z \in \mathcal{H}_{\text{ess}}(A_1) :\ & \text{there exists } w \in \mathcal{H}_{\text{ess}}(A) \\ & \text{such that } W_\pm(B,A)w = W_\pm(B,A_1)z\} \end{aligned} \tag{5.7}$$

and the action

$$X_\pm(A, A_1) = W_\pm(B,A)^{-1} W_\pm(B, A_1). \tag{5.8}$$

Recall that the operators $W_\pm(B,A)$, $W_\pm(B,A_1)$ are invertible, so

$$X_\pm(A, A_1)^{-1} = X_\pm(A_1, A). \tag{5.9}$$

Definition 5.3. The closure of the operators $X_\pm(A, A_1)$ in the space $L^2(0,\infty)$ is called *wave operator*:

$$W_\pm(A, A_1) = \overline{X(A, A_1)}. \tag{5.10}$$

Operator A is called **regular** if the function $\mathcal{F}\eta(\tau)$ is rational. Instead of $\Gamma_\pm$ we will write $\Gamma_\pm(A)$ or $\Gamma_\pm(A_1)$.

Theorem 5.4. *Let A, A_1 be regular operators and $\mathfrak{S}(A) = \mathfrak{S}(A_1)$ (see (5.1)), then*

1) *the operators $W_\pm(A, A_1)$, $W_\pm(A, A_1)^{-1}$ are bounded and*

$$W_\pm(A, A_1)^{-1} = W_\pm(A_1, A); \tag{5.11}$$

2) *the subspaces*

$$M_\pm = D(W_\pm(A, A_1)), \quad N_\pm = R(W_\pm(A, A_1))$$

are closed subspaces respectively in $\overline{\mathcal{H}_{\mathrm{ess}}(A_1)}$ and $\overline{\mathcal{H}_{\mathrm{ess}}(A)}$ and have a finite codimension.

Proof. Let us consider the operator $W_+(A, A_1)$ only. In view of (3.21) the equation

$$W_+(B, A)w = W_+(B, A_1)z, \quad \Gamma_+(A_1)z \in H$$

for an unknown element w is equivalent to the equation

$$w - \mathcal{F}^{-1}\Gamma_+(A)w = z - \mathcal{F}^{-1}\Gamma_+(A_1)z, \quad \Gamma_+(A_1)z \in H. \tag{5.12}$$

Note that the equation under the form

$$\Gamma_+(A)w = f, \quad f \in H$$

is equivalent to a system containing a singular integral equation for the function $\psi = \mathcal{F}w$

$$\begin{cases} \dfrac{1}{\delta_+(\tau)}(R_\tau(B)w, \eta)_+ = f(\tau), \quad f \in H \\ (w, w_k) = 0, \quad k = 1, \dots, k_0 \end{cases}$$

where $\{w_k\}$ denotes the set of root elements and functionals corresponding to all eigenvalues and spectral singularities of the operator A (see Definition 3.1). The expression (w, w_k) signifies the scalar product or value of the functional w_k.

Consequently, if $\varphi = \mathcal{F}z$, $\psi = \mathcal{F}w$ then equation (5.12) is equivalent to the system (see (4.1)–(4.3))

$$\begin{cases} \dfrac{1}{\delta_+(\tau)}\left[L_{p,q}\psi(\tau) + R\psi(\tau)\right] = \dfrac{1}{\delta_{1,+}(\tau)}\left[L_{p_1,q_1}\varphi(\tau) + R_1\varphi(\tau)\right] \\ (\psi, \psi_k) = 0, \quad k = 1, \dots, k_0 \end{cases}$$

where $\psi_k = \mathcal{F}w_k$.

Let (see (5.2))

$$p_0(\tau) = \Delta_+(\tau)p_1(\tau), \quad q_0(\tau) = \Delta_+(\tau)q_1(\tau), \quad R_0\varphi(\tau) = \Delta_+(\tau)R_1\varphi(\tau).$$

Then the considered system takes the form

$$L_{p,q}\psi + R\psi = L_{p_0,q_0}\varphi + R_0\varphi \tag{5.13}$$

$$(\psi, \psi_k) = 0, \quad k = 1, \dots, k_0 \tag{5.14}$$

We are looking for a solution of the form

$$\psi = L_{m,n}\varphi + L_{a,b}\theta, \quad \theta \in H \tag{5.15}$$

where (see (5.3))

$$a = p = \frac{1}{2}(\delta_+ + \delta_-), \quad b = -q = \frac{1}{2i}(\delta_+ - \delta_-),$$

and we denote

$$H_{a,b} = \{\varphi \in H : \text{ the solution of (5.13) has the form (5.15)}\}. \tag{5.16}$$

Substituting (5.15) into (5.13) and taking into account the relations $pm - qn = p_0$, $qm + pn = q_0$ and also $pa - qb = p^2 + q^2 = \delta_+\delta_-$, $pb + qa = 0$ we obtain the equation (see Lemma 4.2)

$$\delta_+(\tau)\delta_-(\tau)\theta(\tau) + \int_0^\infty M(s,\tau)\theta(s)ds = f(\tau) \tag{5.17}$$

where the kernel $M(s,\tau)$ is defined by the identity

$$-b(\tau)K_{a,b}\theta(\tau) + RL_{a,b}\theta(\tau) = \int_0^\infty M(s,\tau)\theta(s)ds. \tag{5.18}$$

The right side of the equation (5.17), i.e.,

$$f(\tau) = R_0\varphi(\tau) - RL_{m,n}\varphi(\tau) - q(\tau)K_{m,n}\varphi(\tau), \tag{5.19}$$

is an integral operator on φ:

$$f(\tau) \equiv \int_0^\infty F(s,\tau)\varphi(s)ds. \tag{5.20}$$

Let σ_j, $j = 1, \ldots, j_0$ be roots of the function $\delta_+(\sigma)\delta_-(\sigma)$ of multiplicity k_j and $P(\sigma)$, $Q(\sigma)$ corresponding functions in Definition 5.1. In view of condition (3.19) the function $\xi(\tau)$, defined by the relation

$$\delta_+(\sigma)\delta_-(\sigma) = Q(\tau)\xi(\tau),$$

is bounded, i.e., $C_1 \leqslant |\xi(\tau)| \leqslant C_2$, $\tau > 0$ for some $C_1, C_2 > 0$. Using relations (5.4)–(5.6) we obtain that equation (5.17) is equivalent to the system

$$\xi(\tau)\theta(\tau) + \int_0^\infty (\mathcal{R}M(s,\cdot))(\tau)\theta(s)ds = \mathcal{R}f(\tau), \quad \tau > 0, \tag{5.21}$$

$$\int_0^\infty (\pi_{\mathfrak{G}}M(s,\cdot))(\tau)\theta(s)ds = \pi_{\mathfrak{G}}f(\tau). \tag{5.22}$$

The integral operator in relation (5.21) is compact in view of the next Lemma 5.5 (recall that $\mathcal{F}\eta(\tau)$, $\mathcal{F}\eta_1(\tau)$ are rational functions). So, the solution $\theta(\tau)$ of equation (5.21) exists iff

$$\mathcal{R}f \perp g_1, \ldots, q_\alpha, \quad g_i \in H$$

where $\{g_i\}$ is a finite set of elements. According to (5.19)–(5.20) the last condition signifies that

$$\varphi \perp \varphi_1, \dots, \varphi_\alpha \quad \varphi_i \in H. \tag{5.23}$$

The solution θ of (5.21) as operator on φ represents a bounded operator in H. Consequently, the space $H_{a,b}$ (see (5.16)) is closed and, in view of (5.23),

$$\operatorname{codim} H_{a,b} < \infty. \tag{5.24}$$

The codimension of the domain of values $L_{a,b}\theta$ is finite too. So, under the form (5.15)–(5.16) we describe all solutions ψ except maybe a subspace of finite dimension.

Now we come back to the system (5.13)–(5.14). Let equation (5.13) hold. Then in relation (5.12) we have that $\Gamma_+(A_1)\mathcal{F}^{-1}\varphi \in H \iff \Gamma_+(A)\mathcal{F}^{-1}\psi \in H$. If condition (5.14) where ψ is given by the expression (5.15) holds then relation (5.12) holds too. Therefore $\mathcal{F}^{-1}\varphi \in D(W_+(B, A_1))$, $\mathcal{F}^{-1}\psi \in D(W_+(B, A))$ and, consequently, $\varphi \in D(X_+(A, A_1))$.

Due to Corollary 3.8 we have

$$\overline{\mathcal{F}D(W_+(B, A_1)) \cap H_{a,b}} = H_{a,b}.$$

Therefore, as a result, the estimate

$$\operatorname{codim} \overline{D(X_+(A, A_1))} < \infty$$

is proved. In addition, the operator $X_+(A, A_1) : H \to H$ is bounded. If we repeat the proof for the pair (A_1, A) instead (A, A_1) then we obtain that

$$\operatorname{codim} \overline{R(X_+(A, A_1))} < \infty$$

and the inverse operator $X_+(A, A_1)^{-1} : H \to H$ is bounded too.

Consequently

$$D(\overline{X_+(A, A_1))}\overline{D(X_+(A, A_1))}, \quad R(\overline{X_+(A, A_1))}\overline{R(X_+(A, A_1))}$$

what signifies that the subspaces $M_+ = D(W_+(A, A_1))$ and $N_+ = R(W_+(A, A_1))$ are closed and have finite codimension.

The theorem is proved.

Lemma 5.5. *The following estimate holds* (*see* (5.4), (5.18))

$$\int_0^\infty \int_0^\infty |\mathcal{R}M(s,\tau)|^2 \frac{\rho(\tau)}{\rho(s)} ds d\tau < \infty. \tag{5.25}$$

The proof is based on the properties of the operator $\mathcal{R}_\tau$ in the class of the considered functions (see, e.g., (4.9))

From Definition 5.2 and Theorem 5.4 results the following Corollary.

Corollary 5.6. *If $z \in D(W_+(B, A_1))$ then the equality*

$$W_+(B, A)w = W_+(B, A_1)z \tag{5.26}$$

holds iff $w = W_+(A, A_1)z$, where $z \in D(W_+(A, A_1))$.

Proposition 5.7. *Let $s \in (-\infty, \infty)$. Then the following inclusions*

$$e^{isA}\mathcal{H}_{\text{ess}}(A) \subset \mathcal{H}_{\text{ess}}(A) \tag{5.27}$$

$$e^{isA}D(W_+(B,A)) \subset D(W_+(B,A)) \tag{5.28}$$

and the relation

$$e^{-isB}W_+(B,A)zW_+(B,A)e^{-isA}z,\ z \in D(W_+(B,A)) \tag{5.29}$$

hold.

Proof. It is known that

$$e^{isA}R_\zeta(A) = R_\zeta(A)e^{isA},\ \zeta \in \rho(A).$$

The operator e^{isA} is bounded. Consequently inclusion (5.27) results from Definition 3.1. Consider now the identity

$$e^{i(t+s)B}e^{-i(t+s)A}z = e^{isB}(e^{itB}e^{-itA})e^{-isA}z. \tag{5.30}$$

Suppose $z \in D(A) \cap \mathcal{H}_{\text{ess}}(A)$. As $e^{isA}D(A) \subset D(A)$ then in view of (5.27)

$$e^{-isA}z \in D(A) \cap \mathcal{H}_{\text{ess}}(A). \tag{5.31}$$

If additionally $\Gamma_\pm z \in H_0 \cap H$ then $z \in D(W_+^0(B,A))$ and in the left side of (5.30) there exists the limit if $t \to +\infty$ (see Prop. 3.4). The existence of the limit in the right side of (5.35) and condition (5.31) give (using once more Proposition 3.4) $\Gamma_+(e^{-isA}z) \in H_0 \cap H$. Consequently, $e^{-isA}z \in D(W_+^0(B,A))$ and

$$e^{-isB}W_+^0(B,A)z = W_+^0(B,A)e^{-isA}z.$$

The operators e^{-isB}, e^{-isA} are bounded, so as corollary, the relation is proved.

Theorem 5.8. *If $z \in D(W_+(A,A_1)) \cap D(W_+(B,A_1))$ then $e^{isA_1}z \in D(W_+(A,A_1)) \cap D(W_+(B,A))$ and*

$$e^{isA}W_+(A,A_1)z = W_+(A,A_1)e^{isA_1}z. \tag{5.32}$$

Proof. Proposition 5.7 for the operators A and A_1 respectively gives

$$e^{isB}W_+(B,A)w = W_+(B,A)e^{isA}w,\ w \in D(W_+(B,A)) \tag{5.33}$$

and

$$e^{isB}W_+(B,A_1)z = W_+(B,A_1)e^{isA_1}z,\ z \in D(W_+(B,A_1)). \tag{5.34}$$

Suppose $z \in D(W_+(A,A_1)) \cap D(W_+(B,A_1))$. Then

$$w = W_+(A,A_1)z \in D(W_+(B,A))$$

(see Corollary 5.6). Substituting into (5.33) we obtain

$$e^{isB}W_+(B,A)W_+(A,A_1)z = W_+(B,A)e^{isA}W_+(A,A_1)z.$$

The use of relation (5.26) in the left side gives

$$e^{isB}W_+(B,A_1)z = W_+(B,A)e^{isA}W_+(A,A_1)z.$$

The comparison with (5.34) gives

$$W_+(B,A)e^{isA}W_+(A,A_1)z = W_+(B,A_1)e^{isA_1}z.$$

Due to Corollary 5.6 we obtain that

$$e^{isA_1}z \in D(W_+(A,A_1)) \cap D(W_+(B,A_1))$$

and that relation (5.32) holds.

The theorem is proved.

We will discuss the subspaces (see Theorem 5.4)

$$\mathcal{H}_{\text{ess}}(A_1|A) = D(W_+(A,A_1)),\ \mathcal{H}_{\text{ess}}(A|A_1) = R(W_+(A,A_1)).$$

For this aim we will consider a family of operators A with one spectral singularity $\sigma_0 \in (0,\infty)$ of multiplicity 1. Then $\delta_-(\sigma_0)\delta_+(\sigma_0) = 0$ and from equation (5.17) it follows (if $\tau = \sigma_0$)

$$\int_0^\infty M(s,\sigma_0)\theta(s)ds = f(\sigma_0). \tag{5.35}$$

Using the decompositions of the functions $M(s,\tau)$, $f(\tau)$ according to the general formula (see (5.4))

$$h(\tau) = h(\sigma_0)\frac{\sigma_0+1}{\tau+1} + \frac{\tau-\sigma_0}{\tau+1}\mathcal{R}h(\tau),\ \mathcal{R}h(\tau) = h(\tau) + (\sigma_0+1)\mathcal{R}_{\sigma_0}h(\tau)$$

we obtain (see (5.21))

$$\xi(\tau)\theta(\tau) + \int_0^\infty \mathcal{R}M(s,\tau)\theta(s)ds = \mathcal{R}f(\tau),\ \xi(\tau) = \frac{\tau+1}{\tau-\sigma_0}\delta_-(\tau)\delta_+(\tau).$$

The solution $\theta(\tau)$ defines a bounded operator $\Phi : L^2_\rho(0,\infty) \to L^2_\rho(0,\infty)$, i.e., $\theta = \Phi(\mathcal{R}f)$. Substituting θ into (5.35) we obtain

$$V.p.\int_0^\infty \varphi^*(s)\left[f(s) + (\sigma_0+1)\frac{f(s)}{s-\sigma_0}\right]ds = f(\sigma_0)\left[1 + (\sigma_0+1)V.p.\int_0^\infty \frac{\varphi^*(s)}{s-\sigma_0}ds\right] \tag{5.36}$$

where $\varphi^*(s) = \Phi^*(M(\cdot,\sigma_0))(s)$.

In the concrete case we consider the operator A (see (2.2)) generated by an element $\eta \in L^2(0,\infty)$ such that the transformation $\gamma = \mathcal{F}\eta$ is

$$\gamma(\tau) = \frac{-k}{\tau - (\sqrt{\sigma_0}+ik)^2},\ k>0,\ \sigma_0>0. \tag{5.37}$$

By analogy, an other value k_1 defines the function $\gamma_1(\tau)$ and the operator A_1. Such operators have a unique spectral singularity σ_0 which is not an eigenvalue (because the equation $-y'' = \sigma_0 y$ has no solution in $L^2(0,\infty)$). Obviously $\eta(x) = -ke^{i(\sqrt{\sigma_0}+ik)x}$ (compare (3.18)), so, the element η does not satisfy conditions (2.4)–(2.5). But we need (2.4)–(2.5) in the general case of η only to obtain the differentiability of some auxiliary function. In the concrete case (5.37) we can repeat again the statements of Section 2. Note that

$$\mathcal{R}_\tau(\gamma\rho)(s) = \frac{\gamma(s)\rho(s)-\gamma(\tau)\rho(\tau)}{s-\tau}$$
$$= -\frac{k}{2\pi}\frac{1}{\sqrt{s}+\sqrt{\tau}}\left[\frac{1}{(\sqrt{s}-\sqrt{\sigma_0}-ik)(\sqrt{\tau}-\sqrt{\sigma_0}-ik)}\right.$$
$$\left.+\frac{1}{(\sqrt{s}+\sqrt{\sigma_0}+ik)(\sqrt{\tau}+\sqrt{\sigma_0}+ik)}\right]. \tag{5.38}$$

A simple calculus gives

$$\delta_+(\tau) = \frac{\sqrt{\tau}-\sqrt{\sigma_0}}{\sqrt{\tau}-\sqrt{\sigma_0}+ik},\ \delta_+(\tau) = \frac{\sqrt{\tau}+\sqrt{\sigma_0}}{\sqrt{\tau}+\sqrt{\sigma_0}-ik}$$

and

$$b(\tau) = \frac{1}{2i}(\delta_+(\tau)-\delta_-(\tau)) = \frac{-k\sqrt{\tau}}{\tau-\sigma_0+2ik\sqrt{\sigma_0}+k^2}. \tag{5.39}$$

If $S(\tau) = \delta_-(\tau)/\delta_+(\tau)$ then

$$I(\tau) = \frac{S_1(\tau)}{S(\tau)} = \frac{\sqrt{\tau}-\sqrt{\sigma_0}+ik_1}{\sqrt{\tau}+\sqrt{\sigma_0}-ik_1}\frac{\sqrt{\tau}+\sqrt{\sigma_0}-ik}{\sqrt{\tau}-\sqrt{\sigma_0}+ik} \tag{5.40}$$

and

$$\mathcal{R}_\tau(I)(s) = 2i(k-k_1)\Big[(\sqrt{s}+\sqrt{\tau})(\sqrt{s}+\sqrt{\sigma_0}-ik_1)$$
$$\times(\sqrt{\tau}+\sqrt{\sigma_0}-ik_1)(\sqrt{s}-\sqrt{\sigma_0}+ik)(\sqrt{\tau}-\sqrt{\sigma_0}+ik)\Big]^{-1}. \tag{5.41}$$

Let

$$F_\pm\varphi(\tau) = \frac{1}{2}\varphi(\tau) \pm \frac{1}{2i}\mathcal{H}[\varphi](\tau).$$

Lemma 5.9. *The expression $f(\tau)$ (see (5.19)) takes the form*

$$f(\tau) = -\frac{\sqrt{\tau}-\sqrt{\sigma_0}-ik_1}{\sqrt{\tau}-\sqrt{\sigma_0}+ik}\int_0^\infty \mathcal{R}_\tau(\gamma_1\rho)(s)\varphi(s)ds + \frac{1}{2i}b(\tau)(I(\tau)-1)\int_0^\infty \mathcal{R}_\tau \ln s\varphi(s)ds$$
$$+\int_0^\infty\left[\mathcal{R}_\tau(\gamma\rho)(s)F_-\varphi(s) + \left(\mathcal{R}_\tau(\gamma\rho)(s)I(s)+\frac{1}{\pi}b(\tau)\mathcal{R}_\tau(I)(s)\right)F_+\varphi(s)\right]ds. \tag{5.42}$$

The proof reduces to algebraic transformations and the relation

$$\int_0^\infty \varphi(\tau)\mathcal{H}[\psi](\tau)d\tau = -\int_0^\infty \mathcal{H}[\varphi](\tau)\psi(\tau)d\tau.$$

Obviously, if $\tau = \sigma_0$ then

$$f(\sigma_0) = -\frac{k_1}{k}\int_0^\infty \mathcal{R}_{\sigma_0}(\gamma_1\rho)(s)\varphi(s)ds + \frac{1}{2i}b(\sigma_0)(I(\sigma_0)-1)\int_0^\infty \mathcal{R}_{\sigma_0}\ln s\varphi(s)ds$$

$$+\int_0^\infty \left[\mathcal{R}_{\sigma_0}(\gamma\rho)(s)F_-\varphi(s) + \left(\mathcal{R}_{\sigma_0}(\gamma\rho)(s)I(s) + \frac{1}{\pi}b(\sigma_0)\mathcal{R}_{\sigma_0}(I)(s)\right)F_+\varphi(s)\right]ds. \tag{5.43}$$

Further we suppose that

$$ik_1(\varepsilon) = -2\sqrt{\sigma_0} + i\varepsilon, \quad \varepsilon > 0, \quad \varepsilon \to 0. \tag{5.44}$$

Then in view of (5.40)

$$I(\sigma_0) = \frac{M_1(\varepsilon)}{\varepsilon}, \quad M_1(0) = \frac{2\sqrt{\sigma_0}}{k}(2\sqrt{\sigma_0} - ik) \neq 0$$

and in view of (5.41)

$$\mathcal{R}_{\sigma_0}(I)(s)\frac{M_2(\varepsilon)}{\varepsilon}\frac{Q(s)}{\sqrt{s}-\sqrt{\sigma_0}-i\varepsilon}, \quad M_2(0) = -4i\sqrt{\sigma_0} - 2k \neq 0$$

where

$$Q(s) = \frac{1}{(\sqrt{s}+\sqrt{\sigma_0})(\sqrt{s}-\sqrt{\sigma_0}+ik)}.$$

Now we fix the value k such big that the operator Φ exists and equation (5.36) holds. In this equation the expression $f(\tau)$ only depends on k_1. Taking into account this dependence, condition (5.44) and the relations (5.38)–(5.41) we obtain for the expression (5.43) the following decomposition

$$f(\sigma_0) = f_1(\sigma_0,\varepsilon) + \frac{1}{\varepsilon}f_2(\sigma_0,\varepsilon), \quad f_{1,2}(\sigma_0,\varepsilon) = O(1), \ \varepsilon \to +0. \tag{5.45}$$

Note that

$$f_2(\sigma_0,\varepsilon) = b(\sigma_0)\left[\frac{1}{2i}M_1(\varepsilon)\int_0^\infty \mathcal{R}_{\sigma_0}\ln s\varphi(s)ds\right.$$

$$\left. + \frac{1}{\pi}M_2(\varepsilon)\int_0^\infty \frac{Q(s)}{\sqrt{s}-\sqrt{\sigma_0}-i\varepsilon}F_+\varphi(s)ds\right].$$

Therefore we have a subset N dense in $L^2_\rho(0,\infty)$ (maybe changing k) such that $f_2(\sigma_0, 0) \neq 0$, $\varphi \in N$.

Theorem 5.10. *There exist the operators A, A_1, A_2 of type* (2.2) *with the same set of spectral singularities, $\mathfrak{S}(A) = \mathfrak{S}(A_1) = \mathfrak{S}(A_2)$, such that*

$$\mathcal{H}_{\text{ess}}(A|A_1) \neq \mathcal{H}_{\text{ess}}(A|A_2) \tag{5.46}$$

Proof. It is sufficient to consider the examples stated above with the same unique spectral singularity σ_0. Let the values $k \neq k_1$ be sufficiently big, A, A_1 be corresponding operators. Let φ be a fixed element such that $\varphi \neq 0$, $\varphi \in \mathcal{H}_{\text{ess}}(A|A_1)$. Then equation (5.36) holds. Suppose that k_1 varies according to (5.44), let $k_2 = k_1(\varepsilon)$. If the relation $\mathcal{H}_{\text{ess}}(A|A_1) = \mathcal{H}_{\text{ess}}(A|A_2)$ holds always then equation (5.36) holds too. But this is impossible. Really, due to the form (5.37) the function $\varphi^*(s)$ is differentiable, in view of (5.42) and also (5.38)–(5.41) the left side of the relation (5.36) is bounded if $\varepsilon \to +0$. The right side in (5.36) is unbounded (see (5.45)) for $\varphi \in N$. Obviously for small ε the relation (5.41) is disturbed too for the element $\widetilde{\varphi} \in N \cap B$, where $B = \{\widetilde{\varphi} : ||\widetilde{\varphi} - \varphi|| < r\}$ is some ball. So, $N \cap B \cap \mathcal{H}_{\text{ess}}(A|A_2) = \emptyset$. As the space $\mathcal{H}_{\text{ess}}(A|A_2)$ is closed then $\overline{N \cap B} \cap \mathcal{H}_{\text{ess}}(A|A_2) = \emptyset$ or $B \cap \mathcal{H}_{\text{ess}}(A|A_2) = \emptyset$ what proves (5.46).

The theorem is proved.

Conclusion

In this article we consider the wave operator

$$W_{\pm}(A_2, A_1) = s - \lim_{t \to \pm\infty} e^{itA_2} e^{-itA_1}.$$

Suppose that the operators A_1, A_2 are selfadjoint, and the wave operators

$$W_+(A_1, A_2) \quad \text{and} \quad W_+(A_2, A_1)$$

exist. Then the wave operator

$$W_+(A_2, A_1) : H_{1,ac} \to H_{2,ac}$$

is isometric and transforms the whole of $H_{1,ac}$ onto the whole of $H_{2,ac}$ (see [12], Chap. 10, 12.3).

We study non-selfadjoint case, namely Sturm-Liouville operators with non local boundary conditions. For such operators

1) the wave operator $W_+(A_1, A_1)$ may be defined as bounded operator if the sets of spectral singularities of A_1 and A_2 coincide (Theorem 5.4) and
2) the domain of definition $D(W_+(A_1, A_2))$ in the general case varies simultaneously with the variation of the operator A_2 (Theorem 5.10).

Therefore in the non-selfadjoint case it is expediently to define a so-called relative essential subspace $\mathcal{H}_{\text{ess}}(A_1|A_2)$ instead to define some subspace $\mathcal{H}_{\text{ess}}(A_1)$ for one isolated operator A_1. The idea itself to use the Laplace transform was taken from work [10]. Practically we use the Friedrichs' model. One can find the operators with non-local condition within the Friedrichs' model in [9] too.

References

[1] Birman M.Sh. On perturbation of the spectrum of singular operator under the variation of the boundary condition // Dokl. Akad. SSSR, 1961, n. 137, 761–763 (in Russian).

[2] Sachnovich L.A. Dissipative operators with absolutely continuous spectrum, Trudy Moscow Math. Soc., 1968, v. 19, 211–269 (in Russian).

[3] Pavlov B.S. On the conditions of separability of spectral components for dissipative operators // Izv. Akad. SSSR, ser. math., 1975, v. 39, n.1, 123–148 (in Russian).

[4] Naboko S.N. Functional model of perturbation theory and its applications to the scattering theory // Trudy Moscow Math. Soc., 1980, n. 147, 86–114 (in Russian).

[5] Pavlov B.S. Spectral analysis of differential operators with "spreaded" boundary condition // Probl. Math. Phys., 1973, n. 6, 101–119 (in Russian).

[6] Kato T. Wave operators and similarity for some non-selfadjoint operators // Math. Ann., 1966, n. 162, 258–279.

[7] Kako T., Yajima K. Spectral and scattering theory for a class of non-selfadjoint operators // Sci. Par. Coll. Gen. Educ. Univ. Tokyo, 1976, v. 26, n. 2, 73–89.

[8] Cheremnikh E.V. Nonselfadjoint Friedrichs' model and Weyl function // Dokl. Akad. Ukraine, 2001, n. 8, 22–29 (in Ukrainian).

[9] Cheremnikh E.V. Friedrichs' model and the problems with non-local boundary conditions // Math. Meth. and Phys.–Mech. Fields, 2000, v. 43, n. 3, 146–156 (in Ukrainian).

[10] Alsholm Preben Inverse scattering theory for perturbation rank one // Duke Math. J., 1980, v. 47, n. 2, 11–23.

[11] Diaba F., Cheremnikh E. On asymptotic time for an evolution with non-local boundary condition // J. Dynam. Syst. and Geom. Theories, 2006, v. 4, n. 1, 61–77.

[12] Kato T. Perturbation theory for linear operators, Springer-Verlag, 1966.

[13] Yosida K. Functional analysis, Springer-Verlag, 1965.

E.V. Cheremnikh
Gnatjuka Str. 8, Flat 12A
Lviv
79007, Ukraine
e-mail: echeremn@polynet.lviv.ua

Operator Theory:
Advances and Applications, Vol. 186, 69–87

Semiclassical Results for Ideal Fermion Systems. A Review

Monique Combescure

Abstract. Ideal Fermion systems represent an issue of great recent interest. They consist of N-Fermion gases without interaction but subject to an exterior field or potential. In this paper we review recent studies of such systems in the following situations:

- the low temprerature behavior of confined fermion gases in contact with an exterior reservoir
- the action of a magnetic field on two-dimensional electronic devices
- the transport and dissipation for time-dependent scatterers connected to several leads, known as "quantum pumps".

The response of these different systems under the influence of the exterior potential or field is considered. In the first situation the response is calculated in terms of the "dynamical susceptibility" in the linear response framework, which yields a generalized Kubo formula. In the second situation the response is expressed by the magnetic susceptibility and the magnetization which are estimated semiclassically for various regimes of the temperature as compared with powers of the Planck constant. In the last situation the response is a measurable current estimated in an adiabatic framework.

All these results are contained by a series of papers by D. Robert and myself, and in recent works by Avron, Elgart, Graf and Sadun.

Mathematics Subject Classification (2000). 81-02, 81Q20, 81R30, 82B10.

Keywords. Semiclassical analysis, coherent states methods, fermionic systems, linear response theory.

1. Introduction

The study of ideal fermion gases and of their response in several physical situations has focused a large interest in the recent years. The various situations involved are for example:

(i) external time-dependent perturbations of confined fermion gases in contact with an external reservoir;

(ii) the action of a magnetic field on two-dimensional electronic devices;
(iii) the transport and dissipation for so-called "quantum pumps" modelled by time-dependent scatterers connected to several leads.

The reservoirs (or leads) with which these systems are in contact are at some non-zero temperature T. In the case (i), the "response" is measured via the so-called "dynamical susceptibility" and yields a Generalized Kubo Formula.

In case (ii), the response is the "magnetic susceptibility" which presents interesting and measurable features in the so-called "mesoscopic devices".

In case (iii) the response is a measurable current.

These topics have focused large attention in the physical literature, (see the bibliography), and led to understand the underlying physical phenomena. However until recently, only a few rigorous studies have been attempted so far. (Note [29] where a rigorous semiclassical study of thermodynamic properties of Fermion or Boson gases is performed when the Fermions are confined via a power-law potential; see also [22].)

Recently, some mathematical work has been devoted to the three situations considered above:

- The so-called Linear Response Theory;
- The semiclassical regime where some small parameter (usually the Planck constant) is assumed to tend to zero;
- The adiabatic regime (for case (iii)), where adiabaticity of the external time-dependent perturbation is considered small (and thus can be treated as a "semiclassical parameter").

In this review, we shall present mathematical results for the three cases (i)–(iii). They will be based respectively on [10], [9], [2]. Of course this review is not exhaustive in the sense that many other interesting physical aspects have been considered but have not, to our knowledge, received so far an attention towards their rigorous treatment.

Let us mention a few of them, with bibliographic references, so that interested mathematical-physicists might find a source of inspiration:

- The dielectric response of dilute electron gases in small semiconductor devices ([33], [27]);
- Static semiclassical response of a bounded electron gas ([20], [26]);
- The dynamics of ideal Fermion gases in the presence of a combined time-dependent harmonic potential, and a periodic potential ([6]);
- The Friedel oscillations at zero temperature of the particle density-distribution for ideal fermion gases in one-dimensional harmonic traps ([15]);
- Quasi one-dimensional electron gas in a semiconductor quantum wire in the presence of some random field ([27]);
- A kinetic semiclassical equation for modelling transport in degenerate gases ([24], [23]);

- Ground-state number fluctuations of trapped ideal and interacting fermion gases ([31], [32]);
- the "Riemannium", i.e., a chaotic Fermion gas whose single-particle energy-levels are given by the (imaginary part of) the complex zeros of the Riemann zeta function ([19]);
- The eigenvalue statistics of quantum ideal gases with single-particle energy-levels of the form $E_n = n^\alpha$ ([11]);
- The behavior of a degenerate quantum system as a precursor to a "paired Fermi condensate" analogous the the Bose-Einsten Condensate, at very small temperature ([8]);
- The link of Fermi-Dirac statistics with the number theory ([17]).

The plan of this review article is as follows:

In Section 2 we develop a rigorous semiclassical expansion of the "dynamical susceptibility" for an ideal confined fermion gas, at small temperature, assuming some **chaoticity assumption** of the one-particle motion on the Fermi energy surface [10].

In Section 3, we consider an electron gas in any dimension, subject to an inhomogeneous magnetic field, in a semiclassical regime, and examine various temperature regimes where the magnetic response can be evaluated. This allows a description of the "de Haas-van Alphen" oscillations [9].

In Section 4 we give the results on Transport and Dissipation in "Quantum Pumps" as described by the work of Avron-Elgart-Graf-Sadun [2].

2. Semiclassical results in linear response theory

Let us consider a system of **non-interacting fermions**

- confined by potential,
- in contact with **exterior reservoir** at temperature T.

We assume that a time-varying potential drives the system

- out of EQUILIBRIUM,
- but close to it.

The Response to this external perturbation is measured by the **dynamical susceptibility**.

The complete rigorous analysis is difficult in that it involves the non-equilibrium statistical mechanics.

Here we shall follow a semi-empirical route (small perturbations) and we try to solve this problem "to first order in perturbation", namely in the **Linear Response Framework.**

- We assume some "chaoticity assumption" on the one-body classical dynamics;
- We derive a generalized Kubo Formula;
- We perform semiclassical expansions in $\hbar$ when $\hbar \to 0$;

- We exhibit a **low temperature limit**, where the closed classical trajectories manifest themselves in the Response function (or in the Dynamical Susceptibility).

The model is as follows: we start with a classical Hamiltonian of the form:

$$H(q,p) = \frac{p^2}{2m} + V(q) \quad \text{in } \mathbb{R}^n,$$

where the potential is confining and smooth enough:

$$V(q) \geq c_0(1+q^2)^{s/2}, \qquad c_0, s > 0,$$

$$|\partial^\alpha V(q)| \leq C_\alpha V(q), \quad C_\alpha > 0, \ \forall \alpha \in \mathbb{N}^n,$$

$\hat{H}$ is the Weyl quantization of H; it is a self-adjoint operator on $\mathcal{H} := L^2(\mathbb{R}^n)$

The assumptions imply that $\mathrm{sp}(H) \subset [0,\infty)$ is pure-point:

$$\hat{H}\psi_j = E_j\psi_j, \ j \in \mathbb{N}, \ \psi_j \in \mathcal{H}.$$

The infinite system of fermions is described via states in the Fock Space: $\mathcal{F} := \oplus_{n=0}^{\infty} \otimes_a^n \mathcal{H}$, where $\otimes_a^n \mathcal{H}$ antisymetric tensor product of $\mathcal{H}$. It represents the space of n fermionic states.

The framework is thus the **second quantization**:

$$\hat{\hat{H}} := d\Gamma(\hat{H}) \text{ acting on } \mathcal{F} \quad \text{second quantized Hamiltonian},$$

$$\hat{\hat{N}} := d\Gamma(\mathbb{1}_{\mathcal{H}}) \quad \text{number of particles operator},$$

$n_j \in \{0,1\}$ is the occupation number of any state ψ_j in $\mathcal{F}$ (which expresses the Pauli principle);

$$\widehat{N}_{l,n}(\psi_{j_1} \wedge \cdots \wedge \psi_{j_n}) = n_l(\psi_{j_1} \wedge \cdots \wedge \psi_{j_n}), \ \forall l = j_1, \ldots, j_n,$$

tells whether or not the level E_l is occupied in a given state of $\otimes_a^n \mathcal{H}$.

Define:

$$\widehat{N}_j := \oplus_{n=0}^{\infty} \widehat{N}_{j,n},$$

$$\hat{\hat{N}} = \sum_{j=0}^{\infty} \widehat{N}_j, \quad \hat{\hat{H}} = \sum_{j=0}^{\infty} \widehat{N}_j E_j, \quad \left[\hat{\hat{H}}, \hat{\hat{N}}\right] = 0.$$

The Grand-Canonical Partition function is:

$$Z_G := \mathrm{Tr}\left(e^{\beta\hat{\hat{H}} + \kappa\hat{\hat{N}}}\right),$$

where $\beta = \frac{1}{kT}$, $\kappa\beta\mu$, T is the temperature, and μ is the chemical potential.

The trace is taken in $\mathcal{F}$. Thus Z_G can be rewritten as:

$$Z_G = \prod_{j=0}^{\infty} \left[1 + e^{-\beta(E_j-\mu)}\right]^{-1}.$$

Let f be the Fermi-Dirac function $f(x) := (1+e^x)^{-1}$. The operator in $\mathcal{H}$:

$$\hat{\rho}_{eq} := f\left(\beta(\hat{H}-\mu)\right)$$

is called the Fermi-Dirac one-body operator.

Now assume that the one-body Hamiltonian is slightly **perturbed** in a time-dependent way:

$$\hat{H}_\lambda(t) := \hat{H} + \lambda \hat{A} F(t)$$

λ is a small parameter, $\hat{A}$ is some self-adjoint operator, and F is of the form, with $\alpha > 0$:

$$F(t) = \begin{cases} e^{\alpha t} & t < 0 \\ 1 & t \geq 0\,. \end{cases}$$

The time-dependent density matrix $\hat{\rho}_\lambda(t)$ solves

$$i\hbar \frac{\partial \hat{\rho}_\lambda}{\partial t} = \left[\hat{H}_\lambda\,,\, \hat{\rho}_\lambda\right], \qquad \lim_{t\to-\infty} \widehat{\rho}_\lambda(t) = \widehat{\rho}_{eq}.$$

The problem is:

to which extent will $\widehat{\rho}_\lambda(t)$ wander from $\widehat{\rho}_{eq}$ as the perturbation is switched on?

In the Linear Response Theory the aim is to answer this question "to first order in λ"

Assumption: $A(q)$ is a multiplicative smooth function $|A(q)| \leq C(1+q^2)$.

Denote by $U(t) := e^{-it\hat{H}/\hbar}$ the unitary group of evolution for $\hat{H}$. We know from the work of Yajima [35] that the unitary evolution operator for $\hat{H}_\lambda(t)$ exists and we denote it by

$$V_\lambda(s,t),$$

and define the state ("density matrix") at time t as

$$\hat{\rho}_\lambda(s,t) := V_\lambda(t,s)\hat{\rho}_{eq} V_\lambda(s,t),$$

that equals $\hat{\rho}_{eq}$ at $t = s$.

Proposition 2.1. *The mapping $\lambda \mapsto \widehat{\rho}_\lambda(t,t_0)$ is differentiable near $\lambda = 0$ in the trace-class operator norm sense and we have:*

$$\frac{d}{d\lambda}\widehat{\rho}_\lambda(t,t_0)\bigg|_{\lambda=0} = \frac{1}{i\hbar}\int_{t_0}^{t} dt'\; F(t')\; U(t-t')[\widehat{\rho}_{eq}, \widehat{A}]U(t'-t).$$

Moreover $\widehat{\rho}_\lambda(t,t_0)$ has a limit as $t_0 \to -\infty$, in the trace-class operator norm sense, called $\widehat{\rho}_{eq}(t,\lambda)$ and is also differentiable in λ. Moreover we have:

$$\frac{d}{d\lambda}\widehat{\rho}_{eq}(t,\lambda)\bigg|_{\lambda=0} = \frac{1}{i\hbar}\int_{-\infty}^{t} ds\; F(s)[\widehat{\rho}_{eq}, \widehat{A}_{t-s}],$$

$\widehat{A}_t$ being, by definition, the Heisenberg observable at time t (for the quantum evolution governed by $\widehat{H}$):

$$\widehat{A}_t = U(t)\ \widehat{A}\ U(t)^*.$$

This gives in the trace norm sense:

$$\widehat{\rho}_{eq}(t,\lambda) - \widehat{\rho}_{eq}\frac{\lambda}{i\hbar}\int_{-\infty}^{t} dt' F(t')[\widehat{\rho}_{eq}, \widehat{A}_{t-t'}] + o(\lambda).$$

Assume we want to measure some self-adjoint operator $\hat{B}$ in the "almost-stationary" state $\widehat{\rho}_{eq}(t,\lambda)$:

$$J_\lambda(t) = \mathrm{tr}\left\{\widehat{B}\left(\widehat{\rho}_{eq}(t,\lambda) - \widehat{\rho}_{eq}\right)\right\}.$$

Here tr denotes the trace operation in $\mathcal{H}$ (whereas Tr denoted the trace operation in $\mathcal{F}$).

The first order contribution as $\lambda \to 0$ is:

$$J_L(t) = \lambda \int_{-\infty}^{t} dt'\ F(t')\ \Phi(t-t'),$$

where:

$$\Phi(t) = \frac{1}{i\hbar}\mathrm{tr}\left(\widehat{B}\left[\widehat{\rho}_{eq}, \widehat{A}_t\right]\right) = \frac{1}{i\hbar}\mathrm{tr}\left(\widehat{\rho}_{eq}\left[\widehat{A}, \widehat{B}_{-t}\right]\right)$$

using the cyclicity of the trace.

We take the Fourier transform (in the distributional sense) of $\Phi(t)$; it is the "generalized susceptibility":

$$\chi_{A,B}(\omega) = \int_{-\infty}^{+\infty} \Phi(t)\ e^{i\omega t}\ dt.$$

Our aim is to obtain a semiclassical expansion of $\chi_{A,B}(\omega)$ as $\hbar \to 0$, $\beta \to \infty$, namely a semiclassical expansion at low temperature.

Given g any smooth test function whose Fourier transform $\tilde{g}$ is of exponential decrease at ∞:

$$\int \chi_{A,B}(\omega)g(\omega)d\omega = \frac{1}{i\hbar}\int \mathrm{tr}\left(f_\sigma\left(\frac{\widehat{H}-\mu}{\hbar}\right)[\widehat{A}, \widehat{B}_t]\right)\widetilde{g}(t)dt$$

where σ is a new parameter: $\sigma := \beta\hbar$. f_σ is simply the rescaled Fermi-Dirac function: $f_\sigma(x) := f(\sigma x)$.

However the integral

$$\int \mathrm{tr}\left(f_\sigma\left(\frac{\widehat{H}-\mu}{\hbar}\right)[\widehat{A}, \widehat{B}_t]\right)\widetilde{g}(t)dt$$

suffers from a singularity as $\sigma \to 0$. To overcome this difficulty, we "regularize" it, using instead of f_σ the function

$$f_{\sigma,\eta} := f_\sigma * \eta,$$

where the Fourier Transform $\tilde{\eta}$ has compact support. Thus in distributional sense on the test function g, the **regularized dynamic susceptibility** in the linear response regime can be rewritten as:

$$\frac{1}{2i\pi\hbar}\int\int \mathrm{tr}\left(e^{is(\widehat{H}-\mu)/\hbar}\left[\widehat{A},\widehat{B}_t\right]\right)\widetilde{f_\sigma}(s)\widetilde{\eta}(s)\widetilde{g}(t)dsdt$$

$$:= \int\int dsd\omega\tilde{\eta}(s)g(\omega)\chi_{A,B}(s,\omega).$$

Our aim is to obtain semiclassical estimates of $\chi_{A,B}(s,\omega)$ **in distributional sense**, in terms of the auxiliary parameter $\sigma := \hbar\beta$.

Recall that μ is the chemical potential, which has dimension of energy.

Assumptions:

- The symbol B of $\hat{B}$ satisfies $|\partial_q^\alpha \partial_p^{\alpha'} B| \leq C_{\alpha,\alpha'}$, $\forall\ |\alpha| + |\alpha'| \geq 2$,
- On the classical energy surface $\Sigma_\mu := \{q, p \in \mathbb{R}^n : H(q,p) = \mu\}$ all closed orbits γ are isolated (Gutzwiller assumption).

Let $T_\gamma, P_\gamma, \nu_\gamma$ be respectively the classical period, the Poincaré map and the Maslov index of the orbit γ, and let

$$C_{A,B,\mu}(t) := \int_{[H=\mu]} AB_t \frac{d\sigma_\mu}{|\nabla H|},$$

$$c_\gamma(t) := \int_0^{T_\gamma} A_s(q,p)B_{s+t}(q,p)ds \equiv \sum_{k=-\infty}^{+\infty} c_{\gamma,k}e^{2i\pi kt/T_\gamma}.$$

Then the following result holds true:

Theorem 2.2. *Under the above assumptions, we have, in distributional sense:*

$$\begin{aligned}\chi_{A,B}(s,\omega) = &-h^{-n}\delta_0(s)\otimes\widetilde{C'_{A,B,\mu}}(\omega) + \sum_{j\geq 1}\hbar^{j-n}\mu_j(s,\omega)\\ &+\sum_{\gamma:T_\gamma\neq 0}\frac{\pi e^{i(S_\gamma/\hbar+\nu_\gamma\pi/2)}}{\hbar\sigma\sinh(\pi T_\gamma/\sigma)|det(1-P_\gamma)|^{1/2}}\\ &\qquad\times\left(\delta_{T_\gamma}(s)\otimes\sum_k c_{\gamma*,k}\delta(\omega-\frac{2k\pi}{T_{\gamma*}}) + \sum_{j\geq 1}\hbar^j\nu_{j,\gamma}(s,\omega)\right)\\ &+O(\hbar^{a\gamma_H-\varepsilon-n}),\end{aligned}$$

where

- μ_j *and* $\nu_{j,\gamma}$ *are distributions such that* $\mathrm{Supp}(\mu_j) \subseteq \{0\}\times\mathbb{R}$,
- $\mathrm{Supp}(\nu_{j,\gamma}) \subseteq \{T_\gamma\}\times\mathbb{R}$, γ_H *is a non negative constant depending only on* H *and* μ,
- P_γ *is the Poincaré map of the* **closed** *orbit* γ,
- S_γ *is the classical action along* γ,
- γ^* *is the* **primitive** *closed orbit corresponding to* γ, *and*
- T_{γ^*} *the corresponding period.*

Meaning of the result

It is true in the distributional sense in s, ω on functions $\tilde{\eta}(s)g(\omega)$ where $\tilde{\eta}$ has **compact support**, and $\tilde{g}$ is of **exponential decrease** $\tilde{g}(t) \sim e^{-a|t|}$, as $t \to \pm\infty$. Thus the sum over **closed orbits** γ is **finite** (restricted to $T_\gamma \in \mathrm{Supp}(\tilde{\eta})$).

Idea of proof. We have:

$$\langle \chi(s,\omega)\ ,\ \tilde{\eta}(s)g(\omega)\rangle = \frac{1}{2i\pi\hbar}\int\int \mathrm{Tr}\left(e^{is(\widehat{H}-\mu)/\hbar}\left[\widehat{A},\widehat{B}_t\right]\right)\widetilde{f_\sigma}(s)\tilde{\eta}(s)\tilde{g}(t)dsdt.$$

Write:

$$\mathrm{Tr}(C) \equiv \frac{1}{(2\pi\hbar)^n}\int_{\mathbb{R}^{2n}} dz\langle z|C|z\rangle,$$

where $|z\rangle$ is a **coherent state** located at $z \in \mathbb{R}^{2n}$. We use precise **semiclassical** estimates of $e^{-is\hat{H}/\hbar}|z\rangle$ with respect to $\hbar,\ z,\ s$.

The trace property is ensured via a suitable partition of unity near energy surface $\{H = \mu = \text{Fermi level}\}$.

Important remarks

- The temperature dependence occurs only via the parameter $\sigma \equiv \beta\hbar$, which has dimension of **time**.
- The closed orbits $\gamma \in \{H = \mu\}$ such that $T_\gamma \ll \sigma$ are strongly damped.
- It is a **Gutzwiller-type** formula where
  - The dominant part is a regular one, depending only on $\Sigma_\mu \equiv \{H = \mu\}$,
  - The oscillating part comes from contribution of periodic orbits on Σ_μ,
  - The error terms depend on the test functions considered.

Conclusion

Non-interacting Fermionic Gases slightly driven from equilibrium and in contact with reservoirs at temperature T present interesting **semiclassical** features in terms of the parameters $\hbar$ and $\sigma = \hbar/kT$. The Classical Correlation functions of the Perturbation and of the Observable we want to measure manifest themselves

- via the microcanonical Energy Surface Σ_μ, $\mu =$ Fermi Level,
- via the periodic trajectories lying on this surface and whose length is comparable with σ.

Further details of the proof can be found in the paper by Combescure-Robert [10].

3. Semiclassical results for the magnetic response of an electron gas

The magnetic response theory for a free electron gas is an old problem that goes back to Fock [12], Landau [18], Peierls [25]. The revival of interest in physics arose from the advances of recent experiments that make possible

- to perform very small temperature conditions,
- to make measurements on small two-dimensional electronic devices.

These devices are so "pure" that the classical motion inside them can be considered as "ballistic", namely it is only determined by the confining potential. The bi-dimensional structure of such "electron gases" is realized through semi-conductor heterostructures whose size and shape can be controlled experimentally, together with the number N of "confined" electrons. The system is in contact with a reservoir at temperature T, and submitted to a magnetic field B perpendicular to the surface. The magnetic response, namely the magnetization M or the magnetic susceptibility χ can be measured in terms of N, T, B.

These experiments manifest the sensitivity of the "magnetic response" to the integrable versus chaotic character of the classical dynamics of a single electron in the system. Numerical as well as real experiments on two-dimensional billiards have confirmed this observation, and suggested that the quantum magnetic response is an experimentally accessible criterion for distinguishing classical versus chaotic classical dynamics [21], [1], [28].

A number of theoretical studies have analyzed the magnetic response from a "semiclassical" point of view, namely as properties that manifest themselves in the limit where $\hbar$ (the Planck constant) is small compared to classical actions characterizing the system (say $\hbar \ll a^2 eB/c$ where a is a typical size of the system, e the charge of the electron, c the velocity of light, and B the magnetic field size). In these studies the Coulomb interaction between the electrons is neglected ("free electron gas" to which a thermodynamical formalism can be applied), and the spin is also omitted (only the "orbital magnetism" is considered).

Here we shall present mathematical results for a n-dimensional ideal confined electron gas in the presence of an inhomogeneous magnetic field. The gas is in contact with a reservoir at temperature T. We shall establish different regimes of temperature in which the magnetic response can be obtained semiclassically in the form of asymptotic $\hbar$ expansions. We then show that in the so-called "mesoscopic regime", where the temperature is of the order of $\hbar$, the periodic orbits of the classical single-electron dynamics manifest themselves as highly oscillating contributions to the magnetic response. These oscillations are reminiscent of the well-known "de Haas-van Alphen" oscillations which are usually observed for the so-called Hofstadter model, namely non-interacting (spinless) electrons moving in the plane under the combined action of a magnetic field (perpendicular to the plane), and a periodic potential (see [13], [14], [16] for a semiclassical approach and results).

Our assumption is that the classical single-electron dynamics, with electric as well as magnetic potentials is chaotic. We refer to a similar study by Leboeuf-Monastra [19], where the same assumption is also made on a particular toy-model, called the "Riemannium", and where similar results are derived.

Let us now introduce the model and assumptions in this section.

Let k be the Boltzmann constant and T the temperature. We denote as usual

$$\beta := \frac{1}{kT}. \tag{3.1}$$

Let $\kappa \in \mathbb{R}$ be the size of the magnetic field. We consider the single-electron Hamiltonian of the following form:

$$H_\kappa(q,p) := \frac{1}{2}(p - \kappa a(q))^2 + V(q), \tag{3.2}$$

where $V : \mathbb{R}^n \mapsto \mathbb{R}$ and $a : \mathbb{R}^n \mapsto \mathbb{R}^n$ are $\mathcal{C}^\infty$ functions satisfying the following properties:

(H.1) $$\forall q \in \mathbb{R}^n, \ \ V(q) \geq 1 \ , \ |\partial_q^\alpha V(q)| \leq C_\alpha V(q),$$

(H.2) $$\forall q \in \mathbb{R}^n, \ \ |\partial_q^\alpha a(q)| \leq C_\alpha \ V(q)^{1/2},$$

(H.3) $$\forall q \in \mathbb{R}^n, \ \ V(q) \geq c_0 \left(1 + |q|^2\right)^{s/2} \quad \text{for some } s, \, c_0 > 0$$

(confinement assumption).

Let now $\widehat{H}_\kappa$ be the Weyl quantization of H_κ. The previous assumptions ensure that $\widehat{H}_\kappa$ is self-adjoint and its spectrum $\sigma(\widehat{H}_\kappa) \subset [\varepsilon, \infty)$ is pure point for every $\kappa \in \mathbb{R}$ where $\varepsilon > 0$. Let us call $(E_j)_{j \in \mathbb{N}}$ and $(\varphi_j)_{j\in\mathbb{N}}$ the set of corresponding eigenvalues and eigenstates.

Given $\beta > 0$, we set:

$$F_\beta(x) = -\frac{1}{\beta} \operatorname{Log}\left(1 + e^{-\beta x}\right), \tag{3.3}$$

$$f_\beta(x) = F'_\beta(x) = \left(1 + e^{\beta x}\right)^{-1}, \tag{3.4}$$

f_β is related with the Fermi-Dirac distribution. These functions are meromorphic, with poles (or cuts for F_β) at:

$$x = \frac{2j+1}{\beta} \, i\pi, \qquad j \in \mathbb{Z}.$$

In the grand-canonical ensemble, the thermodynamic potential Ω is given by:

$$\Omega(\beta, \mu, \kappa) = \sum_{j\in\mathbb{N}} F_\beta(E_j - \mu) = \operatorname{tr}\left\{F_\beta\left(\widehat{H}_\kappa - \mu\right)\right\}, \tag{3.5}$$

where $\mu > 0$ is the chemical potential and β is given by (3.1).

Furthermore, the mean-number of particles in the grand-canonical ensemble is given by:

$$N(\beta, \mu, \kappa) = \operatorname{tr}\left(f_\beta\left(\widehat{H}_\kappa - \mu\right)\right). \tag{3.6}$$

In all that follows, we fix the chemical potential μ, which according to (3.6) implies that $N(\beta, \mu, \kappa)$ is large in the semiclassical régime so that we are approaching the thermodynamical limit where canonical and grand canonical ensemble descriptions are equivalent. Using the functional calculus, it is not difficult to see that $F_\beta(\widehat{H} - \mu)$ and $f_\beta(\widehat{H} - \mu)$ are trace-class and that the function : $\kappa \mapsto \Omega(\beta, \mu, \kappa)$ is $\mathcal{C}^\infty$ for $|\kappa| \leq \kappa_0$.

We shall denote $\partial_\kappa = \frac{\partial}{\partial \kappa}$.

Proposition 3.1. *The function:* $\kappa \mapsto \Omega(\beta, \mu, \kappa)$ *is* $\mathcal{C}^\infty$ *on* $\mathbb{R}$. *In particular we have*

$$\partial_\kappa \Omega = \operatorname{tr}\left[f_\beta\left(\widehat{H}_\kappa - \mu\right)\partial_\kappa \widehat{H}_\kappa\right]. \tag{3.7}$$

Now we have the following definitions of magnetization M and magnetic susceptibility χ:

$$M = \partial_\kappa \Omega \;=\; \operatorname{tr}\left[f_\beta\left(\widehat{H}_\kappa - \mu\right)\partial_\kappa \widehat{H}_\kappa\right], \tag{3.8}$$

$$\chi \;=\; \partial_\kappa M. \tag{3.9}$$

We shall now distinguish different Temperature Regimes, where the magnetic response behaves semiclassically in significantly different ways:

- When T is larger than $\hbar$ we have asymptotic expansions in $\hbar$ of the Thermodynamic Grand-Canonical Function, and thus of the Magnetic Response functions. As a Corollary, we recover the **Landau diamagnetic formula** for 2-dimensional ideal electron gases.
- When T is of the order of $\hbar$, we prove that the magnetization and the magnetic susceptibility split into two parts: an average part with a regular asymptotic expansion in $\hbar$, plus an oscillating part in $\hbar$ which represents the contribution of the **classical periodic orbits** that lie on the energy surface at the Fermi level μ.

The first Temperature Regime we consider is $\beta \le \hbar^{-2/3+\varepsilon}$ for some $\varepsilon > 0$:

Theorem 3.2. *For any* $\varepsilon > 0$ *and* $\kappa_0 > 0$, Ω *admits an asymptotic expansion in* $\hbar$, *uniform in* κ *for* $|\kappa| \le \kappa_0$, *and for* $\beta \le \hbar^{\varepsilon - 2/3}$. *More explicitly, for any* $N \in \mathbb{N}$ *we have:*

$$\Omega = h^{-n} \sum_{j=0}^{N} \sum_{k \le \frac{3j}{2}} \frac{(-1)^{k+1}}{k!} \hbar^j \; \Omega_{jk} + O\left(\hbar^{N+1-n} \; \beta^{\frac{3N}{2}+k(n)}\right), \tag{3.10}$$

with

$$\Omega_{jk} = \int_{\mathbb{R}^{2n}} dq \; dp \; d_{jk}(q,p) \; F_\beta^{(k)}(H_\kappa - \mu),$$

d_{jk} *being a suitable linear combination of derivatives of* H_κ *with respect to* q, p *and* $k(n)$ *a constant depending only on the dimension* n $(k(n) \le 2n+1)$. *In particular:*

$$\Omega_{00} = \int_{\mathbb{R}^{2n}} dq \; dp \; F_\beta\left(H_\kappa(q,p) - \mu\right), \tag{3.11}$$

$$\frac{1}{2}\Omega_{22} - \frac{1}{6}\Omega_{23} = -\frac{\beta}{48\pi^2}\int_{\mathbb{R}^{2n}} dq \; dp \frac{\kappa^2 \parallel B(q) \parallel^2 - \sum_{jk} \partial_{jk}^2 V}{\cosh^2\left[\frac{\beta}{2}\left(H_\kappa(q,p) - \mu\right)\right]}, \tag{3.12}$$

where B_{jk} *is the magnetic field*

$$\parallel B \parallel^2 = \sum_{j<k} B_{jk}^2, \qquad B_{jk} = \frac{\partial a_j}{\partial q_k} - \frac{\partial a_k}{\partial q_j}, \; \partial_{jk}^2 V = \frac{\partial^2 V}{\partial q_j \partial q_k}, \tag{3.13}$$

and we have chosen the gauge so that $\partial a/\partial q$ is symmetric. Moreover, the asymptotic expansion can be differentiated term by term with respect to κ and yields an asymptotic expansion of the magnetization and the magnetic susceptibility.

Corollary 3.3 (Landau diamagnetism). *Let χ be defined by (3.9), and μ be non-critical for H_0. Then for $n = 2$ and $\kappa = 0$ we have*

$$\lim_{\hbar\to 0,\beta\to\infty,\beta\leq\hbar^{\varepsilon-2/3}} \chi = -\frac{1}{24\pi^2}\int_{\Sigma_\mu^0} \| B(q) \|^2 \, d\sigma_\mu^0, \tag{3.14}$$

which is nothing but Landau's result of the diamagnetism for a 2-dimensional free electron gas.

For the proofs of these results see [9].

The restriction to the value $\hbar^{-2/3+\varepsilon}$, $\varepsilon > 0$ is rather technical. We can extend the above results to a more natural temperature Regime $\beta \leq \hbar^{-1+\varepsilon}$: Assume in addition to (H1–H3):

(H4) μ is non-critical for H_κ.

Then we have:

Theorem 3.4. *Assume* (H1-4). *Then the magnetization $M = \partial_\kappa \Omega$ has for any temperature T satisfying $\hbar^{1-\varepsilon} \leq T \leq \hbar^{\frac{2}{3}-\varepsilon}$ (some $\varepsilon > 0$) a complete asymptotic expansion in $\hbar$ obtained by taking the derivative in κ of the formal expansion in $\hbar$ for Ω given* (3.10).

We now come to the so-called "mesoscopic regime" where $T \sim \hbar$. We introduce a new parameter σ:

$$\sigma := \beta\hbar, \tag{3.15}$$

as in the previous section. Let Σ_μ^κ be the energy surface at energy μ = Fermi-level (or equivalently the chemical potential):

$$\Sigma_\mu^\kappa := \left\{(q,p) : in\ \mathbb{R}^{2n},\ H_\kappa(q,p) = \mu\right\},$$

and let $d\sigma_\mu^\kappa$ be the Liouville measure on Σ_μ^κ.

Then we assume the following (Gutzwiller Assumption):

(H5): on Σ_μ^κ the closed orbits of period smaller than some prescribed $\tau > 0$ are **nondegenerate**.

We shall define a "regularized" magnetization as follows:

Let $\rho \in C_0^\infty$ be an even function such that:

$$\rho(t) \equiv 1, \qquad \text{if} \quad |t| \leq 1;$$
$$\rho(t) \equiv 0, \qquad \text{if } |t| \geq 2 \text{ and } \int_{\mathbb{R}} \rho(t)dt = 1. \tag{3.16}$$

Let us define for $\tau > 0$,

$$\rho_\tau(t) = \rho(t/\tau), \tag{3.17}$$

and

$$M_\tau = \mathrm{Tr}\left\{(f_\sigma * \widetilde{\rho}_\tau)\left(\frac{\widehat{H}_\kappa - \mu}{\hbar}\right)\partial_\kappa \widehat{H}_\kappa\right\}, \tag{3.18}$$

where $\widetilde{g}$ denotes the inverse Fourier transform of g. Clearly $M_\tau \to M$ when $\tau \to \infty$. However we shall not be able to let $\tau \to \infty$ in this work. This is the same type of restriction we had in the previous Section, where only a "regularized dynamical susceptibility" has been estimated.

Then our result is as follows:

Theorem 3.5. *Let us assume* H-1 *to* H-5 *and* $\sigma = \beta\hbar \in [\sigma_0, \sigma_1]$ *where* $\sigma_1 > \sigma_0 > 0$ *are fixed. For* $\chi_\tau = \chi * \rho_\tau$, $\tau > 0$, *we have the decomposition*

$$\chi_\tau = \overline{\chi} + \chi_{\mathrm{osc}}, \tag{3.19}$$

with

$$\overline{\chi} = -\frac{\hbar^{2-n}}{24\pi^2}\int_{\Sigma_\mu^\kappa} d\sigma_\mu^\kappa \parallel B(q) \parallel^2 + \sum_{k\geq 3-n} c_{\chi,k}(\mu,\kappa,\sigma,T)\hbar^k + O(\hbar^\infty), \tag{3.20}$$

$$\chi_{\mathrm{osc}} = \sum_{\gamma\in(\Gamma_\mu)_\tau} e^{i(S_\gamma/\hbar+\nu_\gamma\frac{\pi}{2})}\left\{\frac{\rho_{1,\tau}(T_\gamma)}{|\det(1-P_\gamma)|^{1/2}}\,\frac{r_\gamma m_\gamma^2/2\sigma}{\sinh(\pi T_\gamma/\sigma)} + \sum_{k\geq 1} d_{\chi,\gamma}^{(k)}\hbar^k\right\} + O(\hbar^\infty), \tag{3.21}$$

where we use the following notations:

- γ *is a periodic orbit in classical phase-space,* S_γ *the classical action along this orbit,*
- T_γ *is the period of the orbit* γ,
- $r_\gamma \in \mathbb{Z}$ *the repetition number of this orbit, and*
- P_γ *the linearized Poincaré map of the orbit* γ *(which doesn't have the eigenvalue* 1 *because of* (H5)). *Furthermore*
- $m_\gamma := \int_0^{T_\gamma^*} dt\, \partial_\kappa H_\kappa(q_t, p_t)$, *where* $T_{\gamma^*} := T_\gamma/r_\gamma$ *is the period of the primitive orbit,*
- m_γ *represents the* **flux** *of the magnetic field through the closed curve* γ,
- $c_{\chi,\gamma}$, $d_{\chi,\gamma}$ *are smooth coefficients depending on the periodic orbit* γ, *on* σ, *and on the function* ρ.

The proof is rather involved (see [9]).

To complete this section let us show that a similar decomposition of the magnetization M into a "smooth" part $\bar{M}$ and a highly oscillating part M_{osc} holds true, and that indeed the dominant term in the semiclassical expansion of $\bar{M}$ is nothing but the well-known diamagnetic Landau term.

Proposition 3.6. *Assume* (H1-4). *Then for any* $\sigma_1 > 0$ *and for* $\kappa_0 > 0$, *uniformly for* $\beta\hbar = \sigma \in]0, \sigma_1[$ *we have,* $\mod O(h^\infty)$,

$$\overline{M} = -\kappa \frac{\hbar^{2-n}}{24\pi^2} \int_{\Sigma_\mu^0} d\sigma_\mu^0 \parallel B(q) \parallel^2 + \sum_{k \geq 3-n} c_k(\mu, 0, \sigma, T)\hbar^k + O(\hbar^\infty), \qquad (3.22)$$

where $d\sigma_\mu^0$ *is the Liouville measure on the energy surface* Σ_μ^0, *i.e., without magnetic field.*

Conclusion

The result of Theorem 3.5 is very similar to the one obtained in Theorem 2.2 for the dynamical susceptibility in the Linear Response Regime for an ideal Fermion gas. The temperature dependence occurs via the **same parameter** $\sigma = \beta\hbar$, and the contribution of closed orbits on the Fermi Energy Surface is strongly damped by the factor

$$\frac{1}{\sinh(\pi T_\gamma/\sigma)}.$$

Natural **classical quantities** appear in the dominant parts of the $\hbar$ expansions, namely here, instead of the Classical Correlation Functions on Energy Surface or along the closed orbits, we have:

- The diamagnetic Landau term $\int_{\Sigma_\mu^0} d\sigma_\mu^0 \|B(q)\|^2$,
- The Flux m_γ of the magnetic field through the closed orbits, the periods of the closed orbits.

4. Transport and dissipations in quantum pumps

This section is devoted to a presentation of the work of [2], (see also [3]). We think that it is pertinent in this review paper, since it presents interesting **scattering** features for the electron gas under consideration, whereas the two previous sections dealt with ideal **confined** fermion gases where the spectrum of the single-particle Hamiltonian was thus pure point. The Hamiltonian of a single-particle is here of spectrally continuous nature. Moreover it is driven via a time-dependent potential, which is assumed adiabatic. That means that there is an adiabatic time-scale called ε^{-1} which is considered large with respect to the dwell time τ of the particles into the scatterer. Thus the parameter ε plays the rôle of a "semiclassical parameter", although here $\hbar$ is taken equal to 1.

An adiabatic quantum pump is a slowly time-varying scatterer connected to several leads. Each lead may have several channels. The total number of channels in each lead will be denoted by n. Each channel is represented by a semi-infinite, one-dimensional, single mode ideal wire. It is assumed that the particles propagating in each channel are non-interacting and have all energy $k^2/2$; along each lead, electrons can enter or leave the pump.

Then the question is:

How electrons are transported from one lead to another due to the time-dependent potential of the scatterer (hence the name of "pump")?

This question was first addressed in [7] in the framework of Linear Response Theory. They established that the expected current is a function of the scattering matrix and of its time-derivative. The interest in the subject was piqued by the experiments ([30], although it is not completely clear that the experiment is actually represented by an adiabatic pump).

The mathematical framework is as follows:

• The single-particle Hilbert space is given by

$$\mathcal{H} = \mathcal{H}_0 \oplus L^2(\mathbb{R}_+, \mathbb{C}^n),$$

where states in $L^2(\mathbb{R}_+, \mathbb{C}^n) := \oplus_{j=1}^n L^2(\mathbb{R}_+)$ (resp. $\mathcal{H}_0$) describe electrons in the leads (resp. in the pump). Denote by $\Pi_j,\ j = 0, 1, \ldots, n,\ \mathcal{H} \to \mathcal{H}$ the projection onto $\mathcal{H}_0$ if $j = 0$, and onto the jth copy of $L^2(\mathbb{R}_+)$ if $j = 1, \ldots, n$.

• The Hamiltonian is a slowly varying time-dependent self-adjoint operator in $\mathcal{H}$, namely given by $\hat{H}_\varepsilon(s) \equiv \hat{H}(\varepsilon^{-1}s)$ where ε is the small adiabatic parameter. The assumptions on $\hat{H}(t)$ will ensure the existence of a unitary propagator in $\mathcal{H}$, called $U_\varepsilon(s, s')$ satisfying the Schrödinger equation:

$$i\partial_s U_\varepsilon(s, s') = \varepsilon^{-1}\hat{H}_\varepsilon(s)U_\varepsilon(s, s').$$

• The assumptions on $\hat{H}_\epsilon(s)$ are as follows:

(H1): $\hat{H}_\varepsilon(s) - \hat{H}_\varepsilon(s')$ is bounded in $\mathcal{H}$ and smooth in $s,\ s'$,

(H2): $\|\Pi_0(\hat{H}_\varepsilon(s) + i)^{-m}\|_1 < C$ for all $s,\ \varepsilon$, and some $m \in \mathbb{N}$,

(H3) $\hat{H}_\varepsilon(s)\psi = -d^2\psi/dx^2$ for $\psi \in C_0^\infty(\mathbb{R}_+, \mathbb{C}^n)$,

(H4) $\sigma_{pp}(\hat{H}_\varepsilon(s)) \cap (0, \infty)$ is empty,

(H5) $\hat{H}_\varepsilon(s) \equiv \hat{H}_-,\ \forall s \leq 0$.

(Here $\|\cdot\|_1$ denotes the trace class norm on $\mathcal{H}$.)

• The electrons in the incoming state should be in an equilibrium state. We take it as a density matrix $\hat{\rho} := \rho(\hat{H}_-)$, where ρ is a function of bounded variation with $\mathrm{Supp}(\rho) \subset (0, \infty)$. A good example for it is the zero-temperature equilibrium state at Fermi-level μ: $\rho(\lambda) := \Theta(\mu - \lambda)$, where Θ is the Heaviside function.

• Units are chosen so that $k = \hbar = m = e = 1$ where k is the Boltzmann constant, and $m,\ e$ the mass and charge of the electron.

• One can now define the "charge current" operator. Consider the following operator in $\mathcal{H}$:

$$A := 0 \oplus \frac{1}{2i}\Big(\frac{d}{dx}v(x) + v(x)\frac{d}{dx}\Big),$$

where $v : [0, \infty) \to \mathbb{R}$ is smooth, with $v'(x) \geq 0$, $\mathrm{Supp}(v) \subset [b, \infty)$, some $b > 0$, and $v(x) = x$ for $x > c > b$. A is self-adjoint and commutes with $\Pi_j,\ \forall j = 0, 1, \ldots, n$, and we denote $A_j := A\Pi_j$.

The operator A allows to distinguish *incoming* and *outgoing* electrons. Take some $a > 0$ and a "switch function" $f \in \mathcal{C}_0^\infty$ where: $f(x) = 0$, for $x < -1$, $f(x) = 1$, for $x > 1$. The *charge current operator* is then defined as

$$I_j(a) := i[\hat{H}_\varepsilon(s)\ ,\ f(A_j - a) + f(-A_j - a)] \equiv I_{j+}(a) + I_{j-}(a).$$

• The expectation value of the current at "epoch" s in the channel j, is then given by

$$\langle I\rangle_j(s, a\varepsilon) := \mathrm{tr}(U_\varepsilon(s,-1)\rho(\hat{H}_-)U_\varepsilon(-1,s)I_j(a)).$$

• In order to define "scattering" by the pump, some reference Hamiltonian $\hat{H}_0$ has to be introduced. Take

$$\hat{H}_0 := h \oplus (-\frac{d^2}{dx^2})_N \oplus \cdots \oplus (-\frac{d^2}{dx^2})_N,$$

where the component of $\hat{H}_0$ on each of the leads is $(-\frac{d^2}{dx^2})_N$ which is simply the Laplacian with Neumann boundary conditions. h for the pump component can be anything, provided that $\hat{H}_0$ obeys (H2). The **time-dependent** S-matrix is given by

$$\mathcal{S}(s) := \lim_{s_\pm \to \pm\infty} U_0(\varepsilon^{-1}(s - s_+))U_\varepsilon(s_+, s_-)U_0(\varepsilon^{-1}(s_- - s)). \qquad (4.23)$$

It is the so-called *frozen scattering operator* usually introduced in adiabatic scattering theory. "Energy" E (the spectral representation for $\hat{H}_0$) is conserved so that the E-fibers $S(s,E)$ of the operator $\mathcal{S}(s)$ are $j\ n \times n$ "scattering matrices".

Then everything is in order to state the result:

Theorem 4.1. *Consider a cut-off function in energy space* $\chi \in \mathcal{C}_0^\infty(0,\infty)$ *with* $\chi = 1$ *on* $\mathrm{Supp}(\rho')$. *Redefine the current operator (with IR and UV cutoff):*

$$I_{,,\pm}(s, a, \varepsilon) := \chi(\hat{H}_\varepsilon(s))i[\hat{H}_\varepsilon(s), f(\pm A_j - a)]\chi(\hat{H}_\varepsilon(s)).$$

Then, under the above assumptions, we have the following expression for the current in the j*th channel:*

$$\lim_{a\to\infty}\lim_{\varepsilon\to 0} \varepsilon^{-1}\langle I\rangle_j(s,a,\varepsilon) = -\frac{i}{2\pi}\int_0^\infty dE\rho'(E)\left(\frac{dS}{ds}S(s)\right)_{j,j}, \qquad (4.24)$$

where $S(s) \equiv S(E,s)$ *(the* E*-fiber of the scattering operator* $\mathcal{S}(s)$*). The double limit is uniform in* $s \in I$, I *being a compact interval.*

The proof is given in [2].

Remark 4.2. *The physical meaning of the limiting procedure of* (4.24) *is that the current measurement is made well outside the scattering region* $(\lim_{a\to\infty})$, *but* **after** *the adiabatic limit* $(\lim_{\varepsilon\to 0})$.

Further results of ref. [3] are:

- The relationship with classical scattering,
- The formulation of charge transport in terms of the Chern character of a natural line bundle.

Conclusion

Semiclassical tools appear useful in the context of an adiabatically driven electron system. Under appropriate mathematical assumptions, it provides the expected physical result: namely a time-dependent scatterer connected to several leads acts as a "quantum pump" that allows charge transfer between the leads. This is the response of the scattering system to a slowly varying potential.

The condition that $\mathrm{Supp}(\rho) \subset (0, \infty)$ might seem strange, since it excludes the natural Fermi-Dirac density matrix at **positive temperature** T:

$$\rho(\hat{H}) := \left(1 + e^{\beta(\hat{H}-\mu)}\right)^{-1}, \tag{4.25}$$

where as in previous sections μ is the chemical potential, and $T \equiv \beta^{-1}$ is the temperature (although this restriction is not present in [3]). It seems rather technical, and the link with physics (exterior leads at any temperature T) should be clearer if it is removed. We think that the introduction of cutoff function in energy in Theorem 4.1 should allow to remove this condition.

Thus this system provides a simple example where the scattering of an electron system leads to a measurable response (whereas studies of previous sections were restricted to bound state systems). If the pump were a "chaotic billiard" the classical dynamics will in general have arbitrarily long orbits, which should give rise to resonances, although this point has not been studied yet mathematically, to our knowledge.

References

[1] Agam O., J Phys. I (France), **4**, 697–730, (1994).

[2] Avron J., Elgart A., Graf G.M., Sadun L. Schnee K., *Adiabatic charge pumping in open quantum systems*, Comm. Pre Appl. Math., **57**, 528, (2004).

[3] Avron J., Elgart A., Graf G.M., Sadun L., *Transport and Dissipation in "Quantum Pumps"*, J. Stat. Phys. **116**, 425–473, (2004).

[4] Avron J., Gutkin B., Oaknin D., *Adiabatic Swimming in an Ideal Quantum Gas*, Phys. Rev. Lett., **96**, 130602, (2006).

[5] Brack M., Murthy M.V., *Harmonically trapped fermion gases: exact and asymptotic results in arbitrary dimensions*, J. Phys. A: Math. Gen. **36**, 1111–1133, (2003).

[6] Bruun G.M., Clark C.W., *Ideal gases in time-dependent traps*, Phys. Rev. A, **61**, 061601(R), (2000).

[7] Büttiker M., Prêtre A., Thomas H., Phys. Rev. Lett. **70**, 4114, (1993).

[8] Butts D.A., Rokhsar D.S., *Trapped Fermi Gases*, Phys. Rev. A, **55**, 4346–4350, (1997).

[9] Combescure M., Robert D., *Rigorous semiclassical results for the magnetic response of an electron gas*, Rev. Math. Phys., **13**, 1055–1073, (2001).

[10] Combescure M., Robert D., *Semiclassical results in the linear response theory*, Ann. Phys., **305**, 45–59, (2003).

[11] Eckhardt B., *Eigenvalue Statistics in Quantum Ideal Gases*, arXiv: chao-dyn/9809005, (1998).

[12] Fock V., Z. Physik, **47**, 446–450, (1928).

[13] Gat O., Avron J., *Semiclassical Analysis and the Magnetization of the Hofstadter Model*, Phys. Rev. Lett. **91**, 186801, (2003).

[14] Gat O., Avron J., *Magnetic fingerprints of fractal spectra and the duality of Hofstadter models*, New J. Phys., **5**, 44.1–44.8, (2003).

[15] Gleisberg F., Wonneberger W., *Noninteracting fermions in a one-dimensional harmonic atom trap: Exact one-particle properties at zero temperature*, Phys. Rev. A **62**, 063602, (2000).

[16] Helffer B., Sjöstrand J., *On diamagnetism and the de Haas-Van Alphen effect*, Ann. I. H.P., Phys. Théor., **52**, 303–352, (1990).

[17] Kubasiak A., Korbicz J., Zakrzewski J., Lewenstein M., *Fermi-Dirac statistics and the number theory*, Europhysics Letters, (2005).

[18] Landau L., Z. Physik, **64**, 629, (1930).

[19] Leboeuf P. Monastra A., *Quantum thermodynamic fluctuations of a chaotic Fermi-gas model*, Nuclear Physics A, **724**, 69–84, (2003).

[20] Lert P.W., Weare J.H., *Static semiclassical response of a bounded electron gas. 1*, J. Chem. Phys., **68**, 2221–2227, (1978).

[21] Levy L.P., Reich D.H., Pfeiffer L., West K., Physica B **189**, 204, (1993).

[22] Li M., Yan Z., Chen J., Chen L., Chen C., *Thermodynamic properties of an ideal Fermi gas in an external potential with* $U = br^t$ *in any dimensional space*, Phys. Rev. A **58**, 1445–1449, (1998).

[23] Molinari V., Sumini M., Rocchi F., *Fermion gases in magnetic fields: a semiclassical treatment*, Eur. Phys. J. D, **12**, 211–217, (2000).

[24] Molinari V., Rocchi F., Sumini M., *Kinetic Description of Rotating Gases in External Magnetic Fields in the Framework of the Thomas-Fermi-Dirac Approach*, Transport Theory and Stat. Phys., **32**, 607–621, (2003).

[25] Peierls R.E., Z. Physik, **80**, 763–791, (1933).

[26] Pezzè L., Pitaevskii L., Smerzi A., Stringari S., *Insulating Behaviour of a Trapped Ideal fermi Gas*, Phys. Rev. Lett. **93**, 120401, (2004).

[27] Quang D.N., Tung N.H., *Semiclassical approach to the density of states of the disordered electron gas in a quantum wire*, Phys. Rev. B **60**, 13648–136358, (1999).

[28] Richter K., Ullmo D., Jalabert R., *Orbital magnetism in the ballistic regime: geometrical effects*, Phys. Rep. **276**, 1–83, (1996).

[29] Salasnich L., *Ideal quantum gases in D-dimensional space and power-law potentials*, Journ. Math. Phys., **41**, 8016–8024, (2000).

[30] Switkes M., Marcus C.M., Campman K., Gossard A.G., Science, **283**, 1907, (1999).

[31] Tran M.N., *Exact ground-state number fluctuations of trapped ideal and interacting fermions*, J. Phys. A: Math. Gen. **36**, 961–973, (2003).

[32] Tran M.N., Murthy M.V., Bhaduri R.K., *Ground-state fluctuations in finite Fermi systems*, Phys. Rev. E, **63** 031105, (2001).

[33] van Faassen E., *Dielectric response of a nondegenerate electron gas in semiconductor nanocrystallites*, Phys. Rev. B, **58**, 15729–15735, (1998).

[34] van Zyl B.P., *Analytical expression for the first-order density matrix of a d-dimensional harmonically confined Fermi gas at finite temperature*, Phys. Rev. A, **68**, 033601, (2003)

[35] Yajima K., *Existence of solutions for Schrödinger evolution equations*, Commun. Math. Phys., **110**, 415–426, (1987).

Monique Combescure
IPNL, Bâtiment Paul Dirac
4 rue Enrico Fermi,Université Lyon-1
F-69622 Villeurbanne Cedex, France
e-mail: monique.combescure@ipnl.in2p3.fr

Operator Theory:
Advances and Applications, Vol. 186, 89–102

Quadratic Hamiltonians and Their Renormalization

Jan Dereziński

Abstract. Quadratic bosonic Hamiltonians in Fock spaces with an infinite number of degrees of freedom have a surprisingly rich mathematical theory. In this article I review recent results about two classes of such operators: the van Hove Hamiltonians and the Bogoliubov Hamiltonians.

Mathematics Subject Classification (2000). 47B99, 81T99.

Keywords. Second quantization, Fock spaces, quadratic Hamiltonians.

1. Introduction

In this article I would like to review the results of two recent papers [D, BD] on quadratic bosonic Hamiltonians with an infinite number of degrees of freedom. I would like to convince the reader that their theory is surprisingly rich.

Let a_ξ^*/a_ξ denote *creation/annihilation operators* satisfying

$$[a_\xi, a_{\xi'}] = [a_\xi^*, a_{\xi'}^*] = 0, \quad [a_\xi, a_{\xi'}^*] = \delta_{\xi,\xi'}, \tag{1}$$

and acting on a *bosonic Fock space*. (Above, $\delta_{\xi,\xi'}$ denotes the delta function. Strictly speaking, a_ξ^*/a_ξ are operator-valued measures and they acquire the meaning of an operator only after smearing out with appropriate test functions.)

The first class of Hamiltonians that I would like to discuss was studied in [D] and is given by a formal expression of the form

$$H = \int h(\xi) a_\xi^* a_\xi \mathrm{d}\xi + \int \overline{z}(\xi) a_\xi \mathrm{d}\xi + \int z(\xi) a_\xi^* \mathrm{d}\xi + c, \tag{2}$$

where $h(\xi) \geq 0$. Note that in the above expression the constant c can be infinite. Following [Sch], operators of the form (2) will be called *van Hove Hamiltonians.*

The second class of operators was recently studied by L. Bruneau together with myself in [BD]. Operators from this class are given by a formal expression of the form

$$H = \int h(\xi) a_\xi^* a_\xi \mathrm{d}\xi + \frac{1}{2}\int g(\xi,\xi') a_\xi^* a_{\xi'}^* \mathrm{d}\xi + \frac{1}{2}\int \overline{g}(\xi,\xi') a_\xi a_{\xi'} \mathrm{d}\xi + c. \qquad (3)$$

We will call them *Bogoliubov Hamiltonians*. This name is justified by the famous application of such Hamiltonians in the study of the Bose gas due to Bogoliubov [Bog].

There are several questions that one can pose about these operators. They include: When the above formal expressions defines a self-adjoint operator? When they are bounded from below? When they have a ground state? What is their scattering theory? Rather complete answers to these questions exist in the case of van Hove Hamiltonians. For Bogoliubov Hamiltonians, the answers are not so complete, but still we have a number of interesting results about them.

Van Hove and Bogoliubov Hamiltonians are used in quantum physics very often. A lot of interesting physical phenomena can be explained just with help of quadratic Hamiltonians.

In my paper I would like to convince the reader that also from the mathematical point of view they are interesting objects and illustrate various curious properties of unbounded operators.

Quadratic Hamiltonians are also useful, because they help to understand properties of more complicated Hamiltonians used in quantum theory such as those studied in [Fr, DG1, BFS, DJ].

2. Notation

Let us briefly review the notation for bosonic Fock spaces that we will use in our paper [Be, RS2, BR, GJ, DG1, D1]. Suppose that $\mathcal{Z}$ is a Hilbert space. The bosonic Fock space over the one-particle space $\mathcal{Z}$ is defined as

$$\Gamma_{\mathrm{s}}(\mathcal{Z}) := \mathop{\oplus}_{n=0}^{\infty} \otimes_{\mathrm{s}}^n \mathcal{Z}.$$

It has a distinguished vector called the vacuum vector $\Omega = 1 \in \otimes_{\mathrm{s}}^0 \mathcal{Z} = \mathbb{C}$.

The bosonic Fock space can be viewed as a commutative algebra with the product defined as follows: if $\Psi \in \otimes_{\mathrm{s}}^n \mathcal{Z}$, $\Phi \in \otimes_{\mathrm{s}}^m \mathcal{Z}$, then

$$\Psi \otimes_{\mathrm{s}} \Phi := \Theta_{\mathrm{s}}\ \Psi \otimes \Phi \in \otimes_{\mathrm{s}}^{n+m} \mathcal{Z},$$

where Θ_{s} is the symmetrizing operator.

For $z \in \mathcal{Z}$ we define the creation operator

$$a^*(z)\Psi := \sqrt{n+1}\, z \otimes_{\mathrm{s}} \Psi, \quad \Psi \in \otimes_{\mathrm{s}}^n \mathcal{Z},$$

and the annihilation operator $a(z) := (a^*(z))^*$.

In a large part of the literature one assumes that $\mathcal{Z}$ equals $L^2(\Xi)$ for some measure space $(\Xi, \mathrm{d}\xi)$. One introduces "operator-valued measures" a_ξ/a_ξ^* satisfying (1). If $\xi \mapsto z(\xi)$ is a square integrable function then

$$a^*(z) = \int z(\xi)a_\xi^*\mathrm{d}\xi, \quad a(z) = \int \overline{z}(\xi)a_\xi\mathrm{d}\xi.$$

We will use both notations. The notation involving the operator-valued measures will be called "traditional" – it is lengthy and depends on an arbitrary identification $\mathcal{Z} = L^2(\Xi)$, but is perhaps more familiar to some readers and often convenient.

For an operator q on $\mathcal{Z}$ we define the operator $\Gamma(q)$ on $\Gamma_{\mathrm{s}}(\mathcal{Z})$ by

$$\Gamma(q)\Big|_{\otimes_{\mathrm{s}}^n\mathcal{Z}} = q\otimes\cdots\otimes q.$$

For an operator h on $\mathcal{Z}$ we define the operator $\mathrm{d}\Gamma(h)$ on $\Gamma_{\mathrm{s}}(\mathcal{Z})$ by

$$\mathrm{d}\Gamma(h)\Big|_{\otimes_{\mathrm{s}}^n\mathcal{Z}} = h\otimes 1^{(n-1)\otimes} + \cdots 1^{(n-1)\otimes}\otimes h.$$

If h is the multiplication operator by $h(\xi)$, then in the traditional notation we have

$$\mathrm{d}\Gamma(h) = \int h(\xi)a_\xi^*a_\xi\mathrm{d}\xi.$$

Note the identity $\Gamma(\mathrm{e}^{\mathrm{i}th}) = \mathrm{e}^{\mathrm{i}t\mathrm{d}\Gamma(h)}$.

For $g \in \otimes_{\mathrm{s}}^2\mathcal{Z}$ we define the 2-particle creation operator

$$a^*(g)\Psi := \sqrt{(n+2)(n+1)}g\otimes_{\mathrm{s}}\Psi, \quad \Psi\in\otimes_{\mathrm{s}}^n\mathcal{Z},$$

and the annihilation operator $a(g) = a^*(g)^*$.
In the traditional notation, if g equals the function $g(\xi,\xi')$, then we have

$$a^*(g) = \int g(\xi,\xi')a_\xi^*a_{\xi'}^*\mathrm{d}\xi\mathrm{d}\xi', \quad a(g) = \int \overline{g}(\xi,\xi')a_\xi a_{\xi'}\mathrm{d}\xi\mathrm{d}\xi'.$$

3. Van Hove Hamiltonians

In this section we summarize properties of van Hove Hamiltonians, following [D].

Let $\mathcal{Z} = L^2(\Xi)$. Let $\Xi \ni \xi \mapsto h(\xi)$ be a positive function. Let $\xi \mapsto z(\xi)$ be a function on Ξ such that

$$\int_{h<1}|z(\xi)|^2\mathrm{d}\xi + \int_{h\geq 1}\frac{|z(\xi)|^2}{h(\xi)^2}\mathrm{d}\xi < \infty.$$

Then we can define a family of unitary operators on $\Gamma_{\mathrm{s}}(\mathcal{Z})$

$$V(t) := \Gamma(\mathrm{e}^{\mathrm{i}th})\exp\left(a^*((1-\mathrm{e}^{-\mathrm{i}th})h^{-1}z) - \mathrm{hc}\right).$$

One can easily check that

$$V(t_1)V(t_2) = c(t_1,t_2)V(t_1+t_2)$$

for some complex numbers $c(t_1,t_2)$.

For an operator $B \in B(\Gamma_s(\mathcal{Z}))$ we define

$$\beta_t(B) := V(t)BV(t)^*.$$

Then β is a 1-parameter group of $*$-automorphisms of the algebra of bounded operators on the Fock space, pointwise continuous in the strong operator topology.

By a general theorem [BR], there exists a self-adjoint operator H such that

$$\beta_t(B) = \mathrm{e}^{\mathrm{i}tH}\, B\, \mathrm{e}^{-\mathrm{i}tH}.$$

H is defined uniquely up to an additive constant. We call it a *van Hove Hamiltonian.* It is easy to see that formally it is given by (2), which contains an arbitrary constant c. One can ask if there is a natural choice of c. It turns out that there exist two such natural choices. To describe them it is convenient (especially, if we want to be rigorous) to use the unitary groups generated by van Hove Hamiltonians.

The following theorem describes the unitary group generated by van Hove Hamiltonians of the first kind:

Theorem 1. *Let*

$$\int_{h(\xi)<1} |z(\xi)|^2 \mathrm{d}\xi + \int_{h(\xi)\geq 1} \frac{|z(\xi)|^2}{h(\xi)} \mathrm{d}\xi < \infty.$$

Then

$$U_{\mathrm{I}}(t) := \exp\left(\mathrm{i}\int |z(\xi)|^2 \frac{\sin th(\xi) - th(\xi)}{h^2(\xi)} \mathrm{d}\xi\right) V(t) \tag{4}$$

is a strongly continuous unitary group.

We define the *type* I *van Hove Hamiltonian* H_{I} to be the self-adjoint generator of (4), that is $U_{\mathrm{I}}(t) = \mathrm{e}^{\mathrm{i}tH_{\mathrm{I}}}$. Formally,

$$H_{\mathrm{I}} = \int h(\xi) a_\xi^* a_\xi \mathrm{d}\xi + \int \overline{z}(\xi) a_\xi \mathrm{d}\xi + \int z(\xi) a_\xi^* \mathrm{d}\xi.$$

It satisfies $\Omega \in \mathrm{Dom} H_{\mathrm{I}}$, $(\Omega | H_{\mathrm{I}} \Omega) = 0$.

Note that

$$\inf \mathrm{sp} H_{\mathrm{I}} = -\int \frac{|z(\xi)|^2}{h(\xi)} \mathrm{d}\xi,$$

(which can be $-\infty$). The linear perturbation contained in (2) is an operator iff $\int |z(\xi)|^2 \mathrm{d}\xi < \infty$, otherwise it is a quadratic form.

Another natural class of van Hove Hamiltonians is described in the following theorem:

Theorem 2. *Let*

$$\int_{h(\xi)<1} \frac{|z(\xi)|^2}{h(\xi)} \mathrm{d}\xi + \int_{h(\xi)\geq 1} \frac{|z(\xi)|^2}{h^2(\xi)} \mathrm{d}\xi < \infty. \tag{5}$$

Then

$$U_{\mathrm{II}}(t) := \exp\left(\mathrm{i}\int |z(\xi)|^2 \frac{\sin th(\xi)}{h^2(\xi)} \mathrm{d}\xi\right) V(t) \tag{6}$$

is a strongly continuous unitary group.

We define the *type* II *van Hove Hamiltonian* H_{II} to be the self-adjoint generator of (6), that is $U_{\mathrm{II}}(t) = \mathrm{e}^{\mathrm{i}tH_{\mathrm{II}}}$. Formally,

$$H_{\mathrm{II}} = \int h(\xi)\left(a_\xi^* + \frac{\overline{z}(\xi)}{h(\xi)}\right)\left(a_\xi + \frac{z(\xi)}{h(\xi)}\right)\mathrm{d}\xi.$$

It satisfies $\inf \mathrm{sp} H_{\mathrm{II}} = 0$.

Let us introduce the following unitary operator called sometimes the *dressing operator*:

$$U := \exp\left(-a^*(\frac{z}{h}) + a(\frac{z}{h})\right). \tag{7}$$

Note that (7) is well defined iff

$$\int \frac{|z(\xi)|^2}{h^2(\xi)}\mathrm{d}\xi < \infty.$$

It intertwines H_{II} and the free van Hove Hamiltonian:

$$H_{\mathrm{II}} = U \int h(\xi) a_\xi^* a_\xi \mathrm{d}\xi \; U^*.$$

Hence, in this case H_{II} has a ground state. Otherwise H_{II} has no ground state.

Both H_{I} and H_{II} are well defined iff

$$\int \frac{|z(\xi)|^2}{h(\xi)}\mathrm{d}\xi < \infty,$$

and then

$$H_{\mathrm{II}} = H_{\mathrm{I}} + \int \frac{|z(\xi)|^2}{h(\xi)}\mathrm{d}\xi < \infty.$$

If

$$\int_{h(\xi)<1} \frac{|z(\xi)|^2}{h(\xi)}\mathrm{d}\xi = \int_{h(\xi)\geq 1} \frac{|z(\xi)|^2}{h(\xi)}\mathrm{d}\xi = \infty,$$

then neither H_{I} nor H_{II} is well defined.

Altogether we have 3 kinds of situations that lead to different infrared behaviors of the van Hove Hamiltonians. Likewise, we have 3 possible ultraviolet behaviors. Thus, altogether we have $3 \times 3 = 9$ situations that lead to van Hove Hamiltonians with distinct properties.

They are summarized in the Table on page 94.

In the literature, the analysis of the ultraviolet problem of van hove Hamiltonians can be found in [Be, Sch], following earlier treatments [vH, EP, To, GS]. The understanding of the infrared problem can be traced back to [BN], and then was discussed in a series of papers [Ki]. Closely related problems of coherent representations was discussed already in [Frie]. Nevertheless, it seems that [D] gives the first complete treatment of this subject in the literature.

	$\int_{h>1}\|z\|^2<\infty$	$\int_{h>1}\|z\|^2=\infty$ $\int_{h>1}\frac{\|z\|^2}{h}<\infty$	$\int_{h>1}\frac{\|z\|^2}{h}=\infty$ $\int_{h>1}\frac{\|z\|^2}{h^2}<\infty$	
$\int_{h<1}\frac{\|z\|^2}{h^2}<\infty$				H_{II} defined gr. st. exists
$\int_{h<1}\frac{\|z\|^2}{h^2}=\infty$ $\int_{h<1}\frac{\|z\|^2}{h}<\infty$				H_{II} defined no gr. st.
$\int_{h<1}\frac{\|z\|^2}{h}=\infty$ $\int_{h<1}\|z\|^2<\infty$				unbounded from below
	H_{I} defined pert. is an operator	H_{I} defined pert. is not an operator	infinite renormalization	

4. Scattering theory of van Hove Hamiltonians

The main goal of this section is a description of scattering theory for van Hove Hamiltonians. It is based on [D]

Let us start with some remarks about scattering theory in an abstract setting (see, e.g., [Ya, Kato, RS3]). Suppose we are given two self-adjoint operators: H_0 and H.

In the standard approach to scattering theory, which works, e.g., for 2-body Schrödinger operators, the *wave operators* are defined by

$$\Omega^{\pm} := \text{s-}\lim_{t\to\pm\infty} \mathrm{e}^{\mathrm{i}tH}\,\mathrm{e}^{-\mathrm{i}tH_0}\,. \tag{8}$$

They satisfy $\Omega^{\pm}H_0 = H\Omega^{\pm}$ and are isometric. If $\mathrm{Ran}\Omega^{+} = \mathrm{Ran}\Omega^{-}$, then the *scattering operator*

$$S = \Omega^{+*}\Omega^{-}$$

is unitary and $H_0S = SH_0$.

Unfortunately, this approach does not work for van Hove Hamiltonians. Let us describe an alternative, less-known approach to scattering theory.

Again, we start from two self-adjoint operators: H_0 and H. We introduce the *unrenormalized Abelian wave operators*:

$$\Omega^{\pm}_{\mathrm{ur}} := \text{s-}\lim_{\epsilon\searrow 0} 2\epsilon \int_0^\infty \mathrm{e}^{-2\epsilon t}\,\mathrm{e}^{\pm\mathrm{i}tH}\,\mathrm{e}^{\mp\mathrm{i}tH_0}\,\mathrm{d}t. \tag{9}$$

They satisfy $\Omega^{\pm}_{\mathrm{ur}}H_0 = H\Omega^{\pm}_{\mathrm{ur}}$ but do not have to be isometric. Note that if the usual wave operators $\Omega^{\pm}$ defined by (8) exist, then so do the unrenormalized Abelian ones, and they coincide. However, (9) may exist even if the usual wave operators do not.

Assume that the operators $Z^{\pm} := \Omega_{\rm ur}^{\pm *}\Omega_{\rm ur}^{\pm}$ have a zero kernel. Then we can define the *renormalized Abelian wave operators*

$$\Omega_{\rm rn}^{\pm} := \Omega_{\rm ur}^{\pm}(Z^{\pm})^{-1/2}.$$

They also satisfy $\Omega_{\rm rn}^{\pm}H_0 = H\Omega_{\rm rn}^{\pm}$ and are isometric.

If $\mathrm{Ran}\Omega_{\rm rn}^{+} = \mathrm{Ran}\Omega_{\rm rn}^{-}$, then the *renormalized scattering operator*

$$S_{\rm rn} = \Omega_{\rm rn}^{+*}\Omega_{\rm rn}^{-}$$

is unitary and $H_0 S_{\rm rn} = S_{\rm rn} H_0$.

Note that the alternative approach is more suitable for quantum field theory than the standard. In particular, as we will see, it works well in the case of van Hove Hamiltonians.

Let

$$H_0 = \int h(\xi) a_\xi^* a_\xi \mathrm{d}\xi,$$

$$H = \int h(\xi) a_\xi^* a_\xi \mathrm{d}\xi + \int \overline{z}(\xi) a_\xi \mathrm{d}\xi + \int z(\xi) a_\xi^* \mathrm{d}\xi + \int \frac{|z(\xi)|^2}{h(\xi)} \mathrm{d}\xi.$$

(In other words, H is a van Hove Hamiltonian of type II). Suppose that h has an absolutely continuous spectrum and the assumption (5) is satisfied. Then it is not difficult to show that the unrenormalized Abelian wave operators exist. One can compute explicitly the wave and scattering operators:

$$\Omega_{\rm ur}^{\pm} = Z^{1/2}U, \quad \Omega_{\rm rn}^{\pm} = U, \quad S_{\rm rn} = 1.$$

where U is the dressing operator and

$$Z = \exp \int \frac{|z(\xi)|^2}{h^2(\xi)} \mathrm{d}\xi.$$

Unfortunately, the scattering operator is trivial.

Note in parenthesis that scattering theory for operators similar but more complicated than van Hove Hamiltonians can be quite interesting [DG1, DG2].

5. Bogoliubov Hamiltonians

In this section we describe mathematical theory of Bogoliubov Hamiltonians following [BD]. Again, it is not obvious how to define those Hamiltonians. The expression (3) is not very convenient for their rigorous definition. In order to formulate a definition that is natural and as general as possible, it is convenient to think in terms of the *classical phase space* underlying the given bosonic Fock space. To this end we need to recall some notions from linear algebra and the formalism of second quantization.

Let $\mathcal{Z}$ be a complex Hilbert space. We will write $\overline{\mathcal{Z}}$ for the space complex conjugate to $\mathcal{Z}$. The real vector space

$$\mathcal{Y} := \{(z, \overline{z}) \ : \ z \in \mathcal{Z}\} \subset \mathcal{Z} \oplus \overline{\mathcal{Z}}$$

equipped with a natural symplectic form

$$(z_1, \overline{z}_1)\omega(z_2, \overline{z}_2) := \mathrm{Im}(z_1|z_2).$$

has the meaning of the *dual of the classical phase space* of the quantum system described by the bosonic Fock space $\Gamma_\mathrm{s}(\mathcal{Z})$.

For $y = (z, \overline{z}) \in \mathcal{Y}$ we define the corresponding *Weyl operator*

$$W(y) := \mathrm{e}^{\mathrm{i}a^*(z)+\mathrm{i}a(z)}.$$

Note that $W(y_1)W(y_2) = \mathrm{e}^{-\frac{\mathrm{i}}{2}y_1\omega y_2}W(y_1 + y_2)$.

A map r on $\mathcal{Y}$ is called *symplectic* if

$$(ry_1)\omega(ry_2) = y_1\omega y_2.$$

For such r,

$$W(ry_1)W(ry_2) = \mathrm{e}^{-\frac{\mathrm{i}}{2}y_1\omega y_2}W(r(y_1 + y_2)),$$

and thus the commutation relations of Weyl operators are preserved.

Every linear map r on $\mathcal{Y}$ can be uniquely extended to a complex linear map on $\mathcal{Z} \oplus \overline{\mathcal{Z}}$ and written as

$$r = \begin{bmatrix} p & q \\ \overline{q} & \overline{p} \end{bmatrix}.$$

r is symplectic iff

$$\begin{aligned} p^*p - \overline{q}^*\overline{q} &= 1, \quad -\overline{p}^*\overline{q} + q^*p = 0, \\ pp^* - qq^* &= 1, \quad \overline{q}p^* - \overline{p}q^* = 0. \end{aligned}$$

We have the decomposition

$$r = \begin{bmatrix} 1 & 0 \\ d^* & 1 \end{bmatrix} \begin{bmatrix} p & 0 \\ 0 & \overline{p}^{*-1} \end{bmatrix} \begin{bmatrix} 1 & c \\ 0 & 1 \end{bmatrix},$$

with symmetric operators $d := q\overline{p}^{-1}$, $c := p^{-1}q$. (We say that d is a *symmetric* operator iff $d = \overline{d}^*$.)

Theorem 3 (Shale Theorem). [Sh] *Let r be symplectic. There exists a unitary U, which we call a* Bogoliubov implementer, *such that*

$$UW(y)U^* = W(ry), \quad y \in \mathcal{Y},$$

iff $\mathrm{Tr}q^*q < \infty$.

The map $B(\Gamma_\mathrm{s}(\mathcal{Z})) \ni A \mapsto UAU^*$, where U is a Bogoliubov implementer, will be called a *Bogoliubov automorphism*. For a given r, a Bogoliubov implementer is determined up to a phase. There exists a distinguished choice, denoted U_nat, satisfying $(\Omega|U_\mathrm{nat}\Omega) > 0$, given by

$$U_\mathrm{nat} := |\det pp^*|^{-\frac{1}{4}}\,\mathrm{e}^{-\frac{1}{2}a^*(d)}\,\Gamma(p^{*-1})\,\mathrm{e}^{\frac{1}{2}a(c)}.$$

An important role in our considerations will be played by *strongly continuous 1-parameter groups of symplectic transformations.* If $\mathbb{R} \ni t \mapsto r(t)$ is such a group, we introduce the maps $t \mapsto p(t), q(t)$ defined by

$$r(t) = \begin{bmatrix} p(t) & q(t) \\ \overline{q}(t) & \overline{p}(t) \end{bmatrix}. \tag{10}$$

If h is a *self-adjoint* operator on $\mathcal{Z}$ and g is a bounded symmetric operator from $\overline{\mathcal{Z}}$ to $\mathcal{Z}$ then

$$r(t) = \exp \mathrm{i}t \begin{bmatrix} h & g \\ \overline{g} & \overline{h} \end{bmatrix} \tag{11}$$

is a 1-parameter symplectic group.

Let us consider for a while the case of finite-dimensional $\mathcal{Z}$. In this case theory of Bogoliubov Hamiltonians, while not quite trivial, is well understood. (We still keep the notation $\mathcal{Z} = L^2(\Xi)$, but now Ξ has to be a finite set and the integration over Ξ is just summation.)

Clearly, in finite dimension every continuous 1-parameter symplectic group is of the form (11). Consider a classical quadratic Hamiltonian

$$\begin{aligned} H(\overline{z}, z) = & \int h(\xi)\overline{z}_\xi z_\xi \mathrm{d}\xi \\ & + \frac{1}{2}\int g(\xi,\xi')\overline{z}_\xi\overline{z}_\xi \mathrm{d}\xi\mathrm{d}\xi' + \frac{1}{2}\int \overline{g}(\xi,\xi')z_\xi z_{\xi'}\mathrm{d}\xi\mathrm{d}\xi'. \end{aligned}$$

It is a function on the *classical phase space*

$$\overline{\mathcal{Y}} := \{(\overline{z}, z) \ : \ z \in \mathcal{Z}\} \subset \overline{\mathcal{Z}} \oplus \mathcal{Z}.$$

The *Weyl quantization* of $H(\overline{z}, z)$ equals

$$\begin{aligned} H = & \frac{1}{2}\int h(\xi)a_\xi^* a_\xi \mathrm{d}\xi + \frac{1}{2}\int h(\xi)a_\xi a_\xi^* \mathrm{d}\xi \\ & + \frac{1}{2}\int g(\xi,\xi')a_\xi^* a_{\xi'}^* \mathrm{d}\xi\mathrm{d}\xi' + \frac{1}{2}\int \overline{g}(\xi,\xi')a_\xi a_{\xi'}\mathrm{d}\xi\mathrm{d}\xi' \end{aligned}$$

and corresponds to the choice of c in (3) given by

$$c = \frac{1}{2}\int h(\xi,\xi)\mathrm{d}\xi = \frac{1}{2}\mathrm{Tr}h,$$

H is essentially self-adjoint on finite particle vectors. We have

$$\mathrm{e}^{\mathrm{i}tH} = (\det p(t))^{-\frac{1}{2}}\,\mathrm{e}^{-\frac{1}{2}a^*(d(t))}\,\Gamma(p(t)^{*-1})\,\mathrm{e}^{\frac{1}{2}a(c(t))}.$$

Note that the set of operators of the form

$$(\det p)^{-\frac{1}{2}}\,\mathrm{e}^{-\frac{1}{2}a^*(d)}\,\Gamma(p^{*-1})\,\mathrm{e}^{\frac{1}{2}a(c)} \tag{12}$$

is closed wrt the multiplication. It is called the *metaplectic group* $Mp(\mathcal{Y})$.

Let us now relax the condition $\dim \mathcal{Z} < \infty$ and ask about possible generalizations of the above construction to the case of an arbitrary number of degrees of freedom. Clearly, (12) is well defined provided that $p - 1$ is trace class, or equivalently, $r - 1$ is trace class. The set of operators of this form is also closed wrt multiplication. Thus, as noticed by Lundberg, the metaplectic group can be defined also in the case of an infinite number of degrees of freedom.

We say that a strongly continuous 1-parameter group of symplectic transformations $t \mapsto r(t)$ is *implementable* iff there exists a strongly continuous 1-parameter unitary group $t \mapsto U(t)$, called the *implementing unitary group*, such that

$$U(t)W(y)U^*(t) = W(r(t)y), \quad y \in \mathcal{Y}. \tag{13}$$

Only now, after so much preparation, we introduce the rigorous definition of a Bogoliubov Hamiltonian: A self-adjoint operator H is called a *Bogoliubov Hamiltonian* if there exists a 1-parameter strongly-continuous implementable symplectic group $t \mapsto r(t)$ such that $H := -\mathrm{i}\frac{\mathrm{d}}{\mathrm{d}t}U(t)\Big|_{t=0}$, where $t \mapsto U(t)$ is its implementing unitary group.

The following theorem is proven in [BD]:

Theorem 4. *$t \mapsto r(t)$ is implementable iff* $\mathrm{Tr}q^*(t)q(t) < \infty$ *and*

$$\lim_{t\to 0} \mathrm{Tr}q^*(t)q(t) = 0.$$

Once again, given an implementable Bogoliubov dynamics we have a 1-parameter family of Hamiltonians, formally differing by the constant c. We would like to discuss some of their natural choices.

Let us describe our first choice. Let $t \mapsto r(t)$ be an implementable symplectic group. Let $p(t)$ be defined by (10). We say that $t \mapsto r(t)$ is of *type I* iff $\frac{\mathrm{d}}{\mathrm{d}t}p(t)\Big|_{t=0} = \mathrm{i}h$, $p(t)\,\mathrm{e}^{-\mathrm{i}th} - 1$ is trace class and $\|p(t)\,\mathrm{e}^{-\mathrm{i}th} - 1\|_1 \to 0$.

Theorem 5. *In the type* I *case*

$$U_{\mathrm{I}}(t) := \det(p(t)\,\mathrm{e}^{-\mathrm{i}th})^{-\frac{1}{2}}\,\mathrm{e}^{-\frac{1}{2}a^*(d(t))}\,\Gamma(p(t)^{*-1})\,\mathrm{e}^{\frac{1}{2}a(c(t))}$$

is a strongly continuous 1*-parameter unitary group.*

A *type* I *Bogoliubov Hamiltonian* is defined as

$$H_{\mathrm{I}} := -\mathrm{i}\frac{\mathrm{d}}{\mathrm{d}t}U_{\mathrm{I}}(t)\Big|_{t=0}.$$

Let $t \mapsto r(t)$ be implementable. We say that it is of *type II* iff the implementing 1-parameter group has a generator, which is bounded from below. In this case we define the *type* II *Hamiltonian* to be

$$H_{\mathrm{II}} := -\mathrm{i}\frac{\mathrm{d}}{\mathrm{d}t}U_{\mathrm{II}}(t)\Big|_{t=0}.$$

such that $\inf \mathrm{sp}H_{\mathrm{II}} = 0$ and $U_{\mathrm{II}}(t)$ implements $r(t)$.

For a finite number of degrees of freedom it is easy to see that we have a complete characterization of type I and II Bogoliubov Hamiltonians:

Theorem 6. *Let $\mathcal{Z}$ be finite dimensional. Then*
(1) *$r(t)$ is always type* I *and*

$$H_{\mathrm{I}} = \mathrm{d}\Gamma(h) + \frac{1}{2}a^*(g) + \frac{1}{2}a(g).$$

(2) *$r(t)$ is type* II *iff its classical Hamiltonian is positive definite*

$$\overline{z}hz + \frac{1}{2}\overline{z}g\overline{z} + \frac{1}{2}z\overline{g}z \geq 0,$$

and then

$$H_{\mathrm{II}} = H_{\mathrm{I}} - \frac{1}{4}\mathrm{Tr}\left[\left(\begin{array}{cc} \overline{h}^2 - \overline{g}g & \overline{h}\overline{g} - \overline{g}h \\ hg - g\overline{h} & h^2 - g\overline{g} \end{array}\right)^{1/2} - \left(\begin{array}{cc} \overline{h} & 0 \\ 0 & h \end{array}\right)\right].$$

In the case of an infinite number of degrees of freedom, our results about Bogoliubov Hamiltonians are only partial. Let us give some examples taken from [BD]:

Theorem 7. *Let g be Hilbert-Schmidt. Then*

$$H_{\mathrm{I}} = \mathrm{d}\Gamma(h) + \frac{1}{2}a^*(g) + \frac{1}{2}a(g)$$

is essentially self-adjoint on the algebraic Fock space over $\mathrm{Dom}(h)$ *and* $\mathrm{e}^{\mathrm{i}tH_{\mathrm{I}}}$ *implements $r(t)$ given by* (11).

Theorem 8. *Let h be positive,*

$$\|h^{-1/2} \otimes h^{-1/2}\, g\|_{\Gamma_{\mathrm{s}}^2(\mathcal{Z})} < 1,$$
$$\|h^{-1/2} g\|_{B(\overline{\mathcal{Z}},\mathcal{Z})} < \infty.$$

Then $\frac{1}{2}a^*(g) + \frac{1}{2}a(g)$ is relatively $\mathrm{d}\Gamma(h)$-bounded with the bound less than 1. *Therefore, in this case, both the type* I *and type* II *Bogoliubov Hamiltonians are well defined.*

There is one class of Bogoliubov Hamiltonians, that we were able to analyze rather completely: those satisfying the condition $g\overline{h} = hg$. In this case, without loss of generality we can assume that they are diagonal in a common orthonormal basis $e_1, e_2, \ldots$:

$$he_n = h_n e_n, \; h_n \in \mathbb{R}; \;\; g\overline{e}_n = g_n e_n, \;\; g_n \in \mathbb{C}.$$

We will say that such Hamiltonians are *diagonalizable*.

Theorem 9. [BD] *Suppose h, g are diagonalizable in the above sense.*
(1) *$r(t)$ is well defined iff for some b, $a < 1$, $|g_n| \leq a|h_n| + b$.*
(2) *$r(t)$ is implementable iff* $\sum\limits_n \frac{|g_n|^2}{1+h_n^2} < \infty$.
(3) *$r(t)$ is type* I *iff* $\sum\limits_n \frac{|g_n|^2}{1+|h_n|} < \infty$.
(4) *$r(t)$ is type* II *iff $|g_n| \leq h_n$ and* $\sum\limits_n \frac{|g_n|^2}{h_n+h_n^2} < \infty$.

Theorem 9 shows that there exist implementable 1-parameter symplectic groups, which are not type II, even though their classical Hamiltonian is positive definite. Thus there exist Bogoliubov Hamiltonians unbounded from below with positive classical symbols. This is an example of an interesting infrared behavior of Bogoliubov Hamiltonians.

Theorem 9 shows also that there exist implementable 1-parameter symplectic groups, which are not type I. This means that, in order to express them in terms of creation and annihilation operators, one needs to add an infinite constant – perform an appropriate renormalization. This is an example of an interesting ultraviolet behavior of Bogoliubov Hamiltonians.

There remain various open questions concerning Bogoliubov Hamiltonians. For instance, it would be interesting to give sufficient and necessary conditions for symplectic group $r(t)$ to be of type II in terms of its generator.

Note that Bogoliubov Hamiltonians were studied by various authors, among them Friedrichs [Frie], Berezin [Be], Ruijsenaars [Ru1, Ru2], Araki and his collaborators [A, AY], Matsui and Shimada [MS], Ito and Hiroshima [IH]. Nevertheless, the approach contained [BD], briefly described above, seems to be the most general and flexible.

Acknowledgement

I would like to thank L.Bruneau for a collaboration on the article [BD], which is partly the source of results reviewed in this article.

References

[A] Araki H., On quasifree states of the canonical commutation relations (II), Publ. RIMS, Kyoto Univ. **7**, 121–152 (1971).

[AY] Araki H., Yamagami S., On quasi-equivalence of quasifree states of the canonical commutation relations, Publ. RIMS, Kyoto Univ. **18**, 283–338 (1982).

[Ar] Arai, A.: A note on scattering theory in non-relativistic quantum electrodynamics, J. Phys. A: Math. Gen. 16 (1983) 49–70.

[BFS] Bach V., Fröhlich J., Sigal I, Quantum electrodynamics of confined non-relativistic particles, Adv.Math. **137**, 299–395 (1998).

[Be] Berezin, F.: *The Method of Second Quantization*, second edition, Nauka, 1986.

[Bog] Bogoliubov, N.N.: *J. Phys. (USSR)* **9** (1947) 23; *J. Phys. USSR* **11** (1947) 23, reprinted.

[BN] Bloch, F., Nordsieck, A.: Phys. Rev. 52 (1937) 54.

[BR] Brattelli, O, Robinson D.W.: *Operator Algebras and Quantum Statistical Mechanics, vol. I and II*, Springer, Berlin (1981).

[BD] Dereziński, J., Bruneau, L.: Bogoliubov Hamiltonians and one parameter groups of Bogoliubov transformations, to appear in Journ. Math. Phys.

[D] Dereziński, J.: Van Hove Hamiltonians – exactly solvable models of the infrared and ultraviolet problem, Ann. H. Poincaré 4 (2003) 713–738.

[D1] Dereziński, J.: Introduction to Representations of Canonical Commutation and Anticommutation Relations, in "Large Coulomb Systems – QED", eds. J. Derezinski and H. Siedentop, Lecture Notes in Physics 695, Springer.

[DG1] Dereziński, J., Gérard, C.: Asymptotic completeness in quantum field theory. Massive Pauli-Fierz Hamiltonians, Rev. Math. Phys. 11 (1999) 383.

[DG2] Dereziński, J., C. Gérard) Scattering theory of infrared divergent Pauli-Fierz Hamiltonians, Ann. H. Poincaré 5 (2004) 523–577.

[DJ] Dereziński, J., Jakšić, V.: Spectral theory of Pauli-Fierz operators, Journ. Func. Analysis (2001) 241–327.

[EP] Edwards S., Peierls P.E.: Field equations in functional form, Proc. Roy. Soc. Acad. 224 (1954) 24–33.

[Frie] Friedrichs, K. O. *Mathematical aspects of quantum theory of fields*, New York 1953.

[Fr] Fröhlich, J.: On the infrared problem in a model of scalar electrons an massless, scalar bosons, Ann. Inst. H. Poincaré 19 (1973) 1–103.

[GJ] Glimm J., Jaffe A., *Quantum Physics. A Functional Integral Point of View*, 2nd ed., Springer-Verlag, Berlin, 1987.

[GS] Greenberg O.W., Schweber S.S.: Clothed particle operators in simple models of quantum field theory, Nuovo Cimento 8 (1958) 378–406.

[Ki] Kibble, T.W.B.: J. Math. Phys. 9 (1968) 15, Phys. Rev. 173 (1968) 1527; Phys. Rev. 174 (1968) 1882; Phys. Rev. 175 (1968) 1624.

[Kato] Kato, T.: *Perturbation Theory for Linear Operators*, second edition, Springer-Verlag, Berlin (1976).

[MS] Matsui T., Shimada Y., On quasifree representations of infinite dimensional symplectic group. J. Func. Anal. **215** 67–102 (2004).

[RS2] Reed, M., Simon, B.: *Methods of Modern Mathematical Physics, II. Fourier Analysis, Self-Adjointness*, London, Academic Press (1975).

[RS3] Reed, M., Simon, B.: *Methods of Modern Mathematical Physics, III. Scattering Theory*, London, Academic Press (1978).

[Schwa] Schwarz, A.S.: *Mathematical foundations of quantum field theory*, Atomizdat 1975, Moscow (Russian).

[Sch] Schweber, S.S.: *Introduction to nonrelativistic quantum field theory*, Harper and Row 1962.

[To] Tomonaga, S.: On the effect of the field reactions on the interaction of mesotrons and nuclear particles I. Progr. Theor. Phys. 1 (1946) 83–91.

[vH] van Hove, L.: Les difficultés de divergences pour un modèle particulier de champ quantifié, Physica 18 (1952) 145–152.

[Ya] Yafaev, D.: *Mathematical Scattering Theory*, AMS.

[Sh] Shale D., Linear symmetries of free boson fields, Trans. Amer. Math. Soc. **103**, 149–167 (1962).

[IH] Ito K.R., Hiroshima F., Local exponents and infinitesimal generators of canonical transformations on Boson Fock spaces, Infin. Dimens. Anal. Quantum Probab. Relat. Top. **7**, 547–571 (2004).

[Ru1] Ruijsenaars S.N.M., On Bogoliubov transformations for systems of relativistic charged particles, J. Math. Phys. **18**, 517–526 (1976).

[Ru2] Ruijsenaars S.N.M., On Bogoliubov transforms II. The general case, Ann. Phys. **116**, 105–134 (1978).

Jan Dereziński
Department of Mathematical Methods in Physics
Warsaw University
Hoża 74, 00-682, Warszawa, Poland
e-mail: Jan.Derezinski@fuw.edu.pl

Operator Theory:
Advances and Applications, Vol. 186, 103–113

Jacobi Matrices: Eigenvalues and Spectral Gaps

J. Dombrowski

Abstract. This paper investigates the existence of spectral gaps, and eigenvalues within these spectral gaps, for Jacobi matrices obtained by specific types of oscillating perturbations of unbounded Jacobi matrices with smooth weights. Some results on absolute continuity for such operators are also presented.

Mathematics Subject Classification (2000). 47B36, 47B15, 47B39.

Keywords. Jacobi matrices, spectral gaps. eigenvalues, absolute continuity.

1. Introduction

This paper investigates the spectral properties of a class of infinite Jacobi matrices of the following form:

$$C = \begin{bmatrix} 0 & a_1 & 0 & 0 & \dots \\ a_1 & 0 & a_2 & 0 & \dots \\ 0 & a_2 & 0 & a_3 & \dots \\ 0 & 0 & a_3 & 0 & \dots \\ . & . & . & . & \dots \end{bmatrix} \tag{1.1}$$

$$D_C = \{x = \{x_n\} \in \ell^2 : Cx \in \ell^2\}$$

where it is assumed that for $0 < \alpha \leq 1$ and all $n \in N$, $a_n > 0$, $a_n = n^\alpha + c_n$. Additional conditions are then imposed on the perturbation sequence $\{c_n\}$, specifically on the two difference sequences $\{c_{2n-1} - c_{2n}\}$ and $\{c_{2n+2} - c_{2n}\}$, to create a spectral gap. Under suitable conditions on the sequence $\{a_n\}$ results are presented on the existence of a spectral gap, possible eigenvalues within the gap, and the absolute continuity of the spectral measure outside the spectral gap. The results in this paper are related to those in [3] where the difference sequence $\{c_n\}$ is periodic so that $\{c_{2n-1} - c_{2n}\}$ is a constant sequence and $c_{2n+2} = c_{2n}$. The results below allow the perturbation sequence $\{c_n\}$ to be unbounded.

The sequence $\{c_n\}$ may be bounded or unbounded. However, the conditions imposed on the perturbation sequence $\{c_n\}$ will guarantee that the matrix operator C defined on the indicated maximal domain is a cyclic, self-adjoint operator. Specifically, Carleman's condition $\sum_{n=1}^{\infty} \frac{1}{a_n} = \infty$ will be satisfied. (See [1].) In this case it follows from the spectral theorem that C is unitarily equivalent to a multiplication operator $M_x : D \to L^2(\mu)$ defined on a dense subset D of $L^2(\mu)$ by $M_x : f(x) \to xf(x)$.

If $C = \int \lambda dE_\lambda$ is the spectral decomposition of C then $\mu(\beta) = \|E(\beta)\phi_1\|^2$ where ϕ_1 is the first standard basis vector in ℓ^2. The vector ϕ_1 is a cyclic vector for C since $a_n > 0$. In general, the standard basis vectors $\{\phi_n\}$ in ℓ^2 correspond to a sequence of polynomials $\{P_n(x)\}$ in $L^2(\mu)$ determined by the sequence $\{a_n\}$ as follows:

$$P_1(x) = 1, \qquad P_2(x) = \frac{x}{a_1}$$

$$P_{n+1}(x) = \frac{xP_n(x) - a_{n-1}P_{n-1}(x)}{a_n}, \ n \geq 2.$$

There is a fairly extensive literature related to the study of bounded and unbounded Jacobi operators. See, for example, [9]. It should be noted that every cyclic self-adjoint operator has a tridiagonal matrix representation with respect to the basis generated by the cyclic vector. The diagonal entries vanish when the spectrum is symmetric with respect to the origin. It should be possible, therefore, to model various types of spectral behavior within this class of operators. Results of this type are found, for example, in [3], [6], [7], [8] and [9].

2. Eigenvalues and spectral gaps

This section uses a lemma from [3] to establish the existence of a spectral gap, and to count eigenvalues within the spectral gap, for a class of Jacobi operators of type (1.1). Note that the perturbation sequence $\{c_n\}$ may be unbounded. But the conditions on the difference sequences imply that $\sum_{n=1}^{\infty} \frac{1}{a_n} = \infty$ so that the corresponding Jacobi operator is self-adjoint. (Sufficient conditions for self-adjointness can be found in [1].)

Theorem 2.1. *For $0 < \alpha \leq 1$, let $a_n = n^\alpha + c_n$, $n = 1, 2, \ldots$, where the sequence $\{c_n\}$ is chosen so that $c_n \geq 0$, and the two difference sequences $\{c_{2n} - c_{2n-2}\}$ (with $c_0 = 0$), and $\{c_{2n-1} - c_{2n}\}$ are non-negative, bounded and increasing for $n \geq N \geq 1$. If $\rho = \lim \rho_n$ where $\rho_n = c_{2n} - c_{2n-2}$ and $\omega = \lim \omega_n$, where $\omega_n = c_{2n-1} - c_{2n}$, then $(-\omega - \frac{\rho}{2},\ \omega + \frac{\rho}{2})$ is a gap in the essential spectrum for the operator C defined in (1.1). The interval $(-(\omega_n + \frac{\rho_n}{2}),\ (\omega_n + \frac{\rho_n}{2}))$ for $n \geq N$, contains at most $2n - 1$ eigenvalues.*

Proof. It was shown in [3] that if $C = \int \lambda dE_\lambda$ and $\psi \in D_C$ such that $E(0, \infty)\psi = \psi$ (or $E(-\infty, 0)\psi = \psi$), and if $\psi_n = \langle \psi, \phi_n \rangle$, then it follows that $\sum_{n=1}^{\infty} \psi_{2n}^2 =$

$\sum_{n=1}^{\infty} \psi_{2n-1}^2$. For such a vector ψ, it is also true that $E(0,\infty)C\psi = C\psi$ (or $E(-\infty,0)C\psi = C\psi$). Thus

$$\begin{aligned}
\frac{1}{2}\langle C^2\psi,\psi\rangle &= \frac{1}{2}\langle C\psi, C\psi\rangle \\
&= \sum_{k=1}^{\infty} |a_{2k-1}\psi_{2k-1} + a_{2k}\psi_{2k+1}|^2 \\
&\geq \sum_{k=N}^{\infty} |a_{2k-1}\psi_{2k-1} + a_{2k}\psi_{2k+1}|^2 \\
&= \sum_{k=N}^{\infty} |((2k-1)^\alpha + c_{2k-1})\psi_{2k-1} + ((2k)^\alpha + c_{2k})\psi_{2k+1}|^2 \\
&= \sum_{k=N}^{\infty} |((2k-1)^\alpha + c_{2k})\psi_{2k-1} + ((2k)^\alpha + c_{2k})\psi_{2k+1} \\
&\quad + (c_{2k-1} - c_{2k})\psi_{2k-1}|^2 \\
&= \sum_{k=N}^{\infty} |((2k-1)^\alpha + c_{2k})\psi_{2k-1} + ((2k)^\alpha + c_{2k})\psi_{2k+1}^2|^2 \\
&\quad + \sum_{k=N}^{\infty} 2(c_{2k-1} - c_{2k})[((2k-1)^\alpha + c_{2k})\psi_{2k-1} \\
&\quad + ((2k)^\alpha + c_{2k})\psi_{2k+1}]\psi_{2k-1} + \sum_{k=N}^{\infty} (c_{2k-1} - c_{2k})^2\psi_{2k-1}^2.
\end{aligned}$$

The right-hand side will now be viewed as a sum $S_1 + S_2$, where $S_1 = \sum_{k=N}^{\infty} |((2k-1)^\alpha + c_{2k})\psi_{2k-1} + ((2k)^\alpha + c_{2k})\psi_{2k+1}|^2$. Lower bounds will be given for each of these two pieces. Then

$$\begin{aligned}
S_1 &= \sum_{k=N}^{\infty} |\left((2k-1)^\alpha + \frac{c_{2k} + c_{2k-2}}{2}\right)\psi_{2k-1} + ((2k)^\alpha + c_{2k})\psi_{2k+1} \\
&\quad + \left(\frac{c_{2k} - c_{2k-2}}{2}\right)\psi_{2k-1}|^2 \\
&= \sum_{k=N}^{\infty} \left|\left((2k-1)^\alpha + \frac{c_{2k} + c_{2k-2}}{2}\right)\psi_{2k-1} + ((2k)^\alpha + c_{2k})\psi_{2k+1}\right|^2 \\
&\quad + 2\sum_{k=N}^{\infty} \left(\frac{c_{2k} - c_{2k-2}}{2}\right)\left[\left((2k-1)^\alpha + \frac{c_{2k} + c_{2k-2}}{2}\right)\psi_{2k-1}\right. \\
&\quad \left. + ((2k)^\alpha + c_{2k})\psi_{2k+1}\right]\psi_{2k-1} + \sum_{k=N}^{\infty} \frac{(c_{2k} - c_{2k-2})^2}{4}\psi_{2k-1}^2
\end{aligned}$$

$$\geq 2\sum_{k=N}^{\infty}\left(\frac{c_{2k}-c_{2k-2}}{2}\right)\left[(2k-1)^{\alpha}+\frac{c_{2k}+c_{2k-2}}{2}\right]\psi_{2k-1}^2$$
$$-2\sum_{k=N}^{\infty}\left(\frac{c_{2k}-c_{2k-2}}{2}\right)((2k)^{\alpha}+c_{2k})\left(\frac{1}{2}\psi_{2k+1}^2+\frac{1}{2}\psi_{2k-1}^2\right)$$
$$+\sum_{k=N}^{\infty}\left(\frac{c_{2k}-c_{2k-2}}{2}\right)^2\psi_{2k-1}^2$$
$$\geq 2\sum_{k=N}^{\infty}\left(\frac{c_{2k}-c_{2k-2}}{2}\right)\left((2k-1)^{\alpha}-\frac{1}{2}(2k)^{\alpha}+\frac{c_{2k-2}}{2}\right)\psi_{2k-1}^2$$
$$-2\sum_{k=N+1}^{\infty}\left(\frac{c_{2k-2}-c_{2k-4}}{2}\right)((2k-2)^{\alpha}+c_{2k-2})\frac{1}{2}\psi_{2k-1}^2$$
$$+\sum_{k=N}^{\infty}\left(\frac{c_{2k}-c_{2k-2}}{2}\right)^2\psi_{2k-1}^2.$$

Since it was assumed that $\{c_{2k}-c_{2k-2}\}$, $k\geq N\geq 1$ is an increasing sequence, it follows that

$$S_1\geq\sum_{k=N+1}^{\infty}2\left(\frac{c_{2k}-c_{2k-2}}{2}\right)\left((2k-1)^{\alpha}-\frac{1}{2}(2k)^{\alpha}-\frac{1}{2}(2k-2)^{\alpha}\right)\psi_{2k-1}^2$$
$$+\sum_{k=N}^{\infty}\left(\frac{c_{2k}-c_{2k-2}}{2}\right)^2\psi_{2k-1}^2$$
$$=\sum_{k=N}^{\infty}\left(\frac{c_{2k}-c_{2k-2}}{2}\right)^2\psi_{2k-1}^2.$$

It now remains to determine a lower bound for S_2. The strategy is similar to that used above.

$$S_2=\sum_{k=N}^{\infty}2(c_{2k-1}-c_{2k})[((2k-1)^{\alpha}+c_{2k})\psi_{2k-1}+((2k)^{\alpha}+c_{2k})\psi_{2k+1}]\psi_{2k-1}$$
$$+\sum_{k=N}^{\infty}(c_{2k-1}-c_{2k})^2\psi_{2k-1}^2$$
$$\geq\sum_{k=N}^{\infty}2(c_{2k-1}-c_{2k})((2k-1)^{\alpha}+c_{2k})\psi_{2k-1}^2$$
$$-2\sum_{k=N}^{\infty}(c_{2k-1}-c_{2k})((2k)^{\alpha}+c_{2k})\left[\frac{1}{2}\psi_{2k+1}^2+\frac{1}{2}\psi_{2k-1}^2\right]$$
$$+\sum_{k=N}^{\infty}(c_{2k-1}-c_{2k})^2\psi_{2k-1}^2$$

$$\geq \sum_{k=N}^{\infty} 2(c_{2k-1} - c_{2k}) \left((2k-1)^{\alpha} + c_{2k} - \frac{1}{2}(2k)^{\alpha} - \frac{1}{2}c_{2k} \right) \psi_{2k-1}^2$$
$$- 2 \sum_{k=N+1}^{\infty} (c_{2k-3} - c_{2k-2})((2k-2)^{\alpha} + c_{2k-2})\frac{1}{2}\psi_{2k-1}^2$$
$$+ \sum_{k=N}^{\infty} (c_{2k-1} - c_{2k})^2 \psi_{2k-1}^2.$$

Since it was assumed that $\{c_{2k-1} - c_{2k}\}$, $k \geq N \geq 1$, is an increasing sequence,

$$S_2 \geq \sum_{k=N+1}^{\infty} 2(c_{2k-1} - c_{2k})$$
$$\times \left((2k-1)^{\alpha} - \frac{1}{2}(2k)^{\alpha} - \frac{1}{2}(2k-2)^{\alpha} + \frac{1}{2}c_{2k} - \frac{1}{2}c_{2k-2} \right) \psi_{2k-1}^2$$
$$+ 2(c_{2N-1} - c_{2N})$$
$$\times \left((2N-1)^{\alpha} - \frac{1}{2}(2N)^{\alpha} - \frac{1}{2}(2N-2)^{\alpha} + \frac{1}{2}c_{2N} - \frac{1}{2}c_{2N-2} \right) \psi_{2N-1}^2$$
$$+ \sum_{k=N}^{\infty} (c_{2k-1} - c_{2k})^2 \psi_{2k-1}^2$$
$$\geq \sum_{k=N}^{\infty} (c_{2k-1} - c_{2k})(c_{2k} - c_{2k-2})\psi_{2k-1}^2 + \sum_{k=N}^{\infty} (c_{2k-1} - c_{2k})^2 \psi_{2k-1}^2.$$

Thus

$$\frac{1}{2}\langle C^2\psi, \psi\rangle \geq S_1 + S_2 \geq \sum_{k=N}^{\infty} \left[\frac{(c_{2k} - c_{2k-2})}{2} + (c_{2k-1} - c_{2k}) \right]^2 \psi_{2k-1}^2$$
$$\geq \left(\frac{\rho_N}{2} + \omega_N \right)^2 \sum_{k=N}^{\infty} \psi_{2k-1}^2.$$

Also, if $\langle \psi, \phi_1 \rangle = 0$, $\langle \psi, \phi_3 \rangle = 0, \ldots, \langle \psi, \phi_{2n-1} \rangle = 0$ for $n \geq N$, then

$$\frac{1}{2}\langle C^2\psi, \psi\rangle \geq \left(\frac{\rho_{n+1}}{2} + \omega_{n+1} \right)^2 \sum_{k=n+1}^{\infty} \psi_{2k-1}^2,$$

and it follows that $\langle C^2\psi, \psi\rangle = \|C\psi\|^2 \geq (\frac{\rho_{n+1}}{2} + \omega_{n+1})^2 \|\psi\|^2$. This argument does not apply to an eigenvector corresponding to a zero eigenvalue (when such an eigenvector exists). It does show that for $n \geq N$, C^2 has at most $n-1$ non-zero eigenvalues in the interval $(0, \ (\omega_n + \frac{\rho_n}{2})^2)$ and thus leads to the conclusion that for $n \geq N$, $(-(\omega_n + \frac{\rho_n}{2}), \ (\omega_n + \frac{\rho_n}{2}))$ is a gap in the essential spectrum of the operator C containing at most $2n-1$ eigenvalues.

Corollary 2.2. *For $0 < \alpha \leq 1$, let $a_n = n^{\alpha} + c_n$, $n = 1, 2, \ldots$, where the sequence $\{c_n\}$ is chosen so that $c_n \geq 0$. Suppose that for $n \geq N \geq 1$, $c_{2k} - c_{2k-2} = \rho$ with*

$\rho \geq 0$, and $c_{2k-1} - c_{2k} = \omega$ with $\omega \geq 0$. If $\rho + \omega > 0$, then $(-(\omega + \frac{\varrho}{2}),\ (\omega + \frac{\varrho}{2}))$ is a gap in the essential spectrum containing at most $2N - 1$ eigenvalues.

It is easy to adapt the proof of the previous theorem to establish the following result, which allows c_2 to be negative if $0 < \alpha < 1$.

Theorem 2.3. *For $0 < \alpha \leq 1$, let $a_n = n^\alpha + c_n$, $n = 1, 2, \ldots$, where the sequence $\{c_n\}$ is chosen so that $c_2 \geq -(2-2^\alpha)$, $c_{2n} = c_{2n-2}$ for $n \geq 2$, and $\{c_{2n-1} - c_{2n}\}_{n=1}^\infty$ is non-negative, bounded and increasing. If $\omega_n = \inf_{k \geq n}(c_{2k-1} - c_{2k}) = c_{2n-1} - c_{2n}$, and $\omega = \lim\ \omega_n$, then $(-\omega, \omega)$ is a gap in the essential spectrum. The interval $(-\omega_n,\ \omega_n)$ contains at most $(2n-1)$ eigenvalues.*

Proof. In this case the argument used in the previous proof shows that

$$S_1 = \sum_{k=1}^{\infty} |((2k-1)^\alpha + c_{2k})\psi_{2k-1} + ((2k)^\alpha + c_{2k})\psi_{2k+1}|^2 \geq 0$$

and that

$$S_2 \geq \sum_{k=1}^{\infty} (c_{2k-1} - c_{2k})^2 \psi_{2k-1}^2,$$

and hence that $(-\omega_n,\ \omega_n)$ is a gap in the essential spectrum containing at most $(2n-1)$ eigenvalues.

Corollary 2.4. *If $c_{2k} - c_{2k-2} = 0$, $k \geq 2$, $c_2 \geq -(2 - 2^\alpha)$, and $c_{2k-1} - c_{2k} = \omega > 0$, $k = 1, 2, \ldots$ then $(-\omega,\ \omega)$ is a gap in the essential spectrum containing no nonzero eigenvalues.*

In the following result, a spectral gap is created by choosing the perturbation sequence $\{c_n\}$ so that the difference sequences $\{c_{2n-1} - c_{2n+1}\}$ and $\{c_{2n} - c_{2n-1}\}$ are decreasing.

Theorem 2.5. *For $0 < \alpha \leq 1$, let $a_n = n^\alpha + c_n$, $n = 1, 2, \ldots$, where the sequence $\{c_n\}$ is chosen so that $a_n > 0$ for all n, and the two difference sequences $\{c_{2n-1} - c_{2n+1}\}$ for $n \geq N$, and $\{c_{2n} - c_{2n-1}\}$ for $n \geq N$, are non-negative, bounded and decreasing. If $\rho = \inf_n(c_{2n-1} - c_{2n+1})$ and $\omega = \inf_n(c_{2n} - c_{2n-1})$, and $\rho > 0$ or $\omega > 0$, then $(-(\omega + \frac{\varrho}{2}),\ (\omega + \frac{\varrho}{2}))$ is a gap in the essential spectrum containing at most $2N + 1$ eigenvalues.*

Proof. As in the proof of the previous theorem, choose $\psi \in D_C$ such that

$$E(0, \infty)\psi = \psi \qquad (\text{or } E(-\infty, 0)\psi = \psi).$$

Then if $\psi_n = \langle \psi, \phi_n \rangle$ it follows that

$$\sum_{n=1}^{\infty} \psi_{2n}^2 = \sum_{n=1}^{\infty} \psi_{2n-1}^2.$$

For such a vector ψ it is also true that

$$E(0, \infty)C\psi = C\psi \qquad (\text{or } E(-\infty, 0)C\psi = C\psi).$$

Thus

$$\begin{aligned}
\frac{1}{2}\langle C^2\psi,\psi\rangle &= \frac{1}{2}\langle C\psi, C\psi\rangle \\
&= \sum_{k=1}^{\infty} |a_{2k-1}\psi_{2k-1} + a_{2k}\psi_{2k+1}|^2 \\
&\geq \sum_{k=N}^{\infty} |a_{2k-1}\psi_{2k-1} + a_{2k}\psi_{2k+1}|^2 \\
&= \sum_{k=N}^{\infty} |((2k-1)^\alpha + c_{2k-1})\psi_{2k-1} + ((2k)^\alpha + c_{2k})\psi_{2k+1}|^2 \\
&= \sum_{k=N}^{\infty} |((2k-1)^\alpha + c_{2k-1})\psi_{2k-1} + ((2k)^\alpha + c_{2k-1})\psi_{2k+1} \\
&\qquad + (c_{2k} - c_{2k-1})\psi_{2k+1}|^2 \\
&= \sum_{k=N}^{\infty} |((2k-1)^\alpha + c_{2k-1})\psi_{2k-1} + ((2k)^\alpha + c_{2k-1})\psi_{2k+1}|^2 \\
&\qquad + \sum_{k=N}^{\infty} 2(c_{2k} - c_{2k-1})[((2k-1)^\alpha + c_{2k-1})\psi_{2k-1} \\
&\qquad + ((2k)^\alpha + c_{2k-1})\psi_{2k+1}]\psi_{2k+1} + \sum_{k=N}^{\infty} (c_{2k} - c_{2k-1})^2\psi_{2k+1}^2.
\end{aligned}$$

The right-hand side will now be viewed as a sum $S_1 + S_2$, where $S_1 = \sum_{k=N}^{\infty} |((2k-1)^\alpha + c_{2k-1})\psi_{2k-1} + ((2k)^\alpha + c_{2k-1})\psi_{2k+1}|^2$. Lower bounds will be found for each of these two pieces which will lead to the stated result.

$$\begin{aligned}
S_1 &= \sum_{k=N}^{\infty} \Bigg|((2k-1)^\alpha + c_{2k-1})\psi_{2k-1} + \left((2k)^\alpha + \frac{c_{2k+1} + c_{2k-1}}{2}\right)\psi_{2k+1} \\
&\qquad + \left(\frac{c_{2k-1} - c_{2k+1}}{2}\right)\psi_{2k+1}\Bigg|^2 \\
&= \sum_{k=N}^{\infty} \left|((2k-1)^\alpha + c_{2k-1})\psi_{2k-1} + \left((2k)^\alpha + \frac{c_{2k-1} + c_{2k+1}}{2}\right)\psi_{2k+1}\right|^2 \\
&\qquad + 2\sum_{k=N}^{\infty} \left(\frac{c_{2k-1} - c_{2k+1}}{2}\right)[((2k-1)^\alpha + c_{2k-1})\psi_{2k-1} \\
&\qquad + \left((2k)^\alpha + \frac{c_{2k-1} + c_{2k+1}}{2}\right)\psi_{2k+1}]\psi_{2k+1} \\
&\qquad + \sum_{k=N}^{\infty} \frac{(c_{2k-1} - c_{2k+1})^2}{4}\psi_{2k+1}^2
\end{aligned}$$

$$
\begin{aligned}
&\geq 2\sum_{k=N}^{\infty}\left(\frac{c_{2k-1}-c_{2k+1}}{2}\right)\left[(2k)^{\alpha}+\frac{c_{2k-1}+c_{2k+1}}{2}\right]\psi_{2k+1}^{2}\\
&\quad -2\sum_{k=N}^{\infty}\left(\frac{c_{2k-1}-c_{2k+1}}{2}\right)((2k-1)^{\alpha}+c_{2k-1})\left(\frac{1}{2}\psi_{2k-1}^{2}+\frac{1}{2}\psi_{2k+1}^{2}\right)\\
&\quad +\sum_{k=N}^{\infty}\frac{(c_{2k-1}-c_{2k+1})^{2}}{4}\psi_{2k+1}^{2}\\
&\geq 2\sum_{k=N}^{\infty}\left(\frac{c_{2k-1}-c_{2k+1}}{2}\right)\left[(2k)^{\alpha}-\frac{1}{2}(2k-1)^{\alpha}+\frac{c_{2k+1}}{2}\right]\psi_{2k+1}^{2}\\
&\quad -2\sum_{k=N}^{\infty}\left(\frac{c_{2k-1}-c_{2k+1}}{2}\right)[(2k-1)^{\alpha}+c_{2k-1}]\frac{1}{2}\psi_{2k-1}^{2}\\
&\quad +\sum_{k=N}^{\infty}\frac{(c_{2k-1}-c_{2k+1})^{2}}{4}\psi_{2k+1}^{2}\\
&\geq 2\sum_{k=N}^{\infty}\left(\frac{c_{2k-1}-c_{2k+1}}{2}\right)\left[(2k)^{\alpha}-\frac{1}{2}(2k-1)^{\alpha}+\frac{c_{2k+1}}{2}\right]\psi_{2k+1}^{2}\\
&\quad -2\sum_{k=N-1}^{\infty}\left(\frac{c_{2k+1}-c_{2k+3}}{2}\right)[(2k+1)^{\alpha}+c_{2k+1}]\frac{1}{2}\psi_{2k+1}^{2}\\
&\quad +\sum_{k=N}^{\infty}\frac{(c_{2k-1}-c_{2k+1})^{2}}{4}\psi_{2k+1}^{2}.
\end{aligned}
$$

Using the fact that $\{c_{2n-1}-c_{2n+1}\}$ is a non-negative decreasing sequence, it follows that

$$
\begin{aligned}
S_1 &\geq 2\sum_{k=N}^{\infty}\left(\frac{c_{2k-1}-c_{2k+1}}{2}\right)\left[(2k)^{\alpha}-\frac{1}{2}(2k-1)-\frac{1}{2}(2k+1)^{\alpha}\right]\psi_{2k+1}^{2}\\
&\quad -\frac{1}{2}(c_{2N-1}-c_{2N+1})[(2N-1)^{\alpha}+c_{2N-1}]\psi_{2N\quad 1}^{2}\\
&\quad +\sum_{k=N}^{\infty}\frac{(c_{2k-1}-c_{2k+1})^{2}}{4}\psi_{2k+1}^{2}.
\end{aligned}
$$

The strategy for determining a lower bound for S_2 is similar.

$$
\begin{aligned}
S_2 = \sum_{k=N}^{\infty} 2(c_{2k}-c_{2k-1})[((2k-1)^{\alpha}+c_{2k-1})\psi_{2k-1}+((2k)^{\alpha}+c_{2k-1})\psi_{2k+1}]\psi_{2k+1}\\
+\sum_{k=N}^{\infty}(c_{2k}-c_{2k-1})^{2}\psi_{2k+1}^{2}
\end{aligned}
$$

$$\geq \sum_{k=N}^{\infty} 2(c_{2k} - c_{2k-1})((2k)^\alpha + c_{2k-1})\psi_{2k+1}^2$$
$$- 2\sum_{k=N}^{\infty}(c_{2k} - c_{2k-1})((2k-1)^\alpha + c_{2k-1})\left[\frac{1}{2}\psi_{2k+1}^2 + \frac{1}{2}\psi_{2k-1}^2\right]$$
$$+ \sum_{k=N}^{\infty}(c_{2k} - c_{2k-1})^2\psi_{2k+1}^2$$
$$\geq 2\sum_{k=N}^{\infty}(c_{2k} - c_{2k-1})\left((2k)^\alpha + \frac{1}{2}c_{2k-1} - \frac{1}{2}(2k-1)^\alpha\right)\psi_{2k+1}^2$$
$$- 2\sum_{k=N-1}^{\infty}(c_{2k+2} - c_{2k+1})((2k+1)^\alpha + c_{2k+1})\frac{1}{2}\psi_{2k+1}^2$$
$$+ \sum_{k=N}^{\infty}(c_{2k} - c_{2k-1})^2\psi_{2k+1}^2.$$

Since it was assumed that $\{c_{2n} - c_{2n-1}\}$ is a decreasing sequence for $n \geq N$, it follows that

$$S_2 \geq -(c_{2N} - c_{2N-1})[(2N-1)^\alpha + c_{2N-1}]\psi_{2N-1}^2$$
$$+ 2\sum_{k=N}^{\infty}(c_{2k} - c_{2k-1})$$
$$\times\left((2k)^\alpha + \frac{1}{2}c_{2k-1} - \frac{1}{2}(2k-1)^\alpha - \frac{1}{2}(2k+1)^\alpha - \frac{1}{2}c_{2k+1}\right)\psi_{2k+1}^2$$
$$+ \sum_{k=N}^{\infty}(c_{2k} - c_{2k-1})^2\psi_{2k+1}^2.$$

Combining the two pieces it follows that if $\langle\psi, \phi_1\rangle = 0, \ldots, \langle\psi, \phi_{2N-1}\rangle = 0$ then $\langle C^2\psi, \psi\rangle = \|C\psi\|^2 \geq (\frac{\rho}{2} + \omega)^2\|\psi\|^2$. As in the proof of the previous theorem this argument does not apply to an eigenvector corresponding to a zero eigenvalue (when such an eigenvector exists). Hence it leads to the conclusion that $(-(\omega + \frac{\rho}{2}),\ (\omega + \frac{\rho}{2}))$ is a gap in the essential spectrum of the operator C containing at most $2N + 1$ eigenvalues.

Corollary 2.6. For $0 < \alpha \leq 1$, let $a_n = n^\alpha + c_n$, $n = 1, 2, \ldots$, where the sequence $\{c_n\}$ is chosen so that $a_n > 0$ for all n. Suppose that for $n \geq N$, $c_{2n-1} - c_{2n+1} = \rho$ with $\rho \geq 0$, and $c_{2n} - c_{2n-1} = \omega$ with $\omega \geq 0$. If $\rho > 0$ or $\omega > 0$, then $(-(\omega + \frac{\rho}{2}),\ (\omega + \frac{\rho}{2}))$ is a gap in the essential spectrum containing at most 2N+1 eigenvalues.

3. Absolute continuity

This section presents some results on absolute continuity in the special case $\alpha = 1$.

Theorem 3.1. *Let $a_n = n + c_n$, $n = 1, 2, \ldots$ where the sequence $\{c_n\}$ is chosen so that for all n, $a_n > 0$, for $n \geq 2$, $c_{2n} - c_{2n-2} = \rho$, $\rho \geq 0$, and for $n \geq 1$, $c_{2n-1} - c_{2n} = \omega$, $0 \leq \omega < 1$.*

1. *If $c_2 - \rho + \frac{1}{2}(1-\omega) > 0$, then $(-\omega - \frac{\rho}{2},\ \omega + \frac{\rho}{2})$ is a gap in the essential spectrum and the spectral measure is absolutely continuous on $(-\infty, 0) \cup (0, \infty)$.*
2. *If $1 + c_2 < \frac{\rho}{2}$, then $(-\omega - \frac{\rho}{2},\ \omega + \frac{\rho}{2})$ is a gap in the essential spectrum and the spectral measure is absolutely continuous on $(-\infty, -\omega - \frac{\rho}{2}) \cup (\omega + \frac{\rho}{2}, \infty)$. There exists a non-zero eigenvalue in the spectral gap.*

Proof. Under these assumptions, the existence of the spectral gap follows from Theorem 2.1 above. In addition, using the notation from Theorem 2.1 in [6], $a = a_1 = 1 + c_1 > 0$, and for $n = 1, 2, \ldots$.

$$\delta = a_{2n} - a_{2n-1} = 1 - \omega,$$
$$d = a_{2n+1} - a_{2n} = 1 + c_{2n+1} - c_{2n+2} + c_{2n+2} - c_{2n} = 1 + \rho + \omega.$$

Since $\omega = c_1 - c_2$, it follows that $a_1 = 1 + \omega + c_2$. If $c_2 - \rho + \frac{1}{2}(1-\omega) > 0$ then $a - d + \frac{\delta}{2} > 0$ and it follows from Theorem 2.1 in [6] that the spectral measure is absolutely continuous on $(-\infty, 0) \cup (0, \infty)$. This establishes (i). To prove (ii), let $a = a_1 = r(\omega + \frac{\rho}{2})$, $r > 0$. Then from Theorem 2.1 in [6] it is sufficient to show that $a - d + \frac{\delta}{2} + \frac{d}{2} \cdot \frac{(\omega + \frac{\rho}{2})^2}{a^2} > 0$.

But $a - d + \frac{\delta}{2} + \frac{d}{2} \cdot \frac{(\omega+\frac{\rho}{2})^2}{a^2} = r(\omega + \frac{\rho}{2}) - (1 + \omega + \rho) + \frac{1}{2}(1-\omega) + \frac{1}{2r^2}(1 + \omega + \rho)$. Viewing the right-hand side as a function of r, it can be shown that the minimum occurs when $r = [\frac{1+\rho+\omega}{\omega+\frac{\rho}{2}}]^{\frac{1}{3}} > 1$. If $1 + c_2 < \frac{\rho}{2}$, then $0 < r < 1$, and in this case $a - d + \frac{\delta}{2} + \frac{d}{2} \cdot \frac{(\omega+\frac{\rho}{2})^2}{a^2} > 0$. This establishes absolute continuity outside the spectral gap. It was shown in [6] that zero is an eigenvalue if and only if $\delta > d$. For the case in question zero is not an eigenvalue. Note that $\lambda = a_1 = 1 + c_1 = 1 + \omega + c_2$ is a root for $P_3(\lambda)$. If $s(\lambda) = \{P_n(\lambda)\}$ then the modified Wronskian $W_n(s(\lambda), s(-\lambda)) = (-1)^n 2a_n P_n(\lambda) P_n(-\lambda)$ has at least one node, and it follows from results in [11] that the interval $(-\lambda, \lambda)$ contains a non-zero eigenvalue.

Theorem 3.2. *Let $a_n = n + c_n$, $n = 1, 2, \ldots$ where the sequence $\{c_n\}$ is chosen so that for all n $a_n > 0$, $c_{2n-1} - c_{2n+1} = \rho$ with $\rho \geq 0$, $c_{2n} - c_{2n-1} = \omega$ with $0 \leq \omega \leq 1 - \rho$. If $c_1 + \rho + \frac{3}{2}\omega + \frac{1}{2} > 0$ then $(-\omega - \frac{\rho}{2},\ \omega + \frac{\rho}{2})$ is a gap in the essential spectrum, and the spectral measure is absolutely continuous on $(-\infty, 0) \cup (0, \infty)$.*

Proof. It follows from Theorem 2.5 above that $(-\omega - \frac{\rho}{2},\ \omega + \frac{\rho}{2})$ is a gap in the essential spectrum containing at most 3 eigenvalues. Theorem 2.1 in [6] again leads to a result on absolute continuity. Using the notation from Theorem 2.1

in [6], $a = a_1 = 1 + c_1 > 0$, and for $n = 1, 2, \ldots$.

$$\delta = a_{2n} - a_{2n-1} = 1 + \omega,$$
$$d = a_{2n+1} - a_{2n} = 1 + c_{2n+1} - c_{2n-1} + c_{2n-1} - c_{2n} = 1 - \rho - \omega.$$

It follows that

$$a - d + \frac{\delta}{2} = (1 + c_1) - (1 - \omega - \rho) + \frac{1}{2}(1 + \omega) = c_1 + \rho + \frac{3}{2}\omega + \frac{1}{2} > 0.$$

Thus it follows from Theorem 2.1 in [6] that the spectral measure is absolutely continuous on $(-\infty, 0) \cup (0, \infty)$.

References

[1] J. Berezanskii, *Expansions in Eigenfunctions of Self-adjoint Operators,* Trans. Math. Monographs. **17** Amer. Math. Soc. Providence, RI 1968.

[2] J. Dombrowski, *Cyclic operators, commutators and absolutely continuous measures,* **100** (1987), 457–463.

[3] J. Dombrowski, *Eigenvalues and spectral gaps related to periodic perturbations of Jacobi matrices, Operator Theory: Advances and Applications,* **154** (2004), 91–100.

[4] J. Dombrowski and S. Pedersen, *Orthogonal polynomials, spectral measures and absolute continuity, J. Comput. Appl. Math.,* **65** (1995), 115–124.

[5] J. Dombrowski and S. Pedersen, *Spectral measures and Jacobi matrices related to Laguerre type systems of orthogonal polynomials, Constr. Approx.,* **13** (1997), 421–433.

[6] J. Dombrowski and S. Pedersen, *Spectral Transition parameters for a class of Jacobi matrices, Studia Math.,* **152** (2002), 217–229.

[7] J. Dombrowski, J. Janas. M. Moszynski and S. Pedersen, *Spectral gaps resulting from periodic perturbations of a class of Jacobi operators, Constr. Approx.,* **20** (2004), 585–601.

[8] J. Janas, S. Naboko and G. Stolz, *Spectral theory for a class of periodically perturbed unbounded Jacobi matrices: elementary methods, J. Comput. Appl. Math.,* **171** (2004), 265–276.

[9] M. Moszynski, *Spectral properties of some Jacobi matrices with double weights, J. Math. Anal. Appl.,* **280** (2003), 400–412.

[10] M. Reed and B. Simon, *Methods of Modern Mathematical Physics IV: Analysis of Operators,* Academic Press, New York, 1978.

[11] G. Teschl, *Jacobi Operators and Complete Integrable Nonlinear Lattices,* Amer. Math. Soc., Providence, R.I., 2000.

J. Dombrowski
Department of Mathematics and Statistics
Wright State University
Dayton, OH 45435, USA
e-mail: `joanne.dombrowski@wright.edu`

Operator Theory:
Advances and Applications, Vol. 186, 115–137

Eigenvalue Estimates for the Aharonov-Bohm Operator in a Domain

Rupert L. Frank and Anders M. Hansson

Abstract. We prove semi-classical estimates on moments of eigenvalues of the Aharonov-Bohm operator in bounded two-dimensional domains. Moreover, we present a counterexample to the generalized diamagnetic inequality which was proposed by Erdős, Loss and Vougalter. Numerical studies complement these results.

Mathematics Subject Classification (2000). Primary 35P15; Secondary 35J10, 81Q10.

Keywords. Spectral estimates, Berezin-Li-Yau inequalities, Aharonov-Bohm magnetic field.

1. Introduction

We shall study inequalities for the eigenvalues of the Aharonov-Bohm operator

$$H_\alpha^\Omega := (\mathbf{D} - \alpha \mathbf{A}_0)^2 \qquad \text{in } L_2(\Omega). \tag{1.1}$$

Here $\Omega \subset \mathbb{R}^2$ is a bounded domain, $\mathbf{D} := -i\nabla$ and $\alpha \mathbf{A}_0(x) := \alpha |x|^{-2}(-x_2, x_1)^T$ is a vector potential generating an Aharonov-Bohm magnetic field with flux α through the origin. We shall assume that this point belongs to the interior of the simply-connected hull of Ω and that $\alpha \notin \mathbb{Z}$, for otherwise $\alpha \mathbf{A}_0$ can be gauged away. On the boundary of Ω we impose Dirichlet boundary conditions. More precisely, the operator (1.1) is defined through the closure of the quadratic form $\|(\mathbf{D} - \alpha \mathbf{A}_0)u\|^2$ on $C_0^\infty(\Omega \setminus \{0\})$.

Before stating our main results we would like to recall some well-known semi-classical spectral asymptotics and estimates for the Dirichlet Laplacian $-\Delta^\Omega$ and its magnetic version $(\mathbf{D} - \mathbf{A})_\Omega^2$, $\mathbf{A}$ being an arbitrary vector potential. If Ω is bounded then the spectrum of $-\Delta^\Omega$ is discrete, and by a classical result due to

Weyl (see, e.g., [ReS]) one has, as $\Lambda \to \infty$,

$$\operatorname{tr}(-\Delta^\Omega - \Lambda)_-^\gamma \sim \frac{1}{(2\pi)^2} \iint_{\Omega\times\mathbb{R}^2} (|\xi|^2 - \Lambda)_-^\gamma \, dx\, d\xi = \frac{1}{4\pi(\gamma+1)} |\Omega| \Lambda^{\gamma+1} \tag{1.2}$$

for all $\gamma \geq 0$. Note that the right-hand side involves the symbol $|\xi|^2$ on the phase space $\Omega \times \mathbb{R}^2$. The asymptotics (1.2) are accompanied by the estimate

$$\operatorname{tr}(-\Delta^\Omega - \Lambda)_-^\gamma \leq R_\gamma \frac{1}{(2\pi)^2} \iint_{\Omega\times\mathbb{R}^2} (|\xi|^2 - \Lambda)_-^\gamma \, dx\, d\xi, \qquad \gamma \geq 0, \tag{1.3}$$

with a universal constant R_γ independent of Ω and Λ, and one is interested in the sharp value of this constant R_γ. In view of (1.2) the sharp constant obviously cannot be smaller than 1, and by an argument of Aizenman and Lieb [AL], it is a non-increasing function of γ. Pólya [P] proved the estimate (1.3) for $\gamma = 0$ with constant 1 *under the additional assumption that Ω is a tiling domain.* His famous conjecture that this is true for arbitrary domains is still unproved. Berezin [B1] and independently Li and Yau [LiY] (see also [La]) proved (1.3) for $\gamma \geq 1$ with the sharp constant $R_\gamma = 1$. This also yields the so far best-known bound on the sharp constant for $\gamma = 0$, namely $R_0 \leq 2$. Indeed,

$$\operatorname{tr}(-\Delta^\Omega - \Lambda)_-^0 \leq (\mu - \Lambda)^{-1} \operatorname{tr}(-\Delta^\Omega - \mu)_- \leq (\mu - \Lambda)^{-1} \frac{1}{8\pi} |\Omega| \mu^2, \qquad \mu > \Lambda, \tag{1.4}$$

and the claim follows by optimization with respect to μ. We note that the estimate (1.3) for $\gamma = 0$ and $\gamma = 1$ is closely related to the estimates

$$\lambda_N^\Omega \geq \rho_0 4\pi |\Omega|^{-1} N \qquad \text{and} \qquad \sum_{j=1}^N \lambda_j^\Omega \geq \rho_1 2\pi |\Omega|^{-1} N^2$$

for the eigenvalues λ_j^Ω of the operator $-\Delta^\Omega$. The form (1.3), however, shows the close connection with the Lieb-Thirring inequality, see [LT] and also the review article [LaW2].

We now turn to the 'magnetic' analog of (1.3), i.e., where $-\Delta^\Omega$ is replaced by the Dirichlet realization of the operator $(\mathbf{D} - \mathbf{A})^2_\Omega$ in $L_2(\Omega)$ and $\mathbf{A}$ is a (sufficiently regular) magnetic vector potential. Note that the value of the right-hand side in (1.3) does not change if ξ is replaced by $\xi - \mathbf{A}(x)$. Hence one is interested in the estimate

$$\operatorname{tr}((\mathbf{D} - \mathbf{A})^2_\Omega - \Lambda)_-^\gamma \leq R_\gamma^{\mathrm{mag}} \frac{1}{(2\pi)^2} \iint_{\Omega\times\mathbb{R}^2} (|\xi|^2 - \Lambda)_-^\gamma \, dx\, d\xi, \qquad \gamma \geq 0, \tag{1.5}$$

with a universal constant R_γ^{mag} independent of Ω, Λ and $\mathbf{A}$. It is a consequence of the sharp Lieb-Thirring inequality by Laptev and Weidl [LaW1] that $R_\gamma^{\mathrm{mag}} = 1$ for $\gamma \geq \frac{3}{2}$. Not much is known about (1.5) in the case $\gamma < \frac{3}{2}$. The Laptev-Weidl result and an argument similar to (1.4) yield the (probably non-sharp) estimate $R_\gamma^{\mathrm{mag}} \leq (\frac{5}{3})^{3/2} (\gamma/(\gamma+1))^\gamma$ for $\gamma < \frac{3}{2}$. For $\gamma = 0$ and $\gamma = 1$ in particular one finds the values 2.1517 and 1.0758, respectively. In [ELoV] the estimate (1.5) is shown to hold for $\gamma \geq 1$ with constant 1 *in the special case of a homogeneous magnetic field*

$$\mathbf{A}(x) = \frac{B}{2}(-x_2, x_1)^T. \tag{1.6}$$

It was recently shown in [FLoW] that (1.5) does *not* hold with constant 1 if $0 \leq \gamma < 1$, not even when Ω is tiling. Moreover, the authors determined the optimal constant such that (1.5) holds for all Λ and all Ω under the constraint that $\mathbf{A}$ is given by (1.6).

In this paper we shall consider $\mathbf{A}$ corresponding to an Aharonov-Bohm magnetic field and we shall prove the estimate

$$\operatorname{tr}(H_\alpha^\Omega - \Lambda)_-^\gamma \leq C_\gamma(\alpha) \frac{1}{(2\pi)^2} \iint_{\Omega\times\mathbb{R}^2} (|\xi|^2 - \Lambda)_-^\gamma \, dx\, d\xi, \qquad \gamma \geq 1, \tag{1.7}$$

with a constant $C_\gamma(\alpha)$ given explicitly in terms of Bessel functions. Even though our bound is probably not sharp, it improves upon the previously known estimates. Indeed, numerical evaluation of our constant shows that (1.7) holds for all α with constants $C_0(\alpha) = 1.0540$ and $C_1(\alpha) = 1.0224$ if $\gamma = 0$ and 1, respectively, see Section 4. We complement our analytical results with a numerical study of the eigenvalue of the operator (1.1) for five domains: a disc, a square and three different annuli. In all cases the estimate (1.7) seems to be valid with constant 1. We refer to Section 5 and Figures 1–10 for a detailed account of the outcome of our experiments.

For the proof of our eigenvalue estimate we proceed similarly as in [ELoV]. Indeed, by the Berezin-Lieb inequality, (1.7) is an immediate consequence of the *generalized diamagnetic inequality*

$$\operatorname{tr}\chi_\Omega(H_\alpha - \Lambda)_-^\gamma \chi_\Omega \leq R_\gamma(\alpha) \operatorname{tr}\chi_\Omega(-\Delta - \Lambda)_-^\gamma \chi_\Omega \qquad \text{for all bounded } \Omega \subset \mathbb{R}^2. \tag{1.8}$$

Here $H_\alpha := H_\alpha^{\mathbb{R}^2}$ denotes the Aharonov-Bohm operator in the whole space. In [ELoV] an analogous estimate was proved in the case (1.6) *with constant* 1 when $\gamma \geq 1$. The authors conjectured that such an inequality is not true for an arbitrary magnetic field, but their counterexample contains a gap; the condition 2) on p. 905 can not be satisfied by a non-trivial radial vector field, as it was first pointed out by M. Solomyak. This gap can be removed by a minor change in the argument, since the assumption of radial symmetry is not essential in the proof [ELo]. In the present paper we show that (1.8) does *not* hold with constant 1 in the case of the Aharonov-Bohm operator. We feel that our example is of independent interest and sheds some light on the particularities of the Aharonov-Bohm operator. We even calculate the sharp constant in (1.8), see Theorem 3.1. We establish this by a thorough study of the local spectral density of the operator H_α, see Section 2.

We mention in closing the papers [BaEvLe], [EkF], [H], [MOR], where Lieb-Thirring estimates for the Schrödinger operator $H_\alpha + V$ were obtained. Our estimates can be seen as a refinement of these estimates in the special case where the potential V is equal to a negative constant $-\Lambda$ inside and equal to infinity outside a bounded domain $\Omega \subset \mathbb{R}^2$.

2. The Aharonov-Bohm operator in the whole space

2.1. Diagonalization

In this section we recall some well-known facts about the Aharonov-Bohm operator in the whole space, see, e.g., [AhBo], [Ru]. We denote by H_α the self-adjoint operator in $L_2(\mathbb{R}^2)$ associated with the closure of the quadratic form

$$\int_{\mathbb{R}^2} |(\mathbf{D} - \alpha \mathbf{A}_0)u|^2 \, dx, \qquad u \in C_0^\infty(\mathbb{R}^2 \setminus \{0\}).$$

Here $\mathbf{D} := -i\nabla$, $\mathbf{A}_0(x) := |x|^{-2}(-x_2, x_1)^T$ and $\alpha \in \mathbb{R}$. Moreover, J_ν denotes, as usual, the Bessel function of the first kind of order ν, see [AbSt]. With polar coordinates $x = |x|(\cos\theta_x, \sin\theta_x)$ and similarly for ξ, we define

$$\mathcal{F}_\alpha(\xi, x) := \frac{1}{2\pi} \sum_{n\in\mathbb{Z}} J_{|n-\alpha|}(|\xi||x|) e^{in(\theta_\xi - \theta_x)}, \qquad \xi, x \in \mathbb{R}^2,$$

and put

$$(\mathcal{F}_\alpha u)(\xi) := \int_{\mathbb{R}^d} \mathcal{F}_\alpha(\xi, x) u(x) \, dx, \qquad \xi \in \mathbb{R}^2,$$

for $u \in C_0^\infty(\mathbb{R}^2)$. Note that $\mathcal{F}_0(\xi, x) = (2\pi)^{-1} e^{-i\xi\cdot x}$ [AbSt, 9.1.41], so $\mathcal{F}_0$ is the ordinary Fourier transform, which diagonalizes $H_0 = -\Delta$. Similarly, one has

Lemma 2.1. *For any $\alpha \in \mathbb{R}$, $\mathcal{F}_\alpha$ extends to a unitary operator in $L_2(\mathbb{R}^2)$ and diagonalizes H_α, i.e.,*

$$(\mathcal{F}_\alpha f(H_\alpha)u)(\xi) = f(|\xi|^2)(\mathcal{F}_\alpha u)(\xi), \qquad \xi \in \mathbb{R}^2,$$

for any $u \in L_2(\mathbb{R}^2)$ and $f \in L_\infty(\mathbb{R})$.

We sketch a proof of this assertion for the sake of completeness.

Proof. The orthogonal decomposition

$$L_2(\mathbb{R}^2) = \bigoplus_{n\in\mathbb{Z}} \mathfrak{H}_n, \qquad \mathfrak{H}_n := \{|x|^{-1/2} g(|x|) e^{in\theta_x} : g \in L_2(\mathbb{R}_+)\},$$

reduces H_α. The part of H_α in $\mathfrak{H}_n$ is unitarily equivalent to the operator

$$h_{|n-\alpha|} := -\frac{d^2}{dr^2} + \frac{(n-\alpha)^2 - 1/4}{r^2} \qquad \text{in } L_2(\mathbb{R}_+),$$

which is defined as the Friedrichs extension of the corresponding differential expression on $C_0^\infty(\mathbb{R}_+)$. (We emphasize that in our notation $\mathbb{R}_+$ means the *open* interval $(0, \infty)$.) The operator

$$(\Phi_\nu g)(k) := \int_{\mathbb{R}_+} \sqrt{rk} J_\nu(rk) g(r) \, dr, \qquad k \in \mathbb{R}_+,$$

initially defined for $g \in C_0^\infty(\mathbb{R}_+)$, extends to a unitary operator in $L_2(\mathbb{R}_+)$ and diagonalizes h_ν, i.e.,

$$(\Phi_\nu f(h_\nu) g)(k) = f(k^2)(\Phi_\nu g)(k), \qquad k \in \mathbb{R}_+,$$

for any $g \in L_2(\mathbb{R}_+)$ and $f \in L_\infty(\mathbb{R})$ [Ti, Ch. VIII]. The assertion of the lemma is a simple consequence of these facts. □

The proof of the preceding lemma shows in particular that the operators H_α and $H_{\alpha+m}$ with $m \in \mathbb{Z}$ are unitarily equivalent via multiplication by $e^{im\theta_x}$ (a gauge transformation). Hence, without loss of generality *we shall assume that* $0 \leq \alpha < 1$.

Lemma 2.1 implies that $f(H_\alpha)$, at least formally, is an integral operator with integral kernel

$$\begin{aligned} f(H_\alpha)(x,y) &= \int_{\mathbb{R}^2} \overline{\mathcal{F}_\alpha(\xi,x)} f(|\xi|^2)\mathcal{F}_\alpha(\xi,y)\,d\xi \\ &= \frac{1}{2\pi}\sum_{n\in\mathbb{Z}}\int_0^\infty f(k^2)J_{|n-\alpha|}(k|x|)J_{|n-\alpha|}(k|y|)e^{in(\theta_x-\theta_y)}k\,dk \\ &= \frac{1}{4\pi}\sum_{n\in\mathbb{Z}}\int_0^\infty f(\lambda)J_{|n-\alpha|}(\sqrt{\lambda}|x|)J_{|n-\alpha|}(\sqrt{\lambda}|y|)e^{in(\theta_x-\theta_y)}\,d\lambda. \end{aligned}$$

On the diagonal this is

$$f(H_\alpha)(x,x) = \frac{1}{4\pi}\int_0^\infty f(\lambda)\rho_\alpha(\sqrt{\lambda}|x|)\,d\lambda, \tag{2.1}$$

where

$$\rho_\alpha(t) := \sum_{n\in\mathbb{Z}} J^2_{|n-\alpha|}(t), \qquad t \geq 0. \tag{2.2}$$

In particular, $\frac{1}{4\pi}\rho_\alpha(\sqrt{\lambda}|x|)$ is the *local spectral density* at energy λ. In the following subsection we collect some basic information about this function, and in Subsections 2.3 and 2.5 we prove some results about the precise asymptotic behavior of (2.1) as $|x| \to \infty$ for special choices of functions f. This will allow us to prove that the generalized diamagnetic inequality is violated.

2.2. The spectral density

Our results are based on a detailed study of the function ρ_α from (2.2). We note that if $\alpha = 0$ then $\rho_0 \equiv 1$ by [AbSt, 9.1.76]. An expression in terms of elementary functions is also available [AbSt, 5.2.15] for $\alpha = 1/2$,

$$\rho_{1/2}(t) = \frac{2}{\pi}\int_0^{2t}\frac{\sin s}{s}\,ds. \tag{2.3}$$

As $t \to \infty$, $\rho_{1/2}(t)$ tends to 1 in an oscillating manner. As we shall see, this behavior appears for all $0 < \alpha < 1$. The starting point of our study of ρ_α with non-trivial flux α is the following

Lemma 2.2. *For any* $0 < \alpha < 1$, ρ_α *is a smooth function on* $\mathbb{R}_+$ *with* $\rho_\alpha(0) = 0$, $\rho_\alpha(t) \to 1$ *as* $t \to \infty$ *and*

$$\rho'_\alpha(t) = J_\alpha(t)J_{\alpha-1}(t) + J_{1-\alpha}(t)J_{-\alpha}(t), \qquad t \geq 0. \tag{2.4}$$

Proof. By [Lu, 11.2(10)] and [AbSt, 11.4.42] one has, for all $t \geq 0$,

$$\begin{aligned}\rho_\alpha(t) &= \int_0^t (J_\alpha(s)J_{\alpha-1}(s) + J_{1-\alpha}(s)J_{-\alpha}(s))\, ds \\ &= 1 - \int_t^\infty (J_\alpha(s)J_{\alpha-1}(s) + J_{1-\alpha}(s)J_{-\alpha}(s))\, ds,\end{aligned}$$

which implies the assertion. □

Our next result will not be needed in the sequel, but it helps to clarify the behavior of ρ_α and demonstrates the methods which we shall use later on.

Lemma 2.3. *Let $0 < \alpha < 1$. As $t \to \infty$,*

$$\rho_\alpha(t) = 1 - \frac{\sin \alpha\pi}{\pi} \frac{\cos 2t}{t} + \mathcal{O}(t^{-2}).$$

Proof. The asymptotics [AbSt, 9.2.5]

$$J_\nu(t) = \sqrt{\frac{2}{\pi t}} \left(\cos\left(t - \frac{\pi}{4} - \frac{\nu\pi}{2}\right) - \frac{4\nu^2 - 1}{8t} \sin\left(t - \frac{\pi}{4} - \frac{\nu\pi}{2}\right) + \mathcal{O}(t^{-2}) \right),$$

the formula (2.4) and elementary manipulations show that

$$\rho_\alpha'(t) = \frac{2}{\pi t} \left(\sin \alpha\pi \sin 2t + \frac{(2\alpha - 1)^2 \sin \alpha\pi}{4} \frac{\cos 2t}{t} + \mathcal{O}(t^{-2}) \right).$$

Using that $\rho_\alpha(t) = 1 - \int_t^\infty \rho_\alpha'(s)\, ds$ by Lemma 2.2, we obtain the assertion by repeated integration by parts. □

2.3. Moments of the spectral density

For any $\gamma > -1$ let us define

$$\sigma_{\alpha,\gamma}(r) := \int_0^1 (1-\lambda)^\gamma \rho_\alpha(\sqrt{\lambda} r) d\lambda, \qquad r \geq 0. \tag{2.5}$$

We are interested in the asymptotic behavior of this quantity or, more precisely, in the way it approaches its limit.

Theorem 2.4. *Let $0 < \alpha < 1$ and $\gamma > -1$. As $r \to \infty$,*

$$\sigma_{\alpha,\gamma}(r) = \frac{1}{\gamma + 1} - \Gamma(\gamma + 1) \frac{\sin \alpha\pi}{\pi} \frac{\sin(2r - \frac{1}{2}\gamma\pi)}{r^{2+\gamma}} + \mathcal{O}(r^{-3-\gamma}).$$

For the proof we note that by Lemma 2.2 and dominated convergence

$$\lim_{r \to \infty} \sigma_{\alpha,\gamma}(r) = \int_0^1 (1-\lambda)^\gamma\, d\lambda = \frac{1}{\gamma + 1}.$$

In view of

$$\sigma_{\alpha,\gamma}(r) = \frac{1}{\gamma + 1} - \int_r^\infty \sigma_{\alpha,\gamma}'(s)\, ds,$$

Theorem 2.4 follows via integration by parts from

Proposition 2.5. *Let* $0 < \alpha < 1$ *and* $\gamma > -1$. *As* $r \to \infty$,

$$\sigma'_{\alpha,\gamma}(r) = \frac{\Gamma(\gamma+1)}{r^{\gamma+2}} \frac{\sin \alpha\pi}{\pi} \left(-2\cos\left(2r - \frac{\gamma\pi}{2}\right) - \frac{d_1}{r} \sin\left(2r - \frac{\gamma\pi}{2}\right) + \mathcal{O}(r^{-2}) \right) \tag{2.6}$$

where $d_1 := \frac{1}{8}(1 - 4((\gamma + \frac{3}{2})^2 + (1-2\alpha)^2)$.

We defer the rather technical proof of this proposition to the following subsection. We would like to point out that (2.6) can be proved in an elementary way if $\alpha = \frac{1}{2}$ due to the formula (2.3). Indeed, if we simplify further by taking $\gamma = 1$, then

$$\begin{aligned} \sigma'_{1/2,1}(r) &= \frac{2}{\pi r} \int_0^1 (1-\lambda) \sin(2\sqrt{\lambda} r)\, d\lambda = \frac{4}{\pi r} \int_0^1 k(1-k^2) \sin 2kr\, dk \\ &= -\frac{2}{\pi r} \left(\frac{\sin 2r}{r^2} + \frac{3\cos 2r}{2r^3} + \mathcal{O}(r^{-4}) \right) \end{aligned}$$

by integration by parts. This is (2.6) in this special case.

2.4. Proof of Proposition 2.5

We shall need

Lemma 2.6. *Let* $\nu > -1/2$ *and* $\gamma > -1$. *Then*

$$\int_0^1 J_{\nu+1/2}(kr) J_{\nu-1/2}(kr) k^2 (1-k^2)^\gamma\, dk = c_\gamma \int_0^1 J_{2\nu}(2kr) k (1-k^2)^{\gamma+1/2} dk$$

with

$$c_\gamma := \frac{1}{\sqrt{\pi}} \frac{\Gamma(\gamma+1)}{\Gamma(\gamma+\frac{3}{2})}. \tag{2.7}$$

Proof. By virtue of the series representations [AbSt, 9.1.10, 9.1.14] for $\alpha > 0$,

$$J_{\alpha-1}(t) = \left(\frac{t}{2}\right)^{\alpha-1} \sum_{n=0}^\infty \frac{(-1)^n}{n!\Gamma(\alpha+n)} \left(\frac{t}{2}\right)^{2n},$$

$$J_\alpha(t) J_{\alpha-1}(t) = \left(\frac{t}{2}\right)^{2\alpha-1} \sum_{n=0}^\infty \frac{(-1)^n \Gamma(2\alpha+2n)}{n!\Gamma(\alpha+n)\Gamma(\alpha+n+1)\Gamma(2\alpha+n)} \left(\frac{t}{2}\right)^{2n},$$

the statement is equivalent to

$$\begin{aligned} &\left(\frac{r}{2}\right)^{2\nu-1} \sum_{n=0}^\infty \frac{(-1)^n (r/2)^{2n} \Gamma(2\nu+2n)}{n!\Gamma(2\nu+n)\Gamma(\nu+n)\Gamma(\nu+n+1)} \int_0^1 k^{2\nu+2n+1} (1-k^2)^\gamma dk \\ &= \frac{r^{2\nu-1}}{\sqrt{\pi}} \frac{\Gamma(\gamma+1)}{\Gamma(\gamma+\frac{3}{2})} \sum_{n=0}^\infty \frac{(-1)^n r^{2n}}{n!\Gamma(2\nu+n)} \int_0^1 k^{2\nu+2n} (1-k^2)^{\gamma+1/2} dk, \end{aligned} \tag{2.8}$$

an equality which actually holds termwise. This follows by the beta function identity

$$2\int_0^1 k^{2\alpha-1}(1-k^2)^{\beta-1}dk = \int_0^1 t^{\alpha-1}(1-t)^{\beta-1}dt = B(\alpha,\beta) = \frac{\Gamma(\alpha)\Gamma(\beta)}{\Gamma(\alpha+\beta)}$$

and the duplication formula $\sqrt{\pi}\Gamma(2\alpha) = 2^{2\alpha-1}\Gamma(\alpha)\Gamma\left(\alpha+\frac{1}{2}\right)$ [AbSt, 6.1.18]. □

Lemma 2.7. *Let $\beta > 0$ and $-2 < \nu < 2$. The following decomposition holds:*

$$\int_0^1 J_\nu(kr)k(1-k^2)^{\beta-1}dk = f_{\beta,\nu}(r) + g_{\beta,\nu}(r), \tag{2.9}$$

where $g_{\beta,-\nu}(r) = -g_{\beta,\nu}(r)$ and

$$f_{\beta,\nu}(r) = \frac{\Gamma(\beta)}{2\sqrt{\pi}}\left(\frac{2}{r}\right)^{\beta+1/2}\left[\cos(r-r_0) + \frac{d_1}{r}\sin(r-r_0) + \mathcal{O}(r^{-2})\right], \qquad r\to\infty,$$

with

$$r_0 := \frac{\pi}{2}\left(\beta+\nu+\frac{1}{2}\right) \quad \textit{and} \quad d_1 := \frac{1}{8} - \frac{\beta^2+\nu^2}{2}. \tag{2.10}$$

The function $g_{\beta,\nu}$ has a power-like, non-oscillatory behavior at infinity and dominates $f_{\beta,\nu}$ when $\beta > \frac{3}{2}$. Its odd parity with respect to ν is, however, the property vital to us, since it leads to a useful cancellation.

Proof. The left-hand side in (2.9) can be expressed as a generalized hypergeometric function. Recall that for $p,q \in \mathbb{N}_0$, $p \le q$ and $\alpha_1,\dots,\alpha_p,\beta_1,\dots,\beta_q \in \mathbb{R}\setminus(-\mathbb{N}_0)$,

$${}_pF_q\left(\begin{array}{c}\alpha_1,\alpha_2,\dots,\alpha_p\\ \beta_1,\beta_2,\dots,\beta_q\end{array}\middle| z\right) = \sum_{n=0}^\infty \frac{(\alpha_1)_n(\alpha_2)_n\cdots(\alpha_p)_n}{(\beta_1)_n(\beta_2)_n\cdots(\beta_q)_n}\frac{z^n}{n!}, \qquad z\in\mathbb{C},$$

where $(\alpha)_n = \Gamma(\alpha+n)/\Gamma(\alpha) = \alpha(\alpha+1)\cdots(\alpha+n-1)$. By [Lu, 13.3.2(10)],

$$\begin{aligned}\int_0^1 J_\nu(kr)k(1-k^2)^{\beta-1}dk &= \int_0^{\pi/2} J_\nu(r\sin\theta)\cos^{2\beta-1}\theta\sin\theta\,d\theta\\ &= \frac{r^\nu\Gamma(\beta)\Gamma(\nu/2+1)}{2^{\nu+1}\Gamma(\nu+1)\Gamma(\beta+\nu/2+1)}\,{}_1F_2\left(\begin{array}{c}\nu/2+1\\ \nu+1,\beta+\nu/2+1\end{array}\middle| -\frac{r^2}{4}\right).\end{aligned}$$

Next we use the asymptotics of the generalized hypergeometric function [Lu, 1.3.3(5, 7, 8, 13)],

$$\begin{aligned}&\frac{\Gamma(\nu/2+1)}{\Gamma(\nu+1)\Gamma(\beta+\nu/2+1)}\,{}_1F_2\left(\begin{array}{c}\nu/2+1\\ \nu+1,\beta+\nu/2+1\end{array}\middle| -\frac{r^2}{4}\right)\\ &\sim\frac{(2/r)^{\beta+\nu+1/2}}{2\sqrt{\pi}}\left[e^{i(r-r_0)}\sum_{n=0}^\infty d_n\,(ir)^{-n} + e^{-i(r-r_0)}\sum_{n=0}^\infty d_n\,(-ir)^{-n}\right]\\ &\quad+\frac{\nu(2/r)^{\nu+2}}{2\Gamma(\beta)}\,{}_3F_0\left(\begin{array}{c}1+\nu/2,1-\nu/2,1-\beta\\ -\end{array}\middle| -\frac{4}{r^2}\right), \qquad r\to\infty,\end{aligned}$$

with r_0, d_1 as in (2.10) and $d_0 = 1$. We emphasize that the $\sim$ sign means equality in the sense of asymptotic expansions, and that ${}_3F_0$ is not a well-defined function but denotes an asymptotic expansion. We define $\nu^{-1}r^2 g_{\beta,\nu}(r)$ as a finite approximation to ${}_3F_0$. Namely, let K denote the smallest non-negative integer such that $2K \geq \beta - \frac{3}{2}$ and put

$$g_{\beta,\nu}(r) := \frac{\nu}{r^2}\sum_{n=0}^{K}(1+\nu/2)_n(1-\nu/2)_n(1-\beta)_n\frac{(-1)^n}{n!}\left(\frac{2}{r}\right)^{2n}.$$

This function is antisymmetric in ν and

$$\begin{aligned} f_{\beta,\nu}(r) :=& \int_0^1 J_\nu(kr)k(1-k^2)^{\beta-1}dk - g_{\beta,\nu}(r) \\ =& \frac{r^\nu\Gamma(\beta)}{2^{\nu+1}}\frac{(2/r)^{\beta+\nu+1/2}}{2\sqrt{\pi}}\left[e^{i(r-r_0)}\left(d_0+\frac{d_1}{ir}\right)+e^{-i(r-r_0)}\left(d_0-\frac{d_1}{ir}\right)\right. \\ &\left. + \mathcal{O}(r^{-2})\right] + \mathcal{O}\left(r^{-2(K+2)}\right) \\ =& \frac{\Gamma(\beta)}{2\sqrt{\pi}}\left(\frac{2}{r}\right)^{\beta+1/2}\left[d_0\cos(r-r_0)+\frac{d_1}{r}\sin(r-r_0)+\mathcal{O}(r^{-2})\right] \end{aligned}$$

as claimed. □

Now everything is in place for the

Proof of Proposition 2.5. By Lemmas 2.2 and 2.6,

$$\begin{aligned} \sigma'_{\alpha,\gamma}(r) &= 2\int_0^1 k^2(1-k^2)^\gamma\rho'_\alpha(kr)dk \\ &= 2\int_0^1 (J_\alpha(kr)J_{\alpha-1}(kr)+J_{1-\alpha}(kr)J_{-\alpha}(kr))k^2(1-k^2)^\gamma dk \\ &= 2c_\gamma\int_0^1 (J_{2\alpha-1}(2kr)+J_{1-2\alpha}(2kr))k(1-k^2)^{\gamma+1/2}dk \end{aligned}$$

with c_γ from (2.7). Now we apply Lemma 2.7 with $\beta = \gamma + \frac{3}{2}$ and $\nu = \pm(2\alpha - 1)$. Using the antisymmetry of $g_{\beta,\nu}$ we find that

$$\sigma'_{\alpha,\gamma}(r) = 2c_\gamma\left(f_{\gamma+3/2,2\alpha-1}(r) + f_{\gamma+3/2,1-2\alpha}(r)\right),$$

and the assertion follows after elementary manipulations from the asymptotics of $f_{\beta,\nu}$ given in Lemma 2.7. □

2.5. The Laplace transform of the spectral density

In this subsection we are interested in the quantity

$$\sigma_{\alpha,\infty}(r) := \int_0^\infty e^{-\lambda}\rho_\alpha(\sqrt{\lambda}r)d\lambda, \qquad r \geq 0.$$

Note that this is essentially the Laplace transform of the function $\rho_\alpha(\sqrt{\cdot})$.

Theorem 2.8. *For all $0 < \alpha < 1$ the function $\sigma_{\alpha,\infty}$ is strictly increasing from 0 to 1 on $[0, \infty)$. In particular, $\sigma_{\alpha,\infty}(r) < 1$ for all $r \geq 0$.*

This theorem shows that the oscillations we observed for $\sigma_{\alpha,\gamma}$ are no longer present.

Proof. Since $\rho_\alpha = \rho_{1-\alpha}$ and hence $\sigma_{\alpha,\infty} = \sigma_{1-\alpha,\infty}$, it suffices to treat the case $0 < \alpha \leq 1/2$. By the properties of ρ_α (see Lemma 2.2) and dominated convergence we get $\sigma_{\alpha,\infty}(0) = 0$ and $\sigma_{\alpha,\infty}(r) \to 1$ as $r \to \infty$. Again by Lemma 2.2,

$$\begin{aligned}\sigma'_{\alpha,\infty}(r) &= 2\int_0^\infty k^2 e^{-k^2} \rho'_\alpha(kr)dk \\ &= 2\int_0^\infty k^2 e^{-k^2} \left(J_\alpha(kr)J_{\alpha-1}(kr) + J_{1-\alpha}(kr)J_{-\alpha}(kr)\right) dk,\end{aligned}$$

and hence according to [Lu, 13.4.1(10)],

$$\begin{aligned}\sigma'_{\alpha,\infty}(r) = r^{1-2\alpha}\Bigg[&\frac{2^{1-2\alpha}}{r^{2(1-2\alpha)}\Gamma(\alpha)}\,{}_1F_1\left(\begin{array}{c}\alpha+\frac{1}{2}\\ 2\alpha\end{array}\Bigg| -r^2\right) \\ &+\frac{1}{2^{1-2\alpha}\Gamma(1-\alpha)}\,{}_1F_1\left(\begin{array}{c}\frac{3}{2}-\alpha\\ 2-2\alpha\end{array}\Bigg| -r^2\right)\Bigg].\end{aligned}$$

In the special case $\alpha = 1/2$ we note that

$$\sigma'_{1/2,\infty}(r) = \frac{2}{\sqrt{\pi}}\,{}_1F_1\left(\begin{array}{c}1\\ 1\end{array}\Bigg| -r^2\right) = \frac{2}{\sqrt{\pi}}e^{-r^2}.$$

If $0 < \alpha < \frac{1}{2}$ we can apply the *Kummer transformations* [AbSt, 13.1.27] (note that ${}_1F_1(a, b; z) = M(a, b, z)$ in [AbSt]) to get

$$\begin{aligned}\sigma'_{\alpha,\infty}(r) = r^{1-2\alpha}e^{-r^2}\Bigg[&\frac{2^{1-2\alpha}}{r^{2(1-2\alpha)}\Gamma(\alpha)}\,{}_1F_1\left(\begin{array}{c}\alpha-\frac{1}{2}\\ 2\alpha\end{array}\Bigg| r^2\right) \\ &+\frac{1}{2^{1-2\alpha}\Gamma(1-\alpha)}\,{}_1F_1\left(\begin{array}{c}\frac{1}{2}-\alpha\\ 2-2\alpha\end{array}\Bigg| r^2\right)\Bigg].\end{aligned}$$

By elementary properties of the gamma function this can be rewritten as

$$\sigma'_{\alpha,\infty}(r) = \frac{2\sin\alpha\pi}{\sqrt{\pi}} r^{1-2\alpha}e^{-r^2}U\left(\frac{1}{2}-\alpha, 2-2\alpha, r^2\right)$$

where [AbSt, 13.1.3]

$$\begin{aligned}U\left(\frac{1}{2}-\alpha, 2-2\alpha, r^2\right) = \frac{\pi}{\sin 2\alpha\pi}\frac{1}{r^{2(1-2\alpha)}}\Bigg[&\frac{1}{\Gamma(\frac{1}{2}-\alpha)\Gamma(2\alpha)}\,{}_1F_1\left(\begin{array}{c}\alpha-\frac{1}{2}\\ 2\alpha\end{array}\Bigg| r^2\right) \\ &-\frac{r^{2(1-2\alpha)}}{\Gamma(\alpha-\frac{1}{2})\Gamma(2-2\alpha)}\,{}_1F_1\left(\begin{array}{c}\frac{1}{2}-\alpha\\ 2-2\alpha\end{array}\Bigg| r^2\right)\Bigg].\end{aligned}$$

U is positive by the integral representation [AbSt, 13.2.5], and this proves the theorem. □

3. Counterexample to the generalized diamagnetic inequality

Following [ELoV] we consider the question: *Which non-negative convex functions φ vanishing at infinity satisfy*

$$\operatorname{tr}\chi_\Omega\varphi(H_\alpha)\chi_\Omega \le \operatorname{tr}\chi_\Omega\varphi(-\Delta)\chi_\Omega \quad \textit{for all bounded domains } \Omega\subset\mathbb{R}^2\,? \tag{3.1}$$

By (2.1) the statement (3.1) is equivalent to the pointwise inequality

$$\int_0^\infty \varphi(\lambda)\rho_\alpha(\sqrt{\lambda}r)\,d\lambda \le \int_0^\infty \varphi(\lambda)\,d\lambda, \quad \text{for all } r\in[0,\infty). \tag{3.2}$$

Note that (3.1) is true for the family of functions $\varphi(\lambda)=e^{-t\lambda}$, $t>0$; we shall prove this (even with strict inequality) in Remark 3.2 below. Alternatively, it follows from the 'ordinary' diamagnetic inequality (see, e.g., [HuS]),

$$|\exp(-tH_\alpha)u| \le \exp(-t(-\Delta))|u| \text{ a.e.,} \qquad u\in L_2(\mathbb{R}^2),$$

and [S, Thm. 2.13], since

$$\left\|\chi_\Omega\exp\left(-\frac{t}{2}H_\alpha\right)\right\|_2^2 = \operatorname{tr}\chi_\Omega\exp(-tH_\alpha)\chi_\Omega.$$

We point out that the validity of (3.1) for $\varphi(\lambda)=e^{-t\lambda}$, $t>0$, actually implies that it holds for any function of the form

$$\varphi(\lambda)=\int_0^\infty e^{-t\lambda}w(t)\,dt, \qquad w\ge 0. \tag{3.3}$$

In connection with a Berezin-Li-Yau-type inequality one is particularly interested in the functions

$$\varphi(\lambda)=(\lambda-\Lambda)_-^\gamma, \qquad \gamma\ge 1,\ \Lambda>0. \tag{3.4}$$

Note that these functions cannot be expressed in the form (3.3).

Theorem 3.1. *Let $0<\alpha<1$ and let φ be given by (3.4) for some $\gamma\ge1$, $\Lambda>0$. Then the generalized diamagnetic inequality (3.1) is violated. More precisely, there exist constants $C_1, C_2>0$ (depending on α and γ, but not on Λ) such that for all $|x|\ge C_1\Lambda^{-1/2}$,*

$$\left|\varphi(H_\alpha)(x,x)-\varphi(-\Delta)(x,x)+A_{\alpha,\gamma}(\Lambda)\frac{\sin(2\sqrt{\Lambda}|x|-\frac12\gamma\pi)}{|x|^{2+\gamma}}\right| \le C_2\frac{\Lambda^{(\gamma-1)/2}}{|x|^{3+\gamma}} \tag{3.5}$$

with $A_{\alpha,\gamma}(\Lambda):=(2\pi)^{-2}\Lambda^{\gamma/2}\Gamma(\gamma+1)\sin\alpha\pi$.

Proof. By (2.1) and the scaling $\lambda\mapsto\Lambda\lambda$,

$$\varphi(H_\alpha)(x,x)=\frac{\Lambda^{\gamma+1}}{4\pi}\int_0^1(1-\lambda)^\gamma\rho_\alpha(\sqrt{\Lambda\lambda}|x|)\,d\lambda=\frac{\Lambda^{\gamma+1}}{4\pi}\sigma_{\alpha,\gamma}(\sqrt{\Lambda}|x|), \tag{3.6}$$

where we used the notation (2.5). Note that $\varphi(-\Delta)(x,x)=(4\pi(\gamma+1))^{-1}\Lambda^{\gamma+1}$. The expansion (3.5) is thus a consequence of Theorem 2.4. To prove that the generalized diamagnetic inequality (3.1) is violated, one can consider the domains

$\Omega_n := \{x \in \mathbb{R}^2 : |\sqrt{\Lambda}|x| - r_n| < \varepsilon\}$, $n \in \mathbb{N}$, with $r_n := \pi(n + \frac{1}{4}(\gamma - 1))$ and sufficiently small but fixed $\varepsilon > 0$. It follows easily from (3.5) that (3.1) is violated for all large n. □

Remark 3.2. The analogous statement (with the same proof) is valid for $-1 < \gamma < 1$. Moreover, for exponential functions Theorem 2.8 implies that

$$\exp(-tH_\alpha)(x,x) = \frac{1}{4\pi t}\int_0^\infty \rho_\alpha\left(\sqrt{\lambda}\frac{|x|}{\sqrt{t}}\right) e^{-\lambda} d\lambda = \frac{1}{4\pi t}\sigma_{\alpha,\infty}\left(\frac{|x|}{\sqrt{t}}\right)$$

is *strictly* less than $(4\pi t)^{-1} = \exp(-t(-\Delta))(x,x)$.

The following substitute for (3.1) will be useful later on.

Proposition 3.3. *Let* $0 < \alpha < 1$ *and let* φ *be given by* (3.4) *for some* $\gamma > -1$, $\Lambda > 0$. *Then for all open sets* $\Omega \subset \mathbb{R}^2$

$$\operatorname{tr} \chi_\Omega \varphi(H_\alpha) \chi_\Omega \leq R_\gamma(\alpha) \operatorname{tr} \chi_\Omega \varphi(-\Delta) \chi_\Omega \tag{3.7}$$

with

$$R_\gamma(\alpha) := (\gamma + 1) \sup_{r \geq 0} \int_0^1 (1-\lambda)^\gamma \rho_\alpha(\sqrt{\lambda} r)\, d\lambda. \tag{3.8}$$

Indeed, this is an immediate consequence of (3.6). This shows as well that the right-hand side of (3.8) yields the sharp constant in (3.7).

Remark 3.4. The constant $R_\gamma(\alpha)$ is strictly decreasing with respect to γ. Indeed, following [AL] we write, for $\gamma > \gamma' > -1$ and $0 \leq \lambda \leq 1$,

$$B(\gamma - \gamma', \gamma' + 1)(1-\lambda)^\gamma = \int_0^{1-\lambda} (1 - \lambda - \mu)^{\gamma'} \mu^{\gamma - \gamma' - 1}\, d\mu$$

and find, for any $r \geq 0$,

$$\begin{aligned} B(\gamma - \gamma', \gamma' + 1) &\int_0^1 (1-\lambda)^\gamma \rho_\alpha(\sqrt{\lambda} r)\, d\lambda \\ &= \int_0^1 \left(\int_0^{1-\mu} (1 - \lambda - \mu)^{\gamma'} \rho_\alpha(\sqrt{\lambda} r)\, d\lambda \right) \mu^{\gamma - \gamma' - 1}\, d\mu \\ &= \int_0^1 w(r\sqrt{1-\mu})(1-\mu)^{\gamma'+1} \mu^{\gamma - \gamma' - 1}\, d\mu, \end{aligned}$$

where $w(s) := \int_0^1 (1-\lambda)^{\gamma'} \rho_\alpha(\sqrt{\lambda} s)\, d\lambda$. Since $w(s) \leq R_{\gamma'}(\alpha)(\gamma'+1)^{-1}$ one concludes that

$$\begin{aligned} R_\gamma(\alpha) &\leq R_{\gamma'}(\alpha)(\gamma'+1)^{-1}(\gamma+1)B(\gamma-\gamma',\gamma'+1)^{-1} \int_0^1 (1-\mu)^{\gamma'+1}\mu^{\gamma-\gamma'-1}\, d\mu \\ &= R_{\gamma'}(\alpha)(\gamma'+1)^{-1}(\gamma+1)B(\gamma-\gamma',\gamma'+1)^{-1}B(\gamma-\gamma',\gamma'+2) = R_{\gamma'}(\alpha). \end{aligned}$$

That this inequality is actually strict, follows from the fact that the supremum in (3.8) is attained for some $r_0 \in (0, \infty)$ (see Theorem 2.4) and that $w(r_0\sqrt{1-\mu})$ is non-constant with respect to μ.

We close this section by giving numerical values for $R_\gamma(\alpha)$. Note that

$$\begin{aligned}R_\gamma(\alpha) =&2(\gamma+1)\sup_{r\geq 0}\int_0^1 (1-k^2)^\gamma \rho_\alpha(kr)k\,dk\\ =&\sup_{r\geq 0} r\int_0^1 (1-k^2)^{\gamma+1}\rho'_\alpha(kr)\,dk\\ =&\sup_{r\geq 0} r\int_0^1 (1-k^2)^{\gamma+1}\left(J_\alpha(kr)J_{\alpha-1}(kr)+J_{1-\alpha}(kr)J_{-\alpha}(kr)\right)dk.\end{aligned}$$

The integral can be evaluated by some quadrature algorithm and by Theorem 2.4 the supremum is attained for finite r. This allows us to compute approximate values of $R_\gamma(\alpha)$ $(= R_\gamma(1-\alpha))$,

	$R_\gamma(0.1)$	$R_\gamma(0.2)$	$R_\gamma(0.3)$	$R_\gamma(0.4)$	$R_\gamma(0.5)$
$\gamma=0$	1.01682	1.03262	1.04422	1.05151	1.05397
$\gamma=\frac{1}{2}$	1.01027	1.02050	1.02781	1.03241	1.03395
$\gamma=1$	1.00650	1.01351	1.01833	1.02138	1.02238
$\gamma=\frac{3}{2}$	1.00417	1.00920	1.01250	1.01457	1.01524
$\gamma=2$	1.00267	1.00642	1.00874	1.01019	1.01065

In addition to the monotonicity in γ (see Remark 3.4) the constant $R_\gamma(\alpha)$ seems to be strictly increasing in α, but we have not been able to prove this.

4. A magnetic Berezin-Li-Yau inequality

As in the introduction, let $\Omega\subset\mathbb{R}^2$ be a bounded domain and define H_α^Ω in $L_2(\Omega)$ through the closure of the quadratic form $\|(\mathbf{D}-\alpha\mathbf{A}_0)u\|^2$ on $C_0^\infty(\Omega\setminus\{0\})$. For eigenvalue moments of this operator we shall prove

Theorem 4.1. *Let $0<\alpha<1$, $\gamma\geq 1$ and $\Omega\subset\mathbb{R}^2$ be a bounded domain such that the operator H_α^Ω has discrete spectrum. Then for any $\Lambda>0$,*

$$\operatorname{tr}(H_\alpha^\Omega-\Lambda)_-^\gamma \leq R_\gamma(\alpha)\frac{1}{4\pi(\gamma+1)}|\Omega|\Lambda^{\gamma+1} \tag{4.1}$$

with the constant $R_\gamma(\alpha)$ from (3.8).

As explained in the introduction,

$$\frac{1}{4\pi(\gamma+1)}|\Omega|\Lambda^{\gamma+1}=\frac{1}{(2\pi)^2}\iint_{\mathbb{R}^2\times\Omega}(|\xi|^2-\Lambda)_-^\gamma\,d\xi\,dx$$

is the semi-classical approximation for $\operatorname{tr}(H_\alpha^\Omega-\Lambda)_-^\gamma$. Unfortunately, we can only prove (4.1) with an excess factor $R_\gamma(\alpha)$, which is strictly larger than one by Theorem 2.4. Based on numerical calculations (see next section) we conjecture that (4.1) should be valid with $R_\gamma(\alpha)=1$.

Proof. We first note that

$$\operatorname{tr}(H_\alpha^\Omega - \Lambda)_- \leq \operatorname{tr} \chi_\Omega(\chi_\Omega H_\alpha \chi_\Omega - \Lambda)_- \chi_\Omega. \tag{4.2}$$

Indeed, let (ω_j) be an orthonormal basis of eigenfunctions of H_α^Ω. Then the extension $\tilde\omega_j$ of ω_j by zero belongs to the form domain of $\chi_\Omega H_\alpha \chi_\Omega$ and

$$\begin{aligned}\operatorname{tr}(H_\alpha^\Omega - \Lambda)_- &= \sum_j ((H_\alpha^\Omega - \Lambda)\omega_j, \omega_j)_- = \sum_j ((\chi_\Omega H_\alpha \chi_\Omega - \Lambda)\tilde\omega_j, \tilde\omega_j)_- \\ &\leq \sum_j ((\chi_\Omega H_\alpha \chi_\Omega - \Lambda)_- \tilde\omega_j, \tilde\omega_j) = \operatorname{tr} \chi_\Omega(\chi_\Omega H_\alpha \chi_\Omega - \Lambda)_- \chi_\Omega.\end{aligned}$$

Now let φ be a convex function of the form $\varphi(\lambda) = \int (\lambda - \mu)_- w(\mu)\, d\mu$ for some $w \geq 0$. Then it follows from (4.2) that

$$\operatorname{tr} \varphi(H_\alpha^\Omega) \leq \operatorname{tr} \chi_\Omega \varphi(\chi_\Omega H_\alpha \chi_\Omega)\chi_\Omega,$$

and hence by the Berezin-Lieb inequality (see [B2], [L] and also [LaSa], [La]),

$$\operatorname{tr} \varphi(H_\alpha^\Omega) \leq \operatorname{tr} \chi_\Omega \varphi(\chi_\Omega H_\alpha \chi_\Omega)\chi_\Omega \leq \operatorname{tr} \chi_\Omega \varphi(H_\alpha)\chi_\Omega.$$

As already noted in Remark 3.4, the function $\varphi(\lambda) = (\lambda - \Lambda)_-^\gamma$ is of the considered form. Noting that in this case $\operatorname{tr} \chi_\Omega \varphi(-\Delta)\chi_\Omega = (4\pi(\gamma+1))^{-1}|\Omega|\Lambda^{\gamma+1}$, the assertion follows from Proposition 3.3. □

5. Numerical experiments

5.1. Numerical evaluation of the magnetic Berezin-Li-Yau constant

In this section we present a numerical study, somewhat in the spirit of [LT, Appendix A], of the constant $R_\gamma(\alpha)$ appearing in (4.1) for H_α^Ω on various domains. We consider the unit disc, a square and three annuli:

$$\begin{aligned} A &:= \{x : |x| < 1\}, & C &:= \{x : 1 < |x| < 1.1\}, \\ B &:= \{x : \max(|x_1|, |x_2|) < 1\}, & D &:= \{x : 1 < |x| < 2\}, \\ & & E &:= \{x : 1 < |x| < 11\}. \end{aligned}$$

To expose possible diamagnetic effects we perform identical experiments for $\alpha = 0$ and $\alpha = 0.2$ throughout. If the eigenvalues $\lambda_1, \lambda_2, \ldots$ are known, the task is to determine

$$R_\gamma = \sup_\Lambda r_\gamma(\Lambda), \qquad \text{where } r_\gamma(\Lambda) := \frac{4\pi(\gamma+1)}{|\Omega|\Lambda^{\gamma+1}} \sum_{j:\ \lambda_j < \Lambda} (\Lambda - \lambda_j)^\gamma. \tag{5.1}$$

Since the sum is simply the counting function if $\gamma = 0$, r_0 is decreasing for all Λ but the eigenvalues. In the physically important case $\gamma = 1$, the quotient r_1 is continuous but r_1' has jump increases at the eigenvalues. Figures 1 and 2 show, respectively, the schematic behaviour of r_γ near the bottom of the spectrum and for large Λ in the case where $R_\gamma = 1$ (in fact, the spectrum of H_0^B has been considered). Figure 3 depicts $r_0(\Lambda)$ for an operator with $R_\gamma > 1$ (see [H, Thm. 3.2]), namely the Schrödinger operator $H_\alpha - |x|^{-1}$. The $\times$ symbols on the x axis indicate the eigenvalue loci.

5.2. Computation of the eigenvalues

By separation of variables the spectrum of H_α^A is the union of the spectra of the family of one-dimensional problems

$$-u'' + \frac{(n-\alpha)^2 - 1/4}{r^2} u = \lambda u, \qquad u(1) = 0, \qquad u \in L^2(0,1),$$

parametrized by $n \in \mathbb{Z}$. $\sqrt{r} J_{|n-\alpha|}(\sqrt{\lambda} r)$ is an eigenfunction provided λ is chosen to that $J_{|n-\alpha|}(\sqrt{\lambda}) = 0$, i.e., the eigenvalues are squares of the zeros of the Bessel function of the first kind. Using standard numerical procedures for this task we compute all eigenvalues below 10,000 (approximately 2,500) of H_0^A and all eigenvalues below 50,000 (approx. 12,000) of $H_{0.2}^A$. The lowest eigenvalues of the two operators can be viewed in Figure 4. The reader interested in the dependence of the eigenvalues on the parameter α may find Figure 6 useful, where the lowest eigenvalues of H_α^A as functions of $\alpha \in [0, \frac{1}{2}]$ are shown. We recall that H_α^A and $H_{1-\alpha}^A$ share the same spectrum.

Similarly, the eigenvalue problems for H_α^C, H_α^D and H_α^E are reduced to

$$-u'' + \frac{(n-\alpha)^2 - 1/4}{r^2} u = \lambda u, \qquad u(r_1) = 0 = u(r_2), \qquad u \in L^2(r_1, r_2).$$

In order that the general solution $\sqrt{r}(c_1 J_{|n-\alpha|}(\sqrt{\lambda} r) + c_2 Y_{|n-\alpha|}(\sqrt{\lambda} r))$ satisfy the Dirichlet conditions we need to fix the ratio c_1/c_2 and take λ such that

$$J_{|n-\alpha|}(\sqrt{\lambda} r_1) Y_{|n-\alpha|}(\sqrt{\lambda} r_2) = J_{|n-\alpha|}(\sqrt{\lambda} r_2) Y_{|n-\alpha|}(\sqrt{\lambda} r_1).$$

Equivalently, $\lambda = x^2/r_1^2$ is an eigenvalue whenever x is a zero of the cross-product $J_\nu(t)Y_\nu(\mu t) - J_\nu(\mu t)Y_\nu(t)$ for $\mu = r_2/r_1$, $\nu = |n-\alpha|$. Since the amplitude of the oscillation grows approximately as μ^ν/ν, the computation of the zeros is nontrivial for large orders. By searching in the negative t direction and making use of the simple fact that the asymptotic spacing of the zeros is $\pi/(\mu - 1)$ [AbSt, 9.5.28], we circumvent much of these difficulties in comparison to routines such as MATHEMATICA's *BesselJYJYZeros*. This allows us to compute, for $\alpha = 0$ and 0.2, all eigenvalues below 10,000 (approx. 400) of H_α^C, all eigenvalues below 1,000 (approx. 550) of H_α^D and all eigenvalues below 300 (approx. 9,000) of H_α^E.

H_0^B is simply the Dirichlet Laplacian on $[-1,1]^2$, so for all $k, l \in \mathbb{N}$,

$$\sin\left(k\pi \frac{x+1}{2}\right) \sin\left(l\pi \frac{y+1}{2}\right)$$

is an eigenfunction and $\frac{\pi^2}{4}(k^2 + l^2)$ an eigenvalue. For $\alpha \neq 0$ the absence of radial symmetry prevents us from solving the eigenvalue problem for H_α^B exactly. We therefore resort to the finite element method as implemented in COMSOL MULTIPHYSICS to compute approximate eigenvalues. A 8,800-element mesh, gradually refined near the singularity, is used to compute the 250 lowest eigenvalues, all less than 840. Providing an a priori bound on the computational error would be beyond the scope of this article; it is an encouraging fact, however, that the approximate eigenvalues of $H^A_{0.2}$ do not differ from the exact ones by more than 0.8 %. Figure 5 is the analogue of Figure 4 for H^B_α.

5.3. Outcome of the experiments

We are obviously constrained to computing (5.1) with Λ restricted to a finite interval but provided this interval is sufficiently large it should be possible to tell the behaviors depicted in Figures 2 and 3 apart. Particularly if the function r_γ tends steadily to 1 from below we can conclude with certainty that its supremum will eventually be this number in virtue of the eigenvalue asymptotics. It turns out that *all the operators we consider, whether* $\alpha = 0$ *or* 0.2, *seem to obey a semi-classical eigenvalue estimate with unit constant for both* $\gamma = 0$ *and* 1.

What however differs is the rate of convergence to 1, e.g., r_γ increases slower for H^C_α than for H^E_α since $|C| < |E|$, implying that the point spectrum of the former operator is sparser. To see the influence of α on r_γ we invite the reader to compare Figures 7, 8 and 9. Note that r_γ has the most irregular behavior in the cases $\alpha = 0$ and $\alpha = 0.5$ due to the high degree of degeneracy of the eigenvalues, cf. Figure 6. Moreover we have the impression, from looking at Figures 1, 7 and 10, that the shape of the domain has a limited influence on r_γ. All three plots suggest that $R_0 = 1$ but this has not been proved so far even for H^A_0, the Laplacian on the unit disc!

Acknowledgements

The authors would like to thank A. Laptev for the setting of the problem and helpful remarks. The first author is grateful to E.H. Lieb and R. Seiringer for their hospitality at Princeton University and thanks them, H. Kalf and M. Loss for fruitful discussions.

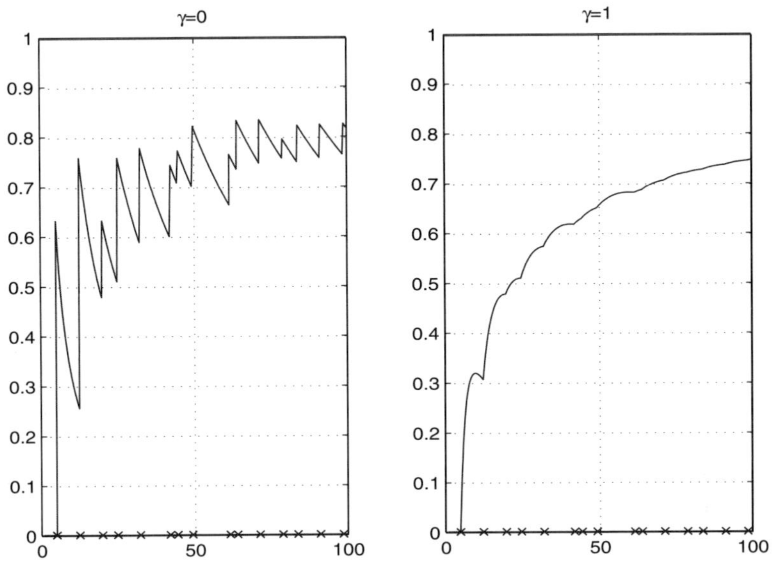

FIGURE 1. r_γ for H_0^B near the bottom of the spectrum

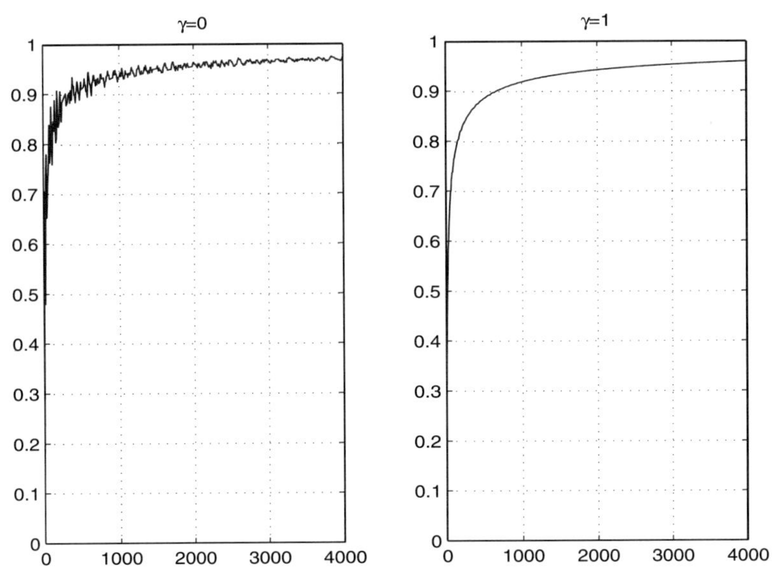

FIGURE 2. $r_\gamma(\Lambda)$ for H_0^B for large Λ

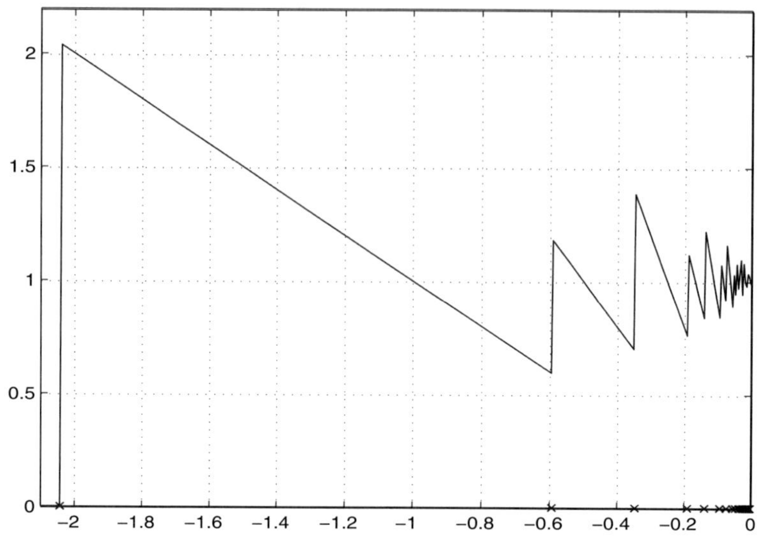

FIGURE 3. r_0 for an operator with $R_0 > 1$

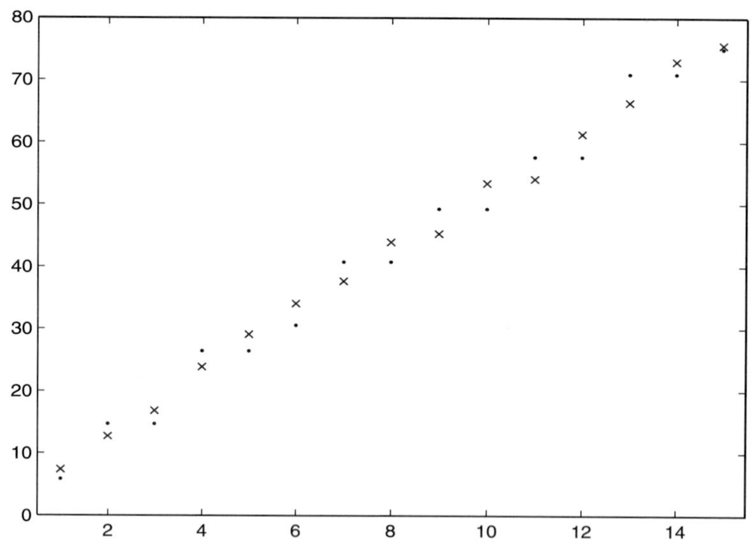

FIGURE 4. Lowest eigenvalues of H_0^A $(\cdot)$ and of $H_{0.2}^A$ $(\times)$

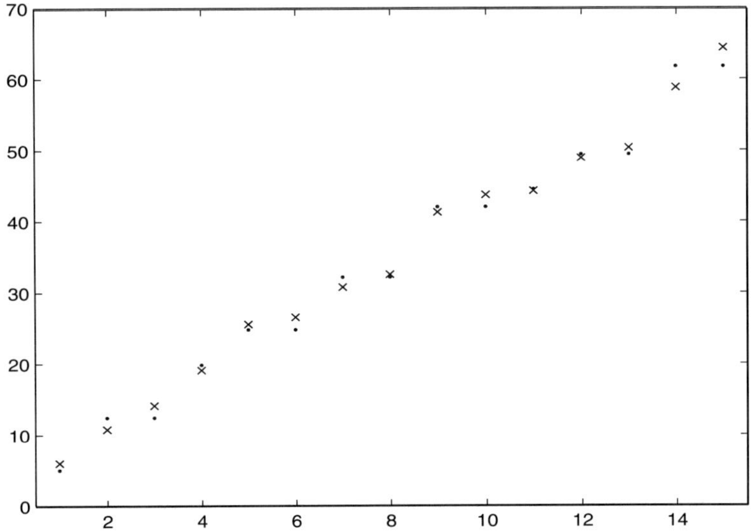

FIGURE 5. Lowest eigenvalues of H_0^B $(\cdot)$ and of $H_{0.2}^B$ $(\times)$

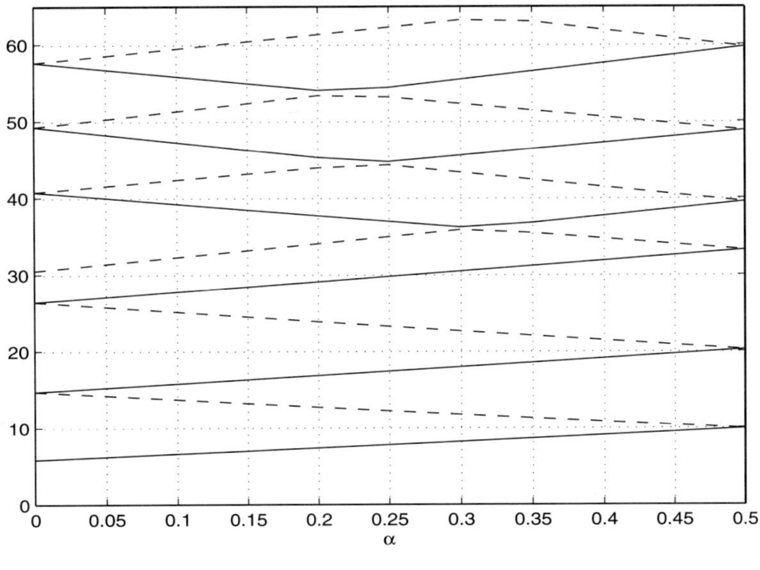

FIGURE 6. Lowest eigenvalues of H_α^A as function of α

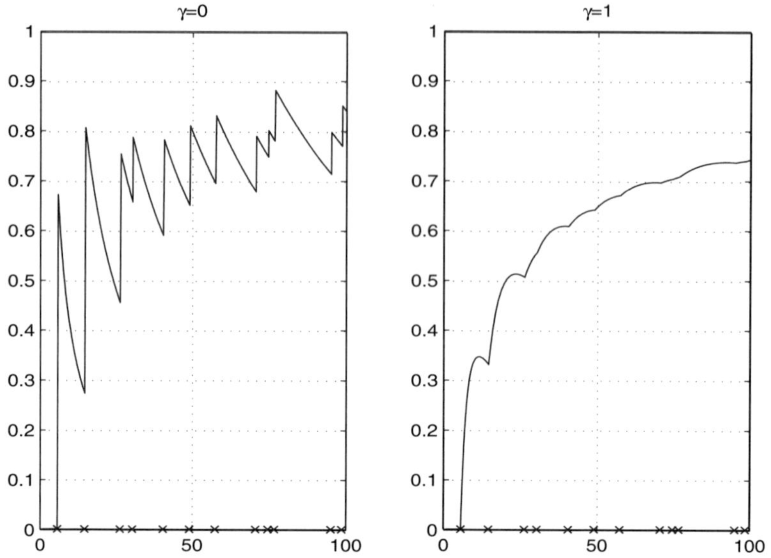

FIGURE 7. r_γ for H_0^A near the bottom of the spectrum

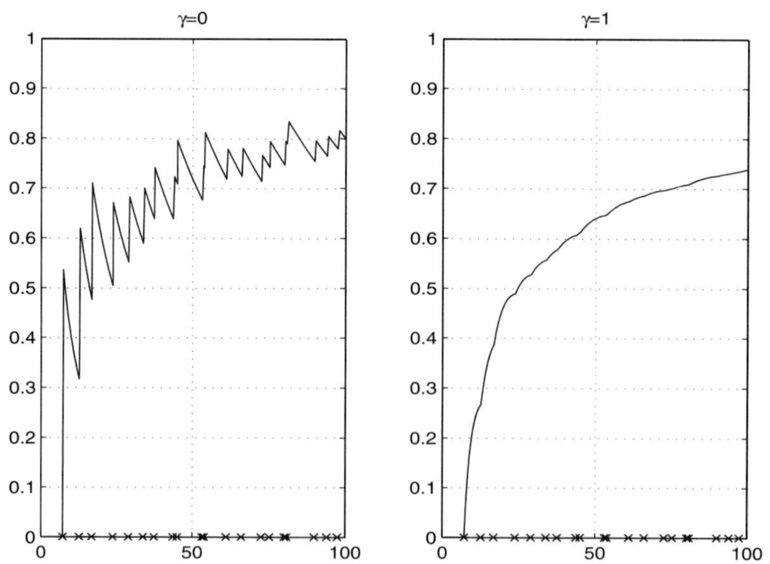

FIGURE 8. r_γ for $H_{0.2}^A$ near the bottom of the spectrum

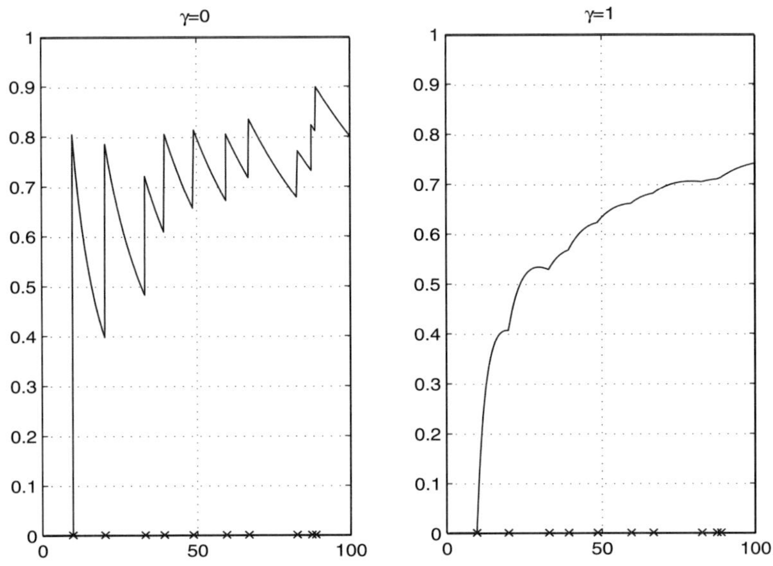

FIGURE 9. r_γ for $H^A_{0.5}$ near the bottom of the spectrum

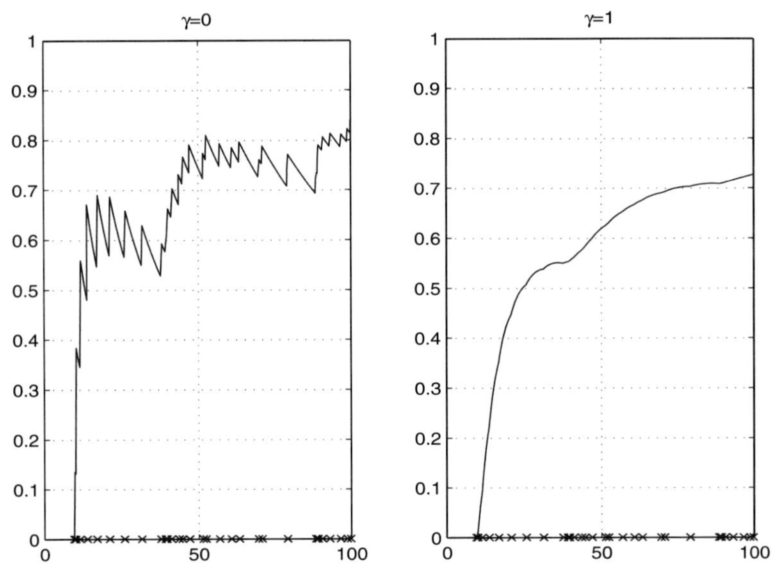

FIGURE 10. r_γ for H^D_0 near the bottom of the spectrum

References

[AhBo] Y. Aharonov, D. Bohm, *Significance of Electromagnetic Potentials in the Quantum Theory.* Phys. Rev. II **115** (1959), 485–491

[AL] M. Aizenman, E. Lieb, *On semiclassical bounds for eigenvalues of Schrödinger operators.* Phys. Lett. A **66** (1978), no. 6, 427–429.

[AbSt] M. Abramowitz, I.A. Stegun, *Handbook of mathematical functions with formulas, graphs, and mathematical tables*, National Bureau of Standards Applied Mathematics Series 55, Washington D.C., 1964.

[BaEvLe] A.A. Balinsky, W.D. Evans, R.T. Lewis, *On the number of negative eigenvalues of Schrödinger operators with an Aharonov-Bohm magnetic field.* Proc. R. Soc. Lond. A **457** (2001) 2481–2489

[B1] F.A. Berezin, *Covariant and contravariant symbols of operators* [Russian]. Math. USSR Izv. **6** (1972), 1117–1151.

[B2] F.A. Berezin, *Convex functions of operators* [Russian]. Mat. Sb. **88** (1972), 268–276.

[EkF] T. Ekholm, R.L. Frank, *On Lieb-Thirring inequalities for Schrödinger operators with virtual level.* Comm. Math. Phys. **264** (2006), no. 3, 725–740.

[ELo] L. Erdős, M. Loss, private communication.

[ELoV] L. Erdős, M. Loss, V. Vougalter, *Diamagnetic behavior of sums of Dirichlet eigenvalues.* Ann. Inst. Fourier **50** (2000), 891–907.

[FLoW] R.L. Frank, M. Loss, T. Weidl, *Pólya's conjecture in the presence of a constant magnetic field.* J. Eur. Math. Soc., to appear.

[H] A.M. Hansson, *On the spectrum and eigenfunctions of the Schrödinger operator with Aharonov-Bohm magnetic field.* Int. J. Math. Math. Sci. **23** (2005), 3751–3766.

[HuS] D. Hundertmark, B. Simon, *A diamagnetic inequality for semigroup differences.* J. Reine Angew. Math. **571** (2004), 107–130.

[La] A. Laptev, *Dirichlet and Neumann eigenvalue problems on domains in Euclidean spaces.* J. Funct. Anal. **151** (1997), no. 2, 531–545.

[LaSa] A. Laptev, Yu. Safarov, *A generalization of the Berezin-Lieb inequality.* Amer. Math. Soc. Transl. (2) **175** (1996), 69–79.

[LaW1] A. Laptev, T. Weidl, *Sharp Lieb-Thirring inequalities in high dimensions.* Acta Math. **184** (2000), no. 1, 87–111.

[LaW2] A. Laptev, T. Weidl, *Recent results on Lieb-Thirring inequalities.* Journées "Équations aux Dérivées Partielles" (La Chapelle sur Erdre, 2000), Exp. No. XX, Univ. Nantes, Nantes, 2000.

[LiY] P. Li, S.-T. Yau, *On the Schrödinger equation and the eigenvalue problem.* Comm. Math. Phys. **88** (1983), 309–318.

[L] E.H. Lieb, *The classical limit of quantum spin systems.* Comm. Math. Phys. **31** (1973), 327–340.

[LT] E.H. Lieb, W. Thirring, *Inequalities for the moments of the eigenvalues of the Schrödinger Hamiltonian and their relation to Sobolev inequalities.* Studies in Mathematical Physics, 269–303. Princeton University Press, Princeton, NJ, 1976.

[Lu] Y.L. Luke, *Integrals of Bessel Functions*, McGraw-Hill, New York-London-Toronto, 1962.

[MOR] M. Melgaard, E.-M. Ouhabaz, G. Rozenblum, *Negative discrete spectrum of perturbed multivortex Aharonov-Bohm Hamiltonians.* Ann. Henri Poincaré 5 (2004), no. 5, 979–1012. Erratum: *ibid.* **6** (2005), no. 2, 397–398.

[P] G. Pólya, *On the eigenvalues of vibrating membranes.* Proc. London Math. Soc. **11** (1961), 419–433.

[ReS] M. Reed, B. Simon, *Methods of Modern Mathematical Physics*, vol. 4. Academic Press, New York-London, 1978.

[Ru] S.N.M. Ruijsenaars, *The Aharonov-Bohm effect and scattering theory*, Ann. Physics **146** (1983), no. 1, 1–34.

[S] B. Simon, *Trace ideals and their applications*, second ed. Mathematical Surveys and Monographs **120**, American Mathematical Society, Providence, RI, 2005.

[Ti] E.C. Titchmarsh, *Theory of Fourier Integrals*, second ed. Oxford University Press, Oxford, 1948.

Rupert L. Frank
Department of Mathematics
School of Engineering Sciences
Royal Institute of Technology
100 44 Stockholm, Sweden
e-mail: `rupert@math.kth.se`

Current address: Department of Mathematics
Princeton University
Princeton, NJ 08544, USA
e-mail: `rlfrank@math.princeton.edu`

Anders M. Hansson
Department of Mathematics
School of Engineering Sciences
Royal Institute of Technology
100 44 Stockholm, Sweden
e-mail: `anhan@math.kth.se`

Operator Theory:
Advances and Applications, Vol. 186, 139–190

The Ablowitz–Ladik Hierarchy Revisited

Fritz Gesztesy, Helge Holden, Johanna Michor and Gerald Teschl

Abstract. We provide a detailed recursive construction of the Ablowitz–Ladik (AL) hierarchy and its zero-curvature formalism. The two-coefficient AL hierarchy under investigation can be considered a complexified version of the discrete nonlinear Schrödinger equation and its hierarchy of nonlinear evolution equations.

Specifically, we discuss in detail the stationary Ablowitz–Ladik formalism in connection with the underlying hyperelliptic curve and the stationary Baker–Akhiezer function and separately the corresponding time-dependent Ablowitz–Ladik formalism.

Mathematics Subject Classification (2000). Primary 37K15, 37K10; Secondary 39A12, 35Q55.

Keywords. Ablowitz–Ladik hierarchy, discrete NLS.

1. Introduction

The prime example of an integrable nonlinear differential-difference system to be discussed in this paper is the Ablowitz–Ladik system,

$$\begin{aligned} -i\alpha_t - (1-\alpha\beta)(\alpha^- + \alpha^+) + 2\alpha = 0, \\ -i\beta_t + (1-\alpha\beta)(\beta^- + \beta^+) - 2\beta = 0 \end{aligned} \tag{1.1}$$

with $\alpha = \alpha(n,t)$, $\beta = \beta(n,t)$, $(n,t) \in \mathbb{Z} \times \mathbb{R}$. Here we used the notation $f^{\pm}(n) = f(n \pm 1)$, $n \in \mathbb{Z}$, for complex-valued sequences $f = \{f(n)\}_{n\in\mathbb{Z}}$. The system (1.1) arose in the mid-seventies when Ablowitz and Ladik, in a series of papers [3]–[6] (see also [1], [2, Sect. 3.2.2], [7, Ch. 3], [17]), used inverse scattering methods to analyze certain integrable differential-difference systems. In particular, Ablowitz and Ladik [4] (see also [7, Ch. 3]) showed that in the focusing case, where $\beta = -\overline{\alpha}$, and in the defocusing case, where $\beta = \overline{\alpha}$, (1.1) yields the discrete analog of the

Research supported in part by the Research Council of Norway, the US National Science Foundation under Grant No. DMS-0405526, and the Austrian Science Fund (FWF) under Grant No. Y330.

nonlinear Schrödinger equation

$$-i\alpha_t - (1 \pm |\alpha|^2)(\alpha^- + \alpha^+) + 2\alpha = 0. \tag{1.2}$$

We will refer to (1.1) as the Ablowitz–Ladik system. The principal theme of this paper will be to derive a detailed recursive construction of the Ablowitz–Ladik hierarchy, a completely integrable sequence of systems of nonlinear evolution equations on the lattice $\mathbb{Z}$ whose first nonlinear member is the Ablowitz–Ladik system (1.1). In addition, we discuss the zero-curvature formalism for the Ablowitz–Ladik (AL) hierarchy in detail.

Since the original discovery of Ablowitz and Ladik in the mid-seventies, there has been great interest in the area of integrable differential-difference equations. Two principal directions of research are responsible for this development: Originally, the development was driven by the theory of completely integrable systems and its applications to fields such as nonlinear optics, and more recently, it gained additional momentum due to its intimate connections with the theory of orthogonal polynomials. In this paper we will not discuss the connection with orthogonal polynomials (see, however, the introduction of [31]) and instead refer to the recent references [13], [20], [37], [38], [42], [43], [44], [47], [48], [49], and the literature cited therein.

The first systematic discussion of the Ablowitz–Ladik (AL) hierarchy appears to be due to Schilling [45] (cf. also [51], [55], [58]); infinitely many conservation laws are derived, for instance, by Ding, Sun, and Xu [21]; the bi-Hamiltonian structure of the AL hierarchy is considered by Ercolani and Lozano [23]; connections between the AL hierarchy and the motion of a piecewise linear curve have been established by Doliwa and Santini [22]; Bäcklund and Darboux transformations were studied by Geng [26] and Vekslerchik [56]; the Hirota bilinear formalism, AL τ-functions, etc., were considered by Vekslerchik [55]. The initial value problem for half-infinite AL systems was discussed by Common [19], for an application of the inverse scattering method to (1.2) we refer to Vekslerchik and Konotop [57]. This just scratches the surface of these developments and the interested reader will find much more material in the references cited in these papers and the ones discussed below. Algebro-geometric (and periodic) solutions of the AL system (1.1) have briefly been studied by Ahmad and Chowdhury [8], [9], Bogolyubov, Prikarpatskii, and Samoilenko [14], Bogolyubov and Prikarpatskii [15], Chow, Conte, and Xu [18], Geng, Dai, and Cao [27], and Vaninsky [53].

In an effort to analyze models describing oscillations in nonlinear dispersive wave systems, Miller, Ercolani, Krichever, and Levermore [40] (see also [39]) gave a detailed analysis of algebro-geometric solutions of the AL system (1.1). Introducing

$$U(z) = \begin{pmatrix} z & \alpha \\ \beta z & 1 \end{pmatrix}, \quad V(z) = i \begin{pmatrix} z - 1 - \alpha\beta^- & \alpha - \alpha^- z^{-1} \\ \beta^- z - \beta & 1 + \alpha^-\beta - z^{-1} \end{pmatrix} \tag{1.3}$$

with $z \in \mathbb{C} \setminus \{0\}$ a spectral parameter, the authors in [40] relied on the fact that the Ablowitz–Ladik system (1.1) is equivalent to the zero-curvature equations

$$U_t + UV - V^+U = 0. \tag{1.4}$$

Miller, Ercolani, Krichever, and Levermore [40] then derived the theta function representations of α, β satisfying the AL system (1.1). Vekslerchik [54] also studied finite-genus solutions for the AL hierarchy by establishing connections with Fay's identity for theta functions. Recently, a detailed study of algebro-geometric solutions for the entire AL hierarchy has been provided in [31]. The latter reference also contains an extensive discussion of the connection between the Ablowitz–Ladik system (1.1) and orthogonal polynomials on the unit circle. The algebro-geometric initial value problem for the Ablowitz–Ladik hierarchy with complex-valued initial data, that is, the construction of α and β by starting from a set of initial data (nonspecial divisors) of full measure, will be presented in [32]. The Hamiltonian and Lax formalisms for the AL hierarchy will be revisited in [33].

In addition to these recent developments on the AL system and the AL hierarchy, we offer a variety of results in this paper apparently not covered before. These include:

- An effective recursive construction of the AL hierarchy using Laurent polynomials.
- The detailed connection between the AL hierarchy and a "complexified" version of transfer matrices first introduced by Baxter [11], [12].
- A detailed treatment of the stationary and time-dependent Ablowitz–Ladik formalism.

The structure of this paper is as follows: In Section 2 we describe our zero-curvature formalism for the Ablowitz–Ladik (AL) hierarchy. Extending a recursive polynomial approach discussed in great detail in [29] in the continuous case and in [16], [30, Ch. 4], [52, Chs. 6, 12] in the discrete context to the case of Laurent polynomials with respect to the spectral parameter, we derive the AL hierarchy of systems of nonlinear evolution equations whose first nonlinear member is the Ablowitz–Ladik system (1.1). Section 3 is devoted to a detailed study of the stationary AL hierarchy. We employ the recursive Laurent polynomial formalism of Section 2 to describe nonnegative divisors of degree p on a hyperelliptic curve $\mathcal{K}_p$ of genus p associated with the $\underline{p}$th system in the stationary AL hierarchy. The corresponding time-dependent results for the AL hierarchy are presented in detail in Section 4. Finally, Appendix A is of a technical nature and summarizes expansions of various key quantities related to the Laurent polynomial recursion formalism as the spectral parameter tends to zero or to infinity.

2. The Ablowitz–Ladik hierarchy, recursion relations, zero-curvature pairs, and hyperelliptic curves

In this section we provide the construction of the Ablowitz–Ladik hierarchy employing a polynomial recursion formalism and derive the associated sequence of Ablowitz–Ladik zero-curvature pairs. Moreover, we discuss the hyperelliptic curve underlying the stationary Ablowitz–Ladik hierarchy.

We denote by $\mathbb{C}^{\mathbb{Z}}$ the set of complex-valued sequences indexed by $\mathbb{Z}$.

Throughout this section we suppose the following hypothesis.

Hypothesis 2.1. *In the stationary case we assume that α, β satisfy*

$$\alpha, \beta \in \mathbb{C}^{\mathbb{Z}}, \quad \alpha(n)\beta(n) \notin \{0,1\}, \; n \in \mathbb{Z}. \tag{2.1}$$

In the time-dependent case we assume that α, β satisfy

$$\begin{aligned} &\alpha(\,\cdot\,,t), \beta(\,\cdot\,,t) \in \mathbb{C}^{\mathbb{Z}}, \; t \in \mathbb{R}, \quad \alpha(n,\,\cdot\,), \beta(n,\,\cdot\,) \in C^1(\mathbb{R}), \; n \in \mathbb{Z}, \\ &\alpha(n,t)\beta(n,t) \notin \{0,1\}, \; (n,t) \in \mathbb{Z} \times \mathbb{R}. \end{aligned} \tag{2.2}$$

Actually, up to Remark 2.11 our analysis will be time-independent and hence only the lattice variations of α and β will matter.

We denote by $S^{\pm}$ the shift operators acting on complex-valued sequences $f = \{f(n)\}_{n\in\mathbb{Z}} \in \mathbb{C}^{\mathbb{Z}}$ according to

$$(S^{\pm} f)(n) = f(n \pm 1), \quad n \in \mathbb{Z}. \tag{2.3}$$

Moreover, we will frequently use the notation

$$f^{\pm} = S^{\pm} f, \quad f \in \mathbb{C}^{\mathbb{Z}}. \tag{2.4}$$

To construct the Ablowitz–Ladik hierarchy we will try to generalize (1.3) by considering the 2×2 matrix

$$U(z) = \begin{pmatrix} z & \alpha \\ z\beta & 1 \end{pmatrix}, \quad z \in \mathbb{C}, \tag{2.5}$$

and making the ansatz

$$V_{\underline{p}}(z) = i \begin{pmatrix} G_{\underline{p}}^-(z) & -F_{\underline{p}}^-(z) \\ H_{\underline{p}}^-(z) & -K_{\underline{p}}^-(z) \end{pmatrix}, \quad \underline{p} = (p_-, p_+) \in \mathbb{N}_0^2, \tag{2.6}$$

where $G_{\underline{p}}$, $K_{\underline{p}}$, $F_{\underline{p}}$, and $H_{\underline{p}}$ are chosen as Laurent polynomials[1] (suggested by the appearance of z^{-1} in the matrix V in (1.3))

$$\begin{aligned} G_{\underline{p}}(z) &= \sum_{\ell=1}^{p_-} z^{-\ell} g_{p_- -\ell,-} + \sum_{\ell=0}^{p_+} z^{\ell} g_{p_+ -\ell,+}, \\ F_{\underline{p}}(z) &= \sum_{\ell=1}^{p_-} z^{-\ell} f_{p_- -\ell,-} + \sum_{\ell=0}^{p_+} z^{\ell} f_{p_+ -\ell,+}, \\ H_{\underline{p}}(z) &= \sum_{\ell=1}^{p_-} z^{-\ell} h_{p_- -\ell,-} + \sum_{\ell=0}^{p_+} z^{\ell} h_{p_+ -\ell,+}, \\ K_{\underline{p}}(z) &= \sum_{\ell=1}^{p_-} z^{-\ell} k_{p_- -\ell,-} + \sum_{\ell=0}^{p_+} z^{\ell} k_{p_+ -\ell,+}. \end{aligned} \tag{2.7}$$

[1]In this paper, a sum is interpreted as zero whenever the upper limit in the sum is strictly less than its lower limit.

Without loss of generality we will only look at the time-independent case and add time later on. Then the stationary zero-curvature equation,

$$0 = UV_{\underline{p}} - V_{\underline{p}}^{+}U, \tag{2.8}$$

is equivalent to the following relationships between the Laurent polynomials

$$UV_{\underline{p}} - V_{\underline{p}}^{+}U = i\begin{pmatrix} z(G_{\underline{p}}^{-} - G_{\underline{p}}) + z\beta F_{\underline{p}} + \alpha H_{\underline{p}}^{-} & F_{\underline{p}} - zF_{\underline{p}}^{-} - \alpha(G_{\underline{p}} + K_{\underline{p}}^{-}) \\ z\beta(G_{\underline{p}}^{-} + K_{\underline{p}}) - zH_{\underline{p}} + H_{\underline{p}}^{-} & -z\beta F_{\underline{p}}^{-} - \alpha H_{\underline{p}} + K_{\underline{p}} - K_{\underline{p}}^{-} \end{pmatrix}, \tag{2.9}$$

respectively, to

$$z(G_{\underline{p}}^{-} - G_{\underline{p}}) + z\beta F_{\underline{p}} + \alpha H_{\underline{p}}^{-} = 0, \tag{2.10}$$

$$z\beta F_{\underline{p}}^{-} + \alpha H_{\underline{p}} - K_{\underline{p}} + K_{\underline{p}}^{-} = 0, \tag{2.11}$$

$$-F_{\underline{p}} + zF_{\underline{p}}^{-} + \alpha(G_{\underline{p}} + K_{\underline{p}}^{-}) = 0, \tag{2.12}$$

$$z\beta(G_{\underline{p}}^{-} + K_{\underline{p}}) - zH_{\underline{p}} + H_{\underline{p}}^{-} = 0. \tag{2.13}$$

Lemma 2.2. *Suppose the Laurent polynomials defined in* (2.7) *satisfy the zero-curvature equation* (2.8), *then*

$$f_{0,+} = 0, \quad h_{0,-} = 0, \quad g_{0,\pm} = g_{0,\pm}^{-}, \quad k_{0,\pm} = k_{0,\pm}^{-}, \tag{2.14}$$

$$k_{\ell,\pm} - k_{\ell,\pm}^{-} = g_{\ell,\pm} - g_{\ell,\pm}^{-}, \; \ell = 0,\dots,p_{\pm}-1, \quad g_{p_+,+} - g_{p_+,+}^{-} = k_{p_+,+} - k_{p_+,+}^{-}. \tag{2.15}$$

Proof. Comparing coefficients at the highest order of z in (2.11) and the lowest in (2.10) immediately yields $f_{0,+} = 0$, $h_{0,-} = 0$. Then $g_{0,+} = g_{0,+}^{-}$, $k_{0,-} = k_{0,-}^{-}$ are necessarily lattice constants by (2.10), (2.11). Since $\det(U(z)) \neq 0$ for $z \in \mathbb{C}\setminus\{0\}$ by (2.1), (2.8) yields $\operatorname{tr}(V_{\underline{p}}^{+}) = \operatorname{tr}(UV_{\underline{p}}U^{-1}) = \operatorname{tr}(V_{\underline{p}})$ and hence

$$G_{\underline{p}} - G_{\underline{p}}^{-} = K_{\underline{p}} - K_{\underline{p}}^{-}, \tag{2.16}$$

implying (2.15). Taking $\ell = 0$ in (2.15) then yields $g_{0,-} = g_{0,-}^{-}$ and $k_{0,+} = k_{0,+}^{-}$. □

In particular, this lemma shows that we can choose

$$k_{\ell,\pm} = g_{\ell,\pm}, \; 0 \le \ell \le p_{\pm}-1, \quad k_{p_+,+} = g_{p_+,+} \tag{2.17}$$

without loss of generality (since this can always be achieved by adding a Laurent polynomial times the identity to $V_{\underline{p}}$, which does not affect the zero-curvature equation). Hence the ansatz (2.7) can be refined as follows (it is more convenient in the following to re-label $h_{p_+,+} = h_{p_- -1,-}$ and $k_{p_+,+} = g_{p_-,-}$, and hence, $g_{p_-,-} = g_{p_+,+}$),

$$F_{\underline{p}}(z) = \sum_{\ell=1}^{p_-} f_{p_- -\ell,-} z^{-\ell} + \sum_{\ell=0}^{p_+ -1} f_{p_+ -1-\ell,+} z^{\ell}, \tag{2.18}$$

$$G_{\underline{p}}(z) = \sum_{\ell=1}^{p_-} g_{p_- -\ell,-} z^{-\ell} + \sum_{\ell=0}^{p_+} g_{p_+ -\ell,+} z^{\ell}, \tag{2.19}$$

$$H_{\underline{p}}(z) = \sum_{\ell=0}^{p_- -1} h_{p_- -1-\ell,-} z^{-\ell} + \sum_{\ell=1}^{p_+} h_{p_+-\ell,+} z^{\ell}, \tag{2.20}$$

$$K_{\underline{p}}(z) = G_{\underline{p}}(z) \text{ since } g_{p_-,-} = g_{p_+,+}. \tag{2.21}$$

In particular, (2.21) renders $V_{\underline{p}}$ in (2.6) traceless in the stationary context. We emphasize, however, that equation (2.21) ceases to be valid in the time-dependent context: In the latter case (2.21) needs to be replaced by

$$K_{\underline{p}}(z) = \sum_{\ell=0}^{p_-} g_{p_- -\ell,-} z^{-\ell} + \sum_{\ell=1}^{p_+} g_{p_+-\ell,+} z^{\ell} = G_{\underline{p}}(z) + g_{p_-,-} - g_{p_+,+}. \tag{2.22}$$

Plugging the refined ansatz (2.18)–(2.21) into the zero-curvature equation (2.8) and comparing coefficients then yields the following result.

Lemma 2.3. *Suppose that U and $V_{\underline{p}}$ satisfy the zero-curvature equation* (2.8). *Then the coefficients $\{f_{\ell,\pm}\}_{\ell=0,\dots,p_\pm-1}$, $\{g_{\ell,\pm}\}_{\ell=0,\dots,p_\pm}$, and $\{h_{\ell,\pm}\}_{\ell=0,\dots,p_\pm-1}$ of $F_{\underline{p}}$, $G_{\underline{p}}$, $H_{\underline{p}}$, and $K_{\underline{p}}$ in* (2.18)–(2.21) *satisfy the following relations*

$$g_{0,+} = \tfrac{1}{2}c_{0,+}, \quad f_{0,+} = -c_{0,+}\alpha^+, \quad h_{0,+} = c_{0,+}\beta, \tag{2.23}$$

$$g_{\ell+1,+} - g^-_{\ell+1,+} = \alpha h^-_{\ell,+} + \beta f_{\ell,+}, \quad 0 \le \ell \le p_+ - 1, \tag{2.24}$$

$$f^-_{\ell+1,+} = f_{\ell,+} - \alpha(g_{\ell+1,+} + g^-_{\ell+1,+}), \quad 0 \le \ell \le p_+ - 2, \tag{2.25}$$

$$h_{\ell+1,+} = h^-_{\ell,+} + \beta(g_{\ell+1,+} + g^-_{\ell+1,+}), \quad 0 \le \ell \le p_+ - 2, \tag{2.26}$$

and

$$g_{0,-} = \tfrac{1}{2}c_{0,-}, \quad f_{0,-} = c_{0,-}\alpha, \quad h_{0,-} = -c_{0,-}\beta^+, \tag{2.27}$$

$$g_{\ell+1,-} - g^-_{\ell+1,-} = \alpha h_{\ell,-} + \beta f^-_{\ell,-}, \quad 0 \le \ell \le p_- - 1, \tag{2.28}$$

$$f_{\ell+1,-} = f^-_{\ell,-} + \alpha(g_{\ell+1,-} + g^-_{\ell+1,-}), \quad 0 \le \ell \le p_- - 2, \tag{2.29}$$

$$h^-_{\ell+1,-} = h_{\ell,-} - \beta(g_{\ell+1,-} + g^-_{\ell+1,-}), \quad 0 \le \ell \le p_- - 2. \tag{2.30}$$

Here $c_{0,\pm} \in \mathbb{C}$ are given constants. In addition, (2.8) *reads*

$$\begin{aligned} 0 &= UV_{\underline{p}} - V^+_{\underline{p}}U \\ &= i \begin{pmatrix} 0 & \begin{matrix} -\alpha(g_{p_+,+} + g^-_{p_-,-}) \\ +f_{p_+-1,+} - f^-_{p_--1,-} \end{matrix} \\ \begin{matrix} z(\beta(g^-_{p_+,+} + g_{p_-,-}) \\ -h_{p_--1,-} + h^-_{p_+-1,+}) \end{matrix} & 0 \end{pmatrix}. \end{aligned} \tag{2.31}$$

Given Lemma 2.3, we now introduce the sequences $\{f_{\ell,\pm}\}_{\ell\in\mathbb{N}_0}$, $\{g_{\ell,\pm}\}_{\ell\in\mathbb{N}_0}$, and $\{h_{\ell,\pm}\}_{\ell\in\mathbb{N}_0}$ recursively by

$$g_{0,+} = \tfrac{1}{2}c_{0,+}, \quad f_{0,+} = -c_{0,+}\alpha^+, \quad h_{0,+} = c_{0,+}\beta, \tag{2.32}$$

$$g_{\ell+1,+} - g^-_{\ell+1,+} = \alpha h^-_{\ell,+} + \beta f_{\ell,+}, \quad \ell \in \mathbb{N}_0, \tag{2.33}$$

$$f^-_{\ell+1,+} = f_{\ell,+} - \alpha(g_{\ell+1,+} + g^-_{\ell+1,+}), \quad \ell \in \mathbb{N}_0, \tag{2.34}$$

$$h_{\ell+1,+} = h^-_{\ell,+} + \beta(g_{\ell+1,+} + g^-_{\ell+1,+}), \quad \ell \in \mathbb{N}_0, \tag{2.35}$$

and

$$g_{0,-} = \tfrac{1}{2}c_{0,-}, \quad f_{0,-} = c_{0,-}\alpha, \quad h_{0,-} = -c_{0,-}\beta^+, \tag{2.36}$$

$$g_{\ell+1,-} - g^-_{\ell+1,-} = \alpha h_{\ell,-} + \beta f^-_{\ell,-}, \quad \ell \in \mathbb{N}_0, \tag{2.37}$$

$$f_{\ell+1,-} = f^-_{\ell,-} + \alpha(g_{\ell+1,-} + g^-_{\ell+1,-}), \quad \ell \in \mathbb{N}_0, \tag{2.38}$$

$$h^-_{\ell+1,-} = h_{\ell,-} - \beta(g_{\ell+1,-} + g^-_{\ell+1,-}), \quad \ell \in \mathbb{N}_0. \tag{2.39}$$

For later use we also introduce

$$f_{-1,\pm} = h_{-1,\pm} = 0. \tag{2.40}$$

Remark 2.4. The sequences $\{f_{\ell,+}\}_{\ell\in\mathbb{N}_0}$, $\{g_{\ell,+}\}_{\ell\in\mathbb{N}_0}$, and $\{h_{\ell,+}\}_{\ell\in\mathbb{N}_0}$ can be computed recursively as follows: Assume that $f_{\ell,+}$, $g_{\ell,+}$, and $h_{\ell,+}$ are known. Equation (2.33) is a first-order difference equation in $g_{\ell+1,+}$ that can be solved directly and yields a local lattice function that is determined up to a new constant denoted by $c_{\ell+1,+} \in \mathbb{C}$. Relations (2.34) and (2.35) then determine $f_{\ell+1,+}$ and $h_{\ell+1,+}$, etc. The sequences $\{f_{\ell,-}\}_{\ell\in\mathbb{N}_0}$, $\{g_{\ell,-}\}_{\ell\in\mathbb{N}_0}$, and $\{h_{\ell,-}\}_{\ell\in\mathbb{N}_0}$ are determined similarly.

Upon setting

$$\gamma = 1 - \alpha\beta, \tag{2.41}$$

one explicitly obtains

$$\begin{aligned}
f_{0,+} &= c_{0,+}(-\alpha^+),\\
f_{1,+} &= c_{0,+}\big(-\gamma^+\alpha^{++} + (\alpha^+)^2\beta\big) + c_{1,+}(-\alpha^+),\\
g_{0,+} &= \tfrac{1}{2}c_{0,+},\\
g_{1,+} &= c_{0,+}(-\alpha^+\beta) + \tfrac{1}{2}c_{1,+},\\
g_{2,+} &= c_{0,+}\big((\alpha^+\beta)^2 - \gamma^+\alpha^{++}\beta - \gamma\alpha^+\beta^-\big) + c_{1,+}(-\alpha^+\beta) + \tfrac{1}{2}c_{2,+},\\
h_{0,+} &= c_{0,+}\beta,\\
h_{1,+} &= c_{0,+}\big(\gamma\beta^- - \alpha^+\beta^2\big) + c_{1,+}\beta,\\
f_{0,-} &= c_{0,-}\alpha,\\
f_{1,-} &= c_{0,-}\big(\gamma\alpha^- - \alpha^2\beta^+\big) + c_{1,-}\alpha,\\
g_{0,-} &= \tfrac{1}{2}c_{0,-},\\
g_{1,-} &= c_{0,-}(-\alpha\beta^+) + \tfrac{1}{2}c_{1,-},\\
g_{2,-} &= c_{0,-}\big((\alpha\beta^+)^2 - \gamma^+\alpha\beta^{++} - \gamma\alpha^-\beta^+\big) + c_{1,-}(-\alpha\beta^+) + \tfrac{1}{2}c_{2,-},\\
h_{0,-} &= c_{0,-}(-\beta^+),\\
h_{1,-} &= c_{0,-}\big(-\gamma^+\beta^{++} + \alpha(\beta^+)^2\big) + c_{1,-}(-\beta^+), \quad \text{etc.}
\end{aligned} \tag{2.42}$$

Here $\{c_{\ell,\pm}\}_{\ell\in\mathbb{N}}$ denote summation constants which naturally arise when solving the difference equations for $g_{\ell,\pm}$ in (2.33), (2.37).

In particular, by (2.31), the stationary zero-curvature relation (2.8), $0 = UV_{\underline{p}} - V_{\underline{p}}^+ U$, is equivalent to

$$-\alpha(g_{p_+,+} + g_{p_-,-}^-) + f_{p_+-1,+} - f_{p_--1,-}^- = 0, \tag{2.43}$$

$$\beta(g_{p_+,+}^- + g_{p_-,-}) + h_{p_+-1,+}^- - h_{p_--1,-} = 0. \tag{2.44}$$

Thus, varying $p_\pm \in \mathbb{N}_0$, equations (2.43) and (2.44) give rise to the stationary Ablowitz–Ladik (AL) hierarchy which we introduce as follows

$$\begin{aligned} \text{s-AL}_{\underline{p}}(\alpha,\beta) = \begin{pmatrix} -\alpha(g_{p_+,+} + g_{p_-,-}^-) + f_{p_+-1,+} - f_{p_--1,-}^- \\ \beta(g_{p_+,+}^- + g_{p_-,-}) + h_{p_+-1,+}^- - h_{p_--1,-} \end{pmatrix} = 0, \\ \underline{p} = (p_-, p_+) \in \mathbb{N}_0^2. \end{aligned} \tag{2.45}$$

Explicitly (recalling $\gamma = 1 - \alpha\beta$ and taking $p_- = p_+$ for simplicity),

$$\begin{aligned}
&\text{s-AL}_{(0,0)}(\alpha,\beta) = \begin{pmatrix} -c_{(0,0)}\alpha \\ c_{(0,0)}\beta \end{pmatrix} = 0, \\
&\text{s-AL}_{(1,1)}(\alpha,\beta) = \begin{pmatrix} -\gamma(c_{0,-}\alpha^- + c_{0,+}\alpha^+) - c_{(1,1)}\alpha \\ \gamma(c_{0,+}\beta^- + c_{0,-}\beta^+) + c_{(1,1)}\beta \end{pmatrix} = 0, \\
&\text{s-AL}_{(2,2)}(\alpha,\beta) = \begin{pmatrix} -\gamma\big(c_{0,+}\alpha^{++}\gamma^+ + c_{0,-}\alpha^{--}\gamma^- - \alpha(c_{0,+}\alpha^+\beta^- + c_{0,-}\alpha^-\beta^+) \\ -\beta(c_{0,-}(\alpha^-)^2 + c_{0,+}(\alpha^+)^2)\big) \\ \gamma\big(c_{0,-}\beta^{++}\gamma^+ + c_{0,+}\beta^{--}\gamma^- - \beta(c_{0,+}\alpha^+\beta^- + c_{0,-}\alpha^-\beta^+) \\ -\alpha(c_{0,+}(\beta^-)^2 + c_{0,-}(\beta^+)^2)\big) \end{pmatrix} \\
&\qquad + \begin{pmatrix} -\gamma(c_{1,-}\alpha^- + c_{1,+}\alpha^+) - c_{(2,2)}\alpha \\ \gamma(c_{1,+}\beta^- + c_{1,-}\beta^+) + c_{(2,2)}\beta \end{pmatrix} = 0, \quad \text{etc.},
\end{aligned} \tag{2.46}$$

represent the first few equations of the stationary Ablowitz–Ladik hierarchy. Here we introduced

$$c_{\underline{p}} = (c_{p,-} + c_{p,+})/2, \quad p_\pm \in \mathbb{N}_0. \tag{2.47}$$

By definition, the set of solutions of (2.45), with $p_\pm$ ranging in $\mathbb{N}_0$ and $c_{\ell,\pm} \in \mathbb{C}$, $\ell \in \mathbb{N}_0$, represents the class of algebro-geometric Ablowitz–Ladik solutions.

In the special case $\underline{p} = (1,1)$, $c_{0,\pm} = 1$, and $c_{(1,1)} = -2$, one obtains the stationary version of the Ablowitz–Ladik system (1.1)

$$\begin{pmatrix} -\gamma(\alpha^- + \alpha^+) + 2\alpha \\ \gamma(\beta^- + \beta^+) - 2\beta \end{pmatrix} = 0. \tag{2.48}$$

Subsequently, it will also be useful to work with the corresponding homogeneous coefficients $\hat{f}_{\ell,\pm}$, $\hat{g}_{\ell,\pm}$, and $\hat{h}_{\ell,\pm}$, defined by the vanishing of all summation constants $c_{k,\pm}$ for $k = 1, \dots, \ell$, and choosing $c_{0,\pm} = 1$,

$$\hat{f}_{0,+} = -\alpha^+, \quad \hat{f}_{0,-} = \alpha, \quad \hat{f}_{\ell,\pm} = f_{\ell,\pm}|_{c_{0,\pm}=1,\, c_{j,\pm}=0, j=1,\dots,\ell}, \quad \ell \in \mathbb{N}, \tag{2.49}$$

$$\hat{g}_{0,\pm} = \tfrac{1}{2}, \quad \hat{g}_{\ell,\pm} = g_{\ell,\pm}|_{c_{0,\pm}=1,\, c_{j,\pm}=0, j=1,\dots,\ell}, \quad \ell \in \mathbb{N}, \tag{2.50}$$

$$\hat{h}_{0,+} = \beta, \quad \hat{h}_{0,-} = -\beta^+, \quad \hat{h}_{\ell,\pm} = h_{\ell,\pm}|_{c_{0,\pm}=1,\, c_{j,\pm}=0, j=1,\dots,\ell}, \quad \ell \in \mathbb{N}. \tag{2.51}$$

By induction one infers that

$$f_{\ell,\pm} = \sum_{k=0}^{\ell} c_{\ell-k,\pm}\hat{f}_{k,\pm}, \quad g_{\ell,\pm} = \sum_{k=0}^{\ell} c_{\ell-k,\pm}\hat{g}_{k,\pm}, \quad h_{\ell,\pm} = \sum_{k=0}^{\ell} c_{\ell-k,\pm}\hat{h}_{k,\pm}. \tag{2.52}$$

In a slight abuse of notation we will occasionally stress the dependence of $f_{\ell,\pm}$, $g_{\ell,\pm}$, and $h_{\ell,\pm}$ on α, β by writing $f_{\ell,\pm}(\alpha,\beta)$, $g_{\ell,\pm}(\alpha,\beta)$, and $h_{\ell,\pm}(\alpha,\beta)$.

Remark 2.5. Using the nonlinear recursion relations (A.29)–(A.34) recorded in Theorem A.1, one infers inductively that all homogeneous elements $\hat{f}_{\ell,\pm}$, $\hat{g}_{\ell,\pm}$, and $\hat{h}_{\ell,\pm}$, $\ell \in \mathbb{N}_0$, are polynomials in α, β, and some of their shifts. (Alternatively, one can prove directly by induction that the nonlinear recursion relations (A.29)–(A.34) are equivalent to that in (2.32)–(2.39) with all summation constants put equal to zero, $c_{\ell,\pm} = 0$, $\ell \in \mathbb{N}$.)

Remark 2.6. As an efficient tool to later distinguish between nonhomogeneous and homogeneous quantities $f_{\ell,\pm}$, $g_{\ell,\pm}$, $h_{\ell,\pm}$, and $\hat{f}_{\ell,\pm}$, $\hat{g}_{\ell,\pm}$, $\hat{h}_{\ell,\pm}$, respectively, we now introduce the notion of degree as follows. Denote

$$f^{(r)} = S^{(r)} f, \quad f = \{f(n)\}_{n\in\mathbb{Z}} \in \mathbb{C}^{\mathbb{Z}}, \quad S^{(r)} = \begin{cases} (S^+)^r, & r \geq 0, \\ (S^-)^{-r}, & r < 0, \end{cases} \quad r \in \mathbb{Z}, \tag{2.53}$$

and define

$$\deg\big(\alpha^{(r)}\big) = r, \quad \deg\big(\beta^{(r)}\big) = -r, \quad r \in \mathbb{Z}. \tag{2.54}$$

This then results in

$$\begin{aligned} &\deg\big(\hat{f}_{\ell,+}^{(r)}\big) = \ell + 1 + r, \quad \deg\big(\hat{f}_{\ell,-}^{(r)}\big) = -\ell + r, \quad \deg\big(\hat{g}_{\ell,\pm}^{(r)}\big) = \pm\ell, \\ &\deg\big(\hat{h}_{\ell,+}^{(r)}\big) = \ell - r, \quad \deg\big(\hat{h}_{\ell,-}^{(r)}\big) = -\ell - 1 - r, \quad \ell \in \mathbb{N}_0, \ r \in \mathbb{Z}, \end{aligned} \tag{2.55}$$

using induction in the linear recursion relations (2.32)–(2.39).

In accordance with our notation introduced in (2.49)–(2.51), the corresponding homogeneous stationary Ablowitz–Ladik equations are defined by

$$\text{s-}\widehat{\text{AL}}_{\underline{p}}(\alpha,\beta) = \text{s-AL}_{\underline{p}}(\alpha,\beta)\big|_{c_{0,\pm}=1,\, c_{\ell,\pm}=0,\, \ell=1,\dots,p_\pm}, \quad \underline{p} = (p_-, p_+) \in \mathbb{N}_0^2. \tag{2.56}$$

We also note the following useful result.

Lemma 2.7. *The coefficients $f_{\ell,\pm}$, $g_{\ell,\pm}$, and $h_{\ell,\pm}$ satisfy the relations*

$$\begin{aligned} g_{\ell,+} - g_{\ell,+}^- &= \alpha h_{\ell,+} + \beta f_{\ell,+}^-, \quad \ell \in \mathbb{N}_0, \\ g_{\ell,-} - g_{\ell,-}^- &= \alpha h_{\ell,-}^- + \beta f_{\ell,-}, \quad \ell \in \mathbb{N}_0. \end{aligned} \tag{2.57}$$

Moreover, we record the following symmetries,

$$\hat{f}_{\ell,\pm}(c_{0,\pm},\alpha,\beta) = \hat{h}_{\ell,\mp}(c_{0,\mp},\beta,\alpha), \quad \hat{g}_{\ell,\pm}(c_{0,\pm},\alpha,\beta) = \hat{g}_{\ell,\mp}(c_{0,\mp},\beta,\alpha), \quad \ell \in \mathbb{N}_0. \tag{2.58}$$

Proof. The relations (2.57) are derived as follows:

$$\begin{aligned}
\alpha h_{\ell+1,+} + \beta f^{-}_{\ell+1,+} &= \alpha h^{-}_{\ell,+} + \alpha\beta(g_{\ell+1,+} + g^{-}_{\ell+1,+}) + \beta f_{\ell,+} - \alpha\beta(g_{\ell+1,+} + g^{-}_{\ell+1,+}) \\
&= \alpha h^{-}_{\ell,+} + \beta f_{\ell,+} \\
&= g_{\ell+1,+} - g^{-}_{\ell+1,+},
\end{aligned} \tag{2.59}$$

and

$$\begin{aligned}
\alpha h^{-}_{\ell+1,-} + \beta f_{\ell+1,-} &= \alpha h_{\ell,-} - \alpha\beta(g_{\ell+1,-} + g^{-}_{\ell+1,-}) + \beta f^{-}_{\ell,-} + \alpha\beta(g_{\ell+1,-} + g^{-}_{\ell+1,-}) \\
&= \alpha h_{\ell,-} + \beta f^{-}_{\ell,-} \\
&= g_{\ell+1,-} + g^{-}_{\ell+1,-}.
\end{aligned} \tag{2.60}$$

The statement (2.58) follows by showing that $\hat{h}_{\ell,\mp}(\beta,\alpha)$ and $\hat{g}_{\ell,\mp}(\beta,\alpha)$ satisfy the same recursion relations as those of $\hat{f}_{\ell,\pm}(\alpha,\beta)$ and $\hat{g}_{\ell,\pm}(\alpha,\beta)$, respectively. That the recursion constants are the same, follows from the observation that the corresponding coefficients have the proper degree. □

Next we turn to the Laurent polynomials $F_{\underline{p}}$, $G_{\underline{p}}$, $H_{\underline{p}}$, and $K_{\underline{p}}$ defined in (2.18)–(2.20) and (2.22). Explicitly, one obtains

$$\begin{aligned}
F_{(0,0)} &= 0, \\
F_{(1,1)} &= c_{0,-}\alpha z^{-1} + c_{0,+}(-\alpha^{+}), \\
F_{(2,2)} &= c_{0,-}\alpha z^{-2} + \big(c_{0,-}(\gamma\alpha^{-} - \alpha^{2}\beta^{+}) + c_{1,-}\alpha\big)z^{-1} \\
&\quad + c_{0,+}\big(-\gamma^{+}\alpha^{++} + (\alpha^{+})^{2}\beta\big) + c_{1,+}(-\alpha^{+}) + c_{0,+}(-\alpha^{+})z, \\
G_{(0,0)} &= \tfrac{1}{2}c_{0,+}, \\
G_{(1,1)} &= \tfrac{1}{2}c_{0,-}z^{-1} + c_{0,+}(-\alpha^{+}\beta) + \tfrac{1}{2}c_{1,+} + \tfrac{1}{2}c_{0,+}z, \\
G_{(2,2)} &= \tfrac{1}{2}c_{0,-}z^{-2} + \big(c_{0,-}(-\alpha\beta^{+}) + \tfrac{1}{2}c_{1,-}\big)z^{-1} \\
&\quad + c_{0,+}\big((\alpha^{+}\beta)^{2} - \gamma^{+}\alpha^{++}\beta - \gamma\alpha^{+}\beta^{-}\big) + c_{1,+}(-\alpha^{+}\beta) + \tfrac{1}{2}c_{2,+} \\
&\quad + \big(c_{0,+}(-\alpha^{+}\beta) + \tfrac{1}{2}c_{1,+}\big)z + \tfrac{1}{2}c_{0,+}z^{2}, \\
H_{(0,0)} &= 0, \\
H_{(1,1)} &= c_{0,-}(-\beta^{+}) + c_{0,+}\beta z, \\
H_{(2,2)} &= c_{0,-}(-\beta^{+})z^{-1} + c_{0,-}\big(-\gamma^{+}\beta^{++} + \alpha(\beta^{+})^{2}\big) + c_{1,-}(-\beta^{+}) \\
&\quad + \big(c_{0,+}(\gamma\beta^{-} - \alpha^{+}\beta^{2}) + c_{1,+}\beta\big)z + c_{0,+}\beta z^{2}, \\
K_{(0,0)} &= \tfrac{1}{2}c_{0,-}, \\
K_{(1,1)} &= \tfrac{1}{2}c_{0,-}z^{-1} + c_{0,-}(-\alpha\beta^{+}) + \tfrac{1}{2}c_{1,-} + \tfrac{1}{2}c_{0,+}z, \\
K_{(2,2)} &= \tfrac{1}{2}c_{0,-}z^{-2} + \big(c_{0,-}(-\alpha\beta^{+}) + \tfrac{1}{2}c_{1,-}\big)z^{-1} \\
&\quad + c_{0,-}\big((\alpha\beta^{+})^{2} - \gamma^{+}\alpha\beta^{++} - \gamma\alpha^{-}\beta^{+}\big) + c_{1,-}(-\alpha\beta^{+}) + \tfrac{1}{2}c_{2,-} \\
&\quad + \big(c_{0,+}(-\alpha^{+}\beta) + \tfrac{1}{2}c_{1,+}\big)z + \tfrac{1}{2}c_{0,+}z^{2}, \quad \text{etc.}
\end{aligned} \tag{2.61}$$

The corresponding homogeneous quantities are defined by ($\ell \in \mathbb{N}_0$)

$$
\begin{aligned}
&\widehat{F}_{0,\mp}(z) = 0, \quad \widehat{F}_{\ell,-}(z) = \sum_{k=1}^{\ell} \hat{f}_{\ell-k,-} z^{-k}, \quad \widehat{F}_{\ell,+}(z) = \sum_{k=0}^{\ell-1} \hat{f}_{\ell-1-k,+} z^{k},\\
&\widehat{G}_{0,-}(z) = 0, \quad \widehat{G}_{\ell,-}(z) = \sum_{k=1}^{\ell} \hat{g}_{\ell-k,-} z^{-k},\\
&\widehat{G}_{0,+}(z) = \frac{1}{2}, \quad \widehat{G}_{\ell,+}(z) = \sum_{k=0}^{\ell} \hat{g}_{\ell-k,+} z^{k},\\
&\widehat{H}_{0,\mp}(z) = 0, \quad \widehat{H}_{\ell,-}(z) = \sum_{k=0}^{\ell-1} \hat{h}_{\ell-1-k,-} z^{-k}, \quad \widehat{H}_{\ell,+}(z) = \sum_{k=1}^{\ell} \hat{h}_{\ell-k,+} z^{k},\\
&\widehat{K}_{0,-}(z) = \frac{1}{2}, \quad \widehat{K}_{\ell,-}(z) = \sum_{k=0}^{\ell} \hat{g}_{\ell-k,-} z^{-k} = \widehat{G}_{\ell,-}(z) + \hat{g}_{\ell,-},\\
&\widehat{K}_{0,+}(z) = 0, \quad \widehat{K}_{\ell,+}(z) = \sum_{k=1}^{\ell} \hat{g}_{\ell-k,+} z^{k} = \widehat{G}_{\ell,+}(z) - \hat{g}_{\ell,+}.
\end{aligned}
\tag{2.62}
$$

Similarly, with $F_{\ell_+,+}$, $G_{\ell_+,+}$, $H_{\ell_+,+}$, and $K_{\ell_+,+}$ denoting the polynomial parts of $F_{\underline{\ell}}$, $G_{\underline{\ell}}$, $H_{\underline{\ell}}$, and $K_{\underline{\ell}}$, respectively, and $F_{\ell_-,-}$, $G_{\ell_-,-}$, $H_{\ell_-,-}$, and $K_{\ell_-,-}$ denoting the Laurent parts of $F_{\underline{\ell}}$, $G_{\underline{\ell}}$, $H_{\underline{\ell}}$, and $K_{\underline{\ell}}$, $\underline{\ell} = (\ell_-, \ell_+) \in \mathbb{N}_0$, such that

$$
\begin{aligned}
F_{\underline{\ell}}(z) &= F_{\ell_-,-}(z) + F_{\ell_+,+}(z), \quad & G_{\underline{\ell}}(z) &= G_{\ell_-,-}(z) + G_{\ell_+,+}(z),\\
H_{\underline{\ell}}(z) &= H_{\ell_-,-}(z) + H_{\ell_+,+}(z), \quad & K_{\underline{\ell}}(z) &= K_{\ell_-,-}(z) + K_{\ell_+,+}(z),
\end{aligned}
\tag{2.63}
$$

one finds that

$$
\begin{aligned}
F_{\ell_\pm,\pm} &= \sum_{k=1}^{\ell_\pm} c_{\ell_\pm-k,\pm} \widehat{F}_{k,\pm}, \quad & H_{\ell_\pm,\pm} &= \sum_{k=1}^{\ell_\pm} c_{\ell_\pm-k,\pm} \widehat{H}_{k,\pm},\\
G_{\ell_-,-} &= \sum_{k=1}^{\ell_-} c_{\ell_--k,-} \widehat{G}_{k,-}, \quad & G_{\ell_+,+} &= \sum_{k=0}^{\ell_+} c_{\ell_+-k,+} \widehat{G}_{k,+},\\
K_{\ell_-,-} &= \sum_{k=0}^{\ell_-} c_{\ell_--k,-} \widehat{K}_{k,-}, \quad & K_{\ell_+,+} &= \sum_{k=1}^{\ell_+} c_{\ell_+-k,+} \widehat{K}_{k,+}.
\end{aligned}
\tag{2.64}
$$

In addition, one immediately obtains the following relations from (2.58):

Lemma 2.8. *Let* $\ell \in \mathbb{N}_0$. *Then,*

$$\widehat{F}_{\ell,\pm}(\alpha, \beta, z, n) = \widehat{H}_{\ell,\mp}(\beta, \alpha, z^{-1}, n), \tag{2.65}$$

$$\widehat{H}_{\ell,\pm}(\alpha, \beta, z, n) = \widehat{F}_{\ell,\mp}(\beta, \alpha, z^{-1}, n), \tag{2.66}$$

$$\widehat{G}_{\ell,\pm}(\alpha, \beta, z, n) = \widehat{G}_{\ell,\mp}(\beta, \alpha, z^{-1}, n), \tag{2.67}$$

$$\widehat{K}_{\ell,\pm}(\alpha, \beta, z, n) = \widehat{K}_{\ell,\mp}(\beta, \alpha, z^{-1}, n). \tag{2.68}$$

Returning to the stationary Ablowitz–Ladik hierarchy, we will frequently assume in the following that α, β satisfy the $\underline{p}$th stationary Ablowitz–Ladik system s-$\mathrm{AL}_{\underline{p}}(\alpha, \beta) = 0$, supposing a particular choice of summation constants $c_{\ell,\pm} \in \mathbb{C}$, $\ell = 0, \dots, p_\pm$, $p_\pm \in \mathbb{N}_0$, has been made.

Remark 2.9.

(i) The particular choice $c_{0,+} = c_{0,-} = 1$ in (2.45) yields the stationary Ablowitz–Ladik equation. Scaling $c_{0,\pm}$ with the same constant then amounts to scaling $V_{\underline{p}}$ with this constant which drops out in the stationary zero-curvature equation (2.8).

(ii) Different ratios between $c_{0,+}$ and $c_{0,-}$ will lead to different stationary hierarchies. In particular, the choice $c_{0,+} = 2$, $c_{0,-} = \cdots = c_{p_- -1,-} = 0$, $c_{p_-,-} \neq 0$, yields the stationary Baxter–Szegő hierarchy considered in detail in [28]. However, in this case some parts from the recursion relation for the negative coefficients still remain. In fact, (2.39) reduces to $g_{p_-,-} - g_{p_-,-}^- = \alpha h_{p_- -1,-}$, $h_{p_- -1,-} = 0$ and thus requires $g_{p_-,-}$ to be a constant in (2.45) and (2.87). Moreover, $f_{p_- -1,-} = 0$ in (2.45) in this case.

(iii) Finally, by Lemma 2.8, the choice $c_{0,+} = \cdots = c_{p_+ -1,+} = 0$, $c_{p_+,+} \neq 0$, $c_{0,-} = 2$ again yields the Baxter–Szegő hierarchy, but with α and β interchanged.

Next, taking into account (2.21), one infers that the expression $R_{\underline{p}}$, defined as

$$R_{\underline{p}} = G_{\underline{p}}^2 - F_{\underline{p}} H_{\underline{p}}, \tag{2.69}$$

is a lattice constant, that is, $R_{\underline{p}} - R_{\underline{p}}^- = 0$, since taking determinants in the stationary zero-curvature equation (2.8) immediately yields

$$\gamma\big(-(G_{\underline{p}}^-)^2 + F_{\underline{p}}^- H_{\underline{p}}^- + G_{\underline{p}}^2 - F_{\underline{p}} H_{\underline{p}} \big) z = 0. \tag{2.70}$$

Hence, $R_{\underline{p}}(z)$ only depends on z, and assuming in addition to (2.1) that

$$c_{0,\pm} \in \mathbb{C} \setminus \{0\}, \quad \underline{p} = (p_-, p_+) \in \mathbb{N}_0^2 \setminus \{(0,0)\}, \tag{2.71}$$

one may write $R_{\underline{p}}$ as[2]

$$R_{\underline{p}}(z) = \left(\frac{c_{0,+}}{2z^{p_-}} \right)^2 \prod_{m=0}^{2p+1} (z - E_m), \quad \{E_m\}_{m=0}^{2p+1} \subset \mathbb{C} \setminus \{0\},\ p = p_- + p_+ - 1 \in \mathbb{N}_0. \tag{2.72}$$

Moreover, (2.69) also implies

$$\lim_{z \to 0} 4z^{2p_-} R_{\underline{p}}(z) = c_{0,+}^2 \prod_{m=0}^{2p+1} (-E_m) = c_{0,-}^2, \tag{2.73}$$

[2] We use the convention that a product is to be interpreted equal to 1 whenever the upper limit of the product is strictly less than its lower limit.

and hence,

$$\prod_{m=0}^{2p+1} E_m = \frac{c_{0,-}^2}{c_{0,+}^2}. \tag{2.74}$$

Relation (2.69) allows one to introduce a hyperelliptic curve $\mathcal{K}_p$ of (arithmetic) genus $p = p_- + p_+ - 1$ (possibly with a singular affine part), where

$$\begin{aligned} \mathcal{K}_p \colon \mathcal{F}_p(z,y) &= y^2 - 4c_{0,+}^{-2} z^{2p_-} R_{\underline{p}}(z) \\ &= y^2 - \prod_{m=0}^{2p+1} (z - E_m) = 0, \quad p = p_- + p_+ - 1. \end{aligned} \tag{2.75}$$

Remark 2.10. In the special case $p_- = p_+$ and $c_{\ell,+} = c_{\ell,-}$, $\ell = 0,\dots,p_-$, the symmetries of Lemma 2.8 also hold for $F_{\underline{p}}$, $G_{\underline{p}}$, and $H_{\underline{p}}$ and thus $R_{\underline{p}}(1/z) = R_{\underline{p}}(z)$ and hence the numbers E_m, $m = 0,\dots,2p+1$, come in pairs $(E_k, 1/E_k)$, $k = 1,\dots,p+1$.

Equations (2.10)–(2.13) and (2.69) permit one to derive nonlinear difference equations for $F_{\underline{p}}$, $G_{\underline{p}}$, and $H_{\underline{p}}$ separately. One obtains

$$\begin{aligned} &\big((\alpha^+ + z\alpha)^2 F_{\underline{p}} - z(\alpha^+)^2\gamma F_{\underline{p}}^-\big)^2 - 2z\alpha^2\gamma^+\big((\alpha^+ + z\alpha)^2 F_{\underline{p}} + z(\alpha^+)^2\gamma F_{\underline{p}}^-\big)F_{\underline{p}}^+ \\ &+ z^2\alpha^4(\gamma^+)^2(F_{\underline{p}}^+)^2 = 4(\alpha\alpha^+)^2(\alpha^+ + \alpha z)^2 R_{\underline{p}}, \end{aligned} \tag{2.76}$$

$$\begin{aligned} &(\alpha^+ + z\alpha)(\beta + z\beta^+)(z + \alpha^+\beta)(1 + z\alpha\beta^+)G_{\underline{p}}^2 \\ &+ z(\alpha^+\gamma G_{\underline{p}}^- + z\alpha\gamma^+ G_{\underline{p}}^+)(z\beta^+\gamma G_{\underline{p}}^- + \beta\gamma^+ G_{\underline{p}}^+) \\ &- z\gamma\big((\alpha^+\beta + z^2\alpha\beta^+)(2-\gamma^+) + 2z(1-\gamma^+)(2-\gamma)\big)G_{\underline{p}}^- G_{\underline{p}} \\ &- z\gamma^+\big(2z(1-\gamma)(2-\gamma^+) + (\alpha^+\beta + z^2\alpha\beta^+)(2-\gamma)\big)G_{\underline{p}}^+ G_{\underline{p}} \\ &= (\alpha^+\beta - z^2\alpha\beta^+)^2 R_{\underline{p}}, \end{aligned} \tag{2.77}$$

$$\begin{aligned} &z^2\big((\beta^+)^2\gamma H_{\underline{p}}^- - \beta^2\gamma^+ H_{\underline{p}}^+\big)^2 - 2z(\beta + z\beta^+)^2\big((\beta^+)^2\gamma H_{\underline{p}}^- + \beta^2\gamma^+ H_{\underline{p}}^+\big)H_{\underline{p}} \\ &+ (\beta + z\beta^+)^4 H_{\underline{p}}^2 = 4z^2(\beta\beta^+)^2(\beta + \beta^+ z)^2 R_{\underline{p}}. \end{aligned} \tag{2.78}$$

Equations analogous to (2.76)–(2.78) can be used to derive nonlinear recursion relations for the homogeneous coefficients $\hat{f}_{\ell,\pm}$, $\hat{g}_{\ell,\pm}$, and $\hat{h}_{\ell,\pm}$ (i.e., the ones satisfying (2.49)–(2.51) in the case of vanishing summation constants) as proved in Theorem A.1 in Appendix A. This then yields a proof that $\hat{f}_{\ell,\pm}$, $\hat{g}_{\ell,\pm}$, and $\hat{h}_{\ell,\pm}$ are polynomials in α, β, and some of their shifts (cf. Remark 2.5). In addition, as proven in Theorem A.2, (2.76) leads to an explicit determination of the summation constants $c_{1,\pm},\dots,c_{p_\pm,\pm}$ in (2.45) in terms of the zeros $E_0,\dots,E_{2p+1}$ of the associated Laurent polynomial $R_{\underline{p}}$ in (2.72). In fact, one can prove (cf. (A.42))

$$c_{\ell,\pm} = c_{0,\pm} c_\ell\big(\underline{E}^{\pm 1}\big), \quad \ell = 0,\dots,p_\pm, \tag{2.79}$$

where

$$c_0(\underline{E}^{\pm 1}) = 1,$$
$$c_k(\underline{E}^{\pm 1}) \tag{2.80}$$
$$= -\sum_{\substack{j_0,\dots,j_{2p+1}=0\\ j_0+\dots+j_{2p+1}=k}}^{k} \frac{(2j_0)!\cdots(2j_{2p+1})!}{2^{2k}(j_0!)^2\cdots(j_{2p+1}!)^2(2j_0-1)\cdots(2j_{2p+1}-1)} E_0^{\pm j_0}\cdots E_{2p+1}^{\pm j_{2p+1}},$$
$$k\in\mathbb{N},$$

are symmetric functions of $\underline{E}^{\pm 1} = (E_0^{\pm 1},\dots,E_{2p+1}^{\pm 1})$ introduced in (A.5) and (A.6).

Remark 2.11. If α,β satisfy one of the stationary Ablowitz–Ladik equations in (2.45) for a particular value of $\underline{p}$, s-$\mathrm{AL}_{\underline{p}}(\alpha,\beta)=0$, then they satisfy infinitely many such equations of order higher than $\underline{p}$ for certain choices of summation constants $c_{\ell,\pm}$. This can be shown as in [29, Remark I.1.5].

Finally we turn to the time-dependent Ablowitz–Ladik hierarchy. For that purpose the coefficients α and β are now considered as functions of both the lattice point and time. For each system in the hierarchy, that is, for each $\underline{p}$, we introduce a deformation (time) parameter $t_{\underline{p}}\in\mathbb{R}$ in α,β, replacing $\alpha(n),\beta(n)$ by $\alpha(n,t_{\underline{p}}),\beta(n,t_{\underline{p}})$. Moreover, the definitions (2.5), (2.6), and (2.18)–(2.20) of $U, V_{\underline{p}}$, and $F_{\underline{p}}, G_{\underline{p}}, H_{\underline{p}}, K_{\underline{p}}$, respectively, still apply; however, equation (2.21) now needs to be replaced by (2.22) in the time-dependent context.

Imposing the zero-curvature relation

$$U_{t_{\underline{p}}} + UV_{\underline{p}} - V_{\underline{p}}^{+}U = 0, \quad \underline{p}\in\mathbb{N}_0^2, \tag{2.81}$$

then results in the equations

$$\begin{aligned} 0 &= U_{t_{\underline{p}}} + UV_{\underline{p}} - V_{\underline{p}}^{+}U \\ &= i\begin{pmatrix} z(G_{\underline{p}}^- - G_{\underline{p}}) + z\beta F_{\underline{p}} + \alpha H_{\underline{p}}^- & -i\alpha_{t_{\underline{p}}} + F_{\underline{p}} - zF_{\underline{p}}^- - \alpha(G_{\underline{p}} + K_{\underline{p}}^-) \\ -iz\beta_{t_{\underline{p}}} + z\beta(G_{\underline{p}}^- + K_{\underline{p}}) - zH_{\underline{p}} + H_{\underline{p}}^- & -z\beta F_{\underline{p}}^- - \alpha H_{\underline{p}} + K_{\underline{p}} - K_{\underline{p}}^- \end{pmatrix} \\ &= i\begin{pmatrix} 0 & \begin{matrix}-i\alpha_{t_{\underline{p}}} - \alpha(g_{p_+,+} + g_{p_-,-}^-) \\ + f_{p_+-1,+} - f_{p_--1,-}^-\end{matrix} \\ \begin{matrix} z\big(-i\beta_{t_{\underline{p}}} + \beta(g_{p_+,+}^- + g_{p_-,-}) \\ -h_{p_--1,-} + h_{p_+-1,+}^-\big)\end{matrix} & 0 \end{pmatrix}, \end{aligned} \tag{2.82}$$

or equivalently,

$$\alpha_{t_{\underline{p}}} = i\big(zF_{\underline{p}}^- + \alpha(G_{\underline{p}} + K_{\underline{p}}^-) - F_{\underline{p}}\big), \tag{2.83}$$
$$\beta_{t_{\underline{p}}} = -i\big(\beta(G_{\underline{p}}^- + K_{\underline{p}}) - H_{\underline{p}} + z^{-1}H_{\underline{p}}^-\big), \tag{2.84}$$
$$0 = z(G_{\underline{p}}^- - G_{\underline{p}}) + z\beta F_{\underline{p}} + \alpha H_{\underline{p}}^-, \tag{2.85}$$
$$0 = z\beta F_{\underline{p}}^- + \alpha H_{\underline{p}} + K_{\underline{p}}^- - K_{\underline{p}}. \tag{2.86}$$

Varying $\underline{p} \in \mathbb{N}_0^2$, the collection of evolution equations

$$\mathrm{AL}_{\underline{p}}(\alpha,\beta) = \begin{pmatrix} -i\alpha_{t_{\underline{p}}} - \alpha(g_{p_+,+} + g^-_{p_-,-}) + f_{p_+-1,+} - f^-_{p_--1,-} \\ -i\beta_{t_{\underline{p}}} + \beta(g^-_{p_+,+} + g_{p_-,-}) - h_{p_--1,-} + h^-_{p_+-1,+} \end{pmatrix} = 0, \qquad (2.87)$$
$$t_{\underline{p}} \in \mathbb{R},\ \underline{p} = (p_-,p_+) \in \mathbb{N}_0^2,$$

then defines the time-dependent Ablowitz–Ladik hierarchy. Explicitly, taking $p_- = p_+$ for simplicity,

$$\mathrm{AL}_{(0,0)}(\alpha,\beta) = \begin{pmatrix} -i\alpha_{t_{(0,0)}} - c_{(0,0)}\alpha \\ -i\beta_{t_{(0,0)}} + c_{(0,0)}\beta \end{pmatrix} = 0,$$

$$\mathrm{AL}_{(1,1)}(\alpha,\beta) = \begin{pmatrix} -i\alpha_{t_{(1,1)}} - \gamma(c_{0,-}\alpha^- + c_{0,+}\alpha^+) - c_{(1,1)}\alpha \\ -i\beta_{t_{(1,1)}} + \gamma(c_{0,+}\beta^- + c_{0,-}\beta^+) + c_{(1,1)}\beta \end{pmatrix} = 0,$$

$$\mathrm{AL}_{(2,2)}(\alpha,\beta) \qquad (2.88)$$
$$= \begin{pmatrix} -i\alpha_{t_{(2,2)}} - \gamma\big(c_{0,+}\alpha^{++}\gamma^+ + c_{0,-}\alpha^{--}\gamma^- - \alpha(c_{0,+}\alpha^+\beta^- + c_{0,-}\alpha^-\beta^+) \\ -\beta(c_{0,-}(\alpha^-)^2 + c_{0,+}(\alpha^+)^2)\big) \\ -i\beta_{t_{(2,2)}} + \gamma\big(c_{0,-}\beta^{++}\gamma^+ + c_{0,+}\beta^{--}\gamma^- - \beta(c_{0,+}\alpha^+\beta^- + c_{0,-}\alpha^-\beta^+) \\ -\alpha(c_{0,+}(\beta^-)^2 + c_{0,-}(\beta^+)^2)\big) \end{pmatrix}$$
$$+ \begin{pmatrix} -\gamma(c_{1,-}\alpha^- + c_{1,+}\alpha^+) - c_{(2,2)}\alpha \\ \gamma(c_{1,+}\beta^- + c_{1,-}\beta^+) + c_{(2,2)}\beta \end{pmatrix} = 0, \quad \text{etc.},$$

represent the first few equations of the time-dependent Ablowitz–Ladik hierarchy. Here we recall the definition of $c_{\underline{p}}$ in (2.47).

The special case $\underline{p} = (1,1)$, $c_{0,\pm} = 1$, and $c_{(1,1)} = -2$, that is,

$$\begin{pmatrix} -i\alpha_{t_{(1,1)}} - \gamma(\alpha^- + \alpha^+) + 2\alpha \\ -i\beta_{t_{(1,1)}} + \gamma(\beta^- + \beta^+) - 2\beta \end{pmatrix} = 0, \qquad (2.89)$$

represents *the* Ablowitz–Ladik system (1.1).

The corresponding homogeneous equations are then defined by

$$\widehat{\mathrm{AL}}_{\underline{p}}(\alpha,\beta) = \mathrm{AL}_{\underline{p}}(\alpha,\beta)\big|_{c_{0,\pm}=1,\, c_{\ell,\pm}=0,\, \ell=1,\dots,p_\pm} = 0, \quad \underline{p} = (p_-,p_+) \in \mathbb{N}_0^2. \qquad (2.90)$$

By (2.87), (2.33), and (2.37), the time derivative of $\gamma = 1 - \alpha\beta$ is given by

$$\gamma_{t_{\underline{p}}} = i\gamma\big((g_{p_+,+} - g^-_{p_+,+}) - (g_{p_-,-} - g^-_{p_-,-})\big). \qquad (2.91)$$

(Alternatively, this follows from computing the trace of $U_{t_{\underline{p}}}U^{-1} = V_{\underline{p}}^+ - UV_{\underline{p}}U^{-1}$.) For instance, if α, β satisfy $\mathrm{AL}_1(\alpha,\beta) = 0$, then

$$\gamma_{t_1} = i\gamma\big(\alpha(c_{0,-}\beta^+ + c_{0,+}\beta^-) - \beta(c_{0,+}\alpha^+ + c_{0,-}\alpha^-)\big). \qquad (2.92)$$

Remark 2.12. From (2.10)–(2.13) and the explicit computations of the coefficients $f_{\ell,\pm}$, $g_{\ell,\pm}$, and $h_{\ell,\pm}$, one concludes that the zero-curvature equation (2.82) and hence the Ablowitz–Ladik hierarchy is invariant under the scaling transformation

$$\alpha \to \alpha_c = \{c\,\alpha(n)\}_{n\in\mathbb{Z}}, \quad \beta \to \beta_c = \{\beta(n)/c\}_{n\in\mathbb{Z}}, \quad c \in \mathbb{C}\setminus\{0\}. \qquad (2.93)$$

Moreover, $R_{\underline{p}} = G_{\underline{p}}^2 - H_{\underline{p}} F_{\underline{p}}$ and hence $\{E_m\}_{m=0}^{2p+1}$ are invariant under this transformation. Furthermore, choosing $c = e^{ic_{\underline{p}} t}$, one verifies that it is no restriction to assume $c_{\underline{p}} = 0$. This also shows that stationary solutions α, β can only be constructed up to a multiplicative constant.

Remark 2.13.

(i) The special choices $\beta = \pm\overline{\alpha}$, $c_{0,\pm} = 1$ lead to the discrete nonlinear Schrödinger hierarchy. In particular, choosing $c_{(1,1)} = -2$ yields the discrete nonlinear Schrödinger equation in its usual form (see, e.g., [7, Ch. 3] and the references cited therein),

$$-i\alpha_t - (1 \mp |\alpha|^2)(\alpha^- + \alpha^+) + 2\alpha = 0, \tag{2.94}$$

as its first nonlinear element. The choice $\beta = \overline{\alpha}$ is called the *defocusing* case, $\beta = -\overline{\alpha}$ represents the *focusing* case of the discrete nonlinear Schrödinger hierarchy.

(ii) The alternative choice $\beta = \overline{\alpha}$, $c_{0,\pm} = \mp i$, leads to the hierarchy of Schur flows. In particular, choosing $c_{(1,1)} = 0$ yields

$$\alpha_t - (1 - |\alpha|^2)(\alpha^+ - \alpha^-) = 0 \tag{2.95}$$

as the first nonlinear element of this hierarchy (cf. [10], [24], [25], [36], [41], [50]).

3. The stationary Ablowitz–Ladik formalism

This section is devoted to a detailed study of the stationary Ablowitz–Ladik hierarchy. Our principal tools are derived from combining the polynomial recursion formalism introduced in Section 2 and a fundamental meromorphic function ϕ on a hyperelliptic curve $\mathcal{K}_p$. With the help of ϕ we study the Baker–Akhiezer vector Ψ, and trace formulas for α and β.

Unless explicitly stated otherwise, we suppose in this section that

$$\alpha, \beta \in \mathbb{C}^{\mathbb{Z}}, \quad \alpha(n)\beta(n) \notin \{0,1\},\ n \in \mathbb{Z}, \tag{3.1}$$

and assume (2.5), (2.6), (2.8), (2.18)–(2.21), (2.32)–(2.39), (2.40), (2.45), (2.69), (2.72), keeping $p \in \mathbb{N}_0$ fixed.

We recall the hyperelliptic curve

$$\begin{aligned} &\mathcal{K}_p \colon \mathcal{F}_p(z,y) = y^2 - 4c_{0,+}^{-2} z^{2p_-} R_{\underline{p}}(z) = y^2 - \prod_{m=0}^{2p+1} (z - E_m) = 0, \\ &R_{\underline{p}}(z) = \bigg(\frac{c_{0,+}}{2z^{p_-}}\bigg)^2 \prod_{m=0}^{2p+1} (z - E_m), \quad \{E_m\}_{m=0}^{2p+1} \subset \mathbb{C} \setminus \{0\},\ p = p_- + p_+ - 1, \end{aligned} \tag{3.2}$$

as introduced in (2.75). Throughout this section we assume the affine part of $\mathcal{K}_p$ to be nonsingular, that is, we suppose that

$$E_m \neq E_{m'} \text{ for } m \neq m', \quad m, m' = 0, 1, \dots, 2p+1. \tag{3.3}$$

$\mathcal{K}_p$ is compactified by joining two points $P_{\infty_\pm}$, $P_{\infty_+} \neq P_{\infty_-}$, but for notational simplicity the compactification is also denoted by $\mathcal{K}_p$. Points P on $\mathcal{K}_p \backslash \{P_{\infty_+}, P_{\infty_-}\}$ are represented as pairs $P = (z, y)$, where $y(\,\cdot\,)$ is the meromorphic function on $\mathcal{K}_p$ satisfying $\mathcal{F}_p(z, y) = 0$. The complex structure on $\mathcal{K}_p$ is then defined in the usual manner. Hence, $\mathcal{K}_p$ becomes a two-sheeted hyperelliptic Riemann surface of genus p in a standard manner.

We also emphasize that by fixing the curve $\mathcal{K}_p$ (i.e., by fixing $E_0, \dots, E_{2p+1}$), the summation constants $c_{1,\pm}, \dots, c_{p_\pm,\pm}$ in $f_{p_\pm,\pm}$, $g_{p_\pm,\pm}$, and $h_{p_\pm,\pm}$ (and hence in the corresponding stationary s-AL$_{\underline{p}}$ equations) are uniquely determined as is clear from (2.79), (2.80), which establish the summation constants $c_{\ell,\pm}$ as symmetric functions of $E_0^{\pm 1}, \dots, E_{2p+1}^{\pm 1}$.

For notational simplicity we will usually tacitly assume that $p \in \mathbb{N}$ and hence $\underline{p} \in \mathbb{N}_0^2 \backslash \{(0,0), (0,1), (1,0)\}$. (The trivial case $\underline{p} = 0$ is explicitly treated in Example 3.5.)

We denote by $\{\mu_j(n)\}_{j=1,\dots,p}$ and $\{\nu_j(n)\}_{j=1,\dots,p}$ the zeros of $(\,\cdot\,)^{p_-} F_{\underline{p}}(\,\cdot\,, n)$ and $(\,\cdot\,)^{p_- -1} H_{\underline{p}}(\,\cdot\,, n)$, respectively. Thus, we may write

$$F_{\underline{p}}(z) = -c_{0,+}\alpha^+ z^{-p_-} \prod_{j=1}^{p} (z - \mu_j), \tag{3.4}$$

$$H_{\underline{p}}(z) = c_{0,+}\beta z^{-p_- +1} \prod_{j=1}^{p} (z - \nu_j), \tag{3.5}$$

and we recall that (cf. (2.69))

$$R_{\underline{p}} - G_{\underline{p}}^2 = -F_{\underline{p}} H_{\underline{p}}. \tag{3.6}$$

The next step is crucial; it permits us to "lift" the zeros μ_j and ν_j from the complex plane $\mathbb{C}$ to the curve $\mathcal{K}_p$. From (3.6) one infers that

$$R_{\underline{p}}(z) - G_{\underline{p}}(z)^2 = 0, \quad z \in \{\mu_j, \nu_k\}_{j,k=1,\dots,p}. \tag{3.7}$$

We now introduce $\{\hat{\mu}_j\}_{j=1,\dots,p} \subset \mathcal{K}_p$ and $\{\hat{\nu}_j\}_{j=1,\dots,p} \subset \mathcal{K}_p$ by

$$\hat{\mu}_j(n) = (\mu_j(n), (2/c_{0,+})\mu_j(n)^{p_-} G_{\underline{p}}(\mu_j(n), n)), \quad j = 1, \dots, p,\ n \in \mathbb{Z}, \tag{3.8}$$

and

$$\hat{\nu}_j(n) = (\nu_j(n), -(2/c_{0,+})\nu_j(n)^{p_-} G_{\underline{p}}(\nu_j(n), n)), \quad j = 1, \dots, p,\ n \in \mathbb{Z}. \tag{3.9}$$

We also introduce the points $P_{0,\pm}$ by

$$P_{0,\pm} = (0, \pm(c_{0,-}/c_{0,+})) \in \mathcal{K}_p, \quad \frac{c_{0,-}^2}{c_{0,+}^2} = \prod_{m=0}^{2p+1} E_m. \tag{3.10}$$

We emphasize that $P_{0,\pm}$ and $P_{\infty_\pm}$ are not necessarily on the same sheet of $\mathcal{K}_p$.

Next, we briefly recall our conventions used in connection with divisors on $\mathcal{K}_p$. A map, $\mathcal{D}\colon \mathcal{K}_p \to \mathbb{Z}$, is called a divisor on $\mathcal{K}_p$ if $\mathcal{D}(P) \neq 0$ for only finitely many $P \in \mathcal{K}_p$. The set of divisors on $\mathcal{K}_p$ is denoted by $\operatorname{Div}(\mathcal{K}_p)$. We shall employ the following (additive) notation for divisors,

$$\begin{aligned} &\mathcal{D}_{Q_0\underline{Q}} = \mathcal{D}_{Q_0} + \mathcal{D}_{\underline{Q}}, \quad \mathcal{D}_{\underline{Q}} = \mathcal{D}_{Q_1} + \cdots + \mathcal{D}_{Q_m}, \\ &\underline{Q} = \{Q_1, \dots, Q_m\} \in \operatorname{Sym}^m \mathcal{K}_p, \quad Q_0 \in \mathcal{K}_p, \ m \in \mathbb{N}, \end{aligned} \tag{3.11}$$

where for any $Q \in \mathcal{K}_p$,

$$\mathcal{D}_Q \colon \mathcal{K}_p \to \mathbb{N}_0, \quad P \mapsto \mathcal{D}_Q(P) = \begin{cases} 1 & \text{for } P = Q, \\ 0 & \text{for } P \in \mathcal{K}_p \setminus \{Q\}, \end{cases} \tag{3.12}$$

and $\operatorname{Sym}^n \mathcal{K}_p$ denotes the nth symmetric product of $\mathcal{K}_p$. In particular, $\operatorname{Sym}^m \mathcal{K}_p$ can be identified with the set of nonnegative divisors $0 \le \mathcal{D} \in \operatorname{Div}(\mathcal{K}_p)$ of degree m. Moreover, for a nonzero, meromorphic function f on $\mathcal{K}_p$, the divisor of f is denoted by (f). Two divisors $\mathcal{D}$, $\mathcal{E} \in \operatorname{Div}(\mathcal{K}_p)$ are called equivalent, denoted by $\mathcal{D} \sim \mathcal{E}$, if and only if $\mathcal{D} - \mathcal{E} = (f)$ for some $f \in \mathcal{M}(\mathcal{K}_p) \setminus \{0\}$. The divisor class $[\mathcal{D}]$ of $\mathcal{D}$ is then given by $[\mathcal{D}] = \{\mathcal{E} \in \operatorname{Div}(\mathcal{K}_p) \,|\, \mathcal{E} \sim \mathcal{D}\}$. We recall that

$$\deg((f)) = 0, \ f \in \mathcal{M}(\mathcal{K}_p) \setminus \{0\}, \tag{3.13}$$

where the degree $\deg(\mathcal{D})$ of $\mathcal{D}$ is given by $\deg(\mathcal{D}) = \sum_{P \in \mathcal{K}_p} \mathcal{D}(P)$.

Next we introduce the fundamental meromorphic function on $\mathcal{K}_p$ by

$$\begin{aligned} \phi(P,n) &= \frac{(c_{0,+}/2)z^{-p_-}y + G_{\underline{p}}(z,n)}{F_{\underline{p}}(z,n)} && (3.14) \\ &= \frac{-H_{\underline{p}}(z,n)}{(c_{0,+}/2)z^{-p_-}y - G_{\underline{p}}(z,n)}, && (3.15) \\ &\qquad P = (z,y) \in \mathcal{K}_p, \ n \in \mathbb{Z}, \end{aligned}$$

with divisor $(\phi(\,\cdot\,,n))$ of $\phi(\,\cdot\,,n)$ given by

$$(\phi(\,\cdot\,,n)) = \mathcal{D}_{P_{0,-}\underline{\hat{\nu}}(n)} - \mathcal{D}_{P_{\infty_-}\underline{\hat{\mu}}(n)}, \tag{3.16}$$

using (3.4) and (3.5). Here we abbreviated

$$\underline{\hat{\mu}} = \{\hat{\mu}_1, \dots, \hat{\mu}_p\}, \ \underline{\hat{\nu}} = \{\hat{\nu}_1, \dots, \hat{\nu}_p\} \in \operatorname{Sym}^p(\mathcal{K}_p). \tag{3.17}$$

(The function ϕ is closely related to one of the variants of Weyl–Titchmarsh functions discussed in [34], [35], [46] in the special defocusing case $\beta = \bar{\alpha}$.) Given $\phi(\,\cdot\,,n)$, the meromorphic stationary Baker–Akhiezer vector $\Psi(\,\cdot\,,n,n_0)$ on $\mathcal{K}_p$ is then defined by

$$\begin{aligned} &\Psi(P,n,n_0) = \begin{pmatrix} \psi_1(P,n,n_0) \\ \psi_2(P,n,n_0) \end{pmatrix}, \\ &\psi_1(P,n,n_0) = \begin{cases} \prod_{n'=n_0+1}^{n} \big(z + \alpha(n')\phi^-(P,n')\big), & n \ge n_0+1, \\ 1, & n = n_0, \\ \prod_{n'=n+1}^{n_0} \big(z + \alpha(n')\phi^-(P,n')\big)^{-1}, & n \le n_0 - 1, \end{cases} \end{aligned} \tag{3.18}$$

$$\psi_2(P,n,n_0) = \phi(P,n_0) \begin{cases} \prod_{n'=n_0+1}^{n} \big(z\beta(n')\phi^-(P,n')^{-1}+1\big), & n \geq n_0+1, \\ 1, & n = n_0, \\ \prod_{n'=n+1}^{n_0} \big(z\beta(n')\phi^-(P,n')^{-1}+1\big)^{-1}, & n \leq n_0-1. \end{cases} \tag{3.19}$$

Basic properties of ϕ and Ψ are summarized in the following result.

Lemma 3.1. *Suppose that α, β satisfy (3.1) and the $\underline{p}$th stationary Ablowitz–Ladik system (2.45). Moreover, assume (3.2) and (3.3) and let $P = (z,y) \in \mathcal{K}_p \setminus \{P_{\infty_+}, P_{\infty_-}, P_{0,+}, P_{0,-}\}$, $(n,n_0) \in \mathbb{Z}^2$. Then ϕ satisfies the Riccati-type equation*

$$\alpha\phi(P)\phi^-(P) - \phi^-(P) + z\phi(P) = z\beta, \tag{3.20}$$

as well as

$$\phi(P)\phi(P^*) = \frac{H_{\underline{p}}(z)}{F_{\underline{p}}(z)}, \tag{3.21}$$

$$\phi(P) + \phi(P^*) = 2\frac{G_{\underline{p}}(z)}{F_{\underline{p}}(z)}, \tag{3.22}$$

$$\phi(P) - \phi(P^*) = c_{0,+}z^{-p_-}\frac{y(P)}{F_{\underline{p}}(z)}. \tag{3.23}$$

The vector Ψ satisfies

$$U(z)\Psi^-(P) = \Psi(P), \tag{3.24}$$

$$V_{\underline{p}}(z)\Psi^-(P) = -(i/2)c_{0,+}z^{-p_-}y\Psi^-(P), \tag{3.25}$$

$$\psi_2(P,n,n_0) = \phi(P,n)\psi_1(P,n,n_0), \tag{3.26}$$

$$\psi_1(P,n,n_0)\psi_1(P^*,n,n_0) = z^{n-n_0}\frac{F_{\underline{p}}(z,n)}{F_{\underline{p}}(z,n_0)}\Gamma(n,n_0), \tag{3.27}$$

$$\psi_2(P,n,n_0)\psi_2(P^*,n,n_0) = z^{n-n_0}\frac{H_{\underline{p}}(z,n)}{F_{\underline{p}}(z,n_0)}\Gamma(n,n_0), \tag{3.28}$$

$$\psi_1(P,n,n_0)\psi_2(P^*,n,n_0) + \psi_1(P^*,n,n_0)\psi_2(P,n,n_0) \tag{3.29}$$
$$= 2z^{n-n_0}\frac{G_{\underline{p}}(z,n)}{F_{\underline{p}}(z,n_0)}\Gamma(n,n_0),$$

$$\psi_1(P,n,n_0)\psi_2(P^*,n,n_0) - \psi_1(P^*,n,n_0)\psi_2(P,n,n_0) \tag{3.30}$$
$$= -c_{0,+}z^{n-n_0-p_-}\frac{y}{F_{\underline{p}}(z,n_0)}\Gamma(n,n_0),$$

where we used the abbreviation

$$\Gamma(n,n_0) = \begin{cases} \prod_{n'=n_0+1}^{n} \gamma(n'), & n \geq n_0+1, \\ 1, & n = n_0, \\ \prod_{n'=n+1}^{n_0} \gamma(n')^{-1}, & n \leq n_0-1. \end{cases} \tag{3.31}$$

Proof. To prove (3.20) one uses the definition (3.14) of ϕ and equations (2.10), (2.12), and (2.69) to obtain

$$\begin{aligned}
&\alpha\phi(P)\phi^-(P) - \phi(P)^- + z\phi(P) - z\beta \\
&= \frac{1}{F_{\underline{p}}F_{\underline{p}}^-}\Big(\alpha G_{\underline{p}}G_{\underline{p}}^- + (c_{0,+}/2)z^{-p_-}y\alpha(G_{\underline{p}} + G_{\underline{p}}^-) + \alpha R_{\underline{p}} \\
&\quad - (G_{\underline{p}}^- + (c_{0,+}/2)z^{-p_-}y)F_{\underline{p}} + z(G_{\underline{p}} + (c_{0,+}/2)z^{-p_-}y)F_{\underline{p}}^- - z\beta F_{\underline{p}}F_{\underline{p}}^-\Big) \\
&= \frac{1}{F_{\underline{p}}F_{\underline{p}}^-}\Big(\alpha G_{\underline{p}}(G_{\underline{p}} + G_{\underline{p}}^-) + F_{\underline{p}}(-\alpha H_{\underline{p}} - G_{\underline{p}}^- - z\beta F_{\underline{p}}^-) + zF_{\underline{p}}^- G_{\underline{p}}\Big) = 0. \qquad (3.32)
\end{aligned}$$

Equations (3.21)–(3.23) are clear from the definitions of ϕ and y. By definition of ψ, (3.26) holds for $n = n_0$. By induction,

$$\frac{\psi_2(P,n,n_0)}{\psi_1(P,n,n_0)} = \frac{z\beta(n)\phi^-(P,n)^{-1} + 1}{z + \alpha(n)\phi^-(P,n)}\frac{\psi_2^-(P,n,n_0)}{\psi_1^-(P,n,n_0)} = \frac{z\beta(n) + \phi^-(P,n)}{z + \alpha(n)\phi^-(P,n)}, \qquad (3.33)$$

and hence ψ_2/ψ_1 satisfies the Riccati-type equation (3.20)

$$\alpha(n)\phi^-(P,n)\frac{\psi_2(P,n,n_0)}{\psi_1(P,n,n_0)} - \phi^-(P,n) + z\frac{\psi_2(P,n,n_0)}{\psi_1(P,n,n_0)} - z\beta(n) = 0. \qquad (3.34)$$

This proves (3.26).

The definition of ψ implies

$$\begin{aligned}
\psi_1(P,n,n_0) &= (z + \alpha(n)\phi^-(P,n))\psi_1^-(P,n,n_0) \\
&= z\psi_1^-(P,n,n_0) + \alpha(n)\psi_2^-(P,n,n_0), \qquad (3.35)\\
\psi_2(P,n,n_0) &= (z\beta(n)\phi^-(P,n)^{-1} + 1)\psi_2^-(P,n,n_0) \\
&= z\beta(n)\psi_1^-(P,n,n_0) + \psi_2^-(P,n,n_0), \qquad (3.36)
\end{aligned}$$

which proves (3.24). Property (3.25) follows from (3.26) and the definition of ϕ. To prove (3.27) one can use (2.10) and (2.12)

$$\begin{aligned}
\psi_1(P)\psi_1(P^*) &= (z + \alpha\phi^-(P))(z + \alpha\phi^-(P^*))\psi_1^-(P)\psi_1^-(P^*) \\
&= \frac{1}{F_{\underline{p}}^-}(z^2F_{\underline{p}}^- + 2z\alpha G_{\underline{p}}^- + \alpha^2 H_{\underline{p}}^-)\psi_1^-(P)\psi_1^-(P^*) \\
&= \frac{1}{F_{\underline{p}}^-}(z^2F_{\underline{p}}^- - z\alpha\beta F_{\underline{p}} + z\alpha(G_{\underline{p}} + G_{\underline{p}}^-))\psi_1^-(P)\psi_1^-(P^*) \\
&= z\gamma\frac{F_{\underline{p}}}{F_{\underline{p}}^-}\psi_1^-(P)\psi_1^-(P^*). \qquad (3.37)
\end{aligned}$$

Equation (3.28) then follows from (3.22) and (3.24). Finally, equation (3.29) (resp. (3.30)) is proved by combining (3.22) and (3.26) (resp. (3.23) and (3.26)). □

Combining the Laurent polynomial recursion approach of Section 2 with (3.4) and (3.5) readily yields trace formulas for $f_{\ell,\pm}$ and $h_{\ell,\pm}$ in terms of symmetric functions of the zeros μ_j and ν_k of $(\,\cdot\,)^{p_-} F_{\underline{p}}$ and $(\,\cdot\,)^{p_- -1} H_{\underline{p}}$, respectively. For simplicity we just record the simplest cases.

Lemma 3.2. *Suppose that* α, β *satisfy* (3.1) *and the* $\underline{p}$*th stationary Ablowitz–Ladik system* (2.45)*. Then,*

$$\frac{\alpha}{\alpha^+} = (-1)^{p+1} \prod_{j=1}^{p} \mu_j \bigg(\prod_{m=0}^{2p+1} E_m \bigg)^{-1/2}, \tag{3.38}$$

$$\frac{\beta^+}{\beta} = (-1)^{p+1} \prod_{j=1}^{p} \nu_j \bigg(\prod_{m=0}^{2p+1} E_m \bigg)^{-1/2}, \tag{3.39}$$

$$\sum_{j=1}^{p} \mu_j = \alpha^+\beta - \gamma^+ \frac{\alpha^{++}}{\alpha^+} - \frac{c_{1,+}}{c_{0,+}}, \tag{3.40}$$

$$\sum_{j=1}^{p} \nu_j = \alpha^+\beta - \gamma \frac{\beta^-}{\beta} - \frac{c_{1,+}}{c_{0,+}}. \tag{3.41}$$

Proof. We compare coefficients in (2.18) and (3.4)

$$\begin{aligned} z^{p_-} F_{\underline{p}}(z) &= f_{0,-} + \cdots + z^{p_- + p_+ -2} f_{1,+} + z^{p_- + p_+ -1} f_{0,+} \\ &= c_{0,+}\alpha^+ \bigg((-1)^{p+1} \prod_{j=1}^{p} \mu_j + \cdots + z^{p_- + p_+ -2} \sum_{j=1}^{p} \mu_j - z^{p_- + p_+ -1} \bigg) \end{aligned} \tag{3.42}$$

and use $f_{0,-} = c_{0,-}\alpha$ and $f_{1,+} = c_{0,+}\big((\alpha^+)^2\beta - \gamma^+\alpha^{++}\big) - \alpha^+ c_{1,+}$ which yields (3.38) and (3.40). Similarly, one employs $h_{0,-} = -c_{0,-}\beta^+$ and $h_{1,+} = c_{0,+}\big(\gamma\beta^- - \alpha^+\beta^2\big) + \beta c_{1,+}$ for the remaining formulas (3.39) and (3.41). □

Next we turn to asymptotic properties of ϕ and Ψ in a neighborhood of $P_{\infty_\pm}$ and $P_{0,\pm}$.

Lemma 3.3. *Suppose that* α, β *satisfy* (3.1) *and the* $\underline{p}$*th stationary Ablowitz–Ladik system* (2.45)*. Moreover, let* $P = (z, y) \in \mathcal{K}_p \setminus \{P_{\infty_+}, P_{\infty_-}, P_{0,+}, P_{0,-}\}$, $(n, n_0) \in \mathbb{Z}^2$*. Then* ϕ *has the asymptotic behavior*

$$\phi(P) \underset{\zeta\to 0}{=} \begin{cases} \beta + \beta^-\gamma\zeta + O(\zeta^2), & P \to P_{\infty_+}, \\ -(\alpha^+)^{-1}\zeta^{-1} + (\alpha^+)^{-2}\alpha^{++}\gamma^+ + O(\zeta), & P \to P_{\infty_-}, \end{cases} \quad \zeta = 1/z, \tag{3.43}$$

$$\phi(P) \underset{\zeta\to 0}{=} \begin{cases} \alpha^{-1} - \alpha^{-2}\alpha^-\gamma\zeta + O(\zeta^2), & P \to P_{0,+}, \\ -\beta^+\zeta - \beta^{++}\gamma^+\zeta^2 + O(\zeta^3), & P \to P_{0,-}, \end{cases} \quad \zeta = z. \tag{3.44}$$

The components of the Baker–Akhiezer vector Ψ *have the asymptotic behavior*

$$\psi_1(P, n, n_0) \underset{\zeta\to 0}{=} \begin{cases} \zeta^{n_0 - n}(1 + O(\zeta)), & P \to P_{\infty_+}, \\ \frac{\alpha^+(n)}{\alpha^+(n_0)} \Gamma(n, n_0) + O(\zeta), & P \to P_{\infty_-}, \end{cases} \quad \zeta = 1/z, \tag{3.45}$$

$$\psi_1(P,n,n_0) \underset{\zeta\to 0}{=} \begin{cases} \frac{\alpha(n)}{\alpha(n_0)} + O(\zeta), & P\to P_{0,+}, \\ \zeta^{n-n_0}\Gamma(n,n_0)(1+O(\zeta)), & P\to P_{0,-}, \end{cases} \quad \zeta = z, \tag{3.46}$$

$$\psi_2(P,n,n_0) \underset{\zeta\to 0}{=} \begin{cases} \beta(n)\zeta^{n_0-n}(1+O(\zeta)), & P\to P_{\infty_+}, \\ -\frac{1}{\alpha^+(n_0)}\Gamma(n,n_0)\zeta^{-1}(1+O(\zeta)), & P\to P_{\infty_-}, \end{cases} \quad \zeta = 1/z, \tag{3.47}$$

$$\psi_2(P,n,n_0) \underset{\zeta\to 0}{=} \begin{cases} \frac{1}{\alpha(n_0)} + O(\zeta), & P\to P_{0,+}, \\ -\beta^+(n)\Gamma(n,n_0)\zeta^{n+1-n_0}(1+O(\zeta)), & P\to P_{0,-}, \end{cases} \quad \zeta = z. \tag{3.48}$$

The divisors (ψ_j) of ψ_j, $j = 1,2$, are given by

$$(\psi_1(\,\cdot\,,n,n_0)) = \mathcal{D}_{\underline{\hat{\mu}}(n)} - \mathcal{D}_{\underline{\hat{\mu}}(n_0)} + (n-n_0)(\mathcal{D}_{P_{0,-}} - \mathcal{D}_{P_{\infty_+}}), \tag{3.49}$$

$$(\psi_2(\,\cdot\,,n,n_0)) = \mathcal{D}_{\underline{\hat{\nu}}(n)} - \mathcal{D}_{\underline{\hat{\mu}}(n_0)} + (n-n_0)(\mathcal{D}_{P_{0,-}} - \mathcal{D}_{P_{\infty_+}}) + \mathcal{D}_{P_{0,-}} - \mathcal{D}_{P_{\infty_-}}. \tag{3.50}$$

Proof. The existence of the asymptotic expansion of ϕ in terms of the local coordinate $\zeta = 1/z$ near $P_{\infty_\pm}$, respectively, $\zeta = z$ near $P_{0,\pm}$ is clear from the explicit form of ϕ in (3.14) and (3.15). Insertion of the Laurent polynomials $F_{\underline{p}}$ into (3.14) and $H_{\underline{p}}$ into (3.15) then yields the explicit expansion coefficients in (3.43) and (3.44). Alternatively, and more efficiently, one can insert each of the following asymptotic expansions

$$\begin{aligned} \phi(P) &\underset{z\to\infty}{=} \phi_{-1}z + \phi_0 + \phi_1 z^{-1} + O(z^{-2}), \\ \phi(P^*) &\underset{z\to\infty}{=} \phi_0 + \phi_1 z^{-1} + O(z^{-2}), \\ \phi(P) &\underset{z\to 0}{=} \phi_0 + \phi_1 z + O(z^2), \\ \phi(P^*) &\underset{z\to 0}{=} \phi_1 z + \phi_2 z^2 + O(z^3) \end{aligned} \tag{3.51}$$

into the Riccati-type equation (3.20) and, upon comparing coefficients of powers of z, which determines the expansion coefficients ϕ_k in (3.51), one concludes (3.43) and (3.44).

Next we compute the divisor of ψ_1. By (3.18) it suffices to compute the divisor of $z + \alpha\phi^-(P)$. First of all we note that

$$z + \alpha\phi^-(P) = \begin{cases} z + O(1), & P\to P_{\infty_+}, \\ \frac{\alpha^+}{\alpha}\gamma + O(z^{-1}), & P\to P_{\infty_-}, \\ \frac{\alpha}{\alpha^-} + O(z), & P\to P_{0,+}, \\ \gamma z + O(z^2), & P\to P_{0,-}, \end{cases} \tag{3.52}$$

which establishes (3.45) and (3.46). Moreover, the poles of the function $z+\alpha\phi^-(P)$ in $\mathcal{K}_p\setminus\{P_{0,\pm}, P_{\infty_\pm}\}$ coincide with the ones of $\phi^-(P)$, and so it remains to compute the missing p zeros in $\mathcal{K}_p\setminus\{P_{0,\pm}, P_{\infty_\pm}\}$. Using (2.12), (2.21), (2.69), and $y(\hat{\mu}_j) =$

$(2/c_{0,+})\mu_j^{p_-}G_{\underline{p}}(\mu_j)$ (cf. (3.8)) one computes

$$\begin{aligned} z+\alpha\phi^-(P) &= z+\alpha\frac{(c_{0,+}/2)z^{-p_-}y+G_{\underline{p}}^-}{F_{\underline{p}}^-} \\ &= \frac{F_{\underline{p}}+\alpha((c_{0,+}/2)z^{-p_-}y-G_{\underline{p}})}{F_{\underline{p}}^-} \\ &= \frac{F_{\underline{p}}}{F_{\underline{p}}^-}+\alpha\frac{(c_{0,+}/2)^2z^{-2p_-}y^2-G_{\underline{p}}^2}{F_{\underline{p}}^-((c_{0,+}/2)z^{-p_-}y+G_{\underline{p}})} \\ &= \frac{F_{\underline{p}}}{F_{\underline{p}}^-}\bigg(1+\frac{\alpha H_{\underline{p}}}{(c_{0,+}/2)z^{-p_-}y+G_{\underline{p}}}\bigg) \underset{P\to\hat{\mu}_j}{=} \frac{F_{\underline{p}}(z)}{F_{\underline{p}}^-(z)}O(1). \end{aligned} \tag{3.53}$$

Hence the sought after zeros are at $\hat{\mu}_j$, $j=1,\dots,p$ (with the possibility that a zero at $\hat{\mu}_j$ is cancelled by a pole at $\hat{\mu}_j^-$).

Finally, the behavior of ψ_2 follows immediately using $\psi_2=\phi\psi_1$. □

In addition to (3.43), (3.44) one can use the Riccati-type equation (3.20) to derive a convergent expansion of ϕ around $P_{\infty\pm}$ and $P_{0,\pm}$ and recursively determine the coefficients as in Lemma 3.3. Since this is not used later in this section, we omit further details at this point.

Since nonspecial divisors play a fundamental role in the derivation of theta function representations of algebro-geometric solutions of the AL hierarchy in [31], we now take a closer look at them.

Lemma 3.4. *Suppose that α,β satisfy (3.1) and the $\underline{p}$th stationary Ablowitz–Ladik system (2.45). Moreover, assume (3.2) and (3.3) and let $n\in\mathbb{Z}$. Let $\mathcal{D}_{\underline{\hat{\mu}}}$, $\underline{\hat{\mu}}=\{\hat{\mu}_1,\dots,\hat{\mu}_p\}$, and $\mathcal{D}_{\underline{\hat{\nu}}}$, $\underline{\hat{\nu}}=\{\underline{\hat{\nu}}_1,\dots,\underline{\hat{\nu}}_p\}$, be the pole and zero divisors of degree p, respectively, associated with α, β, and ϕ defined according to (3.8) and (3.9), that is,*

$$\begin{aligned} \hat{\mu}_j(n) &= (\mu_j(n),(2/c_{0,+})\mu_j(n)^{p_-}G_{\underline{p}}(\mu_j(n),n)), \quad j=1,\dots,p, \\ \hat{\nu}_j(n) &= (\nu_j(n),-(2/c_{0,+})\nu_j(n)^{p_-}G_{\underline{p}}(\nu_j(n),n)), \quad j=1,\dots,p. \end{aligned} \tag{3.54}$$

Then $\mathcal{D}_{\underline{\hat{\mu}}(n)}$ and $\mathcal{D}_{\underline{\hat{\nu}}(n)}$ are nonspecial for all $n\in\mathbb{Z}$.

Proof. We provide a detailed proof in the case of $\mathcal{D}_{\underline{\hat{\mu}}(n)}$. By [30, Thm. A.31] (see also [29, Thm. A.30]), $\mathcal{D}_{\underline{\hat{\mu}}(n)}$ is special if and only if $\{\hat{\mu}_1(n),\dots,\hat{\mu}_p(n)\}$ contains at least one pair of the type $\{\hat{\mu}(n),\hat{\mu}(n)^*\}$. Hence $\mathcal{D}_{\underline{\hat{\mu}}(n)}$ is certainly nonspecial as long as the projections $\mu_j(n)$ of $\hat{\mu}_j(n)$ are mutually distinct, $\mu_j(n)\neq\mu_k(n)$ for $j\neq k$. On the other hand, if two or more projections coincide for some $n_0\in\mathbb{Z}$, for instance,

$$\mu_{j_1}(n_0)=\cdots=\mu_{j_N}(n_0)=\mu_0, \quad N\in\{2,\dots,p\}, \tag{3.55}$$

then $G_{\underline{p}}(\mu_0,n_0)\neq 0$ as long as $\mu_0\notin\{E_0,\dots,E_{2p+1}\}$. This fact immediately follows from (2.69) since $F_{\underline{p}}(\mu_0,n_0)=0$ but $R_{\underline{p}}(\mu_0)\neq 0$ by hypothesis. In particular,

$\hat{\mu}_{j_1}(n_0), \dots, \hat{\mu}_{j_N}(n_0)$ all meet on the same sheet since

$$\hat{\mu}_{j_r}(n_0) = (\mu_0, (2/c_{0,+})\mu_0^{p_-} G_{\underline{p}}(\mu_0, n_0)), \quad r = 1, \dots, N, \tag{3.56}$$

and hence no special divisor can arise in this manner. Remaining to be studied is the case where two or more projections collide at a branch point, say at $(E_{m_0}, 0)$ for some $n_0 \in \mathbb{Z}$. In this case one concludes $F_{\underline{p}}(z, n_0) \underset{z \to E_{m_0}}{=} O\big((z - E_{m_0})^2\big)$ and

$$G_{\underline{p}}(E_{m_0}, n_0) = 0 \tag{3.57}$$

using again (2.69) and $F_{\underline{p}}(E_{m_0}, n_0) = R_{\underline{p}}(E_{m_0}) = 0$. Since $G_{\underline{p}}(\,\cdot\,, n_0)$ is a Laurent polynomial, (3.57) implies $G_{\underline{p}}(z, n_0) \underset{z \to E_{m_0}}{=} O((z - E_{m_0}))$. Thus, using (2.69) once more, one obtains the contradiction,

$$\begin{aligned} O\big((z - E_{m_0})^2\big) &\underset{z \to E_{m_0}}{=} R_{\underline{p}}(z) \\ &\underset{z \to E_{m_0}}{=} \bigg(\frac{c_{0,+}}{2E_{m_0}^{p_-}}\bigg)^2 (z - E_{m_0}) \Bigg(\prod_{\substack{m=0 \\ m \neq m_0}}^{2p+1} (E_{m_0} - E_m) + O(z - E_{m_0}) \Bigg). \end{aligned} \tag{3.58}$$

Consequently, at most one $\hat{\mu}_j(n)$ can hit a branch point at a time and again no special divisor arises. Finally, by our hypotheses on α, β, $\hat{\mu}_j(n)$ stay finite for fixed $n \in \mathbb{Z}$ and hence never reach the points $P_{\infty_\pm}$. (Alternatively, by (3.43), $\hat{\mu}_j$ never reaches the point P_{∞_+}. Hence, if some $\hat{\mu}_j$ tend to infinity, they all necessarily converge to P_{∞_-}.) Again no special divisor can arise in this manner.

The proof for $\mathcal{D}_{\underline{\hat{\nu}}(n)}$ is analogous (replacing $F_{\underline{p}}$ by $H_{\underline{p}}$ and noticing that by (3.43), ϕ has no zeros near $P_{\infty_\pm}$), thereby completing the proof. □

The results of Sections 2 and 3 have been used extensively in [31] to derive the class of stationary algebro-geometric solutions of the Ablowitz–Ladik hierarchy and the associated theta function representations of α, β, ϕ, and Ψ. These theta function representations also show that $\gamma(n) \notin \{0, 1\}$ for all $n \in \mathbb{Z}$, and hence condition (3.1) is satisfied for the stationary algebro-geometric AL solutions discussed in this section, provided the associated divisors $\mathcal{D}_{\underline{\hat{\mu}}(n)}$ and $\mathcal{D}_{\underline{\hat{\nu}}(n)}$ stay away from $P_{\infty_\pm}, P_{0,\pm}$ for all $n \in \mathbb{Z}$.

We conclude this section with the trivial case $\underline{p} = 0$ excluded thus far.

Example 3.5. *Assume* $\underline{p} = 0$ *and* $c_{0,+} = c_{0,-} = c_0 \neq 0$ *(we recall that* $g_{p_+,+} = g_{p_-,-}$*). Then,*

$$\begin{aligned} &F_{(0,0)} = \widehat{F}_{(0,0)} = H_{(0,0)} = \widehat{H}_{(0,0)} = 0, \quad G_{(0,0)} = K_{(0,0)} = \frac{1}{2}c_0, \\ &\widehat{G}_{(0,0)} = \widehat{K}_{(0,0)} = \frac{1}{2}, \quad R_{(0,0)} = \frac{1}{4}c_0^2, \\ &\alpha = \beta = 0, \\ &U = \begin{pmatrix} z & 0 \\ 0 & 1 \end{pmatrix}, \quad V_{(0,0)} = \frac{ic_0}{2}\begin{pmatrix} 1 & 0 \\ 0 & -1 \end{pmatrix}. \end{aligned} \tag{3.59}$$

Introducing

$$\Psi_+(z,n,n_0) = \begin{pmatrix} z^{n-n_0} \\ 0 \end{pmatrix}, \quad \Psi_-(z,n,n_0) = \begin{pmatrix} 0 \\ 1 \end{pmatrix}, \quad n, n_0 \in \mathbb{Z}, \tag{3.60}$$

one verifies the equations

$$U\Psi_\pm^- = \Psi_\pm, \quad V_{(0,0)}\Psi_\pm^- = \pm\frac{ic_0}{2}\Psi_\pm^-. \tag{3.61}$$

4. The time-dependent Ablowitz–Ladik formalism

In this section we extend the algebro-geometric analysis of Section 3 to the time-dependent Ablowitz–Ladik hierarchy.

For most of this section we assume the following hypothesis.

Hypothesis 4.1.

(i) *Suppose that* α, β *satisfy*

$$\begin{aligned} &\alpha(\,\cdot\,,t), \beta(\,\cdot\,,t) \in \mathbb{C}^{\mathbb{Z}},\ t \in \mathbb{R}, \quad \alpha(n,\,\cdot\,), \beta(n,\,\cdot\,) \in C^1(\mathbb{R}),\ n \in \mathbb{Z}, \\ &\alpha(n,t)\beta(n,t) \notin \{0,1\},\ (n,t) \in \mathbb{Z}\times\mathbb{R}. \end{aligned} \tag{4.1}$$

(ii) *Assume that the hyperelliptic curve* $\mathcal{K}_p$ *satisfies* (3.2) *and* (3.3).

The fundamental problem in the analysis of algebro-geometric solutions of the Ablowitz–Ladik hierarchy consists of solving the time-dependent $\underline{r}$th Ablowitz–Ladik flow with initial data a stationary solution of the $\underline{p}$th system in the hierarchy. More precisely, given $\underline{p} \in \mathbb{N}_0^2 \setminus \{(0,0)\}$ we consider a solution $\alpha^{(0)}$, $\beta^{(0)}$ of the $\underline{p}$th stationary Ablowitz–Ladik system $\text{s-AL}_{\underline{p}}(\alpha^{(0)},\beta^{(0)}) = 0$, associated with the hyperelliptic curve $\mathcal{K}_p$ and a corresponding set of summation constants $\{c_{\ell,\pm}\}_{\ell=1,\dots,p_\pm} \subset \mathbb{C}$. Next, let $\underline{r} = (r_-, r_+) \in \mathbb{N}_0^2$; we intend to construct a solution α, β of the $\underline{r}$th Ablowitz–Ladik flow $\text{AL}_{\underline{r}}(\alpha,\beta) = 0$ with $\alpha(t_{0,\underline{r}}) = \alpha^{(0)}$, $\beta(t_{0,\underline{r}}) = \beta^{(0)}$ for some $t_{0,\underline{r}} \in \mathbb{R}$. To emphasize that the summation constants in the definitions of the stationary and the time-dependent Ablowitz–Ladik equations are independent of each other, we indicate this by adding a tilde on all the time-dependent quantities. Hence we shall employ the notation $\widetilde{V}_{\underline{r}}$, $\widetilde{F}_{\underline{r}}$, $\widetilde{G}_{\underline{r}}$, $\widetilde{H}_{\underline{r}}$, $\widetilde{K}_{\underline{r}}$, $\tilde{f}_{s,\pm}$, $\tilde{g}_{s,\pm}$, $\tilde{h}_{s,\pm}$, $\tilde{c}_{s,\pm}$, in order to distinguish them from $V_{\underline{p}}$, $F_{\underline{p}}$, $G_{\underline{p}}$, $H_{\underline{p}}$, $K_{\underline{p}}$, $f_{\ell,\pm}$, $g_{\ell,\pm}$, $h_{\ell,\pm}$, $c_{\ell,\pm}$, in the following. In addition, we will follow a more elaborate notation inspired by Hirota's τ-function approach and indicate the individual $\underline{r}$th Ablowitz–Ladik flow by a separate time variable $t_{\underline{r}} \in \mathbb{R}$.

Summing up, we are interested in solutions α, β of the time-dependent algebro-geometric initial value problem

$$\begin{aligned} &\widetilde{\text{AL}}_{\underline{r}}(\alpha,\beta) = \begin{pmatrix} -i\alpha_{t_{\underline{r}}} - \alpha(\tilde{g}_{r_+,+} + \tilde{g}_{r_-,-}^-) + \tilde{f}_{r_+-1,+} - \tilde{f}_{r_--1,-}^- \\ -i\beta_{t_{\underline{r}}} + \beta(\tilde{g}_{r_+,+}^- + \tilde{g}_{r_-,-}) - \tilde{h}_{r_--1,-} + \tilde{h}_{r_+-1,+}^- \end{pmatrix} = 0, \\ &(\alpha,\beta)\big|_{t=t_{0,\underline{r}}} = \big(\alpha^{(0)},\beta^{(0)}\big), \end{aligned} \tag{4.2}$$

$$\text{s-AL}_{\underline{p}}\left(\alpha^{(0)},\beta^{(0)}\right)=\begin{pmatrix}-\alpha^{(0)}(g_{p_+,+}+g^-_{p_-,-})+f_{p_+-1,+}-f^-_{p_--1,-}\\ \beta^{(0)}(g^-_{p_+,+}+g_{p_-,-})-h_{p_--1,-}+h^-_{p_+-1,+}\end{pmatrix}=0 \quad (4.3)$$

for some $t_{0,\underline{r}}\in\mathbb{R}$, where $\alpha=\alpha(n,t_{\underline{r}})$, $\beta=\beta(n,t_{\underline{r}})$ satisfy (4.1) and a fixed curve $\mathcal{K}_p$ is associated with the stationary solutions $\alpha^{(0)},\beta^{(0)}$ in (4.3). Here,

$$\underline{p}=(p_-,p_+)\in\mathbb{N}_0^2\setminus\{(0,0)\},\quad \underline{r}=(r_-,r_+)\in\mathbb{N}_0^2,\quad p=p_-+p_+-1. \quad (4.4)$$

In terms of the zero-curvature formulation this amounts to solving

$$U_{t_{\underline{r}}}(z,t_{\underline{r}})+U(z,t_{\underline{r}})\widetilde{V}_{\underline{r}}(z,t_{\underline{r}})-\widetilde{V}_{\underline{r}}^+(z,t_{\underline{r}})U(z,t_{\underline{r}})=0, \quad (4.5)$$

$$U(z,t_{0,\underline{r}})V_{\underline{p}}(z,t_{0,\underline{r}})-V_{\underline{p}}^+(z,t_{0,\underline{r}})U(z,t_{0,\underline{r}})=0. \quad (4.6)$$

One can show (cf. [32]) that the stationary Ablowitz–Ladik system (4.6) is actually satisfied for all times $t_{\underline{r}}\in\mathbb{R}$: Thus, we actually impose

$$U_{t_{\underline{r}}}(z,t_{\underline{r}})+U(z,t_{\underline{r}})\widetilde{V}_{\underline{r}}(z,t_{\underline{r}})-\widetilde{V}_{\underline{r}}^+(z,t_{\underline{r}})U(z,t_{\underline{r}})=0, \quad (4.7)$$

$$U(z,t_{\underline{r}})V_{\underline{p}}(z,t_{\underline{r}})-V_{\underline{p}}^+(z,t_{\underline{r}})U(z,t_{\underline{r}})=0, \quad (4.8)$$

instead of (4.5) and (4.6). For further reference, we recall the relevant quantities here (cf. (2.5), (2.6), (2.18)–(2.22)):

$$\begin{aligned}&U(z)=\begin{pmatrix}z&\alpha\\ z\beta&1\end{pmatrix},\\ &V_{\underline{p}}(z)=i\begin{pmatrix}G^-_{\underline{p}}(z)&-F^-_{\underline{p}}(z)\\ H^-_{\underline{p}}(z)&-G^-_{\underline{p}}(z)\end{pmatrix},\quad \widetilde{V}_{\underline{r}}(z)=i\begin{pmatrix}\widetilde{G}^-_{\underline{r}}(z)&-\widetilde{F}^-_{\underline{r}}(z)\\ \widetilde{H}^-_{\underline{r}}(z)&-\widetilde{K}^-_{\underline{r}}(z)\end{pmatrix},\end{aligned} \quad (4.9)$$

and

$$\begin{aligned}F_{\underline{p}}(z)&=\sum_{\ell=1}^{p_-}f_{p_--\ell,-}z^{-\ell}+\sum_{\ell=0}^{p_+-1}f_{p_+-1-\ell,+}z^{\ell}=-c_{0,+}\alpha^+z^{-p_-}\prod_{j=1}^{p}(z-\mu_j),\\ G_{\underline{p}}(z)&=\sum_{\ell=1}^{p_-}g_{p_--\ell,-}z^{-\ell}+\sum_{\ell=0}^{p_+}g_{p_+-\ell,+}z^{\ell},\\ H_{\underline{p}}(z)&=\sum_{\ell=0}^{p_--1}h_{p_--1-\ell,-}z^{-\ell}+\sum_{\ell=1}^{p_+}h_{p_+-\ell,+}z^{\ell}=c_{0,+}\beta z^{-p_-+1}\prod_{j=1}^{p}(z-\nu_j),\\ \widetilde{F}_{\underline{r}}(z)&=\sum_{s=1}^{r_-}\tilde{f}_{r_--s,-}z^{-s}+\sum_{s=0}^{r_+-1}\tilde{f}_{r_+-1-s,+}z^{s},\\ \widetilde{G}_{\underline{r}}(z)&=\sum_{s=1}^{r_-}\tilde{g}_{r_--s,-}z^{-s}+\sum_{s=0}^{r_+}\tilde{g}_{r_+-s,+}z^{s},\\ \widetilde{H}_{\underline{r}}(z)&=\sum_{s=0}^{r_--1}\tilde{h}_{r_--1-s,-}z^{-s}+\sum_{s=1}^{r_+}\tilde{h}_{r_+-s,+}z^{s},\end{aligned} \quad (4.10)$$

$$\widetilde{K}_{\underline{r}}(z) = \sum_{s=0}^{r_-} \tilde{g}_{r_- -s,-} z^{-s} + \sum_{s=1}^{r_+} \tilde{g}_{r_+ -s,+} z^{s} = \widetilde{G}_{\underline{r}}(z) + \tilde{g}_{r_-,-} - \tilde{g}_{r_+,+} \tag{4.11}$$

for fixed $\underline{p} \in \mathbb{N}_0^2 \setminus \{(0,0)\}$, $\underline{r} \in \mathbb{N}_0^2$. Here $f_{\ell,\pm}$, $\tilde{f}_{s,\pm}$, $g_{\ell,\pm}$, $\tilde{g}_{s,\pm}$, $h_{\ell,\pm}$, and $\tilde{h}_{s,\pm}$ are defined as in (2.32)–(2.39) with appropriate sets of summation constants $c_{\ell,\pm}$, $\ell \in \mathbb{N}_0$, and $\tilde{c}_{k,\pm}$, $k \in \mathbb{N}_0$. Explicitly, (4.7) and (4.8) are equivalent to (cf. (2.10)–(2.13), (2.83)–(2.86)),

$$\alpha_{t_{\underline{r}}} = i\big(z\widetilde{F}_{\underline{r}}^- + \alpha(\widetilde{G}_{\underline{r}} + \widetilde{K}_{\underline{r}}^-) - \widetilde{F}_{\underline{r}}\big), \tag{4.12}$$

$$\beta_{t_{\underline{r}}} = -i\big(\beta(\widetilde{G}_{\underline{r}}^- + \widetilde{K}_{\underline{r}}) - \widetilde{H}_{\underline{r}} + z^{-1}\widetilde{H}_{\underline{r}}^-\big), \tag{4.13}$$

$$0 = z(\widetilde{G}_{\underline{r}}^- - \widetilde{G}_{\underline{r}}) + z\beta\widetilde{F}_{\underline{r}} + \alpha\widetilde{H}_{\underline{r}}^-, \tag{4.14}$$

$$0 = z\beta\widetilde{F}_{\underline{r}}^- + \alpha\widetilde{H}_{\underline{r}} + \widetilde{K}_{\underline{r}}^- - \widetilde{K}_{\underline{r}}, \tag{4.15}$$

$$0 = z(G_{\underline{p}}^- - G_{\underline{p}}) + z\beta F_{\underline{p}} + \alpha H_{\underline{p}}^-, \tag{4.16}$$

$$0 = z\beta F_{\underline{p}}^- + \alpha H_{\underline{p}} - G_{\underline{p}} + G_{\underline{p}}^-, \tag{4.17}$$

$$0 = -F_{\underline{p}} + zF_{\underline{p}}^- + \alpha(G_{\underline{p}} + G_{\underline{p}}^-), \tag{4.18}$$

$$0 = z\beta(G_{\underline{p}} + G_{\underline{p}}^-) - zH_{\underline{p}} + H_{\underline{p}}^-, \tag{4.19}$$

respectively. In particular, (2.69) holds in the present $t_{\underline{r}}$-dependent setting, that is,

$$G_{\underline{p}}^2 - F_{\underline{p}}H_{\underline{p}} = R_{\underline{p}}. \tag{4.20}$$

As in the stationary context (3.8), (3.9) we introduce

$$\hat{\mu}_j(n,t_{\underline{r}}) = (\mu_j(n,t_{\underline{r}}), (2/c_{0,+})\mu_j(n,t_{\underline{r}})^{p_-} G_{\underline{p}}(\mu_j(n,t_{\underline{r}}),n,t_{\underline{r}})) \in \mathcal{K}_p, \\ j = 1,\dots,p,\ (n,t_{\underline{r}}) \in \mathbb{Z}\times\mathbb{R}, \tag{4.21}$$

and

$$\hat{\nu}_j(n,t_{\underline{r}}) = (\nu_j(n,t_{\underline{r}}), -(2/c_{0,+})\nu_j(n,t_{\underline{r}})^{p_-} G_{\underline{p}}(\nu_j(n,t_{\underline{r}}),n,t_{\underline{r}})) \in \mathcal{K}_p, \\ j = 1,\dots,p,\ (n,t_{\underline{r}}) \in \mathbb{Z}\times\mathbb{R}, \tag{4.22}$$

and note that the regularity assumptions (4.1) on α, β imply continuity of μ_j and ν_k with respect to $t_{\underline{r}} \in \mathbb{R}$ (away from collisions of these zeros, μ_j and ν_k are of course C^∞).

In analogy to (3.14), (3.15), one defines the following meromorphic function $\phi(\,\cdot\,,n,t_{\underline{r}})$ on $\mathcal{K}_p$,

$$\phi(P,n,t_{\underline{r}}) = \frac{(c_{0,+}/2)z^{-p_-}y + G_{\underline{p}}(z,n,t_{\underline{r}})}{F_{\underline{p}}(z,n,t_{\underline{r}})} \tag{4.23}$$

$$= \frac{-H_{\underline{p}}(z,n,t_{\underline{r}})}{(c_{0,+}/2)z^{-p_-}y - G_{\underline{p}}(z,n,t_{\underline{r}})}, \tag{4.24}$$

$$P = (z,y) \in \mathcal{K}_p,\ (n,t_{\underline{r}}) \in \mathbb{Z}\times\mathbb{R},$$

with divisor $(\phi(\,\cdot\,,n,t_{\underline{r}}))$ of $\phi(\,\cdot\,,n,t_{\underline{r}})$ given by

$$(\phi(\,\cdot\,,n,t_{\underline{r}})) = \mathcal{D}_{P_{0,-}\underline{\hat{\nu}}(n,t_{\underline{r}})} - \mathcal{D}_{P_{\infty_-}\underline{\hat{\mu}}(n,t_{\underline{r}})}. \tag{4.25}$$

The time-dependent Baker–Akhiezer vector is then defined in terms of ϕ by

$$\Psi(P,n,n_0,t_{\underline{r}},t_{0,\underline{r}}) = \begin{pmatrix} \psi_1(P,n,n_0,t_{\underline{r}},t_{0,\underline{r}}) \\ \psi_2(P,n,n_0,t_{\underline{r}},t_{0,\underline{r}}) \end{pmatrix}, \tag{4.26}$$

$$\psi_1(P,n,n_0,t_{\underline{r}},t_{0,\underline{r}}) = \exp\left(i\int_{t_{0,\underline{r}}}^{t_{\underline{r}}} ds\big(\widetilde{G}_{\underline{r}}(z,n_0,s) - \widetilde{F}_{\underline{r}}(z,n_0,s)\phi(P,n_0,s)\big)\right)$$
$$\times \begin{cases} \prod_{n'=n_0+1}^{n} \big(z+\alpha(n',t_{\underline{r}})\phi^-(P,n',t_{\underline{r}})\big), & n \geq n_0+1, \\ 1, & n = n_0, \\ \prod_{n'=n+1}^{n_0} \big(z+\alpha(n',t_{\underline{r}})\phi^-(P,n',t_{\underline{r}})\big)^{-1}, & n \leq n_0-1, \end{cases} \tag{4.27}$$

$$\psi_2(P,n,n_0,t_{\underline{r}},t_{0,\underline{r}}) = \exp\left(i\int_{t_{0,\underline{r}}}^{t_{\underline{r}}} ds\big(\widetilde{G}_{\underline{r}}(z,n_0,s) - \widetilde{F}_{\underline{r}}(z,n_0,s)\phi(P,n_0,s)\big)\right)$$
$$\times \phi(P,n_0,t_{\underline{r}}) \begin{cases} \prod_{n'=n_0+1}^{n} \big(z\beta(n',t_{\underline{r}})\phi^-(P,n',t_{\underline{r}})^{-1}+1\big), & n \geq n_0+1, \\ 1, & n = n_0, \\ \prod_{n'=n+1}^{n_0} \big(z\beta(n',t_{\underline{r}})\phi^-(P,n',t_{\underline{r}})^{-1}+1\big)^{-1}, & n \leq n_0-1, \end{cases} \tag{4.28}$$
$$P=(z,y)\in\mathcal{K}_p\setminus\{P_{\infty_+},P_{\infty_-},P_{0,+},P_{0,-}\},\ (n,t_{\underline{r}})\in\mathbb{Z}\times\mathbb{R}.$$

One observes that

$$\begin{aligned} &\psi_1(P,n,n_0,t_{\underline{r}},\tilde{t}_{\underline{r}}) = \psi_1(P,n_0,n_0,t_{\underline{r}},\tilde{t}_{\underline{r}})\psi_1(P,n,n_0,t_{\underline{r}},t_{\underline{r}}), \\ &P=(z,y)\in\mathcal{K}_p\setminus\{P_{\infty_+},P_{\infty_-},P_{0,+},P_{0,-}\},\ (n,n_0,t_{\underline{r}},\tilde{t}_{\underline{r}})\in\mathbb{Z}^2\times\mathbb{R}^2. \end{aligned} \tag{4.29}$$

The following lemma records basic properties of ϕ and Ψ in analogy to the stationary case discussed in Lemma 3.1.

Lemma 4.2. *Assume Hypothesis* 4.1 *and suppose that* (4.7), (4.8) *hold. In addition, let* $P=(z,y)\in\mathcal{K}_p\setminus\{P_{\infty_+},P_{\infty_-}\}$, $(n,n_0,t_{\underline{r}},t_{0,\underline{r}})\in\mathbb{Z}^2\times\mathbb{R}^2$. *Then* ϕ *satisfies*

$$\alpha\phi(P)\phi^-(P) - \phi^-(P) + z\phi(P) = z\beta, \tag{4.30}$$

$$\phi_{t_{\underline{r}}}(P) = i\widetilde{F}_{\underline{r}}\phi^2(P) - i\big(\widetilde{G}_{\underline{r}}(z)+\widetilde{K}_{\underline{r}}(z)\big)\phi(P) + i\widetilde{H}_{\underline{r}}(z), \tag{4.31}$$

$$\phi(P)\phi(P^*) = \frac{H_{\underline{p}}(z)}{F_{\underline{p}}(z)}, \tag{4.32}$$

$$\phi(P)+\phi(P^*) = 2\frac{G_{\underline{p}}(z)}{F_{\underline{p}}(z)}, \tag{4.33}$$

$$\phi(P)-\phi(P^*) = c_{0,+}z^{-p_-}\frac{y(P)}{F_{\underline{p}}(z)}. \tag{4.34}$$

Moreover, assuming $P=(z,y)\in\mathcal{K}_p\setminus\{P_{\infty_+},P_{\infty_-},P_{0,+},P_{0,-}\}$, then Ψ satisfies

$$\psi_2(P,n,n_0,t_{\underline{r}},t_{0,\underline{r}}) = \phi(P,n,t_{\underline{r}})\psi_1(P,n,n_0,t_{\underline{r}},t_{0,\underline{r}}), \tag{4.35}$$

$$U(z)\Psi^-(P) = \Psi(P), \tag{4.36}$$

$$V_{\underline{p}}(z)\Psi^-(P) = -(i/2)c_{0,+}z^{-p_-}y\Psi^-(P), \tag{4.37}$$

$$\Psi_{t_{\underline{r}}}(P) = \widetilde{V}_{\underline{r}}^+(z)\Psi(P), \tag{4.38}$$

$$\psi_1(P,n,n_0,t_{\underline{r}},t_{0,\underline{r}})\psi_1(P^*,n,n_0,t_{\underline{r}},t_{0,\underline{r}}) = z^{n-n_0}\frac{F_{\underline{p}}(z,n,t_{\underline{r}})}{F_{\underline{p}}(z,n_0,t_{0,\underline{r}})}\Gamma(n,n_0,t_{\underline{r}}), \tag{4.39}$$

$$\psi_2(P,n,n_0,t_{\underline{r}},t_{0,\underline{r}})\psi_2(P^*,n,n_0,t_{\underline{r}},t_{0,\underline{r}}) = z^{n-n_0}\frac{H_{\underline{p}}(z,n,t_{\underline{r}})}{F_{\underline{p}}(z,n_0,t_{0,\underline{r}})}\Gamma(n,n_0,t_{\underline{r}}), \tag{4.40}$$

$$\begin{aligned}
&\psi_1(P,n,n_0,t_{\underline{r}},t_{0,\underline{r}})\psi_2(P^*,n,n_0,t_{\underline{r}},t_{0,\underline{r}}) + \psi_1(P^*,n,n_0,t_{\underline{r}},t_{0,\underline{r}})\psi_2(P,n,n_0,t_{\underline{r}},t_{0,\underline{r}})\\
&\quad = 2z^{n-n_0}\frac{G_{\underline{p}}(z,n,t_{\underline{r}})}{F_{\underline{p}}(z,n_0,t_{0,\underline{r}})}\Gamma(n,n_0,t_{\underline{r}}),
\end{aligned} \tag{4.41}$$

$$\begin{aligned}
&\psi_1(P,n,n_0,t_{\underline{r}},t_{0,\underline{r}})\psi_2(P^*,n,n_0,t_{\underline{r}},t_{0,\underline{r}}) - \psi_1(P^*,n,n_0,t_{\underline{r}},t_{0,\underline{r}})\psi_2(P,n,n_0,t_{\underline{r}},t_{0,\underline{r}})\\
&\quad = -c_{0,+}z^{n-n_0-p_-}\frac{y}{F_{\underline{p}}(z,n_0,t_{0,\underline{r}})}\Gamma(n,n_0,t_{\underline{r}}),
\end{aligned} \tag{4.42}$$

where

$$\Gamma(n,n_0,t_{\underline{r}}) = \begin{cases} \prod_{n'=n_0+1}^{n}\gamma(n',t_{\underline{r}}), & n\geq n_0+1,\\ 1, & n=n_0,\\ \prod_{n'=n+1}^{n_0}\gamma(n',t_{\underline{r}})^{-1}, & n\leq n_0-1. \end{cases} \tag{4.43}$$

In addition, as long as the zeros $\mu_j(n_0,s)$ of $(\,\cdot\,)^{p_-}F_{\underline{p}}(\,\cdot\,,n_0,s)$ are all simple and distinct from zero for $s\in\mathcal{I}_\mu$, $\mathcal{I}_\mu\subseteq\mathbb{R}$ an open interval, $\Psi(\,\cdot\,,n,n_0,t_{\underline{r}},t_{0,\underline{r}})$ is meromorphic on $\mathcal{K}_p\setminus\{P_{\infty_+},P_{\infty_-},P_{0,+},P_{0,-}\}$ for $(n,t_{\underline{r}},t_{0,\underline{r}})\in\mathbb{Z}\times\mathcal{I}_\mu^2$.

Proof. Equations (4.30), (4.32)–(4.37), and (4.39)–(4.42) are proved as in the stationary case, see Lemma 3.1. Thus, we turn to the proof of (4.31) and (4.38): Differentiating the Riccati-type equation (4.30) yields

$$\begin{aligned}
0 &= \big(\alpha\phi\phi^- - \phi^- + z\phi - z\beta\big)_{t_{\underline{r}}}\\
&= \alpha_{t_{\underline{r}}}\phi\phi^- + (\alpha\phi^- + z)\phi_{t_{\underline{r}}} + (\alpha\phi - 1)\phi^-_{t_{\underline{r}}} - z\beta_{t_{\underline{r}}}\\
&= \big((\alpha\phi^- + z) + (\alpha\phi - 1)S^-\big)\phi_{t_{\underline{r}}} + i\phi\phi^-\big(\alpha(\widetilde{G}_{\underline{r}} + \widetilde{K}^-_{\underline{r}}) + z\widetilde{F}^-_{\underline{r}} - \widetilde{F}_{\underline{r}}\big)\\
&\quad + iz\beta(\widetilde{G}^-_{\underline{r}} + \widetilde{K}_{\underline{r}}) + i(z\widetilde{H}_{\underline{r}} - \widetilde{H}^-_{\underline{r}}),
\end{aligned} \tag{4.44}$$

using (4.12) and (4.13). Next, one employs (3.20) to rewrite

$$(\alpha\phi^- + z) + (\alpha\phi - 1)S^- = \frac{1}{\phi}(z\beta + \phi^-) + \frac{z}{\phi^-}(\beta - \phi)S^-. \tag{4.45}$$

This allows one to calculate the right-hand side of (4.31) using (4.14) and (4.15)

$$\begin{aligned}
&\big((\alpha\phi^- + z) + (\alpha\phi - 1)S^-\big)\big(\widetilde H_{\underline r} + \widetilde F_{\underline r}\phi^2 - (\widetilde G_{\underline r} + \widetilde K_{\underline r})\phi\big)\\
&= (\alpha\phi^- + z)\widetilde H_{\underline r} + (\alpha\phi - 1)\widetilde H_{\underline r}^- + \phi(z\beta + \phi^-)\widetilde F_{\underline r} + z\phi^-(\beta - \phi)\widetilde F_{\underline r}^-\\
&\quad - (z\beta + \phi^-)(\widetilde G_{\underline r} + \widetilde K_{\underline r}) - z(\beta - \phi)(\widetilde G_{\underline r}^- + \widetilde K_{\underline r}^-)\\
&= \phi\phi^-(\widetilde F_{\underline r} - z\widetilde F_{\underline r}^-) + z\widetilde H_{\underline r} - \widetilde H_{\underline r}^- + \phi^-(\alpha\widetilde H_{\underline r} + z\beta\widetilde F_{\underline r}^-) + \phi(\alpha\widetilde H_{\underline r}^- + z\beta\widetilde F_{\underline r})\\
&\quad - z\beta(\widetilde G_{\underline r} + \widetilde K_{\underline r} + \widetilde G_{\underline r}^- + \widetilde K_{\underline r}^-) - z\phi(\widetilde G_{\underline r}^- + \widetilde K_{\underline r}^-) - \phi^-(\widetilde G_{\underline r} + \widetilde K_{\underline r})\\
&= \phi\phi^-(\widetilde F_{\underline r} - z\widetilde F_{\underline r}^-) + z\widetilde H_{\underline r} - \widetilde H_{\underline r}^- - z\beta(\widetilde G_{\underline r}^- + \widetilde K_{\underline r}) + (z\phi - \phi^- - z\beta)(\widetilde G_{\underline r} + \widetilde K_{\underline r}^-)\\
&= \phi\phi^-(\widetilde F_{\underline r} - z\widetilde F_{\underline r}^-) + z\widetilde H_{\underline r} - \widetilde H_{\underline r}^- - z\beta(\widetilde G_{\underline r}^- + \widetilde K_{\underline r}) - \alpha\phi\phi^-(\widetilde G_{\underline r} + \widetilde K_{\underline r}^-).
\end{aligned} \tag{4.46}$$

Hence,

$$\Big(\frac{1}{\phi}(z\beta + \phi^-) + \frac{z}{\phi^-}(\beta - \phi)S^-\Big)\big(\phi_{t_{\underline r}} - i\widetilde H_{\underline r} - i\widetilde F_{\underline r}\phi^2 + i(\widetilde G_{\underline r} + \widetilde K_{\underline r})\phi\big) = 0. \tag{4.47}$$

Solving the first-order difference equation (4.47) then yields

$$\begin{aligned}
&\phi_{t_{\underline r}}(P,n,t_{\underline r}) - i\widetilde F_{\underline r}(z,n,t_{\underline r})\phi(P,n,t_{\underline r})^2\\
&\qquad + i(\widetilde G_{\underline r}(z,n,t_{\underline r}) + \widetilde K_{\underline r}(z,n,t_{\underline r}))\phi(P,n,t_{\underline r}) - i\widetilde H_{\underline r}(z,n,t_{\underline r})\\
&\quad = C(P,t_{\underline r})\begin{cases} \prod_{n'=1}^{n} B(P,n',t_{\underline r})/A(P,n',t_{\underline r}), & n \geq 1,\\ 1, & n = 0,\\ \prod_{n'=n+1}^{0} A(P,n',t_{\underline r})/B(P,n',t_{\underline r}), & n \leq -1 \end{cases}
\end{aligned} \tag{4.48}$$

for some n-independent function $C(\,\cdot\,,t_{\underline r})$ meromorphic on $\mathcal{K}_p$, where

$$A = \phi^{-1}(z\beta + \phi^-), \quad B = -z(\phi^-)^{-1}(\beta - \phi). \tag{4.49}$$

The asymptotic behavior of $\phi(P,n,t_{\underline r})$ in (3.43) then yields (for $t_{\underline r} \in \mathbb{R}$ fixed)

$$\frac{B(P)}{A(P)} \underset{P\to P_{\infty_+}}{=} -(1-\alpha\beta)(\beta^-)^{-1}z^{-1} + O(z^{-2}). \tag{4.50}$$

Since the left-hand side of (4.48) is of order $O(z^{r_+})$ as $P \to P_{\infty_+}$, and C is meromorphic, insertion of (4.50) into (4.48), taking $n \geq 1$ sufficiently large, then yields a contradiction unless $C = 0$. This proves (4.31).

Proving (4.38) is equivalent to showing

$$\psi_{1,t_{\underline r}} = i(\widetilde G_{\underline r} - \phi\widetilde F_{\underline r})\psi_1, \tag{4.51}$$

$$\psi_1\phi_{t_{\underline r}} + \phi\psi_{1,t_{\underline r}} = i(\widetilde H_{\underline r} - \phi\widetilde K_{\underline r})\psi_1, \tag{4.52}$$

using (4.35). Equation (4.52) follows directly from (4.51) and from (4.31),

$$\begin{aligned}
\psi_1\phi_{t_{\underline r}} + \phi\psi_{1,t_{\underline r}} &= \psi_1\big(i\widetilde H_{\underline r} + i\widetilde F_{\underline r}\phi^2 - i(\widetilde G_{\underline r} + \widetilde K_{\underline r})\phi + i(\widetilde G_{\underline r} - \phi\widetilde F_{\underline r})\phi\big)\\
&= i(\widetilde H_{\underline r} - \phi\widetilde K_{\underline r})\psi_1.
\end{aligned} \tag{4.53}$$

To prove (4.51) we start from

$$\begin{aligned}
(z+\alpha\phi^-)_{t_{\underline{r}}} &= \alpha_{t_{\underline{r}}}\phi^- + \alpha\phi^-_{t_{\underline{r}}} \\
&= \phi^- i\big(z\widetilde F^-_{\underline{r}} + \alpha(\widetilde G_{\underline{r}} + \widetilde K^-_{\underline{r}}) - \widetilde F_{\underline{r}}\big) + \alpha i\big(\widetilde H^-_{\underline{r}} + \widetilde F^-_{\underline{r}}(\phi^-)^2 - (\widetilde G^-_{\underline{r}} + \widetilde K^-_{\underline{r}})\phi^-\big) \\
&= i\alpha\phi^-(\widetilde G_{\underline{r}} - \widetilde G^-_{\underline{r}}) + i(z+\alpha\phi^-)\phi^-\widetilde F^-_{\underline{r}} - i\phi^-\widetilde F_{\underline{r}} + i\alpha\widetilde H^-_{\underline{r}} \\
&= i(z+\alpha\phi^-)\big(\widetilde G_{\underline{r}} - \phi\widetilde F_{\underline{r}} - (\widetilde G^-_{\underline{r}} - \phi^-\widetilde F^-_{\underline{r}})\big),
\end{aligned} \tag{4.54}$$

where we used (4.14) and (3.20) to rewrite

$$i\alpha\widetilde H^-_{\underline{r}} - i\phi^-\widetilde F_{\underline{r}} = iz(\widetilde G_{\underline{r}} - \widetilde G^-_{\underline{r}}) - \alpha\phi\phi^-\widetilde F_{\underline{r}} - z\phi\widetilde F_{\underline{r}}. \tag{4.55}$$

Abbreviating

$$\sigma(P,n_0,t_{\underline{r}}) = i\int_0^{t_{\underline{r}}} ds\big(\widetilde G_{\underline{r}}(z,n_0,s) - \widetilde F_{\underline{r}}(z,n_0,s)\phi(P,n_0,s)\big), \tag{4.56}$$

one computes for $n \ge n_0+1$,

$$\begin{aligned}
\psi_{1,t_{\underline{r}}} &= \bigg(\exp(\sigma)\prod_{n'=n_0+1}^{n}(z+\alpha\phi^-)(n')\bigg)_{t_{\underline{r}}} \\
&= \sigma_{t_{\underline{r}}}\psi_1 + \exp(\sigma)\sum_{n'=n_0+1}^{n}(z+\alpha\phi^-)_{t_{\underline{r}}}(n')\prod_{\substack{n''=1\\ n''\neq n'}}^{n}(z+\alpha\phi^-)(n'') \\
&= \psi_1\bigg(\sigma_{t_{\underline{r}}} + i\sum_{n'=n_0+1}^{n}\big((\widetilde G_{\underline{r}} - \widetilde F_{\underline{r}}\phi)(n') - (\widetilde G_{\underline{r}} - \widetilde F_{\underline{r}}\phi)(n'-1)\big)\bigg) \\
&= i(\widetilde G_{\underline{r}} - \widetilde F_{\underline{r}}\phi)\psi_1.
\end{aligned} \tag{4.57}$$

The case $n \le n_0$ is handled analogously establishing (4.51).

That $\Psi(\,\cdot\,,n,n_0,t_{\underline{r}},t_{0,\underline{r}})$ is meromorphic on $\mathcal{K}_p \setminus \{P_{\infty_\pm},P_{0,\pm}\}$ if $F_{\underline{p}}(\,\cdot\,,n_0,t_{\underline{r}})$ has only simple zeros distinct from zero is a consequence of (4.27), (4.28), and of

$$-i\widetilde F_{\underline{r}}(z,n_0,s)\phi(P,n_0,s) \underset{P\to\hat\mu_j(n_0,s)}{=} \partial_s \ln\big(F_{\underline{p}}(z,n_0,s)\big) + O(1), \tag{4.58}$$

using (4.21), (4.25), and (4.59). (Equation (4.59) in Lemma 4.3 follows from (4.31), (4.33), and (4.34) which have already been proven.) □

Next we consider the $t_{\underline{r}}$-dependence of $F_{\underline{p}}$, $G_{\underline{p}}$, and $H_{\underline{p}}$.

Lemma 4.3. *Assume Hypothesis* 4.1 *and suppose that* (4.7), (4.8) *hold. In addition, let* $(z,n,t_{\underline{r}}) \in \mathbb{C}\times\mathbb{Z}\times\mathbb{R}$. *Then,*

$$F_{\underline{p},t_{\underline{r}}} = -2iG_{\underline{p}}\widetilde F_{\underline{r}} + i\big(\widetilde G_{\underline{r}} + \widetilde K_{\underline{r}}\big)F_{\underline{p}}, \tag{4.59}$$

$$G_{\underline{p},t_{\underline{r}}} = iF_{\underline{p}}\widetilde H_{\underline{r}} - iH_{\underline{p}}\widetilde F_{\underline{r}}, \tag{4.60}$$

$$H_{\underline{p},t_{\underline{r}}} = 2iG_{\underline{p}}\widetilde H_{\underline{r}} - i\big(\widetilde G_{\underline{r}} + \widetilde K_{\underline{r}}\big)H_{\underline{p}}. \tag{4.61}$$

In particular, (4.59)–(4.61) *are equivalent to*

$$V_{\underline{p},t_{\underline{r}}} = \big[\widetilde{V}_{\underline{r}}, V_{\underline{p}}\big]. \tag{4.62}$$

Proof. To prove (4.59) one first differentiates equation (4.34)

$$\phi_{t_{\underline{r}}}(P) - \phi_{t_{\underline{r}}}(P^*) = -c_{0,+}z^{-p_-}yF_{\underline{p}}^{-2}F_{\underline{p},t_{\underline{r}}}. \tag{4.63}$$

The time derivative of ϕ given in (4.31) and (4.33) yield

$$\begin{aligned}\phi_{t_{\underline{r}}}(P) - \phi_{t_{\underline{r}}}(P^*) &= i\big(\widetilde{H}_{\underline{r}} + \widetilde{F}_{\underline{r}}\phi(P)^2 - \big(\widetilde{G}_{\underline{r}} + \widetilde{K}_{\underline{r}}\big)\phi(P)\big)\\ &\quad - i\big(\widetilde{H}_{\underline{r}} + \widetilde{F}_{\underline{r}}\phi(P^*)^2 - \big(\widetilde{G}_{\underline{r}} + \widetilde{K}_{\underline{r}}\big)\phi(P^*)\big)\\ &= i\widetilde{F}_{\underline{r}}(\phi(P) + \phi(P^*))(\phi(P) - \phi(P^*))\\ &\quad - i\big(\widetilde{G}_{\underline{r}} + \widetilde{K}_{\underline{r}}\big)(\phi(P) - \phi(P^*))\\ &= 2ic_{0,+}z^{-p_-}\widetilde{F}_{\underline{r}}yG_{\underline{p}}F_{\underline{p}}^{-2} - ic_{0,+}z^{-p_-}\big(\widetilde{G}_{\underline{r}} + \widetilde{K}_{\underline{r}}\big)yF_{\underline{p}}^{-1},\end{aligned} \tag{4.64}$$

and hence

$$F_{\underline{p},t_{\underline{r}}} = -2iG_{\underline{p}}\widetilde{F}_{\underline{r}} + i\big(\widetilde{G}_{\underline{r}} + \widetilde{K}_{\underline{r}}\big)F_{\underline{p}}. \tag{4.65}$$

Similarly, starting from (4.33)

$$\phi_{t_{\underline{r}}}(P) + \phi_{t_{\underline{r}}}(P^*) = 2F_{\underline{p}}^{-2}(F_{\underline{p}}G_{\underline{p},t_{\underline{r}}} - F_{\underline{p},t_{\underline{r}}}G_{\underline{p}}) \tag{4.66}$$

yields (4.60) and

$$0 = R_{\underline{p},t_{\underline{r}}} = 2G_{\underline{p}}G_{\underline{p},t_{\underline{r}}} - F_{\underline{p},t_{\underline{r}}}H_{\underline{p}} - F_{\underline{p}}H_{\underline{p},t_{\underline{r}}} \tag{4.67}$$

proves (4.61). □

Next we turn to the Dubrovin equations for the time variation of the zeros μ_j of $(\,\cdot\,)^{p_-}F_{\underline{p}}$ and ν_j of $(\,\cdot\,)^{p_- -1}H_{\underline{p}}$ governed by the $\widetilde{\mathrm{AL}}_{\underline{r}}$ flow.

Lemma 4.4. *Assume Hypothesis* 4.1 *and suppose that* (4.7), (4.8) *hold on* $\mathbb{Z}\times\mathcal{I}_\mu$ *with* $\mathcal{I}_\mu \subseteq \mathbb{R}$ *an open interval. In addition, assume that the zeros* μ_j, $j = 1,\dots,p$, *of* $(\,\cdot\,)^{p_-}F_{\underline{p}}(\,\cdot\,)$ *remain distinct and nonzero on* $\mathbb{Z}\times\mathcal{I}_\mu$. *Then* $\{\hat{\mu}_j\}_{j=1,\dots,p}$, *defined in* (4.21), *satisfies the following first-order system of differential equations on* $\mathbb{Z}\times\mathcal{I}_\mu$,

$$\mu_{j,t_{\underline{r}}} = -i\widetilde{F}_{\underline{r}}(\mu_j)y(\hat{\mu}_j)(\alpha^+)^{-1}\prod_{\substack{k=1\\k\neq j}}^{p}(\mu_j - \mu_k)^{-1}, \quad j = 1,\dots,p, \tag{4.68}$$

with

$$\hat{\mu}_j(n,\cdot) \in C^\infty(\mathcal{I}_\mu, \mathcal{K}_p), \quad j = 1,\dots,p,\ n\in\mathbb{Z}. \tag{4.69}$$

For the zeros ν_j, $j = 1,\dots,p$, *of* $(\,\cdot\,)^{p_- -1}H_{\underline{p}}(\,\cdot\,)$, *identical statements hold with* μ_j *and* $\mathcal{I}_\mu$ *replaced by* ν_j *and* $\mathcal{I}_\nu$, *etc.* (*with* $\mathcal{I}_\nu \subseteq \mathbb{R}$ *an open interval*). *In particular,* $\{\hat{\nu}_j\}_{j=1,\dots,p}$, *defined in* (4.22), *satisfies the first-order system on* $\mathbb{Z}\times I_\nu$,

$$\nu_{j,t_{\underline{r}}} = i\widetilde{H}_{\underline{r}}(\nu_j)y(\hat{\nu}_j)(\beta\nu_j)^{-1}\prod_{\substack{k=1\\k\neq j}}^{p}(\nu_j - \nu_k)^{-1}, \quad j = 1,\dots,p, \tag{4.70}$$

with

$$\hat{\nu}_j(n,\cdot) \in C^\infty(\mathcal{I}_\nu, \mathcal{K}_p), \quad j=1,\dots,p,\ n\in\mathbb{Z}. \tag{4.71}$$

Proof. It suffices to consider (4.68) for $\mu_{j,t_{\underline{r}}}$. Using the product representation for $F_{\underline{p}}$ in (4.10) and employing (4.21) and (4.59), one computes

$$\begin{aligned} F_{\underline{p},t_{\underline{r}}}(\mu_j) &= \Bigg(c_{0,+}\alpha^+\mu_j^{-p_-}\prod_{\substack{k=1\\k\neq j}}^p(\mu_j-\mu_k)\Bigg)\mu_{j,t_{\underline{r}}} = -2iG_{\underline{p}}(\mu_j)\widetilde{F}_{\underline{r}}(\mu_j) \\ &= -ic_{0,+}\mu_j^{-p_-}y(\hat{\mu}_j)\widetilde{F}_{\underline{r}}(\mu_j), \quad j=1,\dots,p, \end{aligned} \tag{4.72}$$

proving (4.68). The case of (4.70) for $\nu_{j,t_{\underline{r}}}$ is of course analogous using the product representation for $H_{\underline{p}}$ in (4.10) and employing (4.22) and (4.61). □

When attempting to solve the Dubrovin systems (4.68) and (4.70), they must be augmented with appropriate divisors $\mathcal{D}_{\underline{\hat{\mu}}(n_0,t_{0,\underline{r}})} \in \operatorname{Sym}^p\mathcal{K}_p$, $t_{0,\underline{r}}\in\mathcal{I}_\mu$, and $\mathcal{D}_{\underline{\hat{\nu}}(n_0,t_{0,\underline{r}})} \in \operatorname{Sym}^p\mathcal{K}_p$, $t_{0,\underline{r}}\in\mathcal{I}_\nu$, as initial conditions.

Since the stationary trace formulas for $f_{\ell,\pm}$ and $h_{\ell,\pm}$ in terms of symmetric functions of the zeros μ_j and ν_k of $(\,\cdot\,)^{p_-}F_{\underline{p}}$ and $(\,\cdot\,)^{p_--1}H_{\underline{p}}$ in Lemma 3.2 extend line by line to the corresponding time-dependent setting, we next record their $t_{\underline{r}}$-dependent analogs without proof. For simplicity we again confine ourselves to the simplest cases only.

Lemma 4.5. *Assume Hypothesis* 4.1 *and suppose that* (4.7), (4.8) *hold. Then,*

$$\frac{\alpha}{\alpha^+} = (-1)^{p+1}\prod_{j=1}^p\mu_j\bigg(\prod_{m=0}^{2p+1}E_m\bigg)^{-1/2}, \tag{4.73}$$

$$\frac{\beta^+}{\beta} = (-1)^{p+1}\prod_{j=1}^p\nu_j\bigg(\prod_{m=0}^{2p+1}E_m\bigg)^{-1/2}, \tag{4.74}$$

$$\sum_{j=1}^p\mu_j = \alpha^+\beta-\gamma^+\frac{\alpha^{++}}{\alpha^+}-\frac{c_{1,+}}{c_{0,+}}, \tag{4.75}$$

$$\sum_{j=1}^p\nu_j = \alpha^+\beta-\gamma\frac{\beta^-}{\beta}-\frac{c_{1,+}}{c_{0,+}}. \tag{4.76}$$

Next, we turn to the asymptotic expansions of ϕ and Ψ in a neighborhood of $P_{\infty_\pm}$ and $P_{0,\pm}$.

Lemma 4.6. *Assume Hypothesis* 4.1 *and suppose that* (4.7), (4.8) *hold. Moreover, let* $P=(z,y)\in\mathcal{K}_p\setminus\{P_{\infty_+},P_{\infty_-},P_{0,+},P_{0,-}\}$, $(n,n_0,t_{\underline{r}},t_{0,\underline{r}})\in\mathbb{Z}^2\times\mathbb{R}^2$. *Then* ϕ *has the asymptotic behavior*

$$\phi(P) \underset{\zeta\to 0}{=} \begin{cases} \beta+\beta^-\gamma\zeta+O(\zeta^2), & P\to P_{\infty_+}, \\ -(\alpha^+)^{-1}\zeta^{-1}+(\alpha^+)^{-2}\alpha^{++}\gamma^++O(\zeta), & P\to P_{\infty_-}, \end{cases} \quad \zeta=1/z, \tag{4.77}$$

$$\phi(P) \underset{\zeta\to 0}{=} \begin{cases} \alpha^{-1} - \alpha^{-2}\alpha^{-}\gamma\zeta + O(\zeta^2), & P \to P_{0,+}, \\ -\beta^{+}\zeta - \beta^{++}\gamma^{+}\zeta^2 + O(\zeta^3), & P \to P_{0,-}, \end{cases} \quad \zeta = z. \tag{4.78}$$

The component ψ_1 of the Baker–Akhiezer vector Ψ has the asymptotic behavior

$$\psi_1(P,n,n_0,t_{\underline{r}},t_{0,\underline{r}}) \underset{\zeta\to 0}{=} \exp\bigg(\pm\frac{i}{2}(t_{\underline{r}}-t_{0,\underline{r}})\sum_{s=0}^{r_+}\tilde{c}_{r_+-s,+}\zeta^{-s}\bigg)(1+O(\zeta))$$
$$\times \begin{cases} \zeta^{n_0-n}, & P\to P_{\infty_+}, \\ \Gamma(n,n_0,t_{\underline{r}})\frac{\alpha^+(n,t_{\underline{r}})}{\alpha^+(n_0,t_{0,\underline{r}})} & \\ \times\exp\Big(i\int_{t_{0,\underline{r}}}^{t_{\underline{r}}} ds\big(\tilde{g}_{r_+,+}(n_0,s)-\tilde{g}_{r_-,-}(n_0,s)\big)\Big), & P\to P_{\infty_-}, \end{cases} \quad \zeta = 1/z, \tag{4.79}$$

$$\psi_1(P,n,n_0,t_{\underline{r}},t_{0,\underline{r}}) \underset{\zeta\to 0}{=} \exp\bigg(\pm\frac{i}{2}(t_{\underline{r}}-t_{0,\underline{r}})\sum_{s=0}^{r_-}\tilde{c}_{r_--s,-}\zeta^{-s}\bigg)(1+O(\zeta))$$
$$\times \begin{cases} \frac{\alpha(n,t_{\underline{r}})}{\alpha(n_0,t_{0,\underline{r}})}, & P\to P_{0,+}, \\ \Gamma(n,n_0,t_{\underline{r}})\zeta^{n-n_0} & \\ \times\exp\Big(i\int_{t_{0,\underline{r}}}^{t_{\underline{r}}} ds\big(\tilde{g}_{r_+,+}(n_0,s)-\tilde{g}_{r_-,-}(n_0,s)\big)\Big), & P\to P_{0,-}, \end{cases} \quad \zeta = z. \tag{4.80}$$

Proof. Since by the definition of ϕ in (4.23) the time parameter $t_{\underline{r}}$ can be viewed as an additional but fixed parameter, the asymptotic behavior of ϕ remains the same as in Lemma 3.3. Similarly, also the asymptotic behavior of $\psi_1(P,n,n_0,t_{\underline{r}},t_{\underline{r}})$ is derived in an identical fashion to that in Lemma 3.3. This proves (4.79) and (4.80) for $t_{0,\underline{r}} = t_{\underline{r}}$, that is,

$$\psi_1(P,n,n_0,t_{\underline{r}},t_{\underline{r}}) \underset{\zeta\to 0}{=} \begin{cases} \zeta^{n_0-n}(1+O(\zeta)), & P\to P_{\infty_+}, \\ \Gamma(n,n_0,t_{\underline{r}})\frac{\alpha^+(n,t_{\underline{r}})}{\alpha^+(n_0,t_{\underline{r}})} + O(\zeta), & P\to P_{\infty_-}, \end{cases} \quad \zeta = 1/z, \tag{4.81}$$

$$\psi_1(P,n,n_0,t_{\underline{r}},t_{\underline{r}}) \underset{\zeta\to 0}{=} \begin{cases} \frac{\alpha(n,t_{\underline{r}})}{\alpha(n_0,t_{\underline{r}})} + O(\zeta), & P\to P_{0,+}, \\ \Gamma(n,n_0,t_{\underline{r}})\zeta^{n-n_0}(1+O(\zeta)), & P\to P_{0,-}, \end{cases} \quad \zeta = z. \tag{4.82}$$

It remains to investigate

$$\psi_1(P,n_0,n_0,t_{\underline{r}},t_{0,\underline{r}}) = \exp\bigg(i\int_{t_{0,\underline{r}}}^{t_{\underline{r}}} dt\big(\widetilde{G}_{\underline{r}}(z,n_0,t) - \widetilde{F}_{\underline{r}}(z,n_0,t)\phi(P,n_0,t)\big)\bigg). \tag{4.83}$$

The asymptotic expansion of the integrand is derived using Theorem A.2. Focusing on the homogeneous coefficients first, one computes as $P\to P_{\infty_\pm}$,

$$\widehat{G}_{s,+} - \widehat{F}_{s,+}\phi = \widehat{G}_{s,+} - \widehat{F}_{s,+}\frac{G_{\underline{p}} + (c_{0,+}/2)z^{-p_-}y}{F_{\underline{p}}}$$

$$= \widehat{G}_{s,+} - \widehat{F}_{s,+}\bigg(\frac{2z^{p_-}}{c_{0,+}}\frac{G_{\underline{p}}}{y} + 1\bigg)\bigg(\frac{2z^{p_-}}{c_{0,+}}\frac{F_{\underline{p}}}{y}\bigg)^{-1}$$

$$\underset{\zeta\to 0}{=} \pm\frac{1}{2}\zeta^{-s} + \frac{\hat{g}_{0,+} \mp \frac{1}{2}}{\hat{f}_{0,+}}\hat{f}_{s,+} + O(\zeta), \quad P \to P_{\infty_\pm},\ \zeta = 1/z. \tag{4.84}$$

Since

$$\widetilde{F}_{\underline{r}} \underset{\zeta\to 0}{=} \sum_{s=0}^{r_+} \tilde{c}_{r_+-s,+}\widehat{F}_{s,+} + O(\zeta), \quad \widetilde{G}_{\underline{r}} \underset{\zeta\to 0}{=} \sum_{s=0}^{r_+} \tilde{c}_{r_+-s,+}\widehat{G}_{s,+} + O(\zeta), \tag{4.85}$$

one infers from (4.77)

$$\widetilde{G}_{\underline{r}} - \widetilde{F}_{\underline{r}}\phi \underset{\zeta\to 0}{=} \frac{1}{2}\sum_{s=0}^{r_+} \tilde{c}_{r_+-s,+}\zeta^{-s} + O(\zeta), \quad P \to P_{\infty_+},\ \zeta = 1/z. \tag{4.86}$$

Insertion of (4.86) into (4.83) then proves (4.79) as $P \to P_{\infty_+}$.

As $P \to P_{\infty_-}$, we need one additional term in the asymptotic expansion of $\widetilde{F}_{\underline{r}}$, that is, we will use

$$\widetilde{F}_{\underline{r}} \underset{\zeta\to 0}{=} \sum_{s=0}^{r_+} \tilde{c}_{r_+-s,+}\widehat{F}_{s,+} + \sum_{s=0}^{r_-} \tilde{c}_{r_--s,-}\hat{f}_{s-1,-}\zeta + O(\zeta^2). \tag{4.87}$$

This then yields

$$\widetilde{G}_{\underline{r}} - \widetilde{F}_{\underline{r}}\phi \underset{\zeta\to 0}{=} -\frac{1}{2}\sum_{s=0}^{r_+} \tilde{c}_{r_+-s,+}\zeta^{-s} - (\alpha^+)^{-1}(\tilde{f}_{r_+,+} - \tilde{f}_{r_--1,-}) + O(\zeta). \tag{4.88}$$

Invoking (2.34) and (4.2) one concludes that

$$\tilde{f}_{r_--1,-} - \tilde{f}_{r_+,+} = -i\alpha^+_{t_{\underline{r}}} + \alpha^+(\tilde{g}_{r_+,+} - \tilde{g}_{r_-,-}) \tag{4.89}$$

and hence

$$\widetilde{G}_{\underline{r}} - \widetilde{F}_{\underline{r}}\phi \underset{\zeta\to 0}{=} -\frac{1}{2}\sum_{s=0}^{r_+} \tilde{c}_{r_+-s,+}\zeta^{-s} - \frac{i\alpha^+_{t_{\underline{r}}}}{\alpha^+} + \tilde{g}_{r_+,+} - \tilde{g}_{r_-,-} + O(\zeta). \tag{4.90}$$

Insertion of (4.90) into (4.83) then proves (4.79) as $P \to P_{\infty_-}$.

Using Theorem A.2 again, one obtains in the same manner as $P \to P_{0,\pm}$,

$$\widehat{G}_{s,-} - \widehat{F}_{s,-}\phi \underset{\zeta\to 0}{=} \pm\frac{1}{2}\zeta^{-s} - \hat{g}_{s,-} + \frac{\hat{g}_{0,-} \pm \frac{1}{2}}{\hat{f}_{0,-}}\hat{f}_{s,-} + O(\zeta). \tag{4.91}$$

Since

$$\widetilde{F}_{\underline{r}} \underset{\zeta\to 0}{=} \sum_{s=0}^{r_-} \tilde{c}_{r_--s,-}\widehat{F}_{s,-} + \tilde{f}_{r_+-1,+} + O(\zeta), \quad P \to P_{0,\pm},\ \zeta = z, \tag{4.92}$$

$$\widetilde{G}_{\underline{r}} \underset{\zeta\to 0}{=} \sum_{s=0}^{r_-} \tilde{c}_{r_--s,-}\widehat{G}_{s,-} + \tilde{g}_{r_+,+} + O(\zeta), \quad P \to P_{0,\pm},\ \zeta = z, \tag{4.93}$$

(4.91)–(4.93) yield

$$\widetilde{G}_{\underline{r}}-\widetilde{F}_{\underline{r}}\phi \underset{\zeta\to 0}{=} \pm\frac{1}{2}\sum_{s=0}^{r_-}\tilde{c}_{r_--s,-}\zeta^{-s}+\tilde{g}_{r_+,+}-\tilde{g}_{r_-,-}-\frac{\hat{g}_{0,-}\pm\frac{1}{2}}{\hat{f}_{0,-}}(\tilde{f}_{r_+-1,+}-\tilde{f}_{r_-,-})+O(\zeta), \tag{4.94}$$

where we again used (4.78), (2.52), and (4.2). As $P \to P_{0,-}$, one thus obtains

$$\widetilde{G}_{\underline{r}}-\widetilde{F}_{\underline{r}}\phi \underset{\zeta\to 0}{=} -\frac{1}{2}\sum_{s=0}^{r_-}\tilde{c}_{r_--s,-}\zeta^{-s}+\tilde{g}_{r_+,+}-\tilde{g}_{r_-,-}, \quad P\to P_{0,-},\ \zeta=z. \tag{4.95}$$

Insertion of (4.95) into (4.83) then proves (4.80) as $P \to P_{0,-}$.

As $P \to P_{0,+}$, one obtains

$$\begin{aligned}\widetilde{G}_{\underline{r}}-\widetilde{F}_{\underline{r}}\phi &\underset{\zeta\to 0}{=} \frac{1}{2}\sum_{s=0}^{r_-}\tilde{c}_{r_--s,-}\zeta^{-s}+\tilde{g}_{r_+,+}-\tilde{g}_{r_-,-}-\frac{1}{\alpha}(\tilde{f}_{r_+-1,+}-\tilde{f}_{r_-,-})+O(\zeta),\\ &\underset{\zeta\to 0}{=} \frac{1}{2}\sum_{s=0}^{r_-}\tilde{c}_{r_--s,-}\zeta^{-s}-\frac{i\alpha_{t_{\underline{r}}}}{\alpha}+O(\zeta), \quad P\to P_{0,+},\ \zeta=z,\end{aligned} \tag{4.96}$$

using $\tilde{f}_{r_-,-}=\tilde{f}^-_{r_--1,-}+\alpha(\tilde{g}_{r_-,-}-\tilde{g}^-_{r_-,-})$ (cf. (2.38)) and (4.2). Insertion of (4.96) into (4.83) then proves (4.80) as $P \to P_{0,+}$. □

Next, we note that Lemma 3.4 on nonspecial divisors in the stationary context extends to the present time-dependent situation without a change. Indeed, since $t_{\underline{r}} \in \mathbb{R}$ just plays the role of a parameter, the proof of Lemma 3.4 extends line by line and is hence omitted.

Lemma 4.7. *Assume Hypothesis 4.1 and suppose that (4.7), (4.8) hold. Moreover, let $(n,t_{\underline{r}}) \in \mathbb{Z}\times\mathbb{R}$. Denote by $\mathcal{D}_{\underline{\hat{\mu}}}$, $\underline{\hat{\mu}}=\{\hat{\mu}_1,\dots,\hat{\mu}_p\}$ and $\mathcal{D}_{\underline{\hat{\nu}}}$, $\underline{\hat{\nu}}=\{\hat{\nu}_1,\dots,\hat{\nu}_p\}$, the pole and zero divisors of degree p, respectively, associated with α, β, and ϕ defined according to (4.21) and (4.22), that is,*

$$\hat{\mu}_j(n,t_{\underline{r}})=(\mu_j(n,t_{\underline{r}}),(2/c_{0,+})\mu_j(n,t_{\underline{r}})^{p_-}G_{\underline{p}}(\mu_j(n,t_{\underline{r}}),n,t_{\underline{r}})), \quad j=1,\dots,p, \tag{4.97}$$

$$\hat{\nu}_j(n,t_{\underline{r}})=(\nu_j(n,t_{\underline{r}}),-(2/c_{0,+})\nu_j(n,t_{\underline{r}})^{p_-}G_{\underline{p}}(\nu_j(n,t_{\underline{r}}),n,t_{\underline{r}})), \quad j=1,\dots,p. \tag{4.98}$$

Then $\mathcal{D}_{\underline{\hat{\mu}}(n,t_{\underline{r}})}$ and $\mathcal{D}_{\underline{\hat{\nu}}(n,t_{\underline{r}})}$ are nonspecial for all $(n,t_{\underline{r}}) \in \mathbb{Z}\times\mathbb{R}$.

Finally, we note that

$$\begin{aligned}\Gamma(n,n_0,t_{\underline{r}})&=\Gamma(n,n_0,t_{0,\underline{r}})\\ &\times\exp\bigg(i\int_{t_{0,\underline{r}}}^{t_{\underline{r}}} ds\big(\tilde{g}_{r_+,+}(n,s)-\tilde{g}_{r_+,+}(n_0,s)-\tilde{g}_{r_-,-}(n,s)+\tilde{g}_{r_-,-}(n_0,s)\big)\bigg),\end{aligned} \tag{4.99}$$

which follows from (2.91), (3.31), and from

$$\Gamma(n,n_0,t_{\underline{r}})_{t_{\underline{r}}} = \sum_{j=n_0+1}^{n} \gamma(j,t_{\underline{r}})_{t_{\underline{r}}} \prod_{\substack{k=n_0+1\\ k\neq j}}^{n} \gamma(j,t_{\underline{r}}) \tag{4.100}$$
$$= i\big(\tilde g_{r_+,+}(n,t_{\underline{r}}) - \tilde g_{r_+,+}(n_0,t_{\underline{r}}) - \tilde g_{r_-,-}(n,t_{\underline{r}}) + \tilde g_{r_-,-}(n_0,t_{\underline{r}})\big)\Gamma(n,n_0,t_{\underline{r}})$$

after integration with respect to $t_{\underline{r}}$.

The results of Sections 2–4 have been used extensively in [31] to derive the class of time-dependent algebro-geometric solutions of the Ablowitz–Ladik hierarchy and the associated theta function representations of α, β, ϕ, and Ψ. These theta function representations also show that $\gamma(n,t_{\underline{r}}) \notin \{0,1\}$ for all $(n,t_{\underline{r}}) \in \mathbb{Z}\times\mathbb{R}$, and hence condition (4.1) is satisfied for the time-dependent algebro-geometric AL solutions discussed in this section, provided the associated divisors $\mathcal{D}_{\underline{\hat\mu}(n,t_{\underline{r}})}$ and $\mathcal{D}_{\underline{\hat\nu}(n,t_{\underline{r}})}$ stay away from $P_{\infty_\pm}, P_{0,\pm}$ for all $(n,t_{\underline{r}}) \in \mathbb{Z}\times\mathbb{R}$.

Appendix A. Asymptotic spectral parameter expansions and nonlinear recursion relations

In this appendix we consider asymptotic spectral parameter expansions of $F_{\underline{p}}/y$, $G_{\underline{p}}/y$, and $H_{\underline{p}}/y$ in a neighborhood of $P_{\infty\pm}$ and $P_{0,\pm}$, the resulting recursion relations for the homogeneous coefficients $\hat f_\ell$, $\hat g_\ell$, and $\hat h_\ell$, their connection with the nonhomogeneous coefficients f_ℓ, g_ℓ, and h_ℓ, and the connection between $c_{\ell,\pm}$ and $c_\ell(\underline{E}^{\pm1})$. We will employ the notation

$$\underline{E}^{\pm1} = \big(E_0^{\pm1},\dots,E_{2p+1}^{\pm1}\big). \tag{A.1}$$

We start with the following elementary results (consequences of the binomial expansion) assuming $\eta\in\mathbb{C}$ such that $|\eta| < \min\{|E_0|^{-1},\dots,|E_{2p+1}|^{-1}\}$:

$$\bigg(\prod_{m=0}^{2p+1}(1-E_m\eta)\bigg)^{-1/2} = \sum_{k=0}^{\infty}\hat c_k(\underline{E})\eta^k, \tag{A.2}$$

where

$$\hat c_0(\underline{E}) = 1,$$
$$\hat c_k(\underline{E}) = \sum_{\substack{j_0,\dots,j_{2p+1}=0\\ j_0+\dots+j_{2p+1}=k}}^{k} \frac{(2j_0)!\cdots(2j_{2p+1})!}{2^{2k}(j_0!)^2\cdots(j_{2p+1}!)^2} E_0^{j_0}\cdots E_{2p+1}^{j_{2p+1}}, \quad k\in\mathbb{N}. \tag{A.3}$$

The first few coefficients explicitly read

$$\hat c_0(\underline{E}) = 1,\ \hat c_1(\underline{E}) = \frac{1}{2}\sum_{m=0}^{2p+1} E_m,\ \hat c_2(\underline{E}) = \frac{1}{4}\sum_{\substack{m_1,m_2=0\\ m_1<m_2}}^{2p+1} E_{m_1}E_{m_2} + \frac{3}{8}\sum_{m=0}^{2p+1}E_m^2, \quad \text{etc.} \tag{A.4}$$

Similarly,

$$\left(\prod_{m=0}^{2p+1}(1-E_m\eta)\right)^{1/2}=\sum_{k=0}^{\infty}c_k(\underline{E})\eta^k, \tag{A.5}$$

where

$$\begin{aligned}&c_0(\underline{E})=1,\\&c_k(\underline{E})=\sum_{\substack{j_0,\dots,j_{2p+1}=0\\ j_0+\cdots+j_{2p+1}=k}}^{k}\frac{(2j_0)!\cdots(2j_{2p+1})!\,E_0^{j_0}\cdots E_{2p+1}^{j_{2p+1}}}{2^{2k}(j_0!)^2\cdots(j_{2p+1}!)^2(2j_0-1)\cdots(2j_{2p+1}-1)},\quad k\in\mathbb{N}.\end{aligned} \tag{A.6}$$

The first few coefficients explicitly are given by

$$c_0(\underline{E})=1,\ c_1(\underline{E})=-\frac{1}{2}\sum_{m=0}^{2p+1}E_m,\ c_2(\underline{E})=\frac{1}{4}\sum_{\substack{m_1,m_2=0\\ m_1<m_2}}^{2p+1}E_{m_1}E_{m_2}-\frac{1}{8}\sum_{m=0}^{2p+1}E_m^2,\quad\text{etc.} \tag{A.7}$$

Multiplying (A.2) and (A.5) and comparing coefficients of η^k one finds

$$\sum_{\ell=0}^{k}\hat{c}_{k-\ell}(\underline{E})c_\ell(\underline{E})=\delta_{k,0},\quad k\in\mathbb{N}_0. \tag{A.8}$$

Next, we turn to asymptotic expansions of various quantities in the case of the Ablowitz–Ladik hierarchy assuming $\alpha,\beta\in\mathbb{C}^{\mathbb{Z}}$, $\alpha(n)\beta(n)\notin\{0,1\}$, $n\in\mathbb{Z}$. Consider a fundamental system of solutions $\Psi_\pm(z,\cdot\,)=(\psi_{1,\pm}(z,\cdot\,),\psi_{2,\pm}(z,\cdot\,))^\top$ of $U(z)\Psi_\pm^-(z)=\Psi_\pm(z)$ for $z\in\mathbb{C}$ (or in some subdomain of $\mathbb{C}$), with U given by (2.5), such that

$$\det(\Psi_-(z),\Psi_+(z))\neq 0. \tag{A.9}$$

Introducing

$$\phi_\pm(z,n)=\frac{\psi_{2,\pm}(z,n)}{\psi_{1,\pm}(z,n)},\quad z\in\mathbb{C},\ n\in\mathbb{N}, \tag{A.10}$$

then $\phi_\pm$ satisfy the Riccati-type equation

$$\alpha\phi_\pm\phi_\pm^- - \phi_\pm^- + z\phi_\pm = z\beta, \tag{A.11}$$

and one introduces in addition,

$$\mathfrak{f}=\frac{2}{\phi_+-\phi_-}, \tag{A.12}$$

$$\mathfrak{g}=\frac{\phi_++\phi_-}{\phi_+-\phi_-}, \tag{A.13}$$

$$\mathfrak{h}=\frac{2\phi_+\phi_-}{\phi_+-\phi_-}. \tag{A.14}$$

Using the Riccati-type equation (A.11) and its consequences,

$$\alpha(\phi_+\phi_+^- - \phi_-\phi_-^-) - (\phi_+^- - \phi_-^-) + z(\phi_+ - \phi_-) = 0, \tag{A.15}$$
$$\alpha(\phi_+\phi_+^- + \phi_-\phi_-^-) - (\phi_+^- + \phi_-^-) + z(\phi_+ + \phi_-) = 2z\beta, \tag{A.16}$$

one derives the identities

$$z(\mathfrak{g}^- - \mathfrak{g}) + z\beta\mathfrak{f} + \alpha\mathfrak{h}^- = 0, \tag{A.17}$$
$$z\beta\mathfrak{f}^- + \alpha\mathfrak{h} - \mathfrak{g} + \mathfrak{g}^- = 0, \tag{A.18}$$
$$-\mathfrak{f} + z\mathfrak{f}^- + \alpha(\mathfrak{g} + \mathfrak{g}^-) = 0, \tag{A.19}$$
$$z\beta(\mathfrak{g}^- + \mathfrak{g}) - z\mathfrak{h} + \mathfrak{h}^- = 0, \tag{A.20}$$
$$\mathfrak{g}^2 - \mathfrak{f}\mathfrak{h} = 1. \tag{A.21}$$

Moreover, (A.17)–(A.20) and (A.21) also permit one to derive nonlinear difference equations for $\mathfrak{f}$, $\mathfrak{g}$, and $\mathfrak{h}$ separately, and one obtains

$$\begin{aligned}&\big((\alpha^+ + z\alpha)^2\mathfrak{f} - z(\alpha^+)^2\gamma\mathfrak{f}^-\big)^2 - 2z\alpha^2\gamma^+\big((\alpha^+ + z\alpha)^2\mathfrak{f} + z(\alpha^+)^2\gamma\mathfrak{f}^-\big)\mathfrak{f}^+\\&\quad + z^2\alpha^4(\gamma^+)^2(\mathfrak{f}^+)^2 = 4(\alpha\alpha^+)^2(\alpha^+ + \alpha z)^2,\end{aligned} \tag{A.22}$$

$$\begin{aligned}&(\alpha^+ + z\alpha)(\beta + z\beta^+)(z + \alpha^+\beta)(1 + z\alpha\beta^+)\mathfrak{g}^2\\&\quad + z(\alpha^+\gamma\mathfrak{g}^- + z\alpha\gamma^+\mathfrak{g}^+)(z\beta^+\gamma\mathfrak{g}^- + \beta\gamma^+\mathfrak{g}^+)\\&\quad - z\gamma\big((\alpha^+\beta + z^2\alpha\beta^+)(2-\gamma^+) + 2z(1-\gamma^+)(2-\gamma)\big)\mathfrak{g}^-\mathfrak{g}\\&\quad - z\gamma^+\big(2z(1-\gamma)(2-\gamma^+) + (\alpha^+\beta + z^2\alpha\beta^+)(2-\gamma)\big)\mathfrak{g}^+\mathfrak{g}\\&= (\alpha^+\beta - z^2\alpha\beta^+)^2,\end{aligned} \tag{A.23}$$

$$\begin{aligned}&z^2\big((\beta^+)^2\gamma\mathfrak{h}^- - \beta^2\gamma^+\mathfrak{h}^+\big)^2 - 2z(\beta + z\beta^+)^2\big((\beta^+)^2\gamma\mathfrak{h}^- + \beta^2\gamma^+\mathfrak{h}^+\big)\mathfrak{h}\\&\quad + (\beta + z\beta^+)^4\mathfrak{h}^2 = 4z^2(\beta\beta^+)^2(\beta + \beta^+ z)^2.\end{aligned} \tag{A.24}$$

For the precise connection between $\mathfrak{f}$, $\mathfrak{g}$, $\mathfrak{h}$ and the Green's function of the Lax difference expression underlying the AL hierarchy, we refer to [30, App. C], [33].

Next, we assume the existence of the following asymptotic expansions of $\mathfrak{f}$, $\mathfrak{g}$, and $\mathfrak{h}$ near $1/z = 0$ and $z = 0$. More precisely, near $1/z = 0$ we assume that

$$\begin{aligned}&\mathfrak{f}(z) \underset{\substack{|z|\to\infty\\ z\in C_R}}{=} -\sum_{\ell=0}^{\infty}\hat{\mathfrak{f}}_{\ell,+}z^{-\ell-1}, \quad \mathfrak{g}(z) \underset{\substack{|z|\to\infty\\ z\in C_R}}{=} -\sum_{\ell=0}^{\infty}\hat{\mathfrak{g}}_{\ell,+}z^{-\ell},\\&\mathfrak{h}(z) \underset{\substack{|z|\to\infty\\ z\in C_R}}{=} -\sum_{\ell=0}^{\infty}\hat{\mathfrak{h}}_{\ell,+}z^{-\ell},\end{aligned} \tag{A.25}$$

for z in some cone C_R with apex at $z = 0$ and some opening angle in $(0, 2\pi]$, exterior to a disk centered at $z = 0$ of sufficiently large radius $R > 0$, for some set of coefficients $\hat{\mathfrak{f}}_{\ell,+}$, $\hat{\mathfrak{g}}_{\ell,+}$, and $\hat{\mathfrak{h}}_{\ell,+}$, $\ell \in \mathbb{N}_0$. Similarly, near $z=0$ we assume that

$$\begin{aligned}
&\mathfrak{f}(z) \underset{\substack{|z|\to 0\\ z\in C_r}}{=} -\sum_{\ell=0}^{\infty} \hat{\mathfrak{f}}_{\ell,-} z^{\ell}, \quad \mathfrak{g}(z) \underset{\substack{|z|\to 0\\ z\in C_r}}{=} -\sum_{\ell=0}^{\infty} \hat{\mathfrak{g}}_{\ell,-} z^{\ell}, \\
&\mathfrak{h}(z) \underset{\substack{|z|\to 0\\ z\in C_r}}{=} -\sum_{\ell=0}^{\infty} \hat{\mathfrak{h}}_{\ell,-} z^{\ell+1},
\end{aligned} \tag{A.26}$$

for z in some cone C_r with apex at $z=0$ and some opening angle in $(0,2\pi]$, interior to a disk centered at $z=0$ of sufficiently small radius $r>0$, for some set of coefficients $\hat{\mathfrak{f}}_{\ell,-}$, $\hat{\mathfrak{g}}_{\ell,-}$, and $\hat{\mathfrak{h}}_{\ell,-}$, $\ell \in \mathbb{N}_0$. Then one can prove the following result.

Theorem A.1. *Assume $\alpha,\beta \in \mathbb{C}^{\mathbb{Z}}$, $\alpha(n)\beta(n) \notin \{0,1\}$, $n\in\mathbb{Z}$, and the existence of the asymptotic expansions* (A.25) *and* (A.26)*. Then $\mathfrak{f}$, $\mathfrak{g}$, and $\mathfrak{h}$ have the following asymptotic expansions as $|z|\to\infty$, $z\in C_R$, respectively, $|z|\to 0$, $z\in C_r$,*

$$\begin{aligned}
&\mathfrak{f}(z) \underset{\substack{|z|\to \infty\\ z\in C_R}}{=} -\sum_{\ell=0}^{\infty} \hat{f}_{\ell,+} z^{-\ell-1}, \quad \mathfrak{g}(z) \underset{\substack{|z|\to \infty\\ z\in C_R}}{=} -\sum_{\ell=0}^{\infty} \hat{g}_{\ell,+} z^{-\ell}, \\
&\mathfrak{h}(z) \underset{\substack{|z|\to \infty\\ z\in C_R}}{=} -\sum_{\ell=0}^{\infty} \hat{h}_{\ell,+} z^{-\ell},
\end{aligned} \tag{A.27}$$

and

$$\begin{aligned}
&\mathfrak{f}(z) \underset{\substack{|z|\to 0\\ z\in C_r}}{=} -\sum_{\ell=0}^{\infty} \hat{f}_{\ell,-} z^{\ell}, \quad \mathfrak{g}(z) \underset{\substack{|z|\to 0\\ z\in C_r}}{=} -\sum_{\ell=0}^{\infty} \hat{g}_{\ell,-} z^{\ell}, \\
&\mathfrak{h}(z) \underset{\substack{|z|\to 0\\ z\in C_r}}{=} -\sum_{\ell=0}^{\infty} \hat{h}_{\ell,-} z^{\ell+1},
\end{aligned} \tag{A.28}$$

where $\hat{f}_{\ell,\pm}$, $\hat{g}_{\ell,\pm}$, and $\hat{h}_{\ell,\pm}$ are the homogeneous versions of the coefficients $f_{\ell,\pm}$, $g_{\ell,\pm}$, and $h_{\ell,\pm}$ defined in (2.49)–(2.51)*. In particular, $\hat{f}_{\ell,\pm}$, $\hat{g}_{\ell,\pm}$, and $\hat{h}_{\ell,\pm}$ can be computed from the following nonlinear recursion relations*[3]

$$\begin{aligned}
&\hat{f}_{0,+} = -\alpha^+, \quad \hat{f}_{1,+} = (\alpha^+)^2\beta - \gamma^+\alpha^{++}, \\
&\hat{f}_{2,+} = -(\alpha^+)^3\beta^2 + \gamma(\alpha^+)^2\beta^- + \gamma^+\big((\alpha^{++})^2\beta^+ - \gamma^{++}\alpha^{+++} + 2\alpha^+\alpha^{++}\beta\big), \\
&\alpha^4\alpha^+\hat{f}_{\ell,+} = \frac{1}{2}\bigg((\alpha^+)^4 \sum_{m=0}^{\ell-4} \hat{f}_{m,+}\hat{f}_{\ell-m-4,+} + \alpha^4 \sum_{m=1}^{\ell-1} \hat{f}_{m,+}\hat{f}_{\ell-m,+} \\
&\qquad - 2(\alpha^+)^2 \sum_{m=0}^{\ell-3} \hat{f}_{m,+}\big(-2\alpha\alpha^+\hat{f}_{\ell-m-3,+} + (\alpha^+)^2\gamma\hat{f}^-_{\ell-m-3,+} + \alpha^2\gamma^+\hat{f}^+_{\ell-m-3,+}\big)
\end{aligned}$$

[3]We recall, a sum is interpreted as zero whenever the upper limit in the sum is strictly less than its lower limit.

$$
\begin{aligned}
&+\sum_{m=0}^{\ell-2}\big(\alpha^4(\gamma^+)^2\hat f^+_{m,+}\hat f^+_{\ell-m-2,+}+(\alpha^+)^2\gamma\hat f^-_{m,+}((\alpha^+)^2\gamma\hat f^-_{\ell-m-2,+}\\
&\qquad -2\alpha^2\gamma^+\hat f^+_{\ell-m-2,+})\\
&\quad -2\alpha\alpha^+\hat f_{m,+}(-3\alpha\alpha^+\hat f_{\ell-m-2,+}+2(\alpha^+)^2\gamma\hat f^-_{\ell-m-2,+}+2\alpha^2\gamma^+\hat f^+_{\ell-m-2,+})\big)\\
&-2\alpha^2\sum_{m=0}^{\ell-1}\hat f_{m,+}\big(-2\alpha\alpha^+\hat f_{\ell-m-1,+}+(\alpha^+)^2\gamma\hat f^-_{\ell-m-1,+}+\alpha^2\gamma^+\hat f^+_{\ell-m-1,+}\big)\Bigg),\\
&\hfill \ell\geq 3, \quad \text{(A.29)}
\end{aligned}
$$

$$
\begin{aligned}
&\hat f_{0,-}=\alpha, \quad \hat f_{1,-}=\gamma\alpha^- -\alpha^2\beta^+,\\
&\hat f_{2,-}=\alpha^3(\beta^+)^2-\gamma^+\alpha^2\beta^{++}-\gamma\big((\alpha^-)^2\beta-\gamma^-\alpha^{--}+2\alpha^-\alpha\beta^+\big),\\
&\alpha(\alpha^+)^4\hat f_{\ell,-}=-\frac{1}{2}\Bigg(\alpha^4\sum_{m=0}^{\ell-4}\hat f_{m,-}\hat f_{\ell-m-4,-}+(\alpha^+)^4\sum_{m=1}^{\ell-1}\hat f_{m,-}\hat f_{\ell-m,-}\\
&-2\alpha^2\sum_{m=0}^{\ell-3}\hat f_{m,-}\big(-2\alpha\alpha^+\hat f_{\ell-m-3,-}+(\alpha^+)^2\gamma\hat f^-_{\ell-m-3,-}+\alpha^2\gamma^+\hat f^+_{\ell-m-3,-}\big)\\
&+\sum_{m=0}^{\ell-2}\big(\alpha^4(\gamma^+)^2\hat f^+_{m,-}\hat f^+_{\ell-m-2,-}\\
&\qquad +(\alpha^+)^2\gamma\hat f^-_{m,-}((\alpha^+)^2\gamma\hat f^-_{\ell-m-2,-}-2\alpha^2\gamma^+\hat f^+_{\ell-m-2,-})\\
&\quad -2\alpha\alpha^+\hat f_{m,-}(-3\alpha\alpha^+\hat f_{\ell-m-2,-}+2(\alpha^+)^2\gamma\hat f^-_{\ell-m-2,-}+2\alpha^2\gamma^+\hat f^+_{\ell-m-2,-})\big)\\
&-2(\alpha^+)^2\sum_{m=0}^{\ell-1}\hat f_{m,-}\big(-2\alpha\alpha^+\hat f_{\ell-m-1,-}+(\alpha^+)^2\gamma\hat f^-_{\ell-m-1,-}+\alpha^2\gamma^+\hat f^+_{\ell-m-1,-}\big)\Bigg),\\
&\hfill \ell\geq 3, \quad \text{(A.30)}
\end{aligned}
$$

$$
\begin{aligned}
&\hat g_{0,+}=\frac{1}{2}, \quad \hat g_{1,+}=-\alpha^+\beta,\\
&\hat g_{2,+}=(\alpha^+\beta)^2-\gamma^+\alpha^{++}\beta-\gamma\alpha^+\beta^+,\\
&(\alpha\beta^+)^2\hat g_{\ell,+}=-\Bigg((\alpha^+)^2\beta^2\sum_{m=0}^{\ell-4}\hat g_{m,+}\hat g_{\ell-m-4,+}+\alpha^2(\beta^+)^2\sum_{m=1}^{\ell-1}\hat g_{m,+}\hat g_{\ell-m,+}\\
&+\alpha^+\beta\sum_{m=0}^{\ell-3}\big(\gamma\gamma^+\hat g^-_{m,+}\hat g^+_{\ell-m-3,+}+\hat g_{m,+}((1+\alpha\beta)(1+\alpha^+\beta^+)\hat g_{\ell-m-3,+}\\
&\qquad -(\gamma+\alpha^+\beta^+\gamma)\hat g^-_{\ell-m-3,+}+(-2+\gamma)\gamma^+\hat g^+_{\ell-m-3,+})\big)\\
&+\sum_{m=0}^{\ell-2}\big(\alpha^+\beta^+\gamma^2\hat g^-_{m,+}\hat g^-_{\ell-m-2,+}+\alpha\beta(\gamma^+)^2\hat g^+_{m,+}\hat g^+_{\ell-m-2,+}
\end{aligned}
$$

$$
\begin{aligned}
&\qquad + \hat g_{m,+}((\alpha^+\beta^+ + \alpha^2\alpha^+\beta^2\beta^+ + \alpha\beta(1+\alpha^+\beta^+)^2)\hat g_{\ell-m-2,+}\\
&\qquad - 2(\alpha^+(1+\alpha\beta)\beta^+\gamma\hat g^-_{\ell-m-2,+} + \alpha\beta(1+\alpha^+\beta^+)\gamma^+\hat g^+_{\ell-m-2,+}))\\
&+ \alpha\beta^+\sum_{m=0}^{\ell-1}(\gamma\gamma^+\hat g^-_{m,+}\hat g^+_{\ell-m-1,+} + \hat g_{m,+}((1+\alpha\beta)(1+\alpha^+\beta^+)\hat g_{\ell-m-1,+}\\
&\qquad - (\gamma+\alpha^+\beta^+\gamma)\hat g^-_{\ell-m-1,+} + (-2+\gamma)\gamma^+\hat g^+_{\ell-m-1,+}))\Big), \quad \ell\geq 3,
\end{aligned} \tag{A.31}
$$

$$
\begin{aligned}
&\hat g_{0,-} = \frac{1}{2}, \quad \hat g_{1,-} = -\alpha\beta^+,\\
&\hat g_{2,-} = (\alpha\beta^+)^2 - \gamma^+\alpha\beta^{++} - \gamma\alpha^-\beta^+,\\
&(\alpha^+)^2\beta^2\hat g_{\ell,-} = -\Big(\alpha^2(\beta^+)^2\sum_{m=0}^{\ell-4}\hat g_{m,-}\hat g_{\ell-m-4,-} + (\alpha^+)^2\beta^2\sum_{m=1}^{\ell-1}\hat g_{m,-}\hat g_{\ell-m,-}\\
&\quad + \alpha\beta^+\sum_{m=0}^{\ell-3}(\gamma\gamma^+\hat g^-_{m,-}\hat g^+_{\ell-m-3,-} + \hat g_{m,-}((1+\alpha\beta)(1+\alpha^+\beta^+)\hat g_{\ell-m-3,-}\\
&\qquad - (\gamma+\alpha^+\beta^+\gamma)\hat g^-_{\ell-m-3,-} + (-2+\gamma)\gamma^+\hat g^+_{\ell-m-3,-}))\\
&\quad + \sum_{m=0}^{\ell-2}(\alpha^+\beta^+\gamma^2\hat g^-_{m,-}\hat g^-_{\ell-m-2,-} + \alpha\beta(\gamma^+)^2\hat g^+_{m,-}\hat g^+_{\ell-m-2,-}\\
&\qquad + \hat g_{m,-}((\alpha^+\beta^+ + \alpha^2\alpha^+\beta^2\beta^+ + \alpha\beta(1+\alpha^+\beta^+)^2)\hat g_{\ell-m-2,-}\\
&\qquad - 2(\alpha^+(1+\alpha\beta)\beta^+\gamma\hat g^-_{\ell-m-2,-} + \alpha\beta(1+\alpha^+\beta^+)\gamma^+\hat g^+_{\ell-m-2,-})))\\
&\quad + \alpha^+\beta\sum_{m=0}^{\ell-1}(\gamma\gamma^+\hat g^-_{m,-}\hat g^+_{\ell-m-1,-} + \hat g_{m,-}((1+\alpha\beta)(1+\alpha^+\beta^+)\hat g_{\ell-m-1,-}\\
&\qquad - (\gamma+\alpha^+\beta^+\gamma)\hat g^-_{\ell-m-1,-} + (-2+\gamma)\gamma^+\hat g^+_{\ell-m-1,-}))\Big), \quad \ell\geq 3,
\end{aligned} \tag{A.32}
$$

$$
\begin{aligned}
&\hat h_{0,+} = \beta, \quad \hat h_{1,+} = \gamma\beta^- - \alpha^+\beta^2,\\
&\hat h_{2,+} = (\alpha^+)^2\beta^3 - \gamma^+\alpha^{++}\beta^2 - \gamma\big(\alpha(\beta^-)^2 - \gamma^-\beta^{--} + 2\alpha^+\beta^-\beta\big),\\
&\beta(\beta^+)^4\hat h_{\ell,+} = -\frac{1}{2}\Big(\beta^4\sum_{m=0}^{\ell-4}\hat h_{m,+}\hat h_{\ell-m-4,+} + (\beta^+)^4\sum_{m=1}^{\ell-1}\hat h_{m,+}\hat h_{\ell-m,+}\\
&\quad - 2\beta^2\sum_{m=0}^{\ell-3}\hat h_{m,+}\big(-2\beta\beta^+\hat h_{\ell-m-3,+} + (\beta^+)^2\gamma\hat h^-_{\ell-m-3,+} + \beta^2\gamma^+\hat h^+_{\ell-m-3,+}\big)\\
&\quad + \sum_{m=0}^{\ell-2}(\beta^4(\gamma^+)^2\hat h^+_{m,+}\hat h^+_{\ell-m-2,+}
\end{aligned}
$$

$$
\begin{aligned}
&\quad + (\beta^+)^2\gamma\hat h^-_{m,+}((\beta^+)^2\gamma\hat h^-_{\ell-m-2,+} - 2\beta^2\gamma^+\hat h^+_{\ell-m-2,+})\\
&\quad - 2\beta\beta^+\hat h_{m,+}(-3\beta\beta^+\hat h_{\ell-m-2,+} + 2(\beta^+)^2\gamma\hat h^-_{\ell-m-2,+} + 2\beta^2\gamma^+\hat h^+_{\ell-m-2,+}))\\
&- 2(\beta^+)^2\sum_{m=0}^{\ell-1}\hat h_{m,+}\big(-2\beta\beta^+\hat h_{\ell-m-1,+} + (\beta^+)^2\gamma\hat h^-_{\ell-m-1,+} + \beta^2\gamma^+\hat h^+_{\ell-m-1,+}\big)\Big),\\
&\hspace{22em} \ell\geq 3, \quad \text{(A.33)}
\end{aligned}
$$

$$
\begin{aligned}
&\hat h_{0,-} = -\beta^+, \quad \hat h_{1,-} = -\gamma^+\beta^{++} + \alpha(\beta^+)^2,\\
&\hat h_{2,-} = -\alpha^2(\beta^+)^3 + \gamma\alpha^-(\beta^+)^2 + \gamma\big(\alpha^+(\beta^{++})^2 - \gamma^{++}\beta^{+++} + 2\alpha\beta^+\beta^{++}\big),\\
&\beta^+\beta^4\hat h_{\ell,-} = \frac{1}{2}\Big((\beta^+)^4\sum_{m=0}^{\ell-4}\hat h_{m,-}\hat h_{\ell-m-4,-} + \beta^4\sum_{m=1}^{\ell-1}\hat h_{m,-}\hat h_{\ell-m,-}\\
&\quad - 2(\beta^+)^2\sum_{m=0}^{\ell-3}\hat h_{m,-}\big(-2\beta\beta^+\hat h_{\ell-m-3,-} + (\beta^+)^2\gamma\hat h^-_{\ell-m-3,-} + \beta^2\gamma^+\,\hat h^+_{\ell-m-3,-}\big)\\
&\quad + \sum_{m=0}^{\ell-2}\big(\beta^4(\gamma^+)^2\hat h^+_{m,-}\hat h^+_{\ell-m-2,-}\\
&\qquad + (\beta^+)^2\gamma\hat h^-_{m,-}((\beta^+)^2\gamma\hat h^-_{\ell-m-2,-} - 2\beta^2\gamma^+\hat h^+_{\ell-m-2,-})\\
&\qquad - 2\beta\beta^+\hat h_{m,-}(-3\beta\beta^+\,\hat h_{\ell-m-2,-} + 2(\beta^+)^2\gamma\hat h^-_{\ell-m-2,-} + 2\beta^2\gamma^+\hat h^+_{\ell-m-2,-}))\\
&\quad - 2\beta^2\sum_{m=0}^{\ell-1}\hat h_{m,-}\big(-2\beta\beta^+\hat h_{\ell-m-1,-} + (\beta^+)^2\gamma\hat h^-_{\ell-m-1,-} + \beta^2\gamma^+\hat h^+_{\ell-m-1,-}\big)\Big),\\
&\hspace{22em} \ell\geq 3. \quad \text{(A.34)}
\end{aligned}
$$

Proof. We first consider the expansions (A.27) near $1/z = 0$ and the nonlinear recursion relations (A.29), (A.31), and (A.33) in detail. Inserting expansion (A.25) for $\mathfrak{f}$ into (A.22), the expansion (A.25) for $\mathfrak{g}$ into (A.23), and the expansion (A.25) for $\mathfrak{h}$ into (A.24), then yields the nonlinear recursion relations (A.29), (A.31), and (A.33), but with $\hat f_{\ell,+}$, $\hat g_{\ell,+}$, and $\hat h_{\ell,+}$ replaced by $\hat{\mathfrak{f}}_{\ell,+}$, $\hat{\mathfrak{g}}_{\ell,+}$, and $\hat{\mathfrak{h}}_{\ell,+}$, respectively. From the leading asymptotic behavior one finds that $\hat{\mathfrak{f}}_{0,+} = -\alpha^+$, $\hat{\mathfrak{g}}_{0,+} = \frac{1}{2}$, and $\hat{\mathfrak{h}}_{0,+} = \beta$.

Next, inserting the expansions (A.25) for $\mathfrak{f}$, $\mathfrak{g}$, and $\mathfrak{h}$ into (A.17)–(A.20), and comparing powers of $z^{-\ell}$ as $|z|\to\infty$, $z\in C_R$, one infers that $\mathfrak{f}_{\ell,+}$, $\mathfrak{g}_{\ell,+}$, and $\mathfrak{h}_{\ell,+}$ satisfy the linear recursion relations (2.32)–(2.35). Here we have used (2.21). The coefficients $\hat{\mathfrak{f}}_{0,+}$, $\hat{\mathfrak{g}}_{0,+}$, and $\hat{\mathfrak{h}}_{0,+}$ are consistent with (2.32) for $c_{0,+} = 1$. Hence one concludes that

$$
\hat{\mathfrak{f}}_{\ell,+} = f_{\ell,+}, \quad \hat{\mathfrak{g}}_{\ell,+} = g_{\ell,+}, \quad \hat{\mathfrak{h}}_{\ell,+} = h_{\ell,+}, \quad \ell\in\mathbb{N}_0, \tag{A.35}
$$

for certain values of the summation constants $c_{\ell,+}$. To conclude that actually, $\hat{\mathfrak{f}}_{\ell,+} = \hat f_{\ell,+}$, $\hat{\mathfrak{g}}_{\ell,+} = \hat g_{\ell,+}$, $\hat{\mathfrak{h}}_{\ell,+} = \hat h_{\ell,+}$, $\ell\in\mathbb{N}_0$, and hence all $c_{\ell,+}$, $\ell\in\mathbb{N}$, vanish, we now rely on the notion of degree as introduced in Remark 2.6. To this end we recall that

$$\deg\big(\hat f_{\ell,+}\big) = \ell + 1, \quad \deg\big(\hat g_{\ell,+}\big) = \ell, \quad \deg\big(\hat h_{\ell,+}\big) = \ell, \quad \ell \in \mathbb{N}_0, \tag{A.36}$$

(cf. (2.55)). Similarly, the nonlinear recursion relations (A.29), (A.31), and (A.33) yield inductively that

$$\deg\big(\hat{\mathfrak{f}}_{\ell,+}\big) = \ell + 1, \quad \deg\big(\hat{\mathfrak{g}}_{\ell,+}\big) = \ell, \quad \deg\big(\hat{\mathfrak{h}}_{\ell,+}\big) = \ell, \quad \ell \in \mathbb{N}_0. \tag{A.37}$$

Hence one concludes

$$\hat{\mathfrak{f}}_{\ell,+} = \hat f_{\ell,+}, \quad \hat{\mathfrak{g}}_{\ell,+} = \hat g_{\ell,+}, \quad \hat{\mathfrak{h}}_{\ell,+} = \hat h_{\ell,+}, \quad \ell \in \mathbb{N}_0. \tag{A.38}$$

The proof of the corresponding asymptotic expansion (A.28) and the nonlinear recursion relations (A.30), (A.32), and (A.34) follows precisely the same strategy and is hence omitted. □

Given this general result on asymptotic expansions, we now specialize to the algebro-geometric case at hand. We recall our conventions $y(P) = \mp(\zeta^{-p-1} + O(\zeta^{-p}))$ for P near $P_{\infty_\pm}$ (where $\zeta = 1/z$) and $y(P) = \pm((c_{0,-}/c_{0,+}) + O(\zeta))$ for P near $P_{0,\pm}$ (where $\zeta = z$).

Theorem A.2. *Assume* (3.1), $\text{s-AL}_{\underline{p}}(\alpha,\beta) = 0$, *and suppose* $P = (z,y) \in \mathcal{K}_p \setminus \{P_{\infty_+}, P_{\infty_-}\}$. *Then* $z^{p_-} F_{\underline{p}}/y$, $z^{p_-} G_{\underline{p}}/y$, *and* $z^{p_-} H_{\underline{p}}/y$ *have the following convergent expansions as* $P \to P_{\infty_\pm}$, *respectively,* $P \to P_{0,\pm}$,

$$\frac{z^{p_-}}{c_{0,+}} \frac{F_{\underline{p}}(z)}{y} = \begin{cases} \mp \sum_{\ell=0}^\infty \hat f_{\ell,+} \zeta^{\ell+1}, & P \to P_{\infty_\pm}, \quad \zeta = 1/z, \\ \pm \sum_{\ell=0}^\infty \hat f_{\ell,-} \zeta^{\ell}, & P \to P_{0,\pm}, \quad \zeta = z, \end{cases} \tag{A.39}$$

$$\frac{z^{p_-}}{c_{0,+}} \frac{G_{\underline{p}}(z)}{y} = \begin{cases} \mp \sum_{\ell=0}^\infty \hat g_{\ell,+} \zeta^{\ell}, & P \to P_{\infty_\pm}, \quad \zeta = 1/z, \\ \pm \sum_{\ell=0}^\infty \hat g_{\ell,-} \zeta^{\ell}, & P \to P_{0,\pm}, \quad \zeta = z, \end{cases} \tag{A.40}$$

$$\frac{z^{p_-}}{c_{0,+}} \frac{H_{\underline{p}}(z)}{y} = \begin{cases} \mp \sum_{\ell=0}^\infty \hat h_{\ell,+} \zeta^{\ell}, & P \to P_{\infty_\pm}, \quad \zeta = 1/z, \\ \pm \sum_{\ell=0}^\infty \hat h_{\ell,-} \zeta^{\ell+1}, & P \to P_{0,\pm}, \quad \zeta = z, \end{cases} \tag{A.41}$$

where $\zeta = 1/z$ *(resp.,* $\zeta = z$*) is the local coordinate near* $P_{\infty_\pm}$ *(resp.,* $P_{0,\pm}$*) and* $\hat f_{\ell,\pm}$, $\hat g_{\ell,\pm}$, *and* $\hat h_{\ell,\pm}$ *are the homogeneous versions*[4] *of the coefficients* $f_{\ell,\pm}$, $g_{\ell,\pm}$, *and* $h_{\ell,\pm}$ *as introduced in* (2.49)–(2.51). *Moreover, one infers for the* E_m*-dependent summation constants* $c_{\ell,\pm}$, $\ell = 0,\dots,p_\pm$, *in* $F_{\underline{p}}$, $G_{\underline{p}}$, *and* $H_{\underline{p}}$ *that*

$$c_{\ell,\pm} = c_{0,\pm} c_\ell\big(\underline{E}^{\pm 1}\big), \quad \ell = 0,\dots,p_\pm. \tag{A.42}$$

In addition, one has the following relations between the homogeneous and nonhomogeneous recursion coefficients:

$$f_{\ell,\pm} = c_{0,\pm} \sum_{k=0}^{\ell} c_{\ell-k}\big(\underline{E}^{\pm 1}\big) \hat f_{k,\pm}, \quad \ell = 0,\dots,p_\pm, \tag{A.43}$$

[4]Strictly speaking, the coefficients $\hat f_{\ell,\pm}$, $\hat g_{\ell,\pm}$, and $\hat h_{\ell,\pm}$ in (A.39)–(A.41) no longer have a well-defined degree and hence represent a slight abuse of notation since we assumed that $\text{s-AL}_{\underline{p}}(\alpha,\beta) = 0$. At any rate, they are explicitly given by (A.49)–(A.51).

$$g_{\ell,\pm} = c_{0,\pm} \sum_{k=0}^{\ell} c_{\ell-k}\big(\underline{E}^{\pm 1}\big)\hat{g}_{k,\pm}, \quad \ell = 0,\dots,p_\pm, \tag{A.44}$$

$$h_{\ell,\pm} = c_{0,\pm} \sum_{k=0}^{\ell} c_{\ell-k}\big(\underline{E}^{\pm 1}\big)h_{k,\pm}, \quad \ell = 0,\dots,p_\pm. \tag{A.45}$$

Furthermore, one has

$$\begin{aligned} c_{0,\pm}\hat{f}_{\ell,\pm} &= \sum_{k=0}^{\ell} \hat{c}_{\ell-k}\big(\underline{E}^{\pm 1}\big) f_{k,\pm}, \quad \ell = 0,\dots,p_\pm - 1, \\ c_{0,\pm}\hat{f}_{p_\pm,\pm} &= \sum_{k=0}^{p_\pm - 1} \hat{c}_{p_\pm - k}\big(\underline{E}^{\pm 1}\big) f_{k,\pm} + \hat{c}_0(\underline{E}^{\pm 1}) f_{p_\mp - 1,\mp}, \end{aligned} \tag{A.46}$$

$$\begin{aligned} c_{0,\pm}\hat{g}_{\ell,\pm} &= \sum_{k=0}^{\ell} \hat{c}_{\ell-k}\big(\underline{E}^{\pm 1}\big) g_{k,\pm}, \quad \ell = 0,\dots,p_\pm - 1, \\ c_{0,\pm}\hat{g}_{p_\pm,\pm} &= \sum_{k=0}^{p_\pm - 1} \hat{c}_{p_\pm - k}\big(\underline{E}^{\pm 1}\big) g_{k,\pm} + \hat{c}_0(\underline{E}^{\pm 1}) g_{p_\mp,\mp}, \end{aligned} \tag{A.47}$$

$$\begin{aligned} c_{0,\pm}\hat{h}_{\ell,\pm} &= \sum_{k=0}^{\ell} \hat{c}_{\ell-k}\big(\underline{E}^{\pm 1}\big) h_{k,\pm}, \quad \ell = 0,\dots,p_\pm - 1, \\ c_{0,\pm}\hat{h}_{p_\pm,\pm} &= \sum_{k=0}^{p_\pm - 1} \hat{c}_{p_\pm - k}\big(\underline{E}^{\pm 1}\big) h_{k,\pm} + \hat{c}_0(\underline{E}^{\pm 1}) h_{p_\mp - 1,\mp}. \end{aligned} \tag{A.48}$$

For general ℓ (not restricted to $\ell \le p_\pm$) one has[5]

$$c_{0,\pm}\hat{f}_{\ell,\pm} = \begin{cases} \sum_{k=0}^{\ell} \hat{c}_{\ell-k}\big(\underline{E}^{\pm 1}\big) f_{k,\pm}, & \ell = 0,\dots,p_\pm - 1, \\ \sum_{k=0}^{p_\pm - 1} \hat{c}_{\ell-k}\big(\underline{E}^{\pm 1}\big) f_{k,\pm} & \\ \quad + \sum_{k=(p-\ell)\vee 0}^{p_\mp - 1} \hat{c}_{\ell+k-p}\big(\underline{E}^{\pm 1}\big) f_{k,\mp}, & \ell \ge p_\pm, \end{cases} \tag{A.49}$$

$$c_{0,\pm}\hat{g}_{\ell,\pm} = \begin{cases} \sum_{k=0}^{\ell} \hat{c}_{\ell-k}\big(\underline{E}^{\pm 1}\big) g_{k,\pm}, & \ell = 0,\dots,p_\pm - \delta_\pm, \\ \sum_{k=0}^{p_\pm - \delta_\pm} \hat{c}_{\ell-k}\big(\underline{E}^{\pm 1}\big) g_{k,\pm} & \\ \quad + \sum_{k=(p-\ell)\vee 0}^{p_\mp - \delta_\pm} \hat{c}_{\ell+k-p}\big(\underline{E}^{\pm 1}\big) g_{k,\mp}, & \ell \ge p_\pm - \delta_\pm + 1, \end{cases} \tag{A.50}$$

$$c_{0,\pm}\hat{h}_{\ell,\pm} = \begin{cases} \sum_{k=0}^{\ell} \hat{c}_{\ell-k}\big(\underline{E}^{\pm 1}\big) h_{k,\pm}, & \ell = 0,\dots,p_\pm - 1, \\ \sum_{k=0}^{p_\pm - 1} \hat{c}_{\ell-k}\big(\underline{E}^{\pm 1}\big) h_{k,\pm} & \\ \quad + \sum_{k=(p-\ell)\vee 0}^{p_\mp - 1} \hat{c}_{\ell+k-p}\big(\underline{E}^{\pm 1}\big) h_{k,\mp}, & \ell \ge p_\pm. \end{cases} \tag{A.51}$$

[5] $m \vee n = \max\{m, n\}$.

Here we used the convention

$$\delta_\pm = \begin{cases} 0, & +, \\ 1, & -. \end{cases} \tag{A.52}$$

Proof. Identifying

$$\Psi_+(z,\,\cdot\,) \text{ with } \Psi(P,\,\cdot\,,0) \text{ and } \Psi_-(z,\,\cdot\,) \text{ with } \Psi(P^*,\,\cdot\,,0), \tag{A.53}$$

recalling that $W(\Psi(P,\,\cdot\,,0),\Psi(P^*,\,\cdot\,,0)) = -c_{0,+}z^{n-n_0-p_-}yF_{\underline{p}}(z,0)^{-1}\Gamma(n,n_0)$ (cf. (3.30)), and similarly, identifying

$$\phi_+(z,\,\cdot\,) \text{ with } \phi(P,\,\cdot\,) \text{ and } \phi_-(z,\,\cdot\,) \text{ with } \phi(P^*,\,\cdot\,), \tag{A.54}$$

a comparison of (A.10)–(A.14) and the results of Lemmas 3.1 and 3.3 shows that we may also identify

$$\begin{gathered}\mathfrak{f} \text{ with } \mp\frac{2F_{\underline{p}}}{c_{0,+}z^{-p_-}y}, \quad \mathfrak{g} \text{ with } \mp\frac{2G_{\underline{p}}}{c_{0,+}z^{-p_-}y}, \\ \text{and } \mathfrak{h} \text{ with } \mp\frac{2H_{\underline{p}}}{c_{0,+}z^{-p_-}y},\end{gathered} \tag{A.55}$$

the sign depending on whether P tends to $P_{\infty\pm}$ or to $P_{0,\pm}$. In particular, (A.17)–(A.24) then correspond to (2.10)–(2.13), (2.69), (2.76)–(2.78), respectively. Since $z^{p_-}F_{\underline{p}}/y$, $z^{p_-}G_{\underline{p}}/y$, and $z^{p_-}H_{\underline{p}}/y$ clearly have asymptotic (in fact, even convergent) expansions as $|z|\to\infty$ and as $|z|\to 0$, the results of Theorem A.1 apply. Thus, as $P\to P_{\infty\pm}$, one obtains the following expansions using (A.2) and (2.18)–(2.20):

$$\begin{aligned}\frac{z^{p_-}}{c_{0,+}}\frac{F_{\underline{p}}(z)}{y} &\underset{\zeta\to 0}{=} \mp\frac{1}{c_{0,+}}\bigg(\sum_{k=0}^{\infty}\hat{c}_k(\underline{E})\zeta^k\bigg)\bigg(\sum_{\ell=1}^{p_-} f_{p_--\ell,-}\zeta^{p_++\ell} + \sum_{\ell=0}^{p_+-1} f_{p_+-1-\ell,+}\zeta^{p_+-\ell}\bigg), \\ &\underset{\zeta\to 0}{=} \mp\sum_{\ell=0}^{\infty}\hat{f}_{\ell,+}\zeta^{\ell+1},\end{aligned} \tag{A.56}$$

$$\begin{aligned}\frac{z^{p_-}}{c_{0,+}}\frac{G_{\underline{p}}(z)}{y} &\underset{\zeta\to 0}{=} \mp\frac{1}{c_{0,+}}\bigg(\sum_{k=0}^{\infty}\hat{c}_k(\underline{E})\zeta^k\bigg)\bigg(\sum_{\ell=1}^{p_-} g_{p_--\ell,-}\zeta^{p_++\ell} + \sum_{\ell=0}^{p_+} g_{p_+-\ell,+}\zeta^{p_+-\ell}\bigg) \\ &\underset{\zeta\to 0}{=} \mp\sum_{\ell=0}^{\infty}\hat{g}_{\ell,+}\zeta^{\ell},\end{aligned} \tag{A.57}$$

$$\begin{aligned}\frac{z^{p_-}}{c_{0,+}}\frac{H_{\underline{p}}(z)}{y} &\underset{\zeta\to 0}{=} \mp\frac{1}{c_{0,+}}\bigg(\sum_{k=0}^{\infty}\hat{c}_k(\underline{E})\zeta^k\bigg)\bigg(\sum_{\ell=0}^{p_--1} h_{p_--1-\ell,-}\zeta^{p_++\ell} + \sum_{\ell=1}^{p_+} h_{p_+-\ell,+}\zeta^{p_+-\ell}\bigg) \\ &\underset{\zeta\to 0}{=} \mp\sum_{\ell=0}^{\infty}\hat{h}_{\ell,+}\zeta^{\ell}.\end{aligned} \tag{A.58}$$

This implies (A.39)–(A.41) as $P\to P_{\infty\pm}$.

Similarly, as $P \to P_{0,\pm}$, (A.2) and (2.18)–(2.20), and (2.74) imply

$$\begin{aligned}
\frac{z^{p_-}}{c_{0,+}}\frac{F_{\underline{p}}(z)}{y} &\underset{\zeta\to 0}{=} \pm\frac{1}{c_{0,-}}\bigg(\sum_{k=0}^{\infty}\hat{c}_k(\underline{E}^{-1})\zeta^k\bigg) \\
&\qquad\times\bigg(\sum_{\ell=1}^{p_-} f_{p_--\ell,-}\zeta^{p_+-\ell} + \sum_{\ell=0}^{p_+-1} f_{p_+-1-\ell,+}\zeta^{p_++\ell}\bigg) \\
&\underset{\zeta\to 0}{=} \pm\sum_{\ell=0}^{\infty}\hat{f}_{\ell,-}\zeta^{\ell}, \qquad (A.59)\\
\frac{z^{p_-}}{c_{0,+}}\frac{G_{\underline{p}}(z)}{y} &\underset{\zeta\to 0}{=} \pm\frac{1}{c_{0,-}}\bigg(\sum_{k=0}^{\infty}\hat{c}_k(\underline{E}^{-1})\zeta^k\bigg) \\
&\qquad\times\bigg(\sum_{\ell=1}^{p_-} g_{p_--\ell,-}\zeta^{p_+-\ell} + \sum_{\ell=0}^{p_+} g_{p_+-\ell,+}\zeta^{p_++\ell}\bigg) \\
&\underset{\zeta\to 0}{=} \pm\sum_{\ell=0}^{\infty}\hat{g}_{\ell,-}\zeta^{\ell}, \qquad (A.60)\\
\frac{z^{p_-}}{c_{0,+}}\frac{H_{\underline{p}}(z)}{y} &\underset{\zeta\to 0}{=} \pm\frac{1}{c_{0,-}}\bigg(\sum_{k=0}^{\infty}\hat{c}_k(\underline{E}^{-1})\zeta^k\bigg) \\
&\qquad\times\bigg(\sum_{\ell=0}^{p_--1} h_{p_--1-\ell,-}\zeta^{p_+-\ell} + \sum_{\ell=1}^{p_+} h_{p_+-\ell,+}\zeta^{p_++\ell}\bigg) \\
&\underset{\zeta\to 0}{=} \pm\sum_{\ell=0}^{\infty}\hat{h}_{\ell,-}\zeta^{\ell+1}. \qquad (A.61)
\end{aligned}$$

Thus, (A.39)–(A.41) hold as $P \to P_{0,\pm}$.

Next, comparing powers of ζ in the second and third term of (A.56), formula (A.46) follows (and hence (A.49) as well). Formulas (A.47) and (A.48) follow by using (A.57) and (A.58), respectively.

To prove (A.43) one uses (A.8) and finds

$$\begin{aligned}
c_{0,\pm}\sum_{m=0}^{\ell} c_{\ell-m}\big(\underline{E}^{\pm 1}\big)\hat{f}_{m,\pm} &= \sum_{m=0}^{\ell} c_{\ell-m}(\underline{E})\sum_{k=0}^{m}\hat{c}_{m-k}\big(\underline{E}^{\pm 1}\big)f_{k,\pm} \\
&= f_{\ell,\pm}. \qquad (A.62)
\end{aligned}$$

The proofs of (A.44) and (A.45) and those of (A.50) and (A.51) are analogous. □

Finally, we also mention the following system of recursion relations for the homogeneous coefficients $\hat{f}_{\ell,\pm}$, $\hat{g}_{\ell,\pm}$, and $\hat{h}_{\ell,\pm}$.

Lemma A.3. *The homogeneous coefficients $\hat{f}_{\ell,\pm}$, $\hat{g}_{\ell,\pm}$, and $\hat{h}_{\ell,\pm}$ are uniquely defined by the following recursion relations:*

$$\begin{aligned}
\hat{g}_{0,+} &= \frac{1}{2}, \quad \hat{f}_{0,+} = -\alpha^+, \quad \hat{h}_{0,+} = \beta, \\
\hat{g}_{l+1,+} &= \sum_{k=0}^{l} \hat{f}_{l-k,+}\hat{h}_{k,+} - \sum_{k=1}^{l} \hat{g}_{l+1-k,+}\hat{g}_{k,+}, \\
\hat{f}^-_{l+1,+} &= \hat{f}_{l,+} - \alpha(\hat{g}_{l+1,+} + \hat{g}^-_{l+1,+}), \\
\hat{h}_{l+1,+} &= \hat{h}^-_{l,+} + \beta(\hat{g}_{l+1,+} + \hat{g}^-_{l+1,+}),
\end{aligned} \tag{A.63}$$

and

$$\begin{aligned}
\hat{g}_{0,-} &= \frac{1}{2}, \quad \hat{f}_{0,-} = \alpha, \quad \hat{h}_{0,-} = -\beta^+, \\
\hat{g}_{l+1,-} &= \sum_{k=0}^{l} \hat{f}_{l-k,-}\hat{h}_{k,-} - \sum_{k=1}^{l} \hat{g}_{l+1-k,-}\hat{g}_{k,-}, \\
\hat{f}_{l+1,-} &= \hat{f}^-_{l,-} + \alpha(\hat{g}_{l+1,-} + \hat{g}^-_{l+1,-}), \\
\hat{h}^-_{l+1,-} &= \hat{h}_{l,-} - \beta(\hat{g}_{l+1,-} + \hat{g}^-_{l+1,-}).
\end{aligned} \tag{A.64}$$

Proof. One verifies that the coefficients defined via these recursion relations satisfy (2.32)–(2.35) (respectively, (2.36)–(2.39)). Since they are homogeneous of the required degree this completes the proof. ☐

Acknowledgments

F.G., J.M., and G.T. gratefully acknowledge the extraordinary hospitality of the Department of Mathematical Sciences of the Norwegian University of Science and Technology, Trondheim, during extended stays in the summers of 2004–2006, where parts of this paper were written. F.G. and G.T. would like to thank all organizers of the international conference on Operator Theory and Mathematical Physics (OTAMP), Lund, June 2006, and especially, Pavel Kurasov, for their kind invitation and the stimulating atmosphere during the meeting. We are indebted to the anonymous referee for constructive remarks.

References

[1] M.J. Ablowitz. Nonlinear evolution equations – continuous and discrete. *SIAM Rev.* 19:663–684, 1977.

[2] M.J. Ablowitz and P.A. Clarkson. *Solitons, Nonlinear Evolution Equations and Inverse Scattering.* Cambridge University Press, Cambridge, 1991.

[3] M.J. Ablowitz and J.F. Ladik. Nonlinear differential-difference equations. *J. Math. Phys.* 16:598–603, 1975.

[4] M.J. Ablowitz and J.F. Ladik. Nonlinear differential-difference equations and Fourier analysis. *J. Math. Phys.* 17:1011–1018, 1976.

[5] M.J. Ablowitz and J.F. Ladik. A nonlinear difference scheme and inverse scattering. *Studies Appl. Math.* 55:213–229, 1976.

[6] M.J. Ablowitz and J.F. Ladik. On the solution of a class of nonlinear partial difference equations. *Studies Appl. Math.* 57:1–12, 1977.

[7] M.J. Ablowitz, B. Prinari, and A.D. Trubatch. *Discrete and Continuous Nonlinear Schrödinger Systems.* London Mathematical Society Lecture Note Series, Vol. 302, Cambridge University Press, Cambridge, 2004.

[8] S. Ahmad and A. Roy Chowdhury. On the quasi-periodic solutions to the discrete non-linear Schrödinger equation. *J. Phys. A* 20:293–303, 1987.

[9] S. Ahmad and A. Roy Chowdhury. The quasi-periodic solutions to the discrete nonlinear Schrödinger equation. *J. Math. Phys.* 28:134–1137, 1987.

[10] G.S. Ammar and W.B. Gragg. Schur flows for orthogonal Hessenberg matrices. *Hamiltonian and Gradient Flows, Algorithms and Control*, A. Bloch (ed.), Fields Inst. Commun., Vol. 3, Amer. Math. Soc., Providence, RI, 1994, pp. 27–34.

[11] G. Baxter. Polynomials defined by a difference system. *Bull. Amer. Math. Soc.* 66:187–190, 1960.

[12] G. Baxter. Polynomials defined by a difference system. *J. Math. Anal. Appl.* 2:223–263, 1961.

[13] M. Bertola and M. Gekhtman. Biorthogonal Laurent polynomials, Töplitz determinants, minimal Toda orbits and isomonodromic tau functions. *Constr. Approx.*, 26:383–430, 2007.

[14] N.N. Bogolyubov, A.K. Prikarpatskii, and V.G. Samoilenko. Discrete periodic problem for the modified nonlinear Korteweg–de Vries equation. *Sov. Phys. Dokl.* 26:490–492, 1981.

[15] N.N. Bogolyubov and A.K. Prikarpatskii. The inverse periodic problem for a discrete approximation of a nonlinear Schrödinger equation. *Sov. Phys. Dokl.* 27:113–116, 1982.

[16] W. Bulla, F. Gesztesy, H. Holden, and G. Teschl. Algebro-geometric quasi-periodic finite-gap solutions of the Toda and Kac-van Moerbeke hierarchy. *Mem. Amer. Math. Soc.* no 641, 135:1–79, 1998.

[17] S.-C. Chiu and J.F. Ladik. Generating exactly soluble nonlinear discrete evolution equations by a generalized Wronskian technique. *J. Math. Phys.* 18:690–700, 1977.

[18] K.W. Chow, R. Conte, and N. Xu. Analytic doubly periodic wave patterns for the integrable discrete nonlinear Schrödinger (Ablowitz–Ladik) model. *Physics Letters A* 349:422–429, 2006.

[19] A.K. Common. A solution of the initial value problem for half-infinite integrable systems. *Inverse Problems* 8:393–408, 1992.

[20] P. Deift. Riemann–Hilbert methods in the theory of orthogonal polynomials. *Spectral Theory and Mathematical Physics: A Festschrift in Honor of Barry Simon's 60th Birthday. Ergodic Schrödinger Operators, Singular Spectrum, Orthogonal Polynomials, and Inverse Spectral Theory*, F. Gesztesy, P. Deift, C. Galvez, P. Perry, and W. Schlag (eds.), Proceedings of Symposia in Pure Mathematics, Vol. 76/2, Amer. Math. Soc., Providence, RI, 2007, pp. 715–740.

[21] H.-Y. Ding, Y.-P. Sun, and X.-X. Xu. A hierarchy of nonlinear lattice soliton equations, its integrable coupling systems and infinitely many conservation laws. *Chaos, Solitons and Fractals* 30:227–234, 2006.

[22] A. Doliwa and P.M. Santini. Integrable dynamics of a discrete curve and the Ablowitz–Ladik hierarchy. *J. Math. Phys.* 36:1259–1273, 1995.

[23] N.M. Ercolani and G.I. Lozano. A bi-Hamiltonian structure for the integrable, discrete non-linear Schrödinger system. *Physica D* 218:105–121, 2006.

[24] L. Faybusovich and M. Gekhtman. On Schur flows. *J. Phys. A* 32:4671–4680, 1999.

[25] L. Faybusovich and M. Gekhtman. Elementary Toda orbits and integrable lattices. *Inverse Probl.* 41:2905–2921, 2000.

[26] X. Geng. Darboux transformation of the discrete Ablowitz–Ladik eigenvalue problem. *Acta Math. Sci.* 9:21–26, 1989.

[27] X. Geng, H.H. Dai, and C. Cao. Algebro-geometric constructions of the discrete Ablowitz–Ladik flows and applications. *J. Math. Phys.* 44:4573–4588, 2003.

[28] J.S. Geronimo, F. Gesztesy, and H. Holden. Algebro-geometric solutions of the Baxter–Szegő difference equation. *Commun. Math. Phys.* 258:149–177, 2005.

[29] F. Gesztesy and H. Holden. *Soliton Equations and Their Algebro-Geometric Solutions. Volume I:* $(1+1)$*-Dimensional Continuous Models.* Cambridge Studies in Advanced Mathematics, Vol. 79, Cambridge University Press, Cambridge, 2003.

[30] F. Gesztesy, H. Holden, J. Michor, and G. Teschl. *Soliton Equations and Their Algebro-Geometric Solutions. Volume II:* $(1+1)$*-Dimensional Discrete Models.* Cambridge Studies in Advanced Mathematics, Cambridge University Press, Cambridge, to appear.

[31] F. Gesztesy, H. Holden, J. Michor, and G. Teschl. Algebro-geometric finite-band solutions of the Ablowitz–Ladik hierarchy. *Int. Math. Res. Notices*, 2007:1–55, rnm082.

[32] F. Gesztesy, H. Holden, J. Michor, and G. Teschl. The algebro-geometric initial value problem for the Ablowitz–Ladik hierarchy, arXiv:nlin/07063370.

[33] F. Gesztesy, H. Holden, J. Michor, and G. Teschl. Local conservation laws and the Hamiltonian formalism for the Ablowitz–Ladik hierarchy, *Stud. Appl. Math.* 120:361–423, 2008.

[34] F. Gesztesy and M. Zinchenko. Weyl–Titchmarsh theory for CMV operators associated with orthogonal polynomials on the unit circle. *J. Approx. Th.* 139:172–213, 2006.

[35] F. Gesztesy and M. Zinchenko. A Borg-type theorem associated with orthogonal polynomials on the unit circle. *J. London Math. Soc.* 74:757–777, 2006.

[36] L. Golinskii. Schur flows and orthogonal polynomials on the unit circle. *Sbornik Math.* 197:1145–1165, 2006.

[37] R. Killip and I. Nenciu. CMV: The unitary analogue of Jacobi matrices. *Commun. Pure Appl. Math.* 59:1–41, 2006.

[38] L.-C. Li. Some remarks on CMV matrices and dressing orbits. *Int. Math. Res. Notices* 40:2437–2446, 2005.

[39] P.D. Miller. Macroscopic behavior in the Ablowitz–Ladik equations. *Nonlinear Evolution Equations & Dynamical Systems*, V.G. Makhankov, A.R. Bishop, and D.D. Holm (eds.), World Scientific, Singapore, 1995, pp. 158–167.

[40] P.D. Miller, N.M. Ercolani, I.M. Krichever, and C.D. Levermore. Finite genus solutions to the Ablowitz–Ladik equations. *Comm. Pure Appl. Math.* 48:1369–1440, 1995.

[41] A. Mukaihira and Y. Nakamura. Schur flow for orthogonal polynomials on the unit circle and its integrable discretization. *J. Comput. Appl. Math.* 139:75–94, 2002.

[42] I. Nenciu. Lax Pairs for the Ablowitz–Ladik System via Orthogonal Polynomials on the Unit Circle. Ph.D. Thesis, California Institute of Technology, Pasadena, CA, 2005.

[43] I. Nenciu. Lax pairs for the Ablowitz–Ladik system via orthogonal polynomials on the unit circle. *Int. Math. Res. Notices* 11:647–686, 2005.

[44] I. Nenciu. CMV matrices in random matrix theory and integrable systems: a survey. *J. Phys. A* 39:8811–8822, 2006.

[45] R.J. Schilling. A systematic approach to the soliton equations of a discrete eigenvalue problem. *J. Math. Phys.* 30:1487–1501, 1989.

[46] B. Simon. Analogs of the m-function in the theory of orthogonal polynomials on the unit circle. *J. Comp. Appl. Math.* 171:411–424, 2004.

[47] B. Simon. *Orthogonal Polynomials on the Unit Circle, Part* 1*: Classical Theory, Part* 2*: Spectral Theory*, AMS Colloquium Publication Series, Vol. 54, Amer. Math. Soc., Providence, R.I., 2005.

[48] B. Simon. OPUC on one foot. *Bull. Amer. Math. Soc.* 42:431–460, 2005.

[49] B. Simon. CMV matrices: Five years after, *J. Comp. Appl. Math.* 208:120–154, 2007.

[50] B. Simon. Zeros of OPUC and long time asymptotics of Schur and related flows. *Inverse Probl. Imaging* 1:189–215, 2007.

[51] K.M. Tamizhmani and Wen-Xiu Ma. Master symmetries from Lax operators for certain lattice soliton hierarchies. *J. Phys. Soc. Japan* 69:351–361, 2000.

[52] G. Teschl. *Jacobi Operators and Completely Integrable Nonlinear Lattices.* Math. Surveys Monographs, Vol. 72, Amer. Math. Soc., Providence, R.I., 2000.

[53] K.L. Vaninsky. An additional Gibbs' state for the cubic Schrödinger equation on the circle. *Comm. Pure Appl. Math.* 54:537–582, 2001.

[54] V.E. Vekslerchik. Finite genus solutions for the Ablowitz–Ladik hierarchy. *J. Phys. A* 32:4983–4994, 1999.

[55] V.E. Vekslerchik. Functional representation of the Ablowitz–Ladik hierarchy. II. *J. Nonlin. Math. Phys.* 9:157–180, 2002.

[56] V.E. Vekslerchik. Implementation of the Bäcklund transformations for the Ablowitz–Ladik hierarchy. *J. Phys. A* 39:6933–6953, 2006.

[57] V.E. Vekslerchik and V.V. Konotop. Discrete nonlinear Schrödinger equation under non-vanishing boundary conditions. *Inverse Problems* 8:889–909, 1992.

[58] Y. Zeng and S. Rauch-Wojciechowski. Restricted flows of the Ablowitz–Ladik hierarchy and their continuous limits. *J. Phys. A* 28:113–134, 1995.

Fritz Gesztesy
Department of Mathematics, University of Missouri
Columbia, MO 65211, USA
e-mail: fritz@math.missouri.edu
URL: http://www.math.missouri.edu/personnel/faculty/gesztesyf.html

Helge Holden
Department of Mathematical Sciences
Norwegian University of Science and Technology
NO–7491 Trondheim, Norway
e-mail: holden@math.ntnu.no
URL: http://www.math.ntnu.no/~holden/

Johanna Michor
Faculty of Mathematics, University of Vienna
Nordbergstrasse 15, 1090 Wien, Austria
and
International Erwin Schrödinger Institute for Mathematical Physics
Boltzmanngasse 9, 1090 Wien, Austria
e-mail: Johanna.Michor@esi.ac.at
URL: http://www.mat.univie.ac.at/~jmichor/

Gerald Teschl
Faculty of Mathematics, University of Vienna
Nordbergstrasse 15, 1090 Wien, Austria
and
International Erwin Schrödinger Institute for Mathematical Physics
Boltzmanngasse 9, 1090 Wien, Austria
e-mail: Gerald.Teschl@univie.ac.at
URL: http://www.mat.univie.ac.at/~gerald/

Operator Theory:
Advances and Applications, Vol. 186, 191–215

On Dirichlet-to-Neumann Maps and Some Applications to Modified Fredholm Determinants

Fritz Gesztesy, Marius Mitrea, and Maxim Zinchenko

Dedicated with great pleasure to Boris Pavlov
on the occasion of his 70th birthday

Abstract. We consider Dirichlet-to-Neumann maps associated with (not necessarily self-adjoint) Schrödinger operators in $L^2(\Omega; d^n x)$, where $\Omega \subset \mathbb{R}^n$, $n = 2, 3$, are open sets with a compact, nonempty boundary $\partial\Omega$ satisfying certain regularity conditions. As an application we describe a reduction of a certain ratio of modified Fredholm perturbation determinants associated with operators in $L^2(\Omega; d^n x)$ to modified Fredholm perturbation determinants associated with operators in $L^2(\partial\Omega; d^{n-1}\sigma)$, $n = 2, 3$. This leads to a two- and three-dimensional extension of a variant of a celebrated formula due to Jost and Pais, which reduces the Fredholm perturbation determinant associated with a Schrödinger operator on the half-line $(0, \infty)$ to a simple Wronski determinant of appropriate distributional solutions of the underlying Schrödinger equation.

Mathematics Subject Classification (2000). Primary: 47B10, 47G10, Secondary: 34B27, 34L40.

Keywords. Fredholm determinants, non-self-adjoint operators, multi-dimensional Schrödinger operators, Dirichlet-to-Neumann maps.

1. Introduction

To describe the original Fredholm determinant result due to Jost and Pais [34], we need a few preparations. Denoting by $H_{0,+}^D$ and $H_{0,+}^N$ the one-dimensional Dirichlet and Neumann Laplacians in $L^2((0,\infty); dx)$, and assuming

$$V \in L^1((0,\infty); dx), \tag{1.1}$$

Based upon work partially supported by the US National Science Foundation under Grant Nos. DMS-0405526, DMS-0400639, and FRG-0456306.

we introduce the perturbed Schrödinger operators H_+^D and H_+^N in $L^2((0,\infty);dx)$ by

$$\begin{aligned} &H_+^D f = -f'' + Vf, \\ &f \in \operatorname{dom}\big(H_+^D\big) = \{g \in L^2((0,\infty);dx) \,|\, g,g' \in AC([0,R]) \text{ for all } R>0, \\ &\qquad\qquad g(0)=0,\, (-g''+Vg) \in L^2((0,\infty);dx)\}, \end{aligned} \tag{1.2}$$

$$\begin{aligned} &H_+^N f = -f'' + Vf, \\ &f \in \operatorname{dom}\big(H_+^N\big) = \{g \in L^2((0,\infty);dx) \,|\, g,g' \in AC([0,R]) \text{ for all } R>0, \\ &\qquad\qquad g'(0)=0,\, (-g''+Vg) \in L^2((0,\infty);dx)\}. \end{aligned} \tag{1.3}$$

Thus, H_+^D and H_+^N are self-adjoint if and only if V is real-valued, but since the latter restriction plays no special role in our results, we will not assume real-valuedness of V throughout this paper.

A fundamental system of solutions $\phi_+^D(z,\cdot)$, $\theta_+^D(z,\cdot)$, and the Jost solution $f_+(z,\cdot)$ of

$$-\psi''(z,x) + V\psi(z,x) = z\psi(z,x), \quad z \in \mathbb{C}\backslash\{0\},\ x \geq 0, \tag{1.4}$$

are then introduced via the standard Volterra integral equations

$$\phi_+^D(z,x) = z^{-1/2}\sin(z^{1/2}x) + \int_0^x dx'\, z^{-1/2}\sin(z^{1/2}(x-x'))V(x')\phi_+^D(z,x'), \tag{1.5}$$

$$\theta_+^D(z,x) = \cos(z^{1/2}x) + \int_0^x dx'\, z^{-1/2}\sin(z^{1/2}(x-x'))V(x')\theta_+^D(z,x'), \tag{1.6}$$

$$f_+(z,x) = e^{iz^{1/2}x} - \int_x^\infty dx'\, z^{-1/2}\sin(z^{1/2}(x-x'))V(x')f_+(z,x'), \tag{1.7}$$

$$z \in \mathbb{C}\backslash\{0\},\ \operatorname{Im}(z^{1/2}) \geq 0,\ x \geq 0.$$

In addition, we introduce

$$u = \exp(i\arg(V))|V|^{1/2}, \quad v = |V|^{1/2}, \text{ so that } V = uv, \tag{1.8}$$

and denote by I_+ the identity operator in $L^2((0,\infty);dx)$. Moreover, we denote by

$$W(f,g)(x) = f(x)g'(x) - f'(x)g(x), \quad x \geq 0, \tag{1.9}$$

the Wronskian of f and g, where $f,g \in C^1([0,\infty))$. We also use the standard convention to abbreviate (with a slight abuse of notation) the operator of multiplication in $L^2((0,\infty);dx)$ by an element $f \in L^1_{\text{loc}}((0,\infty);dx)$ (and similarly in the higher-dimensional context with $(0,\infty)$ replaced by an appropriate open set $\Omega \subset \mathbb{R}^n$ later) by the same symbol f (rather than M_f, etc.). For additional notational conventions we refer to the paragraph at the end of this introduction.

Then, the following results hold (with $\mathcal{B}_1(\cdot)$ abbreviating the ideal of trace class operators):

Theorem 1.1. *Assume* $V \in L^1((0,\infty);dx)$ *and let* $z \in \mathbb{C}\backslash[0,\infty)$ *with* $\mathrm{Im}(z^{1/2}) > 0$. *Then,*

$$\overline{u\big(H_{0,+}^D - zI_+\big)^{-1}v},\ \overline{u\big(H_{0,+}^N - zI_+\big)^{-1}v} \in \mathcal{B}_1(L^2((0,\infty);dx)) \tag{1.10}$$

and

$$\begin{aligned}\det\Big(I_+ + \overline{u\big(H_{0,+}^D - zI_+\big)^{-1}v}\Big) &= 1 + z^{-1/2}\int_0^\infty dx\, \sin(z^{1/2}x)V(x)f_+(z,x)\\ &= W(f_+(z,\cdot),\phi_+^D(z,\cdot)) = f_+(z,0),\end{aligned} \tag{1.11}$$

$$\begin{aligned}\det\Big(I_+ + \overline{u\big(H_{0,+}^N - zI_+\big)^{-1}v}\Big) &= 1 + iz^{-1/2}\int_0^\infty dx\, \cos(z^{1/2}x)V(x)f_+(z,x)\\ &= -\frac{W(f_+(z,\cdot),\theta_+^D(z,\cdot))}{iz^{1/2}} = \frac{f_+'(z,0)}{iz^{1/2}}.\end{aligned} \tag{1.12}$$

Equation (1.11) is the modern formulation of the celebrated result due to Jost and Pais [34]. Performing calculations similar to Section 4 in [24] for the pair of operators $H_{0,+}^N$ and H_+^N, one obtains the analogous result (1.12).

We emphasize that (1.11) and (1.12) exhibit a spectacular reduction of a Fredholm determinant, that is, an infinite determinant (actually, a symmetrized perturbation determinant), associated with the trace class Birmann–Schwinger kernel of a one-dimensional Schrödinger operator on the half-line $(0,\infty)$, to a simple Wronski determinant of $\mathbb{C}$-valued distributional solutions of (1.4). This fact goes back to Jost and Pais [34] (see also [24], [48], [49], [50, Sect. 12.1.2], [60], [61, Proposition 5.7], and the extensive literature cited in these references). The principal aim of this paper is to explore the extent to which this fact may generalize to higher dimensions. While a direct generalization of (1.11), (1.12) appears to be difficult, we will next derive a formula for the ratio of such determinants which indeed permits a natural extension to higher dimensions.

For this purpose we introduce the boundary trace operators γ_D (Dirichlet trace) and γ_N (Neumann trace) which, in the current one-dimensional half-line situation, are just the functionals,

$$\gamma_D\colon \begin{cases} C([0,\infty)) \to \mathbb{C},\\ \qquad g \mapsto g(0),\end{cases} \qquad \gamma_N\colon \begin{cases} C^1([0,\infty)) \to \mathbb{C},\\ \qquad h \mapsto -h'(0).\end{cases} \tag{1.13}$$

In addition, we denote by $m_{0,+}^D$, m_+^D, $m_{0,+}^N$, and m_+^N the Weyl–Titchmarsh m-functions corresponding to $H_{0,+}^D$, H_+^D, $H_{0,+}^N$, and H_+^N, respectively, that is,

$$m_{0,+}^D(z) = iz^{1/2}, \qquad m_{0,+}^N(z) = -\frac{1}{m_{0,+}^D(z)} = iz^{-1/2}, \tag{1.14}$$

$$m_+^D(z) = \frac{f_+'(z,0)}{f_+(z,0)}, \quad m_+^N(z) = -\frac{1}{m_+^D(z)} = -\frac{f_+(z,0)}{f_+'(z,0)}. \tag{1.15}$$

In the case where V is real-valued, we briefly recall the spectral theoretic significance of m_+^D: It is a Herglotz function (i.e., it maps the open complex upper half-plane $\mathbb{C}_+$ analytically into itself) and the measure $d\rho_+^D$ in its Herglotz representation is then the spectral measure of the operator H_+^D and hence encodes all spectral information of H_+^D. Similarly, m_+^D also encodes all spectral information of H_+^N since $-1/m_+^D = m_+^N$ is also a Herglotz function and the measure $d\rho_+^N$ in its Herglotz representation represents the spectral measure of the operator H_+^N. In particular, $d\rho_+^D$ (respectively, $d\rho_+^N$) uniquely determine V a.e. on $(0,\infty)$ by the inverse spectral approach of Gelfand and Levitan [20] or Simon [59], [26] (see also Remling [56] and Section 6 in the survey [21]).

Then we obtain the following result for the ratio of the perturbation determinants in (1.11) and (1.12):

Theorem 1.2. *Assume* $V \in L^1((0,\infty);dx)$ *and let* $z \in \mathbb{C}\backslash\sigma(H_+^D)$ *with* $\operatorname{Im}(z^{1/2}) > 0$. *Then,*

$$\frac{\det\Big(I_+ + \overline{u\big(H_{0,+}^N - zI_+\big)^{-1}v}\Big)}{\det\Big(I_+ + \overline{u\big(H_{0,+}^D - zI_+\big)^{-1}v}\Big)}$$

$$= 1 - \Big(\overline{\gamma_N(H_+^D - zI_+)^{-1}V\big[\gamma_D(H_{0,+}^N - \overline{z}I_+)^{-1}\big]^*}\Big) \tag{1.16}$$

$$= \frac{W(f_+(z),\phi_+^N(z))}{iz^{1/2}W(f_+(z),\phi_+^D(z))} = \frac{f_+'(z,0)}{iz^{1/2}f_+(z,0)} = \frac{m_+^D(z)}{m_{0,+}^D(z)} = \frac{m_{0,+}^N(z)}{m_+^N(z)}. \tag{1.17}$$

The proper multi-dimensional generalization to Schrödinger operators in $L^2(\Omega;d^nx)$ corresponding to an open set $\Omega\subset\mathbb{R}^n$ with compact, nonempty boundary $\partial\Omega$ then involves the operator-valued generalization of the Weyl–Titchmarsh function $m_+^D(z)$, the Dirichlet-to-Neumann map denoted by $M_\Omega^D(z)$. In particular, we will derive the following multi-dimensional extension of (1.16) and (1.17) in Section 4:

Theorem 1.3. *Assume Hypothesis* 2.6 *and let* $z \in \mathbb{C}\backslash\big(\sigma\big(H_\Omega^D\big)\cup\sigma\big(H_{0,\Omega}^D\big)\cup\sigma\big(H_{0,\Omega}^N\big)\big)$. *Then,*

$$\frac{\det{}_2\Big(I_\Omega + \overline{u\big(H_{0,\Omega}^N - zI_\Omega\big)^{-1}v}\Big)}{\det{}_2\Big(I_\Omega + \overline{u\big(H_{0,\Omega}^D - zI_\Omega\big)^{-1}v}\Big)}$$

$$= \det{}_2\Big(I_{\partial\Omega} - \overline{\gamma_N\big(H_\Omega^D - zI_\Omega\big)^{-1}V\big[\gamma_D(H_{0,\Omega}^N - \overline{z}I_\Omega)^{-1}\big]^*}\Big)e^{\operatorname{tr}(T_2(z))} \tag{1.18}$$

$$= \det{}_2\big(M_\Omega^D(z)M_{0,\Omega}^D(z)^{-1}\big)e^{\operatorname{tr}(T_2(z))}. \tag{1.19}$$

Here, $\det_2(\cdot)$ denotes the modified Fredholm determinant in connection with Hilbert–Schmidt perturbations of the identity, $T_2(z)$ is given by

$$T_2(z) = \overline{\gamma_N\big(H_{0,\Omega}^D - zI_\Omega\big)^{-1}V\big(H_\Omega^D - zI_\Omega\big)^{-1}V\big[\gamma_D\big(H_{0,\Omega}^N - \overline{z}I_\Omega\big)^{-1}\big]^*}, \tag{1.20}$$

and I_Ω and $I_{\partial\Omega}$ represent the identity operators in $L^2(\Omega; d^n x)$ and $L^2(\partial\Omega; d^{n-1}\sigma)$, respectively (with $d^{n-1}\sigma$ the surface measure on $\partial\Omega$).

For pertinent comments on the principal reduction of (a ratio of) modified Fredholm determinants associated with operators in $L^2(\Omega; d^n x)$ on the left-hand side of (1.18) to a modified Fredholm determinant associated with operators in $L^2(\partial\Omega; d^{n-1}\sigma)$ on the right-hand side of (1.18) and especially, in (1.19), we refer to Section 4.

Finally, we briefly list some of the notational conventions used throughout this paper. Let T be a linear operator mapping (a subspace of) a Banach space into another, with $\mathrm{dom}(T)$ and $\mathrm{ran}(T)$ denoting the domain and range of T. The closure of a closable operator S is denoted by $\overline{S}$. The kernel (null space) of T is denoted by $\ker(T)$. The spectrum and resolvent set of a closed linear operator in a separable complex Hilbert space $\mathcal{H}$ (with scalar product denoted by $(\cdot,\cdot)_{\mathcal{H}}$, assumed to be linear in the second factor) will be denoted by $\sigma(\cdot)$ and $\rho(\cdot)$. The Banach spaces of bounded and compact linear operators in $\mathcal{H}$ are denoted by $\mathcal{B}(\mathcal{H})$ and $\mathcal{B}_\infty(\mathcal{H})$, respectively. Similarly, the Schatten–von Neumann (trace) ideals will subsequently be denoted by $\mathcal{B}_p(\mathcal{H})$, $p \in \mathbb{N}$. Analogous notation $\mathcal{B}(\mathcal{H}_1, \mathcal{H}_2)$, $\mathcal{B}_\infty(\mathcal{H}_1, \mathcal{H}_2)$, etc., will be used for bounded, compact, etc., operators between two Hilbert spaces $\mathcal{H}_1$ and $\mathcal{H}_2$. In addition, $\mathrm{tr}(T)$ denotes the trace of a trace class operator $T \in \mathcal{B}_1(\mathcal{H})$ and $\det_p(I_{\mathcal{H}} + S)$ represents the (modified) Fredholm determinant associated with an operator $S \in \mathcal{B}_p(\mathcal{H})$, $p \in \mathbb{N}$ (for $p = 1$ we omit the subscript 1). Moreover, $\mathcal{X}_1 \hookrightarrow \mathcal{X}_2$ denotes the continuous embedding of the Banach space $\mathcal{X}_1$ into the Banach space $\mathcal{X}_2$.

For general references on the theory of modified Fredholm determinants we refer, for instance, to [16, Sect. XI.9], [27, Chs. IX, XI], [28, Sect. IV.2], [58], and [61, Ch. 9].

2. Schrödinger operators with Dirichlet and Neumann boundary conditions

In this section we primarily focus on various properties of Dirichlet, $H_{0,\Omega}^D$, and Neumann, $H_{0,\Omega}^N$, Laplacians in $L^2(\Omega; d^n x)$ associated with open sets $\Omega \subset \mathbb{R}^n$, $n = 2, 3$, introduced in Hypothesis 2.1 below. In particular, we study mapping properties of $\big(H_{0,\Omega}^{D,N} - zI_\Omega\big)^{-q}$, $q \in [0,1]$, (I_Ω the identity operator in $L^2(\Omega; d^n x)$) and trace ideal properties of the maps $f\big(H_{0,\Omega}^{D,N} - zI_\Omega\big)^{-q}$, $f \in L^p(\Omega; d^n x)$, for appropriate $p \geq 2$, and $\gamma_N\big(H_{0,\Omega}^D - zI_\Omega\big)^{-r}$, and $\gamma_D\big(H_{0,\Omega}^N - zI_\Omega\big)^{-s}$, for appropriate $r > 3/4$, $s > 1/4$, with γ_N and γ_D being the Neumann and Dirichlet boundary trace operators defined in (2.2) and (2.3).

At the end of this section we then introduce the Dirichlet and Neumann Schrödinger operators H_Ω^D and H_Ω^N in $L^2(\Omega; d^n x)$, that is, perturbations of the Dirichlet and Neumann Laplacians $H_{0,\Omega}^D$ and $H_{0,\Omega}^N$ by a potential V satisfying Hypothesis 2.6.

We start with introducing our assumptions on the set Ω:

Hypothesis 2.1. *Let $n = 2, 3$ and assume that $\Omega \subset \mathbb{R}^n$ is an open set with a compact, nonempty boundary $\partial\Omega$. In addition, we assume that one of the following three conditions holds:*

(i) *Ω is of class $C^{1,r}$ for some $1/2 < r < 1$;*
(ii) *Ω is convex;*
(iii) *Ω is a Lipschitz domain satisfying a uniform exterior ball condition.*

The class of domains described in Hypothesis 2.1 is a subclass of all Lipschitz domains with compact nonempty boundary. We also note that while $\partial\Omega$ is assumed to be compact, Ω may be unbounded (e.g., an exterior domain) in connection with conditions (i) or (iii). For more details in this context, in particular, for the precise definition of the uniform exterior ball condition, we refer to [23, App. A] and [25, App. A] (and the references cited therein, such as [30, Ch. 1], [31], [32], [35], [40, Ch. 3], [43], [44], [62, p. 189], [63], [68], and [70, Sect. I.4.2]).

First, we introduce the boundary trace operator γ_D^0 (Dirichlet trace) by

$$\gamma_D^0 \colon C(\overline{\Omega}) \to C(\partial\Omega), \quad \gamma_D^0 u = u|_{\partial\Omega}. \tag{2.1}$$

Then there exists a bounded, linear operator γ_D (cf. [40, Theorem 3.38]),

$$\begin{aligned} &\gamma_D \colon H^s(\Omega) \to H^{s-(1/2)}(\partial\Omega) \hookrightarrow L^2(\partial\Omega; d^{n-1}\sigma), \quad 1/2 < s < 3/2, \\ &\gamma_D \colon H^{3/2}(\Omega) \to H^{1-\varepsilon}(\partial\Omega) \hookrightarrow L^2(\partial\Omega; d^{n-1}\sigma), \quad \varepsilon \in (0,1), \end{aligned} \tag{2.2}$$

whose action is compatible with that of γ_D^0. That is, the two Dirichlet trace operators coincide on the intersection of their domains. We recall that $d^{n-1}\sigma$ denotes the surface measure on $\partial\Omega$ and we refer to [23, App. A] for our notation in connection with Sobolev spaces (see also [40, Ch. 3], [68], and [70, Sect. I.4.2]).

Next, we introduce the operator γ_N (Neumann trace) by

$$\gamma_N = \nu \cdot \gamma_D \nabla \colon H^{s+1}(\Omega) \to L^2(\partial\Omega; d^{n-1}\sigma), \quad 1/2 < s < 3/2, \tag{2.3}$$

where ν denotes the outward pointing normal unit vector to $\partial\Omega$. It follows from (2.2) that γ_N is also a bounded operator.

Given Hypothesis 2.1, we introduce the self-adjoint and nonnegative Dirichlet and Neumann Laplacians $H_{0,\Omega}^D$ and $H_{0,\Omega}^N$ associated with the domain Ω as follows,

$$H_{0,\Omega}^D = -\Delta, \quad \operatorname{dom}\big(H_{0,\Omega}^D\big) = \{u \in H^2(\Omega) \,|\, \gamma_D u = 0\}, \tag{2.4}$$

$$H_{0,\Omega}^N = -\Delta, \quad \operatorname{dom}\big(H_{0,\Omega}^N\big) = \{u \in H^2(\Omega) \,|\, \gamma_N u = 0\}. \tag{2.5}$$

A detailed discussion of $H_{0,\Omega}^D$ and $H_{0,\Omega}^N$ is provided in [23, App. A] (cf. also [55, Sects. X.III.14, X.III.15]).

Lemma 2.2. *Assume Hypothesis 2.1. Then the operators $H_{0,\Omega}^D$ and $H_{0,\Omega}^N$ introduced in (2.4) and (2.5) are nonnegative and self-adjoint in $L^2(\Omega; d^n x)$ and the following boundedness properties hold for all $q \in [0,1]$ and $z \in \mathbb{C}\backslash[0,\infty)$,*

$$\big(H_{0,\Omega}^D - zI_\Omega\big)^{-q}, \big(H_{0,\Omega}^N - zI_\Omega\big)^{-q} \in \mathcal{B}\big(L^2(\Omega; d^n x), H^{2q}(\Omega)\big). \tag{2.6}$$

The fractional powers in (2.6) (and in subsequent analogous cases) are defined via the functional calculus implied by the spectral theorem for self-adjoint operators.

As explained in [23, Lemma A.2] (based on results in [33], [40, Thm. 4.4, App. B], [42], [45], [64, Chs. 1, 2], [65, Props. 4.5, 7.9], [66, Sect. 1.3, Thm. 1.18.10, Rem. 4.3.1.2], [69]), the key ingredients in proving Lemma 2.2 are the inclusions

$$\operatorname{dom}\big(H_{0,\Omega}^D\big) \subset H^2(\Omega), \quad \operatorname{dom}\big(H_{0,\Omega}^N\big) \subset H^2(\Omega) \tag{2.7}$$

and real interpolation methods.

The next result is a slight extension of [23, Lemma 6.8] and provides an explicit discussion of the z-dependence of the constant c appearing in estimate (6.48) of [23]. For a proof we refer to [25].

Lemma 2.3 ([25]). *Assume Hypothesis* 2.1 *and let* $2 \leq p$, $n/(2p) < q \leq 1$, $f \in L^p(\Omega; d^n x)$, *and* $z \in \mathbb{C}\backslash[0,\infty)$. *Then,*

$$f\big(H_{0,\Omega}^D - zI_\Omega\big)^{-q},\ f\big(H_{0,\Omega}^N - zI_\Omega\big)^{-q} \in \mathcal{B}_p\big(L^2(\Omega; d^n x)\big), \tag{2.8}$$

and for some $c > 0$ (*independent of* z *and* f)

$$\begin{aligned} &\big\|f\big(H_{0,\Omega}^D - zI_\Omega\big)^{-q}\big\|^2_{\mathcal{B}_p(L^2(\Omega;d^n x))} \\ &\quad \leq c\bigg(1 + \frac{|z|^{2q}+1}{\operatorname{dist}\big(z,\sigma\big(H_{0,\Omega}^D\big)\big)^{2q}}\bigg)\big\|(|\cdot|^2 - z)^{-q}\big\|^2_{L^p(\mathbb{R}^n;d^n x)}\|f\|^2_{L^p(\Omega;d^n x)}, \\ &\big\|f\big(H_{0,\Omega}^N - zI_\Omega\big)^{-q}\big\|^2_{\mathcal{B}_p(L^2(\Omega;d^n x))} \\ &\quad \leq c\bigg(1 + \frac{|z|^{2q}+1}{\operatorname{dist}\big(z,\sigma\big(H_{0,\Omega}^N\big)\big)^{2q}}\bigg)\big\|(|\cdot|^2 - z)^{-q}\big\|^2_{L^p(\mathbb{R}^n;d^n x)}\|f\|^2_{L^p(\Omega;d^n x)}. \end{aligned} \tag{2.9}$$

(Here, in obvious notation, $(|\cdot|^2 - z)^{-q}$ denotes the function $(|x|^2 - z)^{-q}$, $x \in \mathbb{R}^n$.)

Next we recall certain boundedness properties of powers of the resolvents of Dirichlet and Neumann Laplacians multiplied by the Neumann and Dirichlet boundary trace operators, respectively:

Lemma 2.4. *Assume Hypothesis* 2.1 *and let* $\varepsilon > 0$, $z \in \mathbb{C}\backslash[0,\infty)$. *Then,*

$$\gamma_N\big(H_{0,\Omega}^D - zI_\Omega\big)^{-\frac{3+\varepsilon}{4}}, \gamma_D\big(H_{0,\Omega}^N - zI_\Omega\big)^{-\frac{1+\varepsilon}{4}} \in \mathcal{B}\big(L^2(\Omega; d^n x), L^2(\partial\Omega; d^{n-1}\sigma)\big). \tag{2.10}$$

As in [23, Lemma 6.9], Lemma 2.4 follows from Lemma 2.2 and from (2.2) and (2.3).

Corollary 2.5. *Assume Hypothesis* 2.1 *and let* $f_1 \in L^{p_1}(\Omega; d^n x)$, $p_1 \geq 2$, $p_1 > 2n/3$, $f_2 \in L^{p_2}(\Omega; d^n x)$, $p_2 > 2n$, *and* $z \in \mathbb{C}\backslash[0,\infty)$. *Then, denoting by* f_1 *and* f_2 *the operators of multiplication by functions* f_1 *and* f_2 *in* $L^2(\Omega; d^n x)$, *respectively, one has*

$$\overline{\gamma_D\big(H_{0,\Omega}^N - zI_\Omega\big)^{-1} f_1} \in \mathcal{B}_{p_1}\big(L^2(\Omega; d^n x), L^2(\partial\Omega; d^{n-1}\sigma)\big), \tag{2.11}$$

$$\overline{\gamma_N\big(H_{0,\Omega}^D - zI_\Omega\big)^{-1}} f_2 \in \mathcal{B}_{p_2}\big(L^2(\Omega;d^nx), L^2(\partial\Omega;d^{n-1}\sigma)\big) \tag{2.12}$$

and for some $c_j(z) > 0$ *(independent of* f_j*),* $j = 1,2$,

$$\Big\| \overline{\gamma_D\big(H_{0,\Omega}^N - zI_\Omega\big)^{-1}} f_1 \Big\|_{\mathcal{B}_{p_1}(L^2(\Omega;d^nx),L^2(\partial\Omega;d^{n-1}\sigma))} \leq c_1(z)\, \|f_1\|_{L^{p_1}(\Omega;d^nx)}, \tag{2.13}$$

$$\Big\| \overline{\gamma_N\big(H_{0,\Omega}^D - zI_\Omega\big)^{-1}} f_2 \Big\|_{\mathcal{B}_{p_2}(L^2(\Omega;d^nx),L^2(\partial\Omega;d^{n-1}\sigma))} \leq c_2(z)\, \|f_2\|_{L^{p_2}(\Omega;d^nx)}. \tag{2.14}$$

As in [23, Corollary 6.10], Corollary 2.5 follows from Lemmas 2.3 and 2.4.

Finally, we turn to our assumptions on the potential V and the corresponding definition of Dirichlet and Neumann Schrödinger operators H_Ω^D and H_Ω^N in $L^2(\Omega;d^nx)$:

Hypothesis 2.6. *Suppose that* Ω *satisfies Hypothesis* 2.1 *and assume that* $V \in L^p(\Omega;d^nx)$ *for some* p *satisfying* $4/3 < p \leq 2$, *in the case* $n = 2$, *and* $3/2 < p \leq 2$, *in the case* $n = 3$.

Assuming Hypothesis 2.6, we next introduce the perturbed operators H_Ω^D and H_Ω^N in $L^2(\Omega;d^nx)$ by alluding to abstract perturbation results due to Kato [36] (see also Konno and Kuroda [37]) as summarized in [23, Sect. 2]: Let V, u, and v denote the operators of multiplication by functions V, $u = \exp(i\arg(V))|V|^{1/2}$, and $v = |V|^{1/2}$ in $L^2(\Omega;d^nx)$, respectively, such that

$$V = uv. \tag{2.15}$$

Since $u, v \in L^{2p}(\Omega;d^nx)$, Lemma 2.3 yields

$$u\big(H_{0,\Omega}^D - zI_\Omega\big)^{-1/2}, \ \overline{\big(H_{0,\Omega}^D - zI_\Omega\big)^{-1/2}v} \in \mathcal{B}_{2p}\big(L^2(\Omega;d^nx)\big), \quad z \in \mathbb{C}\backslash[0,\infty), \tag{2.16}$$

$$u\big(H_{0,\Omega}^N - zI_\Omega\big)^{-1/2}, \ \overline{\big(H_{0,\Omega}^N - zI_\Omega\big)^{-1/2}v} \in \mathcal{B}_{2p}\big(L^2(\Omega;d^nx)\big), \quad z \in \mathbb{C}\backslash[0,\infty), \tag{2.17}$$

and hence, in particular,

$$\mathrm{dom}(u) = \mathrm{dom}(v) \supseteq \mathrm{dom}\Big(\big(H_{0,\Omega}^N\big)^{1/2}\Big) = H^1(\Omega) \supset H^2(\Omega) \supset \mathrm{dom}\big(H_{0,\Omega}^N\big), \tag{2.18}$$

$$\mathrm{dom}(u) = \mathrm{dom}(v) \supseteq H^1(\Omega) \supset H_0^1(\Omega) = \mathrm{dom}\Big(\big(H_{0,\Omega}^D\big)^{1/2}\Big) \supset \mathrm{dom}\big(H_{0,\Omega}^D\big). \tag{2.19}$$

Moreover, (2.16) and (2.17) imply

$$\begin{aligned} \overline{u\big(H_{0,\Omega}^D - zI_\Omega\big)^{-1}v}, \ \overline{u\big(H_{0,\Omega}^N - zI_\Omega\big)^{-1}v} &\in \mathcal{B}_p\big(L^2(\Omega;d^nx)\big) \\ &\subset \mathcal{B}_2\big(L^2(\Omega;d^nx)\big), \quad z \in \mathbb{C}\backslash[0,\infty). \end{aligned} \tag{2.20}$$

Utilizing (2.9) in Lemma 2.3 with $-z > 0$ sufficiently large, such that the $\mathcal{B}_{2p}$-norms of the operators in (2.16) and (2.17) are less than 1, one concludes that the Hilbert–Schmidt norms of the operators in (2.20) are less than 1. Thus, applying [23, Thm. 2.3], one obtains the densely defined, closed operators H_Ω^D and H_Ω^N (which are extensions of $H_{0,\Omega}^D + V$ defined on $\mathrm{dom}\big(H_{0,\Omega}^D\big) \cap \mathrm{dom}(V)$ and $H_{0,\Omega}^N + V$

defined on $\operatorname{dom}(H_{0,\Omega}^N) \cap \operatorname{dom}(V)$, respectively). In particular, the resolvent of H_Ω^D (respectively, H_Ω^N) is explicitly given by

$$\begin{aligned}
\big(H_\Omega^D - zI_\Omega\big)^{-1} &= \big(H_{0,\Omega}^D - zI_\Omega\big)^{-1} - \big(H_{0,\Omega}^D - zI_\Omega\big)^{-1} v \\
&\quad \times \Big[I_\Omega + \overline{u\big(H_{0,\Omega}^D - zI_\Omega\big)^{-1} v}\Big]^{-1} u\big(H_{0,\Omega}^D - zI_\Omega\big)^{-1}, \\
&\qquad\qquad z \in \mathbb{C}\backslash\sigma\big(H_\Omega^D\big),
\end{aligned} \tag{2.21}$$

$$\begin{aligned}
\big(H_\Omega^N - zI_\Omega\big)^{-1} &= \big(H_{0,\Omega}^N - zI_\Omega\big)^{-1} - \big(H_{0,\Omega}^N - zI_\Omega\big)^{-1} v \\
&\quad \times \Big[I_\Omega + \overline{u\big(H_{0,\Omega}^N - zI_\Omega\big)^{-1} v}\Big]^{-1} u\big(H_{0,\Omega}^N - zI_\Omega\big)^{-1}, \\
&\qquad\qquad z \in \mathbb{C}\backslash\sigma\big(H_\Omega^N\big).
\end{aligned} \tag{2.22}$$

Here invertibility of $\Big[I_\Omega + \overline{u\big(H_{0,\Omega}^{D,N} - zI_\Omega\big)^{-1} v}\Big]$ for $z \in \rho\big(H_\Omega^{D,N}\big)$ is guaranteed by arguments discussed, for instance, in [23, Sect. 2] and the literature cited therein.

Although we will not explicitly use the following result in this paper, we feel it is of sufficient independent interest to be included at the end of this section:

Lemma 2.7. *Assume Ω satisfies Hypothesis 2.1 with $n = 2, 3$ replaced by $n \in \mathbb{N}$, $n \geq 2$, suppose that $V \in L^{n/2}(\Omega; d^n x)$, and let*

$$s \in \begin{cases} (0,1) \text{ if } n = 2, \\ [0, \tfrac{3}{2}) \text{ if } n = 3, \\ [0,2) \text{ if } n = 4, \\ [0,2] \text{ if } n \geq 5. \end{cases} \tag{2.23}$$

Then the operator

$$V : H^s(\Omega) \to \big(H^{2-s}(\Omega)\big)^* \tag{2.24}$$

is well defined and bounded, in fact, it is compact.

Proof. The fact that the operator (2.24) is well defined along with the estimate

$$\|V\|_{\mathcal{B}(H^s(\Omega),(H^{2-s}(\Omega))^*)} \leq C(n,\Omega)\|V\|_{L^{n/2}(\Omega)}, \tag{2.25}$$

are direct consequences of standard embedding results (cf. [67, Sect. 3.3.1] for smooth domains and [68] for arbitrary (bounded or unbounded) Lipschitz domains). Once the boundedness of (2.24) has been established, the compactness follows from the fact that if $V_j \in C_0^\infty(\Omega)$ is a sequence of functions with the property that $V_j \underset{j\uparrow\infty}{\to} V$ in $L^{n/2}(\Omega)$, then $V_j \underset{j\uparrow\infty}{\to} V$ in $\mathcal{B}\big(H^s(\Omega), \big(H^{2-s}(\Omega)\big)^*\big)$ by (2.25) and each operator $V_j : H^s(\Omega) \to \big(H^{2-s}(\Omega)\big)^*$ is compact, by Rellich's selection lemma (cf. [17] for smooth domains and [68] for arbitrary Lipschitz domains). Thus, the operator in (2.24) is compact as the operator norm limit of a sequence of compact operators. □

3. Dirichlet and Neumann boundary value problems and Dirichlet-to-Neumann maps

This section is devoted to Dirichlet and Neumann boundary value problems associated with the Helmholtz differential expression $-\Delta - z$ as well as the corresponding differential expression $-\Delta + V - z$ in the presence of a potential V, both in connection with the open set Ω. In addition, we provide a detailed discussion of Dirichlet-to-Neumann, $M_{0,\Omega}^{D}$, M_{Ω}^{D}, and Neumann-to-Dirichlet maps, $M_{0,\Omega}^{N}$, M_{Ω}^{N}, in $L^2(\partial\Omega; d^{n-1}\sigma)$.

Denote by

$$\widetilde{\gamma}_N : \{u \in H^1(\Omega) \,|\, \Delta u \in (H^1(\Omega))^*\} \to H^{-1/2}(\partial\Omega) \tag{3.1}$$

a weak Neumann trace operator defined by

$$\langle \widetilde{\gamma}_N u, \phi \rangle = \int_\Omega d^n x \, \nabla u(x) \cdot \nabla \Phi(x) + \langle \Delta u, \Phi \rangle \tag{3.2}$$

for all $\phi \in H^{1/2}(\partial\Omega)$ and $\Phi \in H^1(\Omega)$ such that $\gamma_D \Phi = \phi$. We note that this definition is independent of the particular extension Φ of ϕ, and that $\widetilde{\gamma}_N$ is an extension of the Neumann trace operator γ_N defined in (2.3). For more details we refer to [23, App. A].

We start with a basic result on the Helmholtz Dirichlet and Neumann boundary value problems:

Theorem 3.1 ([25]). *Assume Hypothesis* 2.1. *Then for every $f \in H^1(\partial\Omega)$ and $z \in \mathbb{C}\backslash\sigma(H_{0,\Omega}^{D})$ the following Dirichlet boundary value problem,*

$$\begin{cases} (-\Delta - z)u_0^D = 0 \text{ on } \Omega, \quad u_0^D \in H^{3/2}(\Omega), \\ \gamma_D u_0^D = f \text{ on } \partial\Omega, \end{cases} \tag{3.3}$$

has a unique solution u_0^D satisfying $\widetilde{\gamma}_N u_0^D \in L^2(\partial\Omega; d^{n-1}\sigma)$. Moreover, there exist constants $C^D = C^D(\Omega, z) > 0$ such that

$$\|u_0^D\|_{H^{3/2}(\Omega)} \leq C^D \|f\|_{H^1(\partial\Omega)}. \tag{3.4}$$

Similarly, for every $g \in L^2(\partial\Omega; d^{n-1}\sigma)$ and $z \in \mathbb{C}\backslash\sigma(H_{0,\Omega}^{N})$ the following Neumann boundary value problem,

$$\begin{cases} (-\Delta - z)u_0^N = 0 \text{ on } \Omega, \quad u_0^N \in H^{3/2}(\Omega), \\ \widetilde{\gamma}_N u_0^N = g \text{ on } \partial\Omega, \end{cases} \tag{3.5}$$

has a unique solution u_0^N satisfying $\gamma_D u_0^N \in H^1(\partial\Omega)$. Moreover, there exist constants $C^N = C^N(\Omega, z) > 0$ such that

$$\|u_0^N\|_{H^{3/2}(\Omega)} \leq C^N \|g\|_{L^2(\partial\Omega; d^{n-1}\sigma)}. \tag{3.6}$$

In addition, (3.3)–(3.6) *imply that the following maps are bounded*

$$\big[\gamma_N\big(\big(H_{0,\Omega}^{D} - zI_\Omega\big)^{-1}\big)^*\big]^* : H^1(\partial\Omega) \to H^{3/2}(\Omega), \quad z \in \mathbb{C}\backslash\sigma(H_{0,\Omega}^{D}), \tag{3.7}$$

$$\big[\gamma_D\big(\big(H_{0,\Omega}^{N} - zI_\Omega\big)^{-1}\big)^*\big]^* : L^2(\partial\Omega; d^{n-1}\sigma) \to H^{3/2}(\Omega), \quad z \in \mathbb{C}\backslash\sigma(H_{0,\Omega}^{N}). \tag{3.8}$$

Finally, the solutions u_0^D *and* u_0^N *are given by the formulas*

$$u_0^D(z) = -\big(\gamma_N\big(H_{0,\Omega}^D - \overline{z}I_\Omega\big)^{-1}\big)^* f, \tag{3.9}$$

$$u_0^N(z) = \big(\gamma_D\big(H_{0,\Omega}^N - \overline{z}I_\Omega\big)^{-1}\big)^* g. \tag{3.10}$$

A detailed proof of Theorem 3.1 will appear in [25].

We temporarily strengthen our hypothesis on V and introduce the following assumption:

Hypothesis 3.2. *Suppose the set* Ω *satisfies Hypothesis* 2.1 *and assume that* $V \in L^2(\Omega; d^n x) \cap L^p(\Omega; d^n x)$ *for some* $p > 2$.

By employing a perturbative approach, we now extend Theorem 3.1 in connection with the Helmholtz differential expression $-\Delta - z$ on Ω to the case of a Schrödinger differential expression $-\Delta + V - z$ on Ω.

Theorem 3.3. *Assume Hypothesis* 3.2. *Then for every* $f \in H^1(\partial\Omega)$ *and* $z \in \mathbb{C}\backslash\sigma\big(H_\Omega^D\big)$ *the following Dirichlet boundary value problem,*

$$\begin{cases} (-\Delta + V - z)u^D = 0 \text{ on } \Omega, \quad u^D \in H^{3/2}(\Omega), \\ \gamma_D u^D = f \text{ on } \partial\Omega, \end{cases} \tag{3.11}$$

has a unique solution u^D *satisfying* $\widetilde{\gamma}_N u^D \in L^2(\partial\Omega; d^{n-1}\sigma)$. *Similarly, for every* $g \in L^2(\partial\Omega; d^{n-1}\sigma)$ *and* $z \in \mathbb{C}\backslash\sigma\big(H_\Omega^N\big)$ *the following Neumann boundary value problem,*

$$\begin{cases} (-\Delta + V - z)u^N = 0 \text{ on } \Omega, \quad u^N \in H^{3/2}(\Omega), \\ \widetilde{\gamma}_N u^N = g \text{ on } \partial\Omega, \end{cases} \tag{3.12}$$

has a unique solution u^N. *Moreover, the solutions* u^D *and* u^N *are given by the formulas*

$$u^D(z) = -\big[\gamma_N\big(\big(H_\Omega^D - zI_\Omega\big)^{-1}\big)^*\big]^* f, \tag{3.13}$$

$$u^N(z) = \big[\gamma_D\big(\big(H_\Omega^N - zI_\Omega\big)^{-1}\big)^*\big]^* g. \tag{3.14}$$

Proof. We temporarily assume that $z \in \mathbb{C}\backslash\big(\sigma\big(H_{0,\Omega}^D\big) \cup \sigma\big(H_\Omega^D\big)\big)$ in the case of the Dirichlet problem and $z \in \mathbb{C}\backslash\big(\sigma\big(H_{0,\Omega}^N\big) \cup \sigma\big(H_\Omega^N\big)\big)$ in the context of the Neumann problem.
Uniqueness follows from the fact that $z \notin \sigma(H_\Omega^D)$ and $z \notin \sigma(H_\Omega^N)$, respectively.

Next, we will show that the functions

$$u^D(z) = u_0^D(z) - \big(H_\Omega^D - zI_\Omega\big)^{-1} V u_0^D(z), \tag{3.15}$$

$$u^N(z) = u_0^N(z) - \big(H_\Omega^N - zI_\Omega\big)^{-1} V u_0^N(z), \tag{3.16}$$

with u_0^D, u_0^N given by Theorem 3.1, satisfy (3.13) and (3.14), respectively. Indeed, it follows from Theorem 3.1 that $u_0^D, u_0^N \in H^{3/2}(\Omega)$ and $\widetilde{\gamma}_N u_0^D \in L^2(\partial\Omega; d^{n-1}\sigma)$. Using the Sobolev embedding theorem $H^{3/2}(\Omega) \hookrightarrow L^q(\Omega; d^n x)$, $q \geq 2$, and the fact that $V \in L^p(\Omega; d^n x)$, $p > 2$, one concludes that $Vu_0^D, Vu_0^N \in L^2(\Omega; d^n x)$, and hence (3.15) and (3.16) are well defined. Since one also has $V \in L^2(\Omega; d^n x)$, it follows from Lemma 2.3 that $V\big(H_{0,\Omega}^D - zI_\Omega\big)^{-1}$ and $V\big(H_{0,\Omega}^N - zI_\Omega\big)^{-1}$ are Hilbert–Schmidt, and hence

$$\big[I + V\big(H_{0,\Omega}^D - zI_\Omega\big)^{-1}\big]^{-1} \in \mathcal{B}\big(L^2(\Omega; d^n x)\big), \quad z \in \mathbb{C}\backslash\big(\sigma\big(H_{0,\Omega}^D\big) \cup \sigma\big(H_\Omega^D\big)\big), \tag{3.17}$$

$$\big[I + V\big(H_{0,\Omega}^N - zI_\Omega\big)^{-1}\big]^{-1} \in \mathcal{B}\big(L^2(\Omega; d^n x)\big), \quad z \in \mathbb{C}\backslash\big(\sigma\big(H_{0,\Omega}^N\big) \cup \sigma\big(H_\Omega^N\big)\big). \tag{3.18}$$

Thus, by (2.4) and (2.5),

$$\big(H_\Omega^D - zI_\Omega\big)^{-1} Vu_0^D = \big(H_{0,\Omega}^D - zI_\Omega\big)^{-1}\big[I + V\big(H_{0,\Omega}^D - zI_\Omega\big)^{-1}\big]^{-1} Vu_0^D \in H^2(\Omega), \tag{3.19}$$

$$\big(H_\Omega^N - zI_\Omega\big)^{-1} Vu_0^N = \big(H_{0,\Omega}^N - zI_\Omega\big)^{-1}\big[I + V\big(H_{0,\Omega}^N - zI_\Omega\big)^{-1}\big]^{-1} Vu_0^N \in H^2(\Omega), \tag{3.20}$$

and hence $u^D, u^N \in H^{3/2}(\Omega)$ and $\widetilde{\gamma}_N u^D \in L^2(\partial\Omega; d^{n-1}\sigma)$. Moreover,

$$\begin{aligned}(-\Delta + V - z)u^D &= (-\Delta - z)u_0^D + Vu_0^D - (-\Delta + V - z)\big(H_\Omega^D - zI_\Omega\big)^{-1} Vu_0^D \\ &= Vu_0^D - I_\Omega V u_0^D = 0, \end{aligned} \tag{3.21}$$

$$\begin{aligned}(-\Delta + V - z)u^N &= (-\Delta - z)u_0^N + Vu_0^N - (-\Delta + V - z)\big(H_\Omega^N - zI_\Omega\big)^{-1} Vu_0^N \\ &= Vu_0^N - I_\Omega V u_0^N = 0, \end{aligned} \tag{3.22}$$

and by (2.4), (2.5) and (3.17), (3.18) one also obtains,

$$\begin{aligned}\gamma_D u^D &= \gamma_D u_0^D - \gamma_D\big(H_\Omega^D - zI_\Omega\big)^{-1} Vu_0^D \\ &= f - \gamma_D\big(H_{0,\Omega}^D - zI_\Omega\big)^{-1}\big[I + V\big(H_{0,\Omega}^D - zI_\Omega\big)^{-1}\big]^{-1} Vu_0^D = f, \end{aligned} \tag{3.23}$$

$$\begin{aligned}\widetilde{\gamma}_N u^N &= \widetilde{\gamma}_N u_0^N - \widetilde{\gamma}_N\big(H_\Omega^N - zI_\Omega\big)^{-1} Vu_0^N \\ &= g - \gamma_N\big(H_{0,\Omega}^N - zI_\Omega\big)^{-1}\big[I + V\big(H_{0,\Omega}^N - zI_\Omega\big)^{-1}\big]^{-1} Vu_0^N = g. \end{aligned} \tag{3.24}$$

Finally, (3.13) and (3.14) follow from (3.9), (3.10), (3.15), (3.16), and the resolvent identity,

$$\begin{aligned}u^D(z) &= \big[I_\Omega - \big(H_\Omega^D - zI_\Omega\big)^{-1} V\big]\big[-\gamma_N\big(\big(H_{0,\Omega}^D - zI_\Omega\big)^{-1}\big)^*\big]^* f \\ &= -\big[\gamma_N\big(\big(H_{0,\Omega}^D - zI_\Omega\big)^{-1}\big)^*\big[I_\Omega - \big(H_\Omega^D - zI_\Omega\big)^{-1} V\big]^*\big]^* f \\ &= -\big[\gamma_N\big(\big(H_\Omega^D - zI_\Omega\big)^{-1}\big)^*\big]^* f, \end{aligned} \tag{3.25}$$

$$\begin{aligned}u^N(z) &= \big[I_\Omega - \big(H_\Omega^N - zI_\Omega\big)^{-1} V\big]\big[\gamma_D\big(\big(H_{0,\Omega}^N - zI_\Omega\big)^{-1}\big)^*\big]^* g \\ &= \big[\gamma_D\big(\big(H_{0,\Omega}^N - zI_\Omega\big)^{-1}\big)^*\big[I_\Omega - \big(H_\Omega^N - zI_\Omega\big)^{-1} V\big]^*\big]^* g \\ &= \big[\gamma_D\big(\big(H_\Omega^N - zI_\Omega\big)^{-1}\big)^*\big]^* g. \end{aligned} \tag{3.26}$$

Analytic continuation with respect to z then permits one to remove the additional condition $z \notin \sigma(H_{0,\Omega}^D)$ in the case of the Dirichlet problem, and the additional condition $z \notin \sigma(H_{0,\Omega}^N)$ in the context of the Neumann problem. $\square$

Assuming Hypothesis 3.2, we now introduce the Dirichlet-to-Neumann maps, $M_{0,\Omega}^D(z)$ and $M_\Omega^D(z)$, associated with $(-\Delta - z)$ and $(-\Delta + V - z)$ on Ω, as follows,

$$M_{0,\Omega}^D(z)\colon \begin{cases} H^1(\partial\Omega) \to L^2(\partial\Omega; d^{n-1}\sigma), \\ \qquad\qquad f \mapsto -\widetilde{\gamma}_N u_0^D, \end{cases} \quad z \in \mathbb{C}\backslash\sigma(H_{0,\Omega}^D), \tag{3.27}$$

where u_0^D is the unique solution of

$$(-\Delta - z)u_0^D = 0 \text{ on } \Omega, \quad u_0^D \in H^{3/2}(\Omega), \quad \gamma_D u_0^D = f \text{ on } \partial\Omega, \tag{3.28}$$

and

$$M_\Omega^D(z)\colon \begin{cases} H^1(\partial\Omega) \to L^2(\partial\Omega; d^{n-1}\sigma), \\ \qquad\qquad f \mapsto -\widetilde{\gamma}_N u^D, \end{cases} \quad z \in \mathbb{C}\backslash\sigma(H_\Omega^D), \tag{3.29}$$

where u^D is the unique solution of

$$(-\Delta + V - z)u^D = 0 \text{ on } \Omega, \quad u^D \in H^{3/2}(\Omega), \quad \gamma_D u^D = f \text{ on } \partial\Omega. \tag{3.30}$$

In addition, still assuming Hypothesis 3.2, we introduce the Neumann-to-Dirichlet maps, $M_{0,\Omega}^N(z)$ and $M_\Omega^N(z)$, associated with $(-\Delta - z)$ and $(-\Delta + V - z)$ on Ω, as follows,

$$M_{0,\Omega}^N(z)\colon \begin{cases} L^2(\partial\Omega; d^{n-1}\sigma) \to H^1(\partial\Omega), \\ \qquad\qquad g \mapsto \gamma_D u_0^N, \end{cases} \quad z \in \mathbb{C}\backslash\sigma(H_{0,\Omega}^N), \tag{3.31}$$

where u_0^N is the unique solution of

$$(-\Delta - z)u_0^N = 0 \text{ on } \Omega, \quad u_0^N \in H^{3/2}(\Omega), \quad \widetilde{\gamma}_N u_0^N = g \text{ on } \partial\Omega, \tag{3.32}$$

and

$$M_\Omega^N(z)\colon \begin{cases} L^2(\partial\Omega; d^{n-1}\sigma) \to H^1(\partial\Omega), \\ \qquad\qquad g \mapsto \gamma_D u^N, \end{cases} \quad z \in \mathbb{C}\backslash\sigma(H_\Omega^N), \tag{3.33}$$

where u^N is the unique solution of

$$(-\Delta + V - z)u^N = 0 \text{ on } \Omega, \quad u^N \in H^{3/2}(\Omega), \quad \widetilde{\gamma}_N u^N = g \text{ on } \partial\Omega. \tag{3.34}$$

It follows from Theorems 3.1 and 3.3, that under the assumption of Hypothesis 3.2, the operators $M_{0,\Omega}^D(z)$, $M_\Omega^D(z)$, $M_{0,\Omega}^N(z)$, and $M_\Omega^N(z)$ are well defined and satisfy the following equalities,

$$M_{0,\Omega}^N(z) = -M_{0,\Omega}^D(z)^{-1}, \quad z \in \mathbb{C}\backslash\big(\sigma(H_{0,\Omega}^D) \cup \sigma(H_{0,\Omega}^N)\big), \tag{3.35}$$

$$M_\Omega^N(z) = -M_\Omega^D(z)^{-1}, \quad z \in \mathbb{C}\backslash\big(\sigma(H_\Omega^D) \cup \sigma(H_\Omega^N)\big), \tag{3.36}$$

and

$$M_{0,\Omega}^D(z) = \widetilde{\gamma}_N \big[\gamma_N \big((H_{0,\Omega}^D - zI_\Omega)^{-1}\big)^*\big]^*, \quad z \in \mathbb{C}\backslash\sigma(H_{0,\Omega}^D), \tag{3.37}$$

$$M_{\Omega}^{D}(z)=\widetilde{\gamma}_N\big[\gamma_N\big(\big(H_{\Omega}^{D}-zI_{\Omega}\big)^{-1}\big)^*\big]^*, \quad z\in\mathbb{C}\backslash\sigma\big(H_{\Omega}^{D}\big), \tag{3.38}$$

$$M_{0,\Omega}^{N}(z)=\gamma_D\big[\gamma_D\big(\big(H_{0,\Omega}^{N}-zI_{\Omega}\big)^{-1}\big)^*\big]^*, \quad z\in\mathbb{C}\backslash\sigma\big(H_{0,\Omega}^{N}\big), \tag{3.39}$$

$$M_{\Omega}^{N}(z)=\gamma_D\big[\gamma_D\big(\big(H_{\Omega}^{N}-zI_{\Omega}\big)^{-1}\big)^*\big]^*, \quad z\in\mathbb{C}\backslash\sigma\big(H_{\Omega}^{N}\big). \tag{3.40}$$

The representations (3.37)–(3.40) provide a convenient point of departure for proving the operator-valued Herglotz property of M_{Ω}^{D} and M_{Ω}^{N}. We will return to this topic in a future paper.

Next, we note that the above formulas (3.37)–(3.40) may be used as alternative definitions of the Dirichlet-to-Neumann and Neumann-to-Dirichlet maps. In particular, we will next use (3.38) and (3.40) to extend the above definition of the operators $M_{\Omega}^{D}(z)$ and $M_{\Omega}^{N}(z)$ to the more general situation governed by Hypothesis 2.6:

Lemma 3.4 ([25]). *Assume Hypothesis* 2.6. *Then the operators $M_{\Omega}^{D}(z)$ and $M_{\Omega}^{N}(z)$ defined by equalities* (3.38) *and* (3.40) *have the following boundedness properties,*

$$M_{\Omega}^{D}(z)\in\mathcal{B}\big(H^1(\partial\Omega),L^2(\partial\Omega;d^{n-1}\sigma)\big), \quad z\in\mathbb{C}\backslash\sigma\big(H_{\Omega}^{D}\big), \tag{3.41}$$

$$M_{\Omega}^{N}(z)\in\mathcal{B}\big(L^2(\partial\Omega;d^{n-1}\sigma),H^1(\partial\Omega)\big), \quad z\in\mathbb{C}\backslash\sigma\big(H_{\Omega}^{N}\big). \tag{3.42}$$

A detailed proof of Lemma 3.4 will be provided in [25].

Weyl–Titchmarsh operators, in a spirit close to ours, have recently been discussed by Amrein and Pearson [2] in connection with the interior and exterior of a ball in $\mathbb{R}^3$ and real-valued potentials $V\in L^\infty(\mathbb{R}^3;d^3x)$. For additional literature on Weyl–Titchmarsh operators, relevant in the context of boundary value spaces (boundary triples, etc.), we refer, for instance, to [1], [3], [4], [5], [6], [7], [14], [15], [22], [29, Ch. 3], [38], [39], [41], [52], [53].

Next, we prove the following auxiliary result, which will play a crucial role in Theorem 4.2, the principal result of this paper.

Lemma 3.5. *Assume Hypothesis* 2.6. *Then the following identities hold,*

$$M_{0,\Omega}^{D}(z)-M_{\Omega}^{D}(z)=\overline{\widetilde{\gamma}_N\big(H_{\Omega}^{D}-zI_{\Omega}\big)^{-1}V\big[\gamma_N\big(\big(H_{0,\Omega}^{D}-zI_{\Omega}\big)^{-1}\big)^*\big]^*}, \\ z\in\mathbb{C}\backslash\big(\sigma\big(H_{0,\Omega}^{D}\big)\cup\sigma\big(H_{\Omega}^{D}\big)\big), \tag{3.43}$$

$$M_{\Omega}^{D}(z)M_{0,\Omega}^{D}(z)^{-1}=I_{\partial\Omega}-\overline{\widetilde{\gamma}_N\big(H_{\Omega}^{D}-zI_{\Omega}\big)^{-1}V\big[\gamma_D\big(\big(H_{0,\Omega}^{N}-zI_{\Omega}\big)^{-1}\big)^*\big]^*}, \\ z\in\mathbb{C}\backslash\big(\sigma\big(H_{0,\Omega}^{D}\big)\cup\sigma\big(H_{\Omega}^{D}\big)\cup\sigma\big(H_{0,\Omega}^{N}\big)\big). \tag{3.44}$$

Proof. Let $z\in\mathbb{C}\backslash\big(\sigma\big(H_{0,\Omega}^{D}\big)\cup\sigma\big(H_{\Omega}^{D}\big)\big)$. Then (3.43) follows from (3.37), (3.38), and the resolvent identity

$$\begin{aligned} M_{0,\Omega}^{D}(z)-M_{\Omega}^{D}(z)&=\widetilde{\gamma}_N\big[\gamma_N\big(\big(H_{0,\Omega}^{D}-zI_{\Omega}\big)^{-1}-\big(H_{\Omega}^{D}-zI_{\Omega}\big)^{-1}\big)^*\big]^* \\ &=\overline{\widetilde{\gamma}_N\big[\gamma_N\big(\big(H_{\Omega}^{D}-zI_{\Omega}\big)^{-1}V\big(H_{0,\Omega}^{D}-zI_{\Omega}\big)^{-1}\big)^*\big]^*} \\ &=\overline{\widetilde{\gamma}_N\big(H_{\Omega}^{D}-zI_{\Omega}\big)^{-1}V\big[\gamma_N\big(\big(H_{0,\Omega}^{D}-zI_{\Omega}\big)^{-1}\big)^*\big]^*}. \end{aligned} \tag{3.45}$$

Next, if $z \in \mathbb{C}\backslash\big(\sigma(H_{0,\Omega}^D) \cup \sigma(H_\Omega^D) \cup \sigma(H_{0,\Omega}^N)\big)$, then it follows from (3.35), (3.39), and (3.43) that

$$\begin{aligned} M_\Omega^D(z) M_{0,\Omega}^D(z)^{-1} &= I_{\partial\Omega} + \big(M_\Omega^D(z) - M_{0,\Omega}^D(z)\big) M_{0,\Omega}^D(z)^{-1} \\ &= I_{\partial\Omega} + \big(M_{0,\Omega}^D(z) - M_\Omega^D(z)\big) M_{0,\Omega}^N(z) \\ &= I_{\partial\Omega} + \overline{\widetilde{\gamma}_N \big(H_\Omega^D - zI_\Omega\big)^{-1} V \big[\gamma_N\big(\big(H_{0,\Omega}^D - zI_\Omega\big)^{-1}\big)^*\big]^*} \\ &\quad \times \gamma_D\big[\gamma_D\big(\big(H_{0,\Omega}^N - zI_\Omega\big)^{-1}\big)^*\big]^*. \end{aligned} \tag{3.46}$$

Let $g \in L^2(\partial\Omega; d^{n-1}\sigma)$. Then by Theorem 3.1,

$$u = \big[\gamma_D\big(\big(H_{0,\Omega}^N - zI_\Omega\big)^{-1}\big)^*\big]^* g \tag{3.47}$$

is the unique solution of

$$(-\Delta - z)u = 0 \text{ on } \Omega, \quad u \in H^{3/2}(\Omega), \quad \widetilde{\gamma}_N u = g \text{ on } \partial\Omega. \tag{3.48}$$

Setting $f = \gamma_D u \in H^1(\partial\Omega)$ and utilizing Theorem 3.1 once again, one obtains

$$\begin{aligned} u &= -\big[\gamma_N\big(H_{0,\Omega}^D - \overline{z}I_\Omega\big)^{-1}\big]^* f \\ &= -\big[\gamma_N\big(\big(H_{0,\Omega}^D - zI_\Omega\big)^{-1}\big)^*\big]^* \gamma_D\big[\gamma_D\big(\big(H_{0,\Omega}^N - zI_\Omega\big)^{-1}\big)^*\big]^* g. \end{aligned} \tag{3.49}$$

Thus, it follows from (3.47) and (3.49) that

$$\begin{aligned} &\big[\gamma_N\big(\big(H_{0,\Omega}^D - zI_\Omega\big)^{-1}\big)^*\big]^* \gamma_D\big[\gamma_D\big(\big(H_{0,\Omega}^N - zI_\Omega\big)^{-1}\big)^*\big]^* \\ &\quad = -\big[\gamma_D\big(\big(H_{0,\Omega}^N - zI_\Omega\big)^{-1}\big)^*\big]^*. \end{aligned} \tag{3.50}$$

Finally, insertion of (3.50) into (3.46) yields (3.44). □

It follows from (4.25)–(4.30) that $\widetilde{\gamma}_N$ can be replaced by γ_N on the right-hand side of (3.43) and (3.44).

4. A multi-dimensional variant of a formula due to Jost and Pais

In this section we prove our multi-dimensional variants of the Jost and Pais formula as discussed in the introduction.

We start with an elementary comment on determinants which, however, lies at the heart of the matter of our multi-dimensional variant of the one-dimensional Jost and Pais result. Suppose $A \in \mathcal{B}(\mathcal{H}_1, \mathcal{H}_2)$, $B \in \mathcal{B}(\mathcal{H}_2, \mathcal{H}_1)$ with $AB \in \mathcal{B}_1(\mathcal{H}_2)$ and $BA \in \mathcal{B}_1(\mathcal{H}_1)$. Then,

$$\det(I_{\mathcal{H}_2} - AB) = \det(I_{\mathcal{H}_1} - BA). \tag{4.1}$$

In particular, $\mathcal{H}_1$ and $\mathcal{H}_2$ may have different dimensions. Especially, one of them may be infinite and the other finite, in which case one of the two determinants in (4.1) reduces to a finite determinant. This case indeed occurs in the original one-dimensional case studied by Jost and Pais [34] as described in detail in [24] and the references therein. In the proof of the next theorem, the role of $\mathcal{H}_1$ and $\mathcal{H}_2$ will be played by $L^2(\Omega; d^n x)$ and $L^2(\partial\Omega; d^{n-1}\sigma)$, respectively.

We start with an extension of a result in [23]:

Theorem 4.1. *Assume Hypothesis* 2.6 *and let* $z \in \mathbb{C}\backslash\big(\sigma\big(H_\Omega^D\big)\cup\sigma\big(H_{0,\Omega}^D\big)\cup\sigma\big(H_{0,\Omega}^N\big)\big)$. *Then,*

$$\overline{\gamma_N\big(H_{0,\Omega}^D - zI_\Omega\big)^{-1}V\big(H_\Omega^D - zI_\Omega\big)^{-1}V\big[\gamma_D\big(H_{0,\Omega}^N - \overline{z}I_\Omega\big)^{-1}\big]^*} \in \mathcal{B}_1\big(L^2(\partial\Omega; d^{n-1}\sigma)\big), \tag{4.2}$$

$$\overline{\gamma_N\big(H_\Omega^D - zI_\Omega\big)^{-1}V\big[\gamma_D\big(H_{0,\Omega}^N - \overline{z}I_\Omega\big)^{-1}\big]^*} \in \mathcal{B}_2\big(L^2(\partial\Omega; d^{n-1}\sigma)\big), \tag{4.3}$$

and

$$\begin{aligned}
&\frac{\det{}_2\Big(I_\Omega + \overline{u\big(H_{0,\Omega}^N - zI_\Omega\big)^{-1}v}\Big)}{\det{}_2\Big(I_\Omega + \overline{u\big(H_{0,\Omega}^D - zI_\Omega\big)^{-1}v}\Big)} \\
&\quad = \det{}_2\Big(I_{\partial\Omega} - \overline{\gamma_N\big(H_\Omega^D - zI_\Omega\big)^{-1}V\big[\gamma_D\big(H_{0,\Omega}^N - \overline{z}I_\Omega\big)^{-1}\big]^*}\Big) \qquad (4.4)\\
&\qquad \times \exp\Big(\operatorname{tr}\Big(\overline{\gamma_N\big(H_{0,\Omega}^D - zI_\Omega\big)^{-1}V\big(H_\Omega^D - zI_\Omega\big)^{-1}V\big[\gamma_D\big(H_{0,\Omega}^N - \overline{z}I_\Omega\big)^{-1}\big]^*}\Big)\Big).
\end{aligned}$$

Proof. From the outset we note that the left-hand side of (4.4) is well defined by (2.20). Let $z \in \mathbb{C}\backslash\big(\sigma\big(H_\Omega^D\big)\cup\sigma\big(H_{0,\Omega}^D\big)\cup\sigma\big(H_{0,\Omega}^N\big)\big)$ and

$$u(x) = \exp(i\arg(V(x)))|V(x)|^{1/2}, \quad v(x) = |V(x)|^{1/2}, \tag{4.5}$$

$$\widetilde{u}(x) = \exp(i\arg(V(x)))|V(x)|^{p/p_1}, \quad \widetilde{v}(x) = |V(x)|^{p/p_2}, \tag{4.6}$$

where

$$p_1 = \begin{cases} 3p/2, & n=2, \\ 4p/3, & n=3, \end{cases} \qquad p_2 = \begin{cases} 3p, & n=2, \\ 4p, & n=3 \end{cases} \tag{4.7}$$

with p as introduced in Hypothesis 2.6. Then it follows that $\frac{1}{p_1} + \frac{1}{p_2} = \frac{1}{p}$, in both cases $n = 2,3$, and hence $V = uv = \widetilde{u}\widetilde{v}$.

Next, we introduce

$$K_D(z) = -\overline{u\big(H_{0,\Omega}^D - zI_\Omega\big)^{-1}v}, \quad K_N(z) = -\overline{u\big(H_{0,\Omega}^N - zI_\Omega\big)^{-1}v} \tag{4.8}$$

and note that

$$[I_\Omega - K_D(z)]^{-1} \in \mathcal{B}\big(L^2(\Omega; d^n x)\big), \quad z \in \mathbb{C}\backslash\big(\sigma\big(H_\Omega^D\big)\cup\sigma\big(H_{0,\Omega}^D\big)\big). \tag{4.9}$$

Thus, utilizing the following facts,

$$[I_\Omega - K_D(z)]^{-1} = I_\Omega + K_D(z)[I_\Omega - K_D(z)]^{-1} \tag{4.10}$$

and

$$\begin{aligned}
1 &= \det{}_2\big([I_\Omega - K_D(z)][I_\Omega - K_D(z)]^{-1}\big) \qquad (4.11)\\
&= \det{}_2\big(I_\Omega - K_D(z)\big)\det{}_2\big([I_\Omega - K_D(z)]^{-1}\big)\exp\big(\operatorname{tr}\big(K_D(z)^2[I_\Omega - K_D(z)]^{-1}\big)\big),
\end{aligned}$$

one obtains

$$\begin{aligned}
&\det{}_2\big([I_\Omega - K_N(z)][I_\Omega - K_D(z)]^{-1}\big)\\
&\quad = \det{}_2\big(I_\Omega - K_N(z)\big)\det{}_2\big([I_\Omega - K_D(z)]^{-1}\big)\\
&\qquad \times \exp\big(\operatorname{tr}\big(K_N(z)K_D(z)[I_\Omega - K_D(z)]^{-1}\big)\big) \qquad (4.12)\\
&\quad = \frac{\det{}_2\big(I_\Omega - K_N(z)\big)}{\det{}_2\big(I_\Omega - K_D(z)\big)}\exp\big(\operatorname{tr}\big((K_N(z)-K_D(z))K_D(z)[I_\Omega - K_D(z)]^{-1}\big)\big).
\end{aligned}$$

At this point, the left-hand side of (4.4) can be rewritten as

$$\begin{aligned}
&\frac{\det{}_2\Big(I_\Omega + \overline{u\big(H_{0,\Omega}^N - zI_\Omega\big)^{-1}v}\Big)}{\det{}_2\Big(I_\Omega + \overline{u\big(H_{0,\Omega}^D - zI_\Omega\big)^{-1}v}\Big)} = \frac{\det{}_2\big(I_\Omega - K_N(z)\big)}{\det{}_2\big(I_\Omega - K_D(z)\big)}\\
&\quad = \det{}_2\big([I_\Omega - K_N(z)][I_\Omega - K_D(z)]^{-1}\big)\\
&\qquad \times \exp\big(\operatorname{tr}\big((K_D(z)-K_N(z))K_D(z)[I_\Omega - K_D(z)]^{-1}\big)\big)\\
&\quad = \det{}_2\big(I_\Omega + (K_D(z)-K_N(z))[I_\Omega - K_D(z)]^{-1}\big) \qquad (4.13)\\
&\qquad \times \exp\big(\operatorname{tr}\big((K_D(z)-K_N(z))K_D(z)[I_\Omega - K_D(z)]^{-1}\big)\big).
\end{aligned}$$

Next, temporarily suppose that $V \in L^p(\Omega; d^nx) \cap L^\infty(\Omega; d^nx)$. Using [23, Lemma A.3] (an extension of a result of Nakamura [47, Lemma 6]) and [23, Remark A.5], one finds

$$\begin{aligned}
K_D(z) - K_N(z) &= -\overline{u\big[\big(H_{0,\Omega}^D - zI_\Omega\big)^{-1} - \big(H_{0,\Omega}^N - zI_\Omega\big)^{-1}\big]v}\\
&= -\overline{u\big[\gamma_D\big(H_{0,\Omega}^N - \overline{z}I_\Omega\big)^{-1}\big]^*}\,\overline{\gamma_N\big(H_{0,\Omega}^D - zI_\Omega\big)^{-1}v} \qquad (4.14)\\
&= -\Big[\overline{\gamma_D\big(H_{0,\Omega}^N - \overline{z}I_\Omega\big)^{-1}\overline{u}}\Big]^*\,\overline{\gamma_N\big(H_{0,\Omega}^D - zI_\Omega\big)^{-1}v}.
\end{aligned}$$

Thus, inserting (4.14) into (4.13) yields,

$$\begin{aligned}
&\frac{\det{}_2\Big(I_\Omega + \overline{u\big(H_{0,\Omega}^N - zI_\Omega\big)^{-1}v}\Big)}{\det{}_2\Big(I_\Omega + \overline{u\big(H_{0,\Omega}^D - zI_\Omega\big)^{-1}v}\Big)}\\
&\quad = \det{}_2\Big(I_\Omega - \Big[\overline{\gamma_D\big(H_{0,\Omega}^N - \overline{z}I_\Omega\big)^{-1}\overline{u}}\Big]^*\,\overline{\gamma_N\big(H_{0,\Omega}^D - zI_\Omega\big)^{-1}v}\\
&\qquad \times \Big[I_\Omega + \overline{u\big(H_{0,\Omega}^D - zI_\Omega\big)^{-1}v}\Big]^{-1}\Big)\\
&\qquad \times \exp\Big(\operatorname{tr}\Big(\Big[\overline{\gamma_D\big(H_{0,\Omega}^N - \overline{z}I_\Omega\big)^{-1}\overline{u}}\Big]^*\,\overline{\gamma_N\big(H_{0,\Omega}^D - zI_\Omega\big)^{-1}v}\\
&\qquad\qquad \times \overline{u\big(H_{0,\Omega}^D - zI_\Omega\big)^{-1}v}\Big[I_\Omega + \overline{u\big(H_{0,\Omega}^D - zI_\Omega\big)^{-1}v}\Big]^{-1}\Big)\Big). \qquad (4.15)
\end{aligned}$$

Then, utilizing Corollary 2.5 with p_1 and p_2 as in (4.7), one finds,

$$\overline{\gamma_D\big(H_{0,\Omega}^N - \overline{z}I_\Omega\big)^{-1}\overline{u}} \in \mathcal{B}_{p_1}\big(L^2(\Omega; d^n x), L^2(\partial\Omega; d^{n-1}\sigma)\big), \tag{4.16}$$

$$\overline{\gamma_N\big(H_{0,\Omega}^D - zI_\Omega\big)^{-1}v} \in \mathcal{B}_{p_2}\big(L^2(\Omega; d^n x), L^2(\partial\Omega; d^{n-1}\sigma)\big), \tag{4.17}$$

and hence,

$$\begin{aligned} \Big[\overline{\gamma_D\big(H_{0,\Omega}^N - \overline{z}I_\Omega\big)^{-1}\overline{u}}\Big]^*\overline{\gamma_N\big(H_{0,\Omega}^D - zI_\Omega\big)^{-1}v} &\in \mathcal{B}_p\big(L^2(\Omega; d^n x)\big) \\ &\subset \mathcal{B}_2\big(L^2(\Omega; d^n x)\big), \end{aligned} \tag{4.18}$$

$$\begin{aligned} \overline{\gamma_N\big(H_{0,\Omega}^D - zI_\Omega\big)^{-1}v}\Big[\overline{\gamma_D\big(H_{0,\Omega}^N - \overline{z}I_\Omega\big)^{-1}\overline{u}}\Big]^* &\in \mathcal{B}_p\big(L^2(\partial\Omega; d^{n-1}\sigma)\big) \\ &\subset \mathcal{B}_2\big(L^2(\partial\Omega; d^{n-1}\sigma)\big). \end{aligned} \tag{4.19}$$

Moreover, using the fact that

$$\Big[I_\Omega + \overline{u\big(H_{0,\Omega}^D - zI_\Omega\big)^{-1}v}\Big]^{-1} \in \mathcal{B}\big(L^2(\Omega; d^n x)\big), \quad z \in \mathbb{C}\backslash\big(\sigma\big(H_\Omega^D\big) \cup \sigma\big(H_{0,\Omega}^D\big)\big), \tag{4.20}$$

one now applies the idea expressed in formula (4.1) and rearranges the terms in (4.15) as follows:

$$\begin{aligned} &\frac{\det{}_2\Big(I_\Omega + \overline{u\big(H_{0,\Omega}^N - zI_\Omega\big)^{-1}v}\Big)}{\det{}_2\Big(I_\Omega + \overline{u\big(H_{0,\Omega}^D - zI_\Omega\big)^{-1}v}\Big)} \\ &\quad = \det{}_2\Big(I_{\partial\Omega} - \overline{\gamma_N\big(H_{0,\Omega}^D - zI_\Omega\big)^{-1}v}\Big[I_\Omega + \overline{u\big(H_{0,\Omega}^D - zI_\Omega\big)^{-1}v}\Big]^{-1} \\ &\qquad\qquad \times \Big[\overline{\gamma_D\big(H_{0,\Omega}^N - \overline{z}I_\Omega\big)^{-1}\overline{u}}\Big]^*\Big) \\ &\qquad \times \exp\Big(\operatorname{tr}\Big(\overline{\gamma_N\big(H_{0,\Omega}^D - zI_\Omega\big)^{-1}v}\,\overline{u\big(H_{0,\Omega}^D - zI_\Omega\big)^{-1}v} \\ &\qquad\qquad \times \Big[I_\Omega + \overline{u\big(H_{0,\Omega}^D - zI_\Omega\big)^{-1}v}\Big]^{-1}\Big[\overline{\gamma_D\big(H_{0,\Omega}^N - \overline{z}I_\Omega\big)^{-1}\overline{u}}\Big]^*\Big)\Big) \\ &\quad = \det{}_2\Big(I_{\partial\Omega} - \overline{\gamma_N\big(H_{0,\Omega}^D - zI_\Omega\big)^{-1}\widetilde{v}}\Big[I_\Omega + \overline{\widetilde{u}\big(H_{0,\Omega}^D - zI_\Omega\big)^{-1}\widetilde{v}}\Big]^{-1} \\ &\qquad\qquad \times \Big[\overline{\gamma_D\big(H_{0,\Omega}^N - \overline{z}I_\Omega\big)^{-1}\overline{\widetilde{u}}}\Big]^*\Big) \\ &\qquad \times \exp\Big(\operatorname{tr}\Big(\overline{\gamma_N\big(H_{0,\Omega}^D - zI_\Omega\big)^{-1}\widetilde{v}}\,\overline{\widetilde{u}\big(H_{0,\Omega}^D - zI_\Omega\big)^{-1}\widetilde{v}} \\ &\qquad\qquad \times \Big[I_\Omega + \overline{\widetilde{u}\big(H_{0,\Omega}^D - zI_\Omega\big)^{-1}\widetilde{v}}\Big]^{-1}\Big[\overline{\gamma_D\big(H_{0,\Omega}^N - \overline{z}I_\Omega\big)^{-1}\overline{\widetilde{u}}}\Big]^*\Big)\Big). \end{aligned} \tag{4.21}$$

In the last equality we employed the following simple identities,

$$V = uv = \widetilde{u}\widetilde{v}, \tag{4.22}$$

$$v\Big[I_\Omega + \overline{u\big(H_{0,\Omega}^D - zI_\Omega\big)^{-1}v}\Big]^{-1}u = \widetilde{v}\Big[I + \overline{\widetilde{u}\big(H_{0,\Omega}^D - zI_\Omega\big)^{-1}\widetilde{v}}\Big]^{-1}\widetilde{u}. \tag{4.23}$$

Utilizing (4.21) and the following resolvent identity,

$$\overline{(H_{\Omega}^{D}-zI_{\Omega})^{-1}\widetilde{v}} = \overline{(H_{0,\Omega}^{D}-zI_{\Omega})^{-1}\widetilde{v}}\Big[I_{\Omega}+\overline{\widetilde{u}(H_{0,\Omega}^{D}-zI_{\Omega})^{-1}\widetilde{v}}\Big]^{-1}, \tag{4.24}$$

one arrives at (4.4), subject to the extra assumption $V \in L^p(\Omega; d^n x) \cap L^\infty(\Omega; d^n x)$.

Finally, assuming only $V \in L^p(\Omega; d^n x)$ and utilizing [23, Thm. 3.2], Lemma 2.3, and Corollary 2.5 once again, one obtains

$$\Big[I_{\Omega}+\overline{\widetilde{u}(H_{0,\Omega}^{D}-zI_{\Omega})^{-1}\widetilde{v}}\Big]^{-1} \in \mathcal{B}\big(L^2(\Omega; d^n x)\big), \tag{4.25}$$

$$\widetilde{u}\big(H_{0,\Omega}^{D}-zI_{\Omega}\big)^{-p/p_1} \in \mathcal{B}_{p_1}\big(L^2(\Omega; d^n x)\big), \tag{4.26}$$

$$\widetilde{v}\big(H_{0,\Omega}^{D}-zI_{\Omega}\big)^{-p/p_2} \in \mathcal{B}_{p_2}\big(L^2(\Omega; d^n x)\big), \tag{4.27}$$

$$\overline{\gamma_D\big(H_{0,\Omega}^{N}-zI_{\Omega}\big)^{-1}\widetilde{u}} \in \mathcal{B}_{p_1}\big(L^2(\Omega; d^n x), L^2(\partial\Omega; d^{n-1}\sigma)\big), \tag{4.28}$$

$$\overline{\gamma_N\big(H_{0,\Omega}^{D}-zI_{\Omega}\big)^{-1}\widetilde{v}} \in \mathcal{B}_{p_2}\big(L^2(\Omega; d^n x), L^2(\partial\Omega; d^{n-1}\sigma)\big), \tag{4.29}$$

and hence

$$\overline{\widetilde{u}\big(H_{0,\Omega}^{D}-zI_{\Omega}\big)^{-1}\widetilde{v}} \in \mathcal{B}_p\big(L^2(\Omega; d^n x)\big) \subset \mathcal{B}_2\big(L^2(\Omega; d^n x)\big). \tag{4.30}$$

Relations (4.24)–(4.30) prove (4.2) and (4.3), and hence, the left- and the right-hand sides of (4.4) are well defined for $V \in L^p(\Omega; d^n x)$. Thus, using (2.9), (2.13), (2.14), the continuity of $\det{}_2(\cdot)$ with respect to the Hilbert–Schmidt norm $\|\cdot\|_{\mathcal{B}_2(L^2(\Omega;d^n x))}$, the continuity of $\operatorname{tr}(\cdot)$ with respect to the trace norm $\|\cdot\|_{\mathcal{B}_1(L^2(\Omega;d^n x))}$, and an approximation of $V \in L^p(\Omega; d^n x)$ by a sequence of potentials $V_k \in L^p(\Omega; d^n x) \cap L^\infty(\Omega; d^n x)$, $k \in \mathbb{N}$, in the norm of $L^p(\Omega; d^n x)$ as $k \uparrow \infty$, then extends the result from $V \in L^p(\Omega; d^n x) \cap L^\infty(\Omega; d^n x)$ to $V \in L^p(\Omega; d^n x)$, $n = 2, 3$. □

Given these preparations, we are now ready for the principal result of this paper, the multi-dimensional analog of Theorem 1.2:

Theorem 4.2. *Assume Hypothesis* 2.6 *and let* $z \in \mathbb{C}\backslash\big(\sigma(H_{\Omega}^{D}) \cup \sigma(H_{0,\Omega}^{D}) \cup \sigma(H_{0,\Omega}^{N})\big)$. *Then,*

$$\begin{aligned} M_{\Omega}^{D}(z)M_{0,\Omega}^{D}(z)^{-1} - I_{\partial\Omega} &= -\overline{\gamma_N\big(H_{\Omega}^{D}-zI_{\Omega}\big)^{-1}V\big[\gamma_D(H_{0,\Omega}^{N}-\overline{z}I_{\Omega})^{-1}\big]^*} \\ &\in \mathcal{B}_2\big(L^2(\partial\Omega; d^{n-1}\sigma)\big) \end{aligned} \tag{4.31}$$

and

$$\begin{aligned} &\frac{\det{}_2\Big(I_{\Omega}+\overline{u\big(H_{0,\Omega}^{N}-zI_{\Omega}\big)^{-1}v}\Big)}{\det{}_2\Big(I_{\Omega}+\overline{u\big(H_{0,\Omega}^{D}-zI_{\Omega}\big)^{-1}v}\Big)} \\ &\quad = \det{}_2\Big(I_{\partial\Omega} - \overline{\gamma_N\big(H_{\Omega}^{D}-zI_{\Omega}\big)^{-1}V\big[\gamma_D(H_{0,\Omega}^{N}-\overline{z}I_{\Omega})^{-1}\big]^*}\Big)e^{\operatorname{tr}(T_2(z))} &(4.32)\\ &\quad = \det{}_2\big(M_{\Omega}^{D}(z)M_{0,\Omega}^{D}(z)^{-1}\big)e^{\operatorname{tr}(T_2(z))}, &(4.33) \end{aligned}$$

where

$$T_2(z) = \overline{\gamma_N\big(H_{0,\Omega}^D - zI_\Omega\big)^{-1}V\big(H_\Omega^D - zI_\Omega\big)^{-1}V\big[\gamma_D\big(H_{0,\Omega}^N - \overline{z}I_\Omega\big)^{-1}\big]^*} \in \mathcal{B}_1\big(L^2(\partial\Omega; d^{n-1}\sigma)\big). \tag{4.34}$$

Proof. The result follows from combining Lemma 3.5 and Theorem 4.1. □

A few comments are in order at this point.

The sudden appearance of the exponential term $\exp(\operatorname{tr}(T_2(z)))$ in (4.32) and (4.33), when compared to the one-dimensional case, is due to the necessary use of the modified determinant $\det_p(\cdot)$ in Theorems 4.1 and 4.2.

The multi-dimensional extension (4.32) of (1.16), under the stronger hypothesis $V \in L^2(\Omega; d^n x)$, $n = 2, 3$, first appeared in [23]. However, the present results in Theorem 4.2 go decidedly beyond those in [23] in the sense that the class of domains Ω permitted by Hypothesis 2.1 is greatly enlarged as compared to [23] and the conditions on V satisfying Hypothesis 2.6 are nearly optimal by comparison with the Sobolev inequality (cf. Cheney [13], Reed and Simon [54, Sect. IX.4], Simon [57, Sect. I.1]). Moreover, the multi-dimensional extension (4.33) of (1.17) invoking Dirichlet-to-Neumann maps is a new result.

The principal reduction in Theorem 4.2 reduces (a ratio of) modified Fredholm determinants associated with operators in $L^2(\Omega; d^n x)$ on the left-hand side of (4.32) to modified Fredholm determinants associated with operators in $L^2(\partial\Omega; d^{n-1}\sigma)$ on the right-hand side of (4.32) and especially, in (4.33). This is the analog of the reduction described in the one-dimensional context of Theorem 1.2, where Ω corresponds to the half-line $(0,\infty)$ and its boundary $\partial\Omega$ thus corresponds to the one-point set $\{0\}$.

In the context of elliptic operators on smooth k-dimensional manifolds, the idea of reducing a ratio of zeta-function regularized determinants to a calculation over the $(k-1)$-dimensional boundary has been studied by Forman [18]. He also pointed out that if the manifold consists of an interval, the special case of a pair of boundary points then permits one to reduce the zeta-function regularized determinant to the determinant of a finite-dimensional matrix. The latter case is of course an analog of the one-dimensional Jost and Pais formula mentioned in the introduction (cf. Theorems 1.1 and 1.2). Since then, this topic has been further developed in various directions and we refer, for instance, to Burghelea, Friedlander, and Kappeler [8], [9], [10], [11], Carron [12], Friedlander [19], Müller [46], Park and Wojciechowski [51], and the references therein.

Remark 4.3. The following observation yields a simple application of formula (4.32). Since by the Birman–Schwinger principle (cf., e.g., the discussion in [23, Sect. 3]), for any $z \in \mathbb{C}\backslash\big(\sigma\big(H_\Omega^D\big) \cup \sigma\big(H_{0,\Omega}^D\big) \cup \sigma\big(H_{0,\Omega}^N\big)\big)$, one has $z \in \sigma\big(H_\Omega^N\big)$ if and only if $\det_2\Big(I_\Omega + \overline{u\big(H_{0,\Omega}^N - zI_\Omega\big)^{-1}v}\Big) = 0$, it follows from (4.32) that

$$\begin{aligned} &\text{for all } z \in \mathbb{C}\backslash\big(\sigma\big(H_\Omega^D\big) \cup \sigma\big(H_{0,\Omega}^D\big) \cup \sigma\big(H_{0,\Omega}^N\big)\big), \text{ one has } z \in \sigma\big(H_\Omega^N\big) \\ &\quad \text{if and only if } \det_2\Big(I_{\partial\Omega} - \overline{\gamma_N\big(H_\Omega^D - zI_\Omega\big)^{-1}V\big[\gamma_D\big(H_{0,\Omega}^N - \overline{z}I_\Omega\big)^{-1}\big]^*}\Big) = 0. \end{aligned} \tag{4.35}$$

One can also prove the following analog of (4.32):

$$\begin{aligned}
&\frac{\det{}_2\Big(I_\Omega + \overline{u(H_{0,\Omega}^D - zI_\Omega)^{-1} v}\Big)}{\det{}_2\Big(I_\Omega + \overline{u(H_{0,\Omega}^N - zI_\Omega)^{-1} v}\Big)} \\
&\quad = \det{}_2\Big(I_{\partial\Omega} + \overline{\gamma_N (H_{0,\Omega}^D - zI_\Omega)^{-1} V \big[\gamma_D ((H_\Omega^N - zI_\Omega)^{-1})^*\big]^*}\Big) \qquad (4.36) \\
&\qquad \times \exp\Big(- \operatorname{tr}\Big(\overline{\gamma_N (H_{0,\Omega}^D - zI_\Omega)^{-1} V (H_\Omega^N - zI_\Omega)^{-1} V \big[\gamma_D (H_{0,\Omega}^N - \overline{z}I_\Omega)^{-1}\big]^*}\Big)\Big).
\end{aligned}$$

Then, proceeding as before, one obtains

$$\begin{aligned}
&\text{for all } z \in \mathbb{C}\backslash\big(\sigma(H_\Omega^N) \cup \sigma(H_{0,\Omega}^N) \cup \sigma(H_{0,\Omega}^D)\big), \text{ one has } z \in \sigma(H_\Omega^D) \qquad (4.37) \\
&\quad \text{if and only if } \det{}_2\Big(I_{\partial\Omega} + \overline{\gamma_N (H_{0,\Omega}^D - zI_\Omega)^{-1} V \big[\gamma_D ((H_\Omega^N - zI_\Omega)^{-1})^*\big]^*}\Big) = 0.
\end{aligned}$$

Acknowledgments

We are indebted to Yuri Latushkin and Konstantin A. Makarov for numerous discussions on this topic. We also thank the referee for a careful reading of our manuscript and his constructive comments.

Fritz Gesztesy would like to thank all organizers of the international conference on Operator Theory and Mathematical Physics (OTAMP), and especially, Pavel Kurasov, for their kind invitation, the stimulating atmosphere during the meeting, and the hospitality extended to him during his stay in Lund in June of 2006. He also gratefully acknowledges a research leave for the academic year 2005/06 granted by the Research Council and the Office of Research of the University of Missouri–Columbia.

References

[1] S. Albeverio, J.F. Brasche, M.M. Malamud, and H. Neidhardt, *Inverse spectral theory for symmetric operators with several gaps: scalar-type Weyl functions*, J. Funct. Anal. **228**, 144–188 (2005).

[2] W.O. Amrein and D.B. Pearson, *M operators: a generalisation of Weyl–Titchmarsh theory*, J. Comp. Appl. Math. **171**, 1–26 (2004).

[3] J. Behrndt and M. Langer, *Boundary value problems for partial differential operators on bounded domains*, J. Funct. Anal. **243**, 536–565 (2007).

[4] J. Behrndt, M.M. Malamud, and H. Neidhardt, *Scattering matrices and Weyl functions*, preprint, 2006.

[5] J.F. Brasche, M.M. Malamud, and H. Neidhardt, *Weyl functions and singular continuous spectra of self-adjoint extensions*, in *Stochastic Processes, Physics and Geometry: New Interplays. II. A Volume in Honor of Sergio Albeverio*, F. Gesztesy, H. Holden, J. Jost, S. Paycha, M. Röckner, and S. Scarlatti (eds.), Canadian Mathematical Society Conference Proceedings, Vol. 29, Amer. Math. Soc., Providence, RI, 2000, pp. 75–84.

[6] J.F. Brasche, M.M. Malamud, and H. Neidhardt, *Weyl function and spectral properties of self-adjoint extensions*, Integral Eqs. Operator Theory **43**, 264–289 (2002).

[7] B.M. Brown and M. Marletta, *Spectral inclusion and spectral exactness for PDE's on exterior domains*, IMA J. Numer. Anal. **24**, 21–43 (2004).

[8] D. Burghelea, L. Friedlander, and T. Kappeler, *On the determinant of elliptic differential and finite difference operators in vector bundles over S^1*, Commun. Math. Phys. **138**, 1–18 (1991). Erratum: Commun. Math. Phys. **150**, 431 (1992).

[9] D. Burghelea, L. Friedlander, and T. Kappeler, *Meyer–Vietoris type formula for determinants of elliptic differential operators*, J. Funct. Anal. **107**, 34–65 (1992).

[10] D. Burghelea, L. Friedlander, and T. Kappeler, *Regularized determinants for pseudodifferential operators in vector bundles over S^1*, Integral Eqs. Operator Theory **16**, 496–513 (1993).

[11] D. Burghelea, L. Friedlander, and T. Kappeler, *On the determinant of elliptic boundary value problems on a line segment*, Proc. Amer. Math. Soc. **123**, 3027–3038 (1995).

[12] G. Carron, *Déterminant relatif et la fonction xi*, Amer. J. Math. **124**, 307–352 (2002).

[13] M. Cheney, *Two-dimensional scattering: The number of bound states from scattering data*, J. Math. Phys. **25**, 1449–1455 (1984).

[14] V.A. Derkach and M.M. Malamud, *Generalized resolvents and the boundary value problems for Hermitian operators with gaps*, J. Funct. Anal. **95**, 1–95 (1991).

[15] V.A. Derkach and M.M. Malamud, *The extension theory of Hermitian operators and the moment problem*, J. Math. Sci. **73**, 141–242 (1995).

[16] N. Dunford and J.T. Schwartz, *Linear Operators Part II: Spectral Theory*, Interscience, New York, 1988.

[17] D.E. Edmunds and H. Triebel, *Entropy numbers and approximation numbers in function spaces*, Proc. London Math. Soc. (3) **58**, 137–152 (1989).

[18] R. Forman, *Functional determinants and geometry*, Invent. Math. **88**, 447–493 (1987).

[19] L. Friedlander, *Determinant of the Schrödinger operator on a metric graph*, preprint, 2005.

[20] I.M. Gel'fand and B. M. Levitan, *On the determination of a differential equation from its special function*, Izv. Akad. Nauk SSR. Ser. Mat. **15**, 309–360 (1951) (Russian); Engl. transl. in Amer. Math. Soc. Transl. Ser. 2, **1**, 253–304 (1955).

[21] F. Gesztesy, *Inverse spectral theory as influenced by Barry Simon*, in *Spectral Theory and Mathematical Physics: A Festschrift in Honor of Barry Simon's* 60*th Birthday. Ergodic Schrödinger Operators, Singular Spectrum, Orthogonal Polynomials, and Inverse Spectral Theory*, F. Gesztesy, P. Deift, C. Galvez, P. Perry, and W. Schlag (eds.), Proceedings of Symposia in Pure Mathematics, Vol. 76/2, Amer. Math. Soc., Providence, RI, 2007, pp. 741–820.

[22] F. Gesztesy, N.J. Kalton, K.A. Makarov, and E. Tsekanovskii, *Some Applications of Operator-Valued Herglotz Functions*, in *Operator Theory, System Theory and Related Topics. The Moshe Livšic Anniversary Volume*, D. Alpay and V. Vinnikov (eds.), Operator Theory: Advances and Applications, Vol. 123, Birkhäuser, Basel, 2001, p. 271–321.

[23] F. Gesztesy, Y. Latushkin, M. Mitrea, and M. Zinchenko, *Nonselfadjoint operators, infinite determinants, and some applications*, Russ. J. Math. Phys. **12**, 443–471 (2005).

[24] F. Gesztesy and K.A. Makarov, *(Modified) Fredholm Determinants for Operators with Matrix-Valued Semi-Separable Integral Kernels Revisited*, Integral Eqs. Operator Theory **47**, 457–497 (2003). (See also Erratum **48**, 425–426 (2004) and the corrected electronic only version in **48**, 561–602 (2004).)

[25] F. Gesztesy, M. Mitrea, and M. Zinchenko, *Variations on a theme of Jost and Pais*, J. Funct. Anal. **253**, 399–448 (2007).

[26] F. Gesztesy and B. Simon, *A new approach to inverse spectral theory, II. General real potentials and the connection to the spectral measure*, Ann. Math. **152**, 593–643 (2000).

[27] I. Gohberg, S. Goldberg, and N. Krupnik, *Traces and Determinants for Linear Operators*, Operator Theory: Advances and Applications, Vol. 116, Birkhäuser, Basel, 2000.

[28] I. Gohberg and M.G. Krein, *Introduction to the Theory of Linear Nonselfadjoint Operators*, Translations of Mathematical Monographs, Vol. 18, Amer. Math. Soc., Providence, RI, 1969.

[29] V.I. Gorbachuk and M.L. Gorbachuk, *Boundary Value Problems for Operator Differential Equations*, Kluwer, Dordrecht, 1991.

[30] P. Grisvard, *Elliptic Problems in Nonsmooth Domains*, Pitman, Boston, 1985.

[31] P. Grisvard, G. Iooss, *Problèmes aux limites unilatéraux dans des domaines non réguliers*, Publications des Séminaires de Mathématiques, Université de Rennes **9**, 1–26 (1975).

[32] D.S. Jerison and C.E. Kenig, *The Neumann problem on Lipschitz domains*, Bull. Amer. Math. Soc. (N.S.) **4**, 203–207 (1981).

[33] D. Jerison and C. Kenig, *The inhomogeneous Dirichlet problem in Lipschitz domains*, J. Funct. Anal. **130**, 161–219 (1995).

[34] R. Jost and A. Pais, *On the scattering of a particle by a static potential*, Phys. Rev. **82**, 840–851 (1951).

[35] J. Kadlec, *The regularity of the solution of the Poisson problem in a domain whose boundary is similar to that of a convex domain*, Czechoslovak Math. J. **14**, 386–393 (1964).

[36] T. Kato, *Wave operators and similarity for some non-selfadjoint operators*, Math. Ann. **162**, 258–279 (1966).

[37] R. Konno and S.T. Kuroda, *On the finiteness of perturbed eigenvalues*, J. Fac. Sci., Univ. Tokyo, Sec. I, **13**, 55–63 (1966).

[38] M.M. Malamud and V.I. Mogilevskii, *Krein type formula for canonical resolvents of dual pairs of linear relations*, Methods Funct. Anal. Topology, **8**, No. 4, 72–100 (2002).

[39] M. Marletta, *Eigenvalue problems on exterior domains and Dirichlet to Neumann maps*, J. Comp. Appl. Math. **171**, 367–391 (2004).

[40] W. McLean, *Strongly Elliptic Systems and Boundary Integral Equations*, Cambridge University Press, Cambridge, 2000.

[41] A.B. Mikhailova, B.S. Pavlov, and L.V. Prokhorov, *Intermediate Hamiltonian via Glazman's splitting and analytic perturbation for meromorphic matrix-functions*, Math. Nachr., to appear.

[42] D. Mitrea, M. Mitrea, and M. Taylor, *Layer Potentials, the Hodge Laplacian, and Global Boundary Problems in Nonsmooth Riemannian Manifolds*, Mem. Amer. Math. Soc. **150**, No. 713, (2001).

[43] M. Mitrea, *Boundary value problems and Hardy spaces associated to the Helmholtz equation in Lipschitz domains*, J. Math. Anal. Appl. **202**, 819–842 (1996).

[44] M. Mitrea, *Dirichlet integrals and Gaffney–Friedrichs inequalities in convex domain*, Forum Math. **13**, 531–567 (2001).

[45] M. Mitrea and M. Taylor, *Potential theory on Lipschitz domains in Riemannian manifolds: Sobolev-Besov space results and the Poisson problem*, J. Funct. Anal. **176**, 1–79 (2000).

[46] W. Müller, *Relative zeta functions, relative determinants and scattering theory*, Commun. Math. Phys. **192**, 309–347 (1998).

[47] S. Nakamura, *A remark on the Dirichlet–Neumann decoupling and the integrated density of states*, J. Funct. Anal. **179**, 136–152 (2001).

[48] R.G. Newton, *Relation between the three-dimensional Fredholm determinant and the Jost function*, J. Math. Phys. **13**, 880–883 (1972).

[49] R.G. Newton, *Inverse scattering. I. One dimension*, J. Math. Phys. **21**, 493–505 (1980).

[50] R.G. Newton, *Scattering Theory of Waves and Particles*, 2nd ed., Dover, New York, 2002.

[51] J. Park and K.P. Wojciechowski, *Adiabatic decomposition of the ζ-determinant and Dirichlet to Neumann operator*, J. Geom. Phys. **55**, 241–266 (2005).

[52] B. Pavlov, *The theory of extensions and explicitly-solvable models*, Russ. Math. Surv. **42:6**, 127–168 (1987).

[53] B. Pavlov, *S-matrix and Dirichlet-to-Neumann operators*, Ch. 6.1.6 in *Scattering: Scattering and Inverse Scattering in Pure and Applied Science*, Vol. 2, R. Pike and P. Sabatier (eds.), Academic Press, San Diego, 2002, pp. 1678–1688.

[54] M. Reed and B. Simon, *Methods of Modern Mathematical Physics. II: Fourier Analysis, Self-Adjointness*, Academic Press, New York, 1975.

[55] M. Reed and B. Simon, *Methods of Modern Mathematical Physics. IV: Analysis of Operators,* Academic Press, New York, 1978.

[56] C. Remling, *Inverse spectral theory for one-dimensional Schrödinger operators: The A function*, Math. Z. **245**, 597–617 (2003).

[57] B. Simon, *Quantum Mechanics for Hamiltonians Defined as Quadratic Forms*, Princeton University Press, Princeton, NJ, 1971.

[58] B. Simon, *Notes on infinite determinants of Hilbert space operators*, Adv. Math. **24**, 244–273 (1977).

[59] B. Simon, *A new approach to inverse spectral theory, I. Fundamental formalism*, Ann. Math. **150**, 1029–1057 (1999).

[60] B. Simon, *Resonances in one dimension and Fredholm determinants*, J. Funct. Anal. **178**, 396–420 (2000).

[61] B. Simon, *Trace Ideals and Their Applications*, Mathematical Surveys and Monographs, Vol. 120, 2nd ed., Amer. Math. Soc., Providence, RI, 2005.

[62] E.M. Stein, *Singular Integrals and Differentiability Properties of Functions*, Princeton Mathematical Series, No. 30, Princeton University Press, Princeton, NJ, 1970.

[63] G. Talenti, *Sopra una classe di equazioni ellittiche a coefficienti misurabili*, Ann. Mat. Pura Appl. **69**, 285–304 (1965).

[64] M.E. Taylor, *Pseudodifferential Operators and Nonlinear PDE*, Progress in Mathematics Vol. 100, Birkhäuser, Boston, MA, 1991.

[65] M.E. Taylor, *Partial Differential Equations, Vol. II*, Springer, New York, 1996.

[66] H. Triebel, *Interpolation Theory, Function Spaces, Differential Operators*, North-Holland, Amsterdam, 1978.

[67] H. Triebel, *Theory of Function Spaces*, Monographs in Mathematics, Vol. 78, Birkhäuser, Basel, 1983.

[68] H. Triebel, *Function spaces in Lipschitz domains and on Lipschitz manifolds. Characteristic functions as pointwise multipliers*, Rev. Mat. Complut. **15**, 475–524 (2002).

[69] G. Verchota, *Layer potentials and regularity for the Dirichlet problem for Laplace's equation in Lipschitz domains*, J. Funct. Anal. **59**, 572–611 (1984).

[70] J. Wloka, *Partial Differential Equations*, Cambridge University Press, Cambridge, 1987.

Fritz Gesztesy and Marius Mitrea
Department of Mathematics
University of Missouri
Columbia, MO 65211, USA
e-mail: `fritz@math.missouri.edu`
URL: `http://www.math.missouri.edu/personnel/faculty/gesztesyf.html`
e-mail: `marius@math.missouri.edu`
URL: `http://www.math.missouri.edu/personnel/faculty/mitream.html`

Maxim Zinchenko
Department of Mathematics
California Institute of Technology
Pasadena, CA 91125, USA
e-mail: `maxim@caltech.edu`

Operator Theory:
Advances and Applications, Vol. 186, 217–228

Higher Derivatives of Spectral Functions Associated with One-dimensional Schrödinger Operators

D.J. Gilbert, B.J. Harris and S.M. Riehl

Abstract. We investigate the existence and asymptotic behaviour of higher derivatives of the spectral function, $\rho(\lambda)$, on the positive real axis, in the context of one-dimensional Schrödinger operators on the half-line with integrable potentials. In particular, we identify sufficient conditions on the potential for the existence and continuity of the nth derivative, $\rho^{(n)}(\lambda)$, and outline a systematic procedure for estimating numerical upper bounds for the turning points of such derivatives. The potential relevance of our results to some topical issues in spectral theory is discussed.

Mathematics Subject Classification (2000). 34B20, 34B24, 47B25, 47E05.

Keywords. Sturm-Liouville problems, spectral functions, unbounded selfadjoint operators.

1. Introduction

For the one-dimensional Schrödinger operator H on the half-line, the associated spectral function, $\rho(\lambda)$, and Titchmarsh-Weyl function, $m(z)$, are well-known tools of classical spectral analysis. In particular $\rho(\lambda)$ incorporates key information on the spectrum, such as its location and type, while on the upper half complex plane, $m(z)$ is a Herglotz function which directly reflects the analyticity properties of the resolvent operator, $(H - zI)^{-1}$. These two functions are mutually related through the boundary properties of $m(z)$, as the real axis is approached normally, and through the Stieltjes inversion formula.

The m-function is also connected to solutions of the Schrödinger equation in a number of ways. In the case of integrable potentials, the logarithmic derivative of the Jost solution of the Schrödinger equation can be identified with a generalised Dirichlet m-function, and this analytic function can be continuously extended from the upper half-plane onto the positive real axis [13]. Moreover, given that the one-dimensional Schrödinger equation is of Sturm-Liouville type, the generalised m-function is a particular solution of a non-linear first order Riccati equation, which can sometimes be approximated for large λ using an iterative procedure (see, e.g., [6], [7]).

These intricate relationships underlie the structure of this paper, which we now briefly outline. The selfadjoint operator H on $L_2([0,\infty))$ with Dirichlet boundary condition at 0 is introduced in Section 2, and the main features of the spectrum, $\sigma(H)$, are summarised for the case $q \in L_1([0,\infty))$. We also collect together relevant properties of the spectral function and associated Titchmarsh-Weyl function and briefly discuss the relevance of the first and second derivatives of $\rho(\lambda)$ to spectral concentration and resonances. In Section 3, sufficient conditions for the existence and continuity of the first n derivatives of the spectral density are identified in Theorem 1. Aspects of the relationship between the spectral density and a generalised Dirichlet m-function, $m(x,z)$, are outlined in Section 4, where we also explore the possibility of using the Riccati equation to construct a series representation of $m(x,z)$ which is valid for large $z = \lambda \in \mathbf{R}^+$. In Section 5, we state our main result in Theorem 2. This identifies sufficient conditions on q to ensure that $\rho^{(M+1)}(\lambda) \sim \rho_0^{(M+1)}(\lambda)$ as $\lambda \to \infty$, where $\rho_0(\lambda)$ denotes the Dirichlet spectral function for H when $q \equiv 0$, and it follows that, under the conditions of the theorem, the Mth derivative $\rho^{(M)}(\lambda)$ has no turning points for sufficiently large λ. We conclude with a brief discussion of the results and illustrate their potential application through an explicit example.

The purpose of this paper is to provide an overview of some recent work by the authors on the differentiability of the spectral function. Full proofs of the main results will appear in a separate publication [8].

2. Mathematical background

We consider the time-independent one-dimensional Schrödinger operator H associated with the system

$$Ly := -y'' + q(x)y = zy, \qquad x \in [0,\infty), \;\; z \in \mathbf{C},$$

$$y(0) = 0,$$

where the potential q is real valued and integrable on $[0,\infty)$. In this case L is regular at $x = 0$ and in Weyl's limit point case at infinity, so that for each $z \in \mathbf{C}\backslash\mathbf{R}$ there exists precisely one linearly independent solution of $Lu = zu$ in $L_2([0,\infty))$, and for each $z \in \mathbf{R}$, there exists at most one such solution. The corresponding selfadjoint operator H acting on $\mathcal{H} = L_2([0,\infty))$ is defined by

$$Hf = Lf, f \in D(H),$$

where

$$D(H) = \{f \in \mathcal{H} : Lf \in \mathcal{H}; f, f' \text{ locally a.c.; } f(0) = 0\}.$$

The essential spectrum $\sigma_{\text{ess}}(H)$ of H fills the semi-axis $[0, \infty)$ and is purely absolutely continuous for $\lambda > 0$, while the negative spectrum, if any, consists of isolated eigenvalues, possibly accumulating at $\lambda = 0$. There may also be an eigenvalue at $\lambda = 0$, but only if $xq(x) \notin L_1([0, \infty))$. In the context of this paper we are principally concerned with the absolutely continuous part of the spectrum on $(0, \infty)$.

The spectral function, $\rho(\lambda)$, associated with H is a non-decreasing function on $\mathbf{R}$, which we normalise in the usual way by setting $\rho(0) = 0$. There is a jump discontinuity in the spectral function at each eigenvalue, but otherwise $\rho(\lambda)$ is constant on $\mathbf{R}^-$. For $\lambda > 0$, $\rho(\lambda)$ is strictly increasing and continuously differentiable, with

$$\rho'(\lambda) \sim \frac{1}{\pi}\sqrt{\lambda} \quad \text{as} \quad \lambda \to \infty.$$

(see [2]). The spectrum, $\sigma(H)$, may be defined as the complement in $\mathbf{R}$ of the set of points in a neighbourhood of which $\rho(\lambda)$ is constant, and this is consistent with the more usual definition in terms of the resolvent operator.

The spectrum of H may also be studied through properties of the Titchmarsh-Weyl m-function, $m(z)$, which is closely related to the Green's function for H and hence reflects the analyticity properties of the resolvent operator on $\mathbf{C}^+$. The first derivative of $\rho(\lambda)$ is known as the spectral density and is related to the boundary values of the m-function on $\mathbf{R}$ through the formula

$$\rho'(\lambda) = \lim_{\epsilon \to 0} \frac{1}{\pi} \Im m(\lambda + i\epsilon), \tag{2.1}$$

which holds for all $\lambda \in \mathbf{R}$ for which the respective limits exist.

Key features of the spectrum, such as its decomposition into absolutely continuous, singular continuous and pure point parts, may be extrapolated from knowledge of the spectral density on $\mathbf{R}$, or equivalently, from the boundary behaviour of the m-function as $z \in \mathbf{C}^+$ approaches the real axis normally. The same spectral information may also be inferred from the asymptotic properties of solutions of $Lu = \lambda u$, $\lambda \in \mathbf{R}$, through the theory of subordinacy [9].

The investigation of more subtle features of the spectrum, such as spectral concentration and the associated issue of resonances, can benefit from information on the behaviour of the second derivative of the spectral function, as we now explain. We adopt the following definition of points of spectral concentration (cf. [1]).

Definition 1 The point $\lambda_c \in \mathbf{R}$ is said to be a *point of spectral concentration* of H if

(i) $\rho'(\lambda)$ exists finitely and is continuous in a neighbourhood of λ_c, and
(ii) $\rho'(\lambda)$ has a local maximum at λ_c.

In the present context, where $q \in L_1([0, \infty))$, the definition effectively restricts attention to points of spectral concentration which occur in the interior of the absolutely continuous spectrum, $\sigma_{ac}(H)$, so that points of spectral concentration in the above sense can only arise on $(0, \infty)$. As noted by Eastham [3], a consequence of the definition is that if $\rho''(\lambda)$ exists and has one sign for $\lambda > M > 0$, then

$\rho'(\lambda)$ exists and is absolutely continuous for such λ, but has no local maximum in (M, ∞). Thus knowledge of the large λ behaviour of the second derivative of $\rho(\lambda)$ provides a mechanism for estimating upper bounds for points of spectral concentration (see, e.g., [4], [6], [7]). Points of spectral concentration may be, but need not be, associated with resonances, which occur at poles of the meromorphic continuation of the Green's function on the so-called unphysical sheet. There is an extensive literature on these phenomena, and the relationships between them (see, e.g., [10], [12], [15]).

We will return to the issue of the contribution of higher derivatives of the spectral function to our understanding of the spectrum in Sections 5 and 6.

3. Sufficient conditions for higher derivatives

In this section we identify sufficient conditions on the potential for the existence and continuity of the nth derivative of the spectral density on $\mathbf{R}^+$. The proof is based on classical methods of Titchmarsh and uses the following generalisation of Gronwall's inequality.

Lemma 1. *Let $f(x) \geq 0$, $g(x) \geq 0$, with $f(x)$ continuous and $(1+x)^n g(x) \in L_1([0,\infty))$. If $C > 0$ is constant and*

$$f(x) \leq C(1+x)^n + \int_0^x f(t)g(t)dt$$

for $x \geq 0$, then

$$f(x) \leq C(1+x)^n e^{\int_0^x (1+t)^n g(t)dt}$$

for $x \geq 0$.

Let $\rho^{(n+1)}(\lambda)$ denote the nth derivative of the spectral density, $\rho'(\lambda)$. Theorem 1 demonstrates a neat correlation between the finiteness of the nth moment of q and the existence of the first n derivatives of the spectral density.

Theorem 1. *If $q(x)$ is real valued and satisfies $(x+1)^n q(x) \in L_1([0,\infty))$, then $\rho^{(k+1)}(\lambda)$ exists and is continuous for $\lambda > 0$; $k = 0, 1, \ldots, n$.*

Idea of proof. The proof proceeds by induction and uses the Titchmarsh formula for the spectral density on $\mathbf{R}^+$, viz.

$$\rho'(\lambda) = \frac{1}{\pi\sqrt{\lambda}(a^2(\lambda) + b^2(\lambda))},$$

where

$$\begin{aligned} a(\lambda) &:= \frac{1}{\sqrt{\lambda}} \int_0^\infty \sin(\sqrt{\lambda}y) q(y) \phi(y,\lambda) dy, \\ b(\lambda) &:= \frac{1}{\sqrt{\lambda}} + \frac{1}{\sqrt{\lambda}} \int_0^\infty \cos(\sqrt{\lambda}y) q(y) \phi(y,\lambda) dy, \end{aligned}$$

and $\phi(x,\lambda)$ is the solution of $Lu = \lambda u$ satisfying $\phi(0,\lambda) = 0$, $\phi'(0,\lambda) = 1$. The result is first established for $k = n$, and the remaining cases follow immediately, since $(x+1)^n q(x) \in L_1([0,\infty))$ implies $(x+1)^k q(x) \in L_1([0,\infty))$ for $k = 0,1,2,\dots,n-1$. Note that the Titchmarsh formula (see [14], Chapter V) is valid for all $q \in L_1([0,\infty))$, and hence for all q in the scope of the theorem.

The following corollary to Theorem 1 is immediate.

Corollary 1. *If $q(x) = O(e^{-ax})$ with $a > 0$, then $\rho^{(n)}(\lambda)$ is continuously differentiable for $n = 0,1,2,\dots$.*

We remark that the sufficient conditions of Theorem 1 are not in general necessary to ensure that the corresponding derivatives exist and are continuous for sufficiently large λ. To see this, let $n = 0$ and $q = \sin((x+1)^{\frac{1}{2}})(x+1)^{-\frac{1}{2}}$. Then $(x+1)^n q(x) = q(x) \notin L_1([0,\infty))$, but there exists $\Lambda_0 > 0$ such that $\rho^{(n+1)}(\lambda) = \rho'(\lambda)$ exists and is continuous on (Λ_0,∞) (see [7], Example 2).

4. Series solutions of the Riccati equation

In addition to its relationship to the spectral density, the Titchmarsh-Weyl m-function is also linked to solutions of the differential equation. For $z \in \mathbf{C}^+$, let $\chi(x,z)$ denote the so-called Jost solution of $Lu = zu$, which satisfies $\chi(x,z) \sim \exp(ikx)$ as $x \to \infty$, where $k^2 = z$ and the principal branch is chosen. Then the logarithmic derivative of $\chi(x,z)$ evaluated at $x = 0$ is equal to the value of the Dirichlet m-function at z, i.e.,

$$\left.\frac{\chi'(x,z)}{\chi(x,z)}\right|_{x=0} = m(z),$$

where $'$ denotes differentiation with respect to x. The asymptotic form of $\chi(x,z)$ implies that $\chi(x,z) \in L_2([0,\infty);dx)$, and hence $\chi(x,z)$ cannot vanish at $x = 0$ for $z \in \mathbf{C}^+$, since to do so would imply that H had a non-real eigenvalue at z, and thus contradict the selfadjointness of H. In a similar way, $\chi(x_0,z) \neq 0$ for any $x_0 > 0$, $z \in \mathbf{C}^+$. We may therefore set

$$m(x,z) := \frac{\chi'(x,z)}{\chi(x,z)} \tag{4.1}$$

for $x \geq 0, z \in \mathbf{C}^+$, and it is straightforward to check that for each $x_0 \geq 0$, $m(x_0,z)$ is the Dirichlet m-function associated with the system: $Lu = zu$, $x \geq x_0$, $u(x_0,z) = 0$. We will refer to $m(x,z)$ as the generalised Dirichlet m-function, and note that in terms of our notation in Section 2, $m(z) \equiv m(0,z)$. Moreover, since $q \in L_1([0,\infty))$, the function $m(x,z)$ may be continuously extended onto the positive real axis, $z = \lambda \in \mathbf{R}^+$, for $x \geq 0$. We then have (cf. (4.1)) that $m(x,\lambda)$ is well defined, continuous and non-real for $x \geq 0, \lambda > 0$, and satisfies

$$m(x,\lambda) = \frac{\chi'(x,\lambda)}{\chi(x,\lambda)}, \tag{4.2}$$

where $\chi(x,\lambda) \notin L_2([0,\infty);dx)$ is the pointwise limit as $z \downarrow \lambda$ of the Jost solution, and is itself a solution of $Lu = \lambda u$ [5], [13]. It follows from (4.2) and the well-known relationships between solutions of the Sturm-Liouville and Riccati equations that $m(x,\lambda)$ satisfies the Riccati equation for $x \geq 0$, $\lambda > 0$, so that

$$\frac{\partial}{\partial x} m(x,\lambda) = -\lambda + q(x) - (m(x,\lambda))^2, \tag{4.3}$$

where $m(x,\lambda)$ is the finite non-real limit as $z \downarrow \lambda$ of the generalised Dirichlet m-function, $m(x,z)$.

We now show that we can investigate the behaviour of the spectral density, $\rho_0'(\lambda)$, using the Riccati equation (4.3). For $x \geq 0, \lambda > 0$, we have

$$m(x,\lambda) := \Re m(x,\lambda) + i\Im m(x,\lambda),$$

from which by (2.1),

$$m(0,\lambda) := m(\lambda) = \Re m(\lambda) + i\pi\rho'(\lambda),$$

so that for $\lambda > 0$

$$\rho'(\lambda) = \frac{1}{\pi}\Im m(0,\lambda). \tag{4.4}$$

Thus in principle, $\rho'(\lambda)$ can be obtained for $\lambda > 0$ by finding the appropriate solution of the Riccati equation and evaluating at $x = 0$.

If in addition it can be shown that $m(x,\lambda)$ is differentiable n times with respect to λ for sufficiently large λ, we can also seek conditions under which $\rho^{(n+1)}(\lambda)$ exists and satisfies

$$\rho^{(n+1)}(\lambda) = \frac{1}{\pi}\Im m^{(n)}(x,\lambda)\Big|_{x=0}, \tag{4.5}$$

where $m^{(n)}(x,\lambda)$ denotes the nth partial derivative of $m(x,\lambda)$ with respect to λ. Equations (4.3) and (4.5) will form the basis for our investigation into the existence of upper bounds for turning points of higher derivatives of $\rho(\lambda)$.

The generalised Dirichlet m-function, $m(x,\lambda), x \geq 0, \lambda > 0$, is a particular solution of (4.3) with known asymptotic behaviour as $x,\lambda \to \infty$ for $q \in L_1([0,\infty))$ (see, e.g., [13]). Since (4.3) is not in general solvable in terms of known functions, we proceed by postulating a series solution of the Riccati equation (4.3) of the form

$$m(x,\lambda) = i\sqrt{\lambda} + R(x,\lambda) + g(x,\lambda), \tag{4.6}$$

where

$$g(x,\lambda) := \sum_{n=0}^{\infty} m_n(x,\lambda),$$

with $g(x,\lambda)$, $R(x,\lambda) \in L_1([0,\infty);dx)$, and $g(x,\lambda)$, $R(x,\lambda) \to 0$ as $x,\lambda \to \infty$. These conditions on R and g ensure that the left-hand side of (4.6) has the required asymptotic behaviour and are sufficient to ensure uniqueness of the solution (see, e.g., [6], [13]). The aim is to construct a series representation of $m(x,\lambda)$, which is valid for all $x \geq 0$ and sufficiently large λ, with a view to using this representation to obtain asymptotic estimates of $\rho'(\lambda)$ and its derivatives through the relationships (4.4) and (4.5).

The mechanism through which the function $R(x,\lambda)$ and the series $g(x,\lambda)$ are obtained is as follows. We first substitute $m(x,\lambda)$ from (4.6) into (4.3) to obtain

$$\sum_{k=1}^{\infty}\left(m_k' + 2\left(i\sqrt{\lambda}+R\right)m_k\right) = Q - m_1^2 - \sum_{k=3}^{\infty}\left(m_{k-1}^2 + 2m_{k-1}\sum_{\ell=1}^{k-2} m_\ell\right),$$

where $Q := q - R' - R^2 - 2i\sqrt{\lambda}$ and $'$ denotes differentiation with respect to x. This equation will be satisfied if solutions can be found to the following system of first order ordinary differential equations

$$\begin{aligned} m_1' + 2(i\sqrt{\lambda}+R)m_1 &= Q, \\ m_2' + 2(i\sqrt{\lambda}+R)m_2 &= -m_1^2, \\ m_k' + 2(i\sqrt{\lambda}+R)m_k &= -\left(m_{k-1}^2 + 2m_{k-1}\sum_{\ell=1}^{k-2} m_\ell\right), \qquad k \geq 3, \end{aligned}$$

which can be solved iteratively by quadrature to give

$$m_1(x,\lambda) = -\int_x^\infty e^{2i\sqrt{\lambda}(t-x)+2\int_x^t R(s,\lambda)ds} Q(t,\lambda)dt, \tag{4.7}$$

$$m_2(x,\lambda) = \int_x^\infty e^{2i\sqrt{\lambda}(t-x)+2\int_x^t R(s,\lambda)ds} m_1^2(t,\lambda)dt, \tag{4.8}$$

and for $k \geq 3$,

$$m_k(x,\lambda) = \int_x^\infty e^{2i\sqrt{\lambda}(t-x)+2\int_x^t R(s,\lambda)ds}\left(m_{k-1}^2 + 2m_{k-1}\sum_{\ell=1}^{k-2} m_\ell\right)dt, \tag{4.9}$$

provided the integrals in (4.7), (4.8) and (4.9) converge.

In addition to satisfying the asymptotic and convergence properties noted above, the function $R(x,\lambda)$ is defined in such a way as to ensure that the derivatives $R^{(M)}(x,\lambda)$ and $g^{(M)}(x,\lambda)$ exist and are continuous for some fixed $M \in \mathbf{N}$, where the superfix (M) denotes the Mth derivative with respect to λ. Note that the choice of $R(x,\lambda)$, which determines both $Q(x,\lambda)$ and $g(x,\lambda)$, is not unique, and is also influenced by other considerations, for example a requirement that the series $g(x,\lambda)$ be absolutely and uniformly convergent for all $x \geq 0$ and sufficiently large λ. The process of determining a suitable choice of $R(x,\lambda)$, which is itself dependent on $q(x)$, has the effect of producing a set of conditions on $q(x)$ which are sufficient to ensure that the generation of the series representation can be accomplished. These conditions contribute to the hypothesis of our main theorem in the following section.

5. Upper bounds for the turning points

The main result in this section, Theorem 2, has direct implications for the existence and computability of upper bounds for turning points of higher derivatives of the spectral function, $\rho(\lambda)$, for a significant class of integrable potentials. Note that the differentiability requirements in the hypothesis of the theorem are influenced by the choice of $R(x,\lambda)$ in (5.5), while the inequalities in the hypothesis, together with (4.7) to (4.9) above, are used to establish the intermediate result in Lemma 2.

Theorem 2. *Let $q(x)$ be continuously differentiable $M-1$ times, with the $(M-1)$th derivative $q^{(M-1)}(x)$ absolutely continuous on $[0,\infty)$, for some $M \geq 2$. Suppose that there exists a decreasing function $\phi(x) \in L_1([0,\infty))$, and a continuous non-negative function $\xi(\lambda)$ on $(0,\infty)$ with $\xi(\lambda) \to 0$ as $\lambda \to \infty$, such that for $x \geq 0$,*

$$\begin{aligned}
(x+1)^{\ell+1}\left|q^{(\ell)}(x)\right| &\leq \phi(x) \ \textit{for}\ \ell = 0,1,\ldots,M-1, && (5.1)\\
\int_x^\infty \left|q^{(M)}(t)\right| dt &\leq (x+1)^{-M+1}\phi(x), && (5.2)\\
\left|\int_x^\infty e^{2i\sqrt{\lambda}t} q^{(M)}(t)dt\right| &\leq \xi(\lambda)(x+1)^{-M+1}\phi(x), && (5.3)
\end{aligned}$$

where $q^{(\ell)}(x)$ denotes the ℓth derivative of q. Then there exists $\Lambda_M \geq 0$ such that $\rho^{(M+1)}(\lambda)$ exists, is continuous on (Λ_M,∞), and satisfies

$$\left|\rho^{(M+1)}(\lambda) - \frac{1}{\pi}\left(\sqrt{\lambda} + \Im R(0,\lambda)\right)^{(M)}\right| \leq C\lambda^{-M}\left(\frac{1}{\sqrt{\lambda}} + \xi(\lambda)\right)\int_0^\infty \phi(t)dt \quad (5.4)$$

for all $\lambda > \Lambda_M$, where

$$R(x,\lambda) := \sum_{n=1}^{M} a_n(x)\lambda^{-\frac{n}{2}}, \quad (5.5)$$

with coefficients $\{a_n(x)\}$ defined inductively by

$$\begin{aligned}
a_1(x) &:= -\frac{i}{2}q(x),\\
a_2(x) &:= \frac{q'(x)}{4},\\
a_{n+1}(x) &:= \frac{i}{2}\left(a_n'(x) + \sum_{\ell=1}^{n-1} a_{n-\ell}(x)a_\ell(x)\right), \qquad n = 2,3,\ldots,M-1,
\end{aligned}$$

and the constant C depends on M and q, and is computable.

Idea of proof. The proof of the theorem depends on the following intermediate result.

Lemma 2. *Let $m_j(x,\lambda)$, $j = 1,2,\ldots$, be as in (4.7)–(4.9), and suppose that the hypothesis of Theorem 2 is satisfied. Then for $j = 1,2,\ldots$, there exists $\Lambda_M \geq 0$*

such that for $\lambda > \Lambda_M$, *the derivatives* $m_j^{(M)}(x,\lambda)$ *exist, are continuous and satisfy*

$$\left|m_j^{(M)}(x,\lambda)\right| \le K\frac{1}{2^j}\lambda^{-M}\left(\sqrt{\lambda}+\xi(\lambda)\right)\int_x^\infty \phi(t)dt$$

for all $x \ge 0$, *where the constant* K *depends on* q *and* M, *and is computable.*

It may be inferred from the proof of the lemma that for sufficiently large λ, the series in (4.6) is uniformly and absolutely convergent in x and λ and that (4.6) may be differentiated term by term M times with respect to λ to give

$$m^{(M)}(x,\lambda) = \left(i\sqrt{\lambda}+R(x,\lambda)\right)^{(M)} + \sum_{n=0}^{\infty} m_n^{(M)}(x,\lambda), \tag{5.6}$$

for $\lambda > \Lambda_M$, from which it may be shown that (5.4) follows from (4.5) and Lemma 2.

Let $\rho_0^{(k)}(\lambda)$ denote the kth derivative of the Dirichlet spectral function associated with $Lu = zu$ for the case $q(x) \equiv 0$. The following corollary to Theorem 2 shows that if $q(x)$ satisfies the conditions of the theorem, then $\rho^{(k+1)}(\lambda)$ approaches $\rho_0^{(k+1)}(\lambda)$ asymptotically as $\lambda \to \infty$ for all $k = 0, 1, \ldots, M$.

Corollary 2. *If the hypothesis of Theorem 2 is satisfied, then for* $k = 0, 1, \ldots, M$, *there exists* $\Lambda_k > 0$, *such that* $\rho^{(k+1)}(\lambda)$ *exists for* $\lambda > \Lambda_k$ *and*

$$\rho^{(k+1)}(\lambda) \sim \rho_0^{(k+1)}(\lambda) = \frac{1}{\pi}\left(\sqrt{\lambda}\right)^{(k)} \tag{5.7}$$

as $\lambda \to \infty$. *In particular,* $\rho^{(k+1)}(\lambda)$ *eventually has one sign, namely the sign of the* $(k+1)$*th derivative of* $\rho_0(\lambda)$.

Proof The result for $k = 0$ is already well known (see, e.g., [2]). The first part of the result for $k = M > 0$ follows easily from the asymptotic behaviour:

$$R(0,\lambda)^{(M)} = \left(\sum_{k=1}^{M} a_k(0)\lambda^{-\frac{k}{2}}\right)^{(M)} = O\left(\frac{1}{\sqrt{\lambda}}\right)^{(M)},$$

$$\begin{aligned} \lambda^{-M}\frac{1}{\sqrt{\lambda}} &= O\left(\frac{1}{\sqrt{\lambda}}\right)^{(M)}, \\ \lambda^{-M}\xi(\lambda) &= o\left(\frac{1}{\sqrt{\lambda}}\left(\sqrt{\lambda}\right)^{(M)}\right), \end{aligned}$$

as $\lambda \to \infty$. The second part of the result for $k = M > 0$ follows from the first, taking into account from (5.4) that the sign of $\rho_0^{(M+1)}(\lambda)$ is positive or negative for all λ according as M is odd or even respectively.

To establish the result for $k = 1, 2, \ldots, M-1$, it is only necessary to observe that if the inequalities (5.1) to (5.3) hold for some $M = M_0 \ge 2$, then they also hold for $M = 1, 2, \ldots, M_0 - 1$.

Remark. Note that for given q and M satisfying the conditions of Theorem 2, the constant C is computable. It may then be inferred from the proof of Corollary 2 that for each $k = 0, 1, 2, \ldots, M-1$, there exists a computable constant, B_k, such that $\rho_0^{(k+1)}(\lambda)$ has one sign for $\lambda > B_k$. Such a constant is then an upper bound for the turning points of $\rho_0^{(k)}(\lambda)$.

6. Discussion and example

We first discuss the relevance of our investigation to the long standing conjecture that a close correlation exists between the presence of resonances on the unphysical sheet and points of spectral concentration of $\rho(\lambda)$ on $\mathbf{R}^+$. This conjecture is particularly plausible if q is in $L_1([0,\infty))$, since in this case the well-known Kodaira formula [11], viz.

$$\rho'(\lambda) = \frac{\sqrt{\lambda}}{\pi \mid \chi(0,\lambda) \mid^2}, \qquad \lambda > 0,$$

holds, while resonances occur at zeros of the Jost function $\chi(z) := \chi(0,z)$ on the unphysical sheet, $\pi \leq \arg k < 2\pi$, where $k^2 = z$ as in Section 4. However, evidence for a systematic correlation between the two phenomena is patchy and in the example below we see that even if a resonance lies arbitrarily close to the real axis, it need not give rise to a point of spectral concentration. The possibility remains that resonances which are insufficiently strong to induce points of spectral concentration may nonetheless leave some trace of their existence through points of inflection of $\rho'(\lambda)$, which are themselves turning points of $\rho''(\lambda)$, or in the turning points of still higher derivatives. This is borne out by recent numerical and graphical experiments for the case of exponentially decaying potentials [16], and by the following example, in which the resonance and relevant turning point can be calculated explicitly.

Example Let

$$q = \frac{2a^2}{(1+ax)^2}, \qquad a > 0.$$

Then $q \in L_1([0,\infty))$, and as noted in [5], the Jost solution of $Lu = zu$ is given by

$$\chi(x,z) = e^{ikx}\left(1 - \frac{a}{ik(1+ax)}\right),$$

with $k^2 = z$ as before, so that by our remarks above, there is just one resonance, at $k = -ia$. Using (4.1) and (4.4), and recalling from Section 4 that we have set $z = \lambda$ for z real and positive, yields for $\lambda > 0$,

$$\rho'(\lambda) = \frac{\lambda^{\frac{3}{2}}}{\pi(\lambda + a^2)}, \qquad \text{so that} \qquad \rho''(\lambda) = \frac{\sqrt{\lambda}(\lambda + 3a^2)}{2\pi(\lambda + a^2)^2} > 0, \;\; \lambda > 0,$$

whence it follows from Definition 1 that there are no points of spectral concentration on $(0,\infty)$. However, a further differentiation yields, in the notation of earlier sections,

$$\rho^{(3)}(\lambda) = \frac{3(\lambda + a^2)^2 - 4\lambda(\lambda + 3a^2)}{4\pi\sqrt{\lambda}(\lambda + a^2)^3},$$

from which it may be inferred that $\rho'(\lambda)$ has a point of inflection at $\lambda = (2\sqrt{3} - 3)a^2 \simeq (0.7a)^2$, this being a global maximum of its derivative, $\rho''(\lambda)$, on $(0, \infty)$.

The principal results reported in this paper, Theorems 1 and 2, are concerned with a wider and more systematic investigation into the existence and behaviour of higher derivatives of the spectral function $\rho(\lambda)$, for $q(x) \in L_1([0, \infty))$. In Theorem 1, we identify sufficient conditions on the potential to ensure that the nth derivative of the spectral density exists and is continuous for $\lambda > 0$. From the example above, we see that the conditions are not necessary, since in this case $\rho^{(n)}(\lambda)$ exists and is continuous on $\mathbf{R}^+$ for all n, even though the first moment of q is not finite. We note that Theorem 1 is sharp for $n = 0$, since there exist potentials, for example of von Neumann Wigner type, which just fail to be in $L_1([0, \infty))$, and for which the spectral function has jump discontinuities on $\mathbf{R}^+$. We suspect that the result is also sharp for $n > 0$, but this has not been established.

The scope of Theorem 2 is both more limited and more wide ranging. Evidently, from the statement of the theorem, the conditions on q are quite restrictive and the conclusions only apply when λ is sufficiently large. On the other hand, if the hypothesis is satisfied, then for sufficiently large λ the existence and continuity of the derivatives is assured, the asymptotic behaviour of the derivatives as $\lambda \to \infty$ is identified, and a mechanism for estimating upper bounds for the turning points of the derivatives is available.

The two results are independent in the following sense. Suppose that q and n satisfy the conditions of Theorem 1 for $n = N$; then the conditions of Theorem 2 may fail to be satisfied for the same q with $M = N$, and the converse situation also holds. To see this, consider the examples:

1. $q(x) = \sin(1 + x)^2(1 + x)^{-4}, \quad N = 2,$
2. $q(x) = \sin(1 + x)^{-1}(1 + x)^{-3}, \quad N = 2.$

It is straightforward to check that Theorem 1 is applicable in Example 1, but Theorem 2 is not, while the reverse situation holds in Example 2.

Acknowledgement

The first author wishes to thank the European Commission for partial support under the INTAS programme, grant no. 05-1000008-7883.

References

[1] B.M. Brown, M.S.P. Eastham and D.K.R. McCormack, Spectral concentration and perturbed discrete spectra, *J. Comp. Appl. Math.* **86** (1997), 415–425.

[2] M.S.P. Eastham, The asymptotic nature of the spectral functions in Sturm-Liouville problems with continuous spectrum, *J. Math. Anal. Appl.* **213** (1997), 573–582.

[3] M.S.P. Eastham, On the location of spectral concentration for Sturm-Liouville problems with rapidly decaying potential, *Mathematika* **45** (1998), 25–36.

[4] M.S.P. Eastham, On the location of spectral concentration for perturbed discrete spectra, *Mathematika* **46** (1999), 145–154.

[5] G. Freiling and V. Yurko, *Inverse Sturm-Liouville Problems and their Applications* Nova Science Publishers, New York, 2001.

[6] D.J. Gilbert and B.J. Harris, Bounds for the points of spectral concentration of Sturm-Liouville problems, *Mathematika* **47,** (2000), 327–337 (2002).

[7] D.J. Gilbert, B.J. Harris and S.M. Riehl, The spectral function for Sturm-Liouville problems where the potential is of von Neumann-Wigner type or slowly decaying, *J. Differential Equations* **201,** (2004), 139–159.

[8] D.J. Gilbert, B.J. Harris and S.M. Riehl, to be published.

[9] D.J. Gilbert and D.B. Pearson, On subordinacy and analysis of the spectrum of one-dimensional Schrödinger operators, *J. Math. Anal. Appl.* **128** (1987), 30–56.

[10] P.D. Hislop and I.M. Segal, *Introduction to Spectral Theory, with Applications to Schrödinger Operators*, Springer-Verlag, New York, 1996.

[11] K. Kodaira, The eigenvalue problem for ordinary differential equations of the second order, and Heisenberg's theory of S-matrices, *Amer. J. Math.* **71** (1949), 921–945.

[12] M. Reed and B. Simon, *Methods of Modern Mathematical Physics, IV: Analysis of Operators,* Academic Press, 1978.

[13] A. Rybkin, Some new and old asymptotic representations of the Jost solution and the Weyl m-function for Schrödinger operators on the line, *Bull. London Math. Soc.* **34** (2002), 61–72.

[14] E.C. Titchmarsh, *Eigenfunction Expansions I*, 2nd edition, Oxford University Press, 1962.

[15] E.C. Titchmarsh, *Eigenfunction Expansions II*, Oxford University Press, 1958.

[16] N. White, private communication.

D.J. Gilbert
School of Mathematical Sciences
Dublin Institute of Technology
Kevin Street, Dublin 8, Ireland
e-mail: `Daphne.Gilbert@dit.ie`

B.J. Harris
Department of Mathematical Sciences
Northern Illinois University
Dekalb, IL 60115, USA
e-mail: `harris@math.niu.edu`

S.M. Riehl
Department of Mathematics
University of Northern Iowa
Cedar Falls, IA 50614-0506, USA
e-mail: `riehl@math.uni.edu`

Operator Theory:
Advances and Applications, Vol. 186, 229–239

On the Boundary Value Problem for p-parabolic Equations

Per-Anders Ivert

Abstract. We derive a fundamental existence theorem for the equation

$$u'_t(x,t) - \operatorname{div}(|\nabla_x u|^{p-2}\nabla_x u) = 0$$

with continuous initial and boundary data in a space-time cylinder.

Mathematics Subject Classification (2000). 35K15; 35K20; 35K65.

Keywords. Degenerate parabolic equations, p-parabolic functions.

1. Introduction

For the development of a potential theory for the operator

$$u \mapsto P(u) = u'_t(x,t) - \operatorname{div}(|\nabla_x u|^{p-2}\nabla_x u),$$

it is of course important to have an existence theorem for solutions to the equation $P(u) = 0$ in some class of simple domains with continuous initial and boundary data. To the author's knowledge, existence results published in the literature (see, e.g., [1]) depend on some regularity of these data, beyond continuity. In [6], a potential theory as mentioned above was discussed, and the existence result that we are going to derive here was anticipated but not proved there. The results of [1] do not seem applicable. A natural approach to the problem of securing existence of a solution under the mere assumption of continuity of boundary data, would be to approximate the data uniformly by smooth data and pass to the limit. To make this process work, one needs a preliminary existence theorem under such regularity assumptions on the data that correspond to regularity properties that can be deduced a priori for arbitrary solutions. This preliminary existence theorem is provided by [4]. The purpose of this work is to prove existence of a solution to the equation $P(u) = 0$ in a space-time cylinder, when continuous boundary values on the "parabolic boundary" of the cylinder are prescribed. Throughout the paper, we let Ω be an open subset of $\mathbf{R}^n$ with Lipschitz boundary and T a positive number.

We put $Z = \Omega \times (0,T) \subset \mathbf{R}^{n+1}$ and denote the parabolic boundary of Z by $\partial_P Z$, i.e.,

$$\partial_P Z = (\Omega \times \{0\}) \cup (\partial\Omega \times [0,T]).$$

Finally, the number p is strictly between 1 and ∞.

2. Preparatory estimates

In this section, we adopt the following *General assumptions*:

1. $u \in L^\infty((0,T); L^2(\Omega))$,
2. $f \in L^1(Z; \mathbf{R}^n)$, $g \in L^1(Z; \mathbf{R})$,
3. For all $\varphi \in C_0^\infty(Z)$ the following equality holds:

$$\begin{aligned} &-\iint_Z u(x,t)\,\frac{\partial\varphi}{\partial t}(x,t)\,dxdt \\ &= \iint_Z \big(g(x,t)\varphi(x,t) - f(x,t)\cdot\nabla_x\varphi(x,t)\big)\,dxdt. \end{aligned} \tag{2.1}$$

2.1. Weak continuity in $L^2(\Omega)$

We will show that under the general assumptions the mapping $u : (0,T) \mapsto L^2(\Omega)$ has a canonical representative that is weakly continuous, i.e., continuous from the interval $(0,T)$ into the space $L^2(\Omega)$ with its weak topology. Let η be an increasing C^∞-function on $\mathbf{R}$ with $\eta(\tau) = 0$ for $\tau \le 0$ and $\eta(\tau) = 1$ for $\tau \ge 1$. Put, for $h > 0$ and $\tau \in \mathbf{R}$,

$$\eta_h(\tau) = \eta\big(\frac{\tau}{h}\big), \qquad u_h(x,\tau) = \int_0^T u(x,t)\eta_h'(t-\tau)\,dt,$$

$$g_h(x,\tau) = \int_0^T g(x,t)\eta_h'(t-\tau)\,dt, \quad f_h(x,\tau) = \int_0^T f(x,t)\eta_h'(t-\tau)\,dt.$$

Proposition 2.1. *For every $t \in (0,T)$, the suite $(u_h(\cdot,t))_{h>0}$ converges weakly in $L^2(\Omega)$ as $h \to 0+$, and the limit $u(\cdot,t)$ is continuous as a mapping from $(0,T)$ into $L^2(\Omega)$ with the weak topology.*

Proof. Fix τ_0 and τ with $0 < \tau_0 < \tau < T$. Choose $\zeta \in C_0^\infty(\Omega)$. Choose, in (2.1), $\varphi(x,t) = \big(\eta_h(t-\tau) - \eta_k(t-\tau_0)\big)\zeta(x)$, where $0 < h < T-\tau$ and $0 < k < T-\tau_0$:

$$\begin{aligned} &-\int_\Omega \big(u_h(x,\tau) - u_k(x,\tau_0)\big)\zeta(x)\,dx \\ &= \iint_Z \big(\eta_h(t-\tau) - \eta_k(t-\tau_0)\big)\big(g(x,t)\zeta(x) - f(x,t)\cdot\nabla\zeta(x)\big)\,dxdt \end{aligned} \tag{2.2}$$

As (h,k) tends to $(0,0)$, the member on the right has the limit

$$-\int_{\tau_0}^{\tau}\int_\Omega \big(g(x,t)\zeta(x) - f(x,t)\cdot\nabla\zeta(x)\big)\,dxdt$$

Since the function $t \mapsto \|u(\cdot,t)\|_{L^2(\Omega)}$ is essentially bounded and $L^2(\Omega)$ is weakly complete, we conclude that the suites $(u_h(\cdot,\tau_0))_{h>0}$ and $(u_h(\cdot,\tau))_{h>0}$ converge weakly in $L^2(\Omega)$ as $h \to 0+$. We thus get a canonical representative for u by letting $u(\cdot,t)$, for every $t \in (0,T)$, be the weak limit of $u_h(\cdot,t)$ as $h \to 0+$, and then

$$\int_\Omega (u(x,t_2) - u(x,t_1))\zeta(x)\,dx = \int_{t_1}^{t_2}\int_\Omega (g(x,t)\zeta(x) - f(x,t)\cdot\nabla\zeta(x))\,dxdt$$

whenever $0 < t_1 < t_2 < T$ and $\zeta \in C_0^\infty(\Omega)$. From this equation it is clear, because of the boundedness of $\|u(\cdot,t)\|_{L^2(\Omega)}$, that the mapping $t \to u(\cdot,t)$ is weakly continuous, i.e., continuous as a mapping from $(0,T)$ into $L^2(\Omega)$ with the weak topology. □

2.2. Continuity in $L^2(\Omega)$

We now make the following *Additional assumptions*:

1. $u \in L^p((0,T); W_0^{1,p}(\Omega))$,
2. $f \in L^{p'}(Z;\mathbf{R}^n)$, $g \in L^2(Z;\mathbf{R})$.

We will show that under these Additional assumptions (and the previously made General assumptions) the mapping $t \mapsto u(\cdot,t)$ is continuous also with respect to the norm topology on $L^2(\Omega)$. Let $0 < h < T$, fix t_0 and t in $(0,T-h)$ and let $\zeta \in C_0^\infty(\Omega)$. We return to (2.2), put $k = h$, and differentiate both members with respect to τ:

$$\int_\Omega \frac{\partial u_h}{\partial\tau}(x,\tau)\zeta(x)\,dx = \int_\Omega (g_h(x,\tau)\zeta(x) - f_h(x,\tau)\cdot\nabla\zeta(x))\,dx \tag{2.3}$$

This equality holds whenever $0 < \tau < T - h$, and by approximation we find that it holds for all $\zeta \in L^2(\Omega) \cap W_0^{1,p}(\Omega)$. We will prove

Theorem 2.2. *Under the General assumptions and the Additional assumptions the mapping $t \mapsto u(\cdot,t)$ is continuous from $(0,T)$ to $L^2(\Omega)$.*

Proof. Let $0 < t_1 < t_2 < T$. According to Mazur's lemma there are convex combinations

$$U_\nu = \sum_{j=1}^{N_\nu} c_{\nu,j} u_{h_{\nu,j}}, \quad \nu = 1,2,\ldots,$$

of the functions u_h such that $U_\nu(\cdot,t_1)$ converges to $u(\cdot,t_1)$ and $U_\nu(\cdot,t_2)$ converges to $u(\cdot,t_2)$ in the norm of $L^2(\Omega)$ as $\nu \to \infty$. More precisely,

$$0 < h_{\nu,j} < T - t_2, \quad \lim_{\nu\to\infty}\max_{1\le j\le N_\nu} h_{\nu,j} = 0,$$

$$c_{\nu,j} \ge 0, \quad \sum_{j=1}^{N_\nu} c_{\nu,j} = 1,$$

$$\lim_{\nu\to\infty}\|u(\cdot,t_i) - U_\nu(\cdot,t_i)\|_{L^2(\Omega)} = 0, \quad i = 1,2,$$

where

$$U_\nu(x,\tau) = \int_0^T u(x,t)H'_\nu(t-\tau)\,dt$$

and H_ν is defined by

$$H_\nu = \sum_{j=1}^{N_\nu} c_{\nu,j}\eta_{h_{\nu,j}}, \quad \nu = 1,2,\dots.$$

We also put

$$G_\nu(x,\tau) = \int_0^T g(x,t)H'_\nu(t-\tau)\,dt \quad \text{och} \quad F_\nu(x,\tau) = \int_0^T f(x,t)H'_\nu(t-\tau)\,dt.$$

Then, by linearity, (2.3) holds with u_h, g_h and f_h replaced by U_ν, G_ν and F_ν, respectively. We can, for every fixed t, choose $\zeta(x) = U_\nu(x,\tau)$ and then integrate between t_1 and t_2. This yields

$$\int_\Omega \big(U_\nu(x,t_2)^2 - U_\nu(x,t_1)^2\big)\,dx = \int_\Omega \big(G_\nu(x,\tau)U_\nu(x,\tau) - F_\nu(x,\tau)\cdot\nabla U_\nu(x,\tau)\big)\,dx\,d\tau.$$

As $\nu \to \infty$ we get in the limit

$$\frac{1}{2}\int_\Omega \big(u(x,t_2)^2 - u(x,t_1)^2\big)\,dx = \int_{t_1}^{t_2}\int_\Omega \big(g(x,t)u(x,t) - f(x,t)\cdot\nabla u(x,t)\big)\,dx\,dt. \tag{2.4}$$

This shows that the function $\tau \mapsto \|u(\cdot,\tau)\|_{L^2(\Omega)}$ is continuous, and since the function $\tau \mapsto u(\cdot,\tau)$ is weakly continuous, we find that it is also strongly continuous. □

2.3. The Hilbert transformation on R

For a function f in the Schwarz class $\mathcal{S}(\mathbf{R})$ of rapidly decreasing functions we define its Fourier transform $\hat{f}$ by

$$\hat{f}(\tau) = \int_{-\infty}^{\infty} f(t)e^{-2\pi i t\tau}\,dt$$

and its Hilbert transform $\tilde{f}$ by

$$\tilde{f}(t) = \frac{1}{\pi}\lim_{\varepsilon\searrow 0}\int_{|s-t|>\varepsilon} \frac{f(s)}{t-s}\,ds.$$

In general, $\tilde{f}$ does not belong to $\mathcal{S}(\mathbf{R})$, but if $f \in \mathcal{S}(\mathbf{R})$, then $\tilde{f} \in C^\infty(\mathbf{R})$ and $D^k\tilde{f} \in L^p(\mathbf{R})$ for all $k \in \mathbf{N}$ and all $p \in (1,\infty)$. The Fourier transforms $\hat{f}$ of f and $\hat{\tilde{f}}$ of $\tilde{f}$ are related by

$$\hat{\tilde{f}}(\tau) = -i\,\mathrm{sgn}\tau\,\hat{f}(\tau).$$

Therefore, the Hilbert transformation has a unique isometric extension to $L^2(\mathbf{R})$. It also has a unique continuous extension $L^p(\mathbf{R}) \to L^p(\mathbf{R})$ for each $p \in (1,\infty)$. If $f \in L^2(\mathbf{R})$ and $g \in L^2(\mathbf{R})$, Parseval's formula gives

$$\int_{-\infty}^{\infty} f(t)\overline{\tilde{g}(t)}\,dt = \int_{-\infty}^{\infty} \hat{f}(\tau)\,i\,\mathrm{sgn}\tau\overline{\hat{g}(\tau)}\,d\tau = -\int_{-\infty}^{\infty} \tilde{f}(t)\overline{g(t)}\,dt$$

and, if $f \in \mathcal{S}(\mathbf{R})$ is real valued,

$$\int_{-\infty}^{\infty} f'(t)\tilde{f}(t)\,dt = \int_{-\infty}^{\infty} 2\pi i\tau\hat{f}(\tau)\,i\,\mathrm{sgn}\tau\overline{\hat{f}(\tau)}\,d\tau = -\int_{-\infty}^{\infty} 2\pi|\tau||\hat{f}(\tau)|^2\,d\tau.$$

At the same time

$$\begin{aligned}\int_{-\infty}^{\infty}\int_{-\infty}^{\infty}\Big|\frac{f(s)-f(t)}{s-t}\Big|^2\,ds\,dt &= \int_{-\infty}^{\infty}\int_{-\infty}^{\infty}\Big|\frac{f(t+s)-f(t)}{s}\Big|^2\,dt\,ds\\ &= \int_{-\infty}^{\infty}\int_{-\infty}^{\infty}\Big|\frac{e^{2\pi is\tau}-1}{s}\Big|^2|\hat{f}(\tau)|^2\,d\tau\,ds\\ &= \int_{-\infty}^{\infty}|\hat{f}(\tau)|^2\int_{-\infty}^{\infty}\Big|\frac{e^{2\pi is\tau}-1}{s}\Big|^2\,ds\,d\tau\\ &= \int_{-\infty}^{\infty}4\pi^2|\tau||\hat{f}(\tau)|^2\,d\tau.\end{aligned}$$

Hence,

$$-\int_{-\infty}^{\infty} f'(t)\tilde{f}(t)\,dt = \int_{-\infty}^{\infty} 2\pi|\tau||\hat{f}(\tau)|^2\,d\tau = \frac{1}{2\pi}\int_{-\infty}^{\infty}\int_{-\infty}^{\infty}\Big|\frac{f(s)-f(t)}{s-t}\Big|^2\,ds\,dt,$$

and the latter of these two equalities holds also for all $f \in L^2(\mathbf{R})$.

Definition 2.3. The space $H^{\frac{1}{2}}(\mathbf{R})$ consists of those elements $f \in L^2(\mathbf{R})$ for which

$$\int_{-\infty}^{\infty}\int_{-\infty}^{\infty}\Big|\frac{f(s)-f(t)}{s-t}\Big|^2\,ds\,dt < \infty.$$

This space is normed by

$$\|f\|_{\frac{1}{2}}^2 = \int_{-\infty}^{\infty}(1+4\pi^2\tau^2)^{\frac{1}{2}}|\hat{f}(\tau)|^2\,d\tau.$$

2.4. An estimate in $H^{\frac{1}{2}}(\mathbf{R}; L^2(\Omega))$

Let $0 < h < T$ and let θ be a C^∞-function on $\mathbf{R}$ with $\mathrm{supp}\,\theta \subset (0, T-h)$. Put

$$a_h(t,\tau) = \theta(\tau)\eta_h'(t-\tau), \quad t \in \mathbf{R},\ \tau \in \mathbf{R}.$$

Let $\hat{a}_h(t,\cdot)$ be the Fourier transform and $\tilde{a}_h(t,\cdot)$ the Hilbert transform of $a_h(t,\cdot)$. As a test function ζ in (2.3) we choose, for each fixed τ,

$$\zeta(x) = \theta(\tau)\int_0^T u(x,s)\tilde{a}_h(s,\tau)\,ds.$$

The left-hand member of (2.3) becomes

$$
\begin{aligned}
&-\int_0^T\int_0^T\Big(\int_\Omega u(x,t)u(x,s)\,dx\Big)\eta_h''(t-\tau)\theta(\tau)\tilde{a}_h(s,\tau)\,ds\,dt\\
&\quad=\int_0^T\int_0^T\Big(\int_\Omega u(x,t)u(x,s)\,dx\Big)\frac{\partial a_h}{\partial\tau}(t,\tau)\tilde{a}_h(s,\tau)\,ds\,dt\\
&\quad\quad-\int_0^T\int_\Omega \theta'(\tau)u_h(x,\tau)u(x,s)\tilde{a}_h(s,\tau)\,dx\,ds.
\end{aligned}
$$

From (2.3) we thus get

$$
\begin{aligned}
&\int_0^T\int_0^T\Big(\int_\Omega u(x,t)u(x,s)\,dx\Big)\frac{\partial a_h}{\partial t}(t,\tau)\tilde{a}_h(s,\tau)\,ds\,dt\\
&\quad=\iint_Z \theta(\tau)\big(g_h(x,\tau)u(x,s)-f_h(x,\tau)\cdot\nabla_x u(x,s)\big)\tilde{a}_h(s,\tau)\,dx\,ds\\
&\quad\quad+\int_0^T\int_\Omega \theta'(\tau)u_h(x,\tau)u(x,s)\tilde{a}_h(s,\tau)\,dx\,ds. \qquad (2.5)
\end{aligned}
$$

We will integrate both members of (2.5) with respect to t, but first we note that

$$
\begin{aligned}
&\int_{-\infty}^{\infty}\frac{\partial a_h}{\partial\tau}(t,\tau)\tilde{a}_h(s,\tau)\,d\tau=\int_{-\infty}^{\infty}2\pi i\sigma\hat{a}_h(t,\sigma)\,i\,\mathrm{sgn}\sigma\overline{\hat{a}_h(s,\sigma)}\,d\sigma\\
&\quad=-\int_{-\infty}^{\infty}2\pi|\sigma|\hat{a}_h(t,\sigma)\overline{\hat{a}_h(s,\sigma)}\,d\sigma\\
&\quad=-\frac{1}{2\pi}\int_{-\infty}^{\infty}\int_{-\infty}^{\infty}\Big|\frac{e^{2\pi i\sigma\tau}-1}{\tau}\Big|^2\hat{a}_h(t,\sigma)\overline{\hat{a}_h(s,\sigma)}\,d\tau\,d\sigma\\
&\quad=-\frac{1}{2\pi}\int_{-\infty}^{\infty}\int_{-\infty}^{\infty}\frac{e^{2\pi i\sigma\tau}\hat{a}_h(t,\sigma)-\hat{a}_h(t,\sigma)}{\tau}\cdot\overline{\frac{e^{2\pi i\sigma\tau}\hat{a}_h(s,\sigma)-\hat{a}_h(s,\sigma)}{\tau}}\,d\sigma\,d\tau\\
&\quad=-\frac{1}{2\pi}\int_{-\infty}^{\infty}\int_{-\infty}^{\infty}\frac{a_h(t,\sigma+\tau)-a_h(t,\sigma)}{\tau}\cdot\frac{a_h(s,\sigma+\tau)-a_h(s,\sigma)}{\tau}\,d\sigma\,d\tau.
\end{aligned}
$$

Consequently,

$$
\begin{aligned}
&\int_{-\infty}^{\infty}\int_0^T\int_0^T\Big(\int_\Omega u(x,t)u(x,s)\,dx\Big)\frac{\partial a_h}{\partial t}(t,\tau)\tilde{a}_h(s,\tau)\,ds\,dt\,d\tau\\
&\quad=-\frac{1}{2\pi}\int_\Omega\int_{-\infty}^{\infty}\int_{-\infty}^{\infty}\Big|\int_0^T\frac{u(x,t)\big(a_h(t,\sigma+\tau)-a_h(t,\sigma)\big)}{\tau}\,dt\Big|^2\,d\sigma\,d\tau\,dx\\
&\quad=-\frac{1}{2\pi}\int_\Omega\int_{-\infty}^{\infty}\int_{-\infty}^{\infty}\Big|\frac{\theta(\sigma+\tau)u_h(x,\sigma+\tau)-\theta(\sigma)u_h(x,\sigma)}{\tau}\Big|^2\,d\sigma\,d\tau\,dx.
\end{aligned}
$$

We now put

$$
G_h(x,t)=\int_{-\infty}^{\infty}g_h(x,\tau)\theta(\tau)\tilde{a}_h(t,\tau)\,d\tau,\quad F_h(x,t)=\int_{-\infty}^{\infty}f_h(x,\tau)\theta(\tau)\tilde{a}_h(t,\tau)\,d\tau
$$

and

$$B_h(s,t) = \int_{-\infty}^{\infty} \theta'(\tau)\tilde{a}_h(t,\tau)\eta_h(s-\tau)\,d\tau.$$

When we integrate both members of (2.5) with respect to τ over $\mathbf{R}$, we get

$$\begin{aligned}\frac{1}{2\pi}\int_\Omega\int_{-\infty}^{\infty}\int_{-\infty}^{\infty}\left|\frac{\theta(t)u_h(x,\tau)-\theta(\sigma)u_h(x,\sigma)}{\tau-\sigma}\right|^2 d\sigma\,d\tau\,dx \\ = \iint_Z \big(F_h(x,t)\cdot\nabla u(x,t) - G_h(x,t)u(x,t)\big)\,dxdt \\ + \int_0^T\int_0^T\int_\Omega u(x,s)u(x,t)B_h(s,t)\,dx\,ds\,dt. \qquad (2.6)\end{aligned}$$

Now we let $b_h(x,\cdot)$ be the Hilbert transform of $\theta(\cdot)g_h(x,\cdot)$:

$$\begin{aligned}G_h(x,t) &= \int_{-\infty}^{\infty} g_h(x,\tau)\theta(\tau)\tilde{a}_h(t,\tau)\,d\tau \\ &= -\int_{-\infty}^{\infty} b_h(x,\tau)a_h(t,\tau)\,d\tau = \eta_h' * \big(\theta(\cdot)b_h(x,\cdot)\big)(t).\end{aligned}$$

Since $\eta_h'(\tau)\geq 0$ and $\int_{-\infty}^{\infty}\eta_h'(\tau)\,dt = 1$ we get

$$\begin{aligned}\|G_h(x,\cdot)\|_{L^2(\mathbf{R})} &\leq \|\theta(\cdot)b_h(x,\cdot)\|_{L^2(\mathbf{R})} \leq \max_{\tau\in\mathbf{R}}|\theta(\tau)|\|b_h(x,\cdot)\|_{L^2(\mathbf{R})} \\ &= \max_{\tau\in\mathbf{R}}|\theta(\tau)|\|\theta(\cdot)g_h(x,\cdot)\|_{L^2(\mathbf{R})} \leq \max_{\tau\in\mathbf{R}}|\theta(\tau)|^2\|g(x,\cdot)\|_{L^2(0,T)},\end{aligned}$$

which leads to

$$\|G_h\|_{L^2(Z)} \leq \max_{\tau\in\mathbf{R}}|\theta(\tau)|^2\|g\|_{L^2(Z)}.$$

In the same way we get

$$\|F_h\|_{L^{p'}(\mathbf{R})} \leq C\max_{\tau\in\mathbf{R}}|\theta(\tau)|^2\|f\|_{L^{p'}(Z)},$$

where C is a constant, depending only on p. To estimate the last term in the right-hand member of (2.6), we let $c_h(s,\cdot)$ be the Hilbert transform of $\theta'(\cdot)\eta_h(s-\cdot)$. Then

$$B_h(s,t) = -\int_{-\infty}^{\infty} a_h(t,\tau)c_h(s,\tau)\,dt = \eta_h' * \big(\theta(\cdot)c_h(s,\cdot)\big)(\tau).$$

Hence

$$\begin{aligned}\|B_h(s,\cdot)\|_{L^2(0,T)} &\leq \|\theta(\cdot)c_h(s,\cdot)\|_{L^2(\mathbf{R})} \\ &\leq \max_{\tau\in\mathbf{R}}|\theta(\tau)|\sqrt{\int_{-\infty}^{\infty}|\theta'(\tau)\eta_h(s-\tau)|^2\,d\tau} \\ &\leq \max_{\tau\in\mathbf{R}}|\theta(\tau)|\|\theta'\|_{L^2(\mathbf{R})}.\end{aligned}$$

We arrive at

$$\begin{aligned}
&\Big|\int_0^T\int_\Omega u(x,s)u(x,t)B_h(s,t)\,dx\,dt\Big|\\
&\qquad\le \|B_h(s,\cdot)\|_{L^2(0,T)}\sqrt{\int_0^T\Big(\int_\Omega u(x,t)u(x,s)\,dx\Big)^2\,dt}\\
&\qquad\le \max_{\tau\in\mathbf{R}}|\theta(\tau)|\|\theta'\|_{L^2(\mathbf{R})}\sqrt{\int_0^T \|u(\cdot,t)\|^2_{L^2(\Omega)}\|u(\cdot,s)\|^2_{L^2(\Omega)}\,dt}\\
&\qquad= \max_{\tau\in\mathbf{R}}|\theta(\tau)|\|\theta'\|_{L^2(\mathbf{R})}\|u\|_{L^2(Z)}\|u(\cdot,s)\|_{L^2(\Omega)},
\end{aligned}$$

and finally

$$\Big|\int_0^T\int_0^T\int_\Omega u(x,s)u(x,t)B_h(s,t)\,dx\,dt\,ds\Big|\le\sqrt{T}\max_{\tau\in\mathbf{R}}|\theta(\tau)|\|\theta'\|_{L^2(\mathbf{R})}\|u\|^2_{L^2(Z)}.$$

From (2.6) we now get

$$\begin{aligned}
&\frac{1}{2\pi}\int_\Omega\int_{-\infty}^{\infty}\int_{-\infty}^{\infty}\Big|\frac{\theta(\tau)u_h(x,\tau)-\theta(\sigma)u_h(x,\sigma)}{\tau-\sigma}\Big|^2\,d\sigma\,d\tau\,dx\\
&\quad\le C\max_{\tau\in\mathbf{R}}|\theta(\tau)|^2\|f\|_{L^{p'}(Z)}\|\nabla u\|_{L^p(Z)}+\max_{\tau\in\mathbf{R}}|\theta(\tau)|^2\|g\|_{L^2(Z)}\|u\|_{L^2(Z)}\\
&\qquad+\sqrt{T}\max_{\tau\in\mathbf{R}}|\theta(\tau)|\|\theta'\|_{L^2(\mathbf{R})}\|u\|^2_{L^2(Z)}. \qquad (2.7)
\end{aligned}$$

According to Fatou's lemma, this inequality is preserved when we pass to the limit $h\to 0+$. By suitably choosing θ as a function which equals 1 in the interval $[\delta,T-\delta]$ for some $\delta\in(0,\frac{T}{2})$, we deduce the following

Lemma 2.4. *Let u be a solution to equation* (2.1) *and let the General and Additional assumptions be valid. Let $0<\delta<\frac{T}{2}$. Then*

$$\begin{aligned}
&\frac{1}{2\pi}\int_\Omega\int_\delta^{T-\delta}\int_\delta^{T-\delta}\Big|\frac{u(x,t)-u(x,\sigma)}{t-\sigma}\Big|^2\,d\sigma\,dt\,dx\\
&\qquad\le C\|f\|_{L^{p'}(Z)}\|\nabla u\|_{L^p(Z)}+\|g\|_{L^2(Z)}\|u\|_{L^2(Z)}+\sqrt{\frac{2T}{\delta}}\|u\|^2_{L^2(Z)},
\end{aligned}$$

with a constant C that depends only on p.

3. The boundary value problem

We will be concerned with the solvability of the equation

$$\frac{\partial u}{\partial t}=\operatorname{div}(|\nabla u|^{p-2}\nabla u),\quad (x,t)\in Z \qquad (3.1)$$

under prescribed values on the parabolic boundary $\partial_P Z$.

Definition 3.1. We say that u is a solution to equation (3.1) if the restriction of u to $\Omega' \times (\delta, T-\delta)$ belongs to $L^\infty((\delta, T-\delta), L^2(\Omega')) \cap L^p((\delta, T-\delta); W^{1,p}(\Omega'))$ for every $\delta \in (0, \frac{T}{2})$ and every domain Ω' such that $\overline{\Omega'}$ is a compact subset of Ω, and

$$\iint_Z u(x,t)\,\frac{\partial\varphi}{\partial t}(x,t)\,dxdt = \iint_Z |\nabla_x u(x,t)|^{p-2}\nabla_x u(x,t)\cdot\nabla_x\varphi(x,t)\,dxdt. \quad (3.2)$$

for all $\varphi \in C_0^\infty(Z)$.

Our main result is

Theorem 3.2. *Let ψ be a continuous function on $\partial_P Z$. Then there is a solution u to equation* (3.1)*, continuous on $\overline{Z}$, such that $u(x,t) = \psi(x,t)$ on $\partial_P Z$.*

3.1. The function spaces $B^p(Q_{(+)})$

Following Fontes [4] (with a slight modification of notation), we put

$$B^p_\cdot(Q) = \{u \in L^p(\mathbf{R}; W^{1,p}(\Omega));\ \int_{\mathbf{R}}\int_{\mathbf{R}}\int_\Omega \left|\frac{u(x,s)-u(x,t)}{s-t}\right|^2 dx\,ds\,dt < \infty\}$$

and

$$B^p_0(Q) = \{u \in L^p(\mathbf{R}; W^{1,p}_0(\Omega));\ \int_{\mathbf{R}}\int_{\mathbf{R}}\int_\Omega \left|\frac{u(x,s)-u(x,t)}{s-t}\right|^2 dx\,ds\,dt < \infty\}$$

These two spaces are Banach spaces, normed by

$$\|u\|^2_{B^p} = \Big(\iint_Q (|u(x,t)|^p + |\nabla_x u(x,t)|^p\,dxdt\Big)^{2/p}$$
$$+\frac{1}{2\pi}\int_{\mathbf{R}}\int_{\mathbf{R}}\int_\Omega \left|\frac{u(x,s)-u(x,t)}{s-t}\right|^2 dx\,dsdt.$$

We put $Q_+ = \Omega \times [0,\infty)$. For functions u, defined in Q_+, we define the extension operators E_0 and E_p by

$$E_0u(x,t) = \begin{cases} u(x,t) & \text{if } t \ge 0 \\ 0 & \text{if } t < 0, \end{cases} \qquad E_pu(x,t) = u(x,|t|),$$

and we also introduce the spaces

$$B^p_{\cdot,\cdot}(Q_+) = \{u \in L^p(\mathbf{R}_+; W^{1,p}(\Omega));\ E_pu \in B^p(Q)\},$$
$$B^p_{0,\cdot}(Q_+) = \{u \in L^p(\mathbf{R}_+; W^{1,p}(\Omega));\ E_pu \in B^p_0(Q)\},$$
$$B^p_{*,0}(Q_+) = \{u \in L^p(\mathbf{R}_+; W^{1,p}(\Omega));\ E_0u \in B^p_*(Q)\}.$$

In the last definition, $*$ stands optionally for $\cdot$ or 0. We also put

$$B^p_I(Q_+) = \{u \in B^p_{0,\cdot}(Q_+) \cap C_b([0,\infty), L^2(\Omega));\ \frac{\partial u}{\partial t} \in L^{p'}(\mathbf{R}_+, W^{-1,p'}(\Omega))\}$$

and $X^p(Q_+) = B^p_{\cdot,0}(Q_+) + B^p_I(Q_+)$. We will need the following result

Lemma 3.3. *Let g be a Lipschitz function on $\overline{Z}$. Then g can be extended to Q_+ in such a way that the extension belongs to $X^p(Q_+)$.*

Proof. Extend g to Q_+ in such a way that it satisfies a global Lipschitz condition, vanishes for t large and belongs to $B^p_{\cdot,\cdot}(Q_+)$ (which is obviously possible). Let η be a Lipschitz function on $\overline{\Omega}$, such that $0 \leq \eta(x) \leq 1$, $\eta(x) = 1$ on $\partial\Omega$ and $\eta(x) \leq 1 - \operatorname{dist}(x, \partial\Omega)$ for x close to $\partial\Omega$. Put

$$u_1(x,t) = \eta(x)^{\frac{1}{t}} g(x,t), \quad u_2(x,t) = g(x,t) - u_1(x,t).$$

One can then verify that $u_1 \in B^p_{\cdot,0}(Q_+)$ and $u_2 \in B^p_I(Q_+)$. The details are left to the reader. □

3.2. Known results

A standard result is the following:

Lemma 3.4. *If u is a bounded solution to* (3.1) *and K a compact subset of Z, then $\iint_K |\nabla_x u(x,t)|^p\, dxdt$ can be estimated in terms of $\sup_Z u(x,t)$, the Lebesgue measure of K and the distance from K to the boundary of Z.*

The following result is a special case of Theorem 4.8 in [4]:

Theorem 3.5. *Given $g \in X^p(Q_+)$, there exists a unique solution $u \in X^p(Q_+)$ to equation* (3.1), *such that $u - g \in B^p_{0,0}(Q_+)$.*

From the regularity theory for degenerate and singular parabolic equations (see [2]) it follows

Theorem 3.6. *If the function g in Theorem* 3.5 *is continuous, then the solution u is continuous on $\overline{Q_+}$.*

A standard result is also the following comparison principle.

Theorem 3.7. *If u and v are solutions to equation* (3.1) *in Z, continuous on $\overline{Z}$, and if $u(x,t) \leq v(x,t)$ on $\partial_P Z$, then $u(x,t) \leq v(x,t)$ in Z.*

With the aid of the comparison principle, we can prove

Lemma 3.8. *Suppose that $u, u_1, u_2, \ldots$ are solutions to equation* (3.1) *in Z, continuous on $\overline{Z}$, and assume that the sequence $(u_j|\partial_P Z)_{j=1}^{\infty}$ converges uniformly on $\partial_P Z$ to $u|\partial_P Z$. Then the sequence $(u_j)_{j=1}^{\infty}$ converges uniformly in Z to u.*

Proof. Let $\varepsilon > 0$. For j sufficiently large, $u(x,t) - \varepsilon < u_j(x,t) < u(x,t) + \varepsilon$ for all $(x,t) \in \partial_P Z$. Since $u - \varepsilon$ and $u + \varepsilon$ are also solutions, it follows from the comparison principle that $u(x,t) - \varepsilon < u_j(x,t) < u(x,t) + \varepsilon$ for all $(x,t) \in Z$ if j is sufficiently large. □

4. Proof of the main theorem

We can now prove Theorem 3.2. Let $(\psi_j)_{j=1}^{\infty}$ be a sequence of Lipschitz functions on $\overline{Z}$, converging uniformly on $\partial_P Z$ to ψ. Each ψ_j can, by Lemma 3.3, be extended to a continuous element of $X^p(Q_+)$. By Theorems 3.5 and 3.6, each ψ_j corresponds to a continuous solution u_j to (3.1), taking boundary values ψ_j. By the comparison principle, since

$$\lim_{j,k\to\infty} \max_{(x,t)\in\partial_P Z} |\psi_j(x,t) - \psi_k(x,t)| = 0$$

we have

$$\lim_{j,k\to\infty} \max_{(x,t)\in\overline{Z}} |u_j(x,t) - u_k(x,t)| = 0$$

and thus the sequence $(u_j)_{j=1}^{\infty}$ converges uniformly on $\overline{Z}$ to a continuous function u. It remains to show that u is a solution. We let Ω' be domain such that $\overline{\Omega'}$ is a compact subset of Ω and we let $0 < \delta < \frac{T}{2}$. It follows from Lemma 2.4 and Lemma 3.4 that the quantities

$$\int_{\Omega'}\int_{\delta}^{T-\delta} |\nabla_x u_j(x,t)|^p \, dt \, dx + \frac{1}{2\pi}\int_{\Omega'}\int_{\delta}^{T-\delta}\int_{\delta}^{T-\delta} \left|\frac{u_j(x,\tau) - u_j(x,\sigma)}{\tau - \sigma}\right|^2 d\sigma \, d\tau \, dx$$

are bounded uniformly in j, and thus the corresponding quantity for u is finite. We conclude that the restriction of u to $Z' = \Omega' \times (\delta, T-\delta)$ can be extended to an element of $X^p(\Omega' \times (\delta,\infty))$ and thus there is a continuous solution v to equation (3.1) in Z', taking the values $u(x,t)$ on $\partial_P Z'$. However, on $\partial_P Z'$ the sequence $(u_j)_{j=1}^{\infty}$ converges uniformly to $v(x,t)$ and thus, by Lemma 3.8, it converges uniformly to v in all of Z'. On the other hand, it converges to $u(x,t)$, so u is indeed a solution to (3.1) in Z'. Since Ω' is arbitrary and δ can be chosen arbitrarily small, we conclude that u is a solution to (3.1) in Z', taking the boundary values ψ. This concludes the proof.

References

[1] H.W. Alt and S. Luckhaus, *Quasilinear elliptic-parabolic differential equations.* Math. Z., **183** (1983), 311–341.

[2] E. DiBenedetto, *Degenerate Parabolic Equations.* Springer Verlag, New York, 1993.

[3] M. Fontes, *A Monotone Operator Method for Elliptic-Parabolic Equations.* Comm. PDE, **25**, 3& 4, (2000), 681–702

[4] M. Fontes, *Initial-boundary value problems for parabolic equations.* Institut Mittag-Leffler, Report No. 22 (1999/2000).

[5] M. Fontes, *Parabolic Equations with Low regularity.* Doctoral Thesis 1996:3, Lund University.

[6] T. Kilpeläinen, P. Lindqvist, *On the Dirichlet Boundary Value Problem for a Degenerate Parabolic Equation.* Siam J. Math. Anal., Vol 27, No 3, (1996), 661–683.

Per-Anders Ivert
Centre for Mathematical Sciences
Lund University
S-221 00 LUND, Sweden
e-mail: `Per-Anders.Ivert@math.lu.se`

Operator Theory:
Advances and Applications, Vol. 186, 241–256

On the Rank One Dissipative Operator and the Parseval Formula

Mitsuteru Kadowaki, Hideo Nakazawa and Kazuo Watanabe

Abstract. The generalized Parseval formula concernig a non-selfadjoint operator is studied. In particular, the wave equation with the rank one dissipative term is considered. The existence of scattering and dissipative modes is obtained.

Mathematics Subject Classification (2000). 35P10, 35P25.

Keywords. Wave equation with dissipative term, dissipative mode, scattering mode, Parseval formula, wave operator, spectral singularity.

1. Introduction

The purpose of the present article is to report on our recent results about the spectral problems of wave equations with a dissipative term. Section 2 and 3 were originally intended to be an announcement of results to be presented in [6] with full detail. Our main result is a generalized Parseval formula associated to dissipative wave equation (Theorem 2.2 in Section 2). A sketch of proof of Theorem 2.2 is presented in Section 3. For the sake of completeness, a full proof given in [6] is reproduced in Appendix. In Section 5 we shall take up a Schrödinger model with a dissipative term of rank one and obtain Parseval's formula by the method developed in Sections 2 and 3. In Section 4 we review some results related to ours.

Definition 1.1. *Let $\mathcal{H}$ be a Hilbert space with the inner product $(\cdot,\cdot)$. Let A be a densely defined linear closed operator in $\mathcal{H}$. A is a dissipative operator if and only if for any $u \in D(A)$*

$$\operatorname{Im}(Au, u) \leq 0.$$

The first and second author are supported by Grant-in-Aid for Scientific Research (No.16540161 and No.17740082), Ministry of Education, Culture, Sports, Science and Technology, Japan, respectively.

We remark that the dissipative operator defined by Lax and Phillips [3] is given by $\mathrm{Im}(Au, u) \geq 0$. If A is a maximal dissipative operator (i.e., A has no dissipative extension), it is well known that $\sigma(A)$ (the spectrum of A) is in $\mathbb{C}_0^- = \{z \in \mathbb{C}; \mathrm{Im} z \leq 0\}$. A self-adjoint operator is a maximal dissipative operator.

We consider the situation that A_0 is an absolutely continuous self-adjoint operator with $\sigma(A_0) = \sigma_{ac}(A_0) = \mathbb{R}$ and V is a bounded dissipative operator. We consider the operator $A = A_0 + V$.

1. Selfadjoint case. (cf. [17], [18]) We first make a review on the selfadjonit case. So, V is self-adjoint and small in some sense. Then the following results are known: the wave operator

$$W_- := \text{s-}\lim_{t\to\infty} e^{itA_0} e^{-itA} P_{ac}(A)$$

exists and $\mathrm{Range}(W_-) = P_{ac}(A_0)\mathcal{H}$, where $P_{ac}(A_0)$ is the projection onto the absolutely continuous subspace of A_0, in our case coinciding with $\mathcal{H}$. Furthermore, we know that

$$\mathrm{Ker}(W_-) = P_{pp}(A)\mathcal{H},$$

where $P_{pp}(A)$ is the projection onto the subspace of the pure point of A, and that A has no singular continuous spectrum. Therefore

$$\mathcal{H} = P_{pp}(A)\mathcal{H} \oplus P_{ac}(A)\mathcal{H} \text{ (orthogonal sum)}.$$

As a consequence of these results we have that $e^{-itA}u$ $(u \neq 0)$ is asymptotically free as $t \to \infty$, i.e., there exists $u_0 \in \mathcal{H}$ such that $e^{-itA}u \sim e^{-itA_0}u_0$, $t \to \infty$, if and only if $u \notin \mathrm{ker} W_-$.

2. Dissipative case. Next we assume that V is dissipative. By a similar way to the selfadjoint case, we can define the wave operator by using the evolution semi-group $\{e^{-itA}\}_{t\geq 0}$.(see [18]). However we do not know the property of the spectrum of A exactly even if V is very small in some sense. Naturally, if $\lambda \in \sigma_{pp}(A)$, $\mathrm{Im}\lambda < 0$ and $Au = \lambda u$ $(u \neq 0)$, we know that $e^{-itA}u \to 0$ as $t \to \infty$. But is the converse true? More precisely, if $e^{-itA}u \to 0$ as $t \to \infty$, $u \neq 0$, can we conclude that u belongs to the span of all eigenspaces corresponding to non-real eigenvalues?

Our approach in the present paper follows the Kato smooth perturbation method (cf. [8]). We consider that V is of rank one, i.e., $V = \alpha(\cdot, \varphi)\varphi$, $\alpha \in \mathbb{C}$, $\mathrm{Im}\alpha < 0$, $\varphi \in \mathcal{H}$. Then using the second resolvent equation we see that

$$(A - z)^{-1}u = (A_0 - z)^{-1}u - \alpha \frac{((A_0 - z)^{-1}u, \varphi)}{1 + \alpha((A_0 - z)^{-1}\varphi, \varphi)}(A_0 - z)^{-1}\varphi.$$

As for the singularity of the resolvent of A, for $\mathrm{Im} z < 0$, since $(A_0 - z)^{-1}$ is analytic in $\mathrm{Im} z < 0$, we can neglect the first term of the right-hand side. So the highest singularity w.r.t. z arises from $\Gamma(z) = 1 + \alpha((A_0 - z)^{-1}\varphi, \varphi) = 0$ in $\mathrm{Im} z < 0$. And for $\mathrm{Im} z = 0$, since we know that

$$\|(A_0 - z)^{-1}\| \leq \frac{1}{|\mathrm{Im} z|},$$

the problem is the boundary value of $\Gamma(\lambda - i0)$, $\lambda \in \mathbb{R}$. If A is self-adjoint, we know that

$$\|(A-z)^{-1}\| \leq \frac{1}{|\mathrm{Im} z|}.$$

So the zeros of $\Gamma(z)$ is, at most, of the degree one. On the other hand, if A is dissipative, we do not have similar estimate. Hence, if the zeros of $\Gamma(\lambda - i0)$ is of the degree one, we can see that a representation formula for the wave operator similar to the selfadjoint case is still valid. So, the difficulty in the dissipative case lies the proof of the fact that the zeros of $\Gamma(\lambda - i0)$ is of degree one. This can be done under Assumption (A.2) in Section 2.

We give some comments for the related works. First, the scattering theory for the non-selfadjoint operators has been developed by S.N. Naboko in [11] and then in [12]. Some other results on the subject can be found in T. Kako and K. Yajima [7]. One should note that dissipative Schrödinger operators (in particular, rank one dissipative perturbations) have been extensively studied by Pavlov in a series of works dating back to early 70s. The explicit construction of symmetric functional model suggested by Pavlov (e.g., see [15]) has been also instrumental for the approach later developed by Naboko. One should also note that the wave operators in the situation of rank one operator (although in the selfadjoint case) have been constructed in the same terms as ours in the Yafaev's book [18]. This construction with no significant changes also applies in the dissipative case. As for the Parseval formula, a closely related discussion for the dissipative operators can be found in Pavlov [16].

2. Wave equation in the space 1-dimensional case

We consider the wave equation with a dissipative term in one space dimension:

$$\partial_t^2 u(x,t) + \langle \partial_t u, \varphi\rangle_0 \varphi(x) - \partial_x^2 u(x,t) = 0, \quad (x,t) \in \mathbb{R} \times \mathbb{R}_+ \tag{2.1}$$

where $\langle \cdot,\cdot\rangle_0$ is the usual inner product in $L^2(\mathbb{R})$, and it is assumed that $\varphi \in L^2_s(\mathbb{R}) (= \{\varphi; (1+|x|)^s \varphi \in L^2\})$ for some $s > 1/2$. We regard (2.1) as a perturbation of

$$\partial_t^2 u(x,t) - \partial_x^2 u(x,t) = 0, \quad (x,t) \in \mathbb{R} \times \mathbb{R}. \tag{2.2}$$

Put $f(t) = {}^t(u(x,t), \partial_t u(x,t))$. Then (2.1) and (2.2) can be written respectively as

$$\partial_t f(t) = -iAf(t) \qquad \text{and} \qquad \partial_t f(t) = -iA_0 f(t),$$

where

$$A = i\begin{pmatrix} 0 & 1 \\ \partial_x^2 & -\langle \cdot, \varphi\rangle_0 \varphi \end{pmatrix} \qquad \text{and} \qquad A_0 = i\begin{pmatrix} 0 & 1 \\ \partial_x^2 & 0 \end{pmatrix}.$$

Let $\mathcal{H}$ be the Hilbert space

$$\mathcal{H} = \{f = {}^t(f_1, f_2) \in \mathcal{H}; f_1 \in H^1(\mathbb{R}), f_2 \in L^2(\mathbb{R})\},$$

where $H^s(\mathbb{R})$ denotes the usual Sobolev space, with the inner product of $f = {}^t(f_1, f_2)$ and $g = {}^t(g_1, g_2)$ defined as

$$\langle f, g\rangle = \int_{\mathbb{R}} (\partial_x f_1(x)\overline{\partial_x g_1(x)} + f_2(x)\overline{g_2(x)})dx,$$

The norm of $\mathcal{H}$ is denoted by $\|\cdot\|$. Then we know that with the domain is

$$D(A) = D(A_0) = \{f = {}^t(f_1, f_2) \in \mathcal{H}; f_1 \in H^2(\mathbb{R}), f_2 \in L^2(\mathbb{R})\},$$

A_0 and A are a selfadjoint and a maximal dissipative operator, respectively.

We list some notations. $\hat{\varphi}$ means the usual Fourier transformation of φ. We denote the resolvent of A (resp. A_0) as

$$R(z) = (A - z)^{-1} \text{ for } z \in \rho(A), \quad (\text{resp. } R_0(z) = (A_0 - z)^{-1} \text{ for } z \in \rho(A_0)),$$

where $\rho(A)$ (resp. $\rho(A_0)$) is the resolvent set of the operator A (resp. A_0). We express the resolvent of $-d^2/dx^2$ in $L^2(\mathbb{R})$

$$r_0(z) = (-d^2/dx^2 - z^2)^{-1} \quad z \in \mathbb{C}\setminus\mathbb{R}.$$

We express the solution of (2.1) as follows:

$$f(t) = e^{-itA}f, \ f(0) = f \in \mathcal{H}. \tag{2.3}$$

We say that f_b is *the bound mode*, if f_b is an eigenfunction corresponding to the real eigenvalue of A, and that $f_s(\neq 0)$ is *the scattering mode*, if f_s is asymptotically free, i.e., there exists $f_0 \in D(A_0)$, $f_0 \neq 0$, such that

$$\lim_{t\to\infty} \|e^{-itA}f_s - e^{-itA_0}f_0\| = 0,$$

and that $f_d(\neq 0)$ is *the dissipative mode*, if $e^{-itA}f_d$ is dissipative, i.e.,

$$\lim_{t\to\infty} \|e^{-itA}f_d\| = 0.$$

Theorem 2.1 (cf. Mochizuki [10]). *Let A_0 and A be as above. Then*

1. *A has no real eigenvalues;*
2. *$W = s - \lim_{t\to\infty} e^{itA_0}e^{-itA}$ exists as an operator from $\mathcal{H}$ to $\mathcal{H}$ and $W \neq 0$.*

Theorem 2.1, 1 implies that (2.3) does not have the bound mode. So the following two questions naturally arise:

(Q1) Does (2.3) have the dissipative mode?

(Q2) Is it true that (2.3) is equal to the linear combination of scattering and dissipative modes which are found?

Our aim in the present paper is to answer (Q1) and (Q2). For that purpose we introduce the following assumptions.

Assumptions.

(A1) $$\varphi(x) \in L^2_{1+s}(\mathbb{R}) \quad \text{for some } s > 1/2$$

(A2) $$\text{and} \quad \Phi(\lambda) \leqq \Phi(\mu) \quad (0 \leqq \mu \leqq \lambda),$$
$$\text{where} \quad \Phi(\lambda) = |\hat{\varphi}(\lambda)|^2 + |\hat{\varphi}(-\lambda)|^2.$$

Under the Assumptions (A1) and (A2) we show that the answer to (Q1) and (Q2) are affirmative. In order to see that we investigate some properties of A. First, we describe the spectrum of the operator A (Proposition 2.1) without the proof (cf. [6]) and secondly we derive (generalized) Parseval's formula for the operator A (Theorem 2.2).

We recall the limiting absorption principle for $r_0(z)$. Let $s > 1/2$ and $u \in L^2_s(\mathbb{R})$. Then the resolvent and its extension to the reals can be expressed

$$r_0(\lambda \pm i\kappa)u(x) = \frac{\pm i}{2(\lambda \pm i\kappa)} \int_{\mathbb{R}} e^{\pm i(\lambda \pm i\kappa)|x-y|} u(y)dy$$

for $\lambda \in \mathbb{R}$ and $\kappa \geqq 0$ with $\lambda\kappa \neq 0$. Let $(l, j) = (0, 1)$ or $(1, 0)$. Then a straightforward calculation based on the above expression implies that, for every $\lambda \in \mathbb{R}$, two limits

$$\lambda^l \partial_x^j r_0(\lambda \pm i0) = \lim_{\kappa \downarrow 0} (\lambda \pm i\kappa)^l \partial_x^j r_0(\lambda \pm i\kappa) \tag{2.4}$$

exist in the uniform operator topology of $\mathcal{B}(L^2_s(\mathbb{R}), L^2_{-s}(\mathbb{R}))$.

We fix some $s > 1/2$, denote by $\langle \cdot, \cdot \rangle_0$ the dual coupling between $L^2_s(\mathbb{R})$ and $L^2_{-s}(\mathbb{R})$, and put

$$\Gamma(z) = 1 - iz\langle r_0(z)\varphi, \varphi\rangle_0, \quad \Gamma(\lambda \pm i0) = 1 - i\lambda\langle r_0(\lambda \pm i0)\varphi, \varphi\rangle_0$$

$$\Sigma_\pm = \{z \in \mathbb{C}_\pm; \Gamma(z) = 0\} \quad \text{and} \quad \Sigma^0_\pm = \{\lambda \in \mathbb{R}; \Gamma(\lambda \pm i0) = 0\}.$$

For $z \in \mathbb{C} \setminus \mathbb{R}$ and $f = {}^t(f_1, f_2) \in \mathcal{H}$ the resolvent $R_0(z)$ can be rewritten as

$$R_0(z)f = {}^t(r_0(z)(zf_1 + if_2), i\partial_x r_0(z)\partial_x f_1 + zr_0(z)f_2), \tag{2.5}$$

and for $z = \lambda + i\kappa \in \mathbb{C}_\pm$,

$$\Gamma(z) = 1 + \frac{1}{2}\int_{-\infty}^{\infty} \frac{\kappa}{(r-\lambda)^2 + \kappa^2}\Phi(r)dr - \frac{i}{2}\int_{-\infty}^{\infty} \frac{r-\lambda}{(r-\lambda)^2+\kappa^2}\Phi(r)dr. \tag{2.6}$$

Remark 2.1. *By* (2.6) *we can easily see that* $\Sigma_+ = \Sigma^0_+ = \emptyset$.

Proposition 2.1 (cf.[6]). *Assume* (A1) *and* (A2). *Then*

$$\Sigma_- = \begin{cases} \emptyset, & \text{if } \Gamma(-i0) \geqq 0, \\ \{i\kappa_0\}, & \text{if } \Gamma(-i0) < 0 \end{cases} \tag{2.7}$$

for some $\kappa_0 < 0$ *and*

$$\Sigma^0_- = \begin{cases} \emptyset, & \text{if } \Gamma(-i0) \neq 0, \\ \{0\}, & \text{if } \Gamma(-i0) = 0. \end{cases} \tag{2.8}$$

Moreover, for $\Gamma(-i0) < 0$ *(resp.* $\Gamma(-i0) = 0$*) it holds that* $d\Gamma(z)/dz|_{z=i\kappa_0} \neq 0$ *(resp.* $d\Gamma(\lambda - i0)/d\lambda|_{\lambda=0} \neq 0$*).*

In order to describe our main theorem (Parseval's formula corresponding to A) we need some notations. We put

$$v(z) = \begin{pmatrix} ir_0(z)\varphi \\ zr_0(z)\varphi \end{pmatrix}.$$

Using the second resolvent equation in the case of the rank one perturbation, we see that $z \notin \Sigma_+$ (resp. Σ_-) if and only if $z \in \rho(A) \cap \mathbb{C}_+$ (resp. $z \in \rho(A) \cap \mathbb{C}_-$) and

$$R(z)f = R_0(z)f + \frac{i\langle f, v(\overline{z})\rangle}{\Gamma(z)} v(z) \tag{2.9}$$

for any $f = {}^t(f_1, f_2) \in \mathcal{H}$. Define operators $\mathfrak{F}_0$, $\mathfrak{F}$ and $\mathfrak{G}$ as follows:

$$(\mathfrak{F}_0 f)(\lambda) = \begin{cases} {}^t\left(\dfrac{\lambda\hat{f}_1(\lambda) + i\hat{f}_2(\lambda)}{\sqrt{2}}, \dfrac{\lambda\hat{f}_1(-\lambda) + i\hat{f}_2(-\lambda)}{\sqrt{2}}\right), & \text{if } \lambda > 0 \\ {}^t\left(\dfrac{-\lambda\hat{f}_1(-\lambda) - i\hat{f}_2(-\lambda)}{\sqrt{2}}, \dfrac{-\lambda\hat{f}_1(\lambda) - i\hat{f}_2(\lambda)}{\sqrt{2}}\right), & \text{if } \lambda < 0, \end{cases}$$

and

$$\begin{cases} (\mathfrak{F}f)(\lambda) = (\mathfrak{F}_0 f)(\lambda) + \dfrac{i\langle f, v(\lambda - i0)\rangle}{\Gamma(\lambda + i0)} (\mathfrak{F}_0 \begin{pmatrix} 0 \\ \varphi \end{pmatrix})(\lambda), \\ (\mathfrak{G}f)(\lambda) = (\mathfrak{F}_0 f)(\lambda) - \dfrac{i\langle f, v(\lambda - i0)\rangle}{\overline{\Gamma(\lambda - i0)}} (\mathfrak{F}_0 \begin{pmatrix} 0 \\ \varphi \end{pmatrix})(\lambda). \end{cases}$$

We know that $\mathfrak{F}_0$ is a unitary operator from $\mathcal{H}$ to $L^2(\mathbb{R}, \mathbb{C}^2)$.

Remark 2.2. *The operator* $\mathfrak{F}$ *has the following property:*

$$\int_{-\infty}^{\infty} \langle (\mathfrak{F}Af)(\lambda), \tilde{g}(\lambda)\rangle_{\mathbb{C}^2} d\lambda = \int_{-\infty}^{\infty} \lambda \langle (\mathfrak{F}f)(\lambda), \tilde{g}(\lambda)\rangle_{\mathbb{C}^2} d\lambda. \tag{2.10}$$

for any $f \in D(A)$ *and* $\tilde{g} \in L^2(\mathbb{R}; \mathbb{C}^2)$.

Theorem 2.2. (Parseval formula) *Assume* (A1) *and* (A2). *Let* P *be the projection*

$$Pf = \frac{-1}{2\pi} \frac{\langle f, v(-i\kappa_0)\rangle}{\Gamma'(i\kappa_0)} v(i\kappa_0), \;\; f \in \mathcal{H},$$

and let

$$\mathcal{E} = \{g \in \mathcal{S}(\mathbb{R}) \times \mathcal{S}(\mathbb{R}); \langle v(-i0), g\rangle = 0\}.$$

1. *If* $\Gamma(-i0) \neq 0$, *it holds that*

$$\langle f, g\rangle = \begin{cases} \displaystyle\int_{-\infty}^{\infty} \langle (\mathfrak{F}f)(\lambda), (\mathfrak{G}g)(\lambda)\rangle_{\mathbb{C}^2} d\lambda, & \text{if } \Gamma(-i0) > 0 \\ \displaystyle\int_{-\infty}^{\infty} \langle (\mathfrak{F}f)(\lambda), (\mathfrak{G}g)(\lambda)\rangle_{\mathbb{C}^2} d\lambda + \langle Pf, g\rangle, & \text{if } \Gamma(-i0) < 0 \end{cases}$$

for any $f, g \in \mathcal{H}$.

2. *If* $\Gamma(-i0) = 0$, *it holds that*

$$\langle f, g\rangle = \int_{-\infty}^{\infty} \langle (\mathfrak{F}f)(\lambda), (\mathfrak{G}g)(\lambda)\rangle_{\mathbb{C}^2} d\lambda, \;\; f \in \mathcal{H}, g \in \mathcal{E}$$

Remark 2.3. Gilliam and Schulenberger [2] treated Maxwell's equations with dissipative boundary conditions in half-space in $\mathbb{R}^3$, but they did not deal with perturbed systems of selfadjoint operators.

Now we can answer (Q1) and (Q2). By Theorem 2.2, we see that

$$\mathrm{Ker} W = \begin{cases} \{0\} & \text{if } \Gamma(-i0) \geq 0, \\ \mathrm{Range} P & \text{if } \Gamma(-i0) < 0. \end{cases} \tag{2.11}$$

If $\Gamma(-i0) \geqq 0$, then any $f \in \mathcal{H}$, $f \neq 0$, is the scattering mode. In fact, if $f \neq 0$, then $f_0 := Wf \neq 0$ and

$$\lim_{t \to \infty} \|e^{-itA} f - e^{-itA_0} f_0\| = 0.$$

If $\Gamma(-i0) < 0$, any $f \in \mathcal{H}$ is a linear combination of the scattering and the dissipative modes. In fact, f can be decomposed into

$$f = (f - Pf) + Pf \equiv f_s + f_d,$$

where f_s (resp. f_d) is the scattering (resp. dissipative) mode.

3. Outline of the proof of Theorem 2.2

We give a sketch of a proof. The most difficulty is the proof of Lemma 3.2 in step 4 will be given in the Appendix.

Step 1. Rewrite the time-dependent representation of the wave operator W as follows: (Kato [8])

$$\langle Wf, g \rangle = \lim_{\kappa \downarrow 0} \frac{\kappa}{\pi} \int_{-\infty}^{\infty} \langle R(\lambda + i\kappa) f, R_0(\lambda + i\kappa) g \rangle d\lambda.$$

Using (2.9) (the representation of the resolvent of A), and the definition of $\mathfrak{F}_0$ and $\mathfrak{F}$, we see that

$$\langle Wf, g \rangle = \int_{-\infty}^{\infty} \langle (\mathfrak{F} f)(\lambda), (\mathfrak{F}_0 g)(\lambda) \rangle_{\mathbb{C}^2} d\lambda.$$

Step 2. We put

$$\mathcal{H}_s = \{ f = {}^t (f_1, f_2); \int_{\mathbb{R}} (1 + |x|^2)^s (|\partial_x f_1(x)|^2 + |f_2(x)|^2) dx < \infty \}.$$

By (2.8) in Proposition 2.1 we can prove the following formula: for $s > 1/2$, for $f, g \in \mathcal{H}_s$ and $\lambda \neq 0$

$$\begin{aligned} &\langle (\mathfrak{F} f)(\lambda), (\mathfrak{G} g)(\lambda) \rangle_{\mathbb{C}^2} \\ =& \langle (\mathfrak{F}_0 f)(\lambda), (\mathfrak{F}_0 g)(\lambda) \rangle_{\mathbb{C}^2} + \frac{1}{2\pi} \frac{\langle f, v(\lambda - i0) \rangle \langle v(\lambda + i0), g \rangle}{\Gamma(\lambda + i0)} \\ &- \frac{1}{2\pi} \frac{\langle f, v(\lambda + i0) \rangle \langle v(\lambda - i0), g \rangle}{\Gamma(\lambda - i0)}. \end{aligned} \tag{3.1}$$

Step 3. Using the unitarity of $\mathfrak{F}_0 : \mathcal{H} \longrightarrow L^2(\mathbb{R}, \mathbb{C}^2)$, we see that

$$\begin{aligned} &\int_{-\infty}^{\infty} \langle (\mathfrak{F}f)(\lambda), (\mathfrak{G}g)(\lambda)\rangle_{\mathbb{C}^2} d\lambda \\ =&\langle f, g\rangle + \frac{1}{2\pi}\int_{-\infty}^{\infty} \frac{\langle f, v(\lambda - i0)\rangle\langle v(\lambda + i0), g\rangle}{\Gamma(\lambda + i0)} d\lambda \\ &-\frac{1}{2\pi}\int_{-\infty}^{\infty} \frac{\langle f, v(\lambda + i0)\rangle\langle v(\lambda - i0), g\rangle}{\Gamma(\lambda - i0)} d\lambda. \end{aligned} \tag{3.2}$$

Step 4. In order to calculate the integrals in the right-hand side of (3.2) we use the following two lemmas:

Lemma 3.1 (cf. Lemma 2.4 in [6]). *Assume* (A1), (A2) *and* $\Gamma(-i0) = 0$. *Then it holds that*

1. $\mathfrak{G}g$ *belongs to* $L^2(\mathbb{R}; \mathbb{C}^2)$ *for* $g \in \mathcal{E}$.
2. $\mathcal{E}$ *is dense in* $\mathcal{H}$.

Lemma 3.2. *For any* $f \in \mathcal{S}(\mathbb{R}) \times \mathcal{S}(\mathbb{R})$ ($\mathcal{S}(\mathbb{R})$ *is the set of the rapidly decreasing functions*), *and for any*

$$g \in \begin{cases} \mathcal{S}(\mathbb{R}) \times \mathcal{S}(\mathbb{R}), & \text{if } \Gamma(-i0) \neq 0, \\ \mathcal{E}, \quad \text{if } \Gamma(-i0) = 0, \end{cases}$$

we have

$$\int_{-\infty}^{\infty} \frac{1}{2\pi} \frac{\langle f, v(\lambda + i0)\rangle\langle v(\lambda - i0), g\rangle}{\Gamma(\lambda - i0)} d\lambda = \begin{cases} 0, & \text{if } \Gamma(-i0) \geqq 0, \\ \langle Pf, g\rangle, & \text{if } \Gamma(-i0) < 0, \end{cases} \tag{3.3}$$

and

$$\int_{-\infty}^{\infty} \frac{\langle f, v(\lambda - i0)\rangle\langle v(\lambda + i0), g\rangle}{\Gamma(\lambda + i0)} d\lambda = 0. \tag{3.4}$$

In order to justify formulas (3.3) and (3.4), we need to show that the pole of the integrands lying in the real axis has degree not greater than 1. Actually, by Proposition 2.1 the integrands satisfy such a condition for $f \in \mathcal{S}(\mathbb{R}) \times \mathcal{S}(\mathbb{R})$ and $g \in \mathcal{S}(\mathbb{R}) \times \mathcal{S}(\mathbb{R})$ (resp. $g \in \mathcal{E}$) in the case of $\Gamma(-i0) \neq 0$ (resp. $\Gamma(-i0) = 0$). Using Lemma 3.1 and 3.2 in (3.2), we finish the proof of Theorem 2.2.

4. Some other results

1. Solvable model: (cf. [1], [4], [14],[15])

$$\begin{cases} i\partial_t u(x,t) = -\dfrac{\partial^2}{\partial x^2} u(x,t) + \alpha\langle \delta(\cdot), u(\cdot, t)\rangle \delta(x), \\ u(x,0) = f(x), \end{cases} \tag{4.1}$$

where $\alpha \in \mathbb{C}$, $(\mathrm{Im}\alpha \leq 0)$. We put

$$H_0 = -\frac{d^2}{dx^2}, \quad H_\alpha = H_0 + \alpha\langle\cdot, \delta\rangle\delta$$

$(\mathrm{Im}\alpha \leq 0)$. This operator includes a singular perturbation, and we refer to [1] for the basic information. The operator H_α is given by

$$\begin{aligned} D(H_\alpha) = &\{ U = u + aH_0(H_0^2+1)^{-1}\delta \,|\, u \in H^2(\mathbb{R}),\, a \in \mathbb{C}, \\ &\langle u, \delta\rangle = -a(\alpha^{-1} + \langle\delta, H_0(H_0^2+1)^{-1}\delta\rangle)\}, \end{aligned}$$

$$H_\alpha U = H_0 u - a(H_0^2+1)^{-1}\varphi,\ U \in D(H_\alpha)$$

We can rewrite the equation (4.1) as

$$iu_t(t) = H_\alpha u(t), \quad u(0) = f.$$

Let

$$W(\alpha) = \text{s-}\lim_{t\to\infty} e^{itH_0}e^{-itH_\alpha}.$$

The following Theorems 4.1, 4.2 and 4.3 can be found in [4].

Theorem 4.1. (*Spectral structure of* H_α) *Let* $\alpha = \alpha_1 + i\alpha_2$ *with* $\alpha_1 \leqq 0$, $\alpha_2 \leqq 0$. *Then the spectrum of* H_α *is given by*

$$\sigma(H_\alpha) = \begin{cases} [0,\infty) \cup \{-\frac{\alpha^2}{4}\}, & \text{if } \alpha_1 < 0, \\ [0,\infty), & \text{if } \alpha_1 = 0. \end{cases}$$

Exact classification of the spectrum $\sigma(H_\alpha)$ *is*

$$\sigma_{\mathrm{ess}}(H_\alpha) = \sigma_c(H_\alpha) = [0,\infty), \quad \sigma_r(H_\alpha) = \emptyset$$

and

$$\sigma_p(H_\alpha) = \begin{cases} \sigma_d(H_\alpha) = \{-\frac{\alpha^2}{4}\}, & \text{if } \alpha_1 < 0, \\ \emptyset, & \text{if } \alpha_1 = 0. \end{cases}$$

Moreover the projection associated to the eigenvalue $-\frac{\alpha^2}{4}$ $(\alpha_1 \neq 0)$ *is given by*

$$P_{-\alpha^2/4}f = -\alpha/2\langle f, e^{(\overline{\alpha}|\cdot|)/2}\rangle e^{(\alpha|x|)/2}.$$

Theorem 4.2. (i) *Assume that* $\alpha_1 < 0$ *and* $\alpha_2 < 0$. *Then*

$$\mathrm{Ker}W(\alpha) = \mathrm{Range}P_{-\frac{\alpha^2}{4}}.$$

(ii) *Assume that* $\alpha_1 = 0$ *and* $\alpha_2 < 0$. *Then*

$$\mathrm{Ker}W(i\alpha_2) = \{0\}.$$

This theorem follows from the Parseval formula given in the following Theorem 4.3. We define $\mathcal{F}_\alpha = \mathcal{F}_0 W(\alpha)$ where $\mathcal{F}_0$ is the usual Fourier transformation.

Theorem 4.3. (1) ($\alpha_1 \neq 0$) *For any* $f, g \in \mathcal{H} \cap L^1(\mathbb{R}^1)$ *and* $\alpha \in \{\alpha = \alpha_1 + i\alpha_2; \alpha_1 < 0\} \equiv D$ *we have*

$$\langle \mathcal{F}_\alpha f, \mathcal{F}_{\bar{\alpha}} g\rangle = \langle f, g\rangle + \frac{\alpha}{2}\langle f, e^{(\bar{\alpha}|\cdot|)/2}\rangle\langle e^{(\alpha|\cdot|)/2}, g\rangle.$$

(2) ($\alpha_1 = 0$)

$$\lim_{\varepsilon\to 0}\langle \mathcal{F}_{i\alpha_2} f, \chi_\varepsilon \mathcal{F}_{-i\alpha_2} g\rangle = \langle f, g\rangle + \frac{i\alpha_2}{4}\int_{\mathbb{R}^1} e^{\frac{i\alpha_2}{2}|x|} f(x)dx \int_{\mathbb{R}^1} e^{\frac{i\alpha_2}{2}|y|}\overline{g(y)}dy,$$

where χ_a *is the characteristic function of* $\{k \in \mathbb{R}; a \leqq ||k| + \alpha_2/2|\}$ *for* $a > 0$.

2. Spectrum of the dissipative operator: (cf. [5])

$$\begin{cases} w_{tt}(x,t) - \Delta w(x,t) + b(x)w_t(x,t) = 0, & (t,x) \in (0,\infty)\times\mathbb{R}^N, \\ w(x,0) = w_0(x), \quad w_t(x,0) = w_1(x), \end{cases} \tag{4.2}$$

where $N \geq 1$, $b(\cdot) \in C^1(\mathbb{R}^N \setminus \{0\})$ is a non-negative function. Putting

$$H_b = i\begin{pmatrix} 0 & 1 \\ \Delta & -b \end{pmatrix}, \quad \vec{w}(t) = \begin{pmatrix} w(\cdot,t) \\ w_t(\cdot,t) \end{pmatrix},$$

we can rewrite the above equation as

$$i\vec{w}_t(t) = H_b\vec{w}(t), \quad \vec{w}(0) = \begin{pmatrix} w_0(\cdot) \\ w_1(\cdot) \end{pmatrix}.$$

Theorem 4.4. *Suppose that the coefficient* $b(x)$ *is given by*

$$b_0(x) = \begin{cases} (3-N)|x|^{-1}, & \text{if } N = 1,2, \\ (N-1)|x|^{-1}, & \text{if } N \geq 3, \end{cases}$$

and let the initial data w_0 *and* w_1 *of* (4.2) *be*

$$w_0(x) \equiv \begin{cases} |x|f(|x|), & \text{if } N = 1, \\ f(|x|), & \text{if } N \geq 2 \end{cases}, w_1(x) = \partial_{|x|}\{w_0(|x|)\}.$$

where $f(|x|) = e^{\beta|x|}g(|x|)$, $\beta < 0$, $g \in \mathcal{S}'$ ($\mathcal{S}'$ *is of the class of tempered distributions*). *Then* (4.2) *has an explicit radial solution given by*

$$w(t,x) = \begin{cases} |x|f(|x|+t), & \text{if } N = 1, \\ f(|x|+t), & \text{if } N \geq 2. \end{cases}$$

Therefore if $f \in \mathcal{B}$ ($\mathcal{B}$ *is the set of all the bounded, infinitely differentiable functions*) *then the total energy decays exponentially as* t *tends to infinity.*

Theorem 4.5. *Assume* $N \geq 3$ *and* $|b(x)| \leq b_1|x|^{-1}$ *in* $\mathbb{R}^N$ *for some* $b_1 \in (0, N-2)$. *Then the following inclusion relation holds:*

$$\sigma_p(H_b) \subset \left\{ \kappa = \alpha + i\beta \in \mathbb{C} \mid \beta^2 \leq \frac{b_1^2}{(N-2)^2 - b_1^2}\alpha^2 \right\}.$$

5. Schrödinger model

We consider a 1-dimensional Schrödinger operator with rank 1 perturbation:

$$A = A_0 + \alpha(\cdot, \varphi)\varphi, \ A_0 = -\frac{d^2}{dx^2},$$

where $\alpha \in \mathbb{C}$, $\mathrm{Im}\alpha < 0$ and $(\cdot, \cdot)$ is the usual $L^2(\mathbb{R})$-inner product. We confine ourselves to the case $\varphi(x) = \chi_{[0,1]}(x)$ (characteristic function of the interval $[0, 1]$) and calculate

$$\Gamma(z) = 1 + \alpha((A_0 - z)^{-1}\varphi, \varphi).$$

For $\mathrm{Im}z \neq 0$ using

$$(A_0 - z)^{-1}f(x) = \frac{1}{-2i\sqrt{z}} \int_{\mathbb{R}} e^{i\sqrt{z}|x-y|} f(y)dy$$

we have

$$((A_0 - z)^{-1}\varphi, \varphi) = \frac{1}{w^2} - \frac{1}{w^3}(e^w - 1)$$

where we put $w = i\sqrt{z}$ and for $\mathrm{Im}z > 0$ we take $\mathrm{Im}\sqrt{z} > 0$. The following two lemmas are obtained by explicit calculations:

Lemma 5.1. *Let $z = -i\varepsilon$, $\varepsilon > 0$. Then we have*

$$((A_0 + i\varepsilon)^{-1}\varphi, \varphi) = \frac{-i}{\varepsilon} - \frac{c}{\varepsilon^{3/2}}(e^{-c\sqrt{\varepsilon}} - 1)$$

where $c = e^{i\pi/4}$. Moreover,

$$\lim_{\varepsilon \downarrow 0}((A_0 + i\varepsilon)^{-1}\varphi, \varphi) = 0.$$

Lemma 5.2. *Let $z = \lambda - i0$, $\lambda > 0$. Then we have*

$$((A_0 - \lambda + i0)^{-1}\varphi, \varphi) = -\frac{1}{\lambda} + \frac{1}{\lambda^{3/2}} \sin\sqrt{\lambda} + \frac{i}{\lambda^{3/2}}(\cos\sqrt{\lambda} - 1).$$

Moreover,

$$\lim_{\lambda \downarrow 0}((A_0 - \lambda + i0)^{-1}\varphi, \varphi) = 0.$$

By Lemmas 5.1 and 5.2 we see that $\Gamma(0) \neq 0$.

Next we consider the degree of the zeros of $\Gamma(\lambda - i0)$. For simplicity we put $x = \sqrt{\lambda} > 0$.

Theorem 5.1. *There is no point $x > 0$ such that $\Gamma(x^2 - i0) = \Gamma'(x^2 - i0) = 0$.*

We put $\alpha = a + bi$, $a, b \in \mathbb{R}$, $a \neq 0$ and $b < 0$. We notice that for $x = x_0$

$$\Gamma(x^2 - i0) = 0 \Longleftrightarrow \begin{cases} 1 + a(-x^{-2} + x^{-3}\sin x) - bx^{-3}(\cos x - 1) = 0, \\ ax^3(\cos x - 1) + b(-x^{-2} + x^{-3}\sin x) = 0. \end{cases}$$

We put $f(x) = x - \sin x$. If $\Gamma(x^2 - i0) = \Gamma'(x^2 - i0) = 0$ we see that

$$\Gamma(x_0^2 - i0) = 0 \Longleftrightarrow \begin{cases} x_0^3 + af(x_0) - bf'(x_0) = 0, \\ af'(x_0) + bf(x_0) = 0, \end{cases} \tag{5.1}$$

and

$$\Gamma'(x_0^2 - i0) = 0 \Longleftrightarrow \begin{cases} 3x_0^2 + af'(x_0) - bf''(x_0) = 0, \\ af''(x_0) + bf'(x_0) = 0. \end{cases} \tag{5.2}$$

By (5.1) we obtain

$$f'(x_0) = \frac{b}{a^2+b^2}x_0^3, \quad f(x_0) = -\frac{a}{a^2+b^2}x_0^3, \tag{5.3}$$

and by (5.2)

$$f'(x_0) = -\frac{3a}{a^2+b^2}x_0^2, \quad f''(x_0) = \frac{3b}{a^2+b^2}x_0^2. \tag{5.4}$$

Hence combining (5.3) and (5.4) for $f'(x_0)$ we have $x_0 = -3a/b$. On the other hand, substituting $f''(x) = \sin x = 3bx^2/(a^2+b^2)$ and $\cos x = 1 - f'(x) = 1 + 3ax^2/(a^2+b^2)$ to $1 = \sin^2 x_0 + \cos^2 x_0$, we have $x_0 = \sqrt{-2a/3}$. If $a > 0$, $\sqrt{-2a/3}$ is not real. If $a < 0$, $x_0 = -3a/b < 0$. It implies a contradiction. (In the case $a = 0$ the proof is simpler.)

6. Appendix

We use the following two lemmas and prove Lemma 3.2 given in Section 3. We omit the proofs of Lemmas 6.1 and 6.2.

Lemma 6.1. *Let $s > 1/2$ and $g \in \mathcal{S}(\mathbb{R}) \times \mathcal{S}(\mathbb{R})$. Then one has for $z \in \mathbb{C}$,*

$$\langle v(z), g\rangle = \begin{cases} O(|z|^{-1}) & (|z| \to \infty), \\ O(1) & (|z| \to 0). \end{cases}$$

Lemma 6.2. *It holds that*

$$\inf_{Imz \geqq 0} \mathrm{Re}\Gamma(z) \geqq 1.$$

There exists a positive constant C such that

$$\liminf_{|z|\to\infty, Imz\leqq 0} |\mathrm{Re}\Gamma(z)| \geqq C.$$

We give a proof of (3.3). Let X_1, X_2 and Y be some large positive numbers and ε a small positive number. Put

$$F(z) = -\frac{1}{2\pi}\frac{\langle f, v(\overline{z})\rangle\langle v(z), g\rangle}{\Gamma(z)}$$

for $z \in \mathbb{C}_-$. In the case of $\Gamma(-i0) < 0$, we see by Proposition 2.1 that $i\kappa_0$ is a simple pole of $F(z)$.

According to [17] Theorem XII.5, the projection P corresponding to the eigenvalue $i\kappa_0$ is given by:

$$\langle Pf, g\rangle = \frac{-1}{2\pi i}\int_\gamma \langle R(z)f, g\rangle dz, \quad f, g \in \mathcal{H} \tag{6.1}$$

where γ ($\subset \mathbb{C}_-$) is a contour around $i\kappa_0$. By the residue theorem we know that

$$\langle Pf, g\rangle = \frac{-1}{2\pi}\frac{\langle f, v(-i\kappa_0)\rangle}{\Gamma'(i\kappa_0)}\langle v(i\kappa_0), g\rangle, \; f, g \in \mathcal{H}.$$

Then it follows from Proposition 2.1 and Lemmas 6.1, 6.2 that

$$\sum_{j=1}^{4}\int_{\Gamma_j} F(z)dz = \begin{cases} 0, & \text{if } \Gamma(-i0) \geqq 0, \\ -\langle Pf, g\rangle, & \text{if } \Gamma(-i0) < 0, \end{cases}$$

where

$$\begin{aligned} &\Gamma_1 : z = \lambda - i\varepsilon \quad (\lambda = X_2 \to -X_1), \\ &\Gamma_2 : z = -X_1 + i\kappa \quad (\kappa = -\varepsilon \to -Y), \\ &\Gamma_3 : z = \lambda - iY \quad (\lambda = -X_1 \to X_2) \end{aligned}$$

and

$$\Gamma_4 : z = X_2 + i\kappa \quad (\kappa = -Y \to -\varepsilon).$$

Let us estimate

$$I_j = \int_{\Gamma_j} F(z)dz \quad (j = 2, 3, 4).$$

Note that

$$I_2 = \int_{\varepsilon}^{Y} F(-X_1 + i\kappa)(-i)d\kappa,$$

$$I_3 = \int_{-X_1}^{X_2} F(\lambda - iY)d\lambda$$

and

$$I_4 = \int_{Y}^{\varepsilon} F(X_2 + i\kappa)(-i)d\kappa.$$

Using Proposition 2.1 we have that

$$|I_2| \leqq C\int_{\varepsilon}^{Y}\frac{1}{X_1^2+\kappa^2}d\kappa = \frac{C}{X_1}\left(\tan^{-1}\frac{Y}{X_1} - \tan^{-1}\frac{\varepsilon}{X_1}\right) \leqq \frac{C}{X_1}\frac{\pi}{2},$$

$$|I_3| \leqq C\int_{-X_1}^{X_2}\frac{1}{\lambda^2+Y^2}d\lambda = \frac{C}{Y}\left(\tan^{-1}\frac{X_2}{Y} + \tan^{-1}\frac{X_1}{Y}\right) \leqq \frac{C}{Y}\pi$$

and

$$|I_4| \leqq C\int_{\varepsilon}^{Y}\frac{1}{X_2^2+\kappa^2}d\kappa = \frac{C}{X_2}\left(\tan^{-1}\frac{Y}{X_2} - \tan^{-1}\frac{\varepsilon}{X_2}\right) \leqq \frac{C}{X_2}\frac{\pi}{2},$$

where $C = C(f, g, \varphi) > 0$. Noting that

$$I_1 = \int_{X_2}^{-X_1} F(\lambda - i\varepsilon)d\lambda = -\int_{-X_1}^{X_2} F(\lambda - i\varepsilon)d\lambda,$$

we have

$$C(\frac{1}{X_1}+\frac{1}{X_2}+\frac{1}{Y}) \geqq \begin{cases} |-I_1|, & \text{if } \Gamma(-i0) \geqq 0 \\ |-I_1-\langle Pf,g\rangle|, & \text{if } \Gamma(-i0) < 0. \end{cases}$$

Thus taking $X_1, X_2, Y \to \infty$, we have

$$\int_{-\infty}^{\infty} F(\lambda - i\varepsilon)d\lambda = \begin{cases} 0, & \text{if } \Gamma(-i0) \geqq 0, \\ \langle Pf,g\rangle, & \text{if } \Gamma(-i0) < 0. \end{cases} \tag{6.2}$$

In the case $\Gamma(-i0) = 0$, the following limit exists by L'Hospital theorem and $\Gamma'(-i0) \neq 0$ (cf. Proposition 2.1),

$$\lim_{\lambda\to 0} \frac{\langle v(\lambda - i0), g\rangle}{\Gamma(\lambda - i0)}, \quad g \in \mathcal{E}.$$

Hence, for each λ, the following limit exists:

$$\lim_{\varepsilon\to 0} F(\lambda - i\varepsilon) = -\frac{1}{2\pi}\frac{\langle f, v(\lambda + i0)\rangle\langle v(\lambda - i0), g\rangle}{\Gamma(\lambda - i0)}. \tag{6.3}$$

By keeping (2.7) in mind we see by Lemma 6.1 and 6.2 that for sufficiently small $\varepsilon > 0$ there exist positive numbers C_1 and C_2 independent of ε such that

$$|F(\lambda - i\varepsilon)| \leqq \begin{cases} \dfrac{C_1}{\lambda^2} & (|\lambda| > 1), \\ C_2 & (|\lambda| \leqq 1). \end{cases}$$

Thus by (6.2), (6.3) and by Lebesgue's convergence theorem, we have (3.3). Next we give a sketch of a proof of (3.4). Put

$$G(z) = \frac{1}{2\pi}\frac{\langle f, v(\overline{z})\rangle\langle v(z), g\rangle}{\Gamma(z)}$$

for $z \in \mathbb{C}_+$. Then it follows from Lemma 6.1 and Lemma 6.2 that

$$\sum_{j=1}^{4}\int_{\tilde{\gamma}_j} G(z)dz = 0,$$

where

$$\tilde{\gamma}_1 : z = \lambda + i\varepsilon \quad (\lambda = -X_1 \to X_2),$$
$$\tilde{\gamma}_2 : z = X_2 + i\kappa \quad (\kappa = \varepsilon \to Y),$$
$$\tilde{\gamma}_3 : z = \lambda + iY \quad (\lambda = X_2 \to -X_1)$$

and

$$\tilde{\gamma}_4 : z = -X_1 + i\kappa \quad (\kappa = Y \to \varepsilon).$$

By the same argument as in the proof of 1, we have (3.4).

Acknowledgement

The authors thank the referee for telling us historical remarks and many articles related to this area and for pointing mistakes.

References

[1] S. Albeverio and P. Kurasov, Singular perturbations of differential operators, London Math. Soc. Lect. Note Ser. No. 271, Cambridge Univ. Press, 2000

[2] D.S. Gilliam and J.R. Schulenberger, *Electromagnetic waves in a three-dimensional half-space with a dissipative boundary*, J. Math. Anal. Appl. **89**, 1982, pp. 129–185

[3] P.D. Lax and R.S. Phillips, Scattering Theory, Academic Press, 1967

[4] M. Kadowaki, H. Nakazawa and K. Watanabe, *On the asymptotics of solutions for some Schrödinger equations with dissipative perturbations of rank one*, Hiroshima Math. J., **34**, 2004, pp. 345–369

[5] M. Kadowaki, H. Nakazawa and K. Watanabe, *Exponential decay and spectral structure for wave equation with some dissipations*, Tokyo J. Math, **28**, 2005, pp. 463–470

[6] M. Kadowaki, H. Nakazawa and K. Watanabe, *Parseval formula for wave equations with dissipative term of rank one*, to appear in SUT J. Math.

[7] T. Kako and K. Yajima, *Spectral and scattering theory for a class of non-selfadjoint operators*, Sci. Papers College Gen. Ed. Univ. Tokyo **26** 1976, pp. 73–89

[8] T. Kato, *Wave operators and similarity for some non-self adjoint operators*, Math. Ann. **162**, 1966, pp. 258–279

[9] S.T. Kuroda, An Introduction to Scattering Theory, Lecture Note Series N° 51, Matematisk Institut, Aarhus University, 1980

[10] K. Mochizuki, *Scattering theory for wave equations with dissipative terms*, Publ. RIMS Kyoto Univ., **12**, 1976, pp. 383–390

[11] S.N. Naboko, *A functional model of perturbation theory and its applications to scattering theory* , Proc. Steklov Inst. Math., **147**, 1980, pp. 86–114

[12] S.N. Naboko, *On the conditions for existence of wave operators in the nonselfadjoint case*, Amer. Math. Soc. Transl. Ser. 2, **157**, Amer. Math. Soc., Providence, RI, 1993, pp. 127–149

[13] M.A. Naimark, *Investigation of the spectrum and the expansion in eigenfunction of a non-selfadjoint operator of second order on a semi axis*, Amer. Math. Soc. Transl. Ser, **216**, 1960, pp. 103–193

[14] B.S. Pavlov, *The nonself–adjoint Schrödinger operators*, Topics in Math. Phys. **1**, Consultants Bureau New York, 1967, pp. 87–114

[15] B.S. Pavlov, *Selfadjoint dilation of a dissipative Schrödinger operator, and expansion in its eigenfunction*, (Russian) Math. USSR-Sb. **31**, 1977 pp. 457–478

[16] B.S. Pavlov, *A remark on spectral meaning of the symmetric functional model*, Oper. Theory Adv. Appl., **154**, Birkhäuser, Basel, 2004, pp. 163–177

[17] M. Reed and B. Simon, Methods of Modern Mathematical Physics, IV, Analysis of Operators, Academic Press, 1978

[18] D.R. Yafaev, Mathematical Scattering Theory: General Theory (Translations of Mathematical Monographs), Amer. Math. Soc., 1992

Mitsuteru Kadowaki
Department of Mechanical Engineering
Ehime University
3 Bunkyo–cho Matsuyama
Ehime 790-8577, Japan
e-mail: mkadowaki@eng.ehime-u.ac.jp

Hideo Nakazawa
Department of Mathematics
Chiba Institute of Technology
2–1–1 Shibazono, Narashino
Chiba 275-0023, Japan
e-mail: nakazawa.hideo@it-chiba.ac.jp

Kazuo Watanabe
Department of Mathematics
Gakushuin University
1–5–1 Mejiro Toshima
Tokyo 171-8588, Japan
e-mail: kazuo.watanabe@gakushuin.ac.jp

Operator Theory:
Advances and Applications, Vol. 186, 257–265

On the Schrödinger Operator with Limit-periodic Potential in Dimension Two

Yulia Karpeshina and Young-Ran Lee

Dedicated to Boris S. Pavlov on the occasion of his 70th birthday.

Abstract. This is an announcement of the following results. We consider the Schrödinger operator $H = -\Delta + V(x)$ in dimension two, $V(x)$ being a limit-periodic potential. We prove that the spectrum of H contains a semiaxis and there is a family of generalized eigenfunctions at every point of this semiaxis with the following properties. First, the eigenfunctions are close to plane waves $e^{i\langle \vec{k},\vec{x}\rangle}$ at the high energy region. Second, the isoenergetic curves in the space of momenta $\vec{k}$ corresponding to these eigenfunctions have a form of slightly distorted circles with holes (Cantor type structure). Third, the spectrum corresponding to the eigenfunctions (the semiaxis) is absolutely continuous.

Mathematics Subject Classification (2000). Primary 35J10; Secondary 81Q15.

Keywords. Schrödinger, limit-periodic potential.

1. Main results

We study the operator

$$H = -\Delta + V(x) \tag{1.1}$$

in two dimensions, $V(x)$ being a limit-periodic potential:

$$V(x) = \sum_{r=1}^{\infty} V_r(x), \tag{1.2}$$

where $\{V_r\}_{r=1}^{\infty}$ is a family of periodic potentials with doubling periods and decreasing L_∞-norms, namely, V_r has orthogonal periods $2^{r-1}\vec{\beta}_1$, $2^{r-1}\vec{\beta}_2$ and $\|V_r\|_\infty < \hat{C}exp(-2^{\eta r})$ for some $\eta > \eta_0 > 0$. Without loss of generality we assume that $\hat{C} = 1$ and $\int_{Q_r} V_r(x)dx = 0$, Q_r being the elementary cell of periods corresponding to $V_r(x)$.

Research partially supported by USNSF Grant DMS-0800949.

The one-dimensional analog of (1.1), (1.2) is already thoroughly investigated. It is proven in [1]–[7] that the spectrum of the operator $H_1u = -u'' + Vu$ is generically a Cantor type set. It has a positive Lebesgue measure [1, 6]. The spectrum is absolutely continuous [1, 2], [5]–[9]. Generalized eigenfunctions can be represented in the form of $e^{ikx}u(x)$, $u(x)$ being limit-periodic [5, 6, 7]. The case of a complex-valued potential is studied in [10]. Integrated density of states is investigated in [11]–[14]. Properties of eigenfunctions of the discrete multidimensional limit-periodic Schrödinger operator are studied in [15]. As to the continuum multidimensional case, it is proven [14] that the integrated density of states for (1.1) is the limit of densities of states for periodic operators.

We concentrate here on properties of the spectrum and eigenfunctions of (1.1), (1.2) in the high energy region. We prove the following results for the two-dimensional case.

1. The spectrum of the operator (1.1), (1.2) contains a semiaxis. A proof of an analogous result by different means can be found in in the paper [16]. In [16], the authors consider the operator $H = (-\Delta)^l + V$, $8l > d + 3$, $d \neq 1(\mathrm{mod}4)$, d being the dimension of the space. This obviously includes our case $l = 1$, $d = 2$. However, there is an additional rather strong restrictions on the potential $V(x)$ in [16], which we don't have here: in [16] all the lattices of periods Q_r of periodic potentials V_r need to contain a nonzero vector γ in common, i.e., V(x) is periodic in a direction γ.
2. There are generalized eigenfunctions $\Psi_\infty(\vec{k}, \vec{x})$, corresponding to the semiaxis, which are close to plane waves: for every $\vec{k}$ in an extensive subset $\mathcal{G}_\infty$ of $\mathbb{R}^2$, there is a solution $\Psi_\infty(\vec{k}, \vec{x})$ of the equation $H\Psi_\infty = \lambda_\infty\Psi_\infty$ which can be described by the formula:

$$\Psi_\infty(\vec{k}, \vec{x}) = e^{i\langle \vec{k}, \vec{x}\rangle}\left(1 + u_\infty(\vec{k}, \vec{x})\right), \tag{1.3}$$

$$\|u_\infty\| =_{|\vec{k}|\to\infty} O(|\vec{k}|^{-\gamma_1}), \quad \gamma_1 > 0, \tag{1.4}$$

where $u_\infty(\vec{k}, \vec{x})$ is a limit-periodic function:

$$u_\infty(\vec{k}, \vec{x}) = \sum_{r=1}^{\infty} u_r(\vec{k}, \vec{x}), \tag{1.5}$$

$u_r(\vec{k}, \vec{x})$ being periodic with periods $2^{r-1}\vec{\beta}_1$, $2^{r-1}\vec{\beta}_2$. The eigenvalue $\lambda_\infty(\vec{k})$ corresponding to $\Psi_\infty(\vec{k}, \vec{x})$ is close to $|\vec{k}|^2$:

$$\lambda_\infty(\vec{k}) =_{|\vec{k}|\to\infty} |\vec{k}|^2 + O(|\vec{k}|^{-\gamma_2}), \quad \gamma_2 > 0. \tag{1.6}$$

The "non-resonant" set $\mathcal{G}_\infty$ of the vectors $\vec{k}$, for which (1.3)–(1.6) hold, is an extensive Cantor type set: $\mathcal{G}_\infty = \cap_{n=1}^\infty \mathcal{G}_n$, where $\{\mathcal{G}_n\}_{n=1}^\infty$ is a decreasing sequence of sets in $\mathbb{R}^2$. Each $\mathcal{G}_n$ has a finite number of holes in each bounded region. More and more holes appear when n increases, however holes added at each step are of smaller and smaller size. The set $\mathcal{G}_\infty$ satisfies the estimate:

$$\frac{|(\mathcal{G}_\infty \cap \mathbf{B_R})|}{|\mathbf{B_R}|} =_{R\to\infty} 1 + O(R^{-\gamma_3}), \quad \gamma_3 > 0, \tag{1.7}$$

where $\mathbf{B_R}$ is the disk of radius R centered at the origin, $|\cdot|$ is the Lebesgue measure in $\mathbb{R}^2$.

3. The set $\mathcal{D}_\infty(\lambda)$, defined as a level (isoenergetic) set for $\lambda_\infty(\vec{k})$,

$$\mathcal{D}_\infty(\lambda) = \left\{\vec{k} \in \mathcal{G}_\infty : \lambda_\infty(\vec{k}) = \lambda\right\},$$

is proven to be a slightly distorted circle with infinite number of holes. It can be described by the formula:

$$\mathcal{D}_\infty(\lambda) = \{\vec{k} : \vec{k} = \varkappa_\infty(\lambda, \vec{\nu})\vec{\nu},\ \vec{\nu} \in \mathcal{B}_\infty(\lambda)\}, \tag{1.8}$$

where $\mathcal{B}_\infty(\lambda)$ is a subset of the unit circle S_1. The set $\mathcal{B}_\infty(\lambda)$ can be interpreted as the set of possible directions of propagation for the almost plane waves (1.3). The set $\mathcal{B}_\infty(\lambda)$ has a Cantor type structure and an asymptotically full measure on S_1 as $\lambda \to \infty$:

$$L\left(\mathcal{B}_n(\lambda)\right) =_{\lambda\to\infty} 2\pi + O\left(\lambda^{-\gamma_3/2}\right), \tag{1.9}$$

here and below $L(\cdot)$ is the length of a curve. The value $\varkappa_\infty(\lambda, \vec{\nu})$ in (1.8) is the "radius" of $\mathcal{D}_\infty(\lambda)$ in a direction $\vec{\nu}$. The function $\varkappa_\infty(\lambda, \vec{\nu}) - \lambda^{1/2}$ describes the deviation of $\mathcal{D}_\infty(\lambda)$ from the perfect circle of the radius $\lambda^{1/2}$. It is proven that the deviation is small:

$$\varkappa_\infty(\lambda, \vec{\nu}) =_{\lambda\to\infty} \lambda^{1/2} + O\left(\lambda^{-\gamma_4}\right), \quad \gamma_4 > 0. \tag{1.10}$$

4. Absolute continuity of the branch of the spectrum (the semiaxis) corresponding to $\Psi_\infty(\vec{k}, \vec{x})$ is proven.

To prove the results listed above we develop a modification of the Kolmogorov-Arnold-Moser (KAM) method. This paper is inspired by [17, 18, 19], where the method is used for periodic problems. In [17] KAM method is applied to classical Hamiltonian systems. In [18, 19] the technique developed in [17] is applied to semiclassical approximation for multidimensional periodic Schrödinger operators at high energies.

We consider a sequence of operators

$$H_0 = -\Delta, \qquad H^{(n)} = H_0 + \sum_{r=1}^{M_n} V_r, \quad n \geq 1,\ M_n \to \infty \text{ as } n \to \infty.$$

Obviously, $\|H - H^{(n)}\| \to 0$ as $n \to \infty$ and $H^{(n)} = H^{(n-1)} + W_n$ where $W_n = \sum_{r=M_{n-1}+1}^{M_n} V_r$. We consider each operator $H^{(n)}$, $n \geq 1$, as a perturbation of the previous operator $H^{(n-1)}$. Every operator $H^{(n)}$ is periodic, however the periods go to infinity as $n \to \infty$. We show that there is a λ_*, $\lambda_* = \lambda_*(V)$, such that the semiaxis $[\lambda_*, \infty)$ is contained in the spectra of **all** operators $H^{(n)}$. For every operator $H^{(n)}$ there is a set of eigenfunctions (corresponding to the semiaxis) being

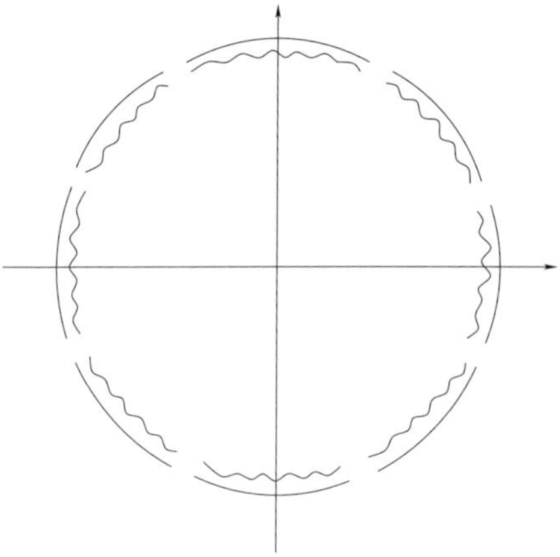

FIGURE 1 Distorted circle with holes, $\mathcal{D}_1(\lambda)$

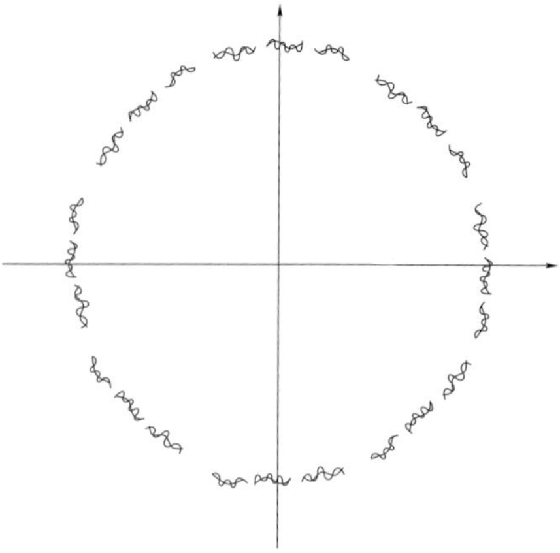

FIGURE 2 Distorted circle with holes, $\mathcal{D}_2(\lambda)$

close to plane waves: for every $\vec{k}$ in an extensive subset $\mathcal{G}_n$ of $\mathbb{R}^2$, there is a solution $\Psi_n(\vec{k},\vec{x})$ of the differential equation $H^{(n)}\Psi_n = \lambda_n\Psi_n$, which can be described by the formula:

$$\Psi_n(\vec{k},\vec{x}) = e^{i\langle\vec{k},\vec{x}\rangle}\left(1+\tilde{u}_n(\vec{k},\vec{x})\right), \quad \|\tilde{u}_n\|_{L_\infty(\mathbb{R}^2)} \underset{|\vec{k}|\to\infty}{=} O(|\vec{k}|^{-\gamma_1}), \quad \gamma_1>0, \tag{1.11}$$

where $\tilde{u}_n(\vec{k},\vec{x})$ has periods $2^{M_n-1}\vec{\beta}_1, 2^{M_n-1}\vec{\beta}_2$.[1] The corresponding eigenvalue $\lambda^{(n)}(\vec{k})$ is close to $|\vec{k}|^2$:

$$\lambda^{(n)}(\vec{k}) =_{|\vec{k}|\to\infty} |\vec{k}|^2 + O\left(|\vec{k}|^{-\gamma_2}\right), \quad \gamma_2>0.$$

The non-resonant set $\mathcal{G}_n$ is proven to be extensive in $\mathbb{R}^2$:

$$\frac{|(\mathcal{G}_\infty \cap \mathbf{B_R})|}{|\mathbf{B_R}|} =_{R\to\infty} 1 + O(R^{-\gamma_3}). \tag{1.12}$$

Estimates (1.11)–(1.12) are uniform in n. The set $\mathcal{D}_n(\lambda)$ is defined as the level (isoenergetic) set for non-resonant eigenvalue $\lambda^{(n)}(\vec{k})$:

$$\mathcal{D}_n(\lambda) = \left\{\vec{k}\in\mathcal{G}_n : \lambda^{(n)}(\vec{k}) = \lambda\right\}.$$

This set is proven to be a slightly distorted circle with a finite number of holes (Fig. 1, 2), the set $\mathcal{D}_1(\lambda)$ being strictly inside the circle of the radius $\lambda^{1/2}$ for sufficiently large λ. The set $\mathcal{D}_n(\lambda)$ can be described by the formula:

$$\mathcal{D}_n(\lambda) = \{\vec{k} : \vec{k} = \varkappa_n(\lambda,\vec{\nu})\vec{\nu},\ \vec{\nu}\in\mathcal{B}_n(\lambda)\}, \tag{1.13}$$

[1]Obviously, $\tilde{u}_n(\vec{k},\vec{x})$ is simply related to functions $u_r(\vec{k},\vec{x})$ used in (1.5):

$$\tilde{u}_n(\vec{k},\vec{x}) = \sum\nolimits_{r=M_{n-1}+1}^{M_n} u_r(\vec{k},\vec{x}).$$

where $\mathcal{B}_n(\lambda)$ is a subset of the unit circle S_1. The set $\mathcal{B}_n(\lambda)$ can be interpreted as the set of possible directions of propagation for almost plane waves (1.11). It has an asymptotically full measure on S_1 as $\lambda \to \infty$:

$$L\left(\mathcal{B}_n(\lambda)\right) =_{\lambda \to \infty} 2\pi + O\left(\lambda^{-\gamma_3/2}\right). \tag{1.14}$$

The set $\mathcal{B}_n$ has only a finite number of holes, however their number is growing with n. More and more holes of a smaller and smaller size are added at each step. The value $\varkappa_n(\lambda, \vec{\nu}) - \lambda^{1/2}$ gives the deviation of $\mathcal{D}_n(\lambda)$ from the perfect circle of the radius $\lambda^{1/2}$ in the direction $\vec{\nu}$. It is proven that the deviation is asymptotically small:

$$\varkappa_n(\lambda, \vec{\nu}) = \lambda^{1/2} + O\left(\lambda^{-\gamma_4}\right), \quad \frac{\partial \varkappa_n(\lambda, \vec{\nu})}{\partial \varphi} = O\left(\lambda^{-\gamma_5}\right), \quad \gamma_4, \gamma_5 > 0, \tag{1.15}$$

φ being an angle variable $\vec{\nu} = (\cos\varphi, \sin\varphi)$.

On each step more and more points are excluded from the non-resonant sets $\mathcal{G}_n$, thus $\{\mathcal{G}_n\}_{n=1}^{\infty}$ is a decreasing sequence of sets. The set $\mathcal{G}_\infty$ is defined as the limit set: $\mathcal{G}_\infty = \cap_{n=1}^{\infty}\mathcal{G}_n$. It has an infinite number of holes at each bounded region, but nevertheless satisfies the relation (1.7). For every $\vec{k} \in \mathcal{G}_\infty$ and every n, there is a generalized eigenfunction of $H^{(n)}$ of the type (1.11). It is proven that the sequence of $\Psi_n(\vec{k}, \vec{x})$ has a limit in $L_\infty(\mathbb{R}^2)$ as $n \to \infty$, when $\vec{k} \in \mathcal{G}_\infty$. The function $\Psi_\infty(\vec{k}, \vec{x}) = \lim_{n\to\infty} \Psi_n(\vec{k}, \vec{x})$ is a generalized eigenfunction of H. It can be written in the form (1.3)–(1.5). Naturally, the corresponding eigenvalue $\lambda_\infty(\vec{k})$ is the limit of $\lambda^{(n)}(\vec{k})$ as $n \to \infty$.

It is shown that $\{\mathcal{B}_n(\lambda)\}_{n=1}^{\infty}$ is a decreasing sequence of sets, since on each step more and more directions are excluded. We consider the limit $\mathcal{B}_\infty(\lambda)$ of $\mathcal{B}_n(\lambda)$:

$$\mathcal{B}_\infty(\lambda) = \bigcap_{n=1}^{\infty} \mathcal{B}_n(\lambda).$$

This set has a Cantor type structure on the unit circle. It is proven that $\mathcal{B}_\infty(\lambda)$ has an asymptotically full measure on the unit circle (see (1.9)). We prove that the sequence $\varkappa_n(\lambda, \vec{\nu})$, $n = 1, 2, \dots$, describing the isoenergetic curves $\mathcal{D}_n$, quickly converges as $n \to \infty$. Hence, $\mathcal{D}_\infty(\lambda)$ can be described as the limit of $\mathcal{D}_n(\lambda)$ in the sense (1.8), where $\varkappa_\infty(\lambda, \vec{\nu}) = \lim_{n\to\infty} \varkappa_n(\lambda, \vec{\nu})$ for every $\vec{\nu} \in \mathcal{B}_\infty(\lambda)$. It is shown that the derivatives of the functions $\varkappa_n(\lambda, \vec{\nu})$ (with respect to the angle variable on the unit circle) have a limit as $n \to \infty$ for every $\vec{\nu} \in \mathcal{B}_\infty(\lambda)$. We denote this limit by $\frac{\partial \varkappa_\infty(\lambda, \vec{\nu})}{\partial \varphi}$. Using (1.15) we prove that

$$\frac{\partial \varkappa_\infty(\lambda, \vec{\nu})}{\partial \varphi} = O\left(\lambda^{-\gamma_5}\right).$$

Thus, the limit curve $\mathcal{D}_\infty(\lambda)$ has a tangent vector in spite of its Cantor type structure, the tangent vector being the limit of the corresponding tangent vectors for $\mathcal{D}_n(\lambda)$ as $n \to \infty$. The curve $\mathcal{D}_\infty(\lambda)$ looks as a slightly distorted circle with infinite number of holes.

Absolute continuity of the branch of the spectrum $[\lambda_*(V), \infty)$, corresponding to the functions $\Psi_\infty(\vec{k}, \vec{x})$, $\vec{k} \in \mathcal{G}_\infty$, follows from continuity properties of level curves $\mathcal{D}_\infty(\lambda)$ with respect to λ, and from convergence of spectral projections corresponding to $\Psi_n(\vec{k}, \vec{x})$, $\vec{k} \in \mathcal{G}_\infty$, to spectral projections of H in the strong sense and uniformly in λ, $\lambda > \lambda_*$.

The main technical difficulty overcome is construction of non-resonance sets $\mathcal{B}_n(\lambda)$ for every fixed sufficiently large λ, $\lambda > \lambda_0(V)$, where $\lambda_0(V)$ is the same for all n. The set $\mathcal{B}_n(\lambda)$ is obtained by deleting a "resonant" part from $\mathcal{B}_{n-1}(\lambda)$. Definition of $\mathcal{B}_{n-1} \setminus \mathcal{B}_n$ includes Bloch eigenvalues of $H^{(n-1)}$. To describe $\mathcal{B}_{n-1} \setminus \mathcal{B}_n$ one has to use not only non-resonant eigenvalues of the type (1.6), but also resonant eigenvalues, for which no suitable formulae are known. Absence of formulae cause difficulties in estimating the size of $\mathcal{B}_n \setminus \mathcal{B}_{n-1}$. To deal with this problem we start with introducing an angle variable φ, $\varphi \in [0, 2\pi)$, $\vec{\nu} = (\cos\varphi, \sin\varphi) \in S_1$ and consider sets $\mathcal{B}_n(\lambda)$ in terms of this variable. Next, we show that the resonant set $\mathcal{B}_{n-1} \setminus \mathcal{B}_n$ can be described as the set of zeros of determinants of the type $Det\big(I + S_n(\varphi)\big)$, $S_n(\varphi)$ being a trace type operator,

$$I + S_n(\varphi) = \Big(H^{(n-1)}\big(\vec{\varkappa}_{n-1}(\varphi) + \vec{b}\big) - \lambda - \epsilon\Big)\Big(H_0\big(\vec{\varkappa}_{n-1}(\varphi) + \vec{b}\big) + \lambda\Big)^{-1},$$

where $\vec{\varkappa}_{n-1}(\varphi)$ is a vector-function describing $\mathcal{D}_{n-1}(\lambda)$, $\vec{\varkappa}_{n-1}(\varphi) = \varkappa_{n-1}(\lambda, \vec{\nu})\vec{\nu}$. To obtain $\mathcal{B}_{n-1} \setminus \mathcal{B}_n$ we take all values of ϵ in a small interval and values of $\vec{b}$ in a finite set, $\vec{b} \neq 0$. Further, we extend our considerations to a complex neighborhood Φ_0 of $[0, 2\pi)$. We show that the determinants are analytic functions of φ in Φ_0, and, by this, reduce the problem of estimating the size of the resonance set to a problem in complex analysis. We use theorems for analytic functions to count zeros of the determinants and to investigate how far the zeros move when ε changes. It enables us to estimate the size of the zero set of the determinants, and, hence, the size of the non-resonance set $\Phi_n \subset \Phi_0$, which is defined as a non-zero set for the determinants. Proving that the non-resonance set Φ_n is sufficiently large, we obtain estimates (1.12) for $\mathcal{G}_n$ and (1.14) for $\mathcal{B}_n$, the set $\mathcal{B}_n$ being the real part of Φ_n. To obtain Φ_n we delete from Φ_0 more and more holes of smaller and smaller radii at each step. Thus, the non-resonance set $\Phi_n \subset \Phi_0$ has a structure of Swiss Cheese (Fig. 3, 4). Deleting resonance set from Φ_0 at each step of the recurrent procedure is called by us a "Swiss Cheese Method". The essential difference of our method from those applied in similar situations before (see, e.g., [17]–[20]) is that we construct a non-resonance set not only on the whole space of a parameter ($\vec{k} \in \mathbb{R}^2$ here), but also on all isoenergetic curves $\mathcal{D}_n(\lambda)$ in the space of parameter, corresponding to sufficiently large λ. Estimates for the size of non-resonance sets on a curve require more subtle technical considerations than those sufficient for description of a non-resonant set in the whole space of the parameter.

The case of the operator $H = (-\Delta)^l + V$, $l \geq 6$, $d = 2$ is considered in [23]. The results proved in [23] are analogous to 1-4 on pages 258, 259. The restriction $l \geq 6$ is technical, it is needed only for the first two steps of the recurrent procedure. We modify the proof in [23] to extend the results to the case $l = 1$. The requirement for

FIGURE 3. The set Φ_2.

FIGURE 4. The set Φ_3.

super exponential decay of $\|V_r\|$ as $r \to \infty$ is essential, since it is needed to ensure convergence of the recurrent procedure. At every step we use the upper bounds on $\|V_r\|$ to prove perturbation formulae for Bloch eigenvalues and eigenfunctions when $\lambda > \lambda_*(V)$, λ_* being the same for all steps. Then, these perturbation formulae are used to prove existence of the limits of $\lambda^{(n)}(\vec{k})$, $\varkappa_n(\lambda, \vec{\nu})$, $\Psi_n(\vec{k}, \vec{x})$ as $n \to \infty$ and, hence, to define $\lambda_\infty(\vec{k})$, $\varkappa_\infty(\lambda, \vec{\nu})$, $\Psi_\infty(\vec{k}, \vec{x})$. It is not particularly important that potentials V_r have doubling periods, in the sense that the periods of the type $q^{r-1}\vec{\beta}_1$, $q^{r-1}\vec{\beta}_2, q \in \mathbb{N}$, can be treated in the same way as doubling.

References

[1] J. Avron, B. Simon *Almost Periodic Schrödinger Operators I: Limit Periodic Potentials.* Commun. Math. Physics, **82** (1981), 101–120.

[2] V.A. Chulaevski *On perturbation of a Schrödinger Operator with Periodic Potential.* Russian Math. Surv., **36(5)**, (1981), 143–144.

[3] J. Moser *An Example of the Schrödinger Operator with Almost-Periodic Potentials and Nowhere Dense Spectrum.* Comment. Math. Helv., **56** (1981), 198–224.

[4] B. Simon *Almost Periodic Schrödinger Operators. A Review.* Advances in Applied Mathematics, **3** (1982), 463–490.

[5] L.A. Pastur, V.A. Tkachenko *On the Spectral Theory of the One-Dimensional Schrödinger Operator with Limit-Periodic Potential.* Dokl. Akad. Nauk SSSR, **279** (1984), 1050–1053; Engl. Transl.: Soviet Math. Dokl., **30** (1984), no. 3, 773–776

[6] L.A. Pastur, V.A. Tkachenko *Spectral Theory of a Class of One-Dimensional Schrödinger Operators with Limit-Periodic Potentials.* Trans. Moscow Math. Soc., **51** (1989), 115–166.

[7] L. Pastur, A. Figotin *Spectra of Random and Almost-Periodic Operators.* Springer-Verlag, 1992, 583pp.

[8] J. Avron, B. Simon *Cantor Sets and Schrödinger Operators: Transient and Recurrent Spectrum.* J. Func. Anal., **43** (1981), 1–31.

[9] S.A. Molchanov and V.A. Chulaevskii *Structure of the Spectrum of Lacunary Limit-Periodic Schrödinger Operator.* Func. Anal. Appl., **18** (1984), 91–92.

[10] L. Zelenko *On a Generic Topological Structure of the Spectrum to One-Dimensional Schrödinger Operators with Complex Limit-Periodic Potentials.* Integral Equations and Operator Theory, **50** (2004), 393–430.

[11] M.A. Shubin *The Density of States for Selfadjoint Elliptic Operators with Almost Periodic Coefficients.* Trudy sem. Petrovskii (Moscow University), **3** (1978), 243–275.

[12] M.A. Shubin *Spectral Theory and Index of Elliptic Operators with Almost Periodic Coefficients.* Russ. Math. Surveys, **34(2)**, (1979), 109–157.

[13] J. Avron, B. Simon *Almost Periodic Schrödinger Operators. II: The Integrated Density of States.* Duke Math. J., **50** (1983), 1, 369–391.

[14] G.V. Rozenblum, M.A. Shubin, M.Z. Solomyak *Spectral Theory of Differential Operators.* Encyclopaedia of Mathematical Sciences, **64**, 1994.

[15] Yu.P. Chuburin *On the Multidimensional Discrete Schrödinger Equation with a Limit Periodic Potential.* Theoretical and Mathematical Physics, **102** (1995), no. 1, 53–59.

[16] M.M. Skriganov, A.V. Sobolev *On the spectrum of a limit-periodic Schrödinger operator* Algebra i Analiz, **17** (2005), 5; English translation: St. Petersburg Math. J. 17 (2006), 815–833.

[17] G. Gallavotti, *Perturbation Theory for Classical Hamiltonian Systems. Scaling and Self-Similarity in Progr. Phys.*, **7**, edited by J. Froehlich, Birkhäuser, Basel, Switzerland, 1983, 359–424.

[18] L.E. Thomas, S.R. Wassel, *Stability of Hamiltonian systems at high energy.* J. Math. Phys. **33(10)**, (1992), 3367–3373.

[19] L.E. Thomas and S.R. Wassel, *Semiclassical Approximation for Schrödinger Operators at High Energy,* Lecture Notes in Physics, **403**, edited by E. Balslev, Springer-Verlag, 1992, 194–210.

[20] J. Bourgain, *Quasiperiodic Solutions of Hamilton Perturbations of 2D Linear Schrödinger Equation.* Ann. of Math. (2), **148** (1998), 2, 363–439.

[21] M. Reed, B. Simon, *Methods of Modern Mathematical Physics.*, Vol IV, Academic Press, 3rd ed., New York – San Francisco – London (1987), 396 pp.

[22] Yu. Karpeshina *Perturbation theory for the Schrödinger operator with a periodic potential.* Lecture Notes in Mathematics, **1663**, Springer-Verlag, 1997, 352 pp.

[23] Yu. Karpeshina, Y.-R. Lee *Spectral Properties of Polyharmonic Operators with Limit-Periodic Potential in Dimension Two.* D'Analyse Mathématique, **102** (2007), pp. 225–310.

Yulia Karpeshina
Department of Mathematics
University of Alabama at Birmingham
1300 University Boulevard
Birmingham, AL, 35294-1170, USA
e-mail: karpeshi@math.uab.edu

Young-Ran Lee
Department of Mathematical Sciences
KAIST (Korea Advanced Institute of Science and Technology)
335 Gwahangno, Yusong-Gu
Daejeon, 305-701, Republic of Korea
e-mail: yrlee4@amath.kaist.ac.kr

Operator Theory:
Advances and Applications, Vol. 186, 267–283

Similarity Problem for Non-self-adjoint Extensions of Symmetric Operators

Alexander V. Kiselev

Abstract. The similarity problem for non-self-adjoint extensions of a symmetric operator having equal deficiency indices is studied. Necessary and sufficient conditions for a wide class of such operators to be similar to self-adjoint ones are obtained. The paper is based on the construction of functional model for the operators of the class considered due to Ryzhov [26, 25] and extends the results of [10] to this class. In addition to solving the similarity problem for the operator itself, we also give necessary and sufficient conditions for similarity of its restrictions to spectral subspaces corresponding to arbitrary Borel sets of the real line. These conditions (together with the technique employed while establishing them) have their direct analogues in the setting of the paper [10].

Mathematics Subject Classification (2000). 47A10, 47A20, 47A45, 47A48, 47A55.

Keywords. Extensions, functional model, non-self-adjoint operators, symmetric operators, similarity theory.

1. Introduction

A non-self-adjoint operator L acting in the Hilbert space H is called similar to a self-adjoint operator A if there exists a bounded, boundedly invertible operator X in H such, that $L = X^{-1}AX$. The similarity problem thus arises, i.e., to ascertain whether a given non-self-adjoint operator L is similar to some self-adjoint operator A or not and to give preferably necessary and sufficient conditions of this.

A criterion for similarity of a general non-self-adjoint operator with real spectrum to a self-adjoint one was established in a form of a pair of integral estimates involving the resolvent of the operator in [17, 5] (see (3.1) below). Unfortunately, this criterion, although given in a very concise form, is hard to verify in applications. This is why it seems reasonable to rewrite it in an equivalent (and suitable

---

The author gratefully appreciates financial support of INTAS (grant no. 05-1000008-7883) and RFBR (grant no. 06-01-00249).

for applications) form under some additional assumptions on the class of operators considered. One such assumption that has yielded rather interesting results in the case of *additive* non-self-adjoint perturbations of self-adjoint operators (see [10]) is that the spectrum of the operator L is absolutely continuous, i.e., the absolutely continuous spectral subspace of the operator coincides with the Hilbert space H (see Section 2 below for the definition and Remark 3.2).

In the present paper we consider the similarity problem for non-self-adjoint extensions of symmetric operators with equal deficiency indices, i.e., we work out the conditions necessary and sufficient for such operators to be similar to self-adjoint ones, extending the results of [10] to the class of non-self-adjoint almost solvable extensions of symmetric operators with absolutely continuous spectrum. Namely, Theorems 3.3 and 3.5 and Corollaries 3.4 and 3.6 below have the same form as the corresponding statements of [10].

The approach employed (based on the functional model of non-self-adjoint operators and quite similar to that of [10]) has also permitted us to study the similarity problem for restrictions of the given non-self-adjoint operator to spectral subspaces of the operator L corresponding to arbitrary Borel sets of the real line. Necessary and sufficient conditions of the named similarity have been obtained. Moreover, direct analogues of these conditions also hold in the setting of the paper [10], although the named paper contains no discussion of this subject. It should be also noted that in [10] the results on the similarity problem have been published without the proof. Results on the similarity problem for restrictions of additive non-self-adjoint perturbations of self-adjoint operators, similar to the ones contained in the present paper, will be published elsewhere.

The conditions of similarity to a self-adjoint operator derived in the present paper differ from ones that have been known previously (that is, the ones of [17, 5]) in that they are formulated in terms of objects defined on the real line only, whereas the existing ones were formulated in terms of half-planes of the complex plane. In a nutshell, using the functional model we have been able to pass to a limit in the Naboko-Van Casteren criterion under the only additional assumption that the operator L has absolutely continuous spectrum.

Since the functional model of a non-self-adjoint operator is of crucial importance for our approach, the proofs of our main results rely heavily upon the recent paper by Ryzhov [25], where the symmetric form of the Nagy-Foiaş functional model has been developed for almost solvable non-self-adjoint extensions of symmetric operators due to Pavlov [21, 20] (see also the paper [16] by Naboko, where this idea was first suggested for additive non-self-adjoint perturbations of self-adjoint operators). The approach of Ryzhov is in its turn based on the technique of the so-called boundary spaces. We continue with a brief introduction to the main concepts and results obtained in this area in Section 2.

Section 3 contains our main result, which is a criterion of similarity of a non-self-adjoint operator of the class considered to a self-adjoint one. This criterion is formulated in two equivalent forms. Depending on the concrete choice of the operator under consideration, one of these forms might prove to be preferable to another from the viewpoint of simplicity of calculations. Further, two concise sufficient conditions of similarity are given that seem to be even more suitable for applications.

Section 4 deals with the similarity problem for the above-mentioned restrictions of non-self-adjoint operator under investigation to spectral subspaces corresponding to arbitrary Borel sets of the real line. We formulate results that fall in line with those of Section 3 and sketch their proof.

Finally, Section 5 demonstrates applicability of our results to analysis of the similarity problem for non-self-adjoint Laplace operator with a potential of zero radius in the case of $\mathbb{R}^3$.

2. The functional model for almost solvable extensions of symmetric operators

The narrative of the present section in the part concerning boundary triples and the theory of almost solvable extensions is based on the works [8, 7, 4, 12], in the part concerning explicit construction of Pavlov-Naboko functional model – on the paper [25], see also [16].

Suppose that A_0 is a symmetric densely defined closed linear operator acting in the Hilbert space H ($D(A_0) \equiv D_{A_0}$ and $R(A_0) \equiv R_{A_0}$ denoting its domain and range respectively; $D(A_0^*) \equiv D_{A_0^*}$, $R(A_0^*) \equiv R_{A_0^*}$ denoting the domain and range of operator A_0^* adjoint to A_0). Assume that A_0 is completely non-self-adjoint, i.e., there exists no subspace H_0 in H such that the restriction $A_0|H_0$ is a self-adjoint operator in H_0. Further assume that the deficiency indices of A_0 (probably being infinite) are equal: $n_+(A_0) = n_-(A_0) \leq \infty$.

Definition 2.1 ([8, 4, 12]). Let Γ_1, Γ_2 be linear mappings of $D_{A_0^*}$ to $\mathcal{H}$ – a separable Hilbert space. The triple $(\mathcal{H}, \Gamma_1, \Gamma_2)$ is called *a boundary triple* for the operator A_0^* if:

1. for all $f, g \in D_{A_0^*}$
$$(A_0^* f, g)_H - (f, A_0^* g)_H = (\Gamma_1 f, \Gamma_2 g)_{\mathcal{H}} - (\Gamma_2 f, \Gamma_1 g)_{\mathcal{H}}.$$
2. the mapping γ defined as $f \longmapsto (\Gamma_1 f; \Gamma_2 f)$, $f \in D_{A_0^*}$ is surjective, i.e., for all $Y_1, Y_2 \in \mathcal{H}$ there exists such $y \in D_{A_0^*}$ that $\Gamma_1 y = Y_1$, $\Gamma_2 y = Y_2$.

A boundary triple can be constructed for any operator A_0 of the class considered. Moreover, the space $\mathcal{H}$ can be chosen in a way such that $\dim \mathcal{H} = n_+ = n_-$.

Definition 2.2 ([8, 7]). A nontrivial extension $\tilde{A}_B$ of the operator A_0 such that $A_0 \subset \tilde{A}_B \subset A_0^*$ is called *almost solvable* if there exists a boundary triple $(\mathcal{H}, \Gamma_1, \Gamma_2)$ for A_0^* and a bounded linear operator B defined everywhere on $\mathcal{H}$ such that for every $f \in D_{A_0^*}$
$$f \in D_{\tilde{A}_B} \text{ if and only if } \Gamma_1 f = B\Gamma_2 f.$$

It can be shown that if an extension $\tilde{A}_B$ of A_0, $A_0 \subset \tilde{A}_B \subset A_0^*$, has regular points (i.e., the points belonging to the resolvent set) in both upper and lower half-planes of the complex plane, then this extension is almost solvable.

The following theorem holds:

Theorem 2.3 ([8, 7]). *Let A_0 be a closed densely defined symmetric operator with $n_+(A_0) = n_-(A_0)$, let $(\mathcal{H}, \Gamma_1, \Gamma_2)$ be a boundary triple of A_0^*. Consider the almost solvable extension $\tilde{A}_B$ of A_0 and the corresponding bounded operator B in $\mathcal{H}$. Then:*

1. $y \in D_{A_0}$ *if and only if* $\Gamma_1 y = \Gamma_2 y = 0$,
2. $\tilde{A}_B$ *is maximal, i.e.,* $\rho(\tilde{A}_B) \neq \emptyset$,
3. $(\tilde{A}_B)^* \subset A_0^*$, $(\tilde{A}_B)^* = \tilde{A}_{(B^*)}$,
4. *operator* $\tilde{A}_B$ *is dissipative if and only if* B *is dissipative,*
5. $(\tilde{A}_B)^* = \tilde{A}_B$ *if and only if* $B^* = B$.

Definition 2.4 ([7, 25]). Let A_0 be a closed densely defined symmetric operator, $n_+(A_0) = n_-(A_0)$, $(\mathcal{H}, \Gamma_1, \Gamma_2)$ is its space of boundary values. The operator-function $M(\lambda)$, defined by

$$M(\lambda)\Gamma_2 f_\lambda = \Gamma_1 f_\lambda, \ \ f_\lambda \in \ker(A_0^* - \lambda I), \ \ \lambda \in \mathbb{C}_\pm, \tag{2.1}$$

is called the Weyl function of a symmetric operator A_0.

The following theorem describing the properties of the Weyl function clarifies its meaning.

Theorem 2.5 ([7]). *Let $M(\lambda)$ be the Weyl function of the symmetric operator A_0 with equal deficiency indices ($n_+(A_0) = n_-(A_0) < \infty$). Let $\tilde{A}_B$ be an almost solvable extension of A_0 corresponding to a bounded operator B. Then for every $\lambda \in \mathbb{C}$:*

1. $M(\lambda)$ *is analytic operator-function when* $\operatorname{Im}\lambda \neq 0$, *its values being bounded linear operators in* $\mathcal{H}$.
2. $(\operatorname{Im}\ M(\lambda)) \operatorname{Im}\ \lambda > 0$ *when* $\operatorname{Im}\lambda \neq 0$.
3. $M(\lambda)^* = M(\overline{\lambda})$ *when* $\operatorname{Im}\lambda \neq 0$.
4. $\lambda \in \sigma_p(\tilde{A}_B)$ *if and only if* $\det(M(\lambda) - B) = 0$.
5. $\lambda \in \rho(\tilde{A}_B)$ *if and only if* $B - M(\lambda)$ *is boundedly invertible.*

Consider now the definition and the properties of the characteristic operator-function of a non-self-adjoint extension of a closed, densely defined symmetric operator [28, 25]. Let A_0 be a symmetric operator, $(\mathcal{H}, \Gamma_1, \Gamma_2)$ its boundary triple and $M(\lambda)$ the corresponding Weyl function. Let $\tilde{A}_B$ be an almost solvable extension of A_0 (we assume for simplicity that it is completely non-self-adjoint). Consider the following definitions:

$$G(\tilde{A}_B) \equiv \{g \in D(\tilde{A}_B) : (\tilde{A}_B f, g) - (f, \tilde{A}_B g) = 0 \text{ for all} f \in D(\tilde{A}_B)\},$$

$$\mathcal{L}(\tilde{A}_B) \equiv D(\tilde{A}_B)/G(\tilde{A}_B).$$

Definition 2.6. ([28, 29]) Hilbert space $\mathcal{L}$ isomorphic to the closure of $\mathcal{L}(\tilde{A}_B)$ w.r.t. the metric $[\cdot,\cdot]$ induced by the non-degenerate but not necessarily positive inner product

$$[\xi_1,\xi_2]_{\mathcal{L}} \equiv \frac{(\tilde{A}_B f_1, f_2) - (f_1, \tilde{A}_B f_2)}{2i}, \ \xi_1,\xi_2 \in \mathcal{L}, \ f_k \in \xi_k, \ k = 1,2,$$

is called the *boundary space* of operator $\tilde{A}_B$.

The mapping $\Gamma : D(\tilde{A}_B) \mapsto \mathcal{L}$, for which

$$[\Gamma f_1, \Gamma f_2]_{\mathcal{L}} = \frac{(\tilde{A}_B f_1, f_2) - (f_1, \tilde{A}_B f_2)}{2i}, \ f_1, f_2 \in D(\tilde{A}_B),$$

is called the *boundary operator* for $\tilde{A}_B$.

Let $\mathcal{L}'$, Γ' be respectively the boundary space and boundary operator for the operator $(-\tilde{A}_B)^*$. Characteristic operator-function $\Theta_{\tilde{A}_B}$ [3] of the operator $\tilde{A}_B$ is defined by the following equality:

$$\Theta_{\tilde{A}_B}(\lambda)\Gamma f = \Gamma' g_\lambda, \tag{2.2}$$

where $\lambda \in \rho(\tilde{A}_B^*)$, $f \in D(\tilde{A}_B)$, $g_\lambda \equiv (\tilde{A}_B^* - \lambda I)^{-1}(\tilde{A}_B - \lambda I)f$.[1]

Considering the polar decomposition of the bounded operator $\operatorname{Im} B \equiv (B - B^*)/2i$ acting in the auxiliary Hilbert space $E \equiv \operatorname{clos}(\operatorname{Ran}(\operatorname{Im}\ B))$, $\operatorname{Im} B = J\alpha^2/2$, where $\alpha \equiv |2\operatorname{Im} B|^{1/2}$, $J \equiv \operatorname{sign}(\operatorname{Im} B|E)$ it can be proved [25] that the space E with the metric $[\cdot,\cdot] \equiv (J\cdot,\cdot)_H$ can be chosen as a boundary space for the operator $\tilde{A}_B$; in this case the boundary operator Γ can be defined as follows [12, 8, 7], see also [25]:

$$\Gamma \equiv J\alpha\Gamma_2; \ D(\Gamma) \equiv D(\tilde{A}_B).$$

and the characteristic function for $\tilde{A}_B$ is given by the following formula:

$$\Theta(\lambda) \equiv I|_E + iJ\alpha(B^* - M(\lambda))^{-1}\alpha|_E, \ \lambda \in \rho(\tilde{A}_B^*). \tag{2.3}$$

Now consider the functional model for a non-self-adjoint extension of symmetric operator A_0 [20, 30, 16, 25]. Alongside with an arbitrary almost solvable extension $\tilde{A}_B$ consider the dissipative [24, 20] extension $\tilde{A}_{B_+}$, corresponding to the bounded operator B_+ defined as follows:

$$B = \Re B + i\alpha J\alpha/2 \longmapsto B_+ \equiv \Re B + i|\operatorname{Im} B| = \Re B + i\alpha^2/2.$$

As stated in [25], the requirement of complete non-self-adjointness of the operator $\tilde{A}_B$ leads to the fact that the dissipative operator $\tilde{A}_{B_+}$ is also completely non-self-adjoint.

Introduce simplified notation: $A \equiv \tilde{A}_B$, $A_+ \equiv \tilde{A}_{B_+}$, and further denote the characteristic function of the operator A_+ by the symbol $S(\lambda)$. It is a contractive in the upper half-plane analytic operator function (this can be quite easily verified based on [8], see, e.g., [25]). Thus $S(\lambda)$ possesses limits almost everywhere on the

[1]For each $\lambda \in \rho(\tilde{A}_B^*)$ the operator-function $\Theta_{\tilde{A}_B}$ acts as a linear bounded operator from $\mathcal{L}$ onto $\mathcal{L}'$ (see [3]).

real line in the strong topology, which we denote by $S(k)$. Consider the model space $\mathfrak{H} = L_2\left(\begin{smallmatrix} I & S^* \\ S & I \end{smallmatrix}\right)$. It is defined in [20] (see also [18] for description of general coordinate-free models) in the following way: consider all two-component vector-functions $(\tilde{g}, g)$ on the axis $(\tilde{g}(k), g(k) \in E, k \in \mathbb{R})$ finite w.r.t. the following metric:

$$\left\langle \begin{pmatrix} \tilde{g} \\ g \end{pmatrix}, \begin{pmatrix} \tilde{g} \\ g \end{pmatrix} \right\rangle = \int_{-\infty}^{\infty} \left\langle \begin{pmatrix} I & S^*(k) \\ S(k) & I \end{pmatrix} \begin{pmatrix} \tilde{g}(k) \\ g(k) \end{pmatrix}, \begin{pmatrix} \tilde{g}(k) \\ g(k) \end{pmatrix} \right\rangle_{E\oplus E} dk.$$

Factoring the set of two-component functions by the set of elements with norm equal to zero and then closing the result w.r.t. the above norm, we finally arrive at the model space $\mathfrak{H}$.

Although we consider $(\tilde{g}, g)$ as a symbol only, the formal expressions $g_- := (\tilde{g} + S^* g)$ and $g_+ := (S\tilde{g} + g)$ (the motivation for the choice of notation is self-evident from what follows) can be shown to represent some true $L_2(E)$-functions on the real line. In what follows we plan to deal mostly with these functions.

Define the following orthogonal subspaces in $\mathfrak{H}$:

$$D_- \equiv \begin{pmatrix} 0 \\ H^2_-(E) \end{pmatrix}, \; D_+ \equiv \begin{pmatrix} H^2_+(E) \\ 0 \end{pmatrix}, \; K \equiv \mathfrak{H} \ominus (D_- \oplus D_+),$$

where $H^2_{+(-)}(E)$ denotes the Hardy class [30] of analytic functions f in the upper (lower) half-plane taking values in the Hilbert space E. These subspaces are "incoming" and "outgoing" subspaces, respectively, in the sense of [13].

The subspace K can be described as $K = \{(\tilde{g}, g) \in \mathfrak{H} : \; g_- \equiv \tilde{g} + S^* g \in H^2_-(E), g_+ \equiv S\tilde{g} + g \in H^2_+(E)\}$. Let P_K be the orthogonal projection of the space $\mathfrak{H}$ onto K, then

$$P_K \begin{pmatrix} \tilde{g} \\ g \end{pmatrix} = \begin{pmatrix} \tilde{g} - P_+(\tilde{g} + S^* g) \\ g - P_-(S\tilde{g} + g) \end{pmatrix},$$

where $P_\pm$ are the orthogonal Riesz projections of the space $L_2(E)$ onto $H^2_\pm(E)$.

The following theorem holds [25], cf. [16, 20]:

Theorem 2.7 (Ryzhov, [25]). *The operator $(A_+ - \lambda_0)^{-1}$ is unitary equivalent to the operator $P_K(k - \lambda_0)^{-1}|_K$ for all $\lambda_0, \operatorname{Im} \lambda_0 < 0$.*

This means, that the operator of multiplication by k serves as the minimal $(\operatorname{clos}_{\operatorname{Im}\lambda \neq 0}(k - \lambda)^{-1} K = \mathfrak{H})$ self-adjoint dilation [30] of the operator A_+.

As it is shown in [25], cf. [16], it is possible to compute the action of the operator A in the model representation of dissipative operator A_+. In particular, the following theorem holds:

Theorem 2.8 (Ryzhov, [25]). *Let $\lambda_0 \in \mathbb{C}_- \cap \rho(A)$; $\mu_0 \in \mathbb{C}_+ \cap \rho(A)$; $(\tilde{g}, g) \in K$. Then*

$$\begin{aligned} (A - \lambda_0 I)^{-1} \begin{pmatrix} \tilde{g} \\ g \end{pmatrix} &= P_K (k - \lambda_0)^{-1} \begin{pmatrix} \tilde{g} \\ g - \mathcal{X}_- \Theta_-^{-1}(\lambda_0)(\tilde{g} + S^* g)(\lambda_0) \end{pmatrix}, \\ (A - \mu_0 I)^{-1} \begin{pmatrix} \tilde{g} \\ g \end{pmatrix} &= P_K (k - \mu_0)^{-1} \begin{pmatrix} \tilde{g} - \mathcal{X}_+ \Theta_+^{-1}(\mu_0)(S\tilde{g} + g)(\mu_0) \\ g \end{pmatrix}, \end{aligned} \tag{2.4}$$

where $\mathcal{X}_\pm \equiv (I \pm J)/2$,

$$\Theta_-(\lambda) \equiv I - i\alpha(B_+ - M(\lambda))^{-1}\alpha\mathcal{X}_- = \mathcal{X}_+ + S^*(\overline{\lambda})\mathcal{X}_-, \quad \text{Im } \lambda < 0,$$
$$\Theta_+(\lambda) \equiv I + i\alpha(B_+^* - M(\lambda))^{-1}\alpha\mathcal{X}_+ = \mathcal{X}_- + S(\lambda)\mathcal{X}_+, \quad \text{Im } \lambda > 0,$$

and $S(\lambda)$ *is as before the characteristic operator-function of the dissipative operator* A_+ *(see* (2.3)*;* [16, 25]*).*

Following [15], we define the linear sets $\hat{N}_\pm$ in $\mathfrak{H}$ as follows:

$$\hat{N}_\pm \equiv \left\{ \begin{pmatrix} \tilde{g} \\ g \end{pmatrix} : \begin{pmatrix} \tilde{g} \\ g \end{pmatrix} \in \mathfrak{H}, P_\pm \left(\mathcal{X}_+ g_+ + \mathcal{X}_- g_-\right) = 0 \right\} \tag{2.5}$$

and introduce the following designation: $N_\pm = \text{clos}\, P_K \hat{N}_\pm$.

Absolutely continuous and singular subspaces of the non-self-adjoint operator A are defined in the same way as in the paper [14]: let $N \equiv \hat{N}_+ \cap \hat{N}_-$, $\tilde{N}_\pm \equiv P_K \hat{N}_\pm$, $\tilde{N}_e \equiv \tilde{N}_+ \cap \tilde{N}_-$. The subspaces $N_\pm(A^*)$ for the adjoint to A operator A^* are defined in an analogous way using the same model representation. Then

$$N_e \equiv \text{clos}\left(\tilde{N}_+ \cap \tilde{N}_-\right) = \text{clos}\, P_K N \equiv \text{clos}\ \tilde{N}_e; \quad N_i \equiv K \ominus N_e(A^*), \tag{2.6}$$

where $N_e(A^*)$ denotes the absolutely continuous subspace of the operator A^*, which can be easily described in similar way in terms of the same model space $\mathfrak{H}$.

An argument quite similar to that of [16] and based on identities established in [25] leads to the following Theorem.

Theorem 2.9. *Let the operator* A *be as above. Then*

(i) *For all* $\text{Im } \lambda < 0$ $(\text{Im } \lambda > 0)$ *and* $(\tilde{g}, g) \in \hat{N}_{-(+)}$, *respectively, one has:*

$$(A - \lambda)^{-1} P_K \begin{pmatrix} \tilde{g} \\ g \end{pmatrix} = P_K \frac{1}{k - \lambda} \begin{pmatrix} \tilde{g} \\ g \end{pmatrix}. \tag{2.7}$$

Conversely, the property (2.7) *for* $\text{Im } \lambda < 0$ $(\text{Im } \lambda > 0)$ *guarantees that the vector* $(\tilde{g}, g)$ *belongs to the set* $\hat{N}_{-(+)}$.

(ii) *The following non-model description of linear sets* $\tilde{N}_\pm \equiv P_K \hat{N}_\pm$ *holds:*

$$\tilde{N}_{+(-)} = \left\{ u \in H : \mathcal{X}_{+(-)}\alpha\Gamma_2(A - \lambda)^{-1}u \in H^2_{+(-)}(E) \right\}. \tag{2.8}$$

Here $\mathcal{X}_{+(-)}\alpha\Gamma_2(A - \lambda)^{-1}u$ *is*[2] *treated as an analytic vector function of* $\lambda \in \mathbb{C}_{+(-)}$ *taking values in the auxiliary Hilbert space* E.

Here and throughout the rest of the paper we use the same notation for the objects of the original Hilbert space H and their model representations. We hope that this convention will not confuse the reader.

We remark that it can be verified [16] that the projections $\mathcal{X}_\pm$ can be dropped altogether in the definition (2.8). The existence of this description gives ground to

[2]That is, analytic continuations of the vector $\mathcal{X}_{+(-)}\alpha\Gamma_2(A-\lambda)^{-1}u$ from the domain of analyticity of the resolvent to the half-plane $\mathbb{C}_{+(-)}$

calling the vectors belonging to the linear sets $\tilde{N}_+, \tilde{N}_-$ and $\tilde{N}_e$ "smooth" (in the corresponding half-planes) vectors of the operator A (see also [16]).

The *proof* of Theorem 2.9 will appear in our forthcoming publication.

3. Similarity problem for absolutely continuous extensions of symmetric operators

Definition 3.1. We call a non-self-adjoint operator A acting in Hilbert space H *an operator with absolutely continuous spectrum* if its absolutely continuous subspace $N_e(A)$ coincides with H.

Remark 3.2. It has to be noted that the requirement $N_e(A) = H$ doesn't actually guarantee that the spectrum of the operator A is *purely* absolutely continuous. Due to the possibility that the absolutely continuous and singular spectral subspaces may intersect, one might face the following situation: $N_e(A) = H$; $N_i(A) \neq \{0\}$ and $N_i(A) \subset N_e(A)$. See [31] for one rather transparent example of this case. Nevertheless, due to the fact that $N_i(A^*) \equiv H \ominus N_e(A)$, one easily sees that in the situation of $N_e(A) = H$ the singular spectral subspace of the *adjoint* operator A^* is trivial.

Let A_0 be as above a closed densely defined linear symmetric operator with equal (finite or infinite) deficiency indices $n_+(A_0) = n_-(A_0) \leq \infty$. Let $(\mathcal{H}, \Gamma_1, \Gamma_2)$ be a boundary triple for A_0^*. Consider operators $\tilde{A}_B \equiv A$, being almost solvable non-self-adjoint extensions of operator A_0. Let the spectrum $\sigma(A)$ of the operator A be a subset of the real axis.[3] Furthermore, without loss of generality we restrict ourselves to consideration of completely non-self-adjoint extensions A only [25, 16].

As above we denote by $\Theta(\lambda)$, $S(\lambda)$ the characteristic operator-functions of operators A, $A_+ \equiv \tilde{A}_{B_+}$, respectively.

Then the following theorem holds:

Theorem 3.3. *Provided that the operator A has absolutely continuous spectrum, the following assertions are equivalent:*

(i) *Operator A is similar to a self-adjoint one*
(ii) *For all $u \in H$ the following estimates hold:*

$$\int ((\Theta(k - i0)J\Theta^*(k - i0) - J)\mathcal{X}_+\alpha\Gamma_2(A_+^* - k - i0)^{-1}u,$$
$$\mathcal{X}_+\alpha\Gamma_2(A_+^* - k - i0)^{-1}u)dk \leq C\|u\|^2$$
$$\int ((J - \Theta^*(k + i0)J\Theta(k + i0))\mathcal{X}_-\alpha\Gamma_2(A_+^* - k - i0)^{-1}u,$$
$$\mathcal{X}_-\alpha\Gamma_2(A_+^* - k - i0)^{-1}u)dk \leq C\|u\|^2.$$

[3]The latter is necessary for operator A to be similar to self-adjoint one.

Proof. It is proved in [17, 5], that a non-self-adjoint operator A acting in Hilbert space H, the spectrum $\sigma(A)$ of which is a subset of real axis, is similar to a self-adjoint operator if and only if the following estimates hold:

$$\begin{aligned} \sup_{\varepsilon>0} \varepsilon \int_{-\infty}^{\infty} \|(A-k-i\varepsilon)^{-1}u\|^2 dk \le C\|u\|^2 \\ \sup_{\varepsilon>0} \varepsilon \int_{-\infty}^{\infty} \|(A^*-k-i\varepsilon)^{-1}u\|^2 dk \le C\|u\|^2 \end{aligned} \tag{3.1}$$

for every $u \in H$.

Our first goal is to rewrite these estimates in terms of the functional model for non-self-adjoint extensions of symmetric operators described in Section 2. We shall see that provided that the spectrum of the operator in question is absolutely continuous it is possible to compute the limits of the left-hand sides in the resulting estimates when $\varepsilon \to 0$ precisely, thus making it possible to replace (3.1) with much more simple conditions, formulated in terms of objects defined on the real line only.

This passage from the uniform integral estimates in the half-plane to the integral estimates on the boundary values simplifies matters simplifies matters in the case of one-dimensional Friedrichs model (see [19, 11], cf. [17]). As shown in [11], our approach reduces the similarity problem to the question of boundedness of certain singular integral operators. The latter could be examined by standard methods of analysis yielding necessary and sufficient conditions for the similarity of the operator of the Friedrichs model to a self-adjoint one. It is also worth mentioning that in the process of this limit procedure we reduce the consideration to analytic functions taking values in Hilbert spaces of (potentially) lower dimensions.

The linear manifold $N \subset \mathfrak{H}$ has been introduced in the previous section as follows: $N \equiv \{(\tilde{g}, g) : \mathcal{X}_-(\tilde{g} + S^*g) + \mathcal{X}_+(S\tilde{g} + g) = 0\}$ (cf. (2.5)) Since A is an operator with absolutely continuous spectrum, $P_K N$ is a dense set in the Hilbert space H. Moreover, for every $(\tilde{g}, g) \in N$ and for all $\lambda_0 \in \mathbb{C}_+$ one has:

$$\begin{aligned} (A-\lambda_0)^{-1} P_K \begin{pmatrix} \tilde{g} \\ g \end{pmatrix} &= P_K \frac{1}{k-\lambda_0} \begin{pmatrix} \tilde{g} \\ g \end{pmatrix} \\ &= \frac{1}{k-\lambda_0} \begin{pmatrix} \tilde{g} \\ g \end{pmatrix} - \begin{pmatrix} P_+ \dfrac{\tilde{g} + S^*g}{k-\lambda_0} \\ P_- \dfrac{S\tilde{g} + g}{k-\lambda_0} \end{pmatrix} \\ &= \frac{1}{k-\lambda_0} P_K \begin{pmatrix} \tilde{g} \\ g \end{pmatrix} + \frac{1}{k-\lambda_0} \begin{pmatrix} P_+(\tilde{g} + S^*g)(\lambda_0) \\ -P_+(S\tilde{g} + g)(\lambda_0) \end{pmatrix}. \end{aligned} \tag{3.2}$$

Note that the first estimate in (3.1) in the model terms can be rewritten as follows:

$$\sup_{\varepsilon>0} \varepsilon \int \left\| (A-k-i\varepsilon)^{-1} P_K \begin{pmatrix} \tilde{g} \\ g \end{pmatrix} \right\|^2 dk \le C \left\| P_K \begin{pmatrix} \tilde{g} \\ g \end{pmatrix} \right\|^2,$$

Since the first summand on the right-hand side of (3.2) obviously satisfies it (due to the self-adjointness of the mapping $(\tilde{g}, g) \mapsto k(\tilde{g}, g)$), is sufficient to consider the second term in (3.2) only.

Taking into account that $S(\lambda)$ is an analytic contractive operator-valued function in $\mathbb{C}_+$, we obtain [9, 30] ($\lambda_0 \equiv x + i\varepsilon$):

$$\begin{aligned}
\operatorname{Im}\lambda_0 &\int \left\| \frac{1}{k-\lambda_0} \begin{pmatrix} P_+(\tilde{g}+S^*g)(\lambda_0) \\ -P_+(S\tilde{g}+g)(\lambda_0) \end{pmatrix} \right\|_{\mathfrak{H}}^2 d\Re\lambda_0 \\
&= \varepsilon \int dx \int dk \frac{1}{(x-k)^2+\varepsilon^2} \\
&\quad\times \left\langle \begin{pmatrix} I & S^* \\ S & I \end{pmatrix} \begin{pmatrix} P_+(\tilde{g}+S^*g)(x+i\varepsilon) \\ -P_+(S\tilde{g}+g)(x+i\varepsilon) \end{pmatrix}, \begin{pmatrix} P_+(\tilde{g}+S^*g)(x+i\varepsilon) \\ -P_+(S\tilde{g}+g)(x+i\varepsilon) \end{pmatrix} \right\rangle \\
&= \pi \int dx \left\langle \begin{pmatrix} I & S^*(x+i\varepsilon) \\ S(x+i\varepsilon) & I \end{pmatrix} \begin{pmatrix} P_+(\tilde{g}+S^*g)(x+i\varepsilon) \\ -P_+(S\tilde{g}+g)(x+i\varepsilon) \end{pmatrix}, \right. \\
&\qquad\qquad \left. \begin{pmatrix} P_+(\tilde{g}+S^*g)(x+i\varepsilon) \\ -P_+(S\tilde{g}+g)(x+i\varepsilon) \end{pmatrix} \right\rangle \\
&= \pi \int dx (\|P_+(\tilde{g}+S^*g)(x+i\varepsilon)\|^2 + \|P_+(S\tilde{g}+g)(x+i\varepsilon)\|^2 \\
&\quad - 2\Re(S(x+i\varepsilon)P_+(S\tilde{g}+g)(x+i\varepsilon), P_+(\tilde{g}+S^*g)(x+i\varepsilon)),
\end{aligned}$$

the latter expression having a limit when $\varepsilon \to 0$ equal [9] to

$$\pi \left\| P_+ \begin{pmatrix} \tilde{g}+S^*g \\ -(S\tilde{g}+g) \end{pmatrix} \right\|_{\mathfrak{H}}^2 .$$

Thus we have proved, that the first inequality in (3.1) is equivalent to the following one:

$$\left\| P_+ \begin{pmatrix} \tilde{g}+S^*g \\ -(S\tilde{g}+g) \end{pmatrix} \right\|_{\mathfrak{H}}^2 \le C \left\| P_K \begin{pmatrix} \tilde{g} \\ g \end{pmatrix} \right\|_{\mathfrak{H}}^2 \quad \text{for all } \begin{pmatrix} \tilde{g} \\ g \end{pmatrix} \in N. \tag{3.3}$$

Analogous computations can be carried out for the second estimate in (3.1). The only difference with the case considered above will arise due to the fact, that smooth vectors for operator A are no longer smooth when the operator A^* is concerned; hence we have to use the general formula for the action of $(A^*-\lambda)^{-1}$ in the model representation (2.4) instead of (2.7). The following assertion holds: the second inequality in (3.1) is equivalent to

$$\left\| \begin{pmatrix} P_+(\tilde{g}+S^*g) - c(k) \\ -P_+(S\tilde{g}+g) + S(k)c(k) \end{pmatrix} \right\|_{\mathfrak{H}}^2 \le C \left\| P_K \begin{pmatrix} \tilde{g} \\ g \end{pmatrix} \right\|_{\mathfrak{H}}^2 , \tag{3.4}$$

where

$$\begin{aligned}
T(\lambda) &\equiv [\mathcal{X}_+ + \mathcal{X}_- S(\lambda)]^{-1}, \\
c(\lambda) &\equiv T(\lambda)(P_+(\tilde{g}+S^*g)(\lambda) + P_+(S\tilde{g}+g)(\lambda)),
\end{aligned}$$

and $c(k) \equiv c(k+i0)$, the boundary values of $c(\lambda)$ almost everywhere on the real line existing due to an argument from [16] for all $(\tilde{g}, g) \in N$.

To complete the proof we now need to rewrite estimates (3.3),(3.4) in terms of objects defined in the "original" Hilbert space H. This can be carried out using another representation of the functional model for the operator A_+, unitarily equivalent to the one used above [25, 16]. Namely, consider the model space $\mathfrak{H}'$ defined as follows:

$$\mathfrak{H}' \equiv \mathcal{D}_- \oplus H \oplus \mathcal{D}_+,$$

where $\mathcal{D}_\pm \equiv L_2(\mathbb{R}_\pm; E)$, $E = \text{clos Ran}(\alpha)$. The explicit formula describing the action of the operator A_+ in $\mathfrak{H}'$ was obtained in [25].

The isometric mapping Φ of $\mathfrak{H}'$ onto $\mathfrak{H}$ has the following form[4] [25]:

$$\begin{aligned} \tilde{g} + S^* g &= -(2\pi)^{-\frac{1}{2}} \alpha\Gamma_2(A_+ - k + i0)^{-1} u + S^*(k)\hat{v}_-(k) + \hat{v}_+(k) \\ S\tilde{g} + g &= -(2\pi)^{-\frac{1}{2}} \alpha\Gamma_2(A_+^* - k - i0)^{-1} u + \hat{v}_-(k) + S(k)\hat{v}_+(k), \end{aligned} \tag{3.5}$$

where $(v_-; u; v_+) \in \mathfrak{H}'$,

$$\hat{v}_\pm(k) = \frac{1}{\sqrt{2\pi}} \int e^{ik\xi} v_\pm(\xi) d\xi$$

is the Fourier transform of the function $v_\pm(\xi)$ extended by zero to the complementary half-line. The boundary values in the strong sense on the real line $\alpha\Gamma_2(A_+^{(*)} - k \pm i0)^{-1} u$ exist and belong to $H_2^\mp(E)$ [25].

The mappings (3.5) together with the absolute continuity of the spectrum of operator A make it possible to complete the proof. We omit here the corresponding purely technical calculations based on identities established in [25] due to the fact that they are similar to those carried out in proving results of [10]. □

Corollary 3.4. *The following estimates are sufficient for the operator A with absolutely continuous spectrum to be similar to a self-adjoint operator:*

$$\begin{aligned} \|\mathcal{X}_+(\Theta(k - i0) J \Theta^*(k - i0) - J)\mathcal{X}_+\| &\le C, \\ \|\mathcal{X}_-(J - \Theta^*(k + i0) J \Theta(k + i0))\mathcal{X}_-\| &\le C \end{aligned}$$

for almost all $k \in \mathbb{R}$.

This assertion immediately follows from the estimates (see [16, 25]):

$$\begin{aligned} \|\alpha\Gamma_2(A_+ - \lambda I)^{-1} u\|_{H_2^-(E)} &\le \sqrt{2\pi}\|u\| \\ \|\alpha\Gamma_2(A_+^* - \lambda I)^{-1} u\|_{H_2^+(E)} &\le \sqrt{2\pi}\|u\| \end{aligned}$$

for all $u \in H$.

Based on the results, formulated in Theorem 3.3 and on the following identity [25]

$$\alpha\Gamma_2(A_+^* - \lambda I)^{-1} = \Theta_+(\lambda)\alpha\Gamma_2(A - \lambda I)^{-1}, \quad \lambda \in \mathbb{C}_+,$$

where as above $\Theta_+(\lambda) \equiv \mathcal{X}_- + S(\lambda)\mathcal{X}_+$, one can rewrite conditions of Theorem 3.3 thus obtaining the following

[4] Under this mapping the subspace $\mathcal{D}_{+(-)}$ is mapped onto $H_2^{+(-)}(E)$ by the Paley-Wiener theorem [9]

Theorem 3.5. *Provided that the operator A has absolutely continuous spectrum, the following assertions are equivalent:*

(i) *Operator A is similar to a self-adjoint one*
(ii) *For all $u \in H$ the following estimates hold:*

$$\int ((I - S^*(k)S(k))\mathcal{X}_+\alpha\Gamma_2(A - k - i0)^{-1}u, \mathcal{X}_+\alpha\Gamma_2(A - k - i0)^{-1}u)dk \le C\|u\|^2$$

$$\int ((I - S^*(k)S(k))\mathcal{X}_-\alpha\Gamma_2(A^* - k - i0)^{-1}u, \mathcal{X}_-\alpha\Gamma_2(A^* - k - i0)^{-1}u)dk \le C\|u\|^2$$

Since the operator-function $S(\lambda)$ is contractive in the upper half-plane, this immediately yields the following

Corollary 3.6. *The following estimates are sufficient for the operator A with absolutely continuous spectrum to be similar to a self-adjoint operator:*

$$\int \left\|\mathcal{X}_+\alpha\Gamma_2(A - k - i0)^{-1}u\right\|^2 dk \le C\|u\|^2$$

$$\int \left\|\mathcal{X}_-\alpha\Gamma_2(A^* - k - i0)^{-1}u\right\|^2 dk \le C\|u\|^2$$

4. Similarity problem for restrictions to invariant spectral subspaces

The spectral projection $\mathcal{P}_\delta$ to the portion δ of the absolutely continuous spectrum was constructed in model terms in [15]. Namely, in the setting of completely non-self-adjoint almost solvable extensions of symmetric operators with equal deficiency indices the following Theorem holds (the proof for additive perturbations of self-adjoint operators can be found in [15]; it is easily extended to the setting of the present paper using results of [25]):

Theorem 4.1. *Suppose that A is a completely non-self-adjoint operator with absolutely continuous spectrum. For any Borel set $\delta \subset \mathbb{R}$ put*

$$\mathcal{P}_\delta P_K \begin{pmatrix} \tilde{g} \\ g \end{pmatrix} = P_K \mathcal{X}_\delta \begin{pmatrix} \tilde{g} \\ g \end{pmatrix}, \tag{4.1}$$

where $\binom{\tilde{g}}{g} \in N$ and $\mathcal{X}_\delta$ is the operator of componentwise multiplication by the characteristic function of the set δ. For the operator $\mathcal{P}_\delta$ defined by (4.1) on the set of smooth vectors $\tilde{N}_e$ the following assertions hold:

(i) $\mathcal{P}_\delta \tilde{N}_e \subset \tilde{N}_e$;
(ii) $(A - \lambda_0)^{-1}\mathcal{P}_\delta = \mathcal{P}_\delta(A - \lambda_0)^{-1}, \quad \operatorname{Im}\lambda_0 \ne 0$;
(iii) $\mathcal{P}_\delta\mathcal{P}_{\delta'} = \mathcal{P}_{\delta\cap\delta'}, \quad \delta, \delta' \subset \mathbb{R}$;
(iv) $\mathcal{P}_\delta u \longrightarrow u$ *as* $\delta \to (-\infty, \infty)$, $\quad u \in \tilde{N}_e$ *(in fact, as* $1 - \mathcal{X}_\delta \to 0$ *in* $L_\infty(\mathbb{R})$*)*;
(v) $\mathcal{P}_\delta u = \lim_{\varepsilon\to+0} \frac{1}{2\pi i}\int_\delta [(A - k - i\varepsilon)^{-1} - (A - k + i\varepsilon)^{-1}]u dk, \quad u \in \tilde{N}_e$.

We remark that the assertion (v) above establishes the connection between the definition of a spectral projection in terms of the functional model with the usual approach to the definition of spectral projections based on the Riesz integral for the resolvent, and thus the term "spectral projection" is justified.

Based on this result, in the present section we obtain conditions, necessary and sufficient for the restrictions of the non-self-adjoint operator A of the class considered in this paper to its spectral subspaces (i.e., to subspaces of the form clos $\mathcal{P}_\delta N_e$) to be similar to self-adjoint operators. We again assume that the operator A is an operator with absolutely continuous spectrum.

In essence, Theorem 4.1 allows to repeat the argument of the previous section in relation to any spectral subspace of the form $\mathcal{X}_\delta \binom{\tilde g}{g}$, where $\binom{\tilde g}{g} \in N$. The characteristic function $\mathcal{X}_\delta$ of the Borel set δ will then manifest itself in the estimates (3.1) rewritten in model terms, see (3.3), (3.4) by changing the interval over which the integral is being taken. Namely, the integral over the real line in the definition of the $\mathfrak{H}$-norm will in all cases be replaced by the integral over the Borel set δ, which together with some obvious modifications immediately yields desired result.

For the sake of completeness, we provide here the corresponding results, directly analogous to the results of Section 3.

Theorem 4.2. *Provided that the operator A has absolutely continuous spectrum, the following assertions are equivalent:*

(i) *The restriction of the operator A to its spectral invariant subspace, corresponding to a Borel set $\delta \subset \mathbb{R}$, is similar to a self-adjoint operator acting in* clos $\mathcal{P}_\delta N_e$.

(ii) *For all $u \in$ clos $\mathcal{P}_\delta N_e$ the following estimates hold:*

$$\int_\delta ((\Theta(k-i0)J\Theta^*(k-i0)-J)\mathcal{X}_+\alpha\Gamma_2(A_+^*-k-i0)^{-1}u,$$
$$\mathcal{X}_+\alpha\Gamma_2(A_+^*-k-i0)^{-1}u)dk \le C\|u\|^2$$

$$\int_\delta ((J-\Theta^*(k+i0)J\Theta(k+i0))\mathcal{X}_-\alpha\Gamma_2(A_+^*-k-i0)^{-1}u,$$
$$\mathcal{X}_-\alpha\Gamma_2(A_+^*-k-i0)^{-1}u)dk \le C\|u\|^2.$$

Theorem 4.3. *Provided that operator A has absolutely continuous spectrum, the following assertions are equivalent:*

(i) *The restriction of the operator A to its spectral invariant subspace, corresponding to a Borel set $\delta \subset \mathbb{R}$, is similar to a self-adjoint operator acting in* clos $\mathcal{P}_\delta N_e$.

(ii) *For all $u \in$ clos $\mathcal{P}_\delta N_e$ the following estimates hold:*

$$\int_\delta ((I-S^*(k)S(k))\mathcal{X}_+\alpha\Gamma_2(A-k-i0)^{-1}u,\mathcal{X}_+\alpha\Gamma_2(A-k-i0)^{-1}u)dk \le C\|u\|^2$$

$$\int_\delta ((I-S^*(k)S(k))\mathcal{X}_-\alpha\Gamma_2(A^*-k-i0)^{-1}u,\mathcal{X}_-\alpha\Gamma_2(A^*-k-i0)^{-1}u)dk \le C\|u\|^2$$

Corollary 4.4. *The following estimates are sufficient for the restriction of the operator A with absolutely continuous spectrum to its spectral invariant subspace, corresponding to a Borel set $\delta \subset \mathbb{R}$, to be similar to a self-adjoint operator:*

$$\|\mathcal{X}_+(\Theta(k-i0)J\Theta^*(k-i0)-J)\mathcal{X}_+\| \le C,$$
$$\|\mathcal{X}_-(J-\Theta^*(k+i0)J\Theta(k+i0))\mathcal{X}_-\| \le C$$

for almost all $k \in \delta$.

Corollary 4.5. *The following estimates are sufficient for the restriction of the operator A with absolutely continuous spectrum to its spectral invariant subspace, corresponding to a Borel set $\delta \subset \mathbb{R}$, to be similar to a self-adjoint operator:*

$$\int_\delta \left\|\mathcal{X}_+\alpha\Gamma_2(A-k-i0)^{-1}u\right\|^2 dk \le C\|u\|^2$$
$$\int_\delta \left\|\mathcal{X}_-\alpha\Gamma_2(A^*-k-i0)^{-1}u\right\|^2 dk \le C\|u\|^2$$

for all $u \in \operatorname{clos} \mathcal{P}_\delta N_e$.

5. Application to non-self-adjoint extensions of the Laplace operator

In this section we demonstrate how the results obtained above apply to the study of the similarity problem for operators that arise when considering Schrödinger equations with potentials of zero radius in quantum mechanics. Namely, following [25], consider a finite set of points $\{x_s\}_{s=1}^n$, $n<\infty$ in $\mathbb{R}^3$ and symmetric operator A_0 being the closure of the Laplace operator $-\Delta$ defined on the manifold $C_0^\infty(\mathbb{R}^3 \setminus \cup x_s)$. It is well known [2, 6, 1] that

$$D(A_0) = \{u \in W_2^2(\mathbb{R}^3):\ u(x_s)=0, s=1,2,\dots,n\}.$$

The deficiency indices of the operator A_0 are equal to (n,n). The following result from [22, 1] makes it possible to describe the domain $D(A_0^*)$ and to construct a boundary triple for the operator considered:

$$D(A_0^*) = \{u \in L_2(\mathbb{R}^3) \cap W_2^2(\mathbb{R}^3 \setminus \{x_s\}_{s=1}^n)$$

with the following asymptotic behaviour when x is close to x_s:

$$u(x) \sim \frac{u_-^s}{|x-x_s|} + u_0^s + O(|x-x_s|^{1/2}), s=1,2,\dots,n\}$$

Moreover, for all $u,v \in D(A_0^*)$ the following equality holds:

$$(A_0^*u,v)_H - (u,A_0^*v)_H = \sum_{s=1}^n (u_0^s\overline{v_-^s} - u_-^s\overline{v_0^s}).$$

It is shown in [25], that $\mathcal{H} \equiv \mathbb{C}^n$ and the mappings Γ_1, Γ_2 defined as

$$\Gamma_1 u \equiv (u_0^1, u_0^2, \dots, u_0^n),\ \Gamma_2 u \equiv (u_-^1, u_-^2, \dots, u_-^n),\ u \in D(A_0^*)$$

can be chosen as a boundary triple for the operator A_0^*. In this case the Weyl function (2.1) $M(\lambda)$ of A_0 is a $(n\times n)$ matrix-function with the matrix elements M_{sj}

$$M_{sj}(\lambda) = \begin{cases} ik, & s=j \\ \frac{e^{ik|x_s-x_j|}}{|x_s-x_j|}, & s\neq j \end{cases}, \quad s,j=1,2,\dots, \tag{5.1}$$

where $k=\sqrt{\lambda}$, Im $k>0$.

In [25] it is proved that the characteristic function of the almost solvable extension A_B corresponding to the n by n matrix B can be calculated as follows:

$$\Theta_{A_B}(\lambda) = I + iJ\alpha(B^* - M(\lambda))^{-1}\alpha$$
$$\Theta_{A_B^*}(\lambda) = I - iJ\alpha(B - M(\lambda))^{-1}\alpha,$$

where as before $\alpha = \sqrt{2|\operatorname{Im} B|}$ and $J = \operatorname{sign}\operatorname{Im} B$. Thus the explicit formula for the Weyl function $M(\lambda)$ given above permits one to immediately verify both conditions of Corollary 3.4. In the case considered this amounts to nothing more than verifying uniform boundedness of two n by n matrix-functions defined on the real line.

Another way to apply the results of Section 3 is the following reformulation of Corollary 3.6. It can be obtained based on the formula from [25] establishing the correspondence between the resolvent of any almost solvable extension A_B and the resolvent of the (self-adjoint) Friedrichs extension A_∞ in terms of the matrix B and the Weyl function $M(\lambda)$.

Theorem 5.1. *Let the operator A_B be as defined in the present section. Let further its spectrum be real. Then if for all $u\in H$*

$$\int \left\|\alpha(B - M(k+i0))^{-1}\Gamma_1(A_\infty - k - i0)^{-1}u\right\|^2 dk \le C\|u\|^2$$
$$\int \left\|\alpha(B^* - M(k+i0))^{-1}\Gamma_1(A_\infty - k - i0)^{-1}u\right\|^2 dk \le C\|u\|^2,$$

where A_∞ is the Friedrichs extension of the operator A_0, i.e., the self-adjoint operator $-\Delta$ defined on the domain $D(A_\infty) = W_2^2(\mathbb{R}^3)$, then *the operator A is similar to a self-adjoint operator.*

Further concise results on the similarity problem for this and other applications, together with the proof of the last theorem, will be a subject of our forthcoming publication.

Acknowledgements

The author is grateful to Prof. S.N. Naboko and Dr. M.M. Faddeev for their constant attention to his work and for valuable discussions of the subject. The author is also grateful to the referee for a number of extremely helpful remarks that hopefully have helped to improve the quality and overall readability of the paper.

References

[1] Albeverio S., Kurasov P., *Singular perturbations of differential operators*, Cambridge Univ. Press, 2000.

[2] Berezin F.A., Faddeev L.D., *A remark on Schrödinger's equation with a singular potential*, Soviet Math. Dokl. 2 (1961).

[3] Brodskij M.S., *Triangular and Jordan representations of linear operators*, English transl. in Amer. Math. Soc., Providence, R.I., 1971.

[4] Bruk V.M., *A certain class of boundary value problems with a spectral parameter in the boundary condition.* Mat. Sb. (N.S.) vol. 100 (142), no. 2 (1976), pp. 210–216.

[5] Van Casteren J., *Boundedness properties of resolvents and semigroups of operators*, Acta Sci. Math. Szeged., vol. 48 (1980), N 1–2.

[6] Demkov Y.N., Ostrovskii V.N., *The use of zero-range potentials in atomic physics*, Nauka, Moskva, 1975 (in Russian).

[7] Derkach V.A. and Malamud M.M., *Generalized resolvents and the boundary value problems for Hermitian operators with gaps,* J. Funct. Anal. vol. 95 (1991), pp. 1–95.

[8] Gorbachuk V.I. and Gorbachuk M.L., *Boundary value problems for operator differential equations.* Translated and revised from the 1984 Russian original. Mathematics and its Applications (Soviet Series), 48. Kluwer Academic Publishers Group, Dordrecht, 1991.

[9] Hoffman K., *Banach spaces of analytic functions*, Prentice-Hall, Englewood Cliffs, N.J., 1962.

[10] Kiselev A.V., Faddeev M.M., *The similarity problem for non-self-adjoint operators with absolutely continuous spectrum*, Func. Anal. App. vol. 34 (2000), no. 2, pp. 140–142.

[11] Kiselev A.V., *Some Spectral Properties of the Non-Self-Adjoint Friedrichs Model Operator*, Math. Proc. of Royal Irish Academy, vol. 105A (2), 2005.

[12] KočubeĭA.N., *On extension of symmetric operators and symmetric binary relations* Math. Notes vol. 17 (1975), pp. 41–48

[13] Lax P.D., Phillips R.S., *Scattering theory*, Pure and Applied Mathematics, 26. Academic Press, Inc., Boston, MA, 1989.

[14] Naboko S.N., *Absolutely continuous spectrum of a nondissipative operator and functional model. I.*, Zapiski Nauchnykh Seminarov LOMI, Vol. 65 (1976), pp. 90–102 (Russian); English translation in J. Sov. Math.

[15] Naboko S.N., *Absolutely continuous spectrum of a nondissipative operator and functional model. II.*, Zapiski Nauchnykh Seminarov LOMI, Vol. 73 (1977), pp. 118–135 (Russian); English translation in J. Sov. Math.

[16] Naboko S.N., *Functional model of perturbation theory and its applications to scattering theory*, Proc. Steklov Inst. Math. (1981), Issue 2., pp. 85–116.

[17] Naboko S.N., *The conditions for similarity to unitary and self-adjoint operators*, Functional Analysis and its applications, vol. 18 (1984), pp. 16–27.

[18] Nikolski N.K., *Operators, Functions and Systems: An Easy Reading*, vol. I, II, AMS, 2002.

[19] Nikolski N., Treil S., *Linear resolvent growth of rank one perturbation of a unitary operator does not imply its similarity to a normal operator,* Journal d'Analyse Mathématique 87 (2002).

[20] Pavlov B.S., *Conditions for separation of the spectral components of a dissipative operator*, Math. USSR-Izv., vol. 39 (1975), pp. 123–148.

[21] Pavlov B.S., *Dilation theory and spectral analysis of non-self-adjoint differential operators*, AMS Transl., vol. 115 (1980), no. 2.

[22] Pavlov B.S., *A model of zero-radius potential with internal structure.* (Russian) Theoret. Math. Phys. vol. 59, no. 3 (1984), pp. 544–550.

[23] Petrov A.M., *Spectral projectors of dissipative operators*, J. Soviet Math., vol. 57 (1991), no. 5, pp. 3440–3443.

[24] Ryzhov V.A., *Absolutely continuous and singular subspaces of a non-self-adjoint operator*, J. Math. Sci., vol. 87 (1997), no. 5, pp. 3886–3911.

[25] Ryzhov V., *Functional model of a class of non-self-adjoint extensions of symmetric operators*, Operator Theory: Adv. Appl. vol. 174 (2007).

[26] Ryzhov V.A., *Functional model of nondissipative operator and the scattering problems*, Ph.D. thesis, St. Petersburg State University, 1996 (in Russian).

[27] Sahnovič L.A., *Nonunitary operators with absolutely continuous spectrum*, Izv. Akad. Nauk SSSR Ser. Mat. vol. 33 (1969), pp. 52–64 (Russian); English translation in Math. USSR-Izv.

[28] Štraus A.V., *Characteristic functions of linear operators.* Izv. Akad. Nauk. SSSR, Ser. Mat., vol. 24, 1 (1960), pp. 43–74; English transl. in AMS Transl. vol. (2) 40 (1964), pp. 1–37.

[29] Štraus A.V., *On extensions and characteristic function of a symmetric operator.* Izv. Akad. Nauk SSSR Ser. Mat. vol. 32 (1968), pp. 186–207; English transl. in Math. USSR - Izvestija, vol. 2 (1968), pp. 181–204.

[30] Sz.-Nagy B. and Foiaş C., *Analyse harmonique des opérateurs de l'espace de Hilbert*, Masson, Paris and Akad. Kiadó, Budapest, 1967.

[31] Veselov V.F., Naboko S.N., *The determinant of the characteristic function and the singular spectrum of a non-self-adjoint operator*, Mat. Sb. vol. 129 (171) (1986), no. 1, pp. 20–39 (Russian); English translation in Math. USSR-Sb.

Alexander V. Kiselev
Dept. of Pure Mathematics and Mathematical Physics
St.Petersburg State University
1 Ulianovskaya st.
198504 Peterhoff, St.Petersburg, Russia
e-mail: alexander.v.kiselev@gmail.com

Operator Theory:
Advances and Applications, Vol. 186, 285–323

Inverse Scattering Problem for a Special Class of Canonical Systems and Non-linear Fourier Integral. Part I: Asymptotics of Eigenfunctions

S. Kupin, F. Peherstorfer, A. Volberg and P. Yuditskii

Abstract. An original approach to the inverse scattering for Jacobi matrices was recently suggested in [20]. The authors considered quite sophisticated spectral sets (including Cantor sets of positive Lebesgue measure), however they did not take into account the mass point spectrum. This paper follows similar lines for the continuous setting with an absolutely continuous spectrum on the half-axis and a pure point spectrum on the negative half-axis satisfying the Blaschke condition. This leads us to the solution of the inverse scattering problem for a class of canonical systems that generalizes the case of Sturm-Liouville (Schrödinger) operator.

Mathematics Subject Classification (2000). Primary: 34L40,42C05; Secondary: 34L25, 81U40.

Keywords. Inverse scattering problem, Schrödinger operator, canonical system, model space, reproducing kernels, asymptotical behavior of the solutions.

1. Faddeev-Marchenko space in Szegő/Blaschke setting

One of the important aspects of the spectral theory of differential operators is the scattering theory [16, 17] and, in particular, the inverse scattering [14]. An original approach to the inverse scattering was recently suggested in [20]. The paper focused on classical Jacobi matrices and connections between the scattering and properties of a special Hilbert transform.

Partially supported by NSF grant DMS-0501067 the Austrian Founds FWF, project number: P16390–N04 and *Marie Curie International Fellowship* within the 6th European Community Framework Programme, Contract MIF1-CT-2005-006966.

In this paper, we carry out the plan of [20] in the continuous situation. Compared with [20], a completely new feature is that the scattering data incorporate the pure point spectrum with infinitely many mass points. Of course, this is a natural and important step in developing the theory. The discussion leads us to the solution of the inverse scattering problem for a class of canonical systems that include the Sturm-Liouville (Schrödinger) equations. At present, though, we are unable to characterize the scattering data corresponding to the last important special case.

The article is quite close to the circle of ideas of [1]–[4], treating the inverse monodromy problem and the inverse spectral problem for a class of canonical systems, and [19], working with a model where the pure point spectrum is not permitted. We also mention an extremely interesting recent paper [8].

The present part of the work is mainly devoted to the asymptotic behavior of certain reproducing kernels (the generalized eigenfunctions). It is organized as follows. Section 1 contains definitions, some general facts and formulations of results on asymptotics. The asymptotic properties of reproducing kernels from certain model spaces are studied in Section 2. Special operator nodes arising from our construction are discussed in Sections 3 and 4. One of the nodes generates a canonical system we are interested in. Its properties and connections to the de Branges spaces of entire functions [5] are also in Section 4. The Sturm-Liouville (Schrödinger) equations are considered in Section 5. An example is given in the first appendix (Section 6). The second appendix (Section 7) relates the whole construction to the matrix A_2 Hunt-Muckenhoupt-Wheeden condition.

We define the L^2-norm on the real axis as

$$||f||^2 = \frac{1}{2\pi}\int_{\mathbb{R}} |f(\lambda)|^2\, d\lambda, \tag{1.1}$$

so that the reproducing kernel of the H^2 subspace is of the form $k(\lambda,\lambda_0) = \frac{i}{\lambda-\bar\lambda_0}$.

The section "Inverse scattering problem on the real axis" in [14, Chap. 3, Sect. 5] begins with a Sturm-Liouville operator

$$-y'' + q(x)y = \lambda^2 y, \quad x\in\mathbb{R}, \tag{1.2}$$

with the potential q satisfying the a priori condition

$$\int_{\mathbb{R}} (1+|x|)|q(x)|\, dx < \infty. \tag{1.3}$$

To such an operator one associates the so-called scattering data

$$\{s_+,\nu_+\}, \tag{1.4}$$

where s_+ is a contractive function on the real axis, $|s_+(\lambda)|\le 1$, $\lambda\in\mathbb{R}$, possessing certain properties and ν_+ is a discrete measure, in fact, supported on a finite number of points $\Lambda=\{\lambda_k\}$ of the imaginary axis, $\frac{\lambda_k}{i}>0$.

We proceed in the opposite direction starting from the scattering data $\{s_+,\nu_+\}$ and going to the potential q. The key point of the construction is that we assume that the scattering data (1.4) satisfy only very natural (and minimal) conditions from the point of view of the function theory. Namely, we suppose that:

- a symmetric on the real axis function s_+, $s_+(\lambda) = \overline{s_+(-\bar{\lambda})}$, satisfies the Szegő condition

$$\int_{\mathbb{R}} \frac{\log(1 - |s_+(\lambda)|^2)}{1+\lambda^2} d\lambda > -\infty, \tag{1.5}$$

- the support Λ of a discrete measure $\nu_+ = \sum_k \nu_+(\lambda_k)\delta_{\lambda_k}$ satisfies the Blaschke condition

$$\sum_{(\lambda_k/i)\le 1} \frac{\lambda_k}{i} < \infty, \quad \sum_{(\lambda_k/i)>1} \frac{i}{\lambda_k} < \infty. \tag{1.6}$$

Let us point out that we did not even assume that the measure ν_+ is finite.

Our plan is to show that already in this case one can associate a certain differential operator of the second order to the given spectral data and then one can prove several specification theorems.

Definition 1.1. An element f of the space $L^2_{\{s_+,\nu_+\}}$ is a function on $\mathbb{R} \cup \Lambda$ such that

$$\begin{aligned} ||f||^2_{\{s_+,\nu_+\}} &= \sum_{\lambda_k\in\Lambda} |f(\lambda_k)|^2 \nu_+(\lambda_k) \\ &+ \frac{1}{4\pi}\int_{\mathbb{R}} \begin{bmatrix} \overline{f(\lambda)} & \overline{-f(-\bar{\lambda})} \end{bmatrix} \begin{bmatrix} 1 & \overline{s_+(\lambda)} \\ s_+(\lambda) & 1 \end{bmatrix} \begin{bmatrix} f(\lambda) \\ -f(-\bar{\lambda}) \end{bmatrix} d\lambda \end{aligned} \tag{1.7}$$

is finite.

Using (1.5) and (1.6) we define the outer in the upper half-plane function s_e as

$$|s_e(\lambda)|^2 + |s_+(\lambda)|^2 = 1 \text{ a.e. on } \mathbb{R}, \tag{1.8}$$

and the Blaschke product

$$B(\lambda) = \prod_k b_{\lambda_k}(\lambda), \tag{1.9}$$

where $b_{\lambda_k}(\lambda) = \dfrac{\lambda - \lambda_k}{\lambda - \bar{\lambda}_k}$ if $\lambda_k/i \le 1$ and $b_{\lambda_k}(\lambda) = -\dfrac{\lambda - \lambda_k}{\lambda - \bar{\lambda}_k}$ if $\lambda_k/i > 1$. We also put

$$S(\lambda) = \begin{bmatrix} s_- & s \\ s & s_+ \end{bmatrix}(\lambda), \ \lambda \in \mathbb{R}, \tag{1.10}$$

where

$$s := \frac{s_e}{B} \quad \text{and} \quad s_- := -\frac{s}{\bar{s}}\bar{s}_+. \tag{1.11}$$

The matrix function S possesses two fundamental properties: $S^*(-\bar{\lambda}) = S(\lambda)$, and it is unitary-valued. The third one is the meromorphic property of the entry s, which has analytic continuation to the upper half-plane as a function of bounded characteristic with a specific nature, that is, it is a ratio of an outer function and a Blaschke product.

The measure ν_- is defined through s_- and ν_+ by

$$\frac{1}{\nu_+(\lambda_k)}\frac{1}{\nu_-(\lambda_k)} = \left|\left(\frac{1}{s}\right)'(\lambda_k)\right|^2. \tag{1.12}$$

A reason for these and the following definitions will be clarified in a moment.

Set

$$\begin{aligned}\begin{bmatrix} sf^+ \\ sf^- \end{bmatrix}(\lambda) &= \begin{bmatrix} s & 0 \\ s_+ & 1 \end{bmatrix}(\lambda)\begin{bmatrix} f^+(\lambda) \\ -f^+(-\bar\lambda) \end{bmatrix} \\ &= \begin{bmatrix} 1 & s_- \\ 0 & s \end{bmatrix}(\lambda)\begin{bmatrix} -f^-(-\bar\lambda) \\ f^-(\lambda) \end{bmatrix}\end{aligned} \tag{1.13}$$

for $\lambda \in \mathbb{R}$ and

$$\begin{aligned} f^-(\lambda_k) &= -i\left(\frac{1}{s}\right)'(\lambda_k)\nu_+(\lambda_k)f^+(\lambda_k), \\ f^+(\lambda_k) &= -i\left(\frac{1}{s}\right)'(\lambda_k)\nu_-(\lambda_k)f^-(\lambda_k). \end{aligned} \tag{1.14}$$

for $\lambda_k \in \Lambda$. These formulas define two mappings inverse to each other, $(.)^+ : L^2_{\{s_-,\nu_-\}} \to L^2_{\{s_+,\nu_+\}}$, and $(.)^- : L^2_{\{s_+,\nu_+\}} \to L^2_{\{s_-,\nu_-\}}$. If $f^+ \in L^2_{\{s_+,\nu_+\}}$, we get $f^- = (f^+)^-$ from the second row of (1.13) and the first row of (1.14). The same applies to $f^+ = (f^-)^+, f^- \in L^2_{\{s_-,\nu_-\}}$ with obvious changes. When it does not cause any ambiguity, *we will write* $f^\pm$ *instead of* $(f^\mp)^\pm$. The corresponding maps will be called $\pm$-mappings. It is not difficult to prove that the $\pm$-mappings are unitary from $L^2_{\{s_+,\nu_+\}}$ to $L^2_{\{s_-,\nu_-\}}$ and vice versa. For example, due to (1.13)

$$\frac{1}{4\pi}\int_{\mathbb{R}}\begin{bmatrix}\overline{f(\lambda)} & \overline{-f(-\bar\lambda)}\end{bmatrix}\begin{bmatrix} 1 & \overline{s_+(\lambda)} \\ s_+(\lambda) & 1 \end{bmatrix}\begin{bmatrix} f(\lambda) \\ -f(-\bar\lambda) \end{bmatrix} d\lambda = \frac{||sf^+||^2 + ||sf^-||^2}{2}, \tag{1.15}$$

where f is f^+ restricted to $\mathbb{R}$ and we have the standard $L^2(\mathbb{R})$-norms in the RHS of the equality. More details on the computation are in [20], Lemma 1.3. It is also important that relations (1.13), (1.14) not only define a duality between these two spaces but, what is more important, a duality between corresponding Hardy subspaces.

Actually we give two versions of definitions of Hardy subspaces (in general, *they are not equivalent*, see an example in Section 6). By the first one, $H^2_{\{s_+,\nu_+\}}$ is basically the closure of H^∞ with respect to the given norm (1.7). More precisely, let $B_{N_1,N_2} = \prod_{k=N_1+1}^{N_2} b_{\lambda_k}$ and $\mathcal{B} = \{B_{N,\infty}\}$. Saying it differently, a Blaschke product B' is in $\mathcal{B}$, if B' is a divisor of B such that B/B' is a finite Blaschke product. Then

$$f = B_{N,\infty}g, \quad g \in H^\infty, \tag{1.16}$$

belongs to $L^2_{\{s_+,\nu_+\}}$. By $H^2_{\{s_+,\nu_+\}}$ we denote the closure in $L^2_{\{s_+,\nu_+\}}$ of functions of the form (1.16). Let us point out that every element f of $H^2_{\{s_+,\nu_+\}}$ is such that $s_e f$

belongs to the standard H^2, see (1.15). Therefore, in fact, $f(\lambda)$ has an analytic continuation from the real axis to the upper half-plane. Moreover, the value $f(\lambda)$ obtained by this continuation, and $f(\lambda_k)$ which is defined for all $\lambda_k \in \Lambda$, since f is a function from $L^2_{\{s_+,\nu_+\}}$, still perfectly coincide.

The second space also consists of functions from $L^2_{\{s_+,\nu_+\}}$ having an analytic continuation to the upper half-plane.

Definition 1.2. A function $f \in L^2_{\{s_+,\nu_+\}}$ belongs to $\hat{H}^2_{\{s_+,\nu_+\}}$ if $g(\lambda) := (s_e f)(\lambda)$, $\lambda \in \mathbb{R}$, belongs to the standard H^2 and

$$f(\lambda_k) = \left(\frac{g}{s_e}\right)(\lambda_k),\ \lambda_k \in \Lambda,$$

where in the RHS g and s_e are defined by their analytic continuation to the upper half-plane.

It turns out that spaces $H^2_{\{s_+,\nu_+\}}$ and $\hat{H}^2_{\{s_+,\nu_+\}}$ are dual in a certain sense.

Theorem 1.3. *Let $f^+ \in L^2_{\{s_+,\nu_+\}} \ominus H^2_{\{s_+,\nu_+\}}$ and let $f^- \in L^2_{\{s_-,\nu_-\}}$ be defined by* (1.13), (1.14). *Then $f^- \in \hat{H}^2_{\{s_-,\nu_-\}}$. In short, we write*

$$(\hat{H}^2_{\{s_-,\nu_-\}})^+ = L^2_{\{s_+,\nu_+\}} \ominus H^2_{\{s_+,\nu_+\}}. \tag{1.17}$$

Proof. We notice that $f^+ \in L^2_{\{s_+,\nu_+\}}$ implies

$$(sf^-)(\lambda) = s_+(\lambda)f^+(\lambda) - f^+(-\bar{\lambda}) \in L^2,\ \lambda \in \mathbb{R}.$$

Since

$$\langle f^+, Bh\rangle_{\{s_+,\nu_+\}} = \langle s_+(\lambda)f^+(\lambda) - f^+(-\bar{\lambda}), -B(-\bar{\lambda})h(-\bar{\lambda})\rangle,\ h \in H^2,$$

it follows from $f^+ \in L^2_{\{s_+,\nu_+\}} \ominus H^2_{\{s_+,\nu_+\}}$ that

$$(s_e f^-)(\lambda) = g(\lambda) := B(\lambda)(s_+(\lambda)f^+(\lambda) - f^+(-\bar{\lambda})) \in H^2.$$

Now we calculate the scalar product

$$\begin{aligned}\langle f^+, \frac{iB(\lambda)}{\lambda-\lambda_k}\rangle_{\{s_+,\nu_+\}} &= f^+(\lambda_k)\overline{iB'(\lambda_k)}\nu_+(\lambda_k) + \langle s_e f^-, \frac{i}{\lambda-\bar{\lambda}_k}\rangle \\ &= f^+(\lambda_k)iB'(\lambda_k)\nu_+(\lambda_k) + g(\lambda_k) = 0.\end{aligned}$$

Therefore, by (1.14) we get

$$f^-(\lambda_k) = \left(\frac{g}{s_e}\right)(\lambda_k),\ \lambda_k \in \Lambda.$$

□

Both $H^2_{\{s_+,\nu_+\}}$ and $\hat{H}^2_{\{s_+,\nu_+\}}$ are spaces of analytic in the upper half-plane functions, so they have reproducing kernels. For $\mu \in \mathbb{C}_+$, we denote them by

$$k_{\{s_+,\nu_+\}}(\lambda,\mu) = k_{\{s_+,\nu_+\}}(.,\mu), \quad \hat{k}_{\{s_+,\nu_+\}}(\lambda,\mu) = \hat{k}_{\{s_+,\nu_+\}}(.,\mu).$$

Recall also that

$$k(\lambda,\mu) = k(.,\mu) = \frac{i}{\lambda-\bar{\mu}}$$

is the reproducing kernel of the standard Hardy space H^2. The first step is to prove asymptotics for the families $\{e^{i\lambda x} k_{\{s_\pm e^{2i\lambda x}, \nu_\pm e^{2i\lambda x}\}}(.,\lambda_0)\}_{x\in\mathbb{R}}$ and $\{e^{i\lambda x} \hat{k}_{\{s_\pm e^{2i\lambda x}, \nu_\pm e^{2i\lambda x}\}}(.,\lambda_0)\}_{x\in\mathbb{R}}$ with $\lambda_0 \in \mathbb{C}_+$.

Theorem 1.4.

i) *The following relations hold true in* $L^2(\mathbb{R})$*:*

$$
\begin{aligned}
s\left\{e^{i\lambda x} k_{\{s_+ e^{2i\lambda x},\, \nu_+ e^{2i\lambda x}\}}(.,\lambda_0)\right\} &= s e^{i\lambda x} k(.,\lambda_0) + o(1),\\
s\left\{e^{i\lambda x} \hat{k}_{\{s_+ e^{2i\lambda x},\, \nu_+ e^{2i\lambda x}\}}(.,\lambda_0)\right\} &= s e^{i\lambda x} k(.,\lambda_0) + o(1)
\end{aligned} \tag{1.18}
$$

as $x \to +\infty$*. Moreover,*

$$
\begin{aligned}
s(-\bar{\lambda}_0) s\left\{e^{i\lambda x} k_{\{s_+ e^{2i\lambda x},\, \nu_+ e^{2i\lambda x}\}}(.,\lambda_0)\right\} &= e^{i\lambda x} k(.,\lambda_0) + s_- e^{-i\lambda x} k(.,-\lambda_0) + o(1),\\
s(-\bar{\lambda}_0) s\left\{e^{i\lambda x} \hat{k}_{\{s_+ e^{2i\lambda x},\, \nu_+ e^{2i\lambda x}\}}(.,\lambda_0)\right\} &= e^{i\lambda x} k(.,\lambda_0) + s_- e^{-i\lambda x} k(.,-\lambda_0) + o(1)
\end{aligned} \tag{1.19}
$$

as $x \to -\infty$.

ii) *On* Λ*, we have:*

$$
e^{i\lambda x} k_{\{s_+ e^{2i\lambda x},\, \nu_+ e^{2i\lambda x}\}}(.,\lambda_0) = o(1), \quad e^{i\lambda x} \hat{k}_{\{s_+ e^{2i\lambda x},\, \nu_+ e^{2i\lambda x}\}}(.,\lambda_0) = o(1) \tag{1.20}
$$

in $L^2_{\nu_+}$*-sense as* $x \to +\infty$*. Furthermore, for a* $\lambda_k \in \Lambda$

$$
\begin{aligned}
\lim_{x\to-\infty} e^{-2\operatorname{Im}\lambda_k x} k_{\{s_+ e^{2i\lambda x},\, \nu_+ e^{2i\lambda x}\}}(\lambda_k,\lambda_k) &= \frac{1}{\nu_+(\lambda_k)},\\
\lim_{x\to-\infty} e^{-2\operatorname{Im}\lambda_k x} \hat{k}_{\{s_+ e^{2i\lambda x},\, \nu_+ e^{2i\lambda x}\}}(\lambda_k,\lambda_k) &= \frac{1}{\nu_+(\lambda_k)}
\end{aligned} \tag{1.21}
$$

It goes without saying that relations (1.18)–(1.20) correspond to scattering "from $+\infty$ to $-\infty$"; compare these formulas to (0.8), (0.25) from [20]. Scattering in the inverse direction ("from $-\infty$ to $+\infty$") is described similarly. We give the formulas for the family $\{e^{-i\lambda x} k_{\{s_- e^{-2i\lambda x}\, \nu_- e^{-2i\lambda x}\}}(.,\lambda_0)\}$ only; asymptotics for $\{e^{-i\lambda x} \hat{k}_{\{s_- e^{-2i\lambda x}\, \nu_- e^{-2i\lambda x}\}}\, (.,\lambda_0)\}$ are the same.

Corollary 1.5. *We have*

$$
\begin{aligned}
s\left\{e^{-i\lambda x} k_{\{s_- e^{-2i\lambda x}\, \nu_- e^{-2i\lambda x}\}}(.,\lambda_0)\right\} &= s e^{-i\lambda x} k(.,\lambda_0) + o(1), \quad x \to -\infty,\\
s(-\bar{\lambda}_0) s\left\{e^{-i\lambda x} k_{\{s_- e^{-2i\lambda x}\, \nu_- e^{-2i\lambda x}\}}(.,\lambda_0)\right\} &= e^{-i\lambda x} k(.,\lambda_0)\\
&\quad + s_+ e^{i\lambda x} k(.,-\lambda_0) + o(1), \quad x \to +\infty
\end{aligned}
$$

in L^2*-sense on the real line. As for* Λ*,*

$$
e^{-i\lambda x} k_{\{s_- e^{-2i\lambda x}\, \nu_- e^{-2i\lambda x}\}}(.,\lambda_0) = o(1)
$$

in $L^2_{\nu_-}$*-sense as* $x \to -\infty$*. As before, for a* $\lambda_k \in \Lambda$*,*

$$
\lim_{x\to+\infty} e^{2\operatorname{Im}\lambda_k x} k_{\{s_- e^{-2i\lambda x}\, \nu_- e^{-2i\lambda x}\}}(\lambda_k,\lambda_k) = \frac{1}{\nu_-(\lambda_k)}.
$$

We set

$$\tilde{k}(.,\lambda_0)=\begin{cases} k(.,\lambda_0), & \lambda\in\mathbb{R},\\ 0, & \lambda\in\Lambda.\end{cases}$$

Theorem 1.4 follows immediately from the following result.

Theorem 1.6. *The following relations hold true:*

$$\lim_{x\to+\infty}||e^{i\lambda x}k_{\{s_+e^{2i\lambda x},\,\nu_+e^{2i\lambda x}\}}(.,\lambda_0)-e^{i\lambda x}\tilde{k}(.,\lambda_0)||_{\{s_+,\nu_+\}} = 0, \quad (1.22)$$

$$\lim_{x\to+\infty}||e^{i\lambda x}\hat{k}_{\{s_+e^{2i\lambda x},\,\nu_+e^{2i\lambda x}\}}(.,\lambda_0)-e^{i\lambda x}\tilde{k}(.,\lambda_0)||_{\{s_+,\nu_+\}} = 0. \quad (1.23)$$

The proof of this theorem is the main purpose of Section 2.

Remark 1.7. Relations (1.18)–(1.20) follow easily from Theorem 1.6.

Indeed, let us have a look at (1.22). Recalling that $\tilde{k}(.,\lambda_0)=0$ on Λ, we see

$$\begin{aligned}||e^{i\lambda x}k_{\{s_+e^{2i\lambda x},\,\nu_+e^{2i\lambda x}\}}(.,\lambda_0) &- e^{i\lambda x}\tilde{k}(.,\lambda_0)||^2_{\{s_+,\nu_+\}}\\ &= ||e^{i\lambda x}k_{\{s_+e^{2i\lambda x},\,\nu_+e^{2i\lambda x}\}}(.,\lambda_0)-e^{i\lambda x}k(.,\lambda_0)||^2_{s_+}\\ &+ ||e^{i\lambda x}k_{\{s_+e^{2i\lambda x},\,\nu_+e^{2i\lambda x}\}}(.,\lambda_0)||^2_{\nu_+}\to 0,\end{aligned}$$

as $x\to+\infty$, so the first relation in (1.20) is proved. Then we notice that

$$\begin{bmatrix}1 & \bar{s}_+\\ s_+ & 1\end{bmatrix}=\begin{bmatrix}|s|^2 & 0\\ 0 & 0\end{bmatrix}+\begin{bmatrix}\bar{s}_+\\ 1\end{bmatrix}\begin{bmatrix}s_+ & 1\end{bmatrix}.$$

This implies that the first summand on the right-hand side of the above equality is

$$\begin{aligned}||\dots||^2_{s_+} &= ||s\left(e^{i\lambda x}k_{\{s_+e^{2i\lambda x},\,\nu_+e^{2i\lambda x}\}}(.,\lambda_0)-e^{i\lambda x}k(.,\lambda_0)\right)||^2_2\\ &+ ||s\left(e^{i\lambda x}k_{\{s_+e^{2i\lambda x},\,\nu_+e^{2i\lambda x}\}}(.,\lambda_0)\right)^-_{s_-}\\ &- \left(e^{-i\lambda x}k(.,-\lambda_0)+s_+e^{i\lambda x}k(.,\lambda_0)\right)||^2_2.\end{aligned}$$

The presence of the first term on the right-hand side shows that we are done with (1.18). To deal with $(e^{i\lambda x}k_{\{s_+e^{2i\lambda x},\,\nu_+e^{2i\lambda x}\}}(.,\lambda_0))^-_{s_-}$, we use Lemma 2.2 and its corollary saying

$$\lim_{x\to+\infty}k_{\{s_-e^{-2i\lambda x}(b_{\lambda_0}b_{-\bar{\lambda}_0})^{-1},\,\nu_-e^{-2i\lambda x}(b_{\lambda_0}b_{-\bar{\lambda}_0})^{-1}\}}(-\bar{\lambda}_0,-\bar{\lambda}_0)=\frac{1}{2\mathrm{Im}\,\lambda_0|s(\lambda_0)|^2}$$

(see also Lemma 2.8). Hence, we come to

$$\begin{aligned}&s(\lambda_0)s\left\{e^{-i\lambda x}\hat{k}_{\{s_-e^{-2i\lambda x}(b_{\lambda_0}b_{-\bar{\lambda}_0})^{-1},\,\nu_-e^{-2i\lambda x}(b_{\lambda_0}b_{-\bar{\lambda}_0})^{-1}\}}\right\}\\ &\qquad= e^{-i\lambda x}k(.,-\bar{\lambda}_0)+s_+(b_{\lambda_0}b_{-\bar{\lambda}_0})e^{i\lambda x}k(.,\bar{\lambda}_0)+o(1)\end{aligned}$$

as $x\to+\infty$. This is the second relation in (1.19) up to changes $x\mapsto -x,\lambda_0\mapsto -\bar{\lambda}_0, s_-(b_{\lambda_0}b_{-\bar{\lambda}_0})^{-1}\mapsto s_+$ and $s_+(b_{\lambda_0}b_{-\bar{\lambda}_0})\mapsto s_-$.

Equalities (1.21) are proved in Corollary 2.5.

2. Asymptotics of reproducing kernels

2.1. Definitions and notation

In this subsection, we prove several propositions concerning special properties of the reproducing kernels introduced in Section 1.

For $\lambda_0 \in \mathbb{C}_+$, let

$$K_{\{s_+,\nu_+\}}(.,\lambda_0) = \frac{k_{\{s_+,\nu_+\}}(.,\lambda_0)}{\sqrt{k_{\{s_+,\nu_+\}}(\lambda_0,\lambda_0)}}, \quad \hat{K}_{\{s_+,\nu_+\}}(.,\lambda_0) = \frac{\hat{k}_{\{s_+,\nu_+\}}(.,\lambda_0)}{\sqrt{\hat{k}_{\{s_+,\nu_+\}}(\lambda_0,\lambda_0)}}$$

be their normalized versions. It is also convenient to put

$$K(\lambda,\lambda_0) = \frac{k(\lambda,\lambda_0)}{||k(.,\lambda_0)||} = \frac{i\,(2\mathrm{Im}\,\lambda_0)^{1/2}}{\lambda - \bar{\lambda}_0}.$$

For a fixed $x \in \mathbb{R}$ we define $H^2_{\{s_+,\nu_+\}}(x)$ as the closure of the functions

$$f(\lambda) = B_{N,\infty}(\lambda) g(\lambda) e^{i\lambda x}, \quad g \in H^\infty, \ B_{N,\infty} \in \mathcal{B}. \tag{2.1}$$

In particular, $H^2_{\{s_+,\nu_+\}} = H^2_{\{s_+,\nu_+\}}(0)$. In the similar way we define the set of spaces $\hat{H}^2_{\{s_+,\nu_+\}}(x)$, so that $\hat{H}^2_{\{s_+,\nu_+\}}$ is related to $x = 0$.

It is easy to see that

$$H^2_{\{s_+,\nu_+\}}(x) = e^{i\lambda x} H^2_{\{s_+ e^{2i\lambda x},\, \nu_+ e^{2i\lambda x}\}}, \quad \hat{H}^2_{\{s_+,\nu_+\}}(x) = e^{i\lambda x} \hat{H}^2_{\{s_+ e^{2i\lambda x},\, \nu_+ e^{2i\lambda x}\}},$$

and

$$\begin{aligned} k_{\{s_+,\nu_+\}}(\lambda,\lambda_0;x) &= e^{ix(\lambda-\bar{\lambda}_0)} k_{\{s_+ e^{2i\lambda x},\, \nu_+ e^{2i\lambda x}\}}(\lambda,\lambda_0), \\ \hat{k}_{\{s_+,\nu_+\}}(\lambda,\lambda_0;x) &= e^{ix(\lambda-\bar{\lambda}_0)} \hat{k}_{\{s_+ e^{2i\lambda x},\, \nu_+ e^{2i\lambda x}\}}(\lambda,\lambda_0) \end{aligned}$$

are the reproducing kernels of these spaces, respectively. We also have their normalized versions

$$\begin{aligned} K_{\{s_+,\nu_+\}}(.,\lambda_0;x) &= e^{ix(\lambda-\mathrm{Re}\,\lambda_0)} K_{\{s_+ e^{2i\lambda x},\, \nu_+ e^{2i\lambda x}\}}(.,\lambda_0), \\ \hat{K}_{\{s_+,\nu_+\}}(.,\lambda_0;x) &= e^{ix(\lambda-\mathrm{Re}\,\lambda_0)} \hat{K}_{\{s_+ e^{2i\lambda x},\, \nu_+ e^{2i\lambda x}\}}(.,\lambda_0). \end{aligned}$$

This section is mainly devoted to the proof of asymptotic formulas for both types of kernels as $x \to +\infty$.

2.2. Some special properties of the reproducing kernels

We comment once again the $\pm$-mappings introduced by formulas (1.13), (1.14). They actually depend on the scattering data $\{s_\pm,\nu_\pm\}$, although we do not indicate this dependence explicitly in most cases.

Lemma 2.1. *Let w be a meromorphic function on $\mathbb{C}$ such that $|w(\lambda)| = 1$ on $\mathbb{R}$ and $w(\lambda_k) \neq 0, \infty$, for all $\lambda_k \in \Lambda$. Put $w_*(\lambda) := \overline{w(-\bar{\lambda})}$. The following diagram is* *commutative*

$$\begin{array}{ccc} L^2_{\{ww_*s_+,ww_*\nu_+\}} & \xrightarrow{w} & L^2_{\{s_+,\nu_+\}} \\ +\Big\uparrow\Big\downarrow- & & +\Big\uparrow\Big\downarrow- \\ L^2_{\{w^{-1}w_*^{-1}s_-,w^{-1}w_*^{-1}\nu_-\}} & \xrightarrow{w_*^{-1}} & L^2_{\{s_-,\nu_-\}} \end{array} \tag{2.2}$$

Here the horizontal arrows are related to the unitary multiplication operators and the vertical arrows are related to two different $\pm$*-duality mappings.*

Proof. Note that both w and w_*^{-1} are well defined on $\mathbb{R}\cup\Lambda$. Evidently, $wf\in L^2_{\{s_+,\nu_+\}}$ means that $f\in L^2_{\{ww_*s_+,ww_*\nu_+\}}$. Since $|w(\lambda)|=1$, $\lambda\in\mathbb{R}$, we have that $\{w^{-1}w_*^{-1}s_-,w^{-1}w_*^{-1}\nu_-\}$ are minus-scattering data for $\{ww_*s_+,ww_*\nu_+\}$ if $\{s_-,\nu_-\}$ corresponds to $\{s_+,\nu_+\}$. In other words, the s-function remains the same for both sets of scattering data. Then we use definitions (1.13), (1.14). □

Let $b_{\lambda_0}=\dfrac{\lambda-\lambda_0}{\lambda-\bar\lambda_0}$. Note that $(b_{\lambda_0})_*=b_{-\bar\lambda_0}$.

Lemma 2.2. *We have*

$$(k_{\{s_+,\nu_+\}}(\lambda,\lambda_0))^-=\frac{1}{s(-\bar\lambda_0)}\frac{b^{-1}_{-\bar\lambda_0}(\lambda)}{2\mathrm{Im}\ \lambda_0}\frac{\hat k_{\{b^{-1}_{\lambda_0}b^{-1}_{-\bar\lambda_0}s_-,b^{-1}_{\lambda_0}b^{-1}_{-\bar\lambda_0}\nu_-\}}(\lambda,-\bar\lambda_0)}{\hat k_{\{b^{-1}_{\lambda_0}b^{-1}_{-\bar\lambda_0}s_-,b^{-1}_{\lambda_0}b^{-1}_{-\bar\lambda_0}\nu_-\}}(-\bar\lambda_0,-\bar\lambda_0)}, \tag{2.3}$$

and, consequently,

$$k_{\{s_+,\nu_+\}}(\lambda_0,\lambda_0)\hat k_{\{b^{-1}_{\lambda_0}b^{-1}_{-\bar\lambda_0}s_-,b^{-1}_{\lambda_0}b^{-1}_{-\bar\lambda_0}\nu_-\}}(-\bar\lambda_0,-\bar\lambda_0)=\frac{1}{s(-\bar\lambda_0)s(\lambda_0)}\frac{1}{(2\mathrm{Im}\ \lambda_0)^2}. \tag{2.4}$$

Proof. First we note that the following one-dimensional spaces coincide

$$\{(k_{\{s_+,\nu_+\}}(\lambda,\lambda_0)\}^-=\{b^{-1}_{-\bar\lambda_0}\hat k_{\{b^{-1}_{\lambda_0}b^{-1}_{-\bar\lambda_0}s_-,b^{-1}_{\lambda_0}b^{-1}_{-\bar\lambda_0}\nu_-\}}(\lambda,-\bar\lambda_0)\}.$$

This follows immediately from Theorem 1.3, but we prefer to give a formal proof. Starting with the orthogonal decomposition

$$\{k_{\{s_+,\nu_+\}}(\lambda,\lambda_0)\}=H^2_{\{s_+,\nu_+\}}\ominus b_{\lambda_0}H^2_{\{b_{\lambda_0}b_{-\bar\lambda_0}s_+,b_{\lambda_0}b_{-\bar\lambda_0}\nu_+\}}$$

we have

$$\{k_{\{s_+,\nu_+\}}(\lambda,\lambda_0)\}^-=(H^2_{\{s_+,\nu_+\}})^-\ominus(b_{\lambda_0}H^2_{\{b_{\lambda_0}b_{-\bar\lambda_0}s_+,b_{\lambda_0}b_{-\bar\lambda_0}\nu_+\}})^-,$$

or, due to (2.2),

$$\{k_{\{s_+,\nu_+\}}(\lambda,\lambda_0)\}^-=(H^2_{\{s_+,\nu_+\}})^-\ominus b^{-1}_{-\bar\lambda_0}(H^2_{\{b_{\lambda_0}b_{-\bar\lambda_0}s_+,b_{\lambda_0}b_{-\bar\lambda_0}\nu_+\}})^-.$$

Now we use Theorem 1.3

$$\begin{aligned}\{k_{\{s_+,\nu_+\}}(\lambda,\lambda_0)\}^-=&(L^2_{\{s_-,\nu_-\}}\ominus\hat H^2_{\{s_-,\nu_-\}})\\ &\ominus b^{-1}_{-\bar\lambda_0}(L^2_{\{b^{-1}_{\lambda_0}b^{-1}_{-\bar\lambda_0}s_-,b^{-1}_{\lambda_0}b^{-1}_{-\bar\lambda_0}\nu_-\}}\ominus\hat H^2_{\{b^{-1}_{\lambda_0}b^{-1}_{-\bar\lambda_0}s_-,b^{-1}_{\lambda_0}b^{-1}_{-\bar\lambda_0}\nu_-\}})\\ =&b^{-1}_{-\bar\lambda_0}(\hat H^2_{\{b^{-1}_{\lambda_0}b^{-1}_{-\bar\lambda_0}s_-,b^{-1}_{\lambda_0}b^{-1}_{-\bar\lambda_0}\nu_-\}}\ominus b_{-\bar\lambda_0}\hat H^2_{\{s_-,\nu_-\}}).\end{aligned}$$

Thus

$$(k_{\{s_+,\nu_+\}}(\lambda,\lambda_0))^- = Cb^{-1}_{-\bar\lambda_0}\hat k_{\{b^{-1}_{\lambda_0}b^{-1}_{-\bar\lambda_0}s_-,b^{-1}_{\lambda_0}b^{-1}_{-\bar\lambda_0}\nu_-\}}(\lambda,-\bar\lambda_0). \tag{2.5}$$

The essential part of the lemma deals with the constant C. We calculate the scalar product

$$\left\langle k_{\{s_+,\nu_+\}}(\lambda,\lambda_0), \frac{iB(\lambda)}{\lambda-\bar\lambda_0}\right\rangle_{\{s_+,\nu_+\}}.$$

On the one hand, since $\frac{iB(\lambda)}{\lambda-\bar\lambda_0}$ belongs to the intersection of $L^2_{\{s_+,\nu_+\}}$ with H^2, we can use the reproducing property of $k_{\{s_+,\nu_+\}}$:

$$\left\langle k_{\{s_+,\nu_+\}}(\lambda,\lambda_0), \frac{iB(\lambda)}{\lambda-\bar\lambda_0}\right\rangle_{\{s_+,\nu_+\}} = \frac{\overline{B(\lambda_0)}}{2\mathrm{Im}\ \lambda_0} = \frac{B(-\bar\lambda_0)}{2\mathrm{Im}\ \lambda_0}. \tag{2.6}$$

On the other hand we can reduce the given scalar product to the scalar product in the standard H^2. Since $B(\lambda_k)=0$, the ν-component disappears and we get

$$\frac{1}{2}\left\langle \begin{bmatrix} 1 & \bar s_+ \\ s_+ & 1 \end{bmatrix}(\lambda)\begin{bmatrix} k_{\{s_+,\nu_+\}}(\lambda,\lambda_0) \\ -k_{\{s_+,\nu_+\}}(-\bar\lambda,\lambda_0)\end{bmatrix}, \begin{bmatrix} \frac{iB(\lambda)}{\lambda-\bar\lambda_0} \\ \frac{iB(-\bar\lambda)}{\lambda+\bar\lambda_0}\end{bmatrix}\right\rangle$$
$$= \left\langle s(\lambda)(k_{\{s_+,\nu_+\}}(\lambda,\lambda_0))^-, \frac{i\overline{B(\lambda)}}{\lambda+\bar\lambda_0}\right\rangle.$$

Substituting here (2.5) and using $s=s_e/B$ we come to

$$C\left\langle s_e(\lambda)\hat k_{\{b^{-1}_{\lambda_0}b^{-1}_{-\bar\lambda_0}s_-,b^{-1}_{\lambda_0}b^{-1}_{-\bar\lambda_0}\nu_-\}}(\lambda,-\bar\lambda_0), b_{-\bar\lambda_0}(\lambda)\frac{i}{\lambda+\bar\lambda_0}\right\rangle.$$

Since $s_e(\lambda)\hat k_{\{b^{-1}_{\lambda_0}b^{-1}_{-\bar\lambda_0}s_-,b^{-1}_{\lambda_0}b^{-1}_{-\bar\lambda_0}\nu_-\}}(\lambda,-\bar\lambda_0)$ belongs to H^2 and $b_{-\bar\lambda_0}(\lambda)\frac{i}{\lambda+\bar\lambda_0} = \frac{i}{\lambda+\lambda_0}$ is the reproducing kernel of H^2, relation (2.6) yields

$$Cs_e(-\bar\lambda_0)\hat k_{\{b^{-1}_{\lambda_0}b^{-1}_{-\bar\lambda_0}s_-,b^{-1}_{\lambda_0}b^{-1}_{-\bar\lambda_0}\nu_-\}}(-\bar\lambda_0,-\bar\lambda_0) = \frac{B(-\bar\lambda_0)}{2\mathrm{Im}\ \lambda_0}.$$

Thus (2.3) is proved. Comparing the norms of these vectors and taking into account that the −-map is an isometry we get (2.4). □

As a consequence of the above lemma, we have

$$b_{-\bar\lambda_0}(e^{i\lambda x}K_{\{s_+e^{2i\lambda x},\,\nu_+e^{2i\lambda x}\}}(.,\lambda_0))^- \in \hat H^2_{\{s_-(b_{\lambda_0}b_{-\bar\lambda_0})^{-1},\,\nu_-(b_{\lambda_0}b_{-\bar\lambda_0})^{-1}\}}, \quad x\ge 0.$$

Indeed, using diagram (2.2), we get for $x\ge 0$

$$(e^{-i\lambda x}K_{\{s_+e^{-2i\lambda x},\,\nu_+e^{-2i\lambda x}\}}(.,\lambda_0))^-_{s_-} = e^{i\lambda x}(K_{\{s_+e^{-2i\lambda x},\,\nu_+e^{-2i\lambda x}\}}(.,\lambda_0))^-_{s_-e^{2i\lambda x}}$$
$$= C(\lambda_0)e^{i\lambda x}\hat K_{\{s_-e^{2i\lambda x}(b_{\lambda_0}b_{-\bar\lambda_0})^{-1},\,\nu_-e^{2i\lambda x}(b_{\lambda_0}b_{-\bar\lambda_0})^{-1}\}}(.,-\bar\lambda_0)$$

by (2.3). So, the latter function is in $\hat H^2_{\{s_-(b_{\lambda_0}b_{-\bar\lambda_0})^{-1},\,\nu_-(b_{\lambda_0}b_{-\bar\lambda_0})^{-1}\}}$.

For discrete measures $\nu_\pm$ (1.6), let $\nu_{\pm N}$ be their truncations

$$\nu_{\pm N} = \sum_{k=1}^{N} \nu_\pm(\lambda_k)\delta_{\lambda_k}.$$

We say few more words about spaces

$$H^2_{\{s_\pm,\nu_\pm\}}, \hat{H}^2_{\{s_\pm,\nu_\pm\}} \quad \text{and} \quad H^2_{\{s_\pm,\nu_{\pm N}\}}, \hat{H}^2_{\{s_\pm,\nu_{\pm N}\}}.$$

Recall that $H^2_{\{s_+,\nu_+\}} \subset \hat{H}^2_{\{s_+,\nu_+\}}$.

Lemma 2.3.

i) *Let* $||s_+||_\infty < 1$ *(or, what is the same,* $\inf_{\mathbb{R}} |s_e| > 0$*). Then*

$$H^2_{\{s_+,\nu_+\}} = \hat{H}^2_{\{s_+,\nu_+\}},$$

and, consequently, $K_{\{s_+,\nu_+\}}(.,\lambda_0) = \hat{K}_{\{s_+,\nu_+\}}(.,\lambda_0)$.

ii) *We always have*

$$K_{\{s_+,\nu_+\}}(\lambda_0,\lambda_0) \le \hat{K}_{\{s_+,\nu_+\}}(\lambda_0,\lambda_0).$$

The equality above takes place if and only if $K_{\{s_+,\nu_+\}}(.,\lambda_0) = \hat{K}_{\{s_+,\nu_+\}}(.,\lambda_0)$.

iii) *Obviously,*

$$H^2_{\{s_+,\nu_+\}} \subset H^2_{\{s_+,\nu_{+N}\}}, \quad \hat{H}^2_{\{s_+,\nu_+\}} \subset \hat{H}^2_{\{s_+,\nu_{+N}\}},$$

and

$$\begin{aligned} K_{\{s_+,\nu_+\}}(\lambda_0,\lambda_0) &\le K_{\{s_+,\nu_{+N}\}}(\lambda_0,\lambda_0), \\ \hat{K}_{\{s_+,\nu_+\}}(\lambda_0,\lambda_0) &\le \hat{K}_{\{s_+,\nu_{+N}\}}(\lambda_0,\lambda_0). \end{aligned}$$

As before, the inequalities become equalities if and only if the corresponding reproducing kernels coincide.

Proof. To prove i), we only have to show the inverse inclusion. Suppose that $f \in \hat{H}^2_{\{s_+,\nu_+\}}$. By Definition 1.2, $f \in L^2_{\{s_+,\nu_+\}}$ and $s_e f \in H^2$. Since $s_e, 1/s_e \in H^\infty$, we see $f \in H^2$ and hence $f \in H^2_{\{s_+,\nu_+\}}$. The claim about the reproducing kernels is trivial.

The inequality in ii) of course follows from inclusion $H^2_{\{s_+,\nu_+\}} \subset \hat{H}^2_{\{s_+,\nu_+\}}$. Consider a system $\{f_n\}_{n\in\mathbb{Z}_+}$, $f_n = b^n_{\lambda_0}\hat{K}_{\{s_+(b_{\lambda_0}b_{-\bar{\lambda}_0})^n,\nu_+(b_{\lambda_0}b_{-\bar{\lambda}_0})^n\}}(.,\lambda_0)$. This is an orthonormal basis in $\hat{H}^2_{\{s_+,\nu_+\}}$. We have $K_{\{s_+,\nu_+\}}(.,\lambda_0) \in \hat{H}^2_{\{s_+,\nu_+\}}$ and $||K_{\{s_+,\nu_+\}}(.,\lambda_0)||_{\{s_+,\nu_+\}} = 1$. So

$$K_{\{s_+,\nu_+\}}(.,\lambda_0) = \sum_n a_n f_n$$

and $a_0 = K_{\{s_+,\nu_+\}}(\lambda_0,\lambda_0)/\hat{K}_{\{s_+,\nu_+\}}(\lambda_0,\lambda_0)$. Obviously, $|a_0|^2 \le 1$ and claim ii) is proved.

Let us have a look at iii). The first inclusion follows from the fact that for $f \in H^2(\mathbb{C}_+)$

$$||B_{N',\infty}f||_{\{s_+,\nu_{+N}\}} \le ||B_{N',\infty}f||_{\{s_+,\nu_+\}}.$$

The second one follows from Definition 1.2 of $\hat{H}^2_{\{s_+,\nu_+\}}$. The inequalities for the reproducing kernels are corollaries of these inclusions; to prove them just argue as in ii). □

Rewriting (2.4), we have

$$K_{\{s_+e^{-2i\lambda x},\, \nu_+e^{-2i\lambda x}\}}(\lambda_0,\lambda_0)K_{\{s_-e^{2i\lambda x}(b_{\lambda_0}b_{-\bar\lambda_0})^{-1},\, \nu_-e^{2i\lambda x}(b_{\lambda_0}b_{-\bar\lambda_0})^{-1}\}}(-\bar\lambda_0,-\bar\lambda_0)$$
$$= \frac{1}{|s(\lambda_0)|(2\mathrm{Im}\,\lambda_0)}.$$

under assumptions i) of the above lemma.

We denote by $P_{\{s_+,\nu_+\}}$ the orthogonal projector from $L^2_{\{s_+,\nu_+\}}$ on $H^2_{\{s_+,\nu_+\}}$. Furthermore, $P_{x,\{s_+,\nu_+\}}$ and $\hat{P}_{x,\{s_+,\nu_+\}}$ are orthogonal projectors on $H^2_{\{s_+,\nu_+\}}(x)$ and $\hat{H}^2_{\{s_+,\nu_+\}}(x)$, correspondingly.

Lemma 2.4. *We have for any $f \in L^2_{\{s_+,\nu_+\}}$:*

$$\text{i)} \qquad \lim_{x\to-\infty} P_{x,\{s_+,\nu_+\}}f = f, \quad \lim_{x\to-\infty} \hat{P}_{x,\{s_+,\nu_+\}}f = f \tag{2.7}$$

$$\text{ii)} \qquad \lim_{x\to+\infty} P_{x,\{s_+,\nu_+\}}f = 0, \quad \lim_{x\to+\infty} \hat{P}_{x,\{s_+,\nu_+\}}f = 0 \tag{2.8}$$

Symbolically, we may say that

$$\text{i)} \lim_{x\to-\infty} e^{i\lambda x}H^2_{\{s_+e^{2i\lambda x},\, \nu_+e^{2i\lambda x}\}} = L^2_{\{s_+,\nu_+\}},$$
$$\lim_{x\to-\infty} e^{i\lambda x}\hat{H}^2_{\{s_+e^{2i\lambda x},\, \nu_+e^{2i\lambda x}\}} = L^2_{\{s_+,\nu_+\}},$$
$$\text{ii)} \lim_{x\to+\infty} e^{i\lambda x}H^2_{\{s_+e^{2i\lambda x},\, \nu_+e^{2i\lambda x}\}} = \{0\},\ \lim_{x\to+\infty} e^{i\lambda x}\hat{H}^2_{\{s_+e^{2i\lambda x},\, \nu_+e^{2i\lambda x}\}} = \{0\}.$$

Proof. We prove the first equality in (2.8); the argument for the second equality is likewise. Relations in (2.7) drop by duality, since

$$L^2_{\{s_+,\nu_+\}} = e^{i\lambda x}H^2_{\{s_+e^{2i\lambda x},\, \nu_+e^{2i\lambda x}\}} \oplus \left(e^{-i\lambda x}\hat{H}^2_{\{s_-e^{-2i\lambda x}\, \nu_-e^{-2i\lambda x}\}}\right)^+.$$

Obviously, $e^{i\lambda x_2}H^2_{\{s_+e^{2i\lambda x_2},\, \nu_+e^{2i\lambda x_2}\}} \subset e^{i\lambda x_1}H^2_{\{s_+e^{2i\lambda x_1},\, \nu_+e^{2i\lambda x_1}\}}$ for $x_1 \le x_2$ and so $k_{\{s_+,\nu_+\}}(\lambda_0,\lambda_0;x) = e^{-2\mathrm{Im}\,\lambda_0 x}k_{\{s_+e^{2i\lambda x},\, \nu_+e^{2i\lambda x}\}}(\lambda_0,\lambda_0)$ is decreasing with respect to $x \in \mathbb{R}$. We have to prove that

$$\lim_{x\to+\infty} e^{-2\mathrm{Im}\,\lambda_0 x}k_{\{s_+e^{2i\lambda x},\, \nu_+e^{2i\lambda x}\}}(\lambda_0,\lambda_0) = 0,$$

which is trivial since the second factor tends to $k(\lambda_0,\lambda_0)$ by Lemma 2.8. □

Corollary 2.5. *We have*

$$\lim_{x\to-\infty} e^{-2\mathrm{Im}\,\lambda_k x}k_{\{s_+e^{2i\lambda x},\, \nu_+e^{2i\lambda x}\}}(\lambda_k,\lambda_k) = \frac{1}{\nu_+(\lambda_k)}.$$

Proof. Let us consider $g_k = (1/\nu_+(\lambda_k))\,\delta_{\lambda_k}$. Recall that

$$k_{\{s_+,\nu_+\}}(.,\lambda_0;x) = e^{ix(\lambda-\bar\lambda_0)}k_{\{s_+e^{2i\lambda x},\,\nu_+e^{2i\lambda x}\}}(.,\lambda_0).$$

Hence, we obtain for a $f \in H^2_{\{s_+,\nu_+\}}(x)$

$$(f,g_k)_{\{s_+,\nu_+\}} = \frac{f(\lambda_k)}{\nu_+(\lambda_k)}\,\nu_+(\lambda_k) = f(\lambda_k).$$

On the other hand,

$$f(\lambda_k) = (f, k_{\{s_+,\nu_+\}}(.,\lambda_k;x))_{\{s_+,\nu_+\}} = (f, P_{x,\{s_+,\nu_+\}}g_k)_{\{s_+,\nu_+\}}.$$

By Lemma 2.4,

$$\lim_{x\to-\infty} ||k_{\{s_+,\nu_+\}}(.,\lambda_k;x)||^2_{\{s_+,\nu_+\}} = ||g_k||^2_{\{s_+,\nu_+\}}$$

which becomes the claim of the corollary if we write the norms explicitly. □

2.3. Proof of Theorem 1.6

Lemma 2.6. *We have*

$$k_{\{s_+,\nu_+\}}(.,\lambda_0) = \lim_{\varepsilon\to 0+}(\varepsilon + I + H_{\{s_+,\nu_+\}})^{-1}k(.,\lambda_0),$$

where $H_{\{s_+,\nu_+\}}$ is the Hankel operator coming from the metric (1.7) *and the limit is understood in $L^2_{\{s_+,\nu_+\}}$-sense.*

The argument follows [20], Lemma 1.2, and is omitted.

Lemma 2.7. *Let $||s_+||_\infty < 1$ and ν_+ be a measure with a finite support. Then*

$$\lim_{x\to+\infty}\frac{K_{\{s_+e^{2i\lambda x},\,\nu_+e^{2i\lambda x}\}}(\lambda_0,\lambda_0)}{K(\lambda_0,\lambda_0)} = 1.$$

Proof. We see that

$$\begin{aligned}
&|k_{\{s_+e^{2i\lambda x},\,\nu_+e^{2i\lambda x}\}}(\lambda_0,\lambda_0) - k(\lambda_0,\lambda_0)|^2\\
&= |(k_{\{s_+e^{2i\lambda x},\,\nu_+e^{2i\lambda x}\}}(.,\lambda_0) - k(.,\lambda_0), k_{\{s_+e^{2i\lambda x},\,\nu_+e^{2i\lambda x}\}}(.,\lambda_0))_{\{s_+e^{2i\lambda x},\,\nu_+e^{2i\lambda x}\}}|^2\\
&= |((I + H_{\{s_+e^{2i\lambda x},\,\nu_+e^{2i\lambda x}\}})\{(I + H_{\{s_+e^{2i\lambda x},\,\nu_+e^{2i\lambda x}\}})^{-1}k(.,\lambda_0) - k(.,\lambda_0)\},\\
&\quad (I + H_{\{s_+e^{2i\lambda x},\,\nu_+e^{2i\lambda x}\}})^{-1}k(.,\lambda_0))|^2\\
&= |(H_{\{s_+e^{2i\lambda x},\,\nu_+e^{2i\lambda x}\}}k(.,\lambda_0), (I + H_{\{s_+e^{2i\lambda x},\,\nu_+e^{2i\lambda x}\}})^{-1}k(.,\lambda_0))|^2\\
&\le C\Big(|(H_{s_+e^{2i\lambda x}}k(.,\lambda_0), (I + H_{\{s_+e^{2i\lambda x},\,\nu_+e^{2i\lambda x}\}})^{-1}k(.,\lambda_0))|^2\\
&\quad + |(H_{\nu_+e^{2i\lambda x}}k(.,\lambda_0), (I + H_{\{s_+e^{2i\lambda x},\,\nu_+e^{2i\lambda x}\}})^{-1}k(.,\lambda_0))|^2\Big)
\end{aligned}$$

The bound for the first term is easy

$$|\dots| \le \frac{1}{1-||s_+||_\infty}||P_+(s_+e^{-2i\bar\lambda x}k(-\bar\lambda,\lambda_0))||_2||k(.,\lambda_0)||_2 \to 0$$

as $x \to +\infty$ by the L^2-Fourier theorem. Since $F = (I + H_{\{s_+ e^{2i\lambda x},\, \nu_+ e^{2i\lambda x}\}})^{-1} k(.,\lambda_0) \in H^2$ satisfies $||F||_2 \le C$, we get $|F(\lambda_k)| \le C/\sqrt{\operatorname{Im}\lambda_k}$ and

$$\begin{aligned} |(H_{\nu_+ e^{2i\lambda x}} k(.,\lambda_0), (I &+ H_{\{s_+ e^{2i\lambda x},\, \nu_+ e^{2i\lambda x}\}})^{-1} k(.,\lambda_0))| \\ &\le C \sum_{k=1}^{N} \nu_+(\lambda_k) e^{-2\operatorname{Im}\lambda_k x} \frac{|k(\lambda_k,\lambda_0)|}{\sqrt{\operatorname{Im}\lambda_k}}. \end{aligned}$$

The right-hand side of the inequality goes to 0 as $x \to +\infty$. □

The following lemma is the main key to the proof of the asymptotics.

Lemma 2.8. *We have*

$$\lim_{x\to+\infty} \frac{K_{\{s_+ e^{2i\lambda x},\, \nu_+ e^{2i\lambda x}\}}(\lambda_0,\lambda_0)}{K(\lambda_0,\lambda_0)} = 1, \tag{2.9}$$

$$\lim_{x\to+\infty} \frac{\hat K_{\{s_+ e^{2i\lambda x},\, \nu_+ e^{2i\lambda x}\}}(\lambda_0,\lambda_0)}{K(\lambda_0,\lambda_0)} = 1. \tag{2.10}$$

Proof. We start with the proof of the first equality. Taking the square root of both sides of (2.4), we see

$$\begin{aligned} &K_{\{s_+ e^{2i\lambda x},\, \nu_+ e^{2i\lambda x}\}}(\lambda_0,\lambda_0) \\ =\ & \frac{1}{2\operatorname{Im}\lambda_0 |s(\lambda_0)|} \frac{1}{\hat K_{\{s_- e^{-2i\lambda x}(b_{\lambda_0} b_{-\bar\lambda_0})^{-1},\, \nu_- e^{-2i\lambda x}(b_{\lambda_0} b_{-\bar\lambda_0})^{-1}\}}(-\bar\lambda_0,-\bar\lambda_0)} \\ \ge\ & \frac{1}{2\operatorname{Im}\lambda_0 |s(\lambda_0)|} \frac{1}{\hat K_{\{s_- e^{-2i\lambda x}(b_{\lambda_0} b_{-\bar\lambda_0})^{-1},\, \nu_{-N} e^{-2i\lambda x}(b_{\lambda_0} b_{-\bar\lambda_0})^{-1}\}}(-\bar\lambda_0,-\bar\lambda_0)} \\ =\ & \frac{|B(\lambda_0)|}{|B_{1,N}(\lambda_0)|} K_{\{s_+ e^{2i\lambda x},\, \nu_{+N} e^{2i\lambda x}\}}(\lambda_0,\lambda_0). \end{aligned} \tag{2.11}$$

Then we continue as

$$\begin{aligned} &K_{\{s_+ e^{2i\lambda x},\, \nu_+ e^{2i\lambda x}\}}(\lambda_0,\lambda_0) \\ \ge\ & |B_{N,\infty}(\lambda_0)| (k_{\{s_+ e^{2i\lambda x},\, \nu_{+N} e^{2i\lambda x}\}}(\lambda_0,\lambda_0))^{1/2} \\ \ge\ & |B_{N,\infty}(\lambda_0)| ((\varepsilon + I + H_{\{s_+ e^{2i\lambda x},\, \nu_{+N} e^{2i\lambda x}\}})^{-1} k(.,\lambda_0))^{1/2}(\lambda_0) \\ =\ & \frac{1}{\sqrt{1+\varepsilon}} |B_{N,\infty}(\lambda_0)| K_{\{\frac{s_+}{1+\varepsilon} e^{2i\lambda x},\, \frac{\nu_{+N}}{1+\varepsilon} e^{2i\lambda x}\}}(\lambda_0,\lambda_0). \end{aligned} \tag{2.12}$$

Let $s_N = s_e/B_{1,N}$. We have

$$\begin{aligned} &K_{\{s_+ e^{2i\lambda x},\, \nu_+ e^{2i\lambda x}\}}(\lambda_0,\lambda_0) \le K_{\{s_+ e^{2i\lambda x},\, \nu_{+N} e^{2i\lambda x}\}}(\lambda_0,\lambda_0) \\ =\ & \frac{1}{2\operatorname{Im}\lambda_0 |s_N(\lambda_0)|} \frac{1}{\hat K_{\{s_- e^{-2i\lambda x}(b_{\lambda_0} b_{-\bar\lambda_0})^{-1},\, \nu_{-N} e^{-2i\lambda x}(b_{\lambda_0} b_{-\bar\lambda_0})^{-1}\}}(-\bar\lambda_0,-\bar\lambda_0)} \\ \le\ & \frac{1}{2\operatorname{Im}\lambda_0 |s_N(\lambda_0)|} \frac{1}{K_{\{s_- e^{-2i\lambda x}(b_{\lambda_0} b_{-\bar\lambda_0})^{-1},\, \nu_{-N} e^{-2i\lambda x}(b_{\lambda_0} b_{-\bar\lambda_0})^{-1}\}}(-\bar\lambda_0,-\bar\lambda_0)} \end{aligned} \tag{2.13}$$

by ii), Lemma 2.3. We set $s_{-,\varepsilon} = s_-/(1+\varepsilon), \nu_{-N,\varepsilon} = \nu_{-N}/(1+\varepsilon)$; the functions $s_{N,\varepsilon}, s_{+,\varepsilon}$ are defined by unitarity of the scattering matrix, and $\nu_{+N,\varepsilon}$ is defined by $\nu_{-N,\varepsilon}$ through relations (1.12). Notice that the support of $\nu_{+N,\varepsilon}$ is the same as the support of ν_{+N} (and equals $\{\lambda_k\}_{k=1,N}$). Since K- and $\hat{K}$-kernels are the same for pairs $\{s_{+,\varepsilon}, \nu_{+N,\varepsilon}\}$ and $\{s_{-,\varepsilon}, \nu_{-N,\varepsilon}\}$ by i), Lemma 2.3, we continue as

$$\begin{aligned}(\dots) &\le \frac{\sqrt{1+\varepsilon}}{2\operatorname{Im}\lambda_0|s_N(\lambda_0)|}\,\frac{1}{K_{\{s_{-,\varepsilon}e^{-2i\lambda x}(b_{\lambda_0}b_{-\bar\lambda_0})^{-1},\,\nu_{-N,\varepsilon}e^{-2i\lambda x}(b_{\lambda_0}b_{-\bar\lambda_0})^{-1}\}}(-\bar\lambda_0,-\bar\lambda_0)}\\ &= \frac{\sqrt{1+\varepsilon}\,|s_{N,\varepsilon}(\lambda_0)|}{|s_N(\lambda_0)|}K_{\{s_{+,\varepsilon}e^{2i\lambda x},\,\nu_{+N,\varepsilon}e^{2i\lambda x}\}}(\lambda_0,\lambda_0).\end{aligned}$$

That is,

$$\begin{aligned}&\frac{|B_{N,\infty}(\lambda_0)|}{\sqrt{1+\varepsilon}}K_{\{\frac{s_+}{1+\varepsilon}e^{2i\lambda x},\,\frac{\nu_{+N}}{1+\varepsilon}e^{2i\lambda x}\}}(\lambda_0,\lambda_0) \le K_{\{s_+e^{2i\lambda x},\,\nu_+e^{2i\lambda x}\}}(\lambda_0,\lambda_0)\\ \le\; &\frac{\sqrt{1+\varepsilon}\,|s_{N,\varepsilon}(\lambda_0)|}{|s_N(\lambda_0)|}K_{\{s_{+,\varepsilon}e^{2i\lambda x},\,\nu_{+N,\varepsilon}e^{2i\lambda x}\}}(\lambda_0,\lambda_0).\end{aligned}$$

The quantities $K_{\{\frac{s_+}{1+\varepsilon}e^{2i\lambda x},\frac{\nu_{+N}}{1+\varepsilon}e^{2i\lambda x}\}}(\lambda_0,\lambda_0)$ and $K_{\{s_{+,\varepsilon}e^{2i\lambda x},\,\nu_{+N,\varepsilon}e^{2i\lambda x}\}}(\lambda_0,\lambda_0)$ tend to $K(\lambda_0,\lambda_0)$ as $x\to+\infty$ by Lemma 2.7. Remaining factors in the left- and right-hand side parts of the inequality go to 1 with $\varepsilon\to+0, N\to+\infty$. Hence, for any $\varepsilon'>0$ we can choose appropriate ε, N to have

$$\begin{aligned}1-\varepsilon' &\le \liminf_{x\to+\infty}\frac{K_{\{s_+e^{2i\lambda x},\,\nu_+e^{2i\lambda x}\}}(\lambda_0,\lambda_0)}{K(\lambda_0,\lambda_0)}\\ &\le \limsup_{x\to+\infty}\frac{K_{\{s_+e^{2i\lambda x},\,\nu_+e^{2i\lambda x}\}}(\lambda_0,\lambda_0)}{K(\lambda_0,\lambda_0)} \le 1+\varepsilon',\end{aligned}$$

and (2.9) is proved.

The proof of (2.10) is almost identical. First of all, to keep the notation we used to, we prove

$$\lim_{x\to+\infty}\frac{\hat{K}_{\{s_-e^{2i\lambda x},\,\nu_-e^{2i\lambda x}\}}(-\bar\lambda_0,-\bar\lambda_0)}{K(-\bar\lambda_0,-\bar\lambda_0)} = 1. \tag{2.14}$$

instead of (2.10). This is obviously the same thing up to changes $-\bar\lambda_0\mapsto\lambda_0$ and $s_-\mapsto s_+$. The second modification is that we estimate the value of a $\hat{K}$-kernel by the values of K-kernels (and not vice versa as we have just done to prove (2.9)).

So, as in (2.12), we have

$$\begin{aligned}\hat{K}_{\{s_-e^{2i\lambda x},\,\nu_-e^{2i\lambda x}\}}(-\bar\lambda_0,-\bar\lambda_0) &\ge K_{\{s_-e^{2i\lambda x},\,\nu_-e^{2i\lambda x}\}}(-\bar\lambda_0,-\bar\lambda_0)\\ &\ge \frac{1}{\sqrt{1+\varepsilon}}K_{\{\frac{s_-}{1+\varepsilon}e^{2i\lambda x},\,\frac{\nu_-}{1+\varepsilon}e^{2i\lambda x}\}}(-\bar\lambda_0,-\bar\lambda_0)\\ &= \frac{|B_{N,\infty}(\lambda_0)|}{\sqrt{1+\varepsilon}}K_{\{\frac{s_-}{1+\varepsilon}e^{2i\lambda x},\,\frac{\nu_{-N}}{1+\varepsilon}e^{2i\lambda x}\}}(-\bar\lambda_0,-\bar\lambda_0).\end{aligned}$$

The first inequality in the above estimate is ii), Lemma 2.3 and the last one repeats computation (2.11). Similarly to (2.13), we get

$$\begin{aligned}
&\hat{K}_{\{s_- e^{2i\lambda x},\, \nu_- e^{2i\lambda x}\}}(-\bar\lambda_0, -\bar\lambda_0) \le \hat{K}_{\{s_- e^{2i\lambda x},\, \nu_{-N} e^{2i\lambda x}\}}(-\bar\lambda_0, -\bar\lambda_0)\\
= &\frac{1}{2\operatorname{Im}\lambda_0 |s_N(\lambda_0)|}\frac{1}{K_{\{s_+ e^{-2i\lambda x}(b_{\lambda_0} b_{-\bar\lambda_0})^{-1},\, \nu_{+N} e^{-2i\lambda x}(b_{\lambda_0} b_{-\bar\lambda_0})^{-1}\}}(\lambda_0,\lambda_0)}\\
\le &\frac{\sqrt{1+\varepsilon}}{2\operatorname{Im}\lambda_0 |s_N(\lambda_0)|}\frac{1}{K_{\{\frac{s_+}{1+\varepsilon} e^{-2i\lambda x}(b_{\lambda_0} b_{-\bar\lambda_0})^{-1},\, \frac{\nu_{+N}}{1+\varepsilon} e^{-2i\lambda x}(b_{\lambda_0} b_{-\bar\lambda_0})^{-1}\}}(\lambda_0,\lambda_0)}\\
= &\frac{\sqrt{1+\varepsilon}\,|s_{N,\varepsilon}(\lambda_0)|}{|s_N(\lambda_0)|}\hat{K}_{\{s_{-,\varepsilon} e^{2i\lambda x},\, \nu_{-N,\varepsilon} e^{2i\lambda x}\}}(-\bar\lambda_0, -\bar\lambda_0)\\
= &\frac{\sqrt{1+\varepsilon}\,|s_{N,\varepsilon}(\lambda_0)|}{|s_N(\lambda_0)|}K_{\{s_{-,\varepsilon} e^{2i\lambda x},\, \nu_{-N,\varepsilon} e^{2i\lambda x}\}}(-\bar\lambda_0, -\bar\lambda_0)
\end{aligned}$$

Above, the pair $\{s_{-,\varepsilon}, \nu_{-N,\varepsilon}\}$ comes from $\{\frac{s_+}{1+\varepsilon}, \frac{\nu_{+N}}{1+\varepsilon}\}$ as explained after (2.13). Hence,

$$\begin{aligned}
&\frac{|B_{N,\infty}(\lambda_0)|}{\sqrt{1+\varepsilon}}K_{\{\frac{s_-}{1+\varepsilon} e^{2i\lambda x},\, \frac{\nu_{-N}}{1+\varepsilon} e^{2i\lambda x}\}}(-\bar\lambda_0, -\bar\lambda_0) \le \hat{K}_{\{s_- e^{2i\lambda x},\, \nu_- e^{2i\lambda x}\}}(-\bar\lambda_0, -\bar\lambda_0)\\
\le &\frac{\sqrt{1+\varepsilon}\,|s_{N,\varepsilon}(\lambda_0)|}{|s_N(\lambda_0)|}K_{\{s_{-,\varepsilon} e^{2i\lambda x},\, \nu_{-N,\varepsilon} e^{2i\lambda x}\}}(-\bar\lambda_0, -\bar\lambda_0).
\end{aligned}$$

Repeating the argument from the first part of the proof, we see that for every $\varepsilon' > 0$

$$\begin{aligned}
1-\varepsilon' \quad &\le \quad \liminf_{x\to+\infty}\frac{\hat{K}_{\{s_- e^{2i\lambda x},\, \nu_- e^{2i\lambda x}\}}(-\bar\lambda_0, -\bar\lambda_0)}{K(-\bar\lambda_0, -\bar\lambda_0)}\\
&\le \quad \limsup_{x\to+\infty}\frac{\hat{K}_{\{s_- e^{2i\lambda x},\, \nu_- e^{2i\lambda x}\}}(-\bar\lambda_0, -\bar\lambda_0)}{K(-\bar\lambda_0, -\bar\lambda_0)} \le 1+\varepsilon',
\end{aligned}$$

and relation (2.14) is proved. □

Proof of Theorem 1.6. At present, the claim of the theorem is an easy consequence of Lemma 2.8. For an arbitrary N, we have

$$\begin{aligned}
||e^{i\lambda x}K_{\{s_+ e^{2i\lambda x},\, \nu_+ e^{2i\lambda x}\}}(.,\lambda_0) \quad &- \quad e^{i\lambda x}\tilde{K}(.,\lambda_0)||_{\{s_+,\nu_+\}}\\
&\le \quad ||e^{i\lambda x}K_{\{s_+ e^{2i\lambda x},\, \nu_+ e^{2i\lambda x}\}}(.,\lambda_0)\\
&- \quad B_{N,\infty}e^{i\lambda x}K(.,\lambda_0)||_{\{s_+,\nu_+\}}\\
&+ \quad ||e^{i\lambda x}\tilde{K}(.,\lambda_0) - B_{N,\infty}e^{i\lambda x}K(.,\lambda_0)||_{\{s_+,\nu_+\}}.
\end{aligned}$$

The claim will then follow if we prove

$$\limsup_{x\to+\infty}||e^{i\lambda x}\tilde{K}(.,\lambda_0) - B_{N,\infty}e^{i\lambda x}K(.,\lambda_0)||_{\{s_+,\nu_+\}} \le C_1|1-B_{N,\infty}(\lambda_0)| \qquad (2.15)$$

$$\begin{aligned}
\limsup_{x\to+\infty}||e^{i\lambda x}K_{\{s_+ e^{2i\lambda x},\, \nu_+ e^{2i\lambda x}\}}(.,\lambda_0) - B_{N,\infty}e^{i\lambda x}K(.,\lambda_0)||_{\{s_+,\nu_+\}}&\\
\le C_2|1-B_{N,\infty}(\lambda_0)| \qquad (2.16)&
\end{aligned}$$

with some constants C_1, C_2. The computation for (2.15) is easy and elementary

$$\begin{aligned} \|e^{i\lambda x}\tilde{K}(.,\lambda_0) &- B_{N,\infty}e^{i\lambda x}K(.,\lambda_0)\|^2_{\{s_+,\nu_+\}} \\ &\leq \|e^{i\lambda x}(1-B_{N,\infty})K(.,\lambda_0)\|^2_{s+} + \|e^{i\lambda x}B_{N,\infty}K(.,\lambda_0)\|^2_{\nu+} \end{aligned}$$

The second term above obviously goes to 0 as $x \to +\infty$; for the first one we have

$$\|e^{i\lambda x}(1-B_{N,\infty})K(.,\lambda_0)\|^2_{s+} \leq 2\|e^{i\lambda x}(1-B_{N,\infty})K(.,\lambda_0)\|^2_2 \leq 4|1-B_{N,\infty}(\lambda_0)|.$$

We pass to (2.16) now. Once again, for an arbitrary N,

$$\begin{aligned} &\|e^{i\lambda x}K_{\{s_+e^{2i\lambda x},\,\nu_+e^{2i\lambda x}\}}(.,\lambda_0) - B_{N,\infty}e^{i\lambda x}K(.,\lambda_0)\|^2_{\{s_+,\nu_+\}} \\ \leq\; &\|e^{i\lambda x}K_{\{s_+e^{2i\lambda x},\,\nu_+e^{2i\lambda x}\}}(.,\lambda_0)\|^2_{\{s_+,\nu_+\}} \\ -\; &2\mathrm{Re}\,(K_{\{s_+e^{2i\lambda x},\,\nu_+e^{2i\lambda x}\}}(.,\lambda_0), B_{N,\infty}K(.,\lambda_0))_{\{s_+e^{2i\lambda x},\,\nu_+e^{2i\lambda x}\}} \\ +\; &\|B_{N,\infty}K(.,\lambda_0)\|^2_{\{s_+e^{2i\lambda x},\,\nu_+e^{2i\lambda x}\}}. \end{aligned}$$

By Lemma 2.8, we get for the second term

$$\mathrm{Re}\,(\ldots) = \frac{B_{N,\infty}(\lambda_0)K(\lambda_0,\lambda_0)}{K_{\{s_+e^{2i\lambda x},\,\nu_+e^{2i\lambda x}\}}(\lambda_0,\lambda_0)} \to B_{N,\infty}(\lambda_0)$$

as $x \to +\infty$. The third term is

$$\begin{aligned} (\ldots) &= \|B_{N,\infty}K(.,\lambda_0)\|^2_{s_+e^{2i\lambda x}} + \|B_{N,\infty}K(.,\lambda_0)\|^2_{\nu_+e^{2i\lambda x}} \\ &\leq \|K(.,\lambda_0)\|^2_2 + \|P_+[s_+e^{2i\lambda x}B_{N,\infty}K(.,\lambda_0)](-\bar{\lambda})\|_2\,\|B_{N,\infty}K(.,\lambda_0)\|_2 \\ &+ \|B_{N,\infty}K(.,\lambda_0)\|^2_{\nu_+e^{2i\lambda x}} \to 1, \end{aligned}$$

since $\|K(.,\lambda_0)\|^2_2 = 1$ and the rest tends to 0 with $x \to +\infty$ (for the second term, this is Fourier L^2-theorem). So, summing up

$$\begin{aligned} \limsup_{x\to+\infty} \|e^{i\lambda x}K_{\{s_+e^{2i\lambda x},\,\nu_+e^{2i\lambda x}\}}(.,\lambda_0) &- B_{N,\infty}e^{i\lambda x}K(.,\lambda_0)\|_{\{s_+,\nu_+\}} \\ &\leq 2\mathrm{Re}\,(1-B_{N,\infty}(\lambda_0)), \end{aligned}$$

and (2.16) is proved.

The proof of (1.23) is likewise, we just have to use (2.10) instead of (2.9). □

3. Unitary node, I

Consider the multiplication operator by $\bar{v}$, $v = \dfrac{\lambda^2-\lambda_0^2}{\lambda^2-\bar{\lambda}_0^2}$, acting in

$$L^2_{\{s_+,\nu_+\}} = (\hat{H}^2_{\{s_-,\nu_-\}})^+ \oplus H^2_{\{s_+,\nu_+\}}. \tag{3.1}$$

Lemma 3.1. *The multiplication operator by* $\bar{v}$ *acts as a unitary operator from*

$$\{\hat{k}^+_{\{s_-,\nu_-\}}(\lambda,\lambda_0)\} \oplus H^2_{\{s_+,\nu_+\}}(x) \tag{3.2}$$

to

$$\{\hat{k}^+_{\{s_-,\nu_-\}}(\lambda,-\bar{\lambda}_0)\} \oplus H^2_{\{s_+,\nu_+\}}(x). \tag{3.3}$$

Proof. It is obvious that the multiplication by $\bar{v} = \frac{b_{-\bar{\lambda}_0}}{b_{\lambda_0}}$ acts from

$$\{f \in \hat{H}^2_{\{s_-,\nu_-\}} : f(\lambda_0) = 0\} = b_{\lambda_0} \hat{H}^2_{\{b_{\lambda_0} b_{-\bar{\lambda}_0} s_-, b_{\lambda_0} b_{-\bar{\lambda}_0} \nu_-\}}$$

to

$$\{f \in \hat{H}^2_{\{s_-,\nu_-\}} : f(-\bar{\lambda}_0) = 0\} = b_{-\bar{\lambda}_0} \hat{H}^2_{\{b_{\lambda_0} b_{-\bar{\lambda}_0} s_-, b_{\lambda_0} b_{-\bar{\lambda}_0} \nu_-\}}.$$

Therefore it acts on their orthogonal complements (3.2), (3.3). □

We now recall the definition of the characteristic function of a unitary node [6, 7] and its functional model. An extensive discussion of the subject and its application to interpolation problems can be found in [11]–[13].

Let K, E_1, E_2 be Hilbert spaces and U be a unitary operator acting from $K \oplus E_1$ to $K \oplus E_2$. We assume that E_1 and E_2 are finite-dimensional ($\dim E_1 = \dim E_2 = 1$ in this section, and $\dim E_1 = \dim E_2 = 2$ in Section 4). The characteristic function is defined by

$$\Theta(\zeta) := P_{E_2} U (I_{K\oplus E_1} - \zeta P_K U)^{-1} | E_1. \tag{3.4}$$

It is a holomorphic in the unit disk $\{\zeta : |\zeta| < 1\}$ contractive-valued operator function. We make a specific assumption that $\Theta(\zeta)$ has an analytic continuation in the exterior of the unite disk through a certain arc $(a, b) \subset \mathbb{T}$ by the symmetry principle

$$\Theta(\zeta) = \Theta^* \left(\frac{1}{\bar{\zeta}}\right)^{-1}.$$

For $f \in K$ define

$$F(\zeta) := P_{E_2} U (I - \zeta P_K U)^{-1} f. \tag{3.5}$$

This E_2-valued holomorphic function belongs to the functional space K_Θ with the following properties.

- $F(\zeta) \in H^2(E_2)$ and it has analytic continuation through the arc (a, b).
- $F_*(\zeta) := \Theta^*(\zeta) F\left(\frac{1}{\bar{\zeta}}\right) \in H^2_-(E_1)$.
- For almost every $\zeta \in \mathbb{T}$ the vector $\begin{bmatrix} F_* \\ F \end{bmatrix}(\zeta)$ belongs to the image of the operator $\begin{bmatrix} I & \Theta^* \\ \Theta & I \end{bmatrix}(\zeta)$, and therefore the scalar product

$$\left\langle \begin{bmatrix} I & \Theta^* \\ \Theta & I \end{bmatrix}^{[-1]} \begin{bmatrix} F_* \\ F \end{bmatrix}, \begin{bmatrix} F_* \\ F \end{bmatrix} \right\rangle_{E_1 \oplus E_2}$$

is well defined and does not depend of the choice of a preimage (the first term in the above scalar product). Moreover,

$$\int_{\mathbb{T}} \left\langle \begin{bmatrix} I & \Theta^* \\ \Theta & I \end{bmatrix}^{[-1]} \begin{bmatrix} F_* \\ F \end{bmatrix}, \begin{bmatrix} F_* \\ F \end{bmatrix} \right\rangle_{E_1 \oplus E_2} dm < \infty. \tag{3.6}$$

One can consider the above items as the definition of the space K_Θ. The integral in (3.6) represents the square of the norm of F in the space.

Note that $P_K U|K$ becomes a certain "standard" operator in the model space

$$f \mapsto F(\zeta) \quad \Longrightarrow \quad P_K U f \mapsto \frac{F(\zeta) - F(0)}{\zeta}, \tag{3.7}$$

see (3.5).

The following simple identity is a convenient tool in the forthcoming calculation.

Lemma 3.2. *For a unitary operator $U : K \oplus E_1 \to K \oplus E_2$*

$$U^* P_{E_2} U (I - \zeta P_K U)^{-1} = I + (\zeta - U^*) P_K U (I - \zeta P_K U)^{-1}. \tag{3.8}$$

Proof. Since $I_{K\oplus E_2} = P_K + P_{E_2}$ and U is unitary we have

$$U^* P_{E_2} U = (I - \zeta P_K U) + (\zeta - U^*) P_K U.$$

Then we multiply this identity by $(I - \zeta P_K U)^{-1}$. □

Theorem 3.3. *Let e_1, e_2 be the normalized vectors in the one-dimensional spaces (3.2) and (3.3),*

$$e_1(\lambda) = \frac{\hat{k}^{+}_{\{s_-,\nu_-\}}(\lambda, \lambda_0)}{\sqrt{\hat{k}_{\{s_-,\nu_-\}}(\lambda_0, \lambda_0)}}, \quad e_2(\lambda) = \frac{\hat{k}^{+}_{\{s_-,\nu_-\}}(\lambda, -\bar{\lambda}_0)}{\sqrt{\hat{k}_{\{s_-,\nu_-\}}(\lambda_0, \lambda_0)}}. \tag{3.9}$$

Then the reproducing kernel of $H^2_{\{s_+,\nu_+\}}$ is of the form

$$k_{\{s_+,\nu_+\}}(\lambda, \mu) = \frac{(ve_2)(\lambda)\overline{(ve_2)(\mu)} - e_1(\lambda)\overline{e_1(\mu)}}{1 - v(\lambda)\overline{v(\mu)}}. \tag{3.10}$$

Proof. First, we are going to find the characteristic function of the multiplication operator by $\bar{v}$ with respect to decompositions (3.2) and (3.3) and the corresponding functional representation of this node.

By (3.9) we fixed "bases" in the one-dimensional spaces. So, instead of the operator we get a scalar function $\theta(\zeta)$:

$$\Theta(\zeta)e_1 := P_{E_2} U (I - \zeta P_K U)^{-1} e_1 = e_2 \theta(\zeta). \tag{3.11}$$

We substitute (3.11) in (3.8)

$$v(\lambda)e_2(\lambda)\theta(\zeta) = e_1(\lambda) + (\zeta - v(\lambda))(P_K U (I - \zeta P_K U)^{-1} e_1)(\lambda). \tag{3.12}$$

Recall an important property of $\hat{k}^{+}_{\{s_-,\nu_-\}}(\lambda, \lambda_0)$: it has analytic continuation in the upper half-plane with the only pole at $-\bar{\lambda}_0$ (see Lemma 2.2). Therefore all terms in (3.12) are analytic in λ and we can choose λ satisfying $v(\lambda) = \zeta$. Then we obtain the characteristic function in terms of the reproducing kernels

$$\theta(v(\lambda)) = \frac{e_1(\lambda)}{v(\lambda)e_2(\lambda)}. \tag{3.13}$$

Similarly for $f \in K = H^2_{\{s_+,\nu_+\}}$ we define the scalar function $F(\zeta)$ by

$$P_{E_2} U(I - \zeta P_K U)^{-1} f = e_2 F(\zeta). \tag{3.14}$$

Using again (3.8) we get

$$v(\lambda) e_2(\lambda) F(\zeta) = f(\lambda) + (\zeta - v(\lambda))(P_K U(I - \zeta P_K U)^{-1} f)(\lambda).$$

Therefore,

$$F(v(\lambda)) = \frac{f(\lambda)}{v(\lambda) e_2(\lambda)}. \tag{3.15}$$

Now we are in a position to get (3.10). Indeed, by (3.14) and (3.15) we proved that the vector

$$P_K(I - \overline{v(\mu)} U^* P_K)^{-1} U^* e_2 \overline{v(\mu) e_2(\mu)}$$

is the reproducing kernel of $K = H^2_{\{s_+,\nu_+\}}$ with respect to μ, $|v(\mu)| < 1$. Using the Darboux identity

$$P_{E_2} U(I - \zeta P_K U)^{-1} P_K (I - \bar{\zeta}_0 U^* P_K)^{-1} U^* | E_2 = \frac{I - \Theta(z)\Theta^*(\zeta_0)}{1 - \zeta \bar{\zeta}_0}$$

(in this setting this is a simple and pleasant exercise) we obtain

$$k_{\{s_+,\nu_+\}}(\lambda, \mu) = v(\lambda) e_2(\lambda) \frac{I - \theta(v(\lambda))\overline{\theta(v(\mu))}}{1 - v(\lambda)\overline{v(\mu)}} \overline{v(\mu) e_2(\mu)}$$

for $|v(\lambda)| < 1$, $|v(\mu)| < 1$. By analyticity and (3.13) we have that relation (3.10) holds for all $\lambda, \mu \in \mathbb{C}_+$. □

Corollary 3.4. *The following Wronskian-type identity is satisfied for the reproducing kernels*

$$\begin{vmatrix} (s e_2^-)(\mu) & (s e_1^-)(\mu) \\ e_2(\mu) & e_1(\mu) \end{vmatrix} = \frac{1}{i} (\log v(\mu))', \ \text{Im } \mu > 0. \tag{3.16}$$

Proof. To be brief, we write $k^-_{\{s_+,\nu_+\}}(.,.)$ instead of $(k_{\{s_+,\nu_+\}}(.,.))^-$. So we multiply $k^-_{\{s_+,\nu_+\}}(\lambda, -\bar{\mu})$ by $b_\mu(\lambda)$ and calculate the resulting function of λ at $\lambda = \mu$. By (2.3) we get

$$\{b_\mu(\lambda) k^-_{\{s_+,\nu_+\}}(\lambda, -\bar{\mu})\}_{\lambda=\mu} = \frac{1}{s(\mu) 2\text{Im } \mu}. \tag{3.17}$$

Now we make the same calculation using representation (3.10). Since

$$k^-_{\{s_+,\nu_+\}}(\lambda, -\bar{\mu}) = \frac{-v(\mu)}{v(\lambda) - v(\mu)} \begin{vmatrix} v(\lambda) e_2^-(\lambda) & \frac{e_1^-(\lambda)}{v(-\bar{\mu}) e_2(-\bar{\mu})} \\ e_1(-\bar{\mu}) & \end{vmatrix},$$

we get in combination with (3.17)

$$-i \frac{v'(\mu)}{v(\mu) s(\mu)} = \begin{vmatrix} v(\mu) e_2^-(\mu) & \frac{e_1^-(\mu)}{v^{-1}(\mu) e_2(-\bar{\mu})} \\ e_1(-\bar{\mu}) & \end{vmatrix}.$$

By the symmetry $\overline{\hat{k}_{\{s_-,\nu_-\}}(\lambda, \lambda_0)} = \hat{k}_{\{s_-,\nu_-\}}(-\bar{\lambda}, -\bar{\lambda}_0)$, we have $\overline{e_2(-\bar{\mu})} = e_1(\mu)$. Thus (3.16) is proved. □

Corollary 3.5. *Let* $\mu \in \mathbb{R}_+$ *and as before* $\operatorname{Re}\lambda_0 > 0$, *then*

$$|e_2(\mu)|^2 - |e_1(\mu)|^2 = \frac{1}{i}(\log v(\mu))'. \tag{3.18}$$

Proof. All terms in (3.16) have boundary values. Recall that on the real axis $(se^{-}_{1,2})(\mu) = (s_- e_{1,2})(\mu) - e_{1,2}(-\bar{\mu})$. Then use again the symmetry of the reproducing kernel. □

We finish this section with a translation of the relation

$$\|f\|^2_{\{s_+,\nu_+\}} = \left\|\frac{f}{ve_2}\right\|^2_{K_\theta}$$

((3.15) is a unitary map from $H^2_{\{s_+,\nu_+\}}$ to K_θ) to the following proposition.

Theorem 3.6. *Let*

$$s_+^\theta(\lambda) := \frac{e_2(-\bar{\lambda})}{e_2(\lambda)}, \quad \lambda \in \mathbb{R}_+, \tag{3.19}$$

extended by the symmetry $s_+^\theta(-\bar{\lambda}) = \overline{s_+^\theta(\lambda)}$ *to the whole* $\mathbb{R}$. *Let* ν_+^θ *be a positive measure on the imaginary half-axis*

$$d\nu_+^\theta(\lambda) := \frac{|dv(\lambda)|}{2\pi|e_2(\lambda)|^2}, \quad \lambda \in i\mathbb{R}_+. \tag{3.20}$$

Then

$$||f||^2_{\{s_+,\nu_+\}} \tag{3.21}$$
$$= \int_{i\mathbb{R}_+} |f(\lambda)|^2 d\nu_+^\theta(\lambda) + \frac{1}{4\pi}\int_{\mathbb{R}} \begin{bmatrix}\overline{f(\lambda)} & \overline{-f(-\bar{\lambda})}\end{bmatrix} \begin{bmatrix} 1 & \overline{s_+^\theta(\lambda)} \\ s_+^\theta(\lambda) & 1\end{bmatrix} \begin{bmatrix} f(\lambda) \\ -f(-\bar{\lambda})\end{bmatrix} d\lambda$$

for all $f \in H^2_{\{s_+,\nu_+\}}$. *In other words*

$$id : H^2_{\{s_+,\nu_+\}} \to H^2_{\{s_+^\theta,\nu_+^\theta\}}$$

is an isometry.

Proof. We use definition of the scalar product in K_θ, relations (3.13), (3.15), and (3.18). □

4. Unitary node, II: a canonical system

In this section we associate a canonical system (see [5, 18]) with the given chain $\{H^2_{\{s_+,\nu_+\}}(x)\}_{x\in\mathbb{R}}$ of subspaces of $L^2_{\{s_+,\nu_+\}}$.

4.1. Characteristic function of a unitary node and transfer matrix. Definitions

This time we consider the unitary multiplication operator by $\bar{v}$,

$$v = \frac{\lambda^2 - \lambda_0^2}{\lambda^2 - \bar{\lambda}_0^2},$$

with respect to the decomposition

$$L^2_{\{s_+,\nu_+\}} = (\hat{H}^2_{\{s_-,\nu_-\}})^+ \oplus K_{\{s_+,\nu_+\}}(x) \oplus H^2_{\{s_+,\nu_+\}}(x). \tag{4.1}$$

Actually this is definition of the space $K_{\{s_+,\nu_+\}}(x)$.

The following lemma is similar to Lemma 3.1.

Lemma 4.1. *The multiplication operator by $\bar{v}$ acts from*

$$\{\hat{k}^+_{\{s_-,\nu_-\}}(\lambda,\lambda_0)\} \oplus K_{\{s_+,\nu_+\}}(x) \oplus \{k_{\{s_+,\nu_+\}}(\lambda,\lambda_0;x)\} \tag{4.2}$$

to

$$\{\hat{k}^+_{\{s_-,\nu_-\}}(\lambda,-\bar{\lambda}_0)\} \oplus K_{\{s_+,\nu_+\}}(x) \oplus \{k_{\{s_+,\nu_+\}}(\lambda,-\bar{\lambda}_0;x)\}. \tag{4.3}$$

We define normalized vectors that form orthonormal bases in E_1 and E_2

$$e_1^{(1)}(\lambda) = \frac{\hat{k}^+_{\{s_-,\nu_-\}}(\lambda,\lambda_0)}{||\hat{k}^+_{\{s_-,\nu_-\}}(\lambda,\lambda_0)||}, \quad e_2^{(1)}(\lambda) = \frac{k_{\{s_+,\nu_+\}}(\lambda,\lambda_0;x)}{||k_{\{s_+,\nu_+\}}(\lambda,\lambda_0;x)||}; \tag{4.4}$$

and

$$e_1^{(2)}(\lambda) = \frac{\hat{k}^+_{\{s_-,\nu_-\}}(\lambda,-\bar{\lambda}_0)}{||\hat{k}^+_{\{s_-,\nu_-\}}(\lambda,-\bar{\lambda}_0)||}, \quad e_2^{(2)}(\lambda) = \frac{k_{\{s_+,\nu_+\}}(\lambda,-\bar{\lambda}_0;x)}{||k_{\{s_+,\nu_+\}}(\lambda,-\bar{\lambda}_0;x)||}. \tag{4.5}$$

We point out that the vectors $e_2^{(i)}(\lambda)$, $i = 1,2$, depend on x and $e_1^{(i)}(\lambda)$, $i = 1,2$, do not.

Generally for an operator $A : H_1 \oplus H_2 \to \tilde{H}_1 \oplus \tilde{H}_2$ its Potapov-Ginzburg transform $\tilde{A} : H_1 \oplus \tilde{H}_2 \to \tilde{H}_1 \oplus H_2$ is defined by [15, 10]

$$\begin{bmatrix} y_1 \\ x_2 \end{bmatrix} = \tilde{A} \begin{bmatrix} x_1 \\ y_2 \end{bmatrix}, \quad \text{where} \quad \begin{bmatrix} y_1 \\ y_2 \end{bmatrix} = A \begin{bmatrix} x_1 \\ x_2 \end{bmatrix}.$$

In terms of the block decomposition of $A = \begin{bmatrix} A_{11} & A_{12} \\ A_{21} & A_{22} \end{bmatrix}$ we have

$$\begin{bmatrix} A_{11} & 0 \\ A_{21} & -I \end{bmatrix} \begin{bmatrix} x_1 \\ y_2 \end{bmatrix} = \begin{bmatrix} I & -A_{12} \\ 0 & -A_{22} \end{bmatrix} \begin{bmatrix} y_1 \\ x_2 \end{bmatrix}.$$

Therefore,

$$\tilde{A} = \begin{bmatrix} I & -A_{12} \\ 0 & -A_{22} \end{bmatrix}^{-1} \begin{bmatrix} A_{11} & 0 \\ A_{21} & -I \end{bmatrix} = \begin{bmatrix} A_{11} - A_{12}A_{22}^{-1}A_{21} & A_{12}A_{22}^{-1} \\ -A_{22}^{-1}A_{21} & A_{22}^{-1} \end{bmatrix}. \tag{4.6}$$

The transformation is well defined if A_{22} is invertible. Note, that if A is unitary,

$$\|y_1\|^2 + \|y_2\|^2 = \|x_1\|^2 + \|x_2\|^2,$$

then $\tilde{A}$ preserves the indefinite metric

$$\|y_1\|^2 - \|x_2\|^2 = \|x_1\|^2 - \|y_2\|^2.$$

For the unitary node given by the multiplication operator by $\bar{v}$ and decompositions (4.2), (4.3):

$$U : (K \oplus \{e_1^{(1)}\}) \oplus \{e_2^{(1)}\} \to (K \oplus \{e_2^{(2)}\}) \oplus \{e_1^{(2)}\} \tag{4.7}$$

we define the j-unitary node

$$\tilde{U} : (K \oplus \{e_1^{(1)}\}) \oplus \{e_1^{(2)}\} \to (K \oplus \{e_2^{(2)}\}) \oplus \{e_2^{(1)}\} \tag{4.8}$$

by (4.6), separating in this way the x-depending "channels".

The characteristic operator-valued function for the node (4.7) is

$$\Theta(\zeta) := P_{E_2} U (I - \zeta P_K U)^{-1} | E_1, \tag{4.9}$$

and its matrix with respect to the chosen bases is

$$\Theta(\zeta) \begin{bmatrix} e_1^{(1)}(\lambda) & e_2^{(1)}(\lambda) \end{bmatrix} \begin{bmatrix} c_1 \\ c_2 \end{bmatrix} = \begin{bmatrix} e_2^{(2)}(\lambda) & e_1^{(2)}(\lambda) \end{bmatrix} \theta(\zeta) \begin{bmatrix} c_1 \\ c_2 \end{bmatrix}, \tag{4.10}$$

where

$$\theta(\zeta) = \begin{bmatrix} \theta_{11} & \theta_{12} \\ \theta_{21} & \theta_{22} \end{bmatrix} (\zeta).$$

Respectively, its functional representation is of the form

$$P_{E_2} U (I - \zeta P_K U)^{-1} f = \begin{bmatrix} e_2^{(2)}(\lambda) & e_1^{(2)}(\lambda) \end{bmatrix} \begin{bmatrix} F_1 \\ F_2 \end{bmatrix} (\zeta), \tag{4.11}$$

for $f \in K_{\{s_+,\nu_+\}}(x)$.

The transfer matrix is actually the characteristic matrix function of the node (4.8). Having (4.9), (4.10), we rewrite (4.7) in the block form as

$$U \begin{bmatrix} \zeta k(\zeta) \\ \begin{bmatrix} c_1 \\ c_2 \end{bmatrix} \end{bmatrix} = \begin{bmatrix} k(\zeta) \\ \begin{bmatrix} \theta_{11} & \theta_{12} \\ \theta_{21} & \theta_{22} \end{bmatrix} (\zeta) \begin{bmatrix} c_1 \\ c_2 \end{bmatrix} \end{bmatrix}. \tag{4.12}$$

Consequently, we get for $\tilde{U}$:

$$\tilde{U} \begin{bmatrix} \zeta k(\zeta) \\ \begin{bmatrix} 1 & 0 \\ \theta_{21}(\zeta) & \theta_{22}(\zeta) \end{bmatrix} \begin{bmatrix} c_1 \\ c_2 \end{bmatrix} \end{bmatrix} = \begin{bmatrix} k(\zeta) \\ \begin{bmatrix} \theta_{11}(\zeta) & \theta_{12}(\zeta) \\ 0 & 1 \end{bmatrix} \begin{bmatrix} c_1 \\ c_2 \end{bmatrix} \end{bmatrix}. \tag{4.13}$$

Therefore the transfer matrix $\mathfrak{A}(\zeta)$ of the j-node is related to $\theta(\zeta)$ by

$$\mathfrak{A}(\zeta) \begin{bmatrix} 1 & 0 \\ \theta_{21}(\zeta) & \theta_{22}(\zeta) \end{bmatrix} = \begin{bmatrix} \theta_{11}(\zeta) & \theta_{12}(\zeta) \\ 0 & 1 \end{bmatrix}.$$

Thus

$$\mathfrak{A}(\zeta) = \begin{bmatrix} \theta_{11}(\zeta) & \theta_{12}(\zeta) \\ 0 & 1 \end{bmatrix} \begin{bmatrix} 1 & 0 \\ \theta_{21}(\zeta) & \theta_{22}(\zeta) \end{bmatrix}^{-1} = \begin{bmatrix} 1 & -\theta_{12}(\zeta) \\ 0 & -\theta_{22}(\zeta) \end{bmatrix}^{-1} \begin{bmatrix} \theta_{11}(\zeta) & 0 \\ \theta_{21}(\zeta) & -1 \end{bmatrix}. \tag{4.14}$$

4.2. Calculating θ and $\mathfrak{A}$

We are following the same lines as in Section 3. Let us substitute (4.10) into (3.8)

$$v(\lambda)\left[e_2^{(2)}(\lambda)\quad e_1^{(2)}(\lambda)\right]\theta(\zeta)=\left[e_1^{(1)}(\lambda)\quad e_2^{(1)}(\lambda)\right] \\ +(\zeta-v(\lambda))\left(P_K U(I-\zeta P_K U)^{-1}\left[e_1^{(1)}\quad e_2^{(1)}\right]\right)(\lambda).$$

All terms here are analytic in λ and we can choose $\lambda\in\mathbb{C}_+$ with the property $v(\lambda)=\zeta$. Then we get

$$v(\lambda)\left[e_2^{(2)}(\lambda)\quad e_1^{(2)}(\lambda)\right]\theta(v(\lambda))=\left[e_1^{(1)}(\lambda)\quad e_2^{(1)}(\lambda)\right]. \tag{4.15}$$

Similarly, by (4.11)

$$v(\lambda)\left[e_2^{(2)}(\lambda)\quad e_1^{(2)}(\lambda)\right]\begin{bmatrix}F_1\\F_2\end{bmatrix}(v(\lambda))=f(\lambda). \tag{4.16}$$

It is tempting to make the change of variable $\lambda\to-\bar\lambda$, $\lambda\in\mathbb{R}$, in (4.15), (4.16) and to write

$$v(\lambda)\left[(e_2^{(2)})^-(\lambda)\quad (e_1^{(2)})^-(\lambda)\right]\theta(v(\lambda))=\left[(e_1^{(1)})^-(\lambda)\quad (e_2^{(1)})^-(\lambda)\right], \tag{4.17}$$

and

$$v(\lambda)\left[(e_2^{(2)})^-(\lambda)\quad (e_1^{(2)})^-(\lambda)\right]\begin{bmatrix}F_1\\F_2\end{bmatrix}(v(\lambda))=f^-(\lambda). \tag{4.18}$$

However, to succeed with this plan, we need to prove that $\theta(\zeta)$ has an analytic continuation in $\mathbb{C}\setminus\mathbb{D}$ etc. That is why we prefer to consider a dual node given by the diagram

$$\begin{array}{ccc} K\oplus E_1 & \xrightarrow{\bar v} & K\oplus E_2\\ \big\downarrow - & & \big\downarrow -\\ K^-\oplus E_1^- & \xrightarrow{\bar v} & K^-\oplus E_2^- \end{array} \tag{4.19}$$

The characteristic matrix-valued function remains the same since we choose basis in $E_{1,2}^-$ as the image of the basis in $E_{1,2}$. Then we obtain (4.17) and (4.18) simply repeating the arguments from (4.15) and (4.16). Hence

$$v(\lambda)\begin{bmatrix}(e_2^{(2)})^- & (e_1^{(2)})^-\\ e_2^{(2)} & e_1^{(2)}\end{bmatrix}(\lambda)\begin{bmatrix}\theta_{11} & \theta_{12}\\ \theta_{21} & \theta_{22}\end{bmatrix}(v(\lambda))=\begin{bmatrix}(e_1^{(1)})^- & (e_2^{(1)})^-\\ e_1^{(1)} & e_2^{(1)}\end{bmatrix}(\lambda), \tag{4.20}$$

and

$$v(\lambda)\begin{bmatrix}(e_2^{(2)})^- & (e_1^{(2)})^-\\ e_2^{(2)} & e_1^{(2)}\end{bmatrix}(\lambda)\begin{bmatrix}F_1\\F_2\end{bmatrix}(v(\lambda))=\begin{bmatrix}f^-\\f\end{bmatrix}(\lambda). \tag{4.21}$$

Lemma 4.2. *Both* $\det\begin{bmatrix}(e_2^{(2)})^- & (e_1^{(2)})^- \\ e_2^{(2)} & e_1^{(2)}\end{bmatrix}(\lambda)$ *and* $\theta_{22}(\lambda)$ *do not vanish identically. Furthermore,*

$$\theta_{22}(\lambda)\det\begin{bmatrix}(e_2^{(2)})^- & (e_1^{(2)})^- \\ e_2^{(2)} & e_1^{(2)}\end{bmatrix}(\lambda) = -\frac{i}{s(\lambda)}\left(\frac{1}{v(\lambda)}\right)'. \tag{4.22}$$

In particular, the characteristic matrix-valued function $\theta(\zeta)$ *and the map* $K_{\{s_+,\nu_+\}}(x) \to K_\theta$ *are well defined by (4.20), (4.21) in terms of the reproducing kernels.*

Proof. This follows from an obvious consequence of (4.20)

$$v(\lambda)\begin{bmatrix}(e_2^{(2)})^- & (e_1^{(2)})^- \\ e_2^{(2)} & e_1^{(2)}\end{bmatrix}(\lambda)\begin{bmatrix}1 & -\theta_{12} \\ 0 & -\theta_{22}\end{bmatrix}(v(\lambda)) = \begin{bmatrix}v(e_2^{(2)})^- & -(e_2^{(1)})^- \\ ve_2^{(2)} & -e_2^{(1)}\end{bmatrix}(\lambda), \tag{4.23}$$

and (3.16) that says

$$\det\begin{bmatrix}v(e_2^{(2)})^- & -(e_2^{(1)})^- \\ ve_2^{(2)} & -e_2^{(1)}\end{bmatrix}(\lambda) = i\frac{v'(\lambda)}{s(\lambda)}.$$

□

By (4.14), we have for the transfer matrix

$$\mathfrak{A}_x(\lambda^2) = \begin{bmatrix}1 & -\theta_{12} \\ 0 & -\theta_{22}\end{bmatrix}^{-1}\begin{bmatrix}\theta_{11} & 0 \\ \theta_{21} & -1\end{bmatrix}(v(\lambda)). \tag{4.24}$$

The map from K_θ to the corresponding de Branges space $\mathcal{H}(\mathfrak{A})$ [5, Sect. 28] is of the form

$$\begin{bmatrix}A_1 \\ A_2\end{bmatrix}(\lambda^2) = \begin{bmatrix}1 & -\theta_{12} \\ 0 & -\theta_{22}\end{bmatrix}^{-1}\begin{bmatrix}F_1 \\ F_2\end{bmatrix}(v(\lambda)). \tag{4.25}$$

Combining (4.20), (4.24) with

$$v(\lambda)\begin{bmatrix}(e_2^{(2)})^- & (e_1^{(2)})^- \\ e_2^{(2)} & e_1^{(2)}\end{bmatrix}(\lambda)\begin{bmatrix}\theta_{11} & 0 \\ \theta_{21} & -1\end{bmatrix}(v(\lambda)) = \begin{bmatrix}(e_1^{(1)})^- & -v(e_1^{(2)})^- \\ e_1^{(1)} & -ve_1^{(2)}\end{bmatrix}(\lambda), \tag{4.26}$$

we have

$$\begin{bmatrix}v(e_2^{(2)})^- & -(e_2^{(1)})^- \\ ve_2^{(2)} & -e_2^{(1)}\end{bmatrix}(\lambda)\mathfrak{A}_x(\lambda^2) = \begin{bmatrix}(e_1^{(1)})^- & -v(e_1^{(2)})^- \\ e_1^{(1)} & -ve_1^{(2)}\end{bmatrix}(\lambda), \tag{4.27}$$

and

$$\begin{bmatrix}v(e_2^{(2)})^- & -(e_2^{(1)})^- \\ ve_2^{(2)} & -e_2^{(1)}\end{bmatrix}(\lambda)\begin{bmatrix}A_1 \\ A_2\end{bmatrix}(\lambda^2) = \begin{bmatrix}f^- \\ f\end{bmatrix}(\lambda). \tag{4.28}$$

Note that the condition $I - \theta\theta^* \geq 0$ is the same as

$$\begin{bmatrix}1 & -\theta_{12} \\ 0 & -\theta_{22}\end{bmatrix}\begin{bmatrix}1 & 0 \\ 0 & -1\end{bmatrix}\begin{bmatrix}1 & -\theta_{12} \\ 0 & -\theta_{22}\end{bmatrix}^* - \begin{bmatrix}\theta_{11} & 0 \\ \theta_{21} & -1\end{bmatrix}\begin{bmatrix}1 & 0 \\ 0 & -1\end{bmatrix}\begin{bmatrix}\theta_{11} & 0 \\ \theta_{21} & -1\end{bmatrix}^* \geq 0.$$

That is, the transformation (4.24) maps contractive matrices onto j-contractions,

$$j - \mathfrak{A} j \mathfrak{A}^* \geq 0, \quad j := \begin{bmatrix} 1 & 0 \\ 0 & -1 \end{bmatrix},$$

although this was clear from the definition of j-node (4.8).

4.3. de Branges' Theorem

Theorem 4.3. *For every $x > 0$,*

$$\mathfrak{A}_x(\lambda^2) = \frac{-is(\lambda)}{v'(\lambda)} \begin{bmatrix} -e_2^{(1)} & (e_2^{(1)})^- \\ -ve_2^{(2)} & v(e_2^{(2)})^- \end{bmatrix} (\lambda) \begin{bmatrix} (e_1^{(1)})^- & -v(e_1^{(2)})^- \\ e_1^{(1)} & -ve_1^{(2)} \end{bmatrix} (\lambda) \tag{4.29}$$

is an entire matrix-valued function of λ^2 and

$$\mathcal{H}(\mathfrak{A}_x) = \left\{ \begin{bmatrix} A_1 \\ A_2 \end{bmatrix} (\lambda^2) = \frac{-is(\lambda)}{v'(\lambda)} \begin{bmatrix} -e_2^{(1)} & (e_2^{(1)})^- \\ -ve_2^{(2)} & v(e_2^{(2)})^- \end{bmatrix} (\lambda) \begin{bmatrix} f^- \\ f \end{bmatrix} (\lambda),\ f \in K_{\{s_+,\nu_+\}}(x) \right\}, \tag{4.30}$$

is the de Branges space of entire functions [5, Sect. 28].

Proof. The entries of the matrix-valued function in RHS of (4.29) are holomorphic in the z-plane, $z = \lambda^2$, cut along the positive half-axis. Moreover, the limits $\mathfrak{A}_x(\xi + i0)$ and $\mathfrak{A}_x(\xi - i0)$ coincide for a.e. $\xi \in \mathbb{R}_+$. We have to prove analyticity of the matrix-valued function on the cut.

Fix $\xi_0 \in \mathbb{R}_+$ and a disk $D_{\xi_0} = D(\xi_0, r)$ of a sufficiently small radius r centered at the point ξ_0, such that $|s(\sqrt{z})|$ is bounded from above and below on its boundary. This is possible because s has non-vanishing boundary values almost everywhere. We would like to apply Morera's Theorem [9, p. 95]. However, the glued functions, generally speaking, do not belong to H^1 in both half-disks $\{z \in D_{\xi_0} : \text{Im } z > 0\}$ and $\{z \in D_{\xi_0} : \text{Im } z < 0\}$. For this reason, first, we have to use the normalization procedure given in Lemma 2.6. Recall that for $\|s_+\| < 1$ every function from $\hat{H}^2$ belongs to the standard H^2 even without the factor s. Hence, for each ϵ we get that the matrix-valued function $\mathfrak{A}_x(z;\epsilon)$ is holomorphic on D_{ξ_0}. On the boundary of this disk $\mathfrak{A}_x(z;\epsilon)$ goes to $\mathfrak{A}_x(z)$ in L^1 with the weight $|s(\sqrt{z})||(z-\xi_0)^2 - r^2|$, that is, actually, in L^1 after multiplication by $(z-\xi_0)^2 - r^2$. Therefore, $\mathfrak{A}_x(z)$ is also holomorphic on the disk. □

We point out that for all x the x-depending matrix in the RHS of (4.29) meets the following normalization condition

$$(ve_2^{(2)})(\lambda_0) = 0, \quad e_2^{(1)}(\lambda_0) > 0. \tag{4.31}$$

As the result we get a family of 2×2 j-contractive matrix-valued functions with a certain normalization at λ_0. The family is monotonic in x, and every matrix is an entire function in λ^2 of the zero mean type (concerning the corollary of the last condition see [5], Theorem 39). According to de Branges' Theorem [5, Sect. 36, 37], Theorem 37, such a family can be included in the chain

$$j\frac{d}{dt}\mathfrak{A}(\lambda^2,t)=\left\{i\lambda^2\begin{bmatrix}\alpha & \beta\\ \bar{\beta} & \alpha\end{bmatrix}(t)+\begin{bmatrix}0 & -\beta\\ \bar{\beta} & 0\end{bmatrix}(t)\right\}\mathfrak{A}(\lambda^2,t),\ |\beta|=\alpha, \tag{4.32}$$

such that $\mathfrak{A}_x(\lambda^2)=\mathfrak{A}(\lambda^2,t_x)$, where $x=x(t)$ is a monotonic function. Here we choose $\lambda_0^2=i$ as the normalization point.

4.4. Parameters of the system in terms of reproducing kernels

Theorem 4.4. *For the system* (4.32)

$$\frac{\beta}{\alpha}=-\frac{s(\lambda_0)}{s(-\bar{\lambda}_0)}\frac{d\,\hat{k}(\lambda_0,-\bar{\lambda}_0)}{d\,\hat{k}(-\bar{\lambda}_0,-\bar{\lambda}_0)}. \tag{4.33}$$

Proof. Set

$$\mathcal{E}=\begin{bmatrix}-e_2^{(1)} & (e_2^{(1)})^-\\ -ve_2^{(2)} & v(e_2^{(2)})^-\end{bmatrix}=\frac{1}{\sqrt{k(\lambda_0,\lambda_0)}}\begin{bmatrix}-k_{\lambda_0} & k_{\bar{\lambda}_0}^-\\ -vk_{-\bar{\lambda}_0} & vk_{-\bar{\lambda}_0}^-\end{bmatrix}. \tag{4.34}$$

The subscripts of the kernels are obvious (see (4.4), (4.5)) and are omitted. We note that due to (4.29) $j\dot{\mathfrak{A}}\mathfrak{A}^{-1}=j\dot{\mathcal{E}}\mathcal{E}^{-1}$. In particular, for $\lambda_0^2=i$

$$j\dot{\mathcal{E}}(\lambda_0)\mathcal{E}^{-1}(\lambda_0)=-\begin{bmatrix}\alpha & 2\beta\\ 0 & \alpha\end{bmatrix}. \tag{4.35}$$

On the other hand, using (2.3), (2.4), we have

$$\mathcal{E}(\lambda_0)=\begin{bmatrix}\tau & a\tau\\ 0 & \tau^{-1}\end{bmatrix}\begin{bmatrix}-1 & 0\\ 0 & C\end{bmatrix} \tag{4.36}$$

with a constant C and

$$\tau:=\sqrt{k(\lambda_0,\lambda_0)},\quad a:=s^2(\lambda_0)(2\mathrm{Im}\ \lambda_0)^2\hat{k}(\lambda_0,-\bar{\lambda}_0). \tag{4.37}$$

In these notations,

$$j\dot{\mathcal{E}}(\lambda_0)\mathcal{E}^{-1}(\lambda_0)=\begin{bmatrix}\frac{\dot{\tau}}{\tau} & \dot{a}\tau^2\\ 0 & \frac{\dot{\tau}}{\tau}\end{bmatrix}.$$

Comparing this with (4.35) we get $\frac{\beta}{\alpha}=-\frac{da}{d(\tau^{-2})}$. Since

$$\tau^{-2}=s(\lambda_0)s(-\bar{\lambda}_0)(2\mathrm{Im}\ \lambda_0)^2\hat{k}(-\bar{\lambda}_0,-\bar{\lambda}_0),$$

we come to (4.33) using (4.37) □

5. de Branges system and Sturm-Liouville equation

In this section, we rewrite the results of the previous sections for a particular case of the Sturm-Liouville equation. Let

$$Ly = -y'' + qy \tag{5.1}$$

be a self-adjoint operator acting on $L^2(\mathbb{R})$, and let

$$u = \frac{L - \bar{\lambda}_0^2}{L - \lambda_0^2}, \quad \text{Re } \lambda_0 > 0, \text{Im } \lambda_0 > 0, \tag{5.2}$$

be its Cayley transform.

Lemma 5.1. *Let* $e^{\pm}(x,\lambda) \in L^2(\mathbb{R}_{\pm})$, $\|e^{\pm}(x,\lambda)\| = 1$, *be such that*

$$-\frac{d^2}{dx^2}e^{\pm}(x,\lambda)+q(x)e^{\pm}(x,\lambda) = \lambda^2 e^{\pm}(x,\lambda), \quad x \in \mathbb{R}_{\pm}, \text{ Im } \lambda > 0, \text{Re } \lambda \neq 0. \tag{5.3}$$

Then for all $f_+ \in L^2(\mathbb{R}_+)$

$$uf_+ = Ce^-(x,\lambda_0)\langle f_+, e^+(x, -\bar{\lambda}_0)\rangle + g_+, \tag{5.4}$$

where $C = C(u)$ *and* $g_+ \in L^2(\mathbb{R}_+)$.

Proof. Let

$$(uf_+)(x) = \begin{cases} g_-(x), & x \in \mathbb{R}_- \\ g_+(x), & x \in \mathbb{R}_+ \end{cases},$$

or, what is the same,

$$\left(\frac{\lambda_0^2 - \bar{\lambda}_0^2}{L - \lambda_0^2} f_+\right)(x) = \begin{cases} g_-(x), & x \in \mathbb{R}_- \\ g_+(x) - f_+(x), & x \in \mathbb{R}_+ \end{cases}.$$

Therefore,

$$-g_-'' + qg_- = \lambda_0^2 g_-, \quad x \in \mathbb{R}_-,$$

and $g_- \in L^2(\mathbb{R}_-)$. That is,

$$g_-(x) = C(f_+)e^-(x,\lambda_0).$$

Thus we get

$$uf_+ = C(f_+)e^-(x,\lambda_0) + g_+.$$

For $C(f_+)$ we have

$$C(f_+) = \langle uf_+, e^-(x,\lambda_0)\rangle = \langle f_+, u^*e^-(x,\lambda_0)\rangle.$$

Now we are looking at

$$u^*e^-(x,\lambda_0) = \left(I + \frac{\bar{\lambda}_0^2 - \lambda_0^2}{L - \bar{\lambda}_0^2}\right) e^-(x,\lambda_0) = \begin{cases} h_-(x), & x \in \mathbb{R}_- \\ h_+(x), & x \in \mathbb{R}_+ \end{cases},$$

or

$$(\bar{\lambda}_0^2 - \lambda_0^2)e^-(x,\lambda_0) = \begin{cases} -\tilde{h}_-'' + q\tilde{h}_- - \bar{\lambda}_0^2\tilde{h}_-, & x \in \mathbb{R}_- \\ -\tilde{h}_+'' + q\tilde{h}_+ - \bar{\lambda}_0^2\tilde{h}_+, & x \in \mathbb{R}_+ \end{cases},$$

where $\tilde{h} = h - e^-(x, \lambda_0)$. This implies

$$\tilde{h} = \begin{cases} -e^-(x,\lambda_0) + C_1 e^-(x, -\bar{\lambda}_0), & x \in \mathbb{R}_- \\ C_2 e^+(x, -\bar{\lambda}_0), & x \in \mathbb{R}_+ \end{cases},$$

where C_1, C_2 are defined by the conditions

$$\tilde{h}(-0) = \tilde{h}(+0), \quad \tilde{h}'(-0) = \tilde{h}'(+0).$$

Notice that the equality $C_2 = 0$ contradicts the linear independence of $e^-(x, \lambda_0)$ and $e^-(x, -\bar{\lambda}_0)$.

Hence,

$$u^* e^-(x, \lambda_0) = \tilde{h} + e^-(x, \lambda_0) = \begin{cases} C_1 e^-(x, -\bar{\lambda}_0), & x \in \mathbb{R}_- \\ C_2 e^+(x, -\bar{\lambda}_0), & x \in \mathbb{R}_+ \end{cases}, \tag{5.5}$$

and (5.4) is proved with

$$C(f_+) = \langle f_+, C_2 e^+(x, -\bar{\lambda}_0)\rangle. \qquad \square$$

Corollary 5.2. *The operator u acts from $L^2(\mathbb{R}_+) \oplus \{e^-(x, -\bar{\lambda}_0)\}$ to $L^2(\mathbb{R}_+) \oplus \{e^-(x, \lambda_0)\}$.*

Proof. Similarly to (5.4),

$$uf_- = Ce^+(x, \lambda_0)\langle f_-, e^-(x, -\bar{\lambda}_0)\rangle + g_-. \tag{5.6}$$

That is, $uf_- = g_- \in L^2(\mathbb{R}_-)$ if $f_- \perp e^-(x, -\bar{\lambda}_0)$. Moreover, (uf_-) is orthogonal to $e^-(x, \lambda_0)$ by (5.5) in this case. $\square$

Corollary 5.3. *For every $x_0 > 0$, the operator u acts from*

$$(L^2[0, x_0] \oplus \{e^-(x, -\bar{\lambda}_0)\}) \oplus \{e^+_{x_0}(x, -\bar{\lambda}_0)\} \tag{5.7}$$

to

$$(L^2[0, x_0] \oplus \{e^+_{x_0}(x, \lambda_0)\}) \oplus \{e^-(x, \lambda_0)\} \tag{5.8}$$

where $e^+_{x_0}(x, \lambda) \in L^2[x_0, \infty)$ is the normalized solution of (5.3).

Below, the dot means the derivative with respect to x.

Theorem 5.4. *The transfer matrix of the unitary node* (5.7), (5.8) *is of the form*

$$\mathfrak{A}_{x_0}(\lambda^2) = \begin{bmatrix} e^+_{x_0}(x_0, \lambda_0) & -e^+_{x_0}(x_0, -\bar{\lambda}_0) \\ \dot{e}^+_{x_0}(x_0, \lambda_0) & -\dot{e}^+_{x_0}(x_0, -\bar{\lambda}_0) \end{bmatrix}^{-1} \mathfrak{B}_{x_0}(\lambda^2) \begin{bmatrix} -e^-(0, -\bar{\lambda}_0) & e^-(0, \lambda_0) \\ -\dot{e}^-(0, -\bar{\lambda}_0) & \dot{e}^-(0, \lambda_0) \end{bmatrix}, \tag{5.9}$$

where

$$\mathfrak{B}_x(\lambda^2) = \begin{bmatrix} c(x, \lambda) & s(x, \lambda) \\ \dot{c}(x, \lambda) & \dot{s}(x, \lambda) \end{bmatrix} \tag{5.10}$$

is the standard transfer matrix for equation (5.1)

$$\frac{d}{dx}\mathfrak{B}_x(\lambda^2) = \begin{bmatrix} 0 & 1 \\ q(x) - \lambda^2 & 0 \end{bmatrix} \mathfrak{B}_x(\lambda^2), \quad \mathfrak{B}_0(\lambda^2) = I. \tag{5.11}$$

Proof. In the block form we have

$$u\begin{bmatrix}\zeta k_\zeta\\ \begin{bmatrix}c_1\\ d_2\end{bmatrix}\end{bmatrix}=\begin{bmatrix}k_\zeta\\ \begin{bmatrix}d_1\\ c_2\end{bmatrix}\end{bmatrix}, \tag{5.12}$$

with

$$\begin{bmatrix}d_1\\ d_2\end{bmatrix}=\mathfrak{A}(\zeta)\begin{bmatrix}c_1\\ c_2\end{bmatrix}. \tag{5.13}$$

In other words,

$$\begin{aligned}\left(I+\frac{\lambda_0^2-\bar\lambda_0^2}{L-\lambda_0^2}\right)&\{\zeta k_\zeta+c_1e^-(x,-\bar\lambda_0)+d_2e^+_{x_0}(x,-\bar\lambda_0)\}\\ &=k_\zeta+d_1e^+_{x_0}(x,\lambda_0)+c_2e^-(x,\lambda_0),\end{aligned}$$

or

$$\begin{aligned}&\frac{\lambda_0^2-\bar\lambda_0^2}{L-\lambda_0^2}\{\zeta k_\zeta+c_1e^-(x,-\bar\lambda_0)+d_2e^+_{x_0}(x,-\bar\lambda_0)\}\\ &\quad=(1-\zeta)k_\zeta+d_1e^+_{x_0}(x,\lambda_0)+c_2e^-(x,\lambda_0)-c_1e^-(x,-\bar\lambda_0)-d_2e^+_{x_0}(x,-\bar\lambda_0).\end{aligned} \tag{5.14}$$

This means that the RHS of (5.14) has the second derivative and we have on the interval $[0,x_0]$

$$-k_\zeta''+qk_\zeta=\lambda^2k_\zeta. \tag{5.15}$$

for the spectral parameter

$$\lambda^2=\lambda_0^2+\frac{\zeta}{1-\zeta}(\lambda_0^2-\bar\lambda_0^2).$$

Above, $\zeta=\dfrac{\lambda^2-\lambda_0^2}{\lambda^2-\bar\lambda_0^2}$.

Let

$$(1-\zeta)k_\zeta=Ac(x,\lambda)+Bs(x,\lambda). \tag{5.16}$$

Then the continuity at $x=0$ implies

$$\begin{bmatrix}A\\ B\end{bmatrix}=\begin{bmatrix}-e^-(0,-\bar\lambda_0) & e^-(0,\lambda_0)\\ -\dot e^-(0,-\bar\lambda_0) & \dot e^-(0,\lambda_0)\end{bmatrix}\begin{bmatrix}c_1\\ c_2\end{bmatrix},$$

and by the continuity at $x=x_0$,

$$\begin{bmatrix}e^+_{x_0}(x_0,\lambda_0) & -e^+_{x_0}(x_0,-\bar\lambda_0)\\ \dot e^+_{x_0}(x_0,\lambda_0) & -\dot e^+_{x_0}(x_0,-\bar\lambda_0)\end{bmatrix}\begin{bmatrix}d_1\\ d_2\end{bmatrix}=\begin{bmatrix}c(x_0,\lambda) & s(x_0,\lambda)\\ \dot c(x_0,\lambda) & \dot s(x_0,\lambda)\end{bmatrix}\begin{bmatrix}A\\ B\end{bmatrix}.$$

The theorem is proved. □

We now compute the parameters of the related canonical system under the chosen normalization.

We start observing that, up to the initial matrix $\mathfrak{A}_0$, the transfer matrix has the same normalization as the transfer matrix (4.24) (or (4.29)) in Section 4.

Corollary 5.5. *The transfer matrix of unitary node* (5.7), (5.8) *is of the form*

$$\mathfrak{A}_x(\lambda^2) = \tilde{\mathfrak{A}}_x(\lambda^2)\mathfrak{A}_0,$$

where

$$\tilde{\mathfrak{A}}_{x_0}(\lambda^2) = \begin{bmatrix} e_{x_0}^+(x_0,\lambda_0) & -e_{x_0}^+(x_0,-\bar{\lambda}_0) \\ \dot{e}_{x_0}^+(x_0,\lambda_0) & -\dot{e}_{x_0}^+(x_0,-\bar{\lambda}_0) \end{bmatrix}^{-1} \mathfrak{B}_{x_0}(\lambda^2) \begin{bmatrix} e_0^+(0,\lambda_0) & -e_0^+(0,-\bar{\lambda}_0) \\ \dot{e}_0^+(0,\lambda_0) & -\dot{e}_0^+(0,-\bar{\lambda}_0) \end{bmatrix}. \tag{5.17}$$

Therefore, $\tilde{\mathfrak{A}}_x(\lambda^2)$ *meets the normalization:*

$$(\tilde{\mathfrak{A}}_x(\lambda_0^2))_{11} > 0, \quad (\tilde{\mathfrak{A}}_x(\lambda_0^2))_{21} = 0$$

for all $x > 0$.

We use the same notation as in (4.36).

Theorem 5.6. *Let* $m_+(\lambda)$ *be the Weyl function of operator* (5.1) *and let*

$$\tilde{\mathfrak{A}}_x(\lambda_0^2) = \begin{bmatrix} \tau & a\tau \\ 0 & \tau^{-1} \end{bmatrix}. \tag{5.18}$$

Then

$$\frac{da}{d(\tau^{-2})} = \frac{\overline{c(x,\lambda_0)}\,\overline{m_+(\lambda_0)} + \overline{s(x,\lambda_0)}}{c(x,\lambda_0)\,m_+(\lambda_0) + s(x,\lambda_0)}. \tag{5.19}$$

Proof. First of all,

$$\begin{bmatrix} e_{x_0}^+(x,\lambda) \\ \dot{e}_{x_0}^+(x,\lambda) \end{bmatrix} = \mathfrak{B}_x \begin{bmatrix} m_+(\lambda) \\ 1 \end{bmatrix} \rho(x_0), \quad x \geq x_0, \tag{5.20}$$

where $\rho(x_0)$ should be found from the condition

$$\int_{x_0}^{\infty} |e_{x_0}^+(x,\lambda)|^2\, dx = 1.$$

Using

$$\frac{d}{dx}\{\mathfrak{B}_x^*(\lambda^2) J \mathfrak{B}_x(\lambda^2)\} = -(\lambda^2 - \bar{\lambda}^2)\mathfrak{B}_x^*(\lambda^2) \begin{bmatrix} 1 \\ 0 \end{bmatrix} \begin{bmatrix} 1 & 0 \end{bmatrix} \mathfrak{B}_x(\lambda^2), \tag{5.21}$$

with $J = \begin{bmatrix} 0 & 1 \\ -1 & 0 \end{bmatrix}$, we obtain

$$\begin{aligned} &\rho^2(x_0) \int_{x_0}^{\infty} \begin{bmatrix} \overline{m_+(\lambda)} & 1 \end{bmatrix} \mathfrak{B}_x^*(\lambda^2) \begin{bmatrix} 1 \\ 0 \end{bmatrix} \begin{bmatrix} 1 & 0 \end{bmatrix} \mathfrak{B}_x(\lambda^2) \begin{bmatrix} m_+(\lambda) \\ 1 \end{bmatrix} dx \\ &= \rho^2(x_0) \frac{\begin{bmatrix} \overline{m_+(\lambda)} & 1 \end{bmatrix} \mathfrak{B}_{x_0}^*(\lambda^2) J \mathfrak{B}_{x_0}(\lambda^2) \begin{bmatrix} m_+(\lambda) \\ 1 \end{bmatrix}}{\lambda^2 - \bar{\lambda}^2}. \end{aligned} \tag{5.22}$$

That is,

$$\rho^2(x_0) = \frac{\lambda^2 - \bar{\lambda}^2}{\begin{bmatrix} \overline{m_+(\lambda)} & 1 \end{bmatrix} \mathfrak{B}_{x_0}^*(\lambda^2) J \mathfrak{B}_{x_0}(\lambda^2) \begin{bmatrix} m_+(\lambda) \\ 1 \end{bmatrix}}. \tag{5.23}$$

In particular

$$\rho^2(0) = -\frac{\lambda^2 - \bar\lambda^2}{m_+(\lambda) - \overline{m_+(\lambda)}}. \tag{5.24}$$

Since for $x \ge x_0$

$$\mathfrak{B}_x(\lambda_0^2)\begin{bmatrix} e_0^+(0,\lambda_0) \\ \dot e_0^+(0,\lambda_0)\end{bmatrix} = \mathfrak{B}_x(\lambda_0^2)\begin{bmatrix} m_+(\lambda_0) \\ 1\end{bmatrix}\rho(0) = \begin{bmatrix} e_{x_0}^+(x,\lambda_0) \\ \dot e_{x_0}^+(x,\lambda_0)\end{bmatrix}\frac{\rho(0)}{\rho(x_0)}, \tag{5.25}$$

we get for the first column of the matrix $\tilde{\mathfrak{A}}_x(\lambda_0^2)$ (5.17)

$$\begin{bmatrix} e_{x_0}^+(x_0,\lambda_0) & -e_{x_0}^+(x_0,-\bar\lambda_0) \\ \dot e_{x_0}^+(x_0,\lambda_0) & -\dot e_{x_0}^+(x_0,-\bar\lambda_0)\end{bmatrix}^{-1}\begin{bmatrix} e_{x_0}^+(x_0,\lambda_0) \\ \dot e_{x_0}^+(x_0,\lambda_0)\end{bmatrix}\frac{\rho(0)}{\rho(x_0)} = \begin{bmatrix} 1 \\ 0\end{bmatrix}\frac{\rho(0)}{\rho(x_0)}.$$

Therefore, we deduce from (5.18) that $\tau = \frac{\rho(0)}{\rho(x_0)}$ and, recalling (5.23), (5.24), we come to

$$\tau^{-2} = -\frac{m_+(\lambda_0) - \overline{m_+(\lambda_0)}}{\begin{bmatrix}\overline{m_+(\lambda_0)} & 1\end{bmatrix}\mathfrak{B}_{x_0}^*(\lambda_0^2)J\mathfrak{B}_{x_0}(\lambda_0^2)\begin{bmatrix} m_+(\lambda_0) \\ 1\end{bmatrix}}. \tag{5.26}$$

To compute $a\tau$, we proceed as

$$\begin{aligned} a\tau\Delta &= -\begin{bmatrix}-\dot e_{x_0}^+(x_0,-\bar\lambda_0) & e_{x_0}^+(x_0,-\bar\lambda_0)\end{bmatrix}\mathfrak{B}_{x_0}(\lambda_0^2)\begin{bmatrix} m_+(\lambda_0) \\ 1\end{bmatrix}\rho(0) \\ &= -\begin{bmatrix} e_{x_0}^+(x_0,-\bar\lambda_0) & \dot e_{x_0}^+(x_0,-\bar\lambda_0)\end{bmatrix}J\mathfrak{B}_{x_0}(\lambda_0^2)\begin{bmatrix} m_+(\lambda_0) \\ 1\end{bmatrix}\rho(0) \\ &= -\rho(x_0)\begin{bmatrix}\overline{m_+(\lambda_0)} & 1\end{bmatrix}\mathfrak{B}_{x_0}^*(\lambda_0^2)J\mathfrak{B}_{x_0}(\lambda_0^2)\begin{bmatrix} m_+(\lambda_0) \\ 1\end{bmatrix}\rho(0), \end{aligned} \tag{5.27}$$

where

$$\begin{aligned} \Delta &= \det\begin{bmatrix} e_{x_0}^+(x_0,\lambda_0) & -e_{x_0}^+(x_0,-\bar\lambda_0) \\ \dot e_{x_0}^+(x_0,\lambda_0) & -\dot e_{x_0}^+(x_0,-\bar\lambda_0)\end{bmatrix} \\ &= \begin{bmatrix} e_{x_0}^+(x_0,-\bar\lambda_0) & \dot e_{x_0}^+(x_0,-\bar\lambda_0)\end{bmatrix}J\begin{bmatrix} e_{x_0}^+(x_0,\lambda_0) \\ \dot e_{x_0}^+(x_0,\lambda_0)\end{bmatrix} \\ &= \rho(x_0)\begin{bmatrix}\overline{m_+(\lambda_0)} & 1\end{bmatrix}\mathfrak{B}_{x_0}^*(\lambda_0^2)J\mathfrak{B}_{x_0}(\lambda_0^2)\begin{bmatrix} m_+(\lambda_0) \\ 1\end{bmatrix}\rho(x_0). \end{aligned} \tag{5.28}$$

Combining (5.27) and (5.28), we obtain

$$a = -\frac{\begin{bmatrix}\overline{m_+(\lambda_0)} & 1\end{bmatrix}\mathfrak{B}_{x_0}^*(\lambda_0^2)J\mathfrak{B}_{x_0}(\lambda_0^2)\begin{bmatrix}\overline{m_+(\lambda_0)} \\ 1\end{bmatrix}}{\begin{bmatrix}\overline{m_+(\lambda_0)} & 1\end{bmatrix}\mathfrak{B}_{x_0}^*(\lambda_0^2)J\mathfrak{B}_{x_0}(\lambda_0^2)\begin{bmatrix} m_+(\lambda_0) \\ 1\end{bmatrix}}. \tag{5.29}$$

Using (5.21) and the Wronskian identity for $\mathfrak{B}_x(\lambda)$ we get (5.19) from (5.26) and (5.29) by a direct computation. □

6. Appendix 1. An example

In this section we give an example which shows that the class of canonical systems discussed in Section 4 is larger than the class of Sturm-Liouville equations from Section 5. We will see that generally

$$H^2_{\{s_+,\nu_+\}} \neq \hat{H}^2_{\{s_+,\nu_+\}}, \tag{6.1}$$

although always $H^2_{\{s_+,\nu_+\}} \subset \hat{H}^2_{\{s_+,\nu_+\}}$ and we will also discuss some other interesting phenomena.

Throughout this section we set $s_+ = \frac{\lambda}{\lambda+i}$ *and* $\nu_+ = 0$.

First, we prove (6.1). Since

$$\langle (s_+ f)(\lambda), -f(-\bar{\lambda})\rangle = 0$$

for all $f \in H^2$, we get $||f||_{\{s_+,\nu_+\}} = ||f||$. Therefore, in this case $H^2_{\{s_+,\nu_+\}}$ coincides with the standard H^2.

On the other hand, we have $s(\lambda) = \frac{i}{\lambda+i}$, so $s \cdot 1 \in H^2$. Let us check that $1 \in L^2_{\{s_+,\nu_+\}}$. This follows from the identity

$$\begin{bmatrix} 1 & -1 \end{bmatrix} \begin{bmatrix} 1 & \overline{\frac{\lambda}{\lambda+i}} \\ \frac{\lambda}{\lambda+i} & 1 \end{bmatrix} \begin{bmatrix} 1 \\ -1 \end{bmatrix} = \frac{2}{|\lambda+i|^2}. \tag{6.2}$$

Hence, by the definition of $\hat{H}^2_{\{s_+,\nu_+\}}$ a constant function belongs to this space, but of course $1 \notin H^2$ and (6.1) is proved.

The above conclusion can be sharpened. Using

$$\frac{1}{2}\left\langle \begin{bmatrix} 1 & \overline{s_+(\lambda)} \\ s_+(\lambda) & 1 \end{bmatrix} \begin{bmatrix} 1 \\ -1 \end{bmatrix}, \begin{bmatrix} f(\lambda) \\ -f(-\bar{\lambda}) \end{bmatrix} \right\rangle = \langle s_+(\lambda) - 1, -f(-\bar{\lambda})\rangle = 0, \tag{6.3}$$

for all $f \in H^2$, we get that 1 is orthogonal to $H^2_{\{s_+,\nu_+\}} \subset \hat{H}^2_{\{s_+,\nu_+\}}$. Actually we have the following orthogonal decomposition

$$\hat{H}^2_{\{s_+,\nu_+\}} = \{1\} \oplus H^2_{\{s_+,\nu_+\}} = \{1\} \oplus H^2. \tag{6.4}$$

This implies that the reproducing kernel of $\hat{H}^2_{\{s_+,\nu_+\}}$ is

$$\hat{k}(\lambda,\lambda_0) = \hat{k}_{\{s_+,\nu_+\}}(\lambda,\lambda_0) = \frac{1}{||1||^2_{\{s_+,\nu_+\}}} + \frac{i}{\lambda - \bar{\lambda}_0}, \tag{6.5}$$

and, by (6.2), $||1||^2_{\{s_+,\nu_+\}} = \frac{1}{2}$.

Now we show that the property $H^2_{\{s_+,\nu_+\}}(x) \neq \hat{H}^2_{\{s_+,\nu_+\}}(x)$ is not x-invariant. Namely, $H^2_{\{s_+,\nu_+\}}(x) = \hat{H}^2_{\{s_+,\nu_+\}}(x)$ for $x > 0$ despite (6.1) for $x = 0$.

Lemma 6.1. *Let* $s_+ = \frac{\lambda}{\lambda+i}$, $\nu_+ = 0$. *Then* $H^2_{\{s_+,\nu_+\}}(x) = \hat{H}^2_{\{s_+,\nu_+\}}(x)$ *for all* $x > 0$.

Proof. Notice that

$$\hat{H}^2_{\{s_+,\nu_+\}}(x) = e^{i\lambda x}\hat{H}^2_{\{s_+e^{2i\lambda x},\nu_+e^{2i\lambda x}\}},$$

and $H^2_{\{s_+,\nu_+\}}(x) = e^{i\lambda x}H^2$. So we have to show that $\hat{H}^2_{\{s_+e^{2i\lambda x},\nu_+e^{2i\lambda x}\}} = H^2$.

By definition $f \in \hat{H}^2_{\{s_+e^{2i\lambda x},\nu_+e^{2i\lambda x}\}}$ means that

$$\begin{aligned} &\frac{i}{\lambda+i}f(\lambda) \in H^2, \\ &e^{2i\lambda x}\frac{\lambda}{\lambda+i}f(\lambda) - f(-\bar{\lambda}) \in L^2. \end{aligned} \tag{6.6}$$

We have to prove that $f \in H^2$.

Let $g(\lambda) = \frac{i}{\lambda+i}f(\lambda)$. Conditions (6.6) can be easily transformed into

$$\lambda\{e^{2i\lambda x}g(\lambda) + g(-\bar{\lambda})\} \in L^2$$

with $g \in H^2$. Let G denote the Fourier transform of g. Obviously, $G \in L^2(\mathbb{R}_+)$ since $g \in H^2$. In these terms we have

$$\{G(2x+t) + G(-t)\}' \in L^2. \tag{6.7}$$

Since the supports of the functions $G(2x+t)$ and $G(-t)$ do not intersect, we get from (6.7) that $G'(t) \in L^2$. Therefore $\lambda g(\lambda) \in L^2$, and, consequently, f belongs to L^2 and, in fact, to H^2. □

Corollary 6.2. *Let* $s_+ = \frac{\lambda}{\lambda+i}$, $\nu_+ = 0$. *Then* $H^2_{\{s_+,\nu_+\}}(x) = \hat{H}^2_{\{s_+,\nu_+\}}(x)$ *for all* $x < 0$.

Proof. We only have to mention that $s_- = s_+$ in our case and to use Theorem 1.3. □

Corollary 6.3. *Let* $s_+ = \frac{\lambda}{\lambda+i}$, $\nu_+ = 0$. *Then*

$$\lim_{x\to -0} H^2_{\{s_+,\nu_+\}}(x) = \hat{H}^2_{\{s_+,\nu_+\}}(0). \tag{6.8}$$

Proof. Obviously, $\lim_{x\to x_0+0} H^2_{\{s_+,\nu_+\}}(x) = H^2_{\{s_+,\nu_+\}}(x_0)$ for $x_0 \geq 0$. Therefore by $s_- = s_+$ and the duality stated in Theorem 1.3,

$$\lim_{x\to x_0-0} \hat{H}^2_{\{s_+,\nu_+\}}(x) = \hat{H}^2_{\{s_+,\nu_+\}}(x_0)$$

for $x_0 \leq 0$. Finally, we use Corollary 6.2

$$\lim_{x\to -0} H^2_{\{s_+,\nu_+\}}(x) = \lim_{x\to -0} \hat{H}^2_{\{s_+,\nu_+\}}(x) = \hat{H}^2_{\{s_+,\nu_+\}}(0).$$ □

This means, in particular, that the canonical system related to the given scattering data is not a Sturm-Liouville equation.

Indeed, let $\mathfrak{A}(x_0, x_1; \lambda^2)$, $x_0 < x_1$, be the transfer matrix (4.24) (or (4.29)). Recalling (4.27), we write

$$\begin{bmatrix} ve_2^{(2)}(x_1) & e_2^{(1)}(x_1) \end{bmatrix} = \begin{bmatrix} e_1^{(1)}(x_0) & ve_1^{(2)}(x_0) \end{bmatrix} \mathfrak{A}(x_0, x_1; \lambda^2).$$

We also introduce $\mathfrak{B}(x_0, x_1; \lambda^2)$ by

$$\begin{bmatrix} ve_2^{(2)}(x_1) & e_2^{(1)}(x_1) \end{bmatrix} = \begin{bmatrix} ve_2^{(2)}(x_0) & e_2^{(1)}(x_0) \end{bmatrix} \mathfrak{B}(x_0, x_1; \lambda^2), \tag{6.9}$$

so that we have the chain rule

$$\mathfrak{A}(x_0, x_2; \lambda^2) = \mathfrak{A}(x_0, x_1; \lambda^2)\mathfrak{B}(x_1, x_2; \lambda^2).$$

Fix $x_0 < 0$, put $x_2 = 0$, and let

$$\mathfrak{B}(\lambda^2) = \lim_{x_1 \to -0} \mathfrak{B}(x_1, 0; \lambda^2).$$

Using (6.9) and (6.8) we have

$$\begin{bmatrix} ve_2^{(2)}(x_1) & e_2^{(1)}(x_1) \end{bmatrix} = \begin{bmatrix} v\hat{e}_2^{(2)}(x_1) & \hat{e}_2^{(1)}(x_1) \end{bmatrix} \mathfrak{B}(\lambda^2), \tag{6.10}$$

where (see (4.4), (4.5))

$$\hat{e}_2^{(1)}(\lambda) = \frac{\hat{k}_{\{s_+,\nu_+\}}(\lambda, \lambda_0; x)}{||\hat{k}_{\{s_+,\nu_+\}}(\lambda, \lambda_0; x)||}, \quad \hat{e}_2^{(2)}(\lambda) = \frac{\hat{k}_{\{s_+,\nu_+\}}(\lambda, -\bar{\lambda}_0; x)}{||\hat{k}_{\{s_+,\nu_+\}}(\lambda, -\bar{\lambda}_0; x)||}.$$

Thus $\mathfrak{B}(\lambda^2)$ is a *non-trivial divisor* of $\mathfrak{A}(x_0, x; \lambda^2)$ for $x \geq 0$. The chain $\mathfrak{A}(x_0, x; \lambda^2)$ is not even continuous in x,

$$\lim_{x \to -0} \mathfrak{A}(x_0, x; \lambda^2) \neq \mathfrak{A}(x_0, 0; \lambda^2).$$

Of course, we can get an explicit formula for $\mathfrak{B}(\lambda^2)$. Since $\mathfrak{B}$ depends on λ^2, (6.10) implies

$$\begin{bmatrix} e_2^{(2)}(\lambda) & e_2^{(1)}(\lambda, \lambda_0) \\ e^{(2)}(-\lambda) & e_2^{(1)}(-\lambda) \end{bmatrix} \begin{bmatrix} v(\lambda) & 0 \\ 0 & 1 \end{bmatrix} = \begin{bmatrix} \hat{e}_2^{(2)}(\lambda) & \hat{e}_2^{(1)}(\lambda) \\ \hat{e}_2^{(2)}(-\lambda) & \hat{e}_2^{(1)}(-\lambda) \end{bmatrix} \begin{bmatrix} v(\lambda) & 0 \\ 0 & 1 \end{bmatrix} \mathfrak{B}(\lambda^2),$$

or, with the help of (6.5),

$$\begin{aligned} &\sqrt{\frac{\hat{k}(\lambda_0, \lambda_0)}{k(\lambda_0, \lambda_0)}} \begin{bmatrix} k(\lambda, -\bar{\lambda}_0) & k(\lambda, \lambda_0) \\ k(-\lambda, -\bar{\lambda}_0) & k(-\lambda, \lambda_0) \end{bmatrix} \begin{bmatrix} v(\lambda) & 0 \\ 0 & 1 \end{bmatrix} \\ &\quad = \begin{bmatrix} \hat{k}(\lambda, -\bar{\lambda}_0) & \hat{k}(\lambda, \lambda_0) \\ \hat{k}(-\lambda, -\bar{\lambda}_0) & \hat{k}(-\lambda, \lambda_0) \end{bmatrix} \begin{bmatrix} v(\lambda) & 0 \\ 0 & 1 \end{bmatrix} \mathfrak{B}(\lambda^2) \\ &\quad = \left\{ 2 \begin{bmatrix} 1 & 1 \\ 1 & 1 \end{bmatrix} + \begin{bmatrix} k(\lambda, -\bar{\lambda}_0) & k(\lambda, \lambda_0) \\ k(-\lambda, -\bar{\lambda}_0) & k(-\lambda, \lambda_0) \end{bmatrix} \right\} \begin{bmatrix} v(\lambda) & 0 \\ 0 & 1 \end{bmatrix} \mathfrak{B}(\lambda^2). \end{aligned} \tag{6.11}$$

Thus, directly,

$$\sqrt{1 + 4\text{Im } \lambda_0} I = \left\{ \frac{i}{\text{Re } \lambda_0} \begin{bmatrix} \lambda^2 - \lambda_0^2 & \lambda^2 - \bar{\lambda}_0^2 \\ -\lambda^2 + \lambda_0^2 & -\lambda^2 + \bar{\lambda}_0^2 \end{bmatrix} + I \right\} \mathfrak{B}(\lambda^2).$$

Note that the determinant of the matrix in curly brackets is $1 + 4\text{Im } \lambda_0$, so $\mathfrak{B}(\lambda^2)$ is indeed an entire function of λ^2 (a linear polynomial),

$$\mathfrak{B}(\lambda^2) = \frac{1}{\sqrt{1 + 4\text{Im } \lambda_0}} \left\{ I + \frac{i}{\text{Re } \lambda_0} \begin{bmatrix} -\lambda^2 + \bar{\lambda}_0^2 & -\lambda^2 + \bar{\lambda}_0^2 \\ \lambda^2 - \lambda_0^2 & \lambda^2 - \lambda_0^2 \end{bmatrix} \right\}.$$

7. Appendix 2. On a certain sufficient condition

7.1. On a certain sufficient condition

The following lemmas are related to attempts to rewrite the A_2 condition for the spectral density [20] directly in terms of the scattering function.

Lemma 7.1. *Let*

$$W = \begin{bmatrix} 1 & \bar{s}_+ \\ s_+ & 1 \end{bmatrix}.$$

The following conditions are equivalent

$$\left\langle W^{-1}P_+W \begin{bmatrix} f(t) \\ \bar{t}f(\bar{t}) \end{bmatrix}, P_+W \begin{bmatrix} f(t) \\ \bar{t}f(\bar{t}) \end{bmatrix} \right\rangle \le Q \left\langle W \begin{bmatrix} f(t) \\ \bar{t}f(\bar{t}) \end{bmatrix}, \begin{bmatrix} f(t) \\ \bar{t}f(\bar{t}) \end{bmatrix} \right\rangle \tag{7.1}$$

for all $f \in L^2_{s_+} \ominus H^2_{s_+}$ *and*

$$||f^-||^2 \le Q||f^-||^2_{s_-} \tag{7.2}$$

for $f^-(t) \in \hat{H}^2_{s_-}, t \in \mathbb{T}$.

Lemma 7.2. *If* (7.2) *holds, then* $\hat{H}^2_{s_-} = H^2_{s_-}$, *moreover the norm in* $\hat{H}^2_{s_-}$ *is equivalent to the standard* H^2*-norm.*

Lemma 7.3. $\hat{H}^2_{s_-} = H^2_{s_-}$ *implies* $\hat{H}^2_{s_+} = H^2_{s_+}$

Proof.

$$\begin{aligned} \hat{H}^2_{s_+} &= (L^2_{s_-} \ominus H^2_{s_-})^+ = (L^2_{s_-} \ominus \hat{H}^2_{s_-})^+ \\ &= L^2_{s_+} \ominus (\hat{H}^2_{s_-})^+ = H^2_{s_+}. \end{aligned}$$

□

Nevertheless, we cannot guarantee that the norm in $H^2_{s_+}$ is equivalent to the H^2-norm. Thus in addition to (7.2) we have to impose the condition

$$\langle f, f \rangle \le Q \left\langle W \begin{bmatrix} f(t) \\ \bar{t}f(\bar{t}) \end{bmatrix}, \begin{bmatrix} f(t) \\ \bar{t}f(\bar{t}) \end{bmatrix} \right\rangle \tag{7.3}$$

for all $f \in H^2$ (this is exactly the condition on equivalence of the norms). Obviously, the last inequality is the same as

$$\left\langle WP_+ \begin{bmatrix} f(t) \\ \bar{t}f(\bar{t}) \end{bmatrix}, P_+ \begin{bmatrix} f(t) \\ \bar{t}f(\bar{t}) \end{bmatrix} \right\rangle \le Q \left\langle W \begin{bmatrix} f(t) \\ \bar{t}f(\bar{t}) \end{bmatrix}, \begin{bmatrix} f(t) \\ \bar{t}f(\bar{t}) \end{bmatrix} \right\rangle, \quad f \in H^2. \tag{7.4}$$

Thus we get

Theorem 7.4. *The combination of the following two conditions*

$$\langle (W^{-\frac{1}{2}}P_+W^{\frac{1}{2}})F, (W^{-\frac{1}{2}}P_+W^{\frac{1}{2}})F \rangle \le Q\langle F, F \rangle \tag{7.5}$$

with $F = W^{\frac{1}{2}} \begin{bmatrix} f(t) \\ \bar{t}f(\bar{t}) \end{bmatrix}$ *and* $f \in L^2_{s_+} \ominus H^2_{s_+}$, *and*

$$\langle (W^{\frac{1}{2}}P_+W^{-\frac{1}{2}})F, (W^{\frac{1}{2}}P_+W^{-\frac{1}{2}})F \rangle \le Q\langle F, F \rangle \tag{7.6}$$

with $F = W^{\frac{1}{2}} \begin{bmatrix} f(t) \\ \bar{t}f(\bar{t}) \end{bmatrix}$ *and* $f \in H^2$ *is equivalent to the first (or the second) condition from* [20], *Theorem* 3.1.

Proof. Relation (7.5) is a slight modification of (7.1) and (7.6) of (7.4), respectively. Observe that if (7.6) holds with $f \in H^2$, then every $f \in H^2_{s_+}$ belongs to H^2. Therefore, in fact, (7.6) has a perfect sense for $f \in H^2_{s_+}$. □

In any case, $W \in A_2$ is a sufficient condition for (7.5), (7.6). Let us transform this matrix condition into a scalar one.

Lemma 7.5. *W is in A_2 if and only if*

$$\sup_I \frac{1}{|I|} \int_I \frac{|s_+ - \langle s_+\rangle_I|^2 + (1 - |\langle s_+\rangle_I|^2)}{1 - |s_+|^2}\, dm < \infty, \tag{7.7}$$

where for an arc $I \subset \mathbb{T}$ we put

$$\langle s_+\rangle_I := \frac{1}{|I|} \int_I s_+\, dm. \tag{7.8}$$

Proof. By definition we have that there exists $Q > 0$ such that

$$\langle W^{-1}\rangle_I \leq Q\langle W\rangle_I^{-1} \tag{7.9}$$

for all $I \subset \mathbb{T}$. Note that

$$\langle W\rangle_I = \begin{bmatrix} 1 & \overline{\langle s_+\rangle_I} \\ \langle s_+\rangle_I & 1 \end{bmatrix} = \begin{bmatrix} 1 & 0 \\ \langle s_+\rangle_I & 1 \end{bmatrix} \begin{bmatrix} 1 & 0 \\ 0 & 1 - |\langle s_+\rangle_I|^2 \end{bmatrix} \begin{bmatrix} 1 & \overline{\langle s_+\rangle_I} \\ 0 & 1 \end{bmatrix}.$$

Therefore (7.9) is equivalent to

$$\begin{bmatrix} 1 & 0 \\ 0 & \sqrt{1 - |\langle s_+\rangle_I|^2} \end{bmatrix} \begin{bmatrix} 1 & \overline{\langle s_+\rangle_I} \\ 0 & 1 \end{bmatrix} \langle W^{-1}\rangle_I \begin{bmatrix} 1 & 0 \\ \langle s_+\rangle_I & 1 \end{bmatrix} \begin{bmatrix} 1 & 0 \\ 0 & \sqrt{1 - |\langle s_+\rangle_I|^2} \end{bmatrix} \leq Q.$$

Since the matrix in the RHS is positive, its boundedness is equivalent to the boundedness of its trace. The last condition with a small effort gives (7.7) and vice versa. □

Acknowledgments. A part of this work was done during the last author's visit to CMI (Centre de Mathématiques et Informatique) at University of Provence. He would like to thank the Department for the hospitality.

References

[1] D. Arov, H. Dym, *J-inner matrix functions, interpolation and inverse problems for canonical systems. I. Foundations*, Integral Equations Operator Theory, **29** (1997), no. 4, 373–454.

[2] D. Arov, H. Dym, *J-inner matrix functions, interpolation and inverse problems for canonical systems. II. The inverse monodromy problem*, Integral Equations Operator Theory, **36** (2000), no. 1, 11–70.

[3] D. Arov, H. Dym, *J-inner matrix functions, interpolation and inverse problems for canonical systems. III. More on the inverse monodromy problem*, Integral Equations Operator Theory, **36** (2000), no. 2, 127–181.

[4] D. Arov, H. Dym, *Strongly regular J-inner matrix-valued functions and inverse problems for canonical systems*, Oper. Theory Adv. Appl., vol. 160, Birkhäuser, Basel, 2005, 101–160.

[5] L. de Branges, *Hilbert spaces of entire functions*, Prentice-Hall, Englewood Cliffs, N.J. 1968.

[6] M. Brodskii, *Characteristic matrix-functions of linear operators*, Mat. Sb., **39** (81) (1956), 176–200 (Russian).

[7] M. Brodskii, M. Livshits, *Spectral analysis of non-selfadjoint operators and open systems*, Uspekhi Mat., **13** 1 (79) (1958), 3–85 (Russian).

[8] D. Damanik, R. Killip, B. Simon, *Necessary and sufficient conditions in the spectral theory of Jacobi matrices and Schrödinger operators*, Int. Math. Res. Not. 2004, no. 22, 1087–1097.

[9] J. Garnett, *Bounded Analytic Functions*, Academic Press, New York, 1981.

[10] V. Katsnelson, *Methods of J-theory in continuous interpolation problems of analysis*, Hokkaido University, Sapporo, 1985.

[11] V. Katsnelson, A. Kheifets, P. Yuditskii, *An abstract interpolation problem and the theory of extensions of isometric operators*, Naukova Dumka, Kiev, 1987, 83–96 (Russian); translation in: Oper. Theory Adv. Appl., **95** (1997), 283–298.

[12] A. Kheifets, *Scattering matrices and Parseval equality in the abstract interpolation problem*, Ph. D. thesis, Kharkov, 1989.

[13] A. Kheifets, P. Yuditskii, *An analysis and extension of V.P. Potapov's approach to interpolation problems with applications to the generalized bi-tangential Schur-Nevanlinna-Pick problem and J-inner-outer factorization*, Oper. Theory Adv. Appl., **72** (1994), 133–161.

[14] V. Marchenko, *Sturm-Liouville operators and applications*, Birkhäuser Verlag, Basel, 1986.

[15] V. Potapov, *The multiplicative structure of J-contractive matrix functions*, Amer. Math. Soc. Transl. (2), vol. 15 (1960), 131–243.

[16] M. Reed, B. Simon, *Methods of modern mathematical physics, I*, Academic Press, New York, 1980.

[17] M. Reed, B. Simon, *Methods of modern mathematical physics, III*, Academic Press, New York, 1979.

[18] Ch. Remling, *Schrödinger operators and de Branges spaces*, J. Funct. Anal., **196** (2002), no. 2, 323–394.

[19] T. Tao, Ch. Thiele, *Nonlinear Fourier Analysis*, IAS/Park City Mathematics Series.

[20] A. Volberg and P. Yuditskii, *On the inverse scattering problem for Jacobi matrices with the spectrum on an interval, a finite system of intervals or a Cantor set of positive length*, Comm. Math. Phys., **226** (2002), 567–605.

S. Kupin
CMI, Université de Provence
39, rue Joliot-Curie
F-13453 Marseille Cedex 13, France
e-mail: `kupin@cmi.univ-mrs.fr`

F. Peherstorfer
Institute for Analysis
Johannes Kepler University Linz
A-4040 Linz, Austria
e-mail: Franz.Peherstorfer@jk.uni-linz.ac.at

A. Volberg
Department of Mathematics
Michigan State University
East Lansing MI 48824, USA
e-mail: volberg@math.msu.edu

P. Yuditskii
Department of Mathematics
Bar Ilan University
Ramat Gan 52900, Israel
e-mail: yuditski@mailhost.cs.biu.ac.il

Operator Theory:
Advances and Applications, Vol. 186, 325–337

Remark on the Compensation of Singularities in Krein's Formula

A.B. Mikhailova and B.S. Pavlov

Abstract. We reduce the spectral problem for an additively perturbed self-adjoint operator $H_V = H_0 - V$, to the dual problem of finding zeros of the operator function

$$\mathrm{Sign}\, V - |V|^{1/2}[H_0 - \lambda]^{-1}|V|^{1/2},$$

and develop the Schmidt perturbation procedure for the resolvent of H_V. Based on Rouche theorem for operator-valued analytic functions, we observe, in the Krein's formula for the perturbed resolvent $[H_V - \lambda I]^{-1}$, the compensation of singularities inherited from H_0, and suggest a convenient algorithm for approximate calculation of the groups of eigenfunctions and eigenvalues of the perturbed operator.

Mathematics Subject Classification (2000). Primary 47N50; Secondary 35Q40.

Keywords. Lippmann-Schwinger equations, eigenvalues, eigenfunctions.

1. Introduction: formal solution of the Lippmann-Schwinger equation and Krein's formula

The non-homogeneous equation with self-adjoint operators H_0, and V in a Hilbert space E

$$[H_0 - V - \lambda I]\, u = f \tag{1.1}$$

can be conveniently reduced, see [1], to the corresponding Lippmann-Schwinger equation

$$u = [H_0 - \lambda I]^{-1} f + [H_0 - \lambda I]^{-1} V u := R^0_\lambda f + R^0_\lambda V u. \tag{1.2}$$

The formal solution of this equation leads to the dual problem of localizing zeros of a relevant operator function. Indeed, assuming that the perturbation $V = V^+$ is bounded, we represent it as $V = |V|^{1/2}\Theta_V|V|^{1/2}$ with a unitary self-adjoint

The authors acknowledge support from the Russian Academy of Sciences, Grant RFBR 03-01-00090.

operator $\Theta_V = [\Theta_V]^+ = \text{Sign}\, V$ acting in $E_V := \overline{|V|^{1/2}E}$. Then we rewrite (1.2) as

$$|V|^{1/2}u = |V|^{1/2}R^0_\lambda f + |V|^{1/2}R^0_\lambda |V|^{1/2}\Theta_V|V|^{1/2}u, \tag{1.3}$$

and find $|V|^{1/2}u$ as

$$|V|^{1/2}u = \left[I_V - |V|^{1/2}R^0_\lambda|V|^{1/2}\Theta_V\right]^{-1}|V|^{1/2}R^0_\lambda f, \tag{1.4}$$

where I_V is the restriction of identity to E_V. This gives a convenient formula for the solution of (1.2)

$$u = \left[R^0_\lambda + R^0_\lambda|V|^{1/2}\frac{I_V}{\Theta_V - |V|^{1/2}R^0_\lambda|V|^{1/2}}|V|^{1/2}R^0_\lambda\right]f. \tag{1.5}$$

A similar formula in operator extension theory was first suggested by M. Krein, see [2], and, in different form, by M. Naimark, [3], as a device for the parametrization (of resolvents) of self-adjoint extensions of a given symmetric operator. Later, the idea of reducing the eigenvalue problem for the additively perturbed operator $H_V = H_0 - V$ to the dual problem of zeros of the denominator $\mathbf{D} = \Theta_V - |V|^{1/2}R^0_\lambda|V|^{1/2}$ appeared in [4, 5], and, since publication [7], is called the Birman-Schwinger principle. All these formulae are based on the reduction of the inversion of an infinite-dimensional operator to the simpler problem of inversion of the corresponding denominator $\mathbf{D}$ with a compact or finite-dimensional $|V|^{1/2}R^0_\lambda|V|^{1/2}$. In this paper, for historical reasons and for sake of brevity, we refer to the corresponding formula for the resolvent

$$R^V_\lambda := [H_0 - V - \lambda I]^{-1} = R^0_\lambda + R^0_\lambda|V|^{1/2}\frac{I_V}{\Theta_V - |V|^{1/2}R^0_\lambda|V|^{1/2}}|V|^{1/2}R^0_\lambda \tag{1.6}$$

as *general Krein's formula*. In the original Krein formula [2] the denominator $\mathbf{D}$ is finite-dimensional. The above derivation of the formula, in case of Schrödinger operators with general bounded self-adjoint perturbations V, which are in a certain way subordinated to H_0 can be found in [6].

Both summands in the right side of (1.6) are singular at the eigenvalues of the non-perturbed operator H_0. For one-dimensional V, the singularities of the first and second summands in the right side of (1.6) inherited from the unperturbed operator H_0, compensate each other, so that only zeros of the denominator $\mathbf{D}$ arise as isolated eigenvalues of the perturbed operator $H_0 - V$. The compensation of the singularity inherited from a single simple eigenvalue of the unperturbed problem was noticed in [8, 9, 10, 11] and used for approximate calculation of resonances in [12]. Though it was commonly expected that the singularities of both summands in the original Krein formula compensate each other in case of isolated singularities, but the proof of the fact for a group of isolated eigenvalues requires a deeper insight into the problem and revealing connections with the Schmidt perturbation procedure, see [1].

The standard analytic perturbation procedure, see [13], is applicable to the problem of perturbation of an isolated eigenvalue λ_0 of the operator H_0, if the perturbation $V : H_0 \to H_V = H_0 - V$ is dominated by the spacing of the non-perturbed operator H_0 at the eigenvalue λ_0 :

$$2 \parallel V \parallel < \min_{\lambda_s \neq \lambda_0} |\lambda_s - \lambda_0| =: \rho(\lambda_0). \tag{1.7}$$

A modified ("Schmidt") analytic perturbation procedure sketched in [1] is based on the conjoint perturbation of the eigenvalue λ_0 and the nearest to it eigenvalue λ_1 to the end of augmentation of the spacing $\rho(\lambda_0)$. Our version of the Schmidt procedure permits to extend the tight bounds (1.7) of application of the standard analytic perturbation procedure to the individual eigenvalues by considering the simultaneous perturbation of a group of eigenvalues, situated on a certain "essential" spectral interval Δ as zeros of the corresponding denominator $\mathbf{D}$, via comparison of it with the relevant model denominator $\mathbf{D}$, based on the operator valued Rouche theorem, [16].

In this paper, based on the announced compensation result for a group of eigenvalues, see Section 2, we obtain in Section 3 convenient formulae for the eigenfunctions and eigenvalues of the perturbed operator. In last section of this paper, based on the compensation of singularities, we reveal a transformation of the intersections of the dispersion curves of a periodic Sturm-Liouvilles problem into quasi-intersections.

2. Compensation of singularities

The original Krein formula permits to reduce the calculation of the resolvent of the perturbed operator to the construction of the inverse of the Birman-Schwinger denominator:

$$\Theta_V - |V|^{1/2} R^0_\lambda |V|^{1/2} := \mathbf{D}. \tag{2.1}$$

In case when the condition $\parallel |V|^{1/2} R^0_\lambda |V|^{1/2} \parallel < 1$ is fulfilled on a real spectral interval Δ, the inverse $\mathbf{D}^{-1}$ can be calculated based on standard analytic perturbation procedure, see for instance [1], resulting in a corresponding Neumann series. We extend the "Schmidt method" suggested by R. Newton, see [1], to treat the perturbation problem for a group of few eigenvalues of H_0 on Δ: $\sigma(H_0) \cap \Delta =: \sigma^\Delta = \{\lambda_1, \lambda_2, \lambda_3, \ldots, \lambda_N\}$. We denote by $\varphi_1, \varphi_2, \varphi_3, \ldots, \varphi_N$ the corresponding eigenfunctions, by E_Δ the linear hull of $\{\varphi_s\}_{s=1}^N$, by P^Δ the orthogonal projection on it in E, and by $H^\Delta = \sum_{s=1}^N \lambda_s \varphi_s\rangle\langle\varphi_s$ the part of H_0 in E^Δ. Assume, that the perturbation V is subordinated to H_0 so that $|V|^{1/2} R^0_\lambda |V|^{1/2}$ is compact on the complement of σ^Δ in the complex neighborhood Ω_Δ of Δ, see in [14, 15] for properties of compact-valued analytic operator functions. Then the denominator $\mathbf{D}$ is a meromorphic function on Ω_Δ with a finite number of real poles λ_s and a finite number of real *vector zeros* λ^V_s, e_s : $\mathbf{D}(\lambda^V_s)\, e_s = 0$, $e_s \in E$. These zeros are eigenvalues, in Ω_Δ, of the perturbed operator, $\cup_s \lambda^V_s = \sigma^\Delta_V$. Hereafter in this paper we assume that the sets σ^Δ, σ^Δ_V, are disjoint, $\sigma^\Delta \cap \sigma^\Delta_V = \emptyset$.

Following [16] the *index* of the vector zero z_s of $\mathbf{D}$ denotes *the total dimension of the corresponding root space.* For instance, for an elementary Blaschke factor

B of degree k in the upper half-plane $\Im\lambda > 0$, with a complex zero z_s and an orthogonal projection P_s onto the null- space $\{\nu : \mathbf{B}(z_s)\nu = 0\}$,

$$\mathbf{B}(\lambda) = \left(\frac{\lambda - z_s}{\lambda - \bar{z}_s}\right)^k P_s + P_s^\perp, \; P_s + P_s^\perp = I,$$

the index is equal dim $P_s \times k$. For poles the index is defined similarly, but with negative k. The total index of a meromorphic function in the domain Ω is the (algebraic) sum of all indices at zeros and poles of the function situated in Ω. The operator version of Rouche's theorem, proved by Gohberg and Sigal, see [16], connects the total indices of two meromorphic operator functions $\mathbf{D}, \mathbf{D}_0$ in a domain Ω, if the difference $\mathbf{D} - \mathbf{D}_0 := \delta\mathbf{D}$ is relatively small on the boundary:

If on the boundary $\partial\Omega$ the inequality

$$\sup_{\partial\Omega} \| I - \mathbf{D}_0^{-1}\mathbf{D} \| = \sup_{\partial\Omega} \| \mathbf{D}_0^{-1} \; \delta\mathbf{D} \| < 1 \tag{2.2}$$

is fulfilled, then the total indices of $\mathbf{D}, \mathbf{D}_0$ in Ω coincide. In particular, if the poles of $\mathbf{D}, \mathbf{D}_0$ coincide and have the same indices, then the total index of zeros of $\mathbf{D}$ in Ω coincides with the total index of zeros of $\mathbf{D}_0$ in Ω.

Combining the idea of Schmidt method, see [1], (2, Chapter 9) with Rouche's theorem for matrix valued functions, [16], we develop a perturbation procedure and localize the isolated eigenvalues λ_s^V of the perturbed operator. This provides a basement for the proof of the compensation of singularities and the derivation, in the next section, a formula for the eigenfunctions of the perturbed operator.

To formulate the main result of this section, we need specific notations. Denote $E \ominus E^\Delta =: E^\perp$ and introduce $P_{E^\perp} =: P^\perp = I - P^\Delta$. The identity operator in E^Δ will be denoted by $I^\Delta = P^\Delta\Big|_{E^\Delta}$. We denote by H^Δ the part of H_0 in E^Δ and by $H^\perp$ the part of H_0 in $E^\perp$, the spectrum of $H^\perp$ will be denoted by $\sigma^\perp$. The resolvents of H^Δ and $H^\perp$ will be denoted by R_λ^Δ and $R_\lambda^\perp$ respectively, $R_\lambda^\Delta = \sum_{\lambda_s \in \Delta} \frac{P_s}{\lambda_s - \lambda}$. A small complex neighborhood of Δ, specified below, will be denoted by Ω_Δ.

In our analysis, when using the operator-valued Rouche theorem, we choose $\mathbf{D}_0 =: \mathbf{D}_{lead} := \Theta^V - |V|^{1/2} R_\lambda^\Delta |V|^{1/2}$ and substitute the condition (1.7) by the condition imposed onto $\delta\mathbf{D} = \mathbf{D}^\perp = |V|^{1/2} R_\lambda^\perp |V|^{1/2}$, on a selected neighborhood Ω_Δ, such that

$$\sup_{\lambda \in \Omega_\Delta} \| \mathbf{D}^\perp(\lambda) \| = \sup_{\lambda \in \Omega_\Delta} \| \, |V|^{1/2} R_\lambda^\perp |V|^{1/2} \| < 1 - \varepsilon, \; \varepsilon > 0. \tag{2.3}$$

Due to the maximum principle the weaker condition

$$\sup_{\lambda \in \partial\Omega_\Delta} \| \mathbf{D}^\perp(\lambda) \| = \sup_{\lambda \in \partial\Omega_\Delta} \| \, |V|^{1/2} R_\lambda^\perp |V|^{1/2} \| < 1 - \varepsilon, \; \varepsilon > 0,$$

is equivalent to (2.3). On the other hand, the following slightly stronger condition also implies (2.3):

$$\sup_{\lambda\in\Omega_\Delta} \parallel |V|^{1/2}P^\perp |V|^{1/2} \parallel \leq (1-\varepsilon) \inf_{\lambda\in\Omega_\Delta,\,\lambda_s\in C\setminus\Omega_\Delta} |\lambda-\lambda_s| = (1-\varepsilon)\mathrm{dist}(\lambda,\sigma^\perp). \tag{2.4}$$

We consider the matrix function acting in E_V

$$\mathcal{D} = \Theta_V - |V|^{1/2} R_\lambda^\perp |V|^{1/2} = \Theta_V - \mathbf{D}^\perp. \tag{2.5}$$

The matrix $\mathcal{D}$ is invertible on Ω_Δ since (2.3) due to the Banach principle, because of the geometrical convergence of the series:

$$\mathcal{D}^{-1} = \Theta^V + \Theta\mathbf{D}^\perp\Theta + \Theta\mathbf{D}^\perp\Theta\mathbf{D}^\perp\Theta + \cdots .$$

Lemma 2.1. *If (2.3) is fulfilled on Ω_Δ, then $\mathcal{D}$ is invertible and*

$$P^\Delta |V|^{1/2}\mathcal{D}^{-1}(\lambda)|V|^{1/2}P^\Delta =: V(\lambda)$$

is analytic on Ω_Δ. It is represented by the geometrically convergent series

$$\begin{aligned} P^\Delta |V|^{1/2}\mathcal{D}^{-1}(\lambda)|V|^{1/2}P^\Delta &= P^\Delta V P^\Delta + P^\Delta |V|^{1/2}[\mathcal{D}^{-1} - \Theta^V]|V|^{1/2}P^\Delta \\ &=: P^\Delta V P^\Delta + P^\Delta V R_\lambda^\perp V P^\Delta + P^\Delta V R_\lambda^\perp V R_\lambda^\perp V P^\Delta + \cdots =: P^\Delta V P^\Delta + \Sigma^\Delta, \end{aligned}$$

where

$$\parallel \Sigma^\Delta(\lambda) \parallel = \parallel P^\Delta |V|^{1/2} \sum_{l\geq 1} \left[\Theta_V \mathbf{D}^\perp\right]^l \Theta_V |V|^{1/2} P^\Delta \parallel \leq \parallel P^\Delta |V|^{1/2} \parallel^2 \frac{1-\varepsilon}{\varepsilon}.$$

Moreover

$$\frac{d\Sigma^\Delta}{d\lambda} > 0.$$

Proof. The proof of the statements can be obtained via a straightforward estimation of the above series. For instance (2.3) implies the above estimation for Σ^Δ on Ω_Δ and

$$\begin{aligned} \Sigma^\Delta &= P^\Delta |V|^{1/2} \sum_{l\geq 1} \left[\Theta_V \mathbf{D}^\perp\right]^l \Theta_V |V|^{1/2} P^\Delta \\ &= P^\Delta V [R_\lambda^\perp + R_\lambda^\perp V R_\lambda^\perp + \cdots] V P^\Delta \\ &= P^\Delta V \frac{I}{H_0^\perp - \lambda I^\perp - P^\perp V P^\perp} V P^\Delta, \end{aligned}$$

hence

$$\frac{d\Sigma^\Delta}{d\lambda} = P^\Delta V \frac{I}{\left[H_0^\perp - \lambda I^\perp - P^\perp V P^\perp\right]^2} V P^\Delta > 0. \qquad \square$$

We will also use the analytic matrix function $\mathbf{d}$ in E^Δ

$$\begin{aligned} \mathbf{d} &= H^\Delta - \lambda I^\Delta - P^\Delta |V|^{1/2}\mathcal{D}^{-1}|V|^{1/2}P^\Delta =: H^\Delta - \lambda I^\Delta - V(\lambda) \\ &= H^\Delta - \lambda I^\Delta - P^\Delta V P^\Delta - \Sigma^\Delta(\lambda). \end{aligned}$$

Theorem 2.1. *If, for given $V = V^+$, there exist a domain Ω_Δ such that the above condition* (2.3) *is fulfilled, then the total multiplicity of eigenvalues of the perturbed operator $H_0 - V$ in Ω_Δ is equal to the total multiplicity of the eigenvalues of H_0 on Δ and the perturbed eigenvalues can be found as zeros of* $\mathbf{d}$. *The resolvent of the perturbed operator is represented on Ω_Δ by the modified Krein formula*

$$R^V_\lambda = R^\perp_\lambda + R^\perp_\lambda P^\perp |V|^{1/2} \mathcal{D}^{-1} |V|^{1/2} P^\perp R^\perp_\lambda \tag{2.6}$$
$$+ \left[P^\Delta + R^\perp_\lambda P^\perp |V|^{1/2} \mathcal{D}^{-1} |V|^{1/2} P^\Delta\right] \frac{I}{\mathbf{d}} \left[P^\Delta + P^\Delta |V|^{1/2} \mathcal{D}^{-1} |V|^{1/2} P^\perp R^\perp_\lambda\right].$$

Proof. The group of leading terms of the denominator on the spectral interval Δ includes Θ_V and the part of the spectral expansion of the unperturbed resolvent in E_Δ: $P^\Delta\, R^0_\lambda\, P^\Delta = R^\Delta_\lambda$, squeezed between the factors $|V|^{1/2}$.

$$\mathbf{D}_{\rm lead} := \Theta_V - |V|^{1/2} R^\Delta_\lambda |V|^{1/2} = \Theta_V + \sum_{s=1}^{N} \frac{|V|^{1/2}\varphi_s\rangle\,\langle |V|^{1/2}\varphi_s}{\lambda - \lambda_s}. \tag{2.7}$$

The rest $\mathbf{D}^\perp(\lambda) := |V|^{1/2} R^\perp_\lambda |V|^{1/2}$ of $\mathbf{D}$, containing the part of the unperturbed resolvent $R^\perp_\lambda$ in the complementary subspace, $E^\perp = E \ominus E^\Delta$ can be estimated on Ω_Δ, due to (2.4), as

$$\| \mathbf{D}^\perp(\lambda) \| \leq \frac{\| \, |V|^{1/2} P^\perp |V|^{1/2} \, \|}{\operatorname{dist}(\sigma^\perp, \lambda)} \leq 1 - \varepsilon. \tag{2.8}$$

Due to the above spectral estimate of $\mathbf{D}^\perp$ and (2.3) $\mathcal{D}$ is invertible for $\lambda \in \Omega_\Delta$.

Note that functions $\mathbf{D} = \mathbf{D}_{\rm lead} + \mathbf{D}^\perp$ and $\mathbf{D}_{\rm lead}$ have the same singularities in Ω_Δ. Then, based on the quoted operator-valued version of Rouche theorem, one can localize the zeros of $\mathbf{D}$ comparing $\mathbf{D}$ with $\mathbf{D}_{\rm lead}$. The inverse of $\mathcal{D}$ is calculated on Ω_Δ via Neumann series. We calculate the inverse of $\mathbf{D}$, solving the equation:

$$\left\{\mathcal{D} - |V|^{1/2} R^\Delta_\lambda |V|^{1/2}\right\} u = |V|^{1/2} R^0_\lambda f, \ f \in E. \tag{2.9}$$

Multiplying it by $\mathcal{D}^{-1}$ and introducing the new variable

$$v = \sum_{s=1}^{N} v_s \varphi_s = R^\Delta_\lambda |V|^{1/2} u,$$

with $v_s = (\lambda_s - \lambda)^{-1} \langle |V|^{1/2}\varphi_s, u\rangle$, we consider the dot-product of the result with $|V|^{1/2}\varphi_t$:

$$(\lambda_t - \lambda) v_t - \sum_{s=1}^{N} \langle |V|^{1/2}\varphi_t, \mathcal{D}^{-1} |V|^{1/2}\varphi_s\rangle v_s = \langle |V|^{1/2}\varphi_t \mathcal{D}^{-1} |V|^{1/2} R^0_\lambda f\rangle. \tag{2.10}$$

This equation for v can be reduced, via multiplication (2.10) by φ_t, and subsequent summation over t, to

$$\left(H^\Delta - \lambda I^\Delta\right) v - P^\Delta |V|^{1/2} \mathcal{D}^{-1} |V|^{1/2} P^\Delta v = P^\Delta |V|^{1/2} \mathcal{D}^{-1} |V|^{1/2} R^0_\lambda f. \tag{2.11}$$

The analytic matrix function $H^{\Delta} - \lambda I^{\Delta} - P^{\Delta}|V|^{1/2}\mathcal{D}^{-1}|V|^{1/2}P^{\Delta} = \mathbf{d}$ is invertible for all $\lambda \in \Omega_{\Delta}$, except the zeros of $\det \mathbf{d}$. Then, on the complement of the set of zeros we have:

$$v = \mathbf{d}^{-1}\, P^{\Delta}|V|^{1/2}\mathcal{D}^{-1}|V|^{1/2}R^0_{\lambda} f.$$

hence (2.11) implies:

$$\begin{aligned} u &= \mathcal{D}^{-1}|V|^{1/2}v + \mathcal{D}^{-1}|V|^{1/2}R^0_{\lambda} f \\ &= \mathcal{D}^{-1}|V|^{1/2}\left[P^{\Delta}\,\mathbf{d}^{-1}\, P^{\Delta}|V|^{1/2}\mathcal{D}^{-1}|V|^{1/2} + I\right] R^0_{\lambda} f. \end{aligned} \tag{2.12}$$

Inserting the above result into the above Krein formula (1.6), we obtain

$$R^V_{\lambda} = R^0_{\lambda} + R^0_{\lambda}|V|^{1/2}\mathcal{D}^{-1}|V|^{1/2}\left[P^{\Delta}\,\mathbf{d}^{-1}\, P^{\Delta}|V|^{1/2}\mathcal{D}^{-1}|V|^{1/2} + I\right] R^0_{\lambda}. \tag{2.13}$$

Both summands in the right side of (2.13) have singularities at the spectrum of H_0. But we are able to observe compensation of the singularities. Let us represent the left and right resolvent factors R^0_{λ} in the right side of (2.13) as

$$R^0_{\lambda} = P^{\Delta}\, R^0_{\lambda} + P^{\perp}\, R^0_{\lambda},$$

and consider the matrix elements of the second term in the right side of (2.13), with respect to the decomposition $E = E^{\Delta} \oplus E^{\perp}$, denoting the square bracket in (2.13) as $\left[P^{\Delta}\,\mathbf{d}^{-1}\, P^{\Delta}|V|^{1/2}\mathcal{D}^{-1}|V|^{1/2} + I\right] = [*]$:

$$\begin{aligned} R^{\Delta}_{\lambda} P^{\Delta}|V|^{1/2}\mathcal{D}^{-1}|V|^{1/2}[*]P^{\Delta}\, R^{\Delta}_{\lambda} &=: {}_{\Delta}\{*\}_{\Delta}, \\ R^{\perp}_{\lambda} P^{\perp}|V|^{1/2}\mathcal{D}^{-1}|V|^{1/2}[*]P^{\Delta}\, R^{\Delta}_{\lambda} &=: {}_{\perp}\{*\}_{\Delta}, \\ R^{\Delta}_{\lambda} P^{\Delta}|V|^{1/2}\mathcal{D}^{-1}|V|^{1/2}[*]P^{\perp}\, R^{\perp}_{\lambda} &=: {}_{\Delta}\{*\}_{\perp}, \\ R^{\perp}_{\lambda} P^{\perp}|V|^{1/2}\mathcal{D}^{-1}|V|^{1/2}[*]P^{\perp}\, R^{\perp}_{\lambda} &=: {}_{\perp}\{*\}_{\perp}. \end{aligned}$$

The square bracket in the first expression can be substituted by

$$[*]P^{\Delta} = P^{\Delta}\,\mathbf{d}^{-1}\, P^{\Delta}|V|^{1/2}\mathcal{D}^{-1}|V|^{1/2}P^{\Delta} + P^{\Delta} = \mathbf{d}^{-1}\,[H^{\Delta} - \lambda I^{\Delta}].$$

Then adding to ${}_{\Delta}\{*\}_{\Delta}$ the component R^{Δ}_{λ} of the first summand R^0_{λ} in E_{Δ}, we obtain

$$\begin{aligned} R^{\Delta} + {}_{\Delta}\{*\}_{\Delta} &= R^{\Delta}_{\lambda}\left[I + P^{\Delta}|V|^{1/2}\mathcal{D}^{-1}|V|^{1/2}P^{\Delta}\,\mathbf{d}^{-1}\, P^{\Delta}\right] \\ &= R^{\Delta}_{\lambda}\,[H^{\Delta}_0 - \lambda I^{\Delta}]\,P^{\Delta}\,\mathbf{d}^{-1}\, P^{\Delta} = P^{\Delta}\,\mathbf{d}^{-1}\, P^{\Delta}, \end{aligned} \tag{2.14}$$

thus, the singularities of first expression at the eigenvalues of H^{Δ}_0 are compensated by the singularities of the first term R^{Δ}_{λ}.

The second expression ${}_{\perp}\{*\}_{\Delta}$ is represented as

$$\begin{aligned} {}_{\perp}\{*\}_{\Delta} &= R^{\perp}_{\lambda} P^{\perp}|V|^{1/2}\mathcal{D}^{-1}|V|^{1/2}[*]P^{\Delta}\, R^{\Delta}_{\lambda} \\ &= R^{\perp}_{\lambda} P^{\perp}|V|^{1/2}\mathcal{D}^{-1}|V|^{1/2}\mathbf{d}^{-1}\,[H^{\Delta} - \lambda I^{\Delta}]\, R^{\Delta}_{\lambda} \\ &= R^{\perp}_{\lambda} P^{\perp}|V|^{1/2}\mathcal{D}^{-1}|V|^{1/2}P^{\Delta}\,\mathbf{d}^{-1}\, P^{\Delta}. \end{aligned} \tag{2.15}$$

Hence it is not singular at the eigenvalues of H_0^Δ. Similarly the third expression is transformed:

$$_\Delta\{*\}_\perp = P^\Delta\, \mathbf{d}^{-1}\, |V|^{1/2}\mathcal{D}^{-1}|V|^{1/2}R_\lambda^\perp.$$

The last expression $_\perp\{*\}_\perp$ does not have singularities at the eigenvalues of H_0 on Δ:

$$\begin{aligned} _\perp\{*\}_\perp = {} & R_\lambda^\perp|V|^{1/2}\mathcal{D}^{-1}|V|^{1/2}P^\Delta\, \mathbf{d}^{-1}P^\Delta|V|^{1/2}\mathcal{D}^{-1}|V|^{1/2}R_\lambda^\perp \\ & - R_\lambda^\perp|V|^{1/2}\mathcal{D}^{-1}|V|^{1/2}R_\lambda^\perp. \end{aligned} \tag{2.16}$$

Summarizing the results derived, we obtain the announced form of the Krein formula in the domain Ω_λ:

$$\begin{aligned} R_\lambda^V = {} & R_\lambda^\perp + R_\lambda^\perp P^\perp|V|^{1/2}\mathcal{D}^{-1}|V|^{1/2}P^\perp R_\lambda^\perp \\ & + \left[P^\Delta + R_\lambda^\perp P^\perp|V|^{1/2}\mathcal{D}^{-1}|V|^{1/2}P^\Delta\right]\mathbf{d}^{-1}\left[P^\Delta + P^\Delta|V|^{1/2}\mathcal{D}^{-1}|V|^{1/2}P^\perp R_\lambda^\perp\right] \end{aligned}$$

The only singularities in the right side of the above expression in Ω_Δ are the zeros $\lambda_t^V \in \Delta$ of the $N \times N$ analytic matrix function $\mathbf{d}$ in agreement with the elementary spectral perturbation result that the shift of the spectrum of H_0 does not exceed the norm of the perturbation. These zeros of $\mathbf{d}$ define the eigenvalues of the perturbed operator $H_0 - V$ in Ω_Δ. Really, the left and right factors

$$P^\perp + R_\lambda^\perp P^\perp|V|^{1/2}\mathcal{D}^{-1}|V|^{1/2}P^\Delta,\quad P^\perp + P^\Delta|V|^{1/2}\mathcal{D}^{-1}|V|^{1/2}P^\perp R_\lambda^\perp \tag{2.17}$$

are invertible analytic functions on Δ, hence none of the vector zeros of $\mathbf{d}$ on Δ can be compensated by the corresponding zero of the factor. □

3. Spectral results

Corollary 3.1. Group localization of eigenvalues. *Assume that, together with condition* (2.3), *the group localization of zeros of the operator function*

$$\mathbf{d}(\lambda) = H^\Delta - \lambda I^\Delta - P^\Delta V(\lambda)P^\Delta = H^\Delta - \lambda I^\Delta - P^\Delta V P^\Delta - \Sigma^\Delta(\lambda)$$

is observed:

1. *The discs B_ρ^s centered at the eigenvalues λ_s^V of the operator $H^\Delta - P^\Delta V P^\Delta$, with radii ρ,*

$$\rho > \frac{\| P^\Delta|V|^{1/2}\Theta_V P^\perp \|^2}{\varepsilon} \tag{3.1}$$

are contained in Ω_Δ.

2. *The union $\cup_s B_\rho^s = \mathbf{B}$ is represented as a sum of non-overlapping connected components $\mathbf{B}_l$: $\mathbf{B} = \cup_l \mathbf{B}_l$.*

Then the total multiplicity of eigenvalues of the perturbed operator $H_0 - V$ in each component $\mathbf{B}_l$ coincides with the multiplicity of eigenvalues of $H_V^\Delta := H^\Delta - P^\Delta V P^\Delta$ in $\mathbf{B}_l$.

Consider, together with the denominator $\mathbf{d}$, the "shortened" denominator

$$H^\Delta - \lambda I^\Delta - P^\Delta |V|^{1/2}\Theta_V |V|^{1/2} P^\Delta =: H^\Delta - \lambda I^\Delta - P^\Delta V P^\Delta =: \mathbf{d}_1(\lambda).$$

Estimate the deviation of the ratio $[\mathbf{d}_1(\lambda)]^{-1}\mathbf{d}(\lambda)$ from the unity. We can estimate the difference $\mathbf{d}(\lambda) - \mathbf{d}_1(\lambda) = \Sigma^\Delta$ based on Lemma 2.1. Then, due to the spectral estimation of the resolvent of H_V^Δ

$$\| \left[H_V^\Delta - \lambda I^\Delta\right]^{-1} \| \leq \rho^{-1},$$

we obtain:

$$\| I - \mathbf{d}(\lambda)\mathbf{d}_1^{-1}(\lambda) \| \leq \| P^\Delta |V|^{1/2} \|^2 \, \frac{1-\varepsilon}{\rho\varepsilon} \leq 1 - \varepsilon.$$

Hence the indices of zeros of $\mathbf{d}$ and $\mathbf{d}_1$ in each connected component $\mathbf{B}_l$ are equal due to the matrix Rouche theorem. Since both functions are analytic in Ω_Δ, the total multiplicity of zeros of $\mathbf{d}$ on each connected component $\mathbf{B}_l$ is equal to the total multiplicity of zeros of $\mathbf{d}_1$.

Choosing ρ, ε properly, we may assume that the first condition $B_\rho^s \in \Omega_\Delta$ is always fulfilled.

Corollary 3.2. The perturbed spectral projections.

The spectral projections at the isolated eigenvalues of a self-adjoint operator coincide with the residues of the corresponding polar terms in the resolvent. Though the operator function $\mathbf{d} = H^\Delta - \lambda I^\Delta - P^\Delta V P^\Delta - \Sigma(\Lambda)$ is self-adjoint if $\lambda \in \Delta$, we should prove that all poles of $\mathbf{d}^{-1}$ are simple. Note first that all zeros of $\mathbf{d}$ are simple, because $\frac{d\mathbf{d}}{d\lambda} = I^\delta + \frac{d\Sigma}{d\lambda} \geq I$. Now, assuming that λ_0^V, e_0 is a simple vector zero of $\mathbf{d}$, and denoting $e_0\rangle\,\langle e_0 =: P_0$, we represent $\mathbf{d}$ in close vicinity of λ_0 as

$$\mathbf{d}(\lambda) = H^\Delta - \lambda I^\Delta - P^\Delta V(\lambda_0) P^\Delta - (\lambda - \lambda_0^V)\frac{d\Sigma^\Delta}{d\lambda} + O\left\{(\lambda - \lambda_0^V)^2\right\}.$$

Neglecting the quadratic error, we represent, in the vicinity of λ_0, the equation

$$\mathbf{d}(\lambda)u = f$$

via the resolvent $\left[H^\Delta - \lambda I^\Delta - P^\Delta V(\lambda_0^V)P^\Delta\right]^{-1} = R_\lambda^0$ as

$$u = R_\lambda^0 f - (\lambda - \lambda_0^V) R_\lambda^0 \frac{d\Sigma^\Delta}{d\lambda}(\lambda_0^V) u \approx R_\lambda^0 f - P_0 \frac{d\Sigma^\Delta}{d\lambda} u,$$

with the projector $P_0 = e_0\rangle\,\langle e_0$. Due to Lemma 2.1 $\langle e_0, \frac{d\Sigma^\Delta}{d\lambda}(\lambda_0^V) e_0\rangle > 0$, hence in a small vicinity of λ_0^V

$$\begin{aligned} u &= R_\lambda^0 f - \frac{1}{1 + \langle e_0, \frac{d\Sigma^\Delta}{d\lambda}(\lambda_0^V) e_0\rangle} P_0 \frac{d\Sigma^\Delta}{d\lambda}(\lambda_0^V) R_\lambda^0 f \\ &\approx \frac{1}{1 + \langle e_0, \frac{d\Sigma^\Delta}{d\lambda}(\lambda_0^V) e_0\rangle} \frac{1}{\lambda_0^V - \lambda} P_0 f + \cdots . \end{aligned}$$

Denoting by $J,\ J^+$ the left and the right factors (2.17), we calculate the polar term of the perturbed resolvent at λ_0

$$\frac{1}{1+\langle e_0,\ \frac{d\Sigma^\Delta}{d\lambda}(\lambda_0^V)e_0\rangle}\,\frac{JP_0J^+}{\lambda_0^V-\lambda}.$$

Then we conclude from the above formula that generally, for the simple vector zeros of $\mathbf{d}$, index one, the following holds

The vector zeros $\lambda_s^V,\ \nu_s^V$ of $\mathbf{d}$ are simple eigenvalues of the perturbed operator H_0-V and the corresponding spectral projections are

$$P_s^V=\frac{JP_0J^+}{1+\langle \nu_s^V,\ \frac{d\Sigma^\Delta}{d\lambda}(\lambda_s^V)\nu_s^V\rangle}$$

The eigenfunctions of the perturbed operator may be obtained from the eigenfunctions $\{\nu_s^V\}$ of the operator $H^\Delta - P^\Delta V(\lambda_s^V)P^\Delta V$ as:

$$\Psi_s = J\nu_s^V,$$

with subsequent normalization.

In case of multiple zeros of $\mathbf{d}$ the factor $\left[1+\langle \nu_s^V,\ \frac{d\Sigma^\Delta}{d\lambda}(\lambda_s^V)\nu_s^V\rangle\right]^{-1}$ is substituted by the corresponding operator-valued function.

4. Example

While the standard analytic perturbation technique is aimed on a description of the behavior of a single eigenvalue under a small perturbation, our version (2.6) of Krein's formula permits to observe the behavior of groups of eigenvalues under perturbations. Note that considering groups of eigenvalues we do not connect the eigenpairs of the perturbed and the ones of the unperturbed operator "one-to-one" (individually), but just establish the correspondence between groups. This additional freedom permits to observe transformation of the intersection of terms into the quasi-intersection and details of the Landau-Zener phenomenon, see for instance [18, 19].

In this section we consider the behavior of the eigenvalues and eigenfunctions of the quasi-periodic Sturm-Liouvilles problem on a finite interval under a perturbation defined by a real continuous potential V. The role of the unperturbed operator in $L_2(0,1)$ is played by

$$H_0(p)u=-u'',\ u\in W_2^2(0,1),\ u(1)=e^{ip}\,u(0),\ \ u'(1)=e^{ip}\,u'(0) \tag{4.1}$$

with quasi-momentum $p, 0<p<2\pi$. The unperturbed eigenpairs are

$$\psi_0^l(x,p)=e^{i(2l\pi+p)x},\ \ \lambda_0^l(p)=(2l\pi+p)^2. \tag{4.2}$$

The perturbed operator, with the potential V is defined as

$$\begin{aligned}H_V(p)u&=-u''+Vu,\ u\in W_2^2(0,1),\\ u(1)&=e^{ip}\,u(0),\ \ u'(1)=e^{ip}\,u'(0),\ \ 0<p<2\pi.\end{aligned} \tag{4.3}$$

It also has discrete spectrum with corresponding eigenpairs

$$\psi_V^l(x,p),\ \lambda_V^l(p). \tag{4.4}$$

The asymptotic behavior of the eigenvalues

$$\lambda_V^l(p) = \lambda_0^l(p) + O(1) \tag{4.5}$$

with uniformly bounded error $O(1)$ and the corresponding asymptotic of eigenfunctions is obtained by WKB methods, see for instance [17]. The low-energy results are obtained for a small potential by the analytic perturbation technique, see for instance [20]. The above Krein formula (2.6) permits to describe the low-energy spectral properties of the quasi-periodic operator for any real continuous potential.

Note, first of all, that the spacing between consecutive eigenvalues $\lambda_0^l(p)$, $\lambda_0^{l+1}(p)$ of the unperturbed operator is growing linearly with l. Hence one always can select, for given continuous real potential V, a spectral interval Δ, such that $\text{dist}\,(\sigma^\Delta(p), \sigma^\perp(p)) > 2\max|V|$. Moreover, one can select Δ such that the parameter ε in the condition (3.1) is large enough, to guarantee that the radius ρ estimated in (3.1) is as small, as we need. Then the search of eigenvalues of the perturbed operator, reduced by Theorem 2.1 to the search of zeros of the corresponding function **d**, is substituted by the much easier search of eigenvalues and eigenfunctions of the finite matrix $H_0 - P^\Delta V P(\Delta)$, with an error ρ. For the calculation of eigenvalues and eigenfunctions of the classical Mathieu equation with $V(x) = \cos 2\pi x$ for $\pi/2 < p < 3\pi/2$ it is sufficient to choose $\Delta = [0, (\frac{7\pi}{2})^2]$. The spectrum $\sigma^\perp$ is situated above $(4\pi)^2$, hence $\text{dist}\,(\sigma^\perp(p), \sigma^\delta(p)) \geq \frac{15\pi^2}{4} >> 1 = \max|V(x)|$, so that $\varepsilon = \frac{15\pi^2}{4} - 1 > 32$. Then ρ can be chosen as 0.03. The original periodic spectral problem in $L_2(R)$ is substituted by the spectral problem with 4×4 matrix:

$$\mathbf{d}_1(\lambda,p) = H_0 - P^\Delta V P^\Delta = \begin{pmatrix} \lambda_0^{-1} & 0 & 0 & 0 \\ 0 & \lambda_0^0 & 0 & 0 \\ 0 & 0 & \lambda_0^1 & 0 \\ 0 & 0 & 0 & \lambda_0^2 \end{pmatrix}$$
$$+ \begin{pmatrix} \langle\psi_0^{-1}, V\psi_0^{-1}\rangle & \langle\psi_0^{-1}, V\psi_0^{0}\rangle & \langle\psi_0^{-1}, V\psi_0^{1}\rangle & \langle\psi_0^{-1}, V\psi_0^{2}\rangle \\ \langle\psi_0^{0}, V\psi_0^{-1}\rangle & \langle\psi_0^{0}, V\psi_0^{0}\rangle & \langle\psi_0^{0}, V\psi_0^{1}\rangle & \langle\psi_0^{0}, V\psi_0^{2}\rangle \\ \langle\psi_0^{1}, V\psi_0^{-1}\rangle & \langle\psi_0^{1}, V\psi_0^{0}\rangle & \langle\psi_0^{1}, V\psi_0^{1}\rangle & \langle\psi_0^{1}, V\psi_0^{2}\rangle \\ \langle\psi_0^{2}, V\psi_0^{-1}\rangle & \langle\psi_0^{2}, V\psi_0^{0}\rangle & \langle\psi_0^{2}, V\psi_0^{1}\rangle & \langle\psi_0^{2}, V\psi_0^{2}\rangle \end{pmatrix}.$$

Here λ_0^{-1}, λ_0^0, λ_0^1, λ_0^2, are the eigenvalues of the unperturbed operator, with zero potential, and ψ_0^{-1}, ψ_0^0, ψ_0^1, ψ_0^2 are the corresponding eigenfunctions. The group of unperturbed eigenvalues is situated on the interval $[0, (7\pi/2)^2]$ and reveal an intersection of terms depending in quasi-momentum p. The perturbed terms λ_V^l, $l = -1,\ 0,\ 1,\ 2$, with non-trivial $V \neq 0$, as functions of p reveal typical quasi-intersections. The first order approximation of the perturbed therm is found as zeros of the determinant det **d**. More precise results can be obtained in course of analysis of the corresponding 4×4 matrix $H_0 - P^\Delta |V|^{1/2} \Theta_V |V|^{1/2} P^\Delta$.

Acknowledgement

The authors are grateful to V. Kruglov and P. Kurasov for important references, and to the referee for useful remarks and suggestions.

References

[1] R. Newton, *Scattering theory of waves and particles* Dover Publications, Inc. Mineola, NY, 2002, 745 p.

[2] M. Krein, *Concerning the resolvents of an Hermitian operator with deficiency index* (m, m), Doklady AN USSR ,**52**, 651 (1946).

[3] M. Naimark, *Self-adjoint extensions of the second kind of a symmetric operators*, Bull. AN USSR, Ser. Math. **4**, 53 (1946).

[4] J. Schwinger, *On bound states of a given potential* Proc. Nat. Acad. Sci. USA, **47** (1961) pp. 122–129.

[5] M.Sh. Birman, *On the spectrum of singular boundary-value problems* Mat. Sbornik (N.S.) **55** (1961) pp. 125–174. English transl. in Amer. Math. Soc. Transl. **53** (1966), pp. 23–80.

[6] D.R. Yafaev, *Mathematical scattering theory. General Theory* Amer. Math. Soc., Providence, RI, (1992) 341 p.

[7] B. Simon, *Trace Ideals and Their Applications* London Math. Soc. Lecture Notes Series, 35. Cambridge Univ. Press, Cambridge, NY (1979) 134 p.

[8] V. Bogevolnov, A. Mikhailova, B. Pavlov, A. Yafyasov, *About Scattering on the Ring* In: "Operator Theory: Advances and Applications", Vol. 124 (Israel Gohberg Anniversary Volume), Ed. A. Dijksma, A.M. Kaashoek, A.C.M. Ran, Birkhäuser, Basel (2001) pp. 155–187.

[9] A. Mikhailova, B. Pavlov, *Resonance Quantum Switch* In: S. Albeverio, N. Elander, W.N. Everitt and P. Kurasov (eds.), Operator Methods in Ordinary and Partial Differential Equations, (S. Kovalevski Symposium, Univ. of Stockholm, June 2000), Birkhäuser, Basel-Boston-Berlin, 2002 (Operator Methods: Advances and Applications 132).

[10] A. Mikhailova, B. Pavlov, *Quantum Domain as a triadic Relay* In: Unconventional models of Computations UMC'2K, ed. I. Antoniou, C. Calude, M.J. Dinneeen, Springer Verlag series for Discrete Mathematics and Theoretical Computer Science (2001) pp. 167–186.

[11] A. Mikhailova, B. Pavlov, L. Prokhorov *Modelling of Quantum Networks* arXiv math-ph/031238, 2004, 69 p.

[12] N. Bagraev, A. Mikhailova, B. Pavlov, L. Prokhorov, A. Yafyasov, *Parameter regime of a resonance quantum switch*, Phys. Rev. B, 165308 (2005).

[13] T. Kato, *Perturbation theory for linear operators* Springer Verlag, Berlin-Heidelberg-NY, second edition (1976).

[14] Y.P. Ginzburg, *Almost invariant properties of analytic operator families* (Russian) Funktsional. Anal. i Prilozhen. 16 (1982), no. 1, 68–69.

[15] Y.P. Ginzburg, L.M. Zemskov, *Multiplicative representations of operator-functions of bounded type* (Russian) Teor. Funktsiĭ Funktsional. Anal. i Prilozhen. No. 53 (1990), 108–119; translation in J. Soviet Math. 58 (1992), no. 6, 569–576.

[16] I.S. Gohberg and E.I. Sigal, *Operator extension of the theorem about logarithmic residue and Rouché theorem* Mat. Sbornik. 84, 607 (1971).

[17] E.C. Titchmarsh, *Eigenfunctions expansions associated with second order differential equations*, Clarendon Press, Oxford, **1** (1962).

[18] C. Zener, *Non-adiabatic crossing of energy-levels* Proc. R. Soc. A, **137**, (1932) 696.

[19] Y.N. Demkov, P.B. Kurasov, V.N. Ostrovski, *Double-periodical in time and energy solvable system with two interacting set of states* Journal of Physics A, Math. and General, **28**, (1995), 434.

[20] V. Arnold, *Remarks on perturbation theory for problems of Mathieu type*, Russian Math. surveys **38**, 4, (1983), 215–233.

A.B. Mikhailova
V.A. Fock Institute for Physics
St. Petersburg State University
Ulyanovskaya 1, Petrodvorets
St. Petersburg, 198504, Russia

B.S. Pavlov
Department of Mathematics
The University of Auckland
Private Bag 92019
Auckland, New Zealand
e-mail: pavlov@math.auckland.ac.nz

Operator Theory:
Advances and Applications, Vol. 186, 339–346

On Some Properties of Infinite-dimensional Elliptic Coordinates

Andrey Osipov

Abstract. A number of the equations in classical mechanics is integrable in Jacobi elliptic coordinates. In recent years a generalization of elliptic coordinates to the infinite case has been offered. We consider this generalization and indicate the connection of infinite-dimensional elliptic coordinates with some inverse spectral problems for infinite Jacobi matrices and Sturm-Liouville operators (two spectra inverse problems). Also a link with the rank one perturbation theory is shown.

Mathematics Subject Classification (2000). 47B36; 49N45, 81Q10.

Keywords. Elliptic coordinates, Inverse spectral problems, Weyl function, Herglotz function, Rank one perturbation.

1. Introduction

One of the most powerful integration methods for equations of motion in classical mechanics is separation of variables. In a number of important cases there exists a substitution known as Jacobi elliptic coordinates which allows one to find such separating variables. For example such problems of classical mechanics as plane motion in the field of two attracting centers, Kepler's problem in the homogeneous force field and geodesic motion on an n-dimensional ellipsoid are integrable in elliptic coordinates [1].

They can be introduced in the following manner. Consider the space $H = \mathbb{C}^p$ (or $\mathbb{R}^p$), a symmetric operator A in H with simple eigenvalues and a vector $x \in H$. Then we can define the elliptic coordinates (e. c.) of the vector x as the set of numbers $\{\gamma_n\}$, which for some fixed $C > 0$ satisfies the equation

$$(R_\lambda x, x) = C, \quad R_\lambda = (A - \lambda E)^{-1}.$$

This work has been supported by the Russian Foundation For Basic Research: project 05-01-00989 and by the Ministry for Science and Technology of Russia: project NSh-5247.2006.1 .

If we take the orthonormal basis $\{e_n\}_{n=1}^p$ in H such that $Ae_n = \lambda_n e_n$, and $\lambda_n < \lambda_{n+1}$, $n = 1, \dots, p-1$ then the above equation can be written as

$$\sum_{n=1}^{p} \frac{|x_n|^2}{\lambda_n - \lambda} = C, \quad x_n = (x, e_n).$$

Assume that $x_n \neq 0$ for all n. One can easily check that the above equation has exactly p roots and $\gamma_1 < \lambda_1$, $\lambda_{n-1} < \gamma_n < \lambda_n$, $n = 2, \dots, p$. For the coordinates x_i we have (see, e.g., [1]):

$$|x_i|^2 = C \frac{\prod_{n=1}^{p}(\lambda_i - \gamma_n)}{\prod_{n \neq i}(\gamma_i - \gamma_n)}, \quad i = 1, \dots, p.$$

V.I. Arnold posed the following question [1]–[2]: Generalize Jacobi elliptic coordinates to the infinite-dimensional setting and find equations of mathematical physics integrable by this method.

The first step in this direction has been made by Kostyuchenko and Stepanov in [3]. Considering a separable Hilbert space H instead of $\mathbb{C}^p$ and applying the above operator approach they found the necessary and sufficient conditions for a sequence $\{\gamma_n\}$ to be the e. c. of a certain vector $x \in H$. Unfortunately these conditions are too restrictive for some cases important for the integration task. In [4] Vaninsky considered a more general concept of e. c. which is operator-free. He has shown that the Camassa-Holm equation (approximation to the Euler equation describing an ideal fluid) is integrable in e. c. introduced in this manner.

The aim of this paper is to indicate the connection between both concepts of e. c. and some two-spectra inverse problems. It allows one to formulate some known results on these inverse problems in terms of e. c. Also our examples show that both concepts of elliptic coordinates can be embedded into the rank one perturbation framework. The first concept corresponds to the bounded perturbations and the second concept to the singular ones.

2. First definition of infinite e. c. and link to inverse problems

Consider a self-adjoint lower semibounded operator A in a separable Hilbert space H with a simple purely discrete spectrum $\{\lambda_n\}_{n=1}^{\infty}$, $\lambda_1 < \lambda_2 < \cdots < \lambda_n \to \infty$ as $n \to \infty$. Let $\{e_n\}_{n=1}^{\infty}$ be its orthonormal eigenbasis. As in [3], define the elliptic coordinates $\{\gamma_n\}_{n=1}^{\infty}$ of a certain cyclic vector $x \in H$ as the ordered roots of the equation (1). If we define a function

$$F(\lambda) = \sum_{n=1}^{\infty} \frac{|x_n|^2}{\lambda_n - \lambda} - C, \quad x_n = (x, e_n),$$

then (1) is equivalent to the equation $F(\lambda) = 0$. This seems to be the most natural way to generalize the above definition of elliptic coordinates to the infinite-dimensional case. $F(\lambda)$ is a Herglotz function, so its zeros and poles interlace:

$$\gamma_1 < \lambda_1 < \gamma_2 < \lambda_2 < \cdots . \tag{2.1}$$

Theorem (Kostyuchenko-Stepanov). *Let the operator A is defined as above. A sequence $\{\gamma_n\}_{n=1}^{\infty}$ satisfying (2.1) represents the elliptic coordinates of some vector $x \in H$ and cyclic for A, iff the following condition holds*

$$\sum_{n=1}^{\infty}(\lambda_n - \gamma_n) < \infty. \tag{2.2}$$

In this case for the coordinates $x_i = (x, e_i)$ of x the following equations hold:

$$|x_i|^2 = C\frac{\prod_{n=1}^{\infty}(\lambda_i - \gamma_n)}{\prod_{n\neq i}(\gamma_i - \gamma_n)}, \quad i \in \mathbb{N} \tag{2.3}$$

$$C\sum_{n=1}^{\infty}(\lambda_n - \gamma_n) = \sum_{i=1}^{\infty}|x_i|^2 \tag{2.4}$$

It is interesting to note (see [4]), that (2.4) is a direct generalization to the infinite-dimensional case of an ancient formula of Boole.

Now we discuss one inverse problem for semi-infinite Jacoby matrices considered in [5]. Namely, consider the matrix

$$J = \begin{pmatrix} b_1 & a_1 & 0 & 0 \\ a_1 & b_2 & a_2 & 0 \\ 0 & \ddots & \ddots & \ddots \end{pmatrix} \quad a_n > 0,\ b_n \in \mathbb{R},\ n \in \mathbb{N},$$

such that the limit point case (see, e.g., [6]) holds, and the spectrum of the corresponding self-adjoint operator J in the space $l^2(\mathbb{N})$ is purely discrete with the accumulation point at $+\infty$. Denote by $\{e_n\}_{n=1}^{\infty}$ is the canonical basis in $l^2(\mathbb{N})$. Let $\{\lambda_n\}_{n=1}^{\infty}$ be the ordered sequence of eigenvalues of J (they are always simple). Also consider the following bounded rank one perturbation of J:

$$J_h = J + h(e_1, \cdot)e_1, \tag{2.5}$$

where h is a negative integer. Denote by $\{\mu_n\}_{n=1}^{\infty}$ its ordered spectrum (which is also purely discrete). The problem studied in [5] is to find J and h for a given $\{\lambda_n\}_{n=1}^{\infty}$ and $\{\mu_n\}_{n=1}^{\infty}$.

First of all, we recall the definitions of the basic objects in the theory of Jacobi operators [6]. Namely, for the operator J one can introduce the right-continuous resolution of identity $E(t)$ such that $J = \int_{\mathbb{R}} t dE(t)$. Then the spectral density function $\rho(t)$ of J is defined as follows:

$$\rho(t) = (E(t)e_1, e_1).$$

Vector e_1 is cyclic for J. Also consider the function

$$m(\lambda) = ((J - \lambda E)^{-1}e_1, e_1) = \int_{\mathbb{R}} \frac{d\rho(t)}{t - \lambda}.$$

which is called the Weyl function of J.

The solution $f = \{f_n\}_{n=1}^{\infty}$ of the equation

$$Jf = \lambda f, \qquad \lambda \in \mathbb{C}$$

is uniquely determined if one gives $f_1 = 1$. For the elements of this solution the following notation is standard: $P_k(\lambda) = f_{k+1}$, $k \in \mathbb{Z}_+$, where the polynomial P_k (of degree k) is referred to as the kth orthogonal polynomial of the first kind associated with J. If for some $\lambda = \lambda_n$, $\sum_{k=0}^{\infty} |P_k(\lambda_n)|^2 < \infty$, then λ_n is the eigenvalue of an operator J and $f = f(\lambda_n)$ is the corresponding eigenvector. The coefficients $\alpha_n = \sum_{k=0}^{\infty} |P_k(\lambda_n)|^2$ are called the normalizing factors.

For the considered operator J the corresponding $\rho(t)$ and $m(\lambda)$ are given by the formulas.

$$\rho(t) = \sum_{\lambda_n \le t} \frac{1}{\alpha_n}, \quad m(\lambda) = \sum_{n=1}^{\infty} \frac{1}{\alpha_n(\lambda_n - \lambda)}. \tag{2.6}$$

Also

$$1 = (e_1, e_1) = \int_{\mathbb{R}} d\rho(t) = \sum_{n=1}^{\infty} \frac{1}{\alpha_n}. \tag{2.7}$$

In general, $\rho(t)$ or $m(\lambda)$ uniquely determines J [6]. In our case it means that the knowledge of $\{\lambda_n\}_{n=1}^{\infty}$ and $\{\alpha_n\}_{n=1}^{\infty}$ is enough for the unique reconstruction of J.

It was shown in [5] that the normalizing factors of J can be calculated by the formula

$$\alpha_i = \frac{-h}{\lambda_i - \mu_i} \frac{\prod_{n \neq i}(\mu_i - \mu_n)}{\prod_{n=1}^{\infty}(\lambda_i - \mu_n)}, \qquad i \in \mathbb{N}, \tag{2.8}$$

and the following trace formula holds

$$\sum_{n=1}^{\infty}(\lambda_n - \mu_n) = -h. \tag{2.9}$$

Therefore (2.8)–(2.9) give us a key to the solution of the above-stated inverse problem. Also in [5] some sufficient conditions for a pair of sequences to be the spectra of a Jacobi operator and its rank one perturbation were given.

On can easily obtain (2.8)–(2.9) by using the Kostyuchenko-Stepanov theorem. Indeed, according to the Aronzajn-Krein formula

$$m_h(\lambda) = \frac{m(\lambda)}{1 + hm(\lambda)}$$

where $m_h(\lambda)$ is the Weyl function of J_h.

Therefore $\{\mu_n\}$ being the poles of $m_h(\lambda)$, satisfy the equation

$$\sum_{n=1}^{\infty} \frac{1}{\alpha_n(\lambda_n - \lambda)} + \frac{1}{h} = 0.$$

For the normalized eigenvectors $\hat{f}_n$ of J we also have

$$(\hat{f}_n, e_1) = \frac{P_0(\lambda_n)}{\sqrt{\alpha_n}} = \frac{1}{\sqrt{\alpha_n}}, \qquad n \in \mathbb{N},$$

so $\{\mu_n\}$ are the elliptic coordinates of the vector e_1 for $C = -\frac{1}{h} > 0$. Then the formulas (2.8)–(2.9) follow directly from (2.3)–(2.4) and (2.7). In view of the above we obtain the following result.

Proposition 2.1. *The spectrum of J and the elliptic coordinates $\{\gamma_n\}_{n=1}^{\infty}$ of e_1 for $C = -\frac{1}{h} > 0$, uniquely determine J and h, and $\{\gamma_n\}$ coincide with the spectrum of J_h defined by* (2.5).

Now consider another vector ϕ, cyclic for J. Define

$$\tilde{J} = J + h(\phi, \cdot)\phi, \quad h < 0$$

and

$$F_h(\lambda) = \int_{\mathbb{R}} \frac{d_h\tau(t)}{t - \tau},$$

where $d_h\tau(t)$ is the spectral measure for ϕ associated to $\tilde{J}_h$. Then applying the Aronzajn-Krein formula we obtain that the eigenvalues of $\tilde{J}_h$ satisfy the equation

$$\sum_{n=1}^{\infty} \frac{\tau_n}{\lambda_n - \lambda} + \frac{1}{h} = 0, \quad \tau_n = (\hat{f}_n, \phi).$$

therefore they are e. c. of ϕ. Conversely, if we take the e. c. $\{\gamma_n\}_{n=1}^{\infty}$ of some ϕ for $C > 0$, then one can check that $\{\gamma_n\}$ is the spectrum of $J - \frac{1}{C}(\phi, \cdot)\phi$

Note that according to the Stone theorem (see, e.g., [6]) the above-defined operators A and J are equivalent. Thus we are coming to the following conclusion.

Theorem 2.2. *Let the operator A in H be defined as above. Then for an arbitrary cyclic $x \in H$ its elliptic coordinates for any $C > 0$ and the spectrum of its rank one perturbation*

$$A - \frac{1}{C}(x, \cdot)x$$

are identical.

3. More general definition and Sturm-Liouville inverse problems

As we see, the condition (2.2) plays an important role in the above constructions. However, as pointed in [4], for the string spectral problem associated to the Camassa-Holm equation, (2.2) does not hold for the spectra of the string with different types of boundary conditions. To overcome this obstacle, in ([4]) the infinite elliptic coordinates were introduced in a more general way.

Namely, pick some $C > 0$ and consider the function

$$F(\lambda) = \sum_{n=1}^{\infty} \tfrac{A_n}{\lambda_n - \lambda} - C \tag{3.1}$$
$$A_n > 0,\ n \in \mathbb{N}; \quad \lambda_n < \lambda_{n+1} \to \infty \quad \text{as} \quad n \to \infty,$$

and satisfying the condition

$$\sum_{n=1}^{\infty} \frac{A_n \lambda_n}{1 + \lambda_n^2} < \infty \tag{3.2}$$

We will consider the roots of $F(\lambda) = 0$ as the elliptic coordinates arising from some sequence $\{A_n\}$ and $C > 0$

Theorem 3.1. *The sequence* $\{\gamma_n\}_{n=1}^{\infty}$ *represents the elliptic coordinates arising from the sequence* $\{A_n\}_{n=1}^{\infty}$ *satisfying* (3.2) *and some* $C > 0$ *iff* $\{\gamma_n\}$ *interlace* $\{\lambda_n\}$*:*

$$\gamma_n < \lambda_n < \gamma_{n+1},\ n \in \mathbb{N}; \qquad \lambda_{N-1} < 0 < \gamma_{N+1},$$

and the following condition holds:

$$\prod_{n=1}^{\infty}{}' \frac{\lambda_n}{\gamma_n} < \infty \tag{3.3}$$

(where the prime in the infinite product means that it does not include the factor $n = N$*).*

Proof. Necessity. Consider the function (3.1), satisfying (3.2). Then it is not hard to see (e.g., by using the Herglotz representation theorem) that $F(\lambda)$ maps the upper half-plane onto itself and its poles and zeros interlace. Also $\lim_{\lambda\to\infty} F(\lambda) = -C$. The $F(\lambda)$ admits the following representation [7]:

$$F(\lambda) = K\left(\frac{\lambda - \gamma_N}{\lambda - \lambda_N}\right)\prod_{n=1}^{\infty}{}'\frac{1-\lambda/\gamma_n}{1-\lambda/\lambda_n}, \qquad K < 0.$$

Therefore $-C = \lim_{\lambda\to-\infty} F(\lambda) = K\prod_{n=1}^{\infty}{}'\frac{\lambda_n}{\gamma_n}$.

Sufficiency. Suppose that the conditions of the theorem are fulfilled. Consider the Herglotz function

$$\tilde{F}(\lambda) = -\prod_{n=1}^{\infty}{}'\frac{1-\lambda/\gamma_n}{1-\lambda/\lambda_n}.$$

From the Herglotz representation theorem we obtain that

$$\tilde{F}(\lambda) = A\lambda + B + \sum_{n=1}^{\infty} A_n\left(\frac{1}{\lambda_n - \lambda} - \frac{\lambda_n}{1+\lambda_n^2}\right)$$

where $A \geq 0,\ B \in \mathbb{R},\ A_n > 0$ and

$$\sum_{n=1}^{\infty}\frac{A_n}{1+\lambda_n^2} < \infty.$$

Since $\lim_{\lambda\to-\infty} \tilde{F}(\lambda) = -\prod_{n=1}^{\infty}{}'\frac{\lambda_n}{\gamma_n} > -\infty$ we have $A = 0$ and

$$\sum_{n=1}^{\infty}\frac{A_n\lambda_n}{1+\lambda_n^2} < \infty.$$

Thus $\tilde{F}(\lambda)$ admits the representation (3.1). □

A similar theorem has been proved in [4] (Theorem 5.2). The above theorem is more suitable for our case.

To see the examples of these e. c. consider the self-adjoint operator $L_h = -\frac{d^2}{dx^2} + q(x)$ in $L^2(0,\pi)$ with boundary conditions

$$y'(0) + hy(0) = 0, \tag{3.4}$$
$$y'(\pi) + Hy(\pi) = 0, \tag{3.5}$$

where $h, H \in \mathbb{R}$ and $q(x) \in C_{\mathbb{R}}[0,\pi]$. Denote by $\{\lambda_n^h\}_{n=1}^{\infty}$ its ordered eigenvalues and by $\phi_n = \phi_n^h$ the corresponding eigenfunctions such that $\phi_n(0) = 1$ and $\phi_n'(0) = -h$. The following asymptotic formulas are well known.

$$\lambda_n^h = (n-1)^2 + \tfrac{2}{\pi}\left(H - h + \tfrac{1}{2}\int_0^\pi q(\tau)d(\tau)\right) + o\left(\tfrac{1}{n}\right), \tag{3.6}$$
$$\alpha_n^h = \int_0^\pi \phi_n^2(x)d(x) = \tfrac{\pi}{2} + o\left(\tfrac{1}{n}\right). \tag{3.7}$$

The spectral density function $\rho_h(t)$ of L_h and its Weyl function $m_h(\lambda)$ are defined similarly to (2.6) ($\alpha_n = \alpha_n^h$). Now pick two values of h : $h_0 = 0$ and $h_1 > 0$. Then, as shown in [8], the eigenvalues of L_{h_1}, being the poles of $m_{h_1}(\lambda)$, satisfy the following equation

$$m_0(\lambda) \equiv \sum_{n=1}^{\infty} \frac{1}{\alpha_n^0(\lambda_n^0 - \lambda)} = \frac{1}{h}.$$

As we see from (3.6–3.7), for the spectrum of L_0 and $\{\gamma_n = \lambda_n^{h_1}\}$ the condition (2.2) is no longer valid, whereas the condition (3.3) is valid. This allows us to formulate the following result.

Proposition 3.2. *For the operator L_0 and any $h > 0$ the spectrum of L_h represents the elliptic coordinates arising from the sequence $\{\frac{1}{\alpha_n^0}\}$ and $C = \frac{1}{h}$.*

The operators L_h can be considered as singular rank one perturbations of L_0 (see, e.g., [9]).

$$L_h = L_0 - h\langle\phi,\cdot\rangle\phi,$$

where $\phi = \delta_0$, the delta function at zero. So for this case, similarly to (1) we can define the elliptic coordinates of the singular vector ϕ for some $C > 0$ as the roots of the equation

$$\langle\phi, (L_0 - \lambda)^{-1}\phi\rangle = C,$$

and the spectrum of L_h coincide with the e. c. introduced in this manner for $h = \frac{1}{C}$.

Finally note that one may obtain a similar results for the Sturm-Liouville operators considered in $L^2(0,\infty)$, having a discrete spectrum and satisfying (3.4) ([8], [10]).

References

[1] V.I. Arnold, *Mathematical methods of classical mechanics* (Russian) Nauka, 1989.

[2] V.I. Arnold, *Arnold's problems.* Springer-Verlag, 2004.

[3] A.G. Kostyuchenko, A.A. Stepanov, *Infinite-dimensional elliptic coordinates.* Funct. Anal. Appl. **33** No 4, (1999), 300–303.

[4] K.L. Vaninsky, *Equations of Camassa-Holm type and Jacobi ellipsoidal coordinates.* Commun. Pure Appl. Math. **58** No 9, (2005), 1149–1187.

[5] R.Z. Khalilova, *On one inverse problem* (Russian). Izv. Akad. Nauk Az. SSR, Ser. Fiz.-Tekh. Mat. Nauk **3-4** (1967), 169–175.

[6] N.I. Akhiezer, *The classical problem and some related problems in analysis.* Hafner Publishing, N. Y., 965.

[7] B.J. Levin, *Distribution of zeros of the entire functions.* Translations of Mathematical Monographs **5**. Amer. Math. Soc., Providence, 1980.

[8] B.M. Levitan, M.G. Gasymov, *On the determination of a Sturm-Liouville equation by two spectra.* Russ. Math. Surv. **19** No. 2, (1964), 1–63.

[9] S. Albeverio, F. Gesztesy, R. Hoegh-Krohn, H. Holden, *Solvable models in quantum mechanics.* Springer-Verlag, 1988.

[10] F. Gesztesy, B. Simon, *Rank one perturbations at infinite coupling.* J. Funct. Anal. **128** No 1, (1995), 245–252.

Andrey Osipov
Scientific-Research Institute for System Studies
Russian Academy of Sciences (NIISI RAN)
Nakhimovskii pr. 36-1
Moscow, 117218, Russia
e-mail: `osipa68@yahoo.com`

Operator Theory:
Advances and Applications, Vol. 186, 347–389

Finite Difference Operators with a Finite-Band Spectrum

F. Peherstorfer and P. Yuditskii

Abstract. We study the correspondence between almost periodic difference operators and algebraic curves (spectral surfaces). An especial role plays the parametrization of the spectral curves in terms of, so-called, branching divisors. The multiplication operator by the covering map with respect to the natural basis in the Hardy space on the surface is the $2d+1$-diagonal matrix; the d-root of the product of the Green functions (counting their multiplicities) with respect to all infinite points on the surface is the symbol of the shift operator. We demonstrate an application of our general construction to the particular covering, which generate almost periodic CMV matrices recently widely discussed. Then we study an important theme: covering of one spectral surface by another one and the related transformations on the set of multidiagonal operators (so-called Renormalization Equations). We prove several new results dealing with Renormalization Equations for periodic Jacobi matrices (polynomial coverings) and for the case of a rational double covering.

Mathematics Subject Classification (2000). 47B36, 30F35, 47B32.

Keywords. Hardy Spaces, Riemann Surfaces, Spectral theory, Jacobi matrices, iterations, expanding polynomials, CMV matrices.

1. Introduction

1.1. Ergodic finite difference operators and associated Riemann surfaces

The standard (three-diagonal) finite-band Jacobi matrices [9, 27] can be defined as almost periodic or even ergodic Jacobi matrices with absolutely continuous spectrum that consists of a finite system of intervals. We wish to find a natural extension of this class of finite difference operators to the multi-diagonal case.

Partially supported by the Austrian Founds FWF, project number: P20413–N18 and *Marie Curie International Fellowship* within the 6th European Community Framework Programme, Contract MIF1-CT-2005-006966.

The correspondence between periodic difference operators and algebraic curves (spectral surfaces) was discussed in detail in [18]. Let us generalize this construction to show (at least on a speculative level) how almost periodic or ergodic operators give rise to a corresponding spectral surface.

Recall the definition of an ergodic operator [7, 19], see also [29]. Let $(\Omega, \mathfrak{A}, d\chi)$ be a separable probability space and let $\mathcal{T} : \Omega \to \Omega$ be an invertible ergodic transformation, i.e., $\mathcal{T}$ is measurable, it preserves $d\chi$, and every measurable $\mathcal{T}$-invariant set has measure 0 or 1. Let $\{q^{(k)}\}_{k=0}^d$ be functions from $L^\infty_{d\chi}$, with $q^{(d)}$ positive-valued and $q^{(0)}$ real-valued. Note that in the periodic case $\Omega = \mathbb{Z}/N\mathbb{Z}$, where N is the period and $\mathcal{T}\{n\} = \{n+1\}$, $\{n\} \in \mathbb{Z}/N\mathbb{Z}$.

Then with almost every $\omega \in \Omega$ we associate a self-adjoint $2d+1$-diagonal operator $J(\omega)$ as follows:

$$(J(\omega)x)_n = \sum_{k=-d}^{d} \overline{q_n^{(k)}(\omega)} x_{n+k}, \quad x = \{x_n\}_{n=-\infty}^{\infty} \in l^2(\mathbb{Z}),$$

where $q_n^{(k)}(\omega) := q^{(k)}(\mathcal{T}^n\omega)$ and $q^{(-k)}(\omega) := \overline{q^{(k)}(\mathcal{T}^{-k}\omega)}$.

Note that the structure of $J(\omega)$ is described by the following identity

$$J(\omega)S = SJ(\mathcal{T}\omega), \tag{1}$$

where S is the shift operator in $l^2(\mathbb{Z})$. The last relation indicates strongly that one can associate with the family of matrices $\{J(\omega)\}_{\omega\in\Omega}$ a natural pair of commuting operators (in the periodic case one just uses the fact that J and S^N commute).

Namely, let $L^2_{d\chi}(l^2(\mathbb{Z}))$ be the space of $l^2(\mathbb{Z})$-valued vector functions, $x(\omega) \in l^2(\mathbb{Z})$, with the norm

$$||x||^2 = \int_\Omega ||x(\omega)||^2 \, d\chi.$$

Define

$$(\widehat{J}x)(\omega) = J(\omega)x(\omega), \quad (\widehat{S}x)(\omega) = Sx(\mathcal{T}\omega), \quad x \in L^2_{d\chi}(l^2(\mathbb{Z})).$$

Then (1) implies

$$(\widehat{J}\widehat{S}x)(\omega) = J(\omega)Sx(\mathcal{T}\omega) = SJ(\mathcal{T}\omega)x(\mathcal{T}\omega) = (\widehat{S}\widehat{J}x)(\omega).$$

Further, $\widehat{S}$ is a unitary operator and $\widehat{J}$ is self-adjoint. The space $L^2_{d\chi}(l^2(\mathbb{Z}_+))$ is an invariant subspace for $\widehat{S}$. It is not invariant with respect to $\widehat{J}$ but it is invariant with respect to $\widehat{J}\widehat{S}^d$. Put

$$\widehat{S}_+ = \widehat{S}|L^2_{d\chi}(l^2(\mathbb{Z}_+)), \quad (\widehat{J}\widehat{S}^d)_+ = \widehat{J}\widehat{S}^d|L^2_{d\chi}(l^2(\mathbb{Z}_+)).$$

Definition 1.1 (local functional model). We say that a pair of commuting operators $A_1 : H \to H$ and $A_2 : H \to H$ has a (local) functional model if there is a unitary embedding $i : H \to H_O$ in a space H_O of functions $F(\zeta)$ holomorphic in some domain O with a reproducing kernel ($F \mapsto F(\zeta_0)$, $\zeta_0 \in O$, is a bounded functional in H_O) such that the operators $i_\star A_1$ and $i_\star A_2$ become a pair of operators of multiplication by holomorphic functions, say

$$A_1 x \mapsto a_1(\zeta)F(\zeta), \quad A_2 x \mapsto a_2(\zeta)F(\zeta).$$

As usual $i_\star A$ is generated by the diagram

$$\begin{array}{ccc} H & \xrightarrow{A} & H \\ i\big\downarrow & & i\big\downarrow \\ H_O & \xrightarrow{i_\star A} & H_O \end{array}$$

Existence of a local functional model implies a number of quite strong consequences. In what follows $a(\zeta)$ and $b(\zeta)$ denote the functions (symbols) related to the operators $(\widehat{J}\widehat{S}^d)_+$ and $\widehat{S}_+$. Let k_ζ be the reproducing kernel in H_O and let $\hat{k}_\zeta$ be its preimage $i^{-1}k_\zeta$ in $L^2_{d\chi}(l^2(\mathbb{Z}_+))$. Then

$$\langle \widehat{S}_+^* \hat{k}_\zeta, x\rangle = \langle \hat{k}_\zeta, \widehat{S}_+ x\rangle = \langle k_\zeta, bF\rangle.$$

By the reproducing property

$$\langle k_\zeta, bF\rangle = \overline{b(\zeta)F(\zeta)} = \langle \overline{b(\zeta)}k_\zeta, F\rangle.$$

Hence,

$$\langle \widehat{S}_+^* \hat{k}_\zeta, x\rangle = \langle \overline{b(\zeta)}\hat{k}_\zeta, x\rangle.$$

That is $\hat{k}_\zeta$ is an eigenvector of $\widehat{S}_+^*$ with the eigenvalue $\overline{b(\zeta)}$. In the same way, $\hat{k}_\zeta$ is an eigenvector of $(\widehat{J}\widehat{S})_+^*$ with the eigenvalue $\overline{a(\zeta)}$.

Thus, if a functional model exists then the spectral problem

$$\begin{cases} \widehat{S}_+^* \hat{k}_\zeta & = \overline{b(\zeta)}\hat{k}_\zeta \\ (\widehat{J}\widehat{S}^d)_+^* \hat{k}_\zeta & = \overline{a(\zeta)}\hat{k}_\zeta \end{cases}, \tag{2}$$

has a solution $\hat{k}_\zeta$ antiholomorphic in ζ. Moreover, linear combinations of all $\hat{k}_\zeta$ are dense in $L^2_{d\chi}(l^2(\mathbb{Z}_+))$. Vice versa, if (2) has a solution of such kind then we define

$$F(\zeta) := \langle x, \hat{k}_\zeta\rangle, \quad ||F||^2 := ||x||^2.$$

This provides a local functional model for the pair $\widehat{S}_+$, $(\widehat{J}\widehat{S}^d)_+$.

The following proposition is evident.

Proposition 1.2. *Let $\mathcal{U} : L^2_{d\chi} \to L^2_{d\chi}$ be the unitary operator associated with the ergodic transformation $\mathcal{T}$: $(\mathcal{U}c)(\omega) = c(\mathcal{T}\omega)$, $c \in L^2_{d\chi}$. We denote by the same letter q both a function $q \in L^\infty_{d\chi}$ and the multiplication operator q (e.g., $(qc)(\omega) := q(\omega)c(\omega)$). Problem (2) is equivalent to the following spectral problem*

$$\left\{\sum_{k=-d}^{d} \mathcal{U}^k \overline{q^{(k)}b^k(\zeta)}\right\} c_\zeta = \overline{z(\zeta)}c_\zeta, \tag{3}$$

where $z(\zeta) := a(\zeta)/b^d(\zeta)$ and c_ζ is an anti-holomorphic $L^2_{d\chi}$-valued vector function. Moreover $\{c_\zeta\}$ is complete in $L^2_{d\chi}$ if and only if $\{\hat{k}_\zeta\}$ is complete in $L^2_{d\chi}(l^2(\mathbb{Z}_+))$.

We may hope to get a *global* functional model, see Section 2, on a Riemann surface

$$X_0 = \{(z, b) : |b| < 1\}$$

with a local coordinates ζ given by (3). We do not claim that the global model always exists (even existence of a local model requires some additional assumptions on the ergodic map and the coefficients functions). But, in particular, in the periodic case, when $L^2_{d\chi} = \mathbb{C}^N$, we have

$$\mathcal{U}\begin{bmatrix} c_0 \\ \vdots \\ c_{N-1} \end{bmatrix} = \begin{bmatrix} c_1 \\ \vdots \\ c_0 \end{bmatrix}, \quad \mathcal{U} = \begin{bmatrix} 0 & 1 & & \\ & \ddots & \ddots & \\ & & \ddots & 1 \\ 1 & & & 0 \end{bmatrix},$$

where $c_n := c(\{n\})$ and the $q^{(k)}$'s became the diagonal matrices. Thus (3) means that the $N \times N$ matrix has a nontrivial annihilating vector and we arrive at the curve X_0 in the form:

$$\det \left[\sum_{k=-d}^{d} \mathcal{U}^k \overline{q^{(k)} b^k} - \bar{z} I \right] = 0$$

and the restriction $|b| < 1$.

The surface X_0 is in generic case of infinite genus. However we can reduce it because X_0 possesses a family of automorphisms. Assume that $X_0 = \mathbb{D}/\Gamma_0$ is its uniformization with a Fuchsian group Γ_0. Let $e_{\{\gamma\}}$ be an eigenvector of $\mathcal{U}$ with an eigenvalue $\bar{\mu}_{\{\gamma\}}$. The systems of eigenfunctions and eigenvalues form both Abelian groups with respect to multiplication. Using (3) we get immediately that $\{\gamma\} : (z, b) \mapsto (z, \mu_{\{\gamma\}} b)$ is an automorphism of X_0. Taking a quotient of X_0 with respect to these automorphisms we obtain a much smaller surface $X = \mathbb{D}/\Gamma$, $\Gamma = \{\gamma\Gamma_0\}$. In periodic case this means that (1.1) is actually a polynomial expression in z, λ and λ^{-1} where $\lambda := b^N$ (since it is invariant with respect to the substitutions $b \mapsto e^{2\pi i \frac{k}{N}} b$, $0 < k < N$), see [18], see also Section 6.

Note that z is still a function on X but b becomes a character automorphic function. Finally, using z we may glue the boundary of X, remove punched points (where $z = \infty$) and get in this way a compact Riemann surface X_c , such that

$$X = (X_c \setminus \{P : z(P) = \infty\}) \setminus E.$$

The simplest assumption is that the boundary E is a finite system of cuts on X_c. In this case we get that the triple $\{X_c, z, E\}$ characterizes the spectrum of a finite difference operator. We came to this triple basically due to heuristic arguments, but the opposite direction is already a solid mathematical fact: every triple of this kind gives rise to a family of ergodic finite difference operators [18, Sect. 6]. In Sections 2 and 3 we give details on constructions of such operators using the theory of Hardy spaces on Riemann surfaces.

Now we would like to note that one can describe all triples of a given type up to a natural equivalence relation. This natural parametrization of the triples is one of the main point of the current paper. It allows us to put consideration from the pure algebraic points of view to a wider setting and to recruit very much analytic tools.

1.2. Parametrization of the spectral curves in terms of branching divisors

Probably the best-known result in spectral theory of the nature we want to discuss deals with the description of the spectrum of periodic Jacobi matrices, see, e.g., [17, 27]. Such a set E should have a form of an inverse polynomial image $E = T^{-1}([-1,1])$. The polynomial T should have all critical points real $\{c_k : T'(c_k) = 0\} \subset \mathbb{R}$, moreover all critical values $t_k := T(c_k)$ should have modulus not less than 1 and their signs should alternate, i.e.: $|T(c_k)| \geq 1$ and

$$T(c_{k-1})T(c_k) < 0 \quad \text{for} \quad \cdots < c_{k-1} < c_k < \cdots .$$

The claim is that the system of numbers $\{t_k : |t_k| \geq 1,\ t_{k-1}t_k < 0\}$, determines a polynomial with the prescribed and ordered critical values t_k uniquely, modulo a change of the independent variable $z \mapsto az + b,\ a > 0, b \in \mathbb{R}$. The proof uses a special representation of T:

$$T(z) = \pm \cos \phi(z),$$

where ϕ is the conformal map of the upper half-plane onto the half-strip with a system of cuts:

$$\Pi = \{w = u + iv : 0 \leq v \leq \pi d\} \setminus \cup_k \{v = \pi k,\ u \leq h_k\},$$

where $d = \deg T$ and $\cosh(h_k) := |t_k|$, $\phi(\infty) = \infty$.

We mention a more general theorem of Maclane [15] and Vinberg [28] on the existence and uniqueness of real polynomials (actually, and entire functions) with prescribed (ordered!) sequences of critical values. In the case of polynomials, this theorem says that there is a one to one correspondence between finite "up-down" real sequences

$$\cdots \leq t_{k-1} \geq t_k \leq t_{k+1} \geq \cdots ,$$

and real polynomials whose all critical points are real, also, modulo a change of the independent variable $z \mapsto az + b,\ a > 0, b \in \mathbb{R}$. The MacLane–Vinberg theorem is based on an explicit description of the Riemann surfaces spread over the plane of the inverse functions T^{-1}.

To be more precise in the general case we start with the following

Definition 1.3. We say that two triples (X_{c1}, z_1, E_1) and (X_{c2}, z_2, E_2) are equivalent if there exists a holomorphic homeomorphism $h : X_{c1} \to X_{c2}$ such that $z_1 = h^*(z_2)$ and $E_2 = h(E_1)$.

Note, that for any triple (X_c, z, E) the holomorphic function $z : X_c \to \overline{\mathbb{C}}$, $\overline{\mathbb{C}} = \mathbb{C} \cup \{\infty\}$, is a ramified covering of $\overline{\mathbb{C}}$. The fact is that it is possible and very convenient to describe equivalence classes of ramified coverings in terms of branching divisor. Namely, a point $P \in X_c$ such that $\frac{dz}{d\zeta}|_P = 0$ where ζ is a local holomorphic coordinate in a neighborhood of P is called *ramification point* (*or, critical point*). Its image $z(P)$ is called *a branching point* (*or, critical value*). The set of all branching points of function z form a branching divisor of z. Note that infinity also can be a branching point. Since in our consideration it plays an exclusive role it is convenient in what follows to denote by $\mathcal{Z} := \{z_i\}_{i=1}^N$ all branching points from the *finite part* of the complex plane $\mathbb{C}$.

Clearly, branching divisors of equivalent functions are the same. Moreover, the compact holomorphic curve X_c is also uniquely determined by the branching divisor and some additional ramification data (of combinatorial type). Namely, assume that z has degree d. Let $w_0 \in \mathbb{C}$ be a non branching point. Fix a system of non-intersecting paths $\gamma = \{\gamma_i\}_{i=1,\dots,N}$. The ith path γ_i connects w_0 and z_i. We want to construct a system of loops $l_i \subset \mathbb{C}$. To construct l_i we start from w_0 and follow first γ_i almost to z_i, then encircle z_i counterclockwise along a small circle and finally go back to w_0 along $-\gamma_i$. Using l_i we associate with each branching point an element of the permutation group $\sigma_i \in \Sigma_d$.

The point w_0 has exactly d preimages. Let us label them by integers $\{1,\dots,d\}$. Let us follow the loop l_i and lift this loop to X_c starting from each of the preimages of w_0. The monodromy along path l_i gives us a permutation $\sigma_i \in \Sigma_d$ of preimages. We add to this system of loops one more l_∞, related to infinity (starting from w_0 we go sufficiently far then make a big circle in the clockwise direction and go back to w_0). This loop gives us one more permutation σ_∞.

Note that the product $\sigma_1 \cdots \cdots \sigma_N$ times σ_∞ is the identity operator. Therefore, the function z, including its behavior at infinity, determines N branching points and N permutations. These permutations are not uniquely defined, they depend on the labeling of preimages of w_0. Therefore, they are determined, up to a conjugacy by the elements of Σ_d.

Given a set of branching points $\mathcal{Z}(z) = \{z_i\}_{i=1,\dots,N} \subset \mathbb{C}$ and a system of permutations $\sigma(z) = (\sigma_1, \dots, \sigma_N) \in \Sigma_d \times \cdots \times \Sigma_d/\Sigma_d$, where the last quotient is taken with respect to diagonal conjugation, we can restore by Riemann theorem the surface X_c and the function z. Actually there is one more topological condition: the surface should be connected. Throughout the paper we assume that the system of permutations guaranteed this condition to be hold.

Hence, the triple (X_c, z, E) is equivalent to the triple $(\mathcal{Z}, \sigma_\gamma, E)$. We use this triple as free parameters determining the spectral surface of a $2d+1$-diagonal matrix.

Summary. Comparably with the case of Jacobi matrices, where we have only system of cuts (spectral intervals) in the complex plane, in the multidiagonal case we have a new additional system of parameters. We have to fix in $\mathbb{C}$ a system of critical points $\mathcal{Z}$, and associate to them a system of permutation σ, which actually depends on the base point $w_0 \in \mathbb{C} \setminus \mathcal{Z}$ and the system of paths γ. They define a Riemann surfaces X_c and a covering z. Then on the set $z^{-1}(\mathbb{R}) \subset X_c$ we chose a system of cuts E, and thus (X_c, z, E) is restored up to the equivalence 1.3.

Remark 1.4. As it was mentioned by the branching divisors language one can extend consideration from the pure algebraic level. In particular, we are very interested in infinite-dimensional generalizations: the point is that starting from the 5-diagonal case we have the branching divisor $\mathcal{Z}$ as a completely new system of parameters characterizing the spectral surface. What is its influence on the properties of the corresponding surface? Consider, indeed, the simplest case of 5-diagonal matrices. Then we have just to specify the point set $\mathcal{Z}$ (all permutations are of the form $\sigma_j = \begin{pmatrix} 1 & 2 \\ 2 & 1 \end{pmatrix}$). What is the speed of accumulation of an *infinite* system of points $\mathcal{Z}$ to, say, *a finite* system of cuts E so that $X_c \setminus E$ is of Widom type, or of Widom type with the Direct Cauchy Theorem (for the definition of these type of surfaces see [12], see also [26] for their role in the spectral theory)? We provide here the following *example.* Let $W(z)$ be the infinite Blaschke product in the upper half-plane with zeros at $\{z_k\}_{k\geq 0}$. Define

$$\mathfrak{R} = \{P = (z, w) : w^2 = W, \quad \Im z \geq 0\}.$$

Then $\sqrt{\frac{z-z_0}{z-\bar{z}_0}}$ is the Green function in $\mathfrak{R}$. Note that $C_k = (z_k, 0)$, $k \geq 1$ are its critical points, and therefore the Widom function is

$$\Delta = \sqrt{W \frac{z-\bar{z}_0}{z-z_0}}.$$

Note that the Carleson condition for this function on $\mathfrak{R}$ is the standard Carleson condition for $\{z_k\}_{k\geq 1}$ in the upper half-plane. Thus the Blaschke and Carleson conditions on the zero set in the half-plane guarantee that $\mathfrak{R}$ is of Widom type and of Widom type with the Direct Cauchy Theorem respectively.

1.3. Structure of the paper and main results

In Section 2, having the spectral surface fixed, we define a system of Hardy spaces on it (natural counterparts of the Hardy space in the unit disk). In that section we assume that there is only one "infinity" $P_0 \in X_c \setminus E$, $z(P_0) = \infty$, that is, the product of permutations $\sigma_1 \cdot \dots \cdot \sigma_N$ is a cycle. There is an intrinsic basis in the Hardy space: each next basis elements has at P_0 a zero of bigger and bigger multiplicity. Of course, this is a counterpart of the standard basis system $\{\zeta^n\}_{n\geq 0}$ in the standard H^2. Extending this basis to the negative integers (the system extends in the direction of functions having a pole at P_0 with growing multiplicity) one gets a basis in the whole L^2. Finally the multiplication operator, by the covering map with respect to this basis, is the $2d+1$-diagonal matrix (to this end it is important to note that z has a pole of multiplicity d at P_0). The complex Green function with zero at P_0 is playing the role of the symbol of the shift operator. Then we study the question of uniqueness of such a model for an ergodic operator (see Theorem 2.3 and the example right after it).

The general case (several "infinities") is considered in Section 3. Now the d-root of the product of the Green functions (counting their multiplicities) with respect to all infinite points on the surface becomes the symbol of the shift. We have the ordering of the infinities as one more parameter, defining the basis system and the corresponding multidiagonal matrix.

In Section 4 we demonstrate, how an application of our general construction to a particular covering, generate the well-known and now widely discussed almost periodic CMV matrices.

Starting from Section 5 we discuss an important theme: covering of one spectral surface by another one and the related transformations on the set of multidiagonal operators (so-called Renormalization Equations). Let $\pi : Y_c \to X_c$ be a d-sheeted covering. For a system of cuts E define $F = \pi^{-1}(E)$. Then we have $\pi : Y_c \setminus F \to X_c \setminus E$. The study is based on the relation between the Hardy spaces on these Riemann surfaces.

The Renormalization Equations generated by polynomial coverings have played an important role in studying of almost periodic Jacobi matrices with a singular continuous spectrum, [5], see also [21], [22]. They act in the most natural way on periodic Jacobi matrices, see Section 6.

In Section 7 we proof several new results dealing with Renormalization Equations for periodic Jacobi matrices: describe the complete set of their solutions; show their relation with the Ruelle operators. Finally, we give a possible generalization of the constructions from Section 6 for a wider class of almost periodic Jacobi matrices with a singular continuous spectrum. In particular, we prove the Lipschitz property of the Darboux transform.

Having in mind importance of the paper [3], where the Renormalization Equation generated by "just" *quadratic polynomial* was used, we investigate in Section 8 the case of *rational double covering* $\pi(v) = \tau v - \frac{\tau-1}{v}$, $\tau > 1$. As usual, the renormalization procedure is simpler to formulate for operators acting on the (integer) half-axis $\mathbb{Z}_+$.

Definition 1.5. Let A be a self-adjoint operator acting in $l^2_+ = l^2(\mathbb{Z}_+)$ with a cyclic vector $|0\rangle$ and the spectrum on $[-1, 1]$. We define its transform $\pi^*(A)$ in the following steps. First we define the upper triangular matrix Φ (with positive diagonal entries) by the condition

$$A^2 + 4\tau(\tau - 1) = \Phi^*\Phi.$$

Then we introduce $A_* := \Phi A \Phi^{-1}$ and define the operator

$$\begin{bmatrix} A & \Phi^* \\ \Phi & A_* \end{bmatrix},$$

acting in $l^2_+ \oplus l^2_+$. Finally, using the unitary operator $U : l^2_+ \to l^2_+ \oplus l^2_+$, such that

$$U|2k\rangle = |k\rangle \oplus 0, \quad U|2k+1\rangle = 0 \oplus |k\rangle$$

we construct

$$\pi^*(A) := \frac{1}{2\tau} U^* \begin{bmatrix} A & \Phi^* \\ \Phi & A_* \end{bmatrix} U : l^2_+ \to l^2_+.$$

We give a theorem (see Theorem 8.8) on the weak convergence of the iterative procedure $A_{n+1} = \pi^*(A_n)$ to an operator with a simple singular continuous spectrum supported on the Julia set of the given expanding mapping and pose here a question on the contractivity of the renormalization operator, at least for large values of τ. The main general conjecture deals with contractivity of all renormalizations, generated by a covering with sufficiently large critical values.

2. The Global Functional Model (single infinity case). Uniqueness Theorem

2.1. Hardy spaces and bases

There are different ways to define Hardy spaces on Riemann surfaces, – the spaces of vector bundles, multivalued functions or forms. These definitions are equivalent. We start from 1-forms, the most natural object with this respect from our point of view.

Let $\pi(\zeta) : \mathbb{D} \to X$ be a uniformization of the surface $X = X_c \setminus E$. Thus there exists a discrete subgroup Γ of the group $SU(1,1)$ consisting of elements of the form

$$\gamma = \begin{bmatrix} \gamma_{11} & \gamma_{12} \\ \gamma_{21} & \gamma_{22} \end{bmatrix}, \ \gamma_{11} = \overline{\gamma_{22}}, \ \gamma_{12} = \overline{\gamma_{21}}, \ \det \gamma = 1,$$

such that $\pi(\zeta)$ is automorphic with respect to Γ, i.e., $\pi(\gamma(\zeta)) = \pi(\zeta)$, $\forall \gamma \in \Gamma$, and any two preimages of $P \in X$ are Γ-equivalent. We normalize

$$Z(\zeta) := (z \circ \pi)(\zeta)$$

by the conditions $Z(0) = \infty$, $(\zeta^d Z)(0) > 0$.

Note that Γ acts dissipatively on $\mathbb{T}$ with respect to the Lebesgue measure dm, that is there exists a measurable (fundamental) set $\mathbb{E}$, which does not contain any two Γ-equivalent points, and the union $\cup_{\gamma \in \Gamma} \gamma(\mathbb{E})$ is a set of full measure. In fact $\mathbb{E}$ can be chosen as a finite union of intervals, – the $\mathbb{T}$-part of the boundary of the fundamental domain. For the space of square summable functions on $\mathbb{E}$ (with respect to dm), we use the notation $L^2_{dm|\mathbb{E}}$.

A character of Γ is a complex-valued function $\alpha : \Gamma \to \mathbb{T}$, satisfying

$$\alpha(\gamma_1 \gamma_2) = \alpha(\gamma_1)\alpha(\gamma_2), \quad \gamma_1, \gamma_2 \in \Gamma.$$

The characters form an Abelian compact group denoted by Γ^*. The further Hardy spaces on X are marked by characters of Γ.

Let f be an analytic function in $\mathbb{D}$, $\gamma \in \Gamma$. Then we put

$$f|[\gamma]_k = \frac{f(\gamma(\zeta))}{(\gamma_{21}\zeta + \gamma_{22})^k} \qquad k = 1, 2.$$

Notice that $f|[\gamma]_2 = f$ for all $\gamma \in \Gamma$, means that the form $f(\zeta)d\zeta$ is invariant with respect to the substitutions $\zeta \to \gamma(\zeta)$ ($f(\zeta)d\zeta$ is an Abelian integral on $\mathbb{D}/\Gamma$). Analogously, $f|[\gamma] = \alpha(\gamma)f$ for all $\gamma \in \Gamma$, $\alpha \in \Gamma^*$, means that the form $|f(\zeta)|^2\,|d\zeta|$ is invariant with respect to these substitutions.

We recall, that a function $f(\zeta)$ is of Smirnov class, if it can be represented as a ratio of two functions from H^∞ with an outer denominator. The following spaces related to the Riemann surface $\mathbb{D}/\Gamma$ are counterparts of the standard Hardy spaces H^2 (H^1) on the unit disk.

Definition 2.1. The space $A^2_1(\Gamma,\alpha)$ ($A^1_2(\Gamma,\alpha)$) is formed by functions f, which are analytic on $\mathbb{D}$ and satisfy the following three conditions

1) f is of Smirnov class
2) $f|[\gamma]=\alpha(\gamma)f \quad (f|[\gamma]_2=\alpha(\gamma)f) \quad \forall\gamma\in\Gamma$
3) $\displaystyle\int_{\mathbb{E}}|f|^2\,dm<\infty \quad \left(\int_{\mathbb{E}}|f|\,dm<\infty\right).$

$A^2_1(\Gamma,\alpha)$ is a Hilbert space with the reproducing kernel $k^\alpha(\zeta,\zeta_0)$, moreover

$$0<\inf_{\alpha\in\Gamma^*}k^\alpha(\zeta_0,\zeta_0)\le\sup_{\alpha\in\Gamma^*}k^\alpha(\zeta_0,\zeta_0)<\infty.$$

Put

$$k^\alpha(\zeta)=k^\alpha(\zeta,0)\quad\text{and}\quad K^\alpha(\zeta)=\overline{K^\alpha_\zeta(0)}=\frac{k^\alpha(\zeta)}{\sqrt{k^\alpha(0)}}.$$

We need one more special function. The Blaschke product

$$b(\zeta)=\zeta\prod_{\gamma\in\Gamma,\gamma\ne 1_2}\frac{\gamma(0)-\zeta}{1-\overline{\gamma(0)}\zeta}\frac{|\gamma(0)|}{\gamma(0)}$$

is called the *Green's function* of Γ with respect to the origin. It is a character-automorphic function, i.e., there exists $\mu\in\Gamma^*$ such that

$$b(\gamma(\zeta))=\mu(\gamma)b(\zeta). \tag{4}$$

Note, if $G(P)=G(P,P_0)$ denotes the Green's function of the surface X, then

$$G(\pi(\zeta))=-\log|b(\zeta)|.$$

We are ready to construct the basis in $A^2_1(\Gamma,\alpha)$. Consider the following subspace of this space

$$\{f\in A^2_1(\Gamma,\alpha):f(0)=0\}.$$

The following two facts are evident

1) $\{f\in A^2_1(\Gamma,\alpha):f(0)=0\}=\{b\tilde f:\tilde f\in A^2_1(\Gamma,\mu^{-1}\alpha)\}=bA^2_1(\Gamma,\mu^{-1}\alpha)$,
2) $A^2_1(\Gamma,\alpha)=\{K^\alpha\}\oplus\{f\in A^2_1(\Gamma,\alpha):f(0)=0\}$.

Thus

$$\begin{aligned}A^2_1(\Gamma,\alpha)=&\{K^\alpha\}\oplus bA^2_1(\Gamma,\mu^{-1}\alpha)\\=&\{K^\alpha\}\oplus\{bK^{\alpha\mu^{-1}}\}\oplus b^2A^2_1(\Gamma,\mu^{-2}\alpha),\end{aligned}$$

and so on.

Basically we proved the following theorem, note, however, that the second statement is not a direct consequence of the first one.

Theorem 2.2. *Given $\alpha \in \Gamma^*$, the system of functions $\{b^n K^{\alpha\mu^{-n}}\}_{n\geq 0}$ forms an orthonormal basis in $A_1^2(\Gamma, \alpha)$; the system $\{b^n K^{\alpha\mu^{-n}}\}_{n\in\mathbb{Z}}$ is an orthonormal basis in $L^2_{dm|\mathbb{E}}$.*

2.2. The Global Functional Model

Of course the constructions in this section and our speculations in Section 1 are closely related and of mutual influence. In this subsection we close the construction from Section 2.1 by proving the Global Functional Model Theorem.

Let $\Gamma_0 := \ker \mu$, μ given by (4), that is $\Gamma_0 = \{\gamma \in \Gamma : \mu(\gamma) = 1\}$. Evidently, $b(\zeta)$ and $(zb^d)(\zeta)$ are holomorphic functions on the surface $X_0 = \mathbb{D}/\Gamma_0$.

Assume that $\alpha_0 \in \Gamma_0$ can be extended to a character on Γ, i.e.,

$$\Omega_{\alpha_0} = \{\alpha \in \Gamma^* : \alpha|\Gamma_0 = \alpha_0\} \neq \emptyset.$$

Note that the set of characters

$$\Omega_\iota = \{\alpha \in \Gamma^* : \alpha|\Gamma_0 = \iota\}$$

where $\iota(\gamma) = 1$ for all $\gamma \in \Gamma_0$ is isomorphic to the set $(\Gamma/\Gamma_0)^*$.

Let us fix an element $\hat{\alpha}_0 \in \Omega_{\alpha_0}$. Since

$$\{\alpha \in \Gamma^* : \alpha|\Gamma_0 = \alpha_0\} = \{\hat{\alpha}_0\beta : \beta \in \Gamma^* \text{ such that } \beta|\Gamma_0 = \iota\}$$

we can define a measure $d\chi_{\alpha_0}(\alpha)$ on Ω_{α_0} by the relation

$$d\chi_{\alpha_0}(\alpha) = d\chi_{\alpha_0}(\hat{\alpha}_0\beta) = d\chi_\iota(\beta),$$

where $d\chi_\iota(\beta)$ is the Haar measure on $(\Gamma/\Gamma_0)^*$ (the measure $d\chi_{\alpha_0}(\alpha)$ does not depend on the choice of the element $\hat{\alpha}_0$).

Obviously, $\mathcal{T}\alpha := \mu^{-1}\alpha$ is an invertible ergodic measure-preserving transformation on $\Omega = \Omega_{\alpha_0}$ with respect to the measure $d\chi = d\chi_{\alpha_0}$.

The following theorem is a slightly modified version of Theorem 2.2 given in [29].

Theorem 2.3. *With respect to the basis from Theorem 2.2, the multiplication operator by z is a $2d+1$-diagonal ergodic finite difference operator with $\Omega = \Omega_{\alpha_0}$, $d\chi = d\chi_{\alpha_0}$, $\mathcal{T}\alpha := \mu^{-1}\alpha$ and $\alpha_0 = \alpha|\Gamma_0$. Moreover, the operators $\widehat{S}_+$ and $(\widehat{J}\widehat{S}^d)_+$ are unitary equivalent to multiplication by b and $(b^d Z)$ in $A_1^2(\Gamma_0, \alpha_0)$ respectively. This unitary map is given by the formula*

$$\sum_{\{\gamma\}\in\Gamma/\Gamma_0} f|[\gamma]\alpha^{-1}(\gamma) = \sum_{n\in\mathbb{Z}_+} x_n(\alpha) b^n K^{\alpha\mu^{-n}}, \quad f \in A_1^2(\Gamma_0, \alpha_0),$$

where the vector function $x(\alpha) := \{x_n(\alpha)\}$ belongs to $L^2_{d\chi}(l^2(\mathbb{Z}_+))$.

2.3. Uniqueness Theorem

The natural question arises: up to which extend is our functional realization unique?

Theorem 2.4. *Assume that a finite difference ergodic operator has a finite band functional model that is there exist a triple* $\{X_c, \tilde{z}, E\}$, *a character* $\alpha_0 \in \Gamma^*$ *and a map* F *from* Ω *to* $\tilde{\Omega} := \Omega_{\alpha_0}$ *such that* $FT\omega = \mu^{-1}F\omega$, $\chi(F^{-1}(A)) = \tilde{\chi}(A)$, $A \subset \tilde{\Omega}$, *with* $d\tilde{\chi} := d\chi_{\alpha_0}$, *here* μ *is the character of the Green's function* $\tilde{b}$ *on* $X_c \setminus E$. *Moreover* $q^{(k)}(\omega) = \tilde{q}^{(k)}(F\omega)$, *where the coefficients* $\tilde{q}^{(k)}(\alpha)$ *are generated by the multiplication operator* $\tilde{z}$ *with respect to the orthonormal basis* $\{\tilde{b}^n K^{\alpha\mu^{-n}}\}_{n\in\mathbb{Z}}$.

If the functions $\tilde{z}$ *and* $\{d\log\tilde{b}/d\tilde{z}\}$ *separate points on* $X_c \setminus E$ *then any local functional model is generated by one of the branches of the function* $\tilde{b}$.

For the proof see [23].

The following example shows that in the case when these two functions $\tilde{z}$ and $\{d\log\tilde{b}/d\tilde{z}\}$ do not separate points on $X_c \setminus E$ one can give different *global* functional realizations for the same ergodic operator.

Example [23]. Let $J = S^d + S^{-d}$. There exist a "trivial" functional model with $X_c \setminus E \sim \mathbb{D}$. In this case J is the multiplication operator by $z = \zeta^d + \zeta^{-d}$ with respect to the standard basis $\{\zeta^l\}$ in $L^2_{\mathbb{T}}$. Note that $b = \zeta$, thus

$$w := \frac{d\log b}{dz} = \frac{1}{\zeta^d - \zeta^{-d}}\frac{1}{d},$$

that is $z^2 + (wd)^{-2} = 4$, $|(wd)^{-1} + z| < 2$.

On the other hand let us fix any polynomial $T(u)$, $\deg T = d$, with real critical values on $\mathbb{R} \setminus [-2,2]$ and define $X_c \setminus E = T^{-1}(\overline{\mathbb{C}} \setminus [-2,2]) \sim \overline{\mathbb{C}} \setminus T^{-1}[-2,2]$. As we discussed the last set is the resolvent set for a d-periodic Jacobi matrix, say J_0. Moreover $T(J_0) = J$, and $-\log|b|$ is just the Green's function of this domain in the complex plane. So, using the standard functional model for J_0 (see Section 7 for details) with the symbols u and b we get a functional model for J with $z = T(u)$ and the same b. Note that as before $z^2 + (wd)^{-2} = 4$, $|(wd)^{-1} + z| < 2$ with $w := \frac{d\log b}{dz}$. □

Remark 2.5. In [8] the identity $T(J_0) = S^d + S^{-d}$ which holds for a periodic Jacobi matrix J_0 with spectrum $T^{-1}[-2,2]$ is called the Magic Formula. There it plays an important role in proving counterparts of Denisov–Rakhmanov and Killip–Simon Theorems for perturbations of periodic Jacbi matrices.

3. Several infinities case. Existence Theorem

Now we examine the situation in which the reduced surface X_c has several "infinities", that is, the covering function z (the symbol of an almost periodic operator) equals infinity at several (distinct) points on X_c.

We start with a simple example.

3.1. A five diagonal matrix of period two

Assume that z is a two sheeted covering with only two branching points, say, $z_1 = -2$, $z_2 = 2$. Thus the corresponding substitutions are $\sigma_1 = \sigma_2 = \begin{pmatrix} 1 & 2 \\ 2 & 1 \end{pmatrix}$. In this case X_c is equivalent to the complex plane $\overline{\mathbb{C}}$, moreover we can put

$$z = v + \frac{1}{v}, \quad v \in \mathbb{C}. \tag{5}$$

Thus, on this surface we have two "infinities" $v = \infty$ and $v = 0$.

Note that $z^{-1}(\mathbb{R}) = \mathbb{T} \cup \mathbb{R}$. We cut $\overline{\mathbb{C}}$ over the interval $[a, b]$, where $0 < a < b < 1$, that is we consider $X_c \setminus E$ of the form $\overline{\mathbb{C}} \setminus [a, b]$. We uniformize $X_c \setminus E$ by

$$v = \frac{a+b}{2} + \frac{b-a}{2}\frac{1/\zeta + \zeta}{2}, \quad \zeta \in \mathbb{D}.$$

For $v = \infty$ we have $\zeta_0 = 0$. Solving

$$\zeta^2 + 2\frac{b+a}{b-a}\zeta + 1 = 0$$

we get the image of the second infinity in $\mathbb{D}$, $\zeta_1 = -\frac{\sqrt{b}-\sqrt{a}}{\sqrt{b}+\sqrt{a}}$. As a result we get a symbol function $z : \mathbb{D} \to \mathbb{C}$ (see (5), (3.1)) for the forthcoming operator with two infinities ζ_0, ζ_1.

Next we derive the symbol b for the shift operator. Recall that the Green function in $\mathbb{D}$ is the Blaschke factor

$$b_{\zeta_0} = \zeta, \quad b_{\zeta_1} = \frac{\zeta - \zeta_1}{1 - \zeta\zeta_1}.$$

The product $b_{\zeta_0} b_{\zeta_1}$ is the smallest unimodular multiplier that cancels poles of z. Since b^2 is of the same nature, $b^2 z$ is holomorphic with a unimodular function b on $\mathbb{T}$, we have $b^2 = b_{\zeta_0} b_{\zeta_1}$.

Finally, we need a certain functional space and an intrinsic basis in it that generalize the construction in Theorem 2.2. Recall b should be related to the shift S, and we are going to define the periodic operator J as the multiplication operator with respect to this basis. To this end we define the following functional spaces. Given $\alpha_k \in \mathbb{T}$, $k = 0, 1$, we associate the space $H^2(\alpha_0, \alpha_1)$ of analytic multivalued functions $f(\zeta)$, $\zeta \in \mathbb{D} \setminus \{\zeta_0, \zeta_1\}$, such that $|f(\zeta)|^2$ has a harmonic majorant and

$$f \circ \gamma_i = \alpha_i f,$$

where γ_i is a small circle around ζ_i. Such a space can be reduced to the standard Hardy space H^2, moreover

$$H^2(\alpha_0, \alpha_1) = b_{\zeta_0}^{\tau_0} b_{\zeta_1}^{\tau_1} H^2, \quad \alpha_k = e^{2\pi i \tau_k}, \ 0 \le \tau_k < 1.$$

Lemma 3.1. *The space $bH^2(-1, 1)$ is a subspace of $H^2(1, -1)$ having a one-dimensional orthogonal complement, moreover*

$$H^2(1, -1) = \{\sqrt{b_{\zeta_1}} k_{\zeta_0}\} \oplus bH^2(-1, 1), \tag{6}$$

where k_{ζ_i} is the reproducing kernel of the standard H^2 with respect to ζ_i.

This lemma allows us to repeat the construction of Subsection 2.1. Iterating, now, the decomposition (6)

$$\begin{aligned} H^2(1,-1) =&\{\sqrt{b_{\zeta_1}}k_{\zeta_0}\}\oplus bH^2(-1,1)\\ =&\{\sqrt{b_{\zeta_1}}k_{\zeta_0}\}\oplus b\{\sqrt{b_{\zeta_0}}k_{\zeta_1}\}\oplus b^2H^2(1,-1)=\cdots, \end{aligned}$$

one gets an orthogonal basis in $H^2(1,-1)$ consisting of vectors of two sorts

$$b^{2m}\{\sqrt{b_{\zeta_1}}k_{\zeta_0}\} \text{ and } b^{2m+1}\{\sqrt{b_{\zeta_0}}k_{\zeta_1}\}.$$

Note that this orthogonal system can be extended to negative integers m so that we obtain a basis in $L^2(1,-1)$.

Theorem 3.2. *With respect to the orthonormal basis*

$$e_n = \begin{cases} b^{2m}\sqrt{b_{\zeta_1}}\frac{k_{\zeta_0}}{||k_{\zeta_0}||}, & n=2m \\ b^{2m+1}\sqrt{b_{\zeta_0}}\frac{k_{\zeta_1}}{||k_{\zeta_1}||}, & n=2m+1 \end{cases}$$

the multiplication operator by z is a 5*-diagonal matrix of period* 2.

3.2. General case

Let $z : X_c \to \mathbb{C}$ be d-sheeted covering with d (distinct) infinities. Further, let $\mathbb{D}/\Gamma$ be a uniformization of $X_c \setminus E$. For a given character $\alpha \in \Gamma^*$ and a *fixed ordering* of infinities $P_1, P_2, \dots, P_d$ we denote by

- $b_j = b_{P_j}$ the Green function with respect to P_j, $b_j \circ \gamma = \mu_j(\gamma) b_j$, $\mu_j \in \Gamma^*$;
- $k_j^\alpha = k_{P_j}^\alpha$ the reproducing kernel of $A_1^2(\Gamma,\alpha)$ with respect to P_j,

$$K_j^\alpha := \frac{k_j^\alpha}{||k_j^\alpha||}.$$

Similar to the previous subsection we get

Theorem 3.3. *Let $b=(b_1\dots b_d)^{\frac{1}{d}}$. With respect to the orthonormal basis*

$$e_n = \begin{cases} b^{dm} b_1^{\frac{1}{d}} \dots b_{d-1}^{\frac{d-1}{d}} K_d^{\alpha(\mu_1\dots\mu_d)^{-m}}, & n = dm \\ b^{dm+1} b_2^{\frac{1}{d}} \dots b_d^{\frac{d-1}{d}} K_1^{\alpha\mu_1^{-1}(\mu_1\dots\mu_d)^{-m}}, & n = dm+1 \\ \dots & \\ b^{dm+d-1} b_d^{\frac{1}{d}} \dots b_{d-2}^{\frac{d-1}{d}} K_{d-1}^{\alpha\mu_1^{-1}\dots\mu_{d-1}^{-1}(\mu_1\dots\mu_d)^{-m}}, & n = dm+d-1 \end{cases} \tag{7}$$

the multiplication operator by z is a $(2d+1)$-diagonal almost periodic matrix.

4. Five-diagonal almost periodic self-adjoint matrices and OPUC

We start again with the two-sheeted covering (5). We have $X_c \simeq \overline{\mathbb{C}}$ and $z^{-1}(\mathbb{R}) = \mathbb{R}\cup\mathbb{T}$, but let us, in this case, cut $\overline{\mathbb{C}}$ on an arbitrary finite union of (necessary non-degenerate) arcs on the unit circle, $E \simeq \{\mathbb{T}\setminus\cup_{j=0}^l(a_j,b_j)\}$.

The domain $X_c \setminus E$ is conformally equivalent to the quotient of the unit disk by the action of a discrete group $\Gamma = \Gamma(E)$. Let

$$v : \mathbb{D} \to \{\overline{\mathbb{C}} \setminus \mathbb{T}\} \cup \{\cup_{j=0}^{l}(a_j, b_j)\} \tag{8}$$

be a covering map, $v \circ \gamma = v$, $\forall \gamma \in \Gamma$. In what follows we assume the following normalization to be hold $v : (-1, 1) \to (a_0, b_0)$, so that one can choose a fundamental domain $\mathfrak{F}$ and a system of generators $\{\gamma_j\}_{j=1}^{l}$ of Γ such that they are symmetric with respect to the complex conjugation:

$$\overline{\mathfrak{F}} = \mathfrak{F}, \quad \overline{\gamma_j} = \gamma_j^{-1}.$$

Denote by $\zeta_0 \in \mathfrak{F}$ the preimage of the origin, $v(\zeta_0) = 0$, then $v(\overline{\zeta_0}) = \infty$. Thus the infinities of z defined by (5) and (8) on $\mathbb{D}/\Gamma$ are trajectories of ζ_0 and $\overline{\zeta_0}$ under the action of the group Γ, $P_0 = \{\gamma(\zeta_0)\}_{\gamma \in \Gamma}$ and $P_1 = \{\gamma(\overline{\zeta_0})\}_{\gamma \in \Gamma}$. There are *two* infinities that we have in the case under consideration.

Thus, to define the function b (the symbol of the shift operator) we have to introduce the Green functions $B(\zeta, \zeta_0)$ and $B(\zeta, \overline{\zeta_0})$. It is convenient to normalize them by $B(\overline{\zeta_0}, \zeta_0) > 0$ and $B(\zeta_0, \overline{\zeta_0}) > 0$. Then

$$v(\zeta) = e^{ic} \frac{B(\zeta, \zeta_0)}{B(\zeta, \overline{\zeta_0})}. \tag{9}$$

Also, we can rotate (if necessary) the v-plane so that $c = 0$. Note that $B(\zeta, \zeta_0)$ is a character-automorphic function

$$B(\gamma(\zeta), \zeta_0) = \mu(\gamma) B(\zeta, \zeta_0), \quad \gamma \in \Gamma,$$

with a certain $\mu \in \Gamma^*$. By (9), $B(\zeta, \overline{\zeta_0})$ has the *same* factor of automorphy,

$$B(\gamma(\zeta), \overline{\zeta_0}) = \mu(\gamma) B(\zeta, \overline{\zeta_0}), \quad \gamma \in \Gamma.$$

By the definition $b^2 = B(\zeta, \zeta_0) B(\zeta, \overline{\zeta_0})$ we get a multivalued analytic function b on the punched surface $\{\mathbb{D}/\Gamma\} \setminus \{P_0, P_1\}$.

In this case we have only two possibilities for ordering of the infinities: $\{P_0, P_1\}$ and $\{P_1, P_0\}$. According to Theorem 3.3, to any of them, say the first one, and to an arbitrary $\alpha \in \Gamma^*$ we can associate the operator $J = J(-1, 1; \alpha)$ by fixing the space $H^2(-1, 1; \alpha) = \sqrt{B(\zeta, \zeta_0)} A_1^2(\alpha)$ and a natural basis in it. Up to a common multiplier $\sqrt{B(\zeta, \zeta_0)}$ it is a basis in $A_1^2(\alpha)$ of the form

$$K^{\alpha}(\zeta, \overline{\zeta_0}),\ B(\zeta, \overline{\zeta_0}) K^{\alpha\mu^{-1}}(\zeta, \zeta_0),\ B(\zeta, \zeta_0) B(\zeta, \overline{\zeta_0}) K^{\alpha\mu^{-2}}(\zeta, \overline{\zeta_0}), \dots .$$

Thus we get the same system of functions that we used describing almost periodic Verblunsky coefficients [20]. The last one we can define by

$$a(\alpha) = a = \frac{K^{\alpha}(\zeta_0, \overline{\zeta_0})}{K^{\alpha}(\zeta_0, \zeta_0)}.$$

In [20] they appear in the following recursion

$$\begin{aligned} K^{\alpha}(\zeta, \overline{\zeta_0}) = &\ a(\alpha) K^{\alpha}(\zeta, \zeta_0) + \rho(\alpha) B(\zeta, \zeta_0) K^{\alpha\mu^{-1}}(\zeta, \overline{\zeta_0}), \\ K^{\alpha}(\zeta, \zeta_0) = &\ \overline{a(\alpha)} K^{\alpha}(\zeta, \overline{\zeta_0}) + \rho(\alpha) B(\zeta, \overline{\zeta_0}) K^{\alpha\mu^{-1}}(\zeta, \zeta_0), \end{aligned} \tag{10}$$

where

$$\rho(\alpha) = \rho = \sqrt{1-|a|^2} = B(\overline{\zeta_0},\zeta_0)\frac{K^{\alpha\mu^{-1}}(\overline{\zeta_0},\overline{\zeta_0})}{K^{\alpha}(\overline{\zeta_0},\overline{\zeta_0})}.$$

The goal of this section is to represent J in terms of Verblunsky coefficients.

Lemma 4.1. *With respect to the basis*

$$\dots,\ K^{\alpha\mu}(\zeta,\zeta_0)/B(\zeta,\zeta_0),\ K^{\alpha}(\zeta,\overline{\zeta_0}),\ B(\zeta,\overline{\zeta_0})K^{\alpha\mu^{-1}}(\zeta,\zeta_0),\ \dots \tag{11}$$

the multiplication operator by v is a matrix having at most two non-vanishing diagonals over the main diagonal. Moreover,

$$v \sim \begin{bmatrix} & \rho(\alpha\mu)\rho(\alpha\mu^2) & 0 & \\ \ddots & -\rho(\alpha\mu)a(\alpha\mu^2) & 0 & \\ & -a(\alpha\mu)\overline{a(\alpha)} & \rho(\alpha)\overline{a(\alpha\mu^{-1})} & \\ & * & -a(\alpha)\overline{a(\alpha\mu^{-1})} & \\ & & & \ddots \end{bmatrix}. \tag{12}$$

Similarly, the multiplication operator by $1/v$ is of the form

$$1/v \sim \begin{bmatrix} & 0 & 0 & \\ \ddots & \rho(\alpha\mu)a(\alpha) & \rho(\alpha)\rho(\alpha\mu) & \\ & -\overline{a(\alpha\mu)}a(\alpha) & -\rho(\alpha)\overline{a(\alpha\mu)} & \\ & * & -a(\alpha\mu^{-1})\overline{a(\alpha)} & \\ & & & \ddots \end{bmatrix}. \tag{13}$$

Proof. We give a proof, say, for (12). Recall (9), from which we can see that the decomposition of the vector $v(\zeta)K^{\alpha}(\zeta,\overline{\zeta_0})$ begins with

$$v(\zeta)K^{\alpha}(\zeta,\overline{\zeta_0}) = c_0\frac{K^{\alpha\mu^2}(\zeta,\overline{\zeta_0})}{B(\zeta,\zeta_0)B(\zeta,\overline{\zeta_0})} + c_1\frac{K^{\alpha\mu}(\zeta,\zeta_0)}{B(\zeta,\zeta_0)} + c_2K^{\alpha}(\zeta,\overline{\zeta_0}) + \cdots.$$

Multiplying by the denominator $B(\zeta,\zeta_0)B(\zeta,\overline{\zeta_0})$ we get

$$\begin{aligned} B^2(\zeta,\zeta_0)K^{\alpha}(\zeta,\overline{\zeta_0}) =& c_0K^{\alpha\mu^2}(\zeta,\overline{\zeta_0}) + c_1K^{\alpha\mu}(\zeta,\zeta_0)B(\zeta,\overline{\zeta_0}) \\ &+ c_2K^{\alpha}(\zeta,\overline{\zeta_0})B(\zeta,\zeta_0)B(\zeta,\overline{\zeta_0}) + \cdots. \end{aligned} \tag{14}$$

First we put $\zeta = \overline{\zeta_0}$. By the definition of $\rho(\alpha)$ we have

$$c_0 = B^2(\overline{\zeta_0},\zeta_0)\frac{K^{\alpha}(\overline{\zeta_0},\overline{\zeta_0})}{K^{\alpha\mu^2}(\overline{\zeta_0},\overline{\zeta_0})} = \rho(\alpha\mu)\rho(\alpha\mu^2).$$

Putting $\zeta = \zeta_0$ in (14) and using the definition of $a(\alpha)$, we have

$$c_1 = -c_0\frac{K^{\alpha\mu^2}(\zeta_0,\overline{\zeta_0})}{K^{\alpha\mu}(\zeta_0,\zeta_0)B(\zeta_0,\overline{\zeta_0})} = -\rho(\alpha\mu)\rho(\alpha\mu^2)\frac{a(\alpha\mu^2)}{\rho(\alpha\mu^2)} = -\rho(\alpha\mu)a(\alpha\mu^2).$$

Doing in the same way we can find a representation for c_2 that would involve derivatives of the reproducing kernels. However, we can find c_2 in terms of a and ρ calculating the scalar product

$$c_2 = \langle B^2(\zeta,\zeta_0)K^\alpha(\zeta,\overline{\zeta_0}), B(\zeta,\zeta_0)B(\zeta,\overline{\zeta_0})K^\alpha(\zeta,\overline{\zeta_0})\rangle.$$

Since $B(\zeta,\zeta_0)$ is unimodular, using (10), we get

$$c_2 = \langle \frac{K^{\alpha\mu}(\zeta,\overline{\zeta_0}) - a(\alpha\mu)K^{\alpha\mu}(\zeta,\zeta_0)}{\rho(\alpha\mu)}, B(\zeta,\overline{\zeta_0})K^\alpha(\zeta,\overline{\zeta_0})\rangle.$$

Recall that $k^\alpha(\zeta,\zeta_0) = K^\alpha(\zeta,\zeta_0)K^\alpha(\zeta_0,\zeta_0)$ is the reproducing kernel. Thus

$$c_2 = -\frac{a(\alpha\mu)}{\rho(\alpha\mu)}\overline{\frac{B(\zeta_0,\overline{\zeta_0})K^\alpha(\zeta_0,\overline{\zeta_0})}{K^{\alpha\mu}(\zeta_0,\zeta_0)}} = -\frac{a(\alpha\mu)}{\rho(\alpha\mu)}\overline{\rho(\alpha\mu)a(\alpha)} = -a(\alpha\mu)\overline{a(\alpha)}.$$

To find the decomposition of the vector $v(\zeta)B(\zeta,\overline{\zeta_0})K^{\alpha\mu^{-1}}(\zeta,\zeta_0)$ is even simpler since only two leading terms are involved. Note that all other columns of the matrix in (12), starting from these two, can be obtain by the character's shift by μ^{-2} along diagonals. □

Now let us remind the CMV representation for operators related to OPUC [24]. For a given sequence of numbers from $\mathbb{D}$

$$\dots, a_{-1}, a_0, a_1, a_2, \dots \tag{15}$$

define unitary matrices

$$A_k = \begin{bmatrix} \overline{a_k} & \rho_k \\ \rho_k & -a_k \end{bmatrix}, \quad \rho_k = \sqrt{1-|a_k|^2},$$

and unitary operators in $l^2(\mathbb{Z})$ given by block-diagonal matrices

$$\mathfrak{A}_0 = \begin{bmatrix} \ddots & & & \\ & A_{-2} & & \\ & & A_0 & \\ & & & \ddots \end{bmatrix}, \quad \mathfrak{A}_1 = S\begin{bmatrix} \ddots & & & \\ & A_{-1} & & \\ & & A_1 & \\ & & & \ddots \end{bmatrix}S^{-1}.$$

The CMV operator $\mathfrak{A}$, related to the sequence (15), is the product

$$\mathfrak{A} = \mathfrak{A}(\{a_k\}) := \mathfrak{A}_0\mathfrak{A}_1. \tag{16}$$

Theorem 4.2. *Define the sequence* $a_k = a(\alpha\mu^{-k})$. *Then* $v \sim \mathfrak{A}(\{a_k\})$, *see* (16).

Proof. Note that the operators in (12) and (13) are mutually conjugated, therefore the under-diagonal entries of both operators are also known. The rest is an easy direct computation. □

Theorem 4.3. *With respect to the basis* (11) *the multiplication operator by* z,

$$z \sim \begin{bmatrix} & q^{(-2)}(\alpha) & 0 & \\ \ddots & q^{(-1)}(\alpha) & q^{(-2)}(\alpha\mu^{-1}) & \\ & q^{(0)}(\alpha) & \overline{\{q^{(-1)}(\alpha\mu^{-1})\}} & \\ & * & q^{(0)}(\alpha\mu^{-1}) & \\ & & & \ddots \end{bmatrix},$$

is defined by the functions

$$q^{(-2)}(\alpha) = \rho(\alpha\mu)\rho(\alpha\mu^2), \quad q^{(-1)}(\alpha) = \rho(\alpha\mu)\{a(\alpha) - a(\alpha\mu^2)\},$$

and

$$q^{(0)}(\alpha) = -2\Re\{a(\alpha)\overline{a(\alpha\mu)}\}.$$

It is worth to mention that the second column is not only the shift by μ^{-1}, there is also conjugation.

Proof. Recall that $z = v + 1/v$ and use the previous lemma. □

Using general constructions from [18] one can define and integrate the flows hierarchy given by

$$\dot{\mathfrak{A}} = [(\mathfrak{A}^n + \mathfrak{A}^{-n})_+, \mathfrak{A}]. \tag{17}$$

Note that for $n = 1$, (17) gives the Schur flow [11].

5. Coverings

Discussing this subject we prefer to use a functional version of presentation of Hardy spaces. First we introduce these spaces and then we will remark how they are related to the spaces of forms.

Let ω_{P_0} denote the harmonic measure on an open surface $X_c \setminus E$ with respect to $P_0 \in X_c \setminus E$. Note that ω_{P_0} is the restriction of the differential $\frac{1}{2\pi i} d \log b(P, P_0)$ on E. By $H^2(\alpha, \omega_{P_0})$ we denote the closure of $H^\infty(\alpha)$ in L^2 with respect to the measure ω_{P_0}. The natural question is how this space is related to the space with another point fixed, say $P_1 \in X_c \setminus E$, or, more generally, with $H^2(\alpha, \omega)$, which denotes the closure of $H^\infty(\alpha)$ with respect to an equivalent norm given by a measure of the form $\omega = \rho\omega_{P_0}$, where $0 < C_1 \le \rho \le C_2 < \infty$. (By the Harnack Theorem ω_{P_1} and ω_{P_0} satisfy this property.)

To answer it, let us define an outer function ϕ, such that $\rho = |\phi|^2$. This function belongs to $H^\infty(\beta)$ with a certain $\beta \in \Gamma^*$. In this case

$$f \mapsto \phi f$$

is a unitary map from $H^2(\alpha, \omega)$ to $H^2(\alpha\beta, \omega_{P_0})$. Then the equality

$$\begin{aligned} \langle (\phi f)(P), k_Q^\alpha(P;\omega)\phi(P)\overline{\phi(Q)}\rangle_{H^2(\alpha\beta,\omega_{P_0})} &= \langle f(P), k_Q^\alpha(P;\omega)\overline{\phi(Q)}\rangle_{H^2(\alpha,\omega)} \\ &= f(Q)\phi(Q) \end{aligned}$$

shows that the reproducing kernels of $H^2(\alpha\beta, \omega_{P_0})$ and $H^2(\alpha, \omega)$ are related by

$$k_Q^{\alpha\beta}(P; \omega_{P_0}) = k_Q^{\alpha}(P; \omega)\phi(P)\overline{\phi(Q)}.$$

For the normalized kernels we have

$$K_Q^{\alpha\beta}(P; \omega_{P_0}) = K_Q^{\alpha}(P; \omega)\frac{\overline{\phi(Q)}}{|\phi(Q)|}\phi(P).$$

Therefore the matrix of a multiplication operator in $H^2(\alpha, \omega)$, with respect to the reproducing kernels based basis, actually can be obtained by a character's shift for the matrix with the same symbol related to the chosen space $H^2(\alpha, \omega_{P_0})$. (Let us mention in brackets a specific normalization of a basis vector given by the unimodular factor $\frac{\overline{\phi(Q)}}{|\phi(Q)|}$.)

The relations between $A_1^2(\alpha)$ and $H^2(\alpha)$ are of the same nature. Indeed, let $\rho : \mathbb{D}/\Gamma \to X_c \setminus E$, $\rho(0) = P_0$, be the uniformization of the given surface. Then $H^2(\alpha)$ is a subspace of the standard H^2 in $\mathbb{D}$: $f(\rho(\zeta)) \in H^2$ for $f \in H^2(\alpha)$, moreover

$$\|f\|^2 = \int_{\mathbb{T}} |f(\rho(t))|^2 \, dm(t),$$

where dm is the Lebesgue measure. Fix a fundamental set $\mathbb{E}$ for the action of Γ on $\mathbb{T}$, $\mathbb{T} = \cup_{\gamma\in\Gamma}\gamma(\mathbb{E})$. Then

$$\|f\|^2 = \int_{\mathbb{E}} |f(\rho(t))|^2 |\psi(t)|^2 \, dm(t),$$

where

$$|\psi(t)|^2 := \sum_{\gamma\in\Gamma} |\gamma'(t)|.$$

Again, we can consider ψ as an outer function and then

$$f \to (f \circ \rho)\psi$$

is the unitary map from $H^2(\alpha)$ to $A_1^2(\alpha\beta)$, where the character β is generated by the 1-form ψ. The β here is a particular character, so when α runs on the whole group Γ^* $\alpha\beta$ covers also all characters, and we have one to one correspondence between two ways of writing Hardy spaces.

But, as we noted above, working with coverings, it will be convenient to use character automorphic H^2-spaces with respect to the following specific measure

$$\omega = \frac{1}{l}\sum_{i=1}^{l} \omega_{P_i}, \tag{18}$$

associated with a system of points $\{P_i\}$ on $X_c \setminus E$. Naturally in what follows P_i's are infinities on $X_c \setminus E$.

Now we can go back to the coverings. Let Γ_X (respectively Γ_Y) be the fundamental group on $X_c \setminus E$ (respectively $Y_c \setminus E$). We have $\pi_* : \Gamma_Y \to \Gamma_X$ ($\pi_*(\gamma)$ is the image of a contour $\gamma \in \Gamma_Y$) and $\pi^* : \Gamma_X \to \Gamma_Y$ ($\pi^*(\gamma)$ is the full preimage of a contour $\gamma \in \Gamma_X$). Note that $\pi^*\pi_* = d\,\mathrm{Id}$. The maps $\pi_* : \Gamma_Y^* \to \Gamma_X^*$ and $\pi^* : \Gamma_X^* \to \Gamma_Y^*$ are defined by duality.

For a system of points $\{P_i\}_{i=1}^l$, $P_i \in X_c \setminus E$, define the measure ω_* on E by (18). Let $\{Q_k^{(i)}\}_{k=1}^d = \pi^{-1}(P_i)$. Define

$$\omega^* = \frac{1}{ld}\sum \omega_{Q_k^{(i)}}.$$

In this case

$$\int_F (f \circ \pi)\omega^* = \int_E f\omega_*. \tag{19}$$

Moreover,

$$\int_F f\omega^* = \int_E (\mathfrak{L}f)\omega_*, \quad \text{where} \quad (\mathfrak{L}f)(P) = \frac{1}{d}\sum_{\pi(Q)=P} f(Q). \tag{20}$$

As a direct consequence of (19), (20) we get

Lemma 5.1. *The map $V : H^2(\alpha, \omega_*) \to H^2(\pi^*\alpha, \omega^*)$, defined by*

$$Vf = f \circ \pi, \quad f \in H^2(\alpha, \omega_*), \tag{21}$$

is an isometry with

$$(V^*f)(P) = (\mathfrak{L}f)(P), \quad f \in H^2(\pi^*\alpha, \omega^*). \tag{22}$$

Also,

$$Vk_{P_0}^{\alpha} = \frac{1}{d}\sum_{\pi(Q_j^{(0)})=P_0} k_{Q_j^{(0)}}^{\pi^*\alpha}.$$

for the reproducing kernel $k_{P_0}^{\alpha} \in H^2(\alpha, \omega_*)$.

Theorem 5.2. *Let $z_* : X_c \setminus E \to \overline{\mathbb{C}}$. Using the notations introduced above, assume that*

$$z_* : E \to \mathbb{R}, \quad z_*^{-1}(\infty) \subset \{P_i\}_{i=1}^l$$

Let $\mathbf{z}_*(\alpha)$ be the multiplication operator by z_* with respect to the basis (7). For an arbitrary ordering of $\{Q_k^{(i)}\}_{1\le k\le d, 1\le i\le l}$ subordinated to the ordering of $\{P_i\}$ consider the operator $\mathbf{z}^*(\pi^*\alpha)$ with the symbol $z^* := z_* \circ \pi$ and the, by (21), related isometry $\mathbf{V}$. Then, the following relations hold

$$\mathbf{V}^*\mathbf{z}^*(\pi^*\alpha) = \mathbf{z}_*(\alpha)\mathbf{V}^*, \quad \mathbf{V}^*S^d = S\mathbf{V}^*. \tag{23}$$

Equations (23) are very close to the so-called Renormalization Equations that we start to discuss now.

6. The renormalization of periodic matrices

We recall some basic facts from the spectral theory of periodic Jacobi matrices. The spectrum E of any periodic matrix J is an inverse polynomial image

$$E = U^{-1}[-1,1] \tag{24}$$

the polynomial U of degree $g+1$ should have all critical points $\{c_U\}$ real and for all critical values $|U(c_U)| \geq 1$. For simplicity we assume that $|U(c_U)| > 1$. Then the spectrum of J consists of g intervals

$$E = [b_0, a_0] \setminus (\cup_{j=1}^{g}(a_j, b_j)).$$

Also it would be convenient for us to normalize U by a linear change of variable such that $b_0 = -1$ and $a_0 = 1$. In this case U is a so-called expanding polynomial.

Having the set E of the above form fixed, let us describe the whole set of periodic Jacobi matrices $J(E)$ with the given spectrum. To this end we associate with U the hyper-elliptic Riemann surface (the surface is given by (1.1) with $\lambda = b^N$, $N = g+1$)

$$X = \{Z = (z, \lambda) : \lambda - 2U(z) + \lambda^{-1} = 0\}.$$

The involution on it is denoted by τ,

$$\tau Z := \left(z, \frac{1}{\lambda}\right) \in X. \tag{25}$$

The set

$$X_+ = \{Z \in X : |\lambda(Z)| < 1\}$$

we call the upper sheet of X. Note $X_+ \simeq \bar{\mathbb{C}} \setminus E$, in fact, $z(Z) \in \bar{\mathbb{C}} \setminus E$ if $Z \in X_+$.

The following well-known theorem describes $J(E)$ in terms of *real* divisors on X. The Jacobian variety of X, $\mathrm{Jac}(X)$, is a g-dimensional complex torus, $\mathrm{Jac}(X) \simeq \mathbb{C}^g/L(X)$, where L is a lattice (that can be chosen in the form $L = \mathbb{Z}^g + \Omega\mathbb{Z}^g$ with $\Im\Omega > 0$). Consider the g-dimensional real subtorus consisting of divisors of the form

$$D(E) = \{D = D_+ - D_C,\ D_+ := \sum_{i=1}^{g} Z_i : Z_i \in X,\ z(Z_i) \in [a_i, b_i]\},$$

here D_C is a point of normalization that we choose of the form

$$D_C := \sum_{i=1}^{g} C_i : C_i \in X,\ z(C_i) = (c_U)_i, |\lambda(C_i)| > 1,$$

– the collections of the points on the lower sheet with the z-coordinates at the critical points. (At least topologically, it is evident $D(E) \simeq \mathbb{R}^g/\mathbb{Z}^g$.)

Theorem 6.1. *For given E of the form (24) there exists a one-to-one correspondence between $J(E)$ and $D(E)$.*

Let now $\tilde{U}$ and T be polynomials of the form described above, and let us define $U = \tilde{U} \circ T$. Then we have a covering π of the Riemann surface $\tilde{X}$ associated to $\tilde{U}$ by the surface X associated to U:

$$\pi(z, \lambda) = (T(z), \lambda), \tag{26}$$

note $\pi : X_+ \to \tilde{X}_+$.

According to the general theory, this covering generates different natural mappings [16], in particular,

$$\pi_* : \mathrm{Jac}(X) \to \mathrm{Jac}(\tilde{X}),$$

and

$$\pi^* : \mathrm{Jac}(\tilde{X}) \to \mathrm{Jac}(X).$$

Thus, in combination with Theorem 6.1, we get the map

$$\begin{array}{ccc} D(\tilde{E}) & \xrightarrow{\pi_T^*} & D(E) \\ \downarrow & & \downarrow \\ J(\tilde{E}) & \xrightarrow{\mathcal{J}_T} & J(E) \end{array}$$

We study this map as described in the previous section.

The differential $\frac{1}{2\pi i} d \log b$, restricted on ∂X_+, is the harmonic measure $d\omega$ of the domain $\bar{\mathbb{C}} \setminus E$ with pole at infinity. The space $L^p(\partial X_+)$, in a sense, is the L^p space with respect to the harmonic measure, but it should be mentioned that $\partial X_+ = (E - i0) \cup (E + i0)$, i.e., an element f of $L^p(\partial X_+)$ may have different values $f(x + i0)$ and $f(x - i0)$, $x \in E$, and $H^p(X_+)$ is the closure of the set of holomorphic functions uniformly bounded on X_+ with respect to this norm.

Since $\tilde{\lambda} \circ \pi = \lambda$ we have the relation

$$\tilde{b} \circ \pi = b^d.$$

Thus, we get

$$\int_{\partial X_+} f \, d\omega = \int_{\partial \tilde{X}_+} \frac{1}{d} \left(\sum_{\pi(Z) = \tilde{Z}} f(Z) \right) (\tilde{Z}) \, d\tilde{\omega} \tag{27}$$

for every $f \in L^1(\partial X_+)$.

As it follows directly from (27), the covering (26) generates an isometric enclosure

$$v_+ : H^2(\tilde{X}) \to H^2(X_+)$$

acting in the natural way

$$(v_+ f)(Z) = f(\pi(Z)).$$

Remark. As it was mentioned, the function b is not single valued but $|b(z)|$ is a single valued function. We define the character $\mu \in \Gamma^*$ by

$$b(\gamma(z)) = \mu(\gamma) b(z).$$

Let γ_j be the contour, that starts at infinity (or any other real point bigger than 1), go in the upper half-plane to the gap (a_j, b_j) and then go back in the lower half-plane to the initial point. Assuming that $b_0 < \cdots < a_j < b_j < a_{j+1} < \cdots < a_0$, we have $\mu(\gamma_j) = e^{-2\pi i \frac{g+1-j}{g+1}}$, equivalently $\omega([b_j, a_0]) = \frac{g+1-j}{g+1}$. Note that the system of the above contours γ_j is a generator of the free group $\Gamma^*(E)$. In other words a character α is uniquely defined by the vector

$$\left[\alpha(\gamma_1),\quad \alpha(\gamma_2),\quad \ldots,\quad \alpha(\gamma_g)\right] \in \mathbb{T}^g.$$

This sets a one-to-one correspondence between $\Gamma^*(E)$ and $\mathbb{T}^g$.

Recall the key role of the reproducing kernels k^α in our construction. In this particular case they are especially well studied [10]. First of all, they have an analytic continuation (as multivalued functions) on the whole X, so we can write $k^\alpha(Z)$.

Theorem 6.2. *For every $\alpha \in \Gamma^*$ the reproducing kernel $k^\alpha(Z)$ has on X exactly g simple poles that do not depend on α and g simple zeros. The divisor $D_+ = \sum_{j=1}^g Z_j$ of zeros*

$$k^\alpha(Z_j) = 0$$

with the divisor of poles form the divisor

$$\operatorname{div}(k^\alpha) = D_+ - D_C \tag{28}$$

that belongs to $D(E)$, moreover (28) sets a one-to-one correspondence between $D(E)$ and $\Gamma^*(E)$.

The functions k^α possess different representations, in particular, in terms of theta-functions [16], and the map $D \mapsto \alpha$ can be written explicitly in terms of Abelian integrals (the Abel map).

Summary. The three objects $J(E)$, $D(E)$ and $\Gamma^*(E)$ are equivalent. Both maps $\Gamma^*(E) \to D(E)$ and $\Gamma^*(E) \to J(E)$ can be defined in terms of the reproducing kernels of the spaces $H^2(X_+, \alpha)$, $\alpha \in \Gamma^*(E)$. The first one is given by (28). It associates to $k^\alpha(Z)$ the sets of its zeros and poles (the poles are fixed and the zeros vary with α). The matrix $J(\alpha) \in J(E)$ is defined as the matrix of the multiplication operator by $z(Z)$

$$z(Z)e_s^\alpha(Z) = p_s^\alpha e_{s-1}^\alpha(Z) + q_s^\alpha e_s^\alpha(Z) + p_{s+1}^\alpha e_{s+1}^\alpha(Z),\ Z \in X,\ s \in \mathbb{Z},$$

with respect to the basis

$$e_s^\alpha(Z) = b^s(Z) K^{\alpha\mu^{-s}}(Z). \tag{29}$$

It is really easy to see that $J(\alpha)$ is periodic: just recall that b^{g+1} is single valued, that is, $\mu^{g+1} = 1$, and therefore the spaces $H^2(X_+, \alpha)$ and $H^2(X_+, \alpha\mu^{-(g+1)})$ (and their reproducing kernels) coincide.

Now we can go back to the Renormalization Equation. Note that π acts on $\Gamma(E)$ in the following natural way:

$$\pi\gamma = \{\pi(Z),\ Z \in \gamma\} \in \Gamma(\tilde{E}), \quad \text{for } \gamma \in \Gamma(E).$$

The map $\pi^* : \Gamma^*(\tilde{E}) \to \Gamma^*(E)$ is defined by duality:

$$(\pi^*\tilde{\alpha})(\gamma) = \tilde{\alpha}(\pi\gamma).$$

Theorem 6.3. *Let T, $T^{-1} : [-1,1] \to [-1,1]$, be an expanding polynomial. Let $\tilde{J}$ be a periodic Jacobi matrix with spectrum $\tilde{E} \subset [-1,1]$, and therefore there exists a polynomial $\tilde{U}$ such that $\tilde{E} = \tilde{U}^{-1}[-1,1]$ and a character $\tilde{\alpha} \in \Gamma^*(\tilde{E})$ such that $\tilde{J} = J(\tilde{\alpha})$. Then*

$$J := J(\pi^*\alpha) = \mathcal{J}_T(\tilde{J})$$

is the periodic Jacobi matrix with spectrum $E = U^{-1}[-1,1]$, $U := \tilde{U} \circ T$, that satisfies the Renormalization Equation

$$V^*(z - J)^{-1}V = \frac{T'(z)}{d}(T(z) - \tilde{J})^{-1},$$

where the isometry matrix V is defined by $V|k\rangle := |kd\rangle$.

Remark 6.4. Let us mention that the Renormalization Equation can be rewritten equivalently in the form of polynomials equations, as it should be since we have the map from one algebraic variety, $\mathrm{Jac}(\tilde{X})$, in the other one, $\mathrm{Jac}(X)$. Equation (6.3) is equivalent to, see [21],

$$V^*T(J) = \tilde{J}V^*,$$

$$V^*\frac{T(z) - T(J)}{z - J}V = T'(z)/d.$$

Proof of Theorem 6.3. First we note, that for the operator multiplication by $z(Z)$ in $L^2(\partial X_+)$, the operator multiplication by $\tilde{z}(\tilde{Z})$ in $L^2(\partial\tilde{X}_+)$, the spectral parameter z_0 and the isometry

$$(vf)(Z) = f(\pi(Z)), \quad v : L^2(\partial\tilde{X}_+) \to L^2(\partial X_+),$$

we have

$$\int_{\partial X_+} \frac{1}{z_0 - z(Z)}|(vf)(Z)|^2\, d\omega = \int_{\partial\tilde{X}_+} \left(\frac{1}{d}\sum_{\pi(Z)=\tilde{Z}} \frac{1}{z_0 - z(Z)}\right)|f(\tilde{Z})|^2\, d\tilde{\omega}. \quad (30)$$

It is evident, that

$$\frac{1}{d}\sum_{T(y)=x}\frac{1}{z_0 - y} = \frac{T'(z_0)/d}{T(z_0) - x}.$$

Thus

$$v^*(z_0 - z(Z))^{-1}v = (T'(z_0)/d)(T(z_0) - \tilde{z}(\tilde{Z}))^{-1}.$$

It remains to show that π transforms the basis vector $\tilde{e}_n^{\tilde{\alpha}} = \tilde{b}^n K^{\tilde{\alpha}\tilde{\mu}^{-n}}$ into

$$e_{nd}^{\pi^*\alpha} = b^{nd}K^{(\pi^*\tilde{\alpha})\mu^{-nd}} = (\tilde{b}^n \circ \pi)K^{\pi^*(\tilde{\alpha}\tilde{\mu}^{-n})}.$$

Or, what is the same, that $K^{\tilde{\alpha}} \circ \pi = K^{\pi^*\alpha}$ for all $\tilde{\alpha} \in \Gamma^*(\tilde{E})$. Note that both functions are of norm one in the same space $H^2(X_+, \pi^*\tilde{\alpha})$, in particular, they have the same character of automorphy $\pi^*\tilde{\alpha} \in \Gamma^*(E)$. Note, finally, that the divisor

$$\operatorname{div}(k^{\tilde{\alpha}} \circ \pi) = \pi^{-1}(\tilde{D}_+) - \pi^{-1}(\tilde{D}_C),$$

where $\operatorname{div}(k^{\tilde{\alpha}}) = \tilde{D}_+ - \tilde{D}_C$, belongs to $D(E)$, therefore $k^{\tilde{\alpha}} \circ \pi$ is the reproducing kernel and the theorem is proved. □

Probably it would be better to call (6.3) the Renormalization Identity in the above theorem. The idea is that one can try to define J as the *solution* of (6.3) with the given $\tilde{J}$. Indeed, in the case of one-sided matrices such an equation has an unique solution. Now we demonstrate that in the two sided case for the given periodic $\tilde{J}$ we can find 2^{d-1} solutions. This proof is based on Theorem 6.2. For an alternative proof see [21].

To find all this solutions of (6.3) let us look a bit more carefully at the proof of Theorem 6.3. Note that the same identity (30) holds for any isometry v of the form

$$vf = v_\theta f = \theta(f \circ \pi),$$

where θ is a unimodular ($|\theta| = 1$) function on ∂X_+.

Concerning the second part of the proof, let us mention that the set of critical points of U splits into two sets:

$$\{c_U\} = T^{-1}\{c_{\tilde{U}}\} \cup \{c_T\}.$$

Correspondingly,

$$\sum (C_U)_j = \sum_k \sum_{\pi(C_U)_{k,j} = (C_{\tilde{U}})_k} (C_U)_{k,j} + \sum (C_T)_j,$$

and the divisor of $k^{\tilde{\alpha}} \circ \pi$ consists of two parts, the one that depends on $\tilde{\alpha}$

$$\pi^{-1}(\tilde{D}),$$

and the part that corresponds to the critical points of the polynomial T

$$\{(C_T)_j\}_{j=1}^{d-1},$$

since

$$D = \operatorname{div}(k^{\tilde{\alpha}} \circ \pi) = \pi^{-1}(\tilde{D}) + \sum_{j=1}^{d-1} (C_T)_j - \pi^{-1}(\tilde{D}_C) - \sum_{j=1}^{d-1} (C_T)_j.$$

Thus we can fix an arbitrary system of points $\{Z_{c,j}\}_{j=1}^{d-1}$ such that $z(Z_{c,j})$ belongs to the same gap in the spectrum E as the critical point $(c_T)_j$. If θ is the canonical product on X with the divisor

$$\operatorname{div}(\theta) = \sum_{j=1}^{d-1} Z_{c,j} - \sum_{j=1}^{d-1} (C_T)_j,$$

then $\theta k^{\tilde{\alpha}} \circ \pi$ is the reproducing kernel simultaneously for all $\tilde{\alpha} \in \Gamma^*(\tilde{E})$. But to make θ unimodular (zeros and poles are symmetric) our choice is restricted just to $Z_{c,j} = (C_T)_j$ or $Z_{c,j} = \tau(C_T)_j$. Note that $\tau(C_T)_j - (C_T)_j$ is the divisor of the complex Green function $b_{(c_T)_j}$. In this way we arrive at

Theorem 6.5. *For an expanding polynomial T, and a periodic Jacobi matrix $\tilde{J} = J(\tilde{\alpha})$, $\tilde{\alpha} \in \Gamma^*(\tilde{E})$ as in Theorem 6.3 there exist 2^{d-1} solutions of the Renormalization Equation (6.3). Denote by $\mu_{(c_T)_j}$ the character generated by the Green's function $b_{(c_T)_j}$, $b_{(c_T)_j} \circ \gamma = \mu_{(c_T)_j}(\gamma) b_{(c_T)_j}$. Then these solutions are of the form*

$$J := J\left(\eta_\delta \pi^* \tilde{\alpha}\right), \quad \eta_\delta := \prod_{j=1}^{d-1} \mu_{(c_T)_j}^{\frac{1}{2}(1+\delta_{(c_T)_j})}, \tag{31}$$

as before

$$\delta = \{\delta_{(c_T)_j}\}, \quad \delta_{(c_T)_j} = \pm 1.$$

Proof. We define the isometry

$$(vf)(Z) = \left(\prod_{j=1}^{d-1} b_{(c_T)_j}^{\frac{1}{2}(1+\delta_{(c_T)_j})}(Z) \right) f(\pi(Z))$$

and then repeat the arguments of the proof of Theorem 6.3. □

7. The Renormalization Equation for two-sided Jacobi matrices – general case

In this section we assume that

$$T(z) = z^d - qdz^{d-1} + \cdots$$

is a *monic* expanding polynomial. Under this normalization $T^{-1} : [-\xi, \xi] \to [-\xi, \xi]$ for a certain $\xi > 0$. It was proved in [21] that the Renormalization Equation has 2^{d-1} solutions for every two sided Jacobi matrix $\tilde{J}$ with the spectrum in $[-\xi, \xi]$, not only for periodic one. Moreover, they are the only possible solutions. For the reader's convenience we formulate these theorems here.

By $l^2_\pm(s)$ we denote the spaces which are formed by $\{|s+k\rangle\}$ with $k \le 0$ and $k \ge 0$ respectively, that is $l^2(\mathbb{Z}) = l^2_-(s) \oplus l^2_+(s+1)$. Correspondingly to these decompositions we set $\tilde{J}_\pm(s) = P_{l^2_\pm(s)} \tilde{J} | l^2_\pm(s)$. Recall that a (finite or infinite) one-sided Jacobi matrix is uniquely determined by its so-called resolvent function

$$\tilde{r}_\pm(z, s) = \langle s | (\tilde{J}_\pm(s) - z)^{-1} | s \rangle. \tag{32}$$

We parametrize this set of solutions by a collections of vectors

$$\delta := \{\delta_c\}_c, \tag{33}$$

where each component δ_c can be chosen as plus or minus one.

Theorem 7.1. *Fix a vector δ of the form (33). For a given $\tilde{J}$ with the spectrum on $[-\xi, \xi]$ define the Jacobi matrix J according to the following algorithm:*

For $s \in \mathbb{Z}$ we put

$$\frac{1}{T^{(s)}(c)} = -\tilde{r}_-(T(c), s), \quad \text{if } \delta_c = -1, \tag{34}$$

and

$$T^{(s)}(c) = -\tilde{p}_{s+1}^2 \tilde{r}_+(T(c), s+1), \quad \text{if } \delta_c = 1,$$

where the functions $\tilde{r}_\pm(z, s)$ are defined by (32). Then define the monic polynomial $T^{(s)}(z)$ of degree d by the interpolation formula

$$T^{(s)}(z) = (z-q)T'(z)/d + \sum_{c:T'(c)=0} \frac{T'(z)}{(z-c)T''(c)} T^{(s)}(c).$$

Define the block

$$J^{(s)} = \begin{bmatrix} q_{sd} & p_{sd+1} & & & \\ p_{sd+1} & q_{sd+1} & p_{sd+2} & & \\ & \ddots & \ddots & \ddots & \\ & & p_{sd+d-2} & q_{sd+d-2} & p_{sd+d-1} \\ & & & p_{sd+d-1} & q_{sd+d-1} \end{bmatrix}$$

by its resolvent function

$$\left\langle 0 \left| (z - J^{(s)})^{-1} \right| 0 \right\rangle = \frac{T'(z)/d}{T^{(s)}(z)},$$

where $T^{(s)}(z)$ is a monic polynomial of degree d. Finally define the entry p_{sd+d} by $p_{sd+1} \cdots p_{sd+d} = \tilde{p}_{s+1}$

We claim that the matrix $J = J(\delta, \tilde{J})$, combined with such blocks and entries over all s, satisfies (6.3).

Theorem 7.2. [21] *Theorem 7.1 describes the whole set of solutions of the Renormalization Equation.*

In [21] we concentrated only on one of the solutions of (6.3), namely that one that is related to the vector

$$\delta_- = \{-1, \ldots, -1\},$$

that is, all $T^{(s)}(c)$ are defined by (34). Note, $J(\delta_-, \tilde{J}) = \mathcal{J}_T(\tilde{J})$ for a periodic $\tilde{J}$. Precisely for this solution we proved (main Theorem 1.1 in [21]):

Theorem 7.3. *Let $\tilde{J}$ be a Jacobi matrix with the spectrum on $[-\xi, \xi]$. Then the Renormalization Equation (6.3) has a solution $J = J(\delta_-, \tilde{J})$ with the spectrum on $T^{-1}([-\xi, \xi])$. Moreover, if $\min_i |t_i| \geq 10\xi$ then*

$$\|J(\delta_-, \tilde{J}_1) - J(\delta_-, \tilde{J}_2)\| \leq \kappa \|\tilde{J}_1 - \tilde{J}_2\| \tag{35}$$

with an absolute constant $\kappa < 1$.

Let us emphasize the special role of the position of the critical values: the transform $\mathcal{J}_T(\tilde{J})$ is a contraction as soon as critical values are sufficiently far away from the spectrum.

In this work we add certain remarks concerning other solutions of (6.3).

7.1. The duality $\delta \mapsto -\delta$

At least for one more solution of the Renormalization Equation is a contraction.

Theorem 7.4. *The dual solution of the Renormalization Equation $J(\tilde{J}, -\delta)$, possesses simultaneously with $J(\tilde{J}, \delta)$ the contractibility property.*

It deals with the following universal involution acting on Jacobi matrices

$$J \to J_\tau := U_\tau J U_\tau, \quad \text{where } U_\tau |l\rangle = |1-l\rangle. \tag{36}$$

Obviously $VU_\tau = U_\tau S^{1-d} V$. Thus, having J as a solution of the renormalization equation corresponding to $\tilde{J}$ we have simultaneously that $S^{d-1} J_\tau S^{1-d}$ solves the equation with the initial $\tilde{J}_\tau$. The following lemma describes which branch corresponds to which in this case.

Lemma 7.5. *Let $J = J(\tilde{J}, \delta)$ then*

$$S^{d-1} J_\tau S^{1-d} = J(\tilde{J}_\tau, -\delta). \tag{37}$$

Proof. We give a proof using the language of Section 5, so formally we prove the claim only for periodic matrices.

Note that the involution (36) is strongly related to the standard involution τ (25) on X. Indeed, the function $K(\tau Z, \alpha)$ has the divisor

$$\tau D_+ - \tau D_C = (\tau D_+ - D_C) - (\tau D_C - D_C),$$

that is,

$$K(\tau Z, \alpha) = \frac{K(Z, \beta)}{b_{c_1}(Z) \dots b_{c_g}(Z)},$$

and $\beta = \nu \alpha^{-1}$, where $\nu = \mu_{c_1} \dots \mu_{c_g}$. Due to this remark and the property $z(\tau Z) = z(Z)$ we have

$$(J(\alpha))_\tau = J(\nu \mu \alpha^{-1}). \tag{38}$$

Now we apply (38) to prove (37). Let $\tilde{J}_\tau = J(\tilde{\alpha})$ with $\tilde{\alpha} \in \Gamma^*(\tilde{X}_+)$. Or, in other words, $\tilde{J} = J(\tilde{\mu}\tilde{\nu}\tilde{\alpha}^{-1})$. Then by (31)

$$J(\tilde{J}, \delta) = J(\eta_\delta \pi^*(\tilde{\mu}\tilde{\nu}\tilde{\alpha}^{-1})), \quad \eta_\delta := \prod_{j=1}^{d-1} \mu_{(c_T)_j}^{\frac{1}{2}(1+\delta_{(c_T)_j})}.$$

But $\pi^* \tilde{\mu} = \mu^d$ and $\pi^*(\tilde{\nu}) = \nu \eta_{\delta_+}^{-1}$ (just to look at the characters of the corresponding Blaschke products). Thus, having in mind that $\eta_\delta \eta_{-\delta} = \eta_{\delta_+}$, we obtain

$$J(\tilde{J}, \delta) = J(\mu^d \nu \eta_{-\delta}^{-1} \pi^*(\tilde{\alpha}^{-1})).$$

Using again (38) we get

$$(J(\tilde{J},\delta))_\tau = J(\mu^{1-d}\eta_{-\delta}\pi^*(\tilde{\alpha})) = S^{1-d}J(\eta_{-\delta}\pi^*(\tilde{\alpha}))S^{d-1},$$

and the lemma and Theorem 7.4 are proved. □

Let Jul(T) be the Julia set of a sufficiently hyperbolic polynomial T. Having two different contractive branches of solutions of the renormalization equation, following [14], to an arbitrary sequence

$$\epsilon = \{\epsilon_0, \epsilon_1, \epsilon_2, \dots\}, \quad \epsilon_j = \delta_\pm.$$

we can associate a limit periodic matrix J with the spectrum on Jul(T). To this end we define J as the limit of

$$J_n := J(\eta_{\epsilon_0}\pi^*\eta_{\epsilon_1}\dots\pi^*\eta_{\epsilon_{n-1}}) \tag{39}$$

and use Theorem 7.4.

7.2. Other solutions of the Renormalization Equation and the Ruelle operators

Iterating the renormalization transform

$$J_{n+1} := J(J_n, \delta_-),$$

we obtain an almost periodic operator $J = \lim J_n$ with spectrum on the Julia set of the sufficiently expanding polynomial T (the key point, here, is of course property (35)). This operator possesses the following structure [21]: it is the direct sum of two (one-sided) Jacobi matrices $J_\pm$. Moreover, the spectral measure of the operator J_+ is the balanced measure μ on the Julia set. That is, it is the eigenmeasure of the Ruelle operator $\mathcal{L}^*$, where

$$(\mathcal{L}f)(x) = \frac{1}{d}\sum_{Tn(y)=x} f(y).$$

The spectral measure ν of J_- is the so-called Bowen–Ruelle measure. It is the eigenmeasure for $\mathcal{L}_2^*$,

$$(\mathcal{L}_2 f)(x) = \sum_{T(y)=x} \frac{f(y)}{T'(x)^2},$$

We conjecture that actually all branches of solutions of the renormalization equation are contractions for sufficiently hyperbolic T. At least the previous subsection looks as a quite strong indication in this direction: considering, instead of initial T, $T^2 = T\circ T$ or its bigger powers, we get, as in (39), several δ's, $\eta_\delta = \eta_{\epsilon_0}\pi^*\eta_{\epsilon_1}\dots\pi^*\eta_{\epsilon_{n-1}}$, possessing the contractibility property with respect to the polynomial T^n and different from $\delta_\pm$ (related to $(\pi^*)^n$).

Similarly to the above statement we formulate

Conjecture. Let $T(z)$ be an expanding polynomial. For a given δ we factorize $T'(z) = A_1(z)A_2(z)$ putting in the first factor all critical points related to $\delta_c = 1$.

Denote by $\sigma_{1,2}$, the (nonnegative) eigenmeasures, corresponding to the Ruelle operators

$$(\mathcal{L}_{A_i} f)(x) = \sum_{T(y)=x} \frac{f(y)}{A_i(y)^2}, \tag{40}$$

i.e., $\mathcal{L}^*_{A_i}\sigma_i = \rho_i\sigma_i$. Finally let $J_{1,2}$ be the one-sided Jacobi matrices associated with $\sigma_{1,2}$. We conjecture that the iterations $J_{n+1} := J(J_n, \delta)$, converges to the block matrix $J = J_- \oplus J_+$ with $J_- = J_1$ and $J_+ = J_2$. In particular this means that all such operators are limit periodic.

In the support of this conjecture let us note that the statement holds true, say, for the polynomial of the form $T^2(z)$, and for $A_1(z) = T'(z)$, $A_2 = T'(T(z))$.

7.3. Shift transformations with the Lipschitz property

We say that the direction $\eta \in \Gamma^*$ has the Lipschitz property with a constant $C(\eta)$ if for all $\alpha, \beta \in \Gamma^*$

$$\|J(\eta\alpha) - J(\eta\beta)\| \le C(\eta)\|J(\alpha) - J(\beta)\|.$$

Then, one can get the contractibility of the map $\eta\pi^*$ in two steps:

$$\begin{aligned}\|J(\eta\pi^*\tilde{\alpha}) - J(\eta\pi^*\tilde{\beta})\| &\le C(\eta)\|J(\pi^*\tilde{\alpha}) - J(\pi^*\tilde{\beta})\| \\ &\le C(\eta)\kappa\|J(\tilde{\alpha}) - J(\tilde{\beta})\|.\end{aligned}$$

Note, that in fact the situation is a bit more involved because we should be able to compare Jacobi matrices with different spectral sets, for example, when $E_i = T^{-1}\tilde{E}_i$, $\tilde{E}_1 \ne \tilde{E}_2$. But we just want to indicate the general idea. Note, in particular, that for directions η_δ of the form (31) such a comparison is possible. Of course, for our goal the constant $C(\eta)$ should be uniformly bounded when we increase the level of hyperbolicity of T making κ smaller.

However the key point of this remark (this way of proof) is that, actually, *we do not need to constrain ourselves by the form of the vector* η. Combining a "Lipschitz" shift by η (the direction is restricted just by this property) with a sufficiently contractive pull-back π^* we arrive at an iterative process that produces a limit periodic Jacobi matrix with the spectrum on the same $\mathrm{Jul}(T)$. In the next subsection we give examples of directions with the required property, see Corollary 7.8.

We do not have a proof of the Lipschitz property of η_δ's, but there is a good chance to generalize the result of the next subsection in a way that at least some of the directions η_δ will be also available.

Finally, we would be very interested to know, whether there is in general a relation between the form of the "weight" vector η and the corresponding weights of the Ruelle operators (if any exists).

7.4. Quadratic polynomials and the Lipschitz property of the Darboux transform

Consider the simplest special case $T(z) = \rho(z^2 - 1) + 1$, $\rho > 2$. Note that the spectral set $E = T^{-1}\tilde{E}$ is symmetric, moreover the matrix related to $H^2(\pi^*\tilde{\alpha})$ has zero main diagonal (as well as a one-sided matrix related to a symmetric measure). Now we introduce a decomposition of $H^2(\pi^*\tilde{\alpha})$ which is very similar to the standard decomposition into even and odd functions.

We define the two-dimensional vector-function representation of $f \in H^2(\pi^*\tilde{\alpha})$

$$f \mapsto \frac{1}{\sqrt{2}} \begin{bmatrix} f(Z_1(\tilde{Z})) \\ f(Z_2(\tilde{Z})) \end{bmatrix} \mapsto \begin{bmatrix} g_1(\tilde{Z}) \\ g_2(\tilde{Z}) \end{bmatrix}, \tag{41}$$

where

$$\begin{bmatrix} g_1(\tilde{Z}) \\ g_2(\tilde{Z}) \end{bmatrix} = \frac{1}{2} \begin{bmatrix} f(Z_1(\tilde{Z})) + f(Z_2(\tilde{Z})) \\ f(Z_1(\tilde{Z})) - f(Z_2(\tilde{Z})) \end{bmatrix},$$

the first component, in a sense, is even and the second is odd. To be more precise, let us describe the analytical properties of this object in $\partial\tilde{X}_+$.

Note that due to

$$\int_{\partial X_+} |f|^2 d\omega = \int_{\partial \tilde{X}_+} \frac{1}{2} \sum_{\pi(Z)=\tilde{Z}} |f|^2 d\tilde{\omega}$$

metrically it is from L^2 with respect to $\tilde{\omega}$, moreover the transformation is norm-preserved.

It is evident that the function g_1 belongs to $H^2(\tilde{X}_+, \tilde{\alpha})$. Consider the second function. Note that the critical points of T are zero and infinity. For a small circle γ around the point $T(0) = -\rho + 1$ we have $g_2 \circ \gamma = -g_2$ and the same property holds for a contour γ that surrounds infinity. Let us introduce

$$\Delta^2 := \tilde{b}_{T(0)}\tilde{b}.$$

Note that for the above contours we have $\Delta \circ \gamma = -\Delta$. We are going to represent g_2 in the form $g_2 = \Delta \hat{g}_2$ and to claim that $\hat{g}_2$ has nice automorphic properties in $\tilde{X}_+$. Let us note that

$$\tilde{b} \frac{\tilde{z} - T(0)}{\tilde{b}_{T(0)}} = \tilde{b}^2 \frac{\tilde{z} - T(0)}{\Delta^2}$$

is an outer function in the domain $\bar{\mathbb{C}} \setminus \tilde{E} \simeq \tilde{X}_+$. So, the square root of this function is well defined. We put

$$\tilde{b}\phi := \sqrt{\tilde{b}^2 \frac{\tilde{z} - T(0)}{\rho\Delta^2}} \tag{42}$$

and denote by $\tilde{\eta}$ the character generated by ϕ, $\phi \circ \gamma = \eta(\gamma)\phi$. Thus (42) reduces the ramification of the function Δ to the function ϕ, which is well defined in the domain, and to the elementary function $\sqrt{\tilde{z} - T(0)}$.

Theorem 7.6. *The transformation $f \mapsto g_1 \oplus \hat{g}_2$ given by (41) is a unitary map from $H^2(\pi^*\tilde{\alpha})$ to $H^2(\tilde{\alpha}) \oplus H^2(\tilde{\alpha}\tilde{\eta})$. Moreover with respect to this representation*

$$zf \mapsto \begin{bmatrix} 0 & \bar{\phi} \\ \phi & 0 \end{bmatrix} \begin{bmatrix} g_1 \\ \hat{g}_2 \end{bmatrix} \tag{43}$$

and

$$v_+ f \mapsto f \oplus 0, \quad f \in H^2(\tilde{\alpha}),$$

where the isometry $v_+ : H^2(\tilde{\alpha}) \to H^2(\pi^*\tilde{\alpha})$ is defined by (6).

Proof. By the definition of Δ we have

$$g_2 = \Delta \hat{g}_2, \text{ where } \hat{g}_2 \in H^2(\tilde{\alpha}\tilde{\eta}).$$

Further, since

$$z_{1,2} = \pm\sqrt{\frac{\tilde{z} - T(0)}{\rho}},$$

we have, say for the second component,

$$\frac{1}{\Delta}\frac{(zf)(Z(\tilde{Z}_1)) - (zf)(Z(\tilde{Z}_2))}{2} = \sqrt{\frac{\tilde{z} - T(0)}{\rho\Delta^2}}\frac{f(Z(\tilde{Z}_1)) + f(Z(\tilde{Z}_2))}{2} = \phi g_1. \tag{44}$$

Since on the boundary of the domain

$$\phi^2\Delta^2 = \frac{\tilde{z} - T(0)}{\rho} = |\phi|^2$$

(the second expression is positive on $\partial\tilde{X}_+$) we have

$$\phi\Delta^2 = \overline{\phi} \quad \text{on } \tilde{E}. \tag{45}$$

Using this relation, similarly to (44), we prove the identity of the first components in (43).

□

Theorem 7.7. *The multiplication operator $\phi : L^2(\partial\tilde{X}_+) \to L^2(\partial\tilde{X}_+)$ with respect to the basis systems (29) related to $\tilde{\alpha}$ and $\tilde{\eta}\tilde{\alpha}$, respectively, is a two diagonal matrix Φ. Moreover,*

$$\Phi^*\Phi = \frac{J(\tilde{\alpha}) - T(0)}{\rho}, \quad \Phi\Phi^* = \frac{J(\tilde{\eta}\tilde{\alpha}) - T(0)}{\rho}. \tag{46}$$

In other words, the transformation $J(\tilde{\alpha}) \mapsto J(\tilde{\eta}\tilde{\alpha})$ is the Darboux transform.

Proof. First of all ϕ is a character-automorphic function with the character $\tilde{\eta}$ with a unique pole at infinity ($\tilde{b}\phi$ is an outer function). Therefore the multiplication operator acts from $\tilde{b}H^2(\tilde{\alpha}\tilde{\mu}^{-1})$ to $H^2(\tilde{\eta}\tilde{\alpha})$. Therefore, the operator Φ has only one non-trivial diagonal above the main diagonal. The adjoint operator has the symbol $\overline{\phi}$. According to (45) it has holomorphic continuation from the boundary inside the domain. Thus Φ^* is a lower triangular matrix. Combining these two facts we get that Φ has only two non-trivial diagonals. Then, just comparing symbols of operators on the left and right parts of (46), we prove these identities. □

Corollary 7.8. *Let* $\tilde{J}_{1,2}$ *be periodic Jacobi matrices with the spectrum on* $[-1,1]$. *Let* $\mathrm{Drb}(\tilde{J}_{1,2},\rho)$ *be their Darboux transforms. Then*

$$\|\mathrm{Drb}(\tilde{J}_1,\rho) - \mathrm{Drb}(\tilde{J}_2,\rho)\| \le C(\rho)\|\tilde{J}_1 - \tilde{J}_2\|. \tag{47}$$

Proof. For the given $\tilde{J}_{1,2}$ we define $J_{1,2}$ via the quadratic polynomial $T(z) = \rho(z^2-1)+1$. Being decomposed into even and odd indexed subspaces they are of the form

$$J_{1,2} = \begin{bmatrix} 0 & \Phi_{1,2}^* \\ \Phi_{1,2} & 0 \end{bmatrix}.$$

Due to the Theorem 7.3, that gives the uniform estimate for $\|J_1 - J_2\|$, we have

$$\|\Phi_1 - \Phi_2\| \le \kappa(\rho)\|\tilde{J}_1 - \tilde{J}_2\|,$$

with $\kappa(\rho) = \frac{C}{\rho-2}$, C is an absolute constant. Using (46) we get (47) with $C(\rho) = \frac{2\rho C}{\rho-2}$. □

8. Double covering $\pi(v) = \tau v - 1/v$

8.1. Over the simply-connected domain

The simplest expanding double coverings are $T_1 = v^2 - \lambda$, $\lambda > 2$, and $T_2 = \tau v - 1/v$, $\tau > 1$. Denote by $\xi_{1,2}$ the fixed points $T_{1,2}(\xi_{1,2}) = \xi_{1,2}$, $\xi_{1,2} > 0$. To start with we give a complete description of finite difference operators related to these coverings over the simply-connected domain $\overline{\mathbb{C}} \setminus E$, where $E = [-\xi_{1,2}, \xi_{1,2}]$.

More precisely, we start with a Jacobi matrix J_0 with constant coefficients. Under the normalization $\sigma(J_0) = E$ we have $J = \frac{\xi_{1,2}}{2}(S + S^{-1})$. That is the symbols of this operator (z_*, b_*) are related by

$$z_* = \frac{\xi_{1,2}}{2}\left(\frac{1}{b_*} + b_*\right), \tag{48}$$

(b_* is the Green function of the domain $\overline{\mathbb{C}} \setminus E$). For $\pi = T_1$ or $\pi = T_2$ we consider the open Riemann surfaces $Y_c \setminus F$ with $Y_c \simeq \overline{\mathbb{C}}$, $F = \pi^{-1}(E)$ and describe operators with the symbols (z^*, b^*):

$$z^* = z_* \circ \pi \quad \text{and} \quad (b^*)^2 = b_* \circ \pi. \tag{49}$$

The main difference between these two cases is that in the first one we have only one infinity on Y_c ($\infty \in \overline{\mathbb{C}}$) and in the second case there are two infinities: $0, \infty \in \overline{\mathbb{C}}$. Correspondingly an intrinsic basis contains the reproducing kernels related only to one given point in the first case and to two specific points in the second case. As result the multiplication operator by v with respect to this basis is a Jacobi matrix in the first case and a five diagonal matrix (of a special form, see Lemma 4.1) in the second.

8.1.1. $\pi = v^2 - \lambda$. Due to (48), (49) we have

$$v^2 - \lambda = \frac{\xi_1}{2}\left(\frac{1}{(b^*)^2} + (b^*)^2\right). \tag{50}$$

Since v is the symbol of a Jacobi matrix,

$$v \sim S\Lambda_1 + \Lambda_0 + \Lambda_1 S^{-1},$$

where $\Lambda_{0,1}$ are diagonal of period two matrices, and $(b^*)^2 \sim S^2$, we have from (50)

$$\begin{aligned} &\lambda_0^{(1)}\lambda_1^{(1)} = \xi_1/2,\\ &\lambda_0^{(0)} + \lambda_1^{(0)} = 0,\\ &(\lambda_0^{(0)})^2 + (\lambda_0^{(1)})^2 + (\lambda_1^{(1)})^2 = \lambda. \end{aligned} \tag{51}$$

8.1.2. $\pi = \tau v - 1/v$. Repeating arguments (49), (48) we get instead of (50)

$$\tau v - 1/v = \frac{\xi_2}{2}\left(\frac{1}{(b^*)^2} + (b^*)^2\right).$$

or

$$\tau v^2 - 1 = \frac{\xi_2}{2}\left(\frac{1}{(b^*)^2} + (b^*)^2\right) v. \tag{52}$$

In this case we have a five diagonal matrix,

$$v \sim S^2\Lambda_2 + S\Lambda_1 + \Lambda_0 + \Lambda_1 S^{-1} + \Lambda_2 S^{-2},$$

but of the specific structure: depending of ordering of infinities one of coefficients $\lambda_0^{(2)}, \lambda_1^{(2)}$ vanishes. Say, $\lambda_1^{(2)} = 0$, respectively $\lambda_0^{(2)} \neq 0$. Using $(b^*)^2 \sim S^2$ we get from (52)

$$\begin{aligned} &\tau S^2\Lambda_2 S^2\Lambda_2 = \frac{\xi_2}{2} S^4\Lambda_2,\\ &\tau(S^2\Lambda_2 S\Lambda_1 + S\Lambda_1 S^2\Lambda_2) = \frac{\xi_2}{2} S^3\Lambda_1,\\ &\tau(S^2\Lambda_2\Lambda_0 + S\Lambda_1 S\Lambda_1 + \Lambda_0 S^2\Lambda_2) = \frac{\xi_2}{2} S^2\Lambda_0,\\ &\tau(S^2\Lambda_2\Lambda_1 S^{-1} + S\Lambda_1\Lambda_0 + \Lambda_0 S\Lambda_1 + \Lambda_1 S\Lambda_2) = \frac{\xi_2}{2} S^2\Lambda_1 S^{-1},\\ &\tau(S^2\Lambda_2^2 S^{-2} + S\Lambda_1^2 S^{-1} + \Lambda_0^2 + \Lambda_1^2 + \Lambda_2^2) - I = \frac{\xi_2}{2}(S^2\Lambda_2 S^{-2} + \Lambda_2). \end{aligned} \tag{53}$$

That is $\tau\lambda_0^{(2)} = \xi_2/2$ and the second relation in (53) is an identity. Further,

$$\tau\lambda_0^{(1)}\lambda_1^{(1)} = -\frac{\xi}{2}\lambda_0^{(0)} = \frac{\xi}{2}\lambda_1^{(0)} \tag{54}$$

and the fourth relation is an identity. Finally, from the last equation we get

$$(\lambda_0^{(0)})^2 + (\lambda_0^{(1)})^2 + (\lambda_1^{(1)})^2 = 1/\tau. \tag{55}$$

Thus (54), (55) are counterparts of (51) in this case.

8.2. The Renormalization Equation in the general case

For $\pi : \overline{\mathbb{C}} \to \overline{\mathbb{C}}$ given by T_2 (that is $X_c \simeq \overline{\mathbb{C}}$ and $Y_c \simeq \overline{\mathbb{C}}$) let E be a system of intervals, a subset of $[-\xi_2, \xi_2]$. As usual $F = \pi^{-1}(E)$. For a system of points on $X_c \setminus E$

$$P_1 = \infty,\ P_2,\ \dots,\ P_l, \tag{56}$$

we define

$$b_*^l = b_{P_1} \dots b_{P_l}.$$

We consider z_* given by the identical map $X_c \to \mathbb{C}$. The finite difference operator $\tilde{J}$ related to this symbol, to a character $\alpha \in \Gamma_X^*$ and to the ordering system of infinities (56), is of the form

$$\tilde{J} = \begin{bmatrix} * & * & \dots & * & & & & \\ * & * & \dots & * & & & & \\ & * & \dots & * & & & & \\ & & \ddots & \vdots & & & & \\ & & & * & * & * & \dots & * \\ & & & & * & * & \dots & * \\ & & & & & * & \dots & * \\ & & & & & & \ddots & \vdots \\ & & & & & & & * \end{bmatrix},$$

because, actually, $z_*(P_j) = \infty$ only for $j = 1$.

The system of infinities on Y_c is given by

$$Q_1^{(1)} = \infty,\ Q_2^{(1)} = 0,\ \dots,\ Q_1^{(l)},\ Q_2^{(l)}, \tag{57}$$

where $\{Q_1^{(j)}, Q_2^{(j)}\} = \pi^{-1}(P_j)$, $j = 1, \dots, l$. Respectively,

$$z^* = z_* \circ \pi = \tau v - \frac{1}{v}, \quad (b^*)^2 = b_* \circ \pi.$$

We are interested to find a relation between the initial operator $\tilde{J}$ and the finite difference operator J with the symbol v related to the character $\pi^*\alpha \in \Gamma_Y^*$ and to the ordering system of infinities (57). Let us point out that the symbol of J is v, not z^*. It gives us an opportunity to iterate the procedure, which appears to be a certain rescaling.

Let us prove the following lemma.

Lemma 8.1. *Let* $T(z) = \frac{Q(z)}{P(z)}$. *Define*

$$(\mathfrak{L}f)(x) = \frac{1}{d} \sum_{T(y)=x} f(y).$$

Then

$$\mathfrak{L}\frac{1}{z-x} = \frac{1}{d}\frac{P'(z)}{P(z)} + \frac{1}{d}\frac{T'(z)}{T(z)-x}.$$

Proof. Note that

$$\sum_{Q(y)-xP(y)=0} \frac{1}{z-y} = \frac{Q'(z)-xP'(z)}{Q(z)-xP(z)}$$

and then collect corresponding terms

$$\frac{Q'(z)-xP'(z)}{Q(z)-xP(z)} = \frac{Q'(z)-Q(z)P'(z)/P(z)+\{Q(z)-xP(z)\}P'(z)/P(z)}{Q(z)-xP(z)}$$
$$= \frac{P'(z)}{P(z)} + \frac{Q'(z)P(z)-Q(z)P'(z)}{P^2(z)}\frac{1}{T(z)-x}. \qquad \square$$

Theorem 8.2. *The following relation holds*

$$\mathbf{V}^*(z-J)^{-1}\mathbf{V} = \frac{1}{d}\frac{P'(z)}{P(z)} + \frac{1}{d}T'(z)(T(z)-\tilde{J})^{-1}, \tag{58}$$

where $\tilde{J} = \mathbf{z}_*(\alpha)$ *and* $J = \mathbf{v}(\pi^*\alpha)$.

Proof. We use the previous lemma and the functional representation of all operators involved in (58). □

8.3. A vector representation of $H^2(\pi^*\alpha)$

Due to

$$\int_F |f|^2\omega^* = \int_E \frac{1}{2}\sum_{\pi(Q)=P}|f|^2\omega_*$$

we have a certain representation of $f \in H^2(\pi^*\alpha)$

$$f \mapsto \frac{1}{\sqrt{2}}\begin{bmatrix} f(Q_1(P)) \\ f(Q_2(P)) \end{bmatrix} \mapsto \begin{bmatrix} g_1(P) \\ g_2(P) \end{bmatrix}, \tag{59}$$

where

$$\begin{bmatrix} g_1(P) \\ g_2(P) \end{bmatrix} = \frac{1}{2}\begin{bmatrix} f(Q_1(P)) + f(Q_2(P)) \\ f(Q_1(P)) - f(Q_2(P)) \end{bmatrix},$$

as a two-dimensional vector-function. Let us describe analytical properties of this object (metrically it is from L^2 with respect to ω_*, moreover the transformation is norm-preserved).

It is evident that the function g_1 belongs to $H^2(\alpha)$. Consider the second function. Let $c_\pm$ be critical points

$$\pi'(c_\pm) = 0 \Rightarrow c_\pm = \pm\frac{i}{\sqrt{\tau}}.$$

For a small circle γ around the point $\pi(c_\pm) = \pm 2i\sqrt{\tau}$ we have $g_2 \circ \gamma = -g_2$. Let us introduce

$$\Delta^2 := b_{\pi(c_+)}b_{\pi(c_-)}.$$

Note that $\Delta \circ \gamma = -\Delta$. Further, since the ratio

$$\frac{b_{\pi(c_+)}}{b_{\pi(c_-)}}(z) = e^{iC}\frac{z-\pi(c_+)}{z-\pi(c_-)}$$

is single-valued in $X_c \setminus E$ both Green functions have the same character of automorphy, which we denote by $\nu \in \Gamma_X^*$. Thus, we get

$$g_2 = \Delta \tilde{g}_2, \quad \text{where} \quad \tilde{g}_2 \in H^2(\alpha\nu^{-1}). \tag{60}$$

We can summarize the result of this subsection as

Theorem 8.3. *The transformation $f \mapsto g_1 \oplus g_2$ given by (59) is a unitary map from $H^2(\pi^*\alpha)$ to $H^2(\alpha) \oplus \Delta H^2(\alpha\nu^{-1})$. Moreover,*

$$Vf \mapsto f \oplus 0, \quad f \in H^2(\alpha),$$

where the isometry $V : H^2(\alpha) \to H^2(\pi^*\alpha)$ is defined by (21). Also,

$$k_{Q_\pm}^{\pi^*\alpha} = k_P^\alpha \oplus (\pm\Delta\overline{\Delta(P)}k_P^{\alpha\nu^{-1}})$$

for the reproducing kernel $k_{Q_\pm}^{\pi^*\alpha} \in H^2(\pi^*\alpha)$, $\pi(Q_\pm) = P$.

Extending this vector representation onto L^2-spaces we get immediately

Theorem 8.4. *The transformation $f \mapsto g_1 \oplus \tilde{g}_2$ given by (59) and (60) is a unitary map from $L^2(\pi^*\alpha)$ to $L^2(\alpha) \oplus L^2(\alpha\nu^{-1})$. With respect to this representation the multiplication operator by $z^* = z_* \circ \pi$ is of the form*

$$\mathbf{z}^*(\pi^*\alpha) \mapsto \begin{bmatrix} \mathbf{z}_*(\alpha) & 0 \\ 0 & \mathbf{z}_*(\alpha\nu^{-1}) \end{bmatrix}. \tag{61}$$

To get a similar representation for the multiplication operator by v we need to introduce the following notations. Let us note that

$$b_\infty^2 \frac{z^2 + 4\tau}{\Delta^2}$$

is an outer function in the domain $\overline{\mathbb{C}} \setminus E$. So, the square root of this function is well defined. We put

$$b_\infty \phi := \sqrt{b_\infty^2 \frac{z^2 + 4\tau}{\Delta^2}}.$$

Since on the boundary of the domain

$$\phi^2 \Delta^2 = z^2 + 4\tau = |\phi|^2$$

(the second expression is positive on E) we have

$$\phi \Delta^2 = \overline{\phi} \quad \text{on } E. \tag{62}$$

Lemma 8.5. *The multiplication operator $\phi : L^2(\alpha) \to L^2(\alpha\nu^{-1})$ with respect to the bases systems related to the infinities $\{P_1, \dots, P_l\}$ has as many diagonals as $\mathbf{z}_*(\alpha)$ and $\mathbf{z}_*(\alpha\nu^{-1})$. Moreover,*

$$\begin{aligned} \boldsymbol{\phi}\mathbf{z}_*(\alpha) &= \mathbf{z}_*(\alpha\nu^{-1})\boldsymbol{\phi}, \\ \boldsymbol{\phi}^*\boldsymbol{\phi} &= \mathbf{z}_*^2(\alpha) + 4\tau, \\ \boldsymbol{\phi}\boldsymbol{\phi}^* &= \mathbf{z}_*^2(\alpha\nu^{-1}) + 4\tau. \end{aligned} \tag{63}$$

Proof. First of all ϕ is a character-automorphic function with the character ν^{-1}, therefore the multiplication operator acts from $L^2(\alpha)$ to $L^2(\alpha\nu^{-1})$. Since $b_\infty \phi$ is an outer function, ϕ has a unique pole at infinity, and, therefore, the operator $\boldsymbol{\phi}$ has the same structure over diagonal as the operator multiplication by z. The adjoint operator has the symbol $\overline{\phi}$. According to (62) it has analytic continuation from the boundary inside the domain with the only pole at infinity. Thus $\boldsymbol{\phi}^*$ is also of the same structure over the main diagonal as $\mathbf{z}_*(\alpha)$ or $\mathbf{z}_*(\alpha\nu^{-1})$. Combining these two facts we get that the whole structure of $\boldsymbol{\phi}$ coincides with the structure of the matrix $\mathbf{z}_*(\alpha)$. Then, just comparing symbols of operators on the left and right parts of (63), we prove these identities. □

Theorem 8.6. *With respect to the decomposition $L^2(\pi^*\alpha) \simeq L^2(\alpha) \oplus L^2(\alpha\nu^{-1})$ the multiplication operator by v is of the form*

$$\mathbf{v}(\pi^*\alpha) \simeq \frac{1}{2\tau}\begin{bmatrix} \mathbf{z}_*(\alpha) & \boldsymbol{\phi}^* \\ \boldsymbol{\phi} & \mathbf{z}_*(\alpha\nu^{-1}) \end{bmatrix}. \tag{64}$$

Let us mention that according to (63), the operators given in (61) and (64) commute and satisfy the identity, which is generated by the symbols identity $z^* = \tau v - 1/v$.

8.4. One sided matrices and the expanding transform $\pi(v) = \tau v - \frac{\tau-1}{v}$

Note that in this normalization $v = 1$ is the positive fixed point, $\pi(1) = 1$. Put $E_0 = [-1, 1]$. For a continuous function f on

$$E_1 = \pi^{-1}([-1,1]) = [-1, -1 + \frac{1}{\tau}] \cup [1 - \frac{1}{\tau}, 1]$$

we define

$$(\mathcal{L}f)(x) = \frac{1}{2}\sum_{\pi(v)=x} f(v).$$

The conjugate operator acts on measures

$$\mathcal{L}^* : C(E_0)^* \to C(E_1)^*.$$

Let $f_0, f_1, f_2, \dots$ be a certain orthonormal system with respect to a (positive) measure $\nu \in C(E_0)^*$ then

$$f_0 \circ \pi, f_1 \circ \pi, f_2 \circ \pi, \dots$$

is orthonormal system with respect to $\mu := \mathcal{L}^*\nu$. Note that if the first system form basis in $L^2_{d\nu}$ the second one form basis in the set of "even" functions from $L^2_{d\mu}$, the functions that are invariant with respect to the substitution $f(-\frac{\tau-1}{\tau v}) = f(v)$.

Example. Let $f_0, f_1, f_2, \dots$ be orthonormal polynomials in $L^2_{d\nu}$. $f_0 \circ \pi, f_1 \circ \pi, f_2 \circ \pi, \dots$ is a certain orthonormal system in $L^2_{d\mu}$ consisting of "polynomials" of v and $1/v$, similarly to the systems that generate CMV matrices:

$$1, v, 1/v, v^2, 1/v^2 \dots.$$

Making a small modification in this procedure, we orthogonalize

$$1, \tau v + \frac{\tau - 1}{v}, \tau v - \frac{\tau - 1}{v}, (\tau v)^2 - \left(\frac{\tau - 1}{v}\right)^2, (\tau v)^2 + \left(\frac{\tau - 1}{v}\right)^2 \dots$$

and denote the orthonormal system by $e_0, e_1, e_2, \dots$.

It is evident that

$$e_{2k} = f_k \circ \pi$$

and

$$e_{2k+1}(v) = \left(\tau v + \frac{\tau - 1}{v}\right) g_k(\pi(v)),$$

where g_k is also orthonormal system of *polynomials* but with respect to the measure $(x^2 + 4\tau(\tau - 1))d\nu(x)$, since

$$\left(\tau v + \frac{\tau - 1}{v}\right)^2 = x^2 + 4\tau(\tau - 1), \quad for\ x = \tau v - \frac{\tau - 1}{v}.$$

Let J be the Jacobi matrix, corresponding to the multiplication operator in $L^2_{d\nu}$ with respect to the basis of the orthonormal polynomials.

The given J we want to describe the multiplication operator in $L^2_{d\mu}$ with respect to $\{e_k\}$.

We decompose $L^2_{d\mu}$ onto even and odd subspaces. Then

$$\tau v - \frac{\tau - 1}{v} \mapsto \begin{bmatrix} J & 0 \\ 0 & J_* \end{bmatrix},$$

where J_* is the Jacobi matrix corresponding to the measure $(x^2 + 4\tau(\tau - 1))d\nu(x)$.

Further,

$$\tau v + \frac{\tau - 1}{v} \mapsto \begin{bmatrix} 0 & \Phi^* \\ \Phi & 0 \end{bmatrix}.$$

It is quite evident that Φ is an upper triangular matrix.

We get that

$$v \mapsto \frac{1}{2\tau} \begin{bmatrix} J & \Phi^* \\ \Phi & J_* \end{bmatrix},$$

and

$$-1/v \mapsto \frac{1}{2(\tau - 1)} \begin{bmatrix} J & -\Phi^* \\ -\Phi & J_* \end{bmatrix}.$$

Therefore

$$\begin{bmatrix} J^2 - \Phi^*\Phi & -J\Phi^* + \Phi^* J_* \\ \Phi J - J_*\Phi & {J_*}^2 - \Phi\Phi^* \end{bmatrix} = \begin{bmatrix} -4\tau(\tau - 1) & 0 \\ 0 & -4\tau(\tau - 1) \end{bmatrix}.$$

Thus Φ can be found in the upper-lower triangular factorization

$$\Phi^*\Phi = J^2 + 4\tau(\tau - 1),$$

and for J_* we have $J_* = \Phi J \Phi^{-1}$.

Thus for

$$J = \begin{bmatrix} 0 & p_1 & & \\ p_1 & 0 & p_2 & \\ & \ddots & \ddots & \ddots \end{bmatrix}$$

we have

$$\pi^*(J) = \frac{1}{2\tau} \begin{bmatrix} 0 & & & & & & & \\ \lambda_0 & 0 & & & & & & \\ p_1 & 0 & 0 & & & & & \\ 0 & p_1\frac{\lambda_1}{\lambda_0} & \lambda_1 & 0 & & & & \\ 0 & 0 & p_2 & 0 & 0 & & & \\ 0 & 0 & 0 & p_2\frac{\lambda_2}{\lambda_1} & \lambda_2 & 0 & & \\ 0 & \frac{p_1p_2}{\lambda_0} & 0 & 0 & p_3 & 0 & 0 & \\ & & \ddots & & & \ddots & \ddots & \end{bmatrix}.$$

The matrix is self-adjoint and λ_n are defined by the recursion

$$\lambda_n^2 = 4\tau(\tau-1) + p_{n+1}^2 + p_n^2 - \frac{p_n^2 p_{n-1}^2}{\lambda_{n-2}^2}.$$

with the initial data

$$\lambda_0^2 = p_1^2 + 4\tau(\tau-1), \quad \lambda_1^2 = p_1^2 + p_2^2 + 4\tau(\tau-1). \qquad \square$$

Theorem 8.7. *Let ν be the spectral measure of A,*

$$\int \frac{d\nu(x)}{x-z} = \langle 0|(A-z)^{-1}|0\rangle. \tag{65}$$

Then $\pi^*(A)$ is a self-adjoint operator with the cyclic vector $|0\rangle$ and the spectral measure $\mu = \mathcal{L}^*\nu$.

Proof. By Definition 1.5 and (65) we have

$$\begin{aligned}\langle 0|(\pi^*(A)-z)^{-1}|0\rangle =& 2\tau \left\langle \begin{bmatrix} A-2\tau z & \Phi^* \\ \Phi & A_* - 2\tau z \end{bmatrix}^{-1} \begin{bmatrix} |0\rangle \\ 0 \end{bmatrix}, \begin{bmatrix} |0\rangle \\ 0 \end{bmatrix} \right\rangle \\ =& 2\tau\langle 0|(A-2\tau z - \Phi^*(A_* - 2\tau z)^{-1}\Phi)^{-1}|0\rangle \\ =& 2\tau\langle 0|(A-2\tau z - \Phi^*\Phi(A-2\tau z)^{-1})^{-1}|0\rangle \\ =& 2\tau\langle 0|(A-2\tau z - (A^2+4\tau(1-\tau))(A-2\tau z)^{-1})^{-1}|0\rangle \\ =& \int \frac{2\tau d\nu(x)}{x-2\tau z - (x^2+4\tau(1-\tau))(x-2\tau z)^{-1}} \\ =& \int \frac{(x-2\tau z)d\nu(x)}{2\tau z^2 - 2xz - 2(1-\tau)}.\end{aligned}$$

Since

$$\left(\mathcal{L}\frac{1}{v-z}\right)(x) = \frac{1}{2}\sum_{\tau v - \frac{\tau-1}{v} = x} \frac{1}{v-z} = \frac{x-2\tau z}{2\tau z^2 - 2xz - 2(1-\tau)},$$

we get

$$\langle 0|(\pi^*(A)-z)^{-1}|0\rangle = \int \left(\mathcal{L}\frac{1}{v-z}\right)(x)d\nu(x) = \int \frac{1}{v-z}d(\mathcal{L}^*\nu)(v)$$

and the theorem is proved. □

Using Ruelle's Theorem with respect to the map $\pi(v)$ we can summarize our considerations by the following theorem.

Theorem 8.8. *The iterative procedure*

$$A_{n+1} = \pi^*(A_n)$$

converges to the operator $A = \lim_{n\to\infty} A_n$ *with the spectral measure* μ *which is the eigenmeasure for the Ruelle operator* $\mathcal{L}^*\mu = \mu$*. The operator* A *is the multiplication operator by the independent variable in* $L^2_{d\mu}$ *with respect to the following basis*

$$e_{2k}(v) = e_k(\pi(v)),$$

$$e_{2k+1}(v) = \left(\tau v + \frac{\tau-1}{v}\right)\sum_{j=0}^{k} c_j^k e_j(\pi(v)), \quad e_0(v) = 1,$$

where the coefficients c_j^k *with* $c_k^k > 0$ *are uniquely determined due to the orthogonality condition* $\langle e_{2k+1}, e_l\rangle = \delta_{2k+1,l}$, $l \le 2k+1$*. Moreover,* $e_k(v)$ *is a rational function of* v *such that* $e_k(A)|0\rangle = |k\rangle$*, and*

$$\begin{bmatrix} c_0^0 & c_0^1 & c_0^2 & \cdots \\ 0 & c_1^1 & c_1^2 & \cdots \\ & \ddots & \ddots & \ddots \end{bmatrix} = \Phi^{-1}.$$

References

[1] J. Avron, B. Simon, *Singular continuous spectrum for a class of almost periodic Jacobi matrices* Bull. AMS **6** (1982), 81–85.

[2] M.F. Barnsley, J.S. Geronimo, A.N. Harrington, *Almost periodic Jacobi matrices associated with Julia sets for polynomials*, Comm. Math. Phys. **99** (1985), no. 3, 303–317.

[3] J. Bellissard, D. Bessis, P. Moussa, *Chaotic states of almost periodic Schrödinger operators*, Phys. Rev. Lett. **49** (1982), no. 10, 701–704.

[4] J. Bellissard, B. Simon, *Cantor spectrum for almost Mathieu equation* J. Funct. Anal., **48** (1982), no. 3, 408–419.

[5] J. Bellissard, *Renormalization group analysis and quasicrystals.* Ideas and methods in quantum and statistical physics (Oslo, 1988), 118–148, Cambridge Univ. Press, Cambridge, 1992.

[6] R. Bowen, *Equilibrium States and the Ergodic Theory of Anosov Diffeomorphisms* Lecture Notes in Mathematics, v. 470, Springer-Verlag, Berlin-Heidelberg-New York, 1975.

[7] H.L. Cycon, R.G. Froese, W. Kirsch, B. Simon, *Schrödinger Operators with Application to Quantum Mechanics and Global Geometry*, Springer-Verlag, Berlin, 1987.

[8] D. Damanik, R. Killip, and B. Simon *Perturbation of orthogonal polynomials with periodic recursion coefficients*, preprint.

[9] B.A. Dubrovin, I.M. Krichever and S.P. Novikov, *Dynamical systems. IV*, 173–280, Springer-Verlag, Berlin, 1990.

[10] J. Fay, *Theta Functions on Riemann Surfaces*, Lecture Notes in Mathematics, Vol. 352, Springer-Verlag, New York/Berlin, 1973.

[11] L. Faybusovich and M. Gekhtman, *On Schur flows*, J. Phys. A **32** (1999), no. 25, 4671–4680.

[12] M. Hasumi, *Hardy Classes on Infinitely Connected Riemann Surfaces*, Lecture Notes in Math. **1027**, Springer-Verlag, Berlin and New York, 1983.

[13] J. Herndon, *Limit periodicity of sequences defined by certain recurrence relations; and Julia sets*, Ph.D. thesis, Georgia Institute of Technology, 1985.

[14] O. Knill, *Isospectral deformations of random Jacobi operators.* Comm. Math. Phys. 151 (1993), no. 2, 403–426.

[15] G. MacLane, Concerning the uniformization of certain Riemann surfaces allied to the inverse-cosine and inverse-gamma surfaces, Trans. AMS, 62 (1947) 99–113.

[16] D. Mumford, *Lectures on theta-functions.* (Russian) Translated from the English by D.Yu. Manin. Translation edited and with a preface by Yu.I. Manin. With appendices by Khirosi Umemura [Hiroshi Umemura] and Takakhiro Shiota [Takahiro Shiota]. "Mir", Moscow, 1988. 448 pp.

[17] P. van Moerbeke, *The spectrum of Jacobi matrices.* Invent. Math. 37 (1976), no. 1, 45–81.

[18] P. van Moerbekke, D. Mumford, *The Spectrum of Difference Operators and Algebraic Curves*, Acta Math. **143** (1979), no. 1-2, 93–154.

[19] L. Pastur and A. Figotin *Spectra of Random and Almost-Periodic Operators*, Springer-Verlag, Berlin, 1986.

[20] F. Peherstorfer and P. Yuditskii, *Almost periodic Verblunsky coefficients and reproducing kernels on Riemann surfaces*, J. Approx. Theory **139** (2006), 91–106

[21] F. Peherstorfer, A. Volberg and P. Yuditskii, *Limit periodic Jacobi matrices with prescribed p-adic hull and singular continuous spectrum*, Math. Res. Lett. **13** (2006), 215–230.

[22] F. Pehersorfer, A. Volberg, P. Yuditskii, *Two weight Hilbert transform and Lipschitz property of Jacobi matrices associated to hyperbolic polynomials,* submitted.

[23] M. Shapiro, V. Vinnikov, P. Yuditskii, *Finite Difference Operators with a Finite-band Spectrum*, Mat. Fiz. Anal. Geom. **11** (2004), no. 3, 331–340.

[24] B. Simon, *Orthogonal Polynomials on the Unit Circle*, AMS Colloquium Series, American Mathematical Society, Providence, RI, expected 2004.

[25] M. Sodin, P. Yuditski, *The limit-periodic finite-difference operator on $l_2(\mathbb{Z})$ associated with iterations of quadratic polynomials.* J. Statist. Phys. 60 (1990), no. 5-6, 863–873.

[26] M. Sodin and P. Yuditskii, *Almost periodic Jacobi matrices with homogeneous spectrum, infinite-dimensional Jacobi inversion, and Hardy spaces of character-automorphic functions*, J. Geom. Anal. **7** (1997), no. 3, 387–435.

[27] G. Teschel, *Jacobi Operators and Completely Integrable Nonlinear Lattices*, Mathematical Surveys and Monographs, **72**, American Mathematical Society, Providence, RI, 2000.

[28] E.B. Vinberg, *Real entire functions with prescribed critical values, Problems of group theory and homological algebra*, Yaroslavl. Gos. U., Yaroslavl, 1989, 127–138.

[29] V. Vinnikov and P. Yuditskii, *Functional models for almost periodic Jacobi matrices and the Toda hierarchy*, MAG **9**(2002), 206–219.

F. Peherstorfer and P. Yuditskii
Abteilung für Dynamische Systeme
und Approximationstheorie
Universität Linz
A-4040 Linz, Austria
e-mail: Franz.Peherstorfer@jku.at
e-mail: Petro.Yudytskiy@jku.at

Operator Theory:
Advances and Applications, Vol. 186, 391–401

Connections Between Dynamical Systems and Crossed Products of Banach Algebras by $\mathbb{Z}$

Christian Svensson, Sergei Silvestrov and Marcel de Jeu

Abstract. Starting with a complex commutative semi-simple regular Banach algebra A and an automorphism σ of A, we form the crossed product of A by the integers, where the latter act on A via iterations of σ. The automorphism induces a topological dynamical system on the character space $\Delta(A)$ of A in a natural way. We prove an equivalence between the property that every non-zero ideal in the crossed product has non-zero intersection with the subalgebra A, maximal commutativity of A in the crossed product, and density of the non-periodic points of the induced system on the character space. We also prove that every non-trivial ideal in the crossed product always intersects the commutant of A non-trivially. Furthermore, under the assumption that A is unital and such that $\Delta(A)$ consists of infinitely many points, we show equivalence between simplicity of the crossed product and minimality of the induced system, and between primeness of the crossed product and topological transitivity of the system.

Mathematics Subject Classification (2000). Primary 47L65 Secondary 16S35, 37B05, 54H20.

Keywords. Crossed product; Banach algebra; ideals, dynamical system; maximal abelian subalgebra.

1. Introduction

A lot of work has been done in the direction of connections between certain topological dynamical systems and crossed product C^*-algebras. In [6] and [7], for example, one starts with a homeomorphism σ of a compact Hausdorff space X and constructs the crossed product C^*-algebra $C(X) \rtimes_\alpha \mathbb{Z}$, where $C(X)$ is the algebra of continuous complex valued functions on X and α is the automorphism of $C(X)$ naturally induced by σ. One of many results obtained is equivalence between simplicity of the algebra and minimality of the system, provided that X consists of infinitely many points, see [1], [3], [6], [7] or, for a more general result, [8]. In [5], a purely algebraic variant of the crossed product is considered, with more general classes of algebras than merely continuous functions on compact Hausdorff spaces serving as "coefficient algebras". For example, it was proved there that, for such crossed products, the analogue of the equivalence between density of non-periodic points of a dynamical system and maximal commutativity of the "coefficient algebra" in the associated crossed product C^*-algebra is true for significantly larger classes of coefficient algebras and associated dynamical systems. In this paper, we go beyond these results and investigate the ideal structure of some of the crossed products considered in [5]. More specifically, we consider crossed products of complex commutative semi-simple regular Banach algebras A with the integers under an automorphism $\sigma : A \to A$.

In Section 2 we give the most general definition of the kind of crossed product that we will use throughout this paper. We also mention the elementary result that the commutant of the coefficient algebra is automatically a commutative subalgebra of the crossed product. The more specific setup that we will be working in is introduced in Section 3. There we also introduce some notation and mention two basic results concerning a canonical isomorphism between certain crossed products, and an explicit description of the commutant of the coefficient algebra in one of them.

According to [7, Theorem 5.4], the following three properties are equivalent:

- The non-periodic points of (X, σ) are dense in X;
- Every non-zero closed ideal I of the crossed product C^*-algebra $C(X) \rtimes_\alpha \mathbb{Z}$ is such that $I \cap C(X) \neq \{0\}$;
- $C(X)$ is a maximal abelian C^*-subalgebra of $C(X) \rtimes_\alpha \mathbb{Z}$.

In Section 4, an analogue of this result is proved for our setup. A reader familiar with the theory of crossed product C^*-algebras will easily recognize that if one chooses $A = C(X)$ for X a compact Hausdorff space in Corollary 4.5 below, then the crossed product is canonically isomorphic to a norm-dense subalgebra of the crossed product C^*-algebra coming from the considered induced dynamical system. We also combine this with a theorem from [5] to conclude a stronger result for the Banach algebra $L_1(G)$, where G is a locally compact abelian group with connected dual group.

In Section 5, we prove the equivalence between algebraic simplicity of the crossed product and minimality of the induced dynamical system in the case when A is unital with its character space consisting of infinitely many points. This is analogous to [7, Theorem 5.3], [1, Theorem VIII 3.9], the main result in [3] and, as a special case of a more general result, [8, Corollary 8.22] for the crossed product C^*-algebra.

In Section 6, the fact that the commutant of A always has non-zero intersection with any non-zero ideal of the crossed product is shown. This should be compared with the fact that A itself may well have zero intersection with such ideals, as Corollary 4.5 shows. The analogue of this result in the context of crossed product C^*-algebras appears to be open.

Finally, in Section 7 we show equivalence between primeness of the crossed product and topological transitivity of the induced system, in the case when A is unital and has an infinite character space. The analogue of this in the context of crossed product C^*-algebras is [7, Theorem 5.5].

2. Definition and a basic result

Let A be an associative commutative complex algebra and let $\Psi : A \to A$ be an algebra automorphism. Consider the set

$$A \rtimes_\Psi \mathbb{Z} = \{f : \mathbb{Z} \to A \,|\, f(n) = 0 \text{ except for a finite number of } n\}.$$

We endow it with the structure of an associative complex algebra by defining scalar multiplication and addition as the usual pointwise operations. Multiplication is defined by *twisted convolution*, $*$, as follows;

$$(f * g)(n) = \sum_{k\in\mathbb{Z}} f(k) \cdot \Psi^k(g(n-k)),$$

where Ψ^k denotes the k-fold composition of Ψ with itself. It is trivially verified that $A \rtimes_\sigma \mathbb{Z}$ *is* an associative $\mathbb{C}$-algebra under these operations. We call it the *crossed product of* A *by* $\mathbb{Z}$ *under* Ψ.

A useful way of working with $A \rtimes_\Psi \mathbb{Z}$ is to write elements $f, g \in A \rtimes_\Psi \mathbb{Z}$ in the form $f = \sum_{n\in\mathbb{Z}} f_n\delta^n$, $g = \sum_{m\in\mathbb{Z}} g_m\delta^m$, where $f_n = f(n)$, $g_m = g(m)$, addition and scalar multiplication are canonically defined, and multiplication is determined by $(f_n\delta^n) * (g_m\delta^m) = f_n \cdot \Psi^n(g_m)\delta^{n+m}$, where $n, m \in \mathbb{Z}$ and $f_n, g_m \in A$ are arbitrary.

Clearly one may canonically view A as an abelian subalgebra of $A \rtimes_\Psi \mathbb{Z}$, namely as $\{f_0\delta^0 \,|\, f_0 \in A\}$. The following elementary result is proved in [5, Proposition 2.1].

Proposition 2.1. *The commutant A' of A in $A \rtimes_\Psi \mathbb{Z}$ is abelian, and thus it is the unique maximal abelian subalgebra containing A.*

3. Setup and two basic results

In what follows, we shall focus on cases when A is a commutative complex Banach algebra, and freely make use of the basic theory for such A, see, e.g., [2]. As conventions tend to differ slightly in the literature, however, we mention that we call a commutative Banach algebra A *semi-simple* if the Gelfand transform on A is injective, and that we call it *regular* if, for every subset F of the character space $\Delta(A)$ that is closed in the Gelfand topology and for every $\phi_0 \in \Delta(A) \setminus F$, there exists an $a \in A$ such that $\widehat{a}(\phi) = 0$ for all $\phi \in F$ and $\widehat{a}(\phi_0) \neq 0$. All topological considerations of the character space $\Delta(A)$ will be done with respect to its Gelfand topology (the weakest topology making all elements in the image of the Gelfand transform of A continuous on $\Delta(A)$).

Now let A be a complex commutative semi-simple regular Banach algebra, and let $\sigma : A \to A$ be an algebra automorphism. As in [5], σ induces a map $\widetilde{\sigma} : \Delta(A) \to \Delta(A)$ (where $\Delta(A)$ denotes the character space of A) defined by $\widetilde{\sigma}(\mu) = \mu \circ \sigma^{-1}, \mu \in \Delta(A)$, which is automatically a homeomorphism when $\Delta(A)$ is endowed with the Gelfand topology. Hence we obtain a topological dynamical system $(\Delta(A), \widetilde{\sigma})$. In turn, $\widetilde{\sigma}$ induces an automorphism $\widehat{\sigma} : \widehat{A} \to \widehat{A}$ (where $\widehat{A}$ denotes the algebra of Gelfand transforms of all elements of A) defined by $\widehat{\sigma}(\widehat{a}) = \widehat{a} \circ \widetilde{\sigma}^{-1} = \widehat{\sigma(a)}$. Therefore we can form the crossed product $\widehat{A} \rtimes_{\widehat{\sigma}} \mathbb{Z}$. We also mention that when speaking of ideals, we will always mean two-sided ideals.

In what follows, we shall make frequent use of the following fact. Its proof consists of a trivial direct verification.

Theorem 3.1. *Let A be a commutative semi-simple Banach algebra and σ be an automorphism, inducing an automorphism $\widehat{\sigma} : \widehat{A} \to \widehat{A}$ as above. Then the map $\Phi : A \rtimes_\sigma \mathbb{Z} \to \widehat{A} \rtimes_{\widehat{\sigma}} \mathbb{Z}$ defined by $\sum_{n\in\mathbb{Z}} a_n \delta^n \mapsto \sum_{n\in\mathbb{Z}} \widehat{a_n} \delta^n$ is an isomorphism of algebras mapping A onto $\widehat{A}$.*

Before stating the next result, we make the following basic definitions.

Definition 3.2. For any nonzero $n \in \mathbb{Z}$ we set

$$\mathrm{Per}^n(\Delta(A)) = \{\mu \in \Delta(A) \mid \mu = \widetilde{\sigma}^n(\mu)\}.$$

Furthermore, we denote the non-periodic points by

$$\mathrm{Per}^\infty(\Delta(A)) = \bigcap_{n\in\mathbb{Z}\setminus\{0\}} (\Delta(A) \setminus \mathrm{Per}^n(\Delta(A)).$$

Finally, for $f \in \widehat{A}$, put

$$\mathrm{supp}(f) = \overline{\{\mu \in \Delta(A) \mid f(\mu) \neq 0\}}.$$

Theorem 3.3. *We have the following explicit description of $\widehat{A}'$ in $\widehat{A} \rtimes_{\widehat{\sigma}} \mathbb{Z}$:*

$$\widehat{A}' = \{\sum_{n\in\mathbb{Z}} f_n \delta^n \mid f_n \in \widehat{A}, \text{ and for all } n \in \mathbb{Z} : \mathrm{supp}(f_n) \subseteq \mathrm{Per}^n(\Delta(A))\}.$$

Proof. This follows from [5, Corollary 3.4], as $\widehat{A}$ trivially separates the points of $\Delta(A)$ and $\mathrm{Per}^n(\Delta(A))$ is a closed set. □

4. Three equivalent properties

In this section we shall conclude that, for certain A, two different algebraic properties of $A \rtimes_\sigma \mathbb{Z}$ are equivalent to density of the non-periodic points of the naturally associated dynamical system on the character space $\Delta(A)$, and hence obtain equivalence of three different properties. The analogue of this result in the context of crossed product C^*-algebras is [7, Theorem 5.4]. We shall also combine this with a theorem from [5] to conclude a stronger result for the Banach algebra $L_1(G)$, where G is a locally compact abelian group with connected dual group.

In [5, Theorem 4.8], the following result is proved.

Theorem 4.1. *Let A be a complex commutative regular semi-simple Banach algebra, $\sigma : A \to A$ an automorphism and $\widetilde{\sigma}$ the homeomorphism of $\Delta(A)$ in the Gelfand topology induced by σ as described above. Then the non-periodic points are dense in $\Delta(A)$ if and only if $\widehat{A}$ is a maximal abelian subalgebra of $\widehat{A} \rtimes_{\widetilde{\sigma}} \mathbb{Z}$. In particular, A is maximal abelian in $A \rtimes_\sigma \mathbb{Z}$ if and only if the non-periodic points are dense in $\Delta(A)$.*

We shall soon prove another algebraic property of the crossed product equivalent to density of the non-periodic points of the induced system on the character space. First, however, we need two easy topological lemmas.

Lemma 4.2. *Let $x \in \Delta(A)$ be such that the points $\widetilde{\sigma}^i(x)$ are distinct for all i such that $-m \leq i \leq n$, where n and m are positive integers. Then there exist an open set U_x containing x such that the sets $\widetilde{\sigma}^i(U_x)$ are pairwise disjoint for all i such that $-m \leq i \leq n$.*

Proof. It is easily checked that any finite set of points in a Hausdorff space can be separated by pairwise disjoint open sets. Separate the points $\widetilde{\sigma}^i(x)$ with disjoint open sets V_i. Then it is readily verified that the set

$$U_x := \widetilde{\sigma}^m(V_{-m}) \cap \widetilde{\sigma}^{m-1}(V_{-m+1}) \cap \cdots \cap V_0 \cap \widetilde{\sigma}^{-1}(V_1) \cap \cdots \cap \widetilde{\sigma}^{-n}(V_n)$$

is an open neighbourhood of x with the required property. □

Lemma 4.3. *The non-periodic points of $(\Delta(A), \widetilde{\sigma})$ are dense if and only if the set* $\mathrm{Per}^n(\Delta(A))$ *has empty interior for all positive integers n.*

Proof. Clearly, if there is a positive integer n_0 such that $\mathrm{Per}^{n_0}(\Delta(A))$ has non-empty interior, the non-periodic points are not dense. For the converse, we recall that $\Delta(A)$ is a Baire space since it is locally compact and Hausdorff, and note that we may write

$$\Delta(A) \setminus \mathrm{Per}^\infty(\Delta(A)) = \bigcup_{n>0} \mathrm{Per}^n(\Delta(A)).$$

If the set of non-periodic points is not dense, its complement has non-empty interior, and as the sets $\mathrm{Per}^n(\Delta(A))$ are clearly all closed, there must exist an integer $n_0 > 0$ such that $\mathrm{Per}^{n_0}(\Delta(A))$ has non-empty interior since $\Delta(A)$ is a Baire space. □

We are now ready to prove the promised result.

Theorem 4.4. *Let A be a complex commutative semi-simple regular Banach algebra, $\sigma : A \to A$ an automorphism and $\widetilde{\sigma}$ the homeomorphism of $\Delta(A)$ in the Gelfand topology induced by σ as described above. Then the non-periodic points are dense in $\Delta(A)$ if and only if every non-zero ideal $I \subseteq A \rtimes_\sigma \mathbb{Z}$ is such that $I \cap A \neq \{0\}$.*

Proof. We first assume that $\overline{\mathrm{Per}^\infty(\Delta(A))} = \Delta(A)$, and work initially in $\widehat{A} \rtimes_{\widetilde{\sigma}} \mathbb{Z}$. Assume that $I \subseteq \widehat{A} \rtimes_{\widetilde{\sigma}} \mathbb{Z}$ is a non-zero ideal, and that $f = \sum_{n\in\mathbb{Z}} f_n \delta^n \in I$. By definition, only finitely many f_n are non-zero. Denote the set of integers n for which $f_n \not\equiv 0$ by $S = \{n_1, \ldots, n_r\}$. Pick a non-periodic point $x \in \Delta(A)$ such that $f_{n_1}(x) \neq 0$ (by density of $\mathrm{Per}^\infty(\Delta(A))$ such x exists). Using the fact that x is not periodic we may, by Lemma 4.2, choose an open neighbourhood U_x of x such that $\widetilde{\sigma}^{-n_i}(U_x) \cap \widetilde{\sigma}^{-n_j}(U_x) = \emptyset$ for $n_i \neq n_j$, $n_i, n_j \in S$. Now by regularity of A we can find a function $g \in \widehat{A}$ that is non-zero in $\widetilde{\sigma}^{-n_1}(x)$, and vanishes outside $\widetilde{\sigma}^{-n_1}(U_x)$. Consider $f * g = \sum_{n\in\mathbb{Z}} f_n \cdot (g \circ \widetilde{\sigma}^{-n})\delta^n$. This is an element in I and clearly the coefficient of δ^{n_1} is the only one that does not vanish on the open set U_x. Again by regularity of A, there is an $h \in \widehat{A}$ that is non-zero in x and vanishes outside U_x. Clearly $h * f * g = [h \cdot (g \circ \widetilde{\sigma}^{-n_1}) f_{n_1}]\delta^{n_1}$ is a non-zero monomial belonging to I. Now any ideal that contains a non-zero monomial automatically contains a non-zero element of $\widehat{A}$. Namely, if $a_i\delta^i \in I$ then $[a_i\delta^i] * [(a_i \circ \widetilde{\sigma}^i)\delta^{-i}] = a_i^2 \in \widehat{A}$. By the canonical isomorphism in Theorem 3.1, the result holds for $A \rtimes_\sigma \mathbb{Z}$ as well.

For the converse, assume that $\overline{\mathrm{Per}^\infty(\Delta(A))} \neq \Delta(A)$. Again we work in $\widehat{A} \rtimes_{\widetilde{\sigma}} \mathbb{Z}$. It follows from Lemma 4.3 that since $\overline{\mathrm{Per}^\infty(\Delta(A))} \neq \Delta(A)$, there exists an integer $n > 0$ such that $\mathrm{Per}^n(\Delta(A))$ has non-empty interior. As A is assumed to be regular, there exists $f \in \widehat{A}$ such that $\mathrm{supp}(f) \subseteq \mathrm{Per}^n(\Delta(A))$. Consider now the ideal $I = (f + f\delta^n)$. Using that f vanishes outside $\mathrm{Per}^n(\Delta(A))$, we have

$$
\begin{aligned}
a_i\delta^i(f + f\delta^n)a_j\delta^j &= [a_i \cdot (a_j \circ \widetilde{\sigma}^{-i})\delta^i] * [f\delta^j + f\delta^{n+j}] \\
&= [a_i \cdot (a_j \circ \widetilde{\sigma}^{-i}) \cdot (f \circ \widetilde{\sigma}^{-i})]\delta^{i+j} + [a_i \cdot (a_j \circ \widetilde{\sigma}^{-i}) \cdot (f \circ \widetilde{\sigma}^{-i})]\delta^{i+j+n}.
\end{aligned}
$$

Therefore any element in I may be written in the form $\sum_i (b_i\delta^i + b_i\delta^{n+i})$. As i runs only through a finite subset of $\mathbb{Z}$, this is not a non-zero monomial. In particular, it is not a non-zero element in $\widehat{A}$. Hence I intersects $\widehat{A}$ trivially. By the canonical isomorphism in Theorem 3.1, the result carries over to $A \rtimes_\sigma \mathbb{Z}$. □

Combining Theorem 4.1 and Theorem 4.4, we now have the following result.

Corollary 4.5. *Let A be a complex commutative semi-simple regular Banach algebra, $\sigma : A \to A$ an automorphism and $\widetilde{\sigma}$ the homeomorphism of $\Delta(A)$ in the Gelfand topology induced by σ as described above. Then the following three properties are equivalent:*

- *The non-periodic points $\mathrm{Per}^\infty(\Delta(A))$ of $(\Delta(A), \widetilde{\sigma})$ are dense in $\Delta(A)$;*
- *Every non-zero ideal $I \subseteq A \rtimes_\sigma \mathbb{Z}$ is such that $I \cap A \neq \{0\}$;*
- *A is a maximal abelian subalgebra of $A \rtimes_\sigma \mathbb{Z}$.*

We shall make use of Corollary 4.5 to conclude a result for a more specific class of Banach algebras. We start by recalling a number of standard results from the theory of Fourier analysis on groups, and refer to [2] and [4] for details. Let G be a locally compact abelian group. Recall that $L_1(G)$ consists of equivalence classes of complex valued Borel measurable functions of G that are integrable with respect to a Haar measure on G, and that $L_1(G)$ equipped with convolution product is a commutative regular semi-simple Banach algebra. A group homomorphism $\gamma : G \to \mathbb{T}$ from a locally compact abelian group to the unit circle is called a *character* of G. The set of all *continuous* characters of G forms a group Γ, the *dual group* of G, if the group operation is defined by

$$(\gamma_1 + \gamma_2)(x) = \gamma_1(x)\gamma_2(x) \ (x \in G; \gamma_1, \gamma_2 \in \Gamma).$$

If $\gamma \in \Gamma$ and if

$$\widehat{f}(\gamma) = \int_G f(x)\gamma(-x)dx \quad (f \in L_1(G)),$$

then the map $f \mapsto \widehat{f}(\gamma)$ is a non-zero complex homomorphism of $L_1(G)$. Conversely, every non-zero complex homomorphism of $L_1(G)$ is obtained in this way, and distinct characters induce distinct homomorpisms. Thus we may identify Γ with $\Delta(L_1(G))$. The function $\widehat{f} : \Gamma \to \mathbb{C}$ defined as above is called the *Fourier transform* of $f \in L_1(G)$, and is precisely the Gelfand transform of f. We denote the set of all such $\widehat{f}$ by $A(\Gamma)$. Furthermore, Γ is a locally compact abelian group in the Gelfand topology.

In [5, Theorem 4.16], the following result is proved.

Theorem 4.6. *Let G be a locally compact abelian group with connected dual group and let $\sigma : L_1(G) \to L_1(G)$ be an automorphism. Then $L_1(G)$ is maximal abelian in $L_1(G) \rtimes_\sigma \mathbb{Z}$ if and only if σ is not of finite order.*

Combining Corollary 4.5 and Theorem 4.6 the following result is immediate.

Corollary 4.7. *Let G be a locally compact abelian group with connected dual group and let $\sigma : L_1(G) \to L_1(G)$ be an automorphism. Then the following three statements are equivalent.*

- *σ is not of finite order;*
- *Every non-zero ideal $I \subseteq L_1(G) \rtimes_\sigma \mathbb{Z}$ is such that $I \cap L_1(G) \neq \{0\}$;*
- *$L_1(G)$ is a maximal abelian subalgebra of $L_1(G) \rtimes_\sigma \mathbb{Z}$.*

5. Minimality versus simplicity

Recall that a topological dynamical system is said to be *minimal* if all of its orbits are dense, and that an algebra is called *simple* if it lacks non-trivial proper ideals.

Theorem 5.1. *Let A be a complex commutative semi-simple regular unital Banach algebra such that $\Delta(A)$ consists of infinitely many points, and let $\sigma : A \to A$ be an algebra automorphism of A. Then $A \rtimes_\sigma \mathbb{Z}$ is simple if and only if the naturally induced system $(\Delta(A), \widetilde{\sigma})$ is minimal.*

Proof. Suppose first that the system is minimal, and assume that I is a proper ideal of $A \rtimes_\sigma \mathbb{Z}$. Note that $I \cap A$ is a proper σ- and σ^{-1}-invariant ideal of A. By basic theory of Banach algebras, $I \cap A$ is contained in a maximal ideal of A (note that $I \cap A \neq A$ as A is unital and I was assumed to be proper), which is the kernel of an element $\mu \in \Delta(A)$. Now $\widehat{I \cap A}$ is a $\widehat{\sigma}$- and $\widehat{\sigma}^{-1}$-invariant proper non-trivial ideal of $\widehat{A}$, all of whose elements vanish in μ. Invariance of this ideal implies that all of its elements even annihilate the whole orbit of μ under $\widetilde{\sigma}$. But by minimality, every such orbit is dense and hence $\widehat{I \cap A} = \{0\}$. By semi-simplicity of A, this means $I \cap A = \{0\}$, so $I = \{0\}$ by Corollary 4.5. For the converse, assume that there is an element $\mu \in \Delta(A)$ whose orbit $O(\mu)$ is not dense. By regularity of A there is a nonzero $g \in \widehat{A}$ that vanishes on $\overline{O(\mu)}$. Then clearly the ideal generated by g in $\widehat{A} \rtimes_{\widetilde{\sigma}} \mathbb{Z}$ consists of finite sums of elements of the form $(f_n\delta^n) * g * (h_m\delta^m) = [f_n \cdot (g \circ \widetilde{\sigma}^{-n}) \cdot (h_m \circ \widetilde{\sigma}^{-n})]\delta^{n+m}$, and hence the coefficient of every power of δ in this ideal must vanish in μ, whence the ideal is proper. Hence by Theorem 3.1, $A \rtimes_\sigma \mathbb{Z}$ is not simple. □

6. Every non-zero ideal has non-zero intersection with A'

We shall now show that any non-zero ideal in $A \rtimes_\sigma \mathbb{Z}$ has non-zero intersection with A'. This should be compared with Corollary 4.5, which says that a non-zero ideal may well intersect A solely in 0. We have no information on the validity of the analogue of this result in the context of crossed product C^*-algebras.

Theorem 6.1. *Let A be a complex commutative semi-simple regular Banach algebra, and $\sigma : A \to A$ an automorphism. Then every non-zero ideal I in $A \rtimes_\sigma \mathbb{Z}$ has non-zero intersection with the commutant A' of A in $A \rtimes_\Psi \mathbb{Z}$, that is $I \cap A' \neq \{0\}$.*

Proof. As usual, we work in $\widehat{A} \rtimes_{\widetilde{\sigma}} \mathbb{Z}$. When $\overline{\mathrm{Per}^\infty(\Delta(A))} = \Delta(A)$, the result follows immediately from Corollary 4.5. We will use induction on the number of non-zero terms in an element $f = \sum_{n\in\mathbb{Z}} f_n\delta^n$ to show that it generates an ideal that intersects $\widehat{A}'$ non-trivially. The starting point for the induction, namely when $f = f_n\delta^n$ with non-zero f_n, is clear since any such element generates an ideal that even intersects $\widehat{A}$ non-trivially, as was shown in the proof of Theorem 4.4. Now assume inductively that the conclusion of the theorem is true for the ideals generated by any element of $\widehat{A} \rtimes_{\widetilde{\sigma}} \mathbb{Z}$ with r non-zero terms for some positive integer r, and consider an element $f = f_{n_1}\delta^{n_1} + \cdots + f_{n_{r+1}}\delta^{n_{r+1}}$. By multiplying from the right with a suitable element we obtain an element in the ideal generated by f of the form $g = \sum_{i=0}^{m_r} g_i\delta^i$ such that $g_0 \not\equiv 0$. If some of the other g_i are zero we are done by induction hypothesis, so we may assume this is not the case. We may also assume that g is not in the commutant of $\widehat{A}$ since otherwise we are of course also done. This means, by Theorem 3.3, that there is such j that $0 < j \leq m_r$ and $\mathrm{supp}(g_j) \not\subseteq \mathrm{Per}^j(\Delta(A))$. Pick an $x \in \mathrm{supp}(g_j)$ such that $x \neq \widetilde{\sigma}^{-j}(x)$ and $g_j(x) \neq 0$. As $\Delta(A)$ is Hausdorff we can choose an open neighbourhood U_x of x such that $U_x \cap \widetilde{\sigma}^{-j}(U_x) = \emptyset$. Regularity of A implies existence of an $h \in \widehat{A}$ such that $h \circ \widetilde{\sigma}^{-j}(x) = 1$ and h vanishes identically outside of $\widetilde{\sigma}^{-j}(U_x)$. Now $g*h = \sum_{i=0}^{m_r} g_i \cdot (h \circ \widetilde{\sigma}^{-i})\delta^i$. Using regularity of A again we pick a function $a \in \widehat{A}$ such that $a(x) = 1$ and a vanishes outside U_x. We have $a*g*h = \sum_{i=0}^{m_r} a \cdot g_i \cdot (h \circ \widetilde{\sigma}^{-i})\delta^i$, which is in the ideal generated by f. Now $a \cdot g_0 \cdot h$ is identically zero since $a \cdot h = 0$. On the other hand, $a \cdot g_j \cdot (h \circ \widetilde{\sigma}^{-j})$ is non-zero in the point x. Hence $a*g*h$ is a non-zero element in the ideal generated by f whose number of non-zero coefficient functions is less than or equal to r. By the induction hypothesis, such an element generates an ideal that intersects the commutant of $\widehat{A}$ non-trivially. By Theorem 3.1 it follows that every non-zero ideal in $A \rtimes_\sigma \mathbb{Z}$ intersects A' non-trivially. □

7. Primeness versus topological transitivity

We shall show that for certain A, $A \rtimes_\sigma \mathbb{Z}$ is prime if and only if the induced system $(\Delta(A), \widetilde{\sigma})$ is topologically transitive. The analogue of this result in the context of crossed product C^*-algebras is in [7, Theorem 5.5].

Definition 7.1. The system $(\Delta(A), \widetilde{\sigma})$ is called *topologically transitive* if for any pair of non-empty open sets U, V of $\Delta(A)$, there exists an integer n such that $\widetilde{\sigma}^n(U) \cap V \neq \emptyset$.

Definition 7.2. The algebra $A \rtimes_\sigma \mathbb{Z}$ is called *prime* if the intersection between any two non-zero ideals I, J is non-zero, that is $I \cap J \neq \{0\}$.

For convenience, we also make the following definition.

Definition 7.3. The map $E : \widehat{A} \rtimes_{\widetilde{\sigma}} \mathbb{Z} \to \widehat{A}$ is defined by $E(\sum_{n\in\mathbb{Z}} f_n \delta^n) = f_0$.

To prove the main theorem of this section, we need the two following topological lemmas.

Lemma 7.4. *If $(\Delta(A), \widetilde{\sigma})$ is not topologically transitive, then there exist two disjoint invariant non-empty open sets O_1 and O_2 such that $\overline{O_1} \cup \overline{O_2} = \Delta(A)$.*

Proof. As the system is not topologically transitive, there exist non-empty open sets $U, V \subseteq \Delta(A)$ such that for any integer n we have $\widetilde{\sigma}^n(U) \cap V = \emptyset$. Now clearly the set $O_1 = \bigcup_{n\in\mathbb{Z}} \widetilde{\sigma}^n(U)$ is an invariant non-empty open set. Then $\overline{O_1}$ is an invariant closed set. It follows that $O_2 = \Delta(A) \backslash \overline{O_1}$ is an invariant open set containing V. Thus we even have that $\overline{O_1} \cup O_2 = \Delta(A)$, and the result follows. □

Lemma 7.5. *If $(\Delta(A), \widetilde{\sigma})$ is topologically transitive and there is an $n_0 > 0$ such that $\Delta(A) = \mathrm{Per}^{n_0}(\Delta(A))$, then $\Delta(A)$ consists of a single orbit and is thus finite.*

Proof. Assume two points $x, y \in \Delta(A)$ are not in the same orbit. As $\Delta(A)$ is Hausdorff we may separate the points $x, \sigma(x), \ldots, \sigma^{n_0-1}(x), y$ by pairwise disjoint open sets $V_0, V_1, \ldots, V_{n_0-1}, V_y$. Now consider the set

$$U_x := V_0 \cap \widetilde{\sigma}^{-1}(V_1) \cap \widetilde{\sigma}^{-2}(V_2) \cap \cdots \cap \widetilde{\sigma}^{-n_0+1}(V_{n_0-1}).$$

Clearly the sets $A_x = \cup_{i=0}^{n_0-1} \widetilde{\sigma}^i(U_x)$ and $A_y = \cup_{i=0}^{n_0-1} \widetilde{\sigma}^i(V_y)$ are disjoint invariant non-empty open sets, which leads us to a contradiction. Hence $\Delta(A)$ consists of one single orbit under $\widetilde{\sigma}$. □

We are now ready for a proof of the following result.

Theorem 7.6. *Let A be a complex commutative semi-simple regular unital Banach algebra such that $\Delta(A)$ consists of infinitely many points, and let σ be an automorphism of A. Then $A \rtimes_\sigma \mathbb{Z}$ is prime if and only if the associated system $(\Delta(A), \widetilde{\sigma})$ on the character space is topologically transitive.*

Proof. Suppose $(\Delta(A), \widetilde{\sigma})$ is not topologically transitive. Then there exists, by Lemma 7.4, two disjoint invariant non-empty open sets O_1 and O_2 such that $\overline{O_1} \cup \overline{O_2} = \Delta(A)$. Let I_1 and I_2 be the ideals generated in $\widehat{A} \rtimes_{\widetilde{\sigma}} \mathbb{Z}$ by $k(\overline{O_1})$ (the set of all functions in $\widehat{A}$ that vanish on $\overline{O_1}$) and $k(\overline{O_2})$ respectively. We have that

$$E(I_1 \cap I_2) \subseteq E(I_1) \cap E(I_2) = k(\overline{O_1}) \cap k(\overline{O_2}) = k(\overline{O_1} \cup \overline{O_2}) = k(\Delta(A)) = \{0\}.$$

It is not difficult to see that if $I \subseteq \widehat{A} \rtimes_{\widetilde{\sigma}} \mathbb{Z}$ is an ideal and $E(I) = \{0\}$, then $I = \{0\}$. Namely, suppose $F = \sum_n f_n \delta^n \in I$ and $f_i \neq 0$ for some integer i. Since A is unital, so is $\widehat{A}$ and thus $\delta^{-1} \in \widehat{A} \rtimes_{\widetilde{\sigma}} \mathbb{Z}$. So $F * \delta^{-i} \in I$ and hence $E(F * \delta^{-i}) = f_i = 0$ which is a contradiction, so $I = \{0\}$. Hence $I_1 \cap I_2 = \{0\}$ and $\widehat{A} \rtimes_{\widetilde{\sigma}} \mathbb{Z}$ is not prime. By Theorem 3.1, neither is $A \rtimes_\sigma \mathbb{Z}$. Next suppose that $(\Delta(A), \widetilde{\sigma})$ is topologically transitive. Assume that $\mathrm{Per}^\infty(\Delta(A))$ is not dense. Then by Lemma 4.3 there is an integer $n_0 > 0$ such that $\mathrm{Per}^{n_0}(\Delta(A))$ has non-empty interior. As $\mathrm{Per}^{n_0}(\Delta(A))$ is invariant and closed, topological transitivity implies that $\Delta(A) = \mathrm{Per}^{n_0}(\Delta(A))$. This, however, is impossible since by Lemma 7.5 it would force $\Delta(A)$ to consist of a single orbit and hence be finite. Thus $\mathrm{Per}^\infty(\Delta(A))$ is dense after all. Now let I and J be two non-zero proper ideals in $A \rtimes_\sigma \mathbb{Z}$. Unitality of A assures us that $I \cap A$ and $J \cap A$ are proper invariant ideals of A and density of $\mathrm{Per}^\infty(\Delta(A))$ assures us that they are non-zero, by Theorem 4.5. Consider $A_I = \{\mu \in \Delta(A) \,|\, \mu(a) = 0 \text{ for all } a \in I \cap A\}$ and $A_J = \{\nu \in \Delta(A) \,|\, \nu(b) = 0 \text{ for all } b \in J \cap A\}$. Now by Banach algebra theory a proper ideal in a commutative unital Banach algebra A is contained in a maximal ideal, and a maximal ideal of A is always precisely the set of zeroes of some $\xi \in \Delta(A)$. This implies that both A_I and A_J are non-empty, and semi-simplicity of A assures us that they are proper subsets of $\Delta(A)$. Clearly they are also closed and invariant under $\widetilde{\sigma}$ and $\widetilde{\sigma}^{-1}$. Hence $\Delta(A) \setminus A_I$ and $\Delta(A) \setminus A_J$ are invariant non-empty open sets. By topological transitivity we must have that these two sets intersect, hence that $A_I \cup A_J \neq \Delta(A)$. This means that there exists $\eta \in \Delta(A)$ and $a \in I \cap A$, $b \in J \cap A$ such that $\eta(a) \neq 0$, $\eta(b) \neq 0$ and hence that $\eta(ab) \neq 0$. Hence $0 \neq ab \in I \cap J$, and we conclude that $A \rtimes_\sigma \mathbb{Z}$ is prime. □

Acknowledgments

This work was supported by a visitor's grant of the *Netherlands Organisation for Scientific Research (NWO), The Swedish Foundation for International Cooperation in Research and Higher Education (STINT), Crafoord Foundation, The Royal Physiographic Society in Lund, The Royal Swedish Academy of Sciences and The Swedish Research Council.*

We are also grateful to Jun Tomiyama and Johan Öinert for useful discussions.

References

[1] Davidson, K.R., *C^*-algebras by example*, American Mathematical Society, Rhode Island, 1996.

[2] Larsen, R., *Banach algebras: an introduction*, Marcel Dekker, Inc., New York, 1973.

[3] Power, S.C., *Simplicity of C^*-algebras of minimal dynamical systems*, J. London Math. Soc. 18, 534–538, 1978.

[4] Rudin, W., *Fourier analysis on groups*, Interscience Publishers, New York-London, 1962.

[5] Svensson, C., Silvestrov, S., de Jeu, M., *Dynamical Systems and Commutants in Crossed Products*, Internat. J. Math., Vol. 18 (2007), no. 4, 455–471.

[6] Tomiyama, J., *Invitation to the C^*-algebras and topological dynamics*, World Sci., Singapore New Jersey Hong Kong, 1987.

[7] Tomiyama, J., *The interplay between topological dynamics and theory of C^*-algebras*, Lecture Note No.2, Global Anal. Research Center, Seoul, 1992.

[8] Williams, D.P., *Crossed Products of C^*-algebras*, Mathematical Surveys and Monographs, American Mathematical Society, Province, R.I., 2007.

Christian Svensson
Mathematical Institute
Leiden University
P.O. Box 9512
2300 RA Leiden, The Netherlands

and

Centre for Mathematical Sciences
Lund University
Box 118
SE-221 00 Lund, Sweden
e-mail: chriss@math.leidenuniv.nl

Sergei Silvestrov
Centre for Mathematical Sciences
Lund University
Box 118
SE-221 00 Lund, Sweden
e-mail: Sergei.Silvestrov@math.lth.se

Marcel de Jeu
Mathematical Institute
Leiden University
P.O. Box 9512
2300 RA Leiden, The Netherlands
e-mail: mdejeu@math.leidenuniv.nl

Operator Theory:
Advances and Applications, Vol. 186, 403–413

Born-Oppenheimer-type Approximations for Degenerate Potentials: Recent Results and a Survey on the Area

Françoise Truc

Abstract. This paper is devoted to the asymptotics of eigenvalues for a Schrödinger operator $H_h = -h^2\Delta + V$ on $L^2(\mathbf{R}^m)$, in the case when the potential V does not fulfill the non degeneracy condition: $V(x) \to +\infty$ as $|x| \to +\infty$. For such a model, the point is that the set defined in the phase space by: $H_h \leq \lambda$ may have an infinite volume, so that the Weyl formula which gives the behaviour of the counting function has to be revisited.

We recall various results in this area, in the classical context ($h = 1$ and $\lambda \to +\infty$), as well as in the semi-classical one ($h \to 0$) and comment the different methods. In Sections 3, 4 we present our joint works with A. Morame (Université de Nantes, e-mail: morame@math.univ-nantes.fr),where we consider a degenerate potential $V(x) = f(y)g(z)$, where g is assumed to be a homogeneous positive function of m variables , smooth outside 0, and f is a smooth and strictly positive function of n variables, with a minimum in 0.

In the case where $f(y) \to +\infty$ as $|y| \to +\infty$, the operator has a compact resolvent and we give the asymptotic behaviour, for small values of h, of the number of eigenvalues less than a fixed energy.

Then, without assumptions on the limit of f, we give a sharp estimate of the low eigenvalues, using a Born Oppenheimer approximation. With a refined approach we localize also higher energies. In the case when the degree of homogeneity is not less than 2, we can even assume that the order of these energies is like the inverse power of the square of h.

Finally we apply the previous methods to a class of potentials in R^d, $d \geq 2$, which vanish on a regular hypersurface.

Mathematics Subject Classification (2000). Primary 35P20.

Keywords. Eigenvalues, semi-classical asymptotics, minimax principle, Born-Oppenheimer approximation, Weyl formula, degenerate potentials.

1. Introduction

Let V be a nonnegative, real and continuous potential on $\mathbf{R}^m$, and h a parameter in $]0,1]$. The spectral asymptotics of the operator $H_h = -h^2\Delta + V$ on $L^2(\mathbf{R}^m)$ have been intensively studied. More precisely it is well known [16] that H_h is essentially selfadjoint with compact resolvent when $V(x) \to +\infty$ as $|x| \to +\infty$ (we shall say that V is non degenerate). Moreover, denoting by $N(\lambda, H_h)$ the number of eigenvalues less than a fixed energy λ, the following semiclassical asymptotics hold, as $h \to 0$:

$$N(\lambda, H_h) \sim h^{-m}(2\pi)^{-m} v_m \int_{\mathbf{R}^m} (\lambda - V(x))_+^{m/2} dx \; . \tag{1.1}$$

In this formula, v_m denotes the volume of the unit ball in $\mathbf{R}^m$, and the notation W_+ means the positive value of W.

Let us note that the classical asymptotics are also given by the formula (1.1), provided we let $h = 1$ and $\lambda \to +\infty$.

In both cases, the result points out the asymptotic correspondence between the number of eigenstates with energy less than λ and the volume in phase space of the set $\{(x,\xi), f(x,\xi) \le \lambda\}$, where $f(x,\xi) = \xi^2 + V(x)$ is the principal symbol of H_h.

In this paper we propose a review of results concerning the degenerate case: the potential V does not tend to infinity with $|x|$, so that the volume in phase space of the previous set may be infinite.

2. The Tauberian approach

Let us explain how the problem of the degenerate case came from the non degenerate one.

In 1950 De Wet and Mandl ([3]) proved the formula (1.1) in its classical version, provided $V(x) \ge 1$ and two more conditions on V:

1) a smoothness condition: V differentiable and $|\nabla V| = \circ(V)$
2) a Tauberian type condition: let $\Phi(V,\lambda) = \int_{\mathbf{R}^m} (\lambda - V(x))_+^{m/2} dx$; it is assumed that there exists c and c' such that: $c\Phi(V,\lambda) \le \lambda\Phi'(V,\lambda) \le c'\Phi(V,\lambda)$.

The first condition is local and the second is global. This last condition was needed to use a Tauberian technique, which consists on studying the asymptotic behavior of the Green's function of the operator H_1 and applying a Tauberian theorem.

Refinements were done by Titchmarsh, Levitan and Kostjucenko,([22], [11], [12]) and then Rosenbljum ([18]) proved that the formula (1.1) holds with "maximal" weakening conditions on V:

1) the smoothness condition is replaced by a condition on the "L^1-modulus of continuity" on unit cubes and by the following assumption: $V(y) \le C'V(x)$ if $|x - y| \le 1$.
2) the Tauberian type condition becomes: $\sigma(2\lambda, V) \le C\sigma(\lambda, V)$ (for large λ), where $\sigma(\lambda, V)$ denotes the volume of the set $\{x \in \mathbf{R}^m; V(x) < \lambda\}$.

Solomyak ([21]) makes the following remark:

Lemma 2.1. *Let V be a positive a-homogeneous potential:*
$V(x) \geq 0$; $V(tx) = t^a V(x)$ *for any* $t \geq 0$ $(a > 0)$.
If moreover $V(x)$ is strictly positive ($V(x) \neq 0$ if $x \neq 0$) the spectrum of H_1 is discrete and the formula (1.1) *takes the form:*

$$N(\lambda, H_1) \sim \gamma_{m,a} \lambda^{\frac{2m+am}{2a}} \int_{S^{m-1}} (V(x))^{-m/a} dx$$

($\gamma_{m,a}$ *is a constant depending only on the parameters m and a*).

From that lemma comes out naturally the idea of investigating the spectrum without the condition of strict positivity (and thus in a case of degeneracy of the potential); the two main results are ([21]):

Theorem 2.2. *The formula of Lemma* 2.1 *still holds for a positive a-homogeneous potential such that $J(V) = \int_{S^{m-1}} (V(x))^{-m/a} dx$ is finite.*

The second result deals with a case where $J(V)$ is infinite: let $V(x) = F(y, z)$, $y \in \mathbf{R}^n$, $z \in \mathbf{R}^p$, $n + p = m$, $m \geq 2$, such that $F(sy, tz) = s^b t^{a-b} F(y, z)$ (with $0 < a < b$) and $F(y, z) > 0$ for $|z||y| \neq 0$. Denote by $\lambda_j(y)$ the eigenvalues of the operator $-\Delta_z + F(y, z)$ in $L^2(\mathbf{R}^p)$ and let $s = \frac{2b}{2+a-b}$, then:

Theorem 2.3.

$$\text{If } \ \frac{n}{b} > \frac{m}{a} \quad N(\lambda, H_1) \sim \gamma_{n,s} \lambda^{\frac{2m+am}{2b}} \int_{S^{m-1}} \Sigma(\lambda_j(y))^{-n/s} dx$$

$$\text{if } \ \frac{n}{b} = \frac{m}{a} \quad N(\lambda, H_1) \sim \frac{a(a+2)}{2b(a-b)} \gamma_{m,a} \lambda^{\frac{2m+am}{2b}} \ln \lambda \int_{S^{n-1} S^{p-1}} F(y, z)^{-m/a} dx.$$

The proof is based on variational techniques and spectral estimates proved in ([18]). But on a heuristic level the result can be understood in the framework of the theory of Schrödinger operators with operator potential.

This last approach can be found in ([17]) where D. Robert extended the theory of pseudodifferential operators in the form developed by Hörmander to pseudodifferential operators with operator symbols. It was thus possible to study cases where the operator has a compact resolvent but the condition $\lim_\infty V(x) = +\infty$ is not fulfilled. As an example it gives the asymptotics of $N(\lambda, H_1)$ for the 2-dimensional potential $V(y, z) = y^{2k}(1 + z^2)^l$, where k et l are strictly positive. The asymptotics are the following:

Theorem 2.4.

$$\text{If } \ k > l \quad N(\lambda, H_1) \sim \gamma_1 \lambda^{\frac{l+k+1}{2l}}$$

$$\text{if } \ k = l \quad N(\lambda, H_1) \sim \gamma_2 \lambda^{\frac{2k+1}{2k}} \ln \lambda$$

$$\text{if } \ k < l \quad N(\lambda, H_1) \sim \gamma_3 \lambda^{\frac{2k+1}{2k}}.$$

The constants γ_i depend only on k and l, but the first one γ_1 takes in account the trace of the operator $(-\Delta_z + z^{2k})^{-(k+1)/2l}$ in $L^2(\mathbf{R})$.

In the 2-dimensional case let us mention the results of B. Simon ([19]). He first recalls Weyl's famous result: let H be the Dirichlet Laplacian in a bounded region Ω in $\mathbf{R}^2$, then the following asymptotics hold:

$$N(\lambda, H) \sim \frac{1}{2}\lambda|\Omega|$$

and then he considers special regions Ω for which the volume (denoted by $|\Omega|$) is infinite but the spectrum of the Laplacian is still discrete. These regions are of the type: $\Omega_\mu = \{(y,z); |y||z|^\mu \leq 1\}$.

Actually the problem can be derived from the study of the asymptotics of Schrödinger operators with the homogeneous potential: $V(y,z) = |y|^\alpha |z|^\beta$.

In order to get these "non-Weyl" asymptotics, he uses the Feynman-Kac formula and the Karamata-Tauberian theorem, but the main tool is what he calls "sliced bread inequalities", which can be seen as a kind of Born-Oppenheimer approximation. More precisely let $H = -\Delta + V(y,z)$ be defined on $\mathbf{R}^{n+p}$, and denote by $\lambda_j(y)$ the eigenvalues of the operator $-\Delta_z + V(y,z)$ in $L^2(\mathbf{R}^p)$. (If the z's are electron coordinates and the y's are nuclear coordinates, the $\lambda_j(y)$ are the Born-Oppenheimer curves.) He proves the following lemma:

$$\mathrm{Tr} e^{-tH} \leq \Sigma_j e^{-t(-\Delta_y + \lambda_j(y))}$$

(when the second term exists).

Thus he gets the two following coupled results:

Theorem 2.5. *If $H = -\Delta + |y|^\alpha |z|^\beta$ and $\alpha < \beta$, then*

$$N(\lambda, H) \sim c_\nu \lambda^{\frac{2\nu+1}{2}} \qquad (\nu = \frac{\beta+2}{2\alpha}).$$

Corollary: *if $H = -\Delta_{\Omega_\mu}$ $(\mu > 1)$, then* $\quad N(\lambda, H) \sim c_\mu \lambda^{\frac{1}{2\mu+1}}$.

Theorem 2.6. *If $H = -\Delta + |y|^\alpha |z|^\alpha$, then* $\quad N(\lambda, H) \sim \frac{1}{\pi}\lambda^{1+\frac{1}{\alpha}} \ln \lambda$.

Corollary: *if $H = -\Delta_{\Omega_\mu}$ $(\mu = 1)$, then* $\quad N(\lambda, H) \sim \frac{1}{\pi}\lambda \ln \lambda$.

The constant c_μ depends only on μ, and the constant c_μ takes in account the trace of the operator $(-\Delta_z + |z|^\beta)^{-\nu}$ in $L^2(\mathbf{R})$.

3. The min-max approach

The result presented in this section is based on the method of Courant and Hilbert, the min-max variational principle. It turns out that this method can be applied to operators in $L^2(\mathbf{R}^m)$ with principal symbols which can degenerate on some non bounded manifold of $T^*(\mathbf{R}^m)$. It is the case for the Schrödinger operator with a magnetic field $H = (D_x - A(x))^2$, which degenerates on $\{(x,\xi) \in T^*(\mathbf{R}^m); \xi = A(x)\}$. If the magnetic field $B = dA$ fulfills the so-called magnetic bottle conditions (mainly: $\lim_\infty \|B(x)\| = \infty$) the spectrum is discrete ([1]) and the classical asymptotics were established by Colin de Verdière ([2]) using the min-max method. The semiclassical version of the result is given in ([23]).

In ([14]), the min-max method is performed to get semiclassical asymptotics for a large class of degenerate potentials, namely potentials of the following form:

$$x = (y,z) \in \mathbf{R}^n \times \mathbf{R}^p,\ n+p=m,\ m \geq 2$$
$$V(x) = f(y)g(z),\ f \in C(\mathbf{R}^n; \mathbf{R}_+^*),$$
$$g \in C(\mathbf{R}^p; \mathbf{R}_+),\ g(tz) = t^a g(z)\ (a>0)\ \forall t>0\ ,\ g(z)>0\ \forall z \neq 0. \tag{3.1}$$

The spectrum of the operator $-\Delta_z + g(z)$ in $L^2(\mathbf{R}^p)$ is discrete and positive. Let us denote by μ_j its eigenvalues. It is easy to make the following remark:

Remark 3.1. *If $f(y) \to +\infty$ as $|y| \to +\infty$ then $H_h = -h^2\Delta + V$ has a compact resolvent.*

Of course if f was supposed to be homogeneous, the asymptotics would be given by Theorem 2.3. Here the assumption on f is only a locally uniform regularity:

$$\exists\, b,c>0 \text{ s.t. } c^{-1} \leq f(y) \quad \text{and} \quad |f(y)-f(y')| \leq cf(y)|y-y'|^b,$$

for any y, y' verifying $|y-y'| \leq 1$.

Theorem 3.2. *Let us assume the previous conditions on f and g. Then there exists $\sigma, \tau \in]0,1[$ such that, for any $\lambda > 0$, one can find $h_0 \in]0,1[$, $C_1, C_2 > 0$ in order to have*

$$(1-h^\sigma C_1) n_{h,f}(\lambda - h^\tau C_2) \leq N(\lambda; H_h) \leq (1+h^\sigma C_1) n_{h,f}(\lambda + h^\tau C_2) \quad \forall h \in]0,h_0[$$

$$\textit{if}\ \ n_{h,f}(\lambda) = h^{-n}(2\pi)^{-n} v_n \int_{\mathbf{R}^n} \Sigma_{j\in\mathbf{N}} [\lambda - h^{2a/(2+a)} f^{2/(2+a)}(y)\mu_j]_+^{n/2} dy\ .$$

Provided some additional conditions on f, the previous result can be refined as follows:

Theorem 3.3. *If moreover one can find a constant C_3 such that, for any $\mu > 1$:*

$$\int_{\{y, f(y) < 2\mu\}} f^{-p/a}(y)dy \leq C_3 \int_{\{y, f(y)<\mu\}} f^{-p/a}(y)dy\ ,$$

then one can take $C_2 = 0$ in Theorem 3.2:

$$(1-h^\sigma C_1) n_{h,f}(\lambda) \leq N(\lambda; H_h) \leq (1+h^\sigma C_1) n_{h,f}(\lambda) \quad \forall h \in]0,h_0[.$$

Remark 3.4. *If moreover $f^{-p/a} \in L^1(\mathbf{R}^n)$ and $g \in C^1(\mathbf{R}^p \backslash \{0\})$, then the formula* (1) *holds.*

The proof of Theorem 3.2 uses a suitable covering of $\mathbf{R}^n$, so that the min-max variational principle allows to deal with Dirichlet and Neumann problems in cylinders for the restrained operator (with a fixed y). The proof of Theorem 3.3 is based on an asymptotic formula of the moment of eigenvalues of $-h^2\Delta_z + g(z)$, which is again obtained using the min-max principle.

As a conclusion, let us notice that if there is some information on the growth of f, then the asymptotics can be computed in terms of power of h:

Remark 3.5. *If there exists $k > 0$ and $C > 0$ such that*

$$\frac{1}{C}|y|^k \leq f(y) \leq C|y|^k \quad \textit{for} \quad |y| > 1,$$

then

$$\begin{aligned} &\textit{if} \ \ k > a \quad N(\lambda, H_h) \approx h^{-m} \\ &\textit{if} \ \ k = a \quad N(\lambda, H_h) \approx h^{-m} \ln \frac{1}{h} \\ &\textit{if} \ \ k < a \quad N(\lambda, H_h) \approx h^{-n-\frac{pa}{k}}. \end{aligned}$$

4. Born-Oppenheimer-type estimates

In last section we have investigated the asymptotic behavior of the number of eigenvalues less then λ of $\widehat{H}_h = -h^2\Delta + f(y)g(z)$.

Theorem 3.2 gives us a hint of what should eigenvalues of $\widehat{H}_h$ look like. This can be done using Born-Oppenheimer-type methods.

We assume as in last section that: $g \in C^\infty(\mathbf{R}^m \setminus \{0\})$ is homogeneous of degree $a > 0$, and assume the following for f:

$$\begin{aligned} f \ \in \ C^\infty(\mathbb{R}^n), \ \forall\alpha \in \mathbb{N}^n, \ (|f(y)| + 1)^{-1}\partial_y^\alpha f(y) \ &\in L^\infty(\mathbb{R}^n) \\ 0 \ < \ f(0) \ &= \ \inf_{y\in\mathbb{R}^n} \ f(y) \\ f(0) \ < \ \liminf_{|y|\to\infty} \ f(y) \ &= \ f(\infty) \\ \partial^2 f(0) \ &> \ 0. \end{aligned} \tag{4.1}$$

$\partial^2 f(0)$ denotes the hessian matrix in 0.

4.1. Using homogeneity

By dividing $\widehat{H}_h$ by $f(0)$, we can change the parameter h and assume that

$$f(0) = 1\,. \tag{4.2}$$

Let us define: $\hbar = h^{2/(2+a)}$ and change z in $z\hbar$; we can use the homogeneity of g (3.1) to get:

$$sp\ (\widehat{H}_h) \ = \ \hbar^a \ sp\ (\widehat{H}^\hbar)\,, \tag{4.3}$$

with $\widehat{H}^\hbar \ = \ \hbar^2 D_y^2 \ + \ D_z^2 \ + \ f(y)g(z)\,.$

Let us denote as usually the increasing sequence of eigenvalues of $D_z^2 + g(z)$, (on $L^2(\mathbb{R}^m)$), by $(\mu_j)_{j>0}$.

The associated eigenfunctions will be denoted by $(\varphi_j)_j$:

By homogeneity (3.1) the eigenvalues of $Q_y(z, D_z) \ = \ D_z^2 \ + \ f(y)g(z)$, on $L^2(\mathbb{R}^m)$), for a fixed y, are given by the sequence $(\lambda_j(y))_{j>0}$, where : $\lambda_j(y) = \mu_j f^{2/(2+a)}(y)$.

So as in [15] we get:

$$\widehat{H}^\hbar \geq \left[\, \hbar^2 D_y^2 \;+\; \mu_1 f^{2/(2+a)}(y) \,\right] . \tag{4.4}$$

This estimate is sharp as we will see below.

Then using the same kind of estimate as (4.4), one can see that

$$\inf\, sp_{\mathrm{ess}}(\widehat{H}^\hbar) \;\geq\; \mu_1 f^{2/(2+a)}(\infty) . \tag{4.5}$$

We are in the Born-Oppenheimer approximation situation described by A. Martinez in [13]: the "effective " potential is given by $\lambda_1(y) = \mu_1 f^{2/(2+a)}(y)$, the first eigenvalue of Q_y, and the assumptions on f ensure that this potential admits one unique and nondegenerate well $U = \{0\}$, with minimal value equal to μ_1. Hence we can apply Theorem 4.1 of [13] and get:

Theorem 4.1. *Under the above assumptions, for any arbitrary $C > 0$, there exists $h_0 > 0$ such that, if $0 < \hbar < h_0$, the operator $(\widehat{H}^\hbar)$ admits a finite number of eigenvalues $E_k(\hbar)$ in $[\mu_1, \mu_1 + C\hbar]$, equal to the number of the eigenvalues e_k of $D_y^2 + \frac{\mu_1}{2+a} < \partial^2 f(0)y,\ y >$ in $[0, +C]$ such that:*

$$E_k(\hbar) = \lambda_k(\widehat{H}^\hbar) \;=\; \lambda_k\left(\hbar^2 D_y^2 + \mu_1 f^{2/(2+a)}(y)\right) \;+\; \mathbf{O}(\hbar^2) . \tag{4.6}$$

More precisely $E_k(\hbar) = \lambda_k(\widehat{H}^\hbar)$ has an asymptotic expansion

$$E_k(\hbar) \;\sim\; \mu_1 \;+\; \hbar\; (\; e_k \;+\; \sum_{j\geq 1} \alpha_{kj} \hbar^{j/2}\;). \tag{4.7}$$

If $E_k(\hbar)$ is asymptotically non degenerate, then there exists a quasimode

$$\phi_k^\hbar(y,z) \;\sim\; \hbar^{-m_k} e^{-\psi(y)/\hbar} \sum_{j\geq 0} \hbar^{j/2} a_{kj}(y,z) , \tag{4.8}$$

satisfying

$$\begin{gathered} C_0^{-1} \leq \|\hbar^{-m_k} e^{-\psi(y)/\hbar} a_{k0}(y,z)\| \leq C_0 \\ \|\hbar^{-m_k} e^{-\psi(y)/\hbar} a_{kj}(y,z)\| \leq C_j \\ \left\| \left(\widehat{H}^\hbar - \mu_1 - \;\hbar e_k - \sum_{1\leq j\leq J} \alpha_{kj}\hbar^{j/2} \right) \right\| \\ \hbar^{-m_k} e^{-\psi(x)/\hbar} \sum_{0\leq j\leq J} \hbar^{j/2} a_{kj}(x,y)\| \;\leq\; C_J \hbar^{(J+1)/2}. \end{gathered} \tag{4.9}$$

The formula (4.7) implies

$$\lambda_k(\widehat{H}^\hbar) \;=\; \mu_1 \;+\; \hbar\lambda_k \left(D_y^2 + \frac{\mu_1}{2+a} < \partial^2 f(0)\; y\; ,\; y\; > \right) \;+\; \mathbf{O}(\hbar^{3/2}) , \tag{4.10}$$

and when $k = 1$, one can improve $\mathbf{O}(\hbar^{3/2})$ into $\mathbf{O}(\hbar^2)$. The function ψ is defined by: $\psi(y) = d(y,0)$, where d denotes the Agmon distance related to the degenerate metric $\mu_1 f^{2/(2+a)}(y) dy^2$.

4.2. Improving Born-Oppenheimer methods

We are interested now with the lower energies of $\widehat{H}^\hbar$. Let us make the change of variables

$$(y,\ z) \rightarrow (y,\ f^{1/(2+a)}(y)z)\ . \tag{4.11}$$

The Jacobian of this diffeomorphism is $f^{m/(2+a)}(y)$, so we perform the change of test functions: $u \rightarrow f^{-m/(4+2a)}(y)u$, to get a unitary transformation.

Thus we get that

$$sp\,(\widehat{H}^\hbar) \;=\; sp\,(\widetilde{H}^\hbar) \tag{4.12}$$

where $\widetilde{H}^\hbar$ is the self-adjoint operator on $L^2(\mathbb{R}^n \times \mathbb{R}^m)$ given by

$$\begin{aligned}\widetilde{H}^\hbar &= \hbar^2 D_y^2 \;+\; f^{2/(2+a)}(y)\left(D_z^2 + g(z)\right)\\ &+ \hbar^2 \frac{2}{(2+a)f(y)}(\nabla f(y) D_y)(zD_z)\\ &+ i\hbar^2 \frac{1}{(2+a)f^2(y)}\left(|\nabla f(y)|^2 - f(y)\Delta f(y)\right)[(zD_z) \;-\; i\frac{m}{2}]\\ &+ \hbar^2 \frac{1}{(2+a)^2 f^2(y)}|\nabla f(y)|^2[(zD_z)^2 \;+\; \frac{m^2}{4}].\end{aligned} \tag{4.13}$$

The only significant role up to order 2 in $\hbar$ will be played actually by the first operator, namely: $\widetilde{H}_1^\hbar \;=\; \hbar^2 D_y^2 \;+\; f^{2/(2+a)}(y)\left(D_z^2 + g(z)\right)$.

This leads to:

Theorem 4.2. . *Under the assumptions* (3.1) *and* (4.1), *for any fixed integer* $N > 0$, *there exists a positive constant* $h_0(N)$ *verifying: for any* $\hbar \in]0, h_0(N)[$, *for any* $k \leq N$ *and any* $j \leq N$ *such that*

$$\mu_j \;<\; \mu_1 f^{2/(2+a)}(\infty)\ ,$$

there exists an eigenvalue $\lambda_{jk} \in sp_d(\widehat{H}^\hbar)$ *such that*

$$|\ \lambda_{jk} \;-\; \lambda_k\left(\hbar^2 D_z^2 + \mu_j f^{2/(2+a)}(z)\right)\ | \;\leq\; \hbar^2 C\ . \tag{4.14}$$

Consequently, when $k = 1$, *we have*

$$|\ \lambda_{j1} \;-\; \left[\mu_j \;+\; \hbar(\mu_j)^{1/2}\frac{tr((\partial^2 f(0))^{1/2})}{(2+a)^{1/2}}\right]\ | \;\leq\; \hbar^2 C\ . \tag{4.15}$$

4.3. Middle energies

We can refine the preceding results when $a \geq 2$, $g \in C^\infty(\mathbb{R}^m)$ and $f(\infty) = \infty$. We get then sharp localization near the μ_j's for much higher values of j's. More precisely we prove:

Theorem 4.3. . *Assume the preceding properties, and consider* j *such that* $\mu_j \leq \hbar^{-2}$; *then for any integer* N, *there exists a constant* C *depending only on* N *such that, for any* $k \leq N$, *there exists an eigenvalue* $\lambda_{jk} \in sp_d(\widehat{H}^\hbar)$ *verifying*

$$|\ \lambda_{jk} \;-\; \lambda_k\left(\hbar^2 D_y^2 + \mu_j f^{2/(2+a)}(y)\right)\ | \;\leq\; C\mu_j\hbar^2\ . \tag{4.16}$$

Consequently, when $k = 1$, we have

$$| \lambda_{j1} - \left[\mu_j + \hbar(\mu_j)^{1/2} \frac{tr((\partial^2 f(0))^{1/2})}{(2+a)^{1/2}} \right] | \leq C\mu_j \hbar^2 . \tag{4.17}$$

4.4. An application

We can apply the previous methods for studying Schrödinger operators on $L^2(\mathbb{R}^d_s)$ with $d \geq 2$,

$$P^h = -h^2 \Delta + V(s) \tag{4.18}$$

with a real and regular potential $V(s)$ satisfying

$$\begin{gathered} V \in C^\infty(\mathbb{R}^d \ ; \ [0, +\infty[) \\ \liminf_{|s| \to \infty} V(s) > 0 \\ \Gamma = V^{-1}(\{0\}) \quad \text{is a regular hypersurface.} \end{gathered} \tag{4.19}$$

Moreover we assume that Γ is connected and that there exist $m \in \mathbb{N}^\star$ and $C_0 > 0$ such that for any s verifying $d(s, \Gamma) < C_0^{-1}$

$$C_0^{-1} d^{2m}(s, \Gamma) \leq V(s) \leq C_0 \, d^{2m}(s, \Gamma) \tag{4.20}$$

($d(E, F)$ denotes the euclidian distance between E and F).

We choose an orientation on Γ and a unit normal vector $N(s)$ on each $s \in \Gamma$, and then, we can define the function on Γ,

$$f(s) = \frac{1}{(2m)!} \left(N(s) \frac{\partial}{\partial s} \right)^{2m} V(s) \, , \quad \forall \, s \in \Gamma \, . \tag{4.21}$$

Then by (4.19) and (4.20), $f(s) > 0$, $\forall s \in \Gamma$.

Finally we assume that the function f achieves its minimum on Γ on a finite number of discrete points:

$$\Sigma_0 = f^{-1}(\{\eta_0\}) = \{s_1, \ldots, s_{\ell_0}\} \, , \quad \text{if} \quad \eta_0 = \min_{s \in \Gamma} f(s) \, , \tag{4.22}$$

and the hessian of f at each point $s_j \in \Sigma_0$ is non degenerate.

Hess $(f)_{s_j}$ has $d - 1$ non negative eigenvalues

$$\rho_1^2(s_j) \leq \cdots \leq \rho_{d-1}^2(s_j) \, , \qquad (\rho_j(s_j) > 0) \, .$$

The eigenvalues $\rho_k^2(s_j)$ do not depend on the choice of coordinates. We denote

$$Tr^+(\text{Hess}\,(f(s_j))) = \sum_{\ell=1}^{d-1} \rho_\ell(s_j) \, . \tag{4.23}$$

We denote by $(\mu_j)_{j \geq 1}$ the increasing sequence of the eigenvalues of the operator $-\dfrac{d^2}{dt^2} + t^{2m}$ on $L^2(\mathbb{R})$.

Theorem 4.4. *Under the above assumptions, for any* $N \in \mathbb{N}^\star$, *there exist* $h_0 \in]0,1]$ *and* $C_0 > 0$ *such that, if* $\mu_j \ll h^{-4m/(m+1)(2m+3)}$, *and if* $\alpha \in \mathbb{N}^{d-1}$ *and* $|\alpha| \leq N$, *then* $\forall s_\ell \in \Sigma_0$, $\exists \lambda^h_{j\ell\alpha} \in sp_d(P^h)$ s.t.

$$\left| \lambda^h_{j\ell\alpha} - h^{2m/(m+1)} \left[\eta_0^{1/(m+1)} \mu_j + h^{1/(m+1)} \mu_j^{1/2} \mathcal{A}_\ell(\alpha) \right] \right|$$

$$\leq h^2 \mu_j^{2+3/2m} C_0 \;;$$

with

$$\mathcal{A}_\ell(\alpha) = \frac{1}{\eta_0^{m/(2m+2)} (m+1)^{1/2}} \left[2\alpha\rho(s_\ell) + Tr^+(\mathrm{Hess}\,(f(s_\ell))) \right].$$

$$(\alpha\rho(s_\ell) = \alpha_1\rho_1(s_\ell) + \cdots + \alpha_{d-1}\rho_{d-1}(s_\ell)).$$

References

[1] J. Avron, I. Herbst, B. Simon, *Duke. Math. J.* **45** (1978), 847–883.

[2] Y. Colin de Verdière, L'asymptotique de Weyl pour les bouteilles magnétiques, *Comm. Math. Phys.* **105** (1986), 327–335.

[3] J.S. de Wet and Mandl, On the asymptotic distribution of eigenvalues, *Proc. Roy. Soc. London Ser.* **200** (1950), 572–580.

[4] B. Helffer: Introduction to the semiclassical analysis for the Schrödinger operator and applications. *Springer Lecture Notes in Math.*, n^0 1336 (1988).

[5] B. Helffer, D. Robert, Propriétés asymptotiques du spectre d'opérateurs pseudo-différentiels sur $\mathbb{R}^n$. Comm. in P.D.E., 7(7), (1982), pp. 795–882.

[6] B. Helffer, D. Robert, Comportement semi-classique du spectre des hamiltoniens quantiques hypoelliptiques. Annales ENS Pise IV, 9,(3), (1982), pp. 405–431.

[7] B. Helffer, J. Sjoestrand, Multiple wells in the semi-classical limit. I. Comm. in P.D.E., 9, (4), (1984), pp. 337–408.

[8] B. Helffer, J. Sjoestrand, Puits multiples en mecanique semi-classique. VI. Ann. Institut H. Poincaré, Phys. Theor., 46,(4), (1987), pp. 353–372.

[9] L. Hörmander, The Weyl calculus of pseudo-differential operators. Comm. Pure Appl. Math., 32, (1979), pp. 359–443.

[10] V. Ivrii, Microlocal analysis and precise spectral asymptotic. Springer, Berlin 1998.

[11] B.M. Levitan, On the asymptotic behavior of Green's function and its expansion in eigenvalues of Schrödinger's equation, *Math. USSR-Sb.41* **83** (1957), 439–458.

[12] A.G. Kostjucenko, Asymptotic distribution of the eigenvalues of elliptic operators, *Soviet Math. Dokl.* **5** (1964), 1171–1175.

[13] A. Martinez, Développement asymptotiques et effet tunnel dans l'approximation de Born-Oppenheimer. Ann. Inst. Henri Poincaré, Vol. 49,(3), (1989), pp. 239–257.

[14] A. Morame, F. Truc, Semiclassical Eigenvalue Asymptotics for a Schrödinger Operator with Degenerate Potential. Asymptotic Anal., 22(1), (2000), pp. 39–49.

[15] A. Morame, F. Truc, Accuracy on eigenvalues for a Schrödinger Operator with a Degenerate Potential in the semi-classical limit. (Submitted).

[16] M. Reed and B. Simon, *Methods of Modern Mathematical Physics* IV, Academic press, New York, 1978.

[17] D. Robert, Comportement asymptotique des valeurs propres d'opérateurs du type de Schrödinger à potentiel dégénéré, *J. Math. Pures Appl.* **61** (1982), 275–300.

[18] G.V. Rosenbljum, Asymptotics of the eigenvalues of the Schrödinger operator, *Math. USSR-Sb.* **22**(3) (1974), 349–371.

[Shu] M.A. Shubin, Pseudodifferential operators and spectral theory. Springer-Verlag, Berlin 1987.

[19] B. Simon, Non-classical eigenvalue asymptotics, *J. Funct. Anal.* **53** (1983), 84–98.

[20] B. Simon: Semi-classical analysis of low lying eigenvalues I. Ann. Inst. H. Poincaré, 38, (1983), pp. 295–307.

[21] M.Z. Solomyak, Asymptotics of the spectrum of the Schrödinger operator with non-regular homogeneous potential, *Math. USSR-Sb.* **55**(1) (1986), 19–37.

[22] E.C. Titchmarsh, On the asymptotic distribution of eigenvalues, *Quart. J. Math. Oxford Ser.*(2) **5** (1954), 228–240.

[23] F. Truc, Semi-classical asymptotics for magnetic bottles, *Asympt. Anal.* **15** (1997), 385–395.

Françoise Truc
Université de Grenoble I
Institut Fourier
UMR 5582 CNRS-UJF
B.P. 74
F-38402 St Martin d'Hères Cedex, FRANCE
e-mail: `Francoise.Truc@ujf-grenoble.fr`

Operator Theory:
Advances and Applications, Vol. 186, 415–435

The Ornstein-Uhlenbeck Semigroup in Bounded and Exterior Lipschitz Domains

Ian Wood

Abstract. We consider bounded Lipschitz domains Ω in $\mathbb{R}^n$. It is shown that the Dirichlet-Laplacian generates an analytic C_0-semigroup on $L^p(\Omega)$ for p in an interval around 2 and that the corresponding Cauchy problem has the maximal L^q-regularity property. We then prove that for bounded or exterior Lipschitz domains Ornstein-Uhlenbeck operators A generate C_0-semigroups in the same p-interval. The method also allows to determine the domain $D(A)$ of A and, if Ω satisfies an outer ball condition, allows to show L^p-L^q-smoothing properties of the semigroups.

Mathematics Subject Classification (2000). 47D06, 47D07.

Keywords. Ornstein-Uhlenbeck operator, Lipschitz domain, exterior domain, semigroup, maximal L^p-regularity.

1. Introduction

We shall discuss the following problem. Let M be a real-valued constant coefficient $n \times n$-matrix and let K be a compact set in $\mathbb{R}^n$ with Lipschitz boundary[1]. Let $\Omega = \mathbb{R}^n \setminus K$ be the exterior domain. On this domain we consider the following equation.

$$\begin{cases} \partial_t u(x,t) = \Delta u(x,t) + Mx \cdot \nabla u(x,t) & \text{in } \Omega \times \mathbb{R}_+, \\ u(x,t) = 0 & \text{on } \partial\Omega \times \mathbb{R}_+, \\ u(x,0) = u_0(x) & \text{for } x \in \Omega. \end{cases} \tag{1.1}$$

Operators of the form $\Delta + Mx \cdot \nabla$ are called *Ornstein-Uhlenbeck operators.*[2]

[1] I.e., every point on the boundary of Ω has a neighbourhood U such that, after an affine change of coordinates, $\partial\Omega \cap U$ is described by the equation $x_n = \varphi(x_1, \dots, x_{n-1})$ where φ is a Lipschitz continuous function and $\Omega \cap U = \{x \in U : x_n > \varphi(x_1, \dots, x_{n-1})\}$

[2] Note that x is actually the space variable, so the notation $\Delta + Mx \cdot \nabla$ is formally bad notation, however it is the standard notation and we will use it in the following.

In recent years, many authors have considered Ornstein-Uhlenbeck operators (see [4], [7], [13], [21], [23], [24]), either from the point of view of analysis or stochastics. All these articles deal with the realization of these operators in certain function spaces over the whole space $\mathbb{R}^n$.

One setting where Ornstein-Uhlenbeck operators arise naturally is in transforming parabolic equations such as the heat equation or the Navier-Stokes equations from rotating domains to a fixed domain Ω (see [16]). Therefore, there is considerable interest in such operators defined in exterior domains of $\mathbb{R}^n$, see, e.g., [16], [15], [11]. In bounded domains, Ornstein-Uhlenbeck operators can be considered as relatively bounded perturbations of the Laplacian. If the domain is not bounded however, the term Mx is unbounded which makes the problem much more difficult to consider.

Before looking at Ornstein-Uhlenbeck operators, we need to study the Dirichlet-Laplacian in bounded Lipschitz domains. Consider Laplace's equation

$$\Delta u = f \text{ in } \Omega, \quad u = 0 \text{ on } \partial\Omega,$$

and the heat equation

$$u' - \Delta u = f \text{ in } \Omega, \quad u = 0 \text{ on } \partial\Omega, \quad u(0) = 0 \text{ in } \Omega,$$

where $\Omega \subseteq \mathbb{R}^n$ is a bounded Lipschitz domain. The usual elliptic regularity results no longer hold for the Laplacian in Lipschitz domains (the domain of the operator is no longer contained in the Sobolev space $W^{2,p}(\Omega)$) and it becomes a difficult problem to determine the domain of the operator. See [29] and the references therein for information on elliptic and parabolic problems in non-smooth domains.

The aim of this paper is to collect the results on the Laplacian in Lipschitz domains and their application to Ornstein-Uhlenbeck operators in one place. The results on the Laplacian have previously appeared in [29] and those on the Ornstein-Uhlenbeck operator in [12] in a very shortened form.

The structure of the paper is the following. In the next section we consider the Dirichlet-Laplacian in Lipschitz domains. We first state known results for the elliptic problem, then we combine these with semigroup theory to show that the Dirichlet-Laplacian generates a C_0-semigroup in $L^p(\Omega)$ for a certain range of p. We then study some of the properties of the semigroup such as analyticity and positivity which finally lead to a maximal L^q-regularity result.

In the last section we study Ornstein-Uhlenbeck operators. First, we look at the well-known $\mathbb{R}^n$-case. We then study the first order drift term separately before looking at Ornstein-Uhlenbeck operators in bounded Lipschitz domains in Section 3.3. Using a cut-off procedure previously used by Hishida ([16], [17] and [18]) we combine the results in $\mathbb{R}^n$ and on bounded domains to obtain results for exterior domains in Section 3.4. Finally, Section 3.5 deals with Lipschitz domains satisfying a uniform outer ball condition which gives additional regularity of the solution when $p \leq 2$. This allows us to show L^p-L^q-smoothing estimates which are useful for the study of semilinear equations.

2. The Laplacian in Lipschitz domains

2.1. Laplace's equation

We first state the main result due to Jerison and Kenig on solutions to the Dirichlet problem in bounded Lipschitz domains. The spaces $L^p_\alpha(\Omega)$ denote the Bessel potential spaces of order α and exponent p over the domain Ω (cf. [19] for the definition and some properties). We remark that for Lipschitz domains Ω and $k \in \mathbb{N}$, we have $L^p_k(\Omega) = W^{k,p}(\Omega)$.

Theorem 2.1. (Jerison, Kenig [19, Theorem 1.1]**).** *Let Ω be a bounded Lipschitz domain in $\mathbb{R}^n$, $n \geq 3$. There exists $\varepsilon \in (0,1]$, depending only on the Lipschitz character*[3] *of Ω such that for every $f \in L^p_{\alpha-2}(\mathbb{R}^n)$ there is a unique solution $u \in L^p_\alpha(\Omega)$ to the inhomogeneous Dirichlet problem*

$$\begin{cases} \Delta u = f & \text{in } \Omega, \\ u = 0 & \text{on } \partial\Omega, \end{cases} \tag{2.1}$$

provided one of the following holds (cf. Figure 1):

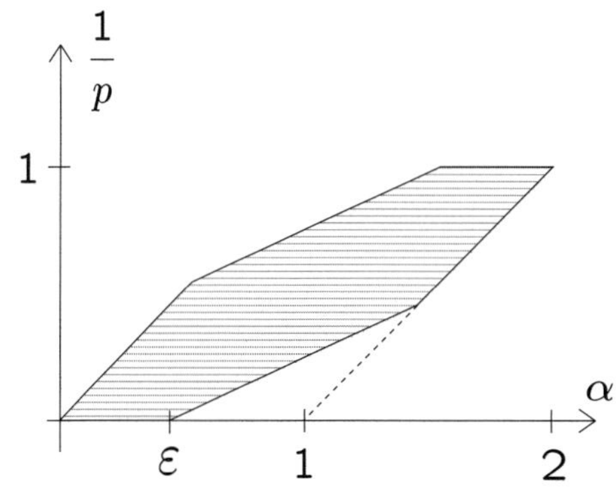

FIGURE 1. $(\alpha, 1/p)$-region in which theorem holds

1. $$1 < p \leq p_0 \quad \text{and} \quad \frac{3}{p} - 1 - \varepsilon < \alpha < 1 + \frac{1}{p},$$
2. $$p_0 < p < p_0' \quad \text{and} \quad \frac{1}{p} < \alpha < 1 + \frac{1}{p},$$
3. $$p_0' \leq p < \infty \quad \text{and} \quad \frac{1}{p} < \alpha < \frac{3}{p} + \varepsilon,$$

where $1/p_0 = 1/2 + \varepsilon/2$ and $1/p_0' = 1/2 - \varepsilon/2$. Moreover, we have the estimate

$$\|u\|_{L^p_\alpha(\Omega)} \leq C \|f\|_{L^p_{\alpha-2}(\mathbb{R}^n)}$$

[3] That is the number of coordinate charts used to cover the boundary $\partial\Omega$ by cylinders such that, inside each cylinder, the domain is the domain above the graph of a Lipschitz function, the radii of these cylinders and the supremum of the norms of the graph functions.

for all $f \in L^p_{\alpha-2}(\mathbb{R}^n)$. *The constant* C *depends on the domain* Ω *only via the Lipschitz character of* Ω. *When the domain is* C^1, p_0 *may be taken to be* 1.

Remark 2.2. In two dimensions there is a similar result with a slightly different range of (α, p) (cf. [19, Theorem 1.3]).

For convex bounded domains $\Omega \subseteq \mathbb{R}^n$ and $1 < p \leq 2$, it is possible to control all second derivatives in $L^p(\Omega)$ by the Laplacian. Note that any bounded convex domain is a Lipschitz domain (cf. [14, Corollary 1.2.2.3]). The following result is due to Fromm.

Theorem 2.3 (Fromm [10]**).** *If* $\Omega \subset \mathbb{R}^n$, $n \geq 2$, *is a bounded and convex domain with diameter* d *and if* $f \in L^p_\alpha(\Omega)$ *then there is a unique* $u \in W^{1,p}_0(\Omega) \cap L^p_{\alpha+2}(\Omega)$ *satisfying* $\Delta u = f$ *in* Ω, *and this solution satisfies the estimate*

$$\|u\|_{\alpha+2,p} \leq C(d) \, \|f\|_{\alpha,p} \tag{2.2}$$

for $-1 \leq \alpha \leq 0$ *and* $1 < p < \frac{2}{\alpha+1}$ (*defining* $\frac{2}{0} = \infty$) *and for* $\alpha = 0$, $p = 2$.

The theorem actually holds in a larger class of domains which are Lipschitz domains that are convex in the neighbourhood of any boundary singularities. This will be important for studying exterior domains later.

Definition 2.4. Let Ω be a domain in $\mathbb{R}^n$. We say that Ω satisfies the outer ball condition if for each $x \in \partial\Omega$, there exists an open ball $B \subseteq \Omega^c$ with $x \in \partial B$. Ω satisfies a uniform outer ball condition if there exists an $R > 0$ such that for all $x \in \partial\Omega$, the ball can be chosen to have radius R.

Remark 2.5. Theorem 2.3 holds in all bounded Lipschitz domains satisfying a uniform outer ball condition (cf. [10, Remarks]). In this case, the constant C in (2.2) depends on more geometric properties of Ω than just the diameter.

2.2. The heat equation

2.2.1. Generation of a semigroup. The aim of this section is to show that if we define the Dirichlet-Laplacian suitably on the Lebesgue spaces $L^p(\Omega)$, for certain domains Ω and a range of exponents p, the Dirichlet-Laplacian is the generator of a C_0-semigroup[4]. We then determine various properties of the generated semigroup. To begin, we introduce the Dirichlet-Laplacian with two different domains of definition.

Definition 2.6. We define the weak Dirichlet-Laplacian Δ^w_p on $L^p(\Omega)$ by

$$D(\Delta^w_p) = \{u \in W^{1,p}_0(\Omega) : \Delta u \in L^p(\Omega)\}, \quad \Delta^w_p u = \Delta u.$$

Here, $\Delta u \in L^p(\Omega)$ is to be understood in the sense of distributions.

[4]cf. , e.g., [2], [9] or [26] for information on C_0-semigroups

In order to obtain results on higher regularity of the solution to the Cauchy problem, we introduce the strong Dirichlet-Laplacian.

Definition 2.7. The strong Dirichlet-Laplacian Δ_p^s on $L^p(\Omega)$ is defined by

$$D(\Delta_p^s) = W^{2,p}(\Omega) \cap W_0^{1,p}(\Omega), \quad \Delta_p^s u = \Delta u.$$

Our aim is to use the Lumer-Phillips Theorem (see, e.g., [2, Theorem 3.4.5]) to prove that the Dirichlet-Laplacian generates a C_0-semigroup of contractions. To apply the theorem, we need to show dissipativity of the operator. An operator A on $L^p(\Omega)$ is dissipative iff $\operatorname{Re}\langle Au, u^*\rangle \leq 0$ for all $u \in D(A)$ with $u^* = |u|^{p-2}\overline{u}\chi_{\{u\neq 0\}}$.

Lemma 2.8. *Let $\Omega \subseteq \mathbb{R}^n$ be a bounded Lipschitz domain and $2 \leq p < \infty$. Then Δ_p^w is dissipative.*

Proof. Let $u \in D(\Delta_p^w)$ and set $u^*(x) = |u(x)|^{p-2}\overline{u}(x)\chi_{\{u\neq 0\}}(x)$. Then $u^* \in L^{p'}(\Omega)$. We need to show $\operatorname{Re}\langle \Delta u, u^*\rangle \leq 0$. The gradient is given by

$$\nabla u^* = \left(|u|^{p-2}\nabla\overline{u} + \frac{p-2}{2}\overline{u}|u|^{p-4}\left(\overline{u}\nabla u + u\nabla\overline{u}\right)\right)\chi_{\{u\neq 0\}}.$$

Using Hölder's inequality, it is easy to check that $\nabla u^* \in L^{p'}(\Omega)$. Moreover $u^*|_{\partial\Omega} = 0$. Therefore, $u^* \in W_0^{1,p'}(\Omega)$ (cf. [19, Proposition 3.3]). Integration by parts and a calculation then yield

$$\begin{aligned}
\operatorname{Re}\langle \Delta u, u^*\rangle &= -\operatorname{Re}\int_\Omega \nabla u \cdot \nabla u^* \\
&= -\operatorname{Re}\int_\Omega |u|^{p-2}|\nabla u|^2\chi_{\{u\neq 0\}} \\
&\qquad - \operatorname{Re}\int_\Omega \frac{p-2}{2}\overline{u}|u|^{p-4}\left(\overline{u}(\nabla u)^2 + u|\nabla u|^2\right)\chi_{\{u\neq 0\}} \\
&= -\int_\Omega \left(|u|^{p-2}|\nabla u|^2 + (p-2)|u|^{p-4}(\operatorname{Re}\overline{u}\nabla u)^2\right)\chi_{\{u\neq 0\}} \\
&= -\int_\Omega |u|^{p-4}\left(\overline{u}\nabla u \cdot u\nabla\overline{u} + (p-2)(\operatorname{Re}\overline{u}\nabla u)^2\right)\chi_{\{u\neq 0\}} \\
&= -\int_\Omega |u|^{p-4}\left[(p-1)(\operatorname{Re}\overline{u}\nabla u)^2 + (\operatorname{Im}\overline{u}\nabla u)^2\right]\chi_{\{u\neq 0\}} \leq 0.
\end{aligned}$$

□

Corollary 2.9. *Let $\Omega \subseteq \mathbb{R}^n$ be a bounded Lipschitz domain and $2 \leq p < \infty$. Then Δ_p^s is dissipative.*

Note that for $p < 2$, the function u^* is not in $W_0^{1,p'}(\Omega)$, so the straightforward integration by parts is not possible.

However, for the strong Dirichlet-Laplacian an approximation procedure yields the desired result:

Lemma 2.10. *Let Ω be a Lipschitz domain, $1 < p < \infty$ and $u^* = |u|^{p-2}\overline{u}\chi_{\{u\neq 0\}}$. Then for $u \in W^{2,p}(\Omega)$ we have*

$$\begin{aligned} \int_\Omega \Delta u\, u^* &= -(p-1)\int_\Omega |u|^{p-4}|\mathrm{Re}(\overline{u}\nabla u)|^2\chi_{\{u\neq 0\}} \\ &\quad -\int_\Omega |u|^{p-4}|\mathrm{Im}(\overline{u}\nabla u)|^2\chi_{\{u\neq 0\}} \\ &\quad -i(p-2)\int_\Omega |u|^{p-4}\mathrm{Re}(\overline{u}\nabla u)\mathrm{Im}(\overline{u}\nabla u)\chi_{\{u\neq 0\}} \\ &\quad +\int_{\partial\Omega} \overline{u}|u|^{p-2}\frac{\partial u}{\partial N}. \end{aligned} \tag{2.3}$$

Proof. This is a special case of Theorem 3.1 from [25] with $\phi = 1$ and $A = \Delta$. Note that the assumptions on the boundary in [25] only require that $C^\infty(\overline{\Omega})$ is dense in $W^{2,p}(\Omega)$ and that traces are well defined. In particular the results hold for Lipschitz domains. □

Corollary 2.11. *Let $\Omega \subseteq \mathbb{R}^n$ be a Lipschitz domain and $1 < p < \infty$. Then Δ_p^s is dissipative.*

We are now in the position to prove one of the main theorems of this section for the strong Dirichlet-Laplacian.

Theorem 2.12. *Let $\Omega \subseteq \mathbb{R}^n$, $n \geq 2$, be a bounded Lipschitz domain satisfying a uniform outer ball condition and $1 < p \leq 2$. Then Δ_p^s generates a C_0-semigroup of contractions on $L^p(\Omega)$.*

Proof. It remains to verify the range condition of the Lumer-Phillips-Theorem, i.e., that

$$(\lambda - \Delta)D(\Delta_p^s) = L^p(\Omega) \text{ for some } \lambda > 0$$

is satisfied. However, from Theorem 2.3, we know that under our assumptions, $0 \in \rho(\Delta_p^s)$ and as the resolvent set is open, we have $\lambda \in \rho(\Delta_p^s)$ for some small $\lambda > 0$ which proves the theorem. □

For the weak Dirichlet-Laplacian we obtain the following result:

Theorem 2.13. *Let Ω be a bounded Lipschitz domain in $\mathbb{R}^n$, where either*

- *$n \geq 3$ and $(3+\delta)' < p < 3+\delta$, where $\delta > 0$ depends only on the Lipschitz character of Ω and $(3+\delta)'$ denotes the conjugate exponent of $3+\delta$,*
- *$n = 2$ and $(4+\delta)' < p < 4+\delta$, where $\delta > 0$ depends only on the Lipschitz character of Ω or*
- *$n \geq 2$ and suppose additionally that Ω satisfies a uniform outer ball condition and $1 < p < \infty$.*

Then the operator Δ_p^w generates a C_0-semigroup of contractions on $L^p(\Omega)$.

Proof. We start with the case $n \geq 3$. Again we verify that the range condition of the Lumer-Phillips-Theorem is satisfied. For $u \in D(\Delta_p^w)$, we have $\Delta u \in L^p(\Omega)$, so $\Delta u \in L^p_{\alpha-2}(\Omega)$ for any $\alpha \leq 2$. By Theorem 2.1, there exists $\delta > 0$ such that whenever $1 < p < 3 + \delta$, we can find $\alpha \in [1, 2]$ and a unique $v \in L^p_\alpha(\Omega)$ such that

$$\begin{cases} \Delta v = \Delta u & \text{in } \Omega, \\ v = 0 & \text{on } \partial\Omega, \end{cases}$$

and $\|v\|_{\alpha,p} \leq C\,\|\Delta u\|_{\alpha-2,p}$. Then $u - v$ is harmonic and by the Maximum Principle we have $u = v$. Thus

$$\|u\|_{1,p} \leq \|u\|_{\alpha,p} \leq C\,\|\Delta u\|_{\alpha-2,p} \leq C\,\|\Delta u\|_p \,.$$

Therefore, $\Delta_p^w : D(\Delta_p^w) \to L^p(\Omega)$ is an injective mapping for our range of exponents p. A similar argument shows that it is also surjective. Thus under our assumptions we have $0 \in \rho(\Delta_p^w)$ for any $1 < p < 3 + \delta$. Moreover, by Lemma 2.8, Δ_p^w is dissipative for $p \geq 2$. This proves the theorem for $2 \leq p < 3 + \delta$.

For the case when $(3+\delta)' < p < 2$ we consider the dual operator. Since $\Delta_{p'}^w$ is m-dissipative, its dual operator $\Delta_{p'}^{w\,\prime}$ is m-dissipative in $L^p(\Omega)$ (cf. [6, Proposition 3.10]). We now claim that $\Delta_p^w \subseteq \Delta_{p'}^{w\,\prime}$, i.e., $D(\Delta_p^w) \subseteq D(\Delta_{p'}^{w\,\prime})$ and both operators coincide on $D(\Delta_p^w)$. To see this, let $v \in D(\Delta_p^w)$, $u \in D(\Delta_{p'}^w)$, $(v_n) \subseteq C_c^\infty(\Omega)$ and $(u_n) \subseteq C_c^\infty(\Omega)$ such that $v_n \to v$ in $W^{1,p}(\Omega)$ and $u_n \to u$ in $W^{1,p'}(\Omega)$. Then

$$\begin{aligned} \langle \Delta u, v\rangle &= \lim_{n\to\infty} \langle \Delta u, v_n\rangle = -\lim_{n\to\infty} \langle \nabla u, \nabla v_n\rangle \\ &= -\langle \nabla u, \nabla v\rangle = -\lim_{n\to\infty} \langle \nabla u_n, \nabla v\rangle = \lim_{n\to\infty} \langle u_n, \Delta v\rangle = \langle u, \Delta v\rangle, \end{aligned}$$

so $v \in D(\Delta_{p'}^{w\,\prime})$ and $\Delta_{p'}^{w\,\prime} v = \Delta v$ as claimed. Therefore Δ_p^w is contained in a dissipative operator and hence is itself dissipative for $(3 + \delta)' < p \leq 2$. Moreover, as we have seen above, for these p the range condition is satisfied. Using the Lumer-Phillips Theorem, this completes the proof for $n \geq 3$.

For $n = 2$, we merely replace Theorem 2.1 by [19, Theorem 1.3], while in the case of domains satisfying a uniform outer ball condition we use Theorem 2.3 and argue in the same way obtaining the larger range of exponents p. □

Remark 2.14. δ is given by $3\varepsilon/(1-\varepsilon)$ where ε is the constant given in Theorem 2.1. Therefore, $(3 + \delta)' = 3/(2 + \varepsilon)$.

Corollary 2.15. *If $\Omega \subseteq \mathbb{R}^n$, $n \geq 2$, is a bounded Lipschitz domain satisfying a uniform outer ball condition and $1 < p \leq 2$, then we have $\Delta_p^w = \Delta_p^s$.*

Proof. Obviously, $\Delta_p^s \subseteq \Delta_p^w$. Since $0 \in \rho(\Delta_p^s) \cap \rho(\Delta_p^w)$, we have $\Delta_p^s = \Delta_p^w$. □

Remark 2.16. To make the statements more concise, in the following we will often only refer to the semigroup generated by the weak Dirichlet-Laplacian, recalling that whenever the strong Dirichlet-Laplacian is a generator, it coincides with the weak Dirichlet-Laplacian.

2.2.2. Properties of the semigroup. We first show that the semigroups generated on $L^p(\Omega)$ are consistent.

Proposition 2.17. *Let Ω and n be as in Theorem 2.13 and suppose p and q satisfy the assumptions on p given there. Then the semigroups T_p generated by Δ_p^w and T_q generated by Δ_q^w are consistent, i.e., if $f \in L^p(\Omega) \cap L^q(\Omega)$ then*

$$T_p(t)f = T_q(t)f \text{ for all } t \geq 0.$$

Proof. W.l.o.g. assume $p < q$. Then, as Ω is bounded, $D(\Delta_q^w) \subseteq D(\Delta_p^w)$. Let $f \in D(\Delta_q^w)$. Then $T_q(t)f$ is the unique classical solution to

$$u' - \Delta u = 0, \; u(0) = f \tag{2.4}$$

with $T_q(t)f \in D(\Delta_q^w)$ for $t \geq 0$. But then $T_q(t)f \in D(\Delta_p^w)$, so it must agree with $T_p(t)f$, the unique classical solution to (2.4) for $f \in D(\Delta_p^w)$. Since $D(\Delta_q^w)$ is dense in $L^q(\Omega)$, we get $T_p(t)f = T_q(t)f$ for all $f \in L^q(\Omega)$. □

This now gives us some further interesting results.

Theorem 2.18. *Let Ω, n and p be as in Theorem 2.13 and let T denote the semigroup generated by Δ_p^w on $L^p(\Omega)$. Then the following holds.*

1. *The spectrum of Δ_p^w is independent of p, i.e., $\sigma(\Delta_p^w) = \sigma(\Delta_2^w)$ for all p in the range.*
2. *T satisfies a Gaussian estimate, i.e., there exist constants $a \geq 0$, $M, b > 0$ such that $|T(t)f| \leq Me^{at}G(bt)|f|$ for $t \geq 0$ where $G(t)f = k_t * f$ is the Gaussian semigroup and k_t is the Gaussian kernel.*
3. *T is analytic.*
4. *T is positive, i.e., $T(t)f \geq 0$ for all $f \geq 0$, $t \geq 0$.*
5. *T is of negative type, i.e., the growth bound $\omega(T)$ of the semigroup satisfies $\omega(T) < 0$. Moreover, $\omega(T)$ is independent of p.*

Proof. We start with 2: On $L^2(\Omega)$, Δ_2^w is identical to the Dirichlet-Laplacian defined via the form in [3, Section 1]. Since the semigroups generated on $L^p(\Omega)$ are consistent, both for the construction of the generator in [3] and for the semigroups generated by Δ_p^w, the semigroups must coincide for all cases. Then by [3, Section 1] the semigroup satisfies a Gaussian estimate.

3 now follows from the Stein Interpolation Theorem, while 1 follows from the Gaussian estimate by a result due to Kunstmann [20].

To see 4, we first deal with with real-valued functions using [6, Corollary 7.15]. In this case, $L^p(\Omega)$ is a Banach lattice, therefore it is sufficient to show that Δ_p^w is dispersive. In the case $p \geq 2$, integration by parts as in the proof of Lemma 2.8 proves this. When dealing with complex-valued functions, the positivity of the semigroup on $L^p(\Omega)$, $p \geq 2$, obviously follows from the real-valued case and the fact that the operator has real-valued coefficients.

Now let $p < 2$. Recall that a semigroup is positive iff the resolvent is positive for sufficiently large $\lambda > 0$ (see, e.g., [6, Proposition 7.1]). Let $f \in L^p(\Omega)$, $f \geq 0$. Then there exist $f_n \in L^2(\Omega)$, $f_n \geq 0$ such that $f_n \to f$ in $L^p(\Omega)$. If Δ_p^w generates a

C_0-semigroup on $L^p(\Omega)$, then from the standard resolvent estimate for generators we obtain that $R(\lambda, \Delta_p^w)f_n \to R(\lambda, \Delta_p^w)f$ in $L^p(\Omega)$. However, positivity of the semigroup on $L^2(\Omega)$ and consistency of the semigroups imply $R(\lambda, \Delta_p^w)f_n \geq 0$ and therefore $R(\lambda, \Delta_p^w)f \geq 0$ almost everywhere. This proves positivity also for the case $p < 2$.

Finally, to prove 5, we use a result due to Weis (cf. [2, Theorem 5.3.6]) which states that for generators A of positive semigroups T on $L^p(\Omega)$, $\omega(T)$ coincides with the spectral bound $s(A)$. By 1, we already know that $\sigma(\Delta_p^w) = \sigma(\Delta_2^w)$, in particular equality holds for the spectral bound. It therefore remains to examine the case $p = 2$.

Using Poincaré's inequality, for $u \in D(\Delta_2^w)$, $u \neq 0$, we have

$$\langle \Delta u, u \rangle = -\|\nabla u\|_2^2 \leq -C\|u\|_2^2 < 0.$$

Then by [26, Theorem 1.3.9], we get that the numerical range and therefore the spectrum of Δ_2^w lie in the half-plane $\{z \in \mathbb{C} : \operatorname{Re} z \leq -C\}$, in particular $s(\Delta_2^w) \leq -C$. □

2.2.3. Maximal L^q-regularity. In the previous section we have proven generator results for the Dirichlet-Laplacian and gathered various properties of the semigroups and their generators. We can now exploit these results to show the maximal regularity property for the Laplacian.

Theorem 2.19. *Let Ω, n and p be as in Theorem 2.13. Then the weak Dirichlet-Laplacian has the maximal regularity property, i.e., for $1 < q < \infty$ and for every $f \in L^q(\mathbb{R}_+, L^p(\Omega))$ there exists a unique solution to*

$$\begin{cases} u'(t) - \Delta u(t) = f(t) & \text{for} \quad t \in \mathbb{R}_+, \\ \quad u(t,x) = 0 & \text{on} \quad \mathbb{R}_+ \times \partial\Omega, \\ \quad u(0) = 0. \end{cases} \tag{2.5}$$

The solution u lies in $L^q(\mathbb{R}_+, D(\Delta_p^w)) \cap W^{1,q}(\mathbb{R}_+, L^p(\Omega))$ and satisfies the estimate

$$\|u\|_{L^q(\mathbb{R}_+, L^p(\Omega))} + \|u'\|_{L^q(\mathbb{R}_+, L^p(\Omega))} + \|\Delta u\|_{L^q(\mathbb{R}_+, L^p(\Omega))} \leq C\|f\|_{L^q(\mathbb{R}_+, L^p(\Omega))}.$$

Proof. We have shown that the semigroup generated by Δ_p^w is contractive (Theorem 2.13), analytic and positive (Theorem 2.18) on $L^p(\Omega)$. Maximal regularity now follows from a result by Weis ([28, Corollary 4d]). Moreover, as the generated semigroup is of negative type (Theorem 2.18), we get $u \in L^q(\mathbb{R}_+, L^p(\Omega))$ (cf. [8]) which leads to the desired estimate using the Closed Graph Theorem. □

Of course, whenever the weak and the strong Laplacian coincide, this also yields maximal regularity for the strong Laplacian. Because of the importance of the result, in particular the better estimate (2.6), we state it here separately.

Theorem 2.20. *Let $1 < p \leq 2$, and $\Omega \subseteq \mathbb{R}^n$, $n \geq 2$, be a bounded Lipschitz domain satisfying a uniform outer ball condition. Then the strong Dirichlet-Laplacian has the maximal L^q-regularity property in $L^p(\Omega)$, i.e., for $1 < q < \infty$ and for every*

$f \in L^q(\mathbb{R}_+, L^p(\Omega))$ there exists a unique solution to (2.5). *The solution u lies in $L^q(\mathbb{R}_+, W^{2,p}(\Omega) \cap W_0^{1,p}(\Omega)) \cap W^{1,q}(\mathbb{R}_+, L^p(\Omega))$ and satisfies the estimate*

$$\|u\|_{L^q(\mathbb{R}_+,W^{2,p}(\Omega))} + \|u'\|_{L^q(\mathbb{R}_+,L^p(\Omega))} \leq C\,\|f\|_{L^q(\mathbb{R}_+,L^p(\Omega))}\,. \tag{2.6}$$

Remark 2.21. The same results can be proven for exterior Lipschitz domains and exterior Lipschitz domains satisfying a uniform outer ball condition. To do this, construct solutions to the resolvent problem in the exterior domain by combining the solutions in bounded domains and the whole space $\mathbb{R}^n$ as is done in the next section for Ornstein-Uhlenbeck operators. Together with dissipativity of the operator this proves generation of a contractive C_0-semigroup via the Lumer-Phillips Theorem. Moreover, we see from the construction that the resolvent operators are consistent on the L^p-spaces, hence the generated semigroups are consistent and they coincide with the positive analytic semigroups gained by the form method in [3]. Invoking Weis' Theorem once more, we obtain maximal L^q-regularity over $L^p(\Omega)$ for the Dirichlet-Laplacian in exterior Lipschitz domains with the same restrictions on the dimension, the domain and the exponent p as in the case of bounded Lipschitz domains.

3. The Ornstein-Uhlenbeck operator in Lipschitz domains

3.1. Ornstein-Uhlenbeck operators on $L^p(\mathbb{R}^n)$

We define the following operator. For $1 < p < \infty$ and $M \in \mathbb{R}^{n\times n}$, let

$$\begin{cases} A_{\mathbb{R}^n} u(x) := \Delta u(x) + Mx \cdot \nabla u(x), \; x \in \mathbb{R}^n, \\ D(A_{\mathbb{R}^n}) := \{u \in W^{2,p}(\mathbb{R}^n) : Mx \cdot \nabla u \in L^p(\mathbb{R}^n)\}. \end{cases} \tag{3.1}$$

We gather the main results on the Ornstein-Uhlenbeck operator $A_{\mathbb{R}^n}$ in a theorem.

Theorem 3.1. *Let $1 < p < \infty$ then the operator $A_{\mathbb{R}^n}$ generates a positive C_0-semigroup on $L^p(\mathbb{R}^n)$ and the semigroup $(e^{tA_{\mathbb{R}^n}})_{t\geq 0}$ has the explicit representation*

$$e^{tA_{\mathbb{R}^n}} f(x) = \frac{1}{(4\pi)^{n/2}(\det Q_t)^{1/2}} \int_{\mathbb{R}^n} f(e^{tM}x - y) e^{-\frac{1}{4}(Q_t^{-1}y,y)} \mathrm{d}y, \; x \in \mathbb{R}^n, \; t > 0,$$

with $Q_t := \int_0^t e^{sM} e^{sM^T}\,\mathrm{d}s$ for $t > 0$. Moreover, there exists $\lambda_0 \in \mathbb{R}$ such that for $\lambda > \lambda_0$ the unique solution of the resolvent problem $(\lambda - A_{\mathbb{R}^n})u = f$ satisfies

$$\|u\|_{W^{2,p}(\mathbb{R}^n)} + \|Mx \cdot \nabla u\|_{L^p(\mathbb{R}^n)} \leq C_\lambda \|f\|_{L^p(\mathbb{R}^n)} \tag{3.2}$$

for some constant C_λ depending on λ, and there exist constants C, ω independent of λ such that the solution to the resolvent problem satisfies the gradient estimate

$$\|\nabla u\|_{L^p(\mathbb{R}^n)} \leq \frac{C}{|\lambda - \omega|^{\frac{1}{2}}} \|f\|_{L^p(\mathbb{R}^n)}\,. \tag{3.3}$$

Furthermore, if $1 < p \leq q \leq \infty$, then there exist constants C, ω such that

$$\left\|e^{tA_{\mathbb{R}^n}} f\right\|_q \leq C t^{-\frac{n}{2}(\frac{1}{p}-\frac{1}{q})} e^{\omega t} \|f\|_p\,, \; t > 0, \; f \in L^p(\mathbb{R}^n), \tag{3.4}$$

and for the gradient we have the estimate

$$\left\|\nabla e^{tA_{\mathbb{R}^n}} f\right\|_q \leq Ct^{-\frac{1}{2}-\frac{n}{2}(\frac{1}{p}-\frac{1}{q})}e^{\omega t}\|f\|_p, \ t>0, \ f \in L^p(\mathbb{R}^n). \tag{3.5}$$

Proof. By [27, Theorem 2.4], we obtain that $A_{\mathbb{R}^n}$ generates a positive C_0-semigroup on $L^p(\mathbb{R}^n)$ and the explicit representation can be found in [23]. The estimate (3.2) then follows from the closed graph theorem. For the gradient estimate (3.3), using [15, Proposition 3.4], we see that the semigroup generated by $A_{\mathbb{R}^n}$ satisfies the estimate

$$\left\|\nabla e^{tA_{\mathbb{R}^n}} f\right\|_p \leq \frac{C}{t^{\frac{1}{2}}} e^{\omega t}\|f\|_p \tag{3.6}$$

for some $\omega \in \mathbb{R}$. For $\lambda > \lambda_0$, the resolvent is given by the Laplace transform of the semigroup (cf. [2]) and we get

$$\begin{aligned}\|\nabla R(\lambda, A_{\mathbb{R}^n})f\|_p &= \left\|\int_0^\infty e^{-\lambda t}\nabla e^{tA_{\mathbb{R}^n}} f \mathrm{d}t\right\|_p \leq C\int_0^\infty t^{-\frac{1}{2}}e^{(\omega-\lambda)t}\mathrm{d}t\,\|f\|_p \\ &\leq C\frac{\Gamma(\frac{1}{2})}{(\lambda-\omega)^{\frac{1}{2}}}\|f\|_p.\end{aligned}$$

Finally, (3.4) and (3.5) follow from [15, Proposition 3.4 and Lemma 3.5]. □

3.2. The drift operator

We have already studied the Laplacian in Lipschitz domains. We now take a closer look at the drift term before combining the results to study the Ornstein-Uhlenbeck operator. We first introduce the drift operator B_Ω on a domain $\Omega \subseteq \mathbb{R}^n$:

$$\begin{cases} B_\Omega u(x) := Mx \cdot \nabla u(x), \ x \in \Omega, \\ D(B_\Omega) := \{u \in W^{1,p}(\Omega) : Mx \cdot \nabla u \in L^p(\Omega)\}. \end{cases} \tag{3.7}$$

Note that the condition $Mx \cdot \nabla u \in L^p(\Omega)$ is trivially satisfied in bounded domains for $u \in W^{1,p}(\Omega)$. Moreover, in this case the drift term is a relatively bounded perturbation of the Laplacian:

Proposition 3.2. *Let D be a bounded Lipschitz domain. Then the drift operator B_D on D is relatively bounded by the weak Dirichlet-Laplacian Δ_p^w in $L^p(D)$ for $(3+\varepsilon)' < p < 3+\varepsilon$ where $\varepsilon > 0$ depends only on the Lipschitz constant of D. The relative bound is given by zero.*

Proof. Obviously we have that $D(B_D) \supseteq D(\Delta_p^w)$. Let $u \in D(\Delta_p^w)$. By Theorem 2.1, we can find $\delta > 0$ such that

$$\|u\|_{1+\delta,p} \leq C\|\Delta u\|_p \quad \text{for} \quad (3+\varepsilon)' < p < 3+\varepsilon.$$

By the resolvent estimate for generators of contractive C_0-semigroups we have

$$\|u\|_p \leq \frac{C}{\lambda}\|(\lambda-\Delta)u\|_p \quad \text{for} \quad \lambda > 0, \quad (3+\varepsilon)' < p < 3+\varepsilon.$$

Set $\Theta = \delta/(1+\delta) \in (0,1)$. Then, using the complex interpolation method (cf. [5]) and Jensen's inequality, we obtain the estimate

$$\begin{aligned}\|u\|_{1,p} &\le \left(C\|\Delta u\|_p\right)^{1-\Theta}\left(\frac{C}{\lambda}\|(\lambda-\Delta)u\|_p\right)^{\Theta}\\ &\le C\lambda^{-\Theta}((1-\Theta)\|\Delta u\|_p + \Theta\|(\lambda-\Delta)u\|_p)\\ &= C\lambda^{-\Theta}\|\Delta u\|_p + C\lambda^{1-\Theta}\|u\|_p\end{aligned}$$

As we can choose $\lambda > 0$ arbitrarily large, this concludes the proof. □

The next step is to prove dissipativity for the drift operator. To do this, we need to integrate by parts. The next result shows that this is possible in our situation.

Proposition 3.3. *Let Ω be a Lipschitz domain in $\mathbb{R}^n$, $u \in W_0^{1,p}(\Omega)$, $\phi \in W^{1,\infty}(\mathbb{R}^n)$ and $1 < p < \infty$. Then we have*

$$\int_\Omega \phi\,\overline{u}|u|^{p-2}\partial_i u = -\int_\Omega \phi\,|u|^{p-2}u\partial_i\overline{u} - \int_\Omega \phi\,\frac{p-2}{2}|u|^{p-2}(\overline{u}\partial_i u + u\partial_i\overline{u}) - \int_\Omega (\partial_i\phi)\,|u|^p. \tag{3.8}$$

Proof. The proof is a simpler version of the proof of Theorem 3.1 in [25], as we only want to integrate by parts once. □

Proposition 3.4. *Let $\Omega \subseteq \mathbb{R}^n$ be a Lipschitz domain and let the drift operator B_Ω be defined as in (3.7). Then $B_\Omega + \frac{\operatorname{tr} M}{p}$ is dissipative in $L^p(\Omega)$ for all $1 < p < \infty$.*

Proof. We need to show that Re $\langle B_\Omega u, u^*\rangle \le -\frac{\operatorname{tr} M}{p}\|u\|_p^p$ where u^* is as in Lemma 2.8. When Ω is unbounded, the term Mx is unbounded which causes some problems. Therefore, we need to introduce a cut-off function. Let $\varphi \in C_c^\infty(\mathbb{R}^n)$, $0 \le \varphi \le 1$ such that $\varphi \equiv 1$ on the unit ball B_1 and supp $\varphi \subseteq B_2$. Let $\varphi_R(x) := \varphi(x/R)$. Then $|\nabla\varphi_R(x)| \le C/R$ and $\nabla\varphi_R$ converges pointwise to 0, while φ_R converges pointwise to 1. We have

$$\begin{aligned}\operatorname{Re}\langle B_\Omega u, u^*\rangle &= \operatorname{Re}\int_\Omega Mx\cdot\nabla u\,|u|^{p-2}\overline{u} = \operatorname{Re}\sum_{i,j=1}^n\int_\Omega m_{ij}x_j(\partial_i u)\,|u|^{p-2}\overline{u}\\ &= \operatorname{Re}\sum_{i,j=1}^n\lim_{R\to\infty}\int_\Omega \varphi_R\,m_{ij}x_j(\partial_i u)\,|u|^{p-2}\overline{u},\end{aligned}$$

where the last equality holds by Lebesgue's Dominated Convergence Theorem. Now we apply Proposition 3.3 with $\phi = \varphi_R m_{ij}x_j$ to obtain

$$\begin{aligned}\int_\Omega \varphi_R\,m_{ij}x_j(\partial_i u)\,|u|^{p-2}\overline{u} &= -\int_\Omega \varphi_R\,m_{ij}x_j\,|u|^{p-2}u\partial_i\overline{u}\\ &\quad -\int_\Omega \varphi_R\,m_{ij}x_j\,\frac{p-2}{2}|u|^{p-2}(\overline{u}\partial_i u + u\partial_i\overline{u})\\ &\quad -\int_\Omega ((\partial_i\varphi_R)\,m_{ij}x_j + \varphi_R\,\delta_{ij}m_{ij})\,|u|^p.\end{aligned}$$

Applying the Dominated Convergence Theorem and collecting our results, we have

$$\begin{aligned}\operatorname{Re}\,\langle B_\Omega u, u^*\rangle &= -\operatorname{Re}\sum_{i,j=1}^n \int_\Omega m_{ij}x_j\,|u|^{p-2}u\partial_i\overline{u} - \operatorname{Re}\sum_{i,j=1}^n \int_\Omega \delta_{ij}m_{ij}\,|u|^p \\ &\quad -\operatorname{Re}\sum_{i,j=1}^n \int_\Omega m_{ij}x_j\,\frac{p-2}{2}|u|^{p-2}(\overline{u}\partial_i u + u\partial_i\overline{u}) \\ &= -(p-1)\operatorname{Re}\,\langle B_\Omega u, u^*\rangle - \operatorname{tr} M\,\|u\|_p^p.\end{aligned}$$

This yields

$$\operatorname{Re}\,\langle B_\Omega u, u^*\rangle = -\frac{\operatorname{tr} M}{p}\,\|u\|_p^p.$$

Therefore, the operator $B_\Omega - \lambda$ is dissipative for $\lambda \geq -\operatorname{tr} M/p$. □

3.3. Ornstein-Uhlenbeck operators on bounded Lipschitz domains

Let D be a bounded Lipschitz domain in $\mathbb{R}^n$, $n \geq 3$. We study the operator

$$\begin{cases} A_D u(x) := \Delta u(x) + Mx\cdot\nabla u(x), \; x \in D, \\ D(A_D) := \{u \in W_0^{1,p}(D) : \Delta u \in L^p(D)\}. \end{cases} \tag{3.9}$$

Our main result for bounded domains is now an easy consequence of perturbation results for C_0-semigroups.

Proposition 3.5. *On bounded Lipschitz domains $D \subseteq \mathbb{R}^n$, the operator A_D generates an analytic quasi-contractive[5] semigroup T for $(3+\varepsilon)' < p < 3+\varepsilon$ where $\varepsilon > 0$ depends on the Lipschitz constant of the domain D. The growth bound of the semigroup can be estimated by $\omega(T) \leq -\frac{\operatorname{tr} M}{p}$. For $\operatorname{Re}\lambda > -\frac{\operatorname{tr} M}{p}$, there exists a constant C_λ depending on λ, such that we have the estimate*

$$\|u\|_{L^p(D)} + \|\Delta u\|_{L^p(D)} + \|Mx\cdot\nabla u\|_{L^p(D)} \leq C_\lambda \|f\|_{L^p(D)} \tag{3.10}$$

for solutions of the resolvent problem $(\lambda - A_D)u = f$. Furthermore, the solution u satisfies the estimate

$$\|u\|_{L^p(D)} \leq \frac{C}{|\lambda + \frac{\operatorname{tr} M}{p}|}\|f\|_{L^p(D)} \tag{3.11}$$

for $\lambda \in -\frac{\operatorname{tr} M}{p} + \Sigma_\varphi$ for some angle φ, where $\Sigma_\varphi := \{\lambda \in \mathbb{C} : |\arg\lambda| < \varphi\}\setminus\{0\}$ and $\arg\lambda$ denotes the argument of λ. For $\lambda > \max\left\{-\frac{\operatorname{tr} M}{p}, 0\right\}$, there exists $\Theta > 0$ such that

$$\|\nabla u\|_{L^p(D)} \leq \frac{C}{\lambda^\Theta}\|f\|_{L^p(D)}. \tag{3.12}$$

---

[5] I.e., there exists $\alpha \in \mathbb{R}$ such that $\left(e^{-\alpha t}T(t)\right)_{t\geq 0}$ is a contractive semigroup

Proof. Since, by Theorem 2.18, the weak Dirichlet-Laplacian generates an analytic semigroup of contractions, Proposition 3.2 implies that A_D generates an analytic C_0-semigroup. The fact that it is quasi-contractive and the growth bound estimate follow from the dissipativity of the operator $B_D + \frac{\operatorname{tr} M}{p}$ (Proposition 3.4) and the Lumer-Phillips-Theorem (cf. [2]). Estimate (3.10) is a consequence of the closed graph theorem, while (3.11) is the standard resolvent estimate for analytic semigroups. For (3.12), we use the representation of the resolvent as

$$R(\lambda, A_D) = R(\lambda, \Delta_p^w) \sum_{n=0}^{\infty} (B_D R(\lambda, \Delta_p^w))^n,$$

where $\lambda > \max\left\{-\frac{\operatorname{tr} M}{p}, 0\right\}$. Due to the relative boundedness, the Neumann series gives a bounded operator, and we only need to show the corresponding estimate for the Laplacian. Now, a similar calculation as in the proof of Proposition 3.2 shows that for $u \in D(A_D)$ we have

$$\|u\|_{1,p} \leq C\lambda^{-\Theta} \|(\lambda - \Delta)u\|_p + C\lambda^{1-\Theta} \|u\|_p .$$

Again using the fact that the Dirichlet-Laplacian generates an analytic semigroup of contractions, we can estimate $\lambda \|u\|_p \leq C \|(\lambda - \Delta)u\|_p$ to obtain

$$\|u\|_{1,p} \leq C\lambda^{-\Theta} \|(\lambda - \Delta)u\|_p .$$

Therefore, $\|\nabla R(\lambda, A_D)\|_{\mathcal{L}(L^p(D))} \leq C\lambda^{-\Theta}$ concluding the proof. □

3.4. Ornstein-Uhlenbeck operators on exterior Lipschitz domains

Let $\Omega = \mathbb{R}^n \setminus \overline{K}$ where K is a bounded Lipschitz domain contained in a ball B_R of radius R. Again we define the Ornstein-Uhlenbeck operator

$$\begin{cases} A_\Omega u(x) := \Delta u(x) + Mx \cdot \nabla u(x), \; x \in \Omega, \\ D(A_\Omega) := \{u \in W_0^{1,p}(\Omega) : \Delta u \in L^p(\Omega) \text{ and } Mx \cdot \nabla u \in L^p(\Omega)\}. \end{cases} \tag{3.13}$$

Using the Lumer-Phillips Theorem, we will see that A_Ω is a closed operator and generates a C_0-semigroup on $L^p(\Omega)$ for $(3+\varepsilon)' < p < 3+\varepsilon$. In a first step, our aim is to show that the range condition $(\lambda - A_\Omega)D(A_\Omega) = L^p(\Omega)$ is satisfied for some $\lambda > 0$ and $(3+\varepsilon)' < p < 3+\varepsilon$.

We use the following notation. The bounded Lipschitz domain $\Omega \cap B_{R+3}$ will be denoted by D. We further introduce a cut-off function $\varphi \in C^\infty(\mathbb{R}^n)$ with $\varphi(x) = 0$ for $x \in B_{R+1}$ and $\varphi(x) = 1$ for $x \in B_{R+2}^c$. Given $f \in L^p(\Omega)$, denote its trivial extension to K by f_0 and its restriction to D by f_*. By u_0 we denote the solution to the resolvent problem $(\lambda - A_{\mathbb{R}^n})u = f_0$ in $\mathbb{R}^n$ given in Theorem 3.1. Let u_* denote the solution to $(\lambda - A_D)u = f_*$ in D given by Proposition 3.5. Note that

$$u_0 \in W^{2,p}(\mathbb{R}^n) \cap \{u \in L^p(\mathbb{R}^n) : Mx \cdot \nabla u \in L^p(\mathbb{R}^n)\}$$
$$u_* \in W_0^{1,p}(D) \text{ and } \Delta u_* \in L^p(D).$$

For the solution of the resolvent problem $(\lambda - A_\Omega)u = f$ we make the ansatz $u = \Theta_\lambda f := \varphi u_0 + (1-\varphi)u_*$. The function u then satisfies

$$\Delta u = \varphi\Delta u_0 + (1-\varphi)\Delta u_* + 2\nabla\varphi\cdot(\nabla u_0 - \nabla u_*) + \Delta\varphi(u_0 - u_*) \quad \text{and}$$

$$Mx\cdot\nabla u = \varphi\, Mx\cdot\nabla u_0 + (1-\varphi)Mx\cdot\nabla u_* + (Mx\cdot\nabla\varphi)(u_0 - u_*).$$

Due to the properties of u_0 and u_*, it is easy to check that $u \in W_0^{1,p}(\Omega)$, $\Delta u \in L^p(\Omega)$ and $Mx\cdot\nabla u \in L^p(\Omega)$. Furthermore, by estimates (3.2) and (3.10), we have

$$\|u\|_{L^p(\Omega)} + \|\Delta u\|_{L^p(\Omega)} + \|Mx\cdot\nabla u\|_{L^p(\Omega)} \le C_{\lambda,\varphi}\,\|f\|_{L^p(\Omega)}\,. \tag{3.14}$$

u satisfies the following equation.

$$\begin{aligned}\lambda u - \Delta u - Mx\cdot\nabla u \;&=\; \varphi f_0 + (1-\varphi)f_* - 2\nabla\varphi\cdot(\nabla u_0 - \nabla u_*)\\ &\quad -\Delta\varphi(u_0 - u_*) - (Mx\cdot\nabla\varphi)(u_0 - u_*) \;=\; (I - T_\lambda)f,\end{aligned}$$

where

$$T_\lambda f := 2\nabla\varphi\cdot(\nabla u_0 - \nabla u_*) + \Delta\varphi(u_0 - u_*) + (Mx\cdot\nabla\varphi)(u_0 - u_*).$$

Our aim is now to invert $I - T_\lambda$ continuously in $\mathcal{L}(L^p(\Omega))$. We need the gradient estimate (3.3) for u_0, and for u_* we use the estimates (3.11) and (3.12). Noting that $\nabla\varphi$ is supported in a compact set, for sufficiently large $\lambda > 0$ this yields

$$\begin{aligned}\|T_\lambda f\|_p \;&\le\; C\,\|\nabla u_0\|_p + C\,\|\nabla u_*\|_p + C\,\|u_0\|_p + C\,\|u_*\|_p\\ &\le\; \frac{C}{|\lambda|^\Theta}\,\|f\|_p + \frac{C}{|\lambda|}\,\|f\|_p \;\le\; \frac{1}{2}\,\|f\|_p\,.\end{aligned}$$

Therefore, for sufficiently large $\lambda > 0$, $I - T_\lambda$ can be continuously inverted in $\mathcal{L}(L^p(\Omega))$ using the Neumann series. Then we can solve the resolvent problem $(\lambda - A_\Omega)u = f$ by setting $u = \Theta_\lambda(I - T_\lambda)^{-1}f$.

Theorem 3.6. *Let Ω be an exterior Lipschitz domain. Then for $(3+\varepsilon)' < p < 3+\varepsilon$ where $\varepsilon > 0$ depends only on the Lipschitz constant of the domain Ω, the operator A_Ω defined as in* (3.13) *generates a quasi-contractive C_0-semigroup T on $L^p(\Omega)$. The growth bound can be estimated by $\omega(T) \le -\frac{\operatorname{tr} M}{p}$.*

Proof. The preceding discussion shows that the range condition of the Lumer-Phillips-Theorem is satisfied. By Proposition 3.4, the operator $B_\Omega + \frac{\operatorname{tr} M}{p}$ is dissipative. Using a smooth cut-off function and the Dominated Convergence Theorem, we can integrate by parts to see that for $p \ge 2$, the weak Dirichlet-Laplacian Δ_p^w is dissipative on exterior Lipschitz domains. This proves the theorem for $2 \le p < 3+\varepsilon$. For $(3+\varepsilon)' < p < 2$ we have so far only shown dissipativity of the weak Laplacian for bounded Lipschitz domains. However, simply setting $M = 0$ in the preceding discussion, we know that Δ_p^w is m-dissipative in exterior Lipschitz domains for $2 \le p < 3+\varepsilon$. The same consideration as in the proof of Theorem 2.13 then proves dissipativity of Δ_p^w and therefore of $A_\Omega + \frac{\operatorname{tr} M}{p}$ also for $(3+\varepsilon)' < p < 2$ in exterior Lipschitz domains. □

3.5. Domains satisfying an outer ball condition

We now want to consider equation (1.1) in domains Ω satisfying a uniform outer ball condition where we can show that the domain of the operator $\Delta + Mx \cdot \nabla$ is contained in $W^{2,p}(\Omega)$. To do this we proceed in the same manner as for general Lipschitz domains using the added regularity of solutions to the heat equation we proved in Theorem 2.20 for these domains.

3.5.1. Bounded domains. Let D be a bounded Lipschitz domain satisfying a uniform outer ball condition. On D, we define the operator

$$\begin{cases} A_D u(x) = \Delta u(x) + Mx \cdot \nabla u(x), \; x \in D, \\ D(A_D) = W^{2,p}(D) \cap W_0^{1,p}(D). \end{cases} \tag{3.15}$$

Proposition 3.7. *The drift operator B_D on D given by*

$$B_D u(x) = Mx \cdot \nabla u(x), \; x \in D, \quad D(B_D) = W^{1,p}(D)$$

is relatively bounded by Δ_p^s in $L^p(D)$ for $1 < p \leq 2$. The relative bound is given by $a_0 = 0$.

Proof. Obviously, $D(B_D) \supseteq D(\Delta_p^s)$. Furthermore, using Ehrling's Lemma (cf. [1, Theorem 4.14]) and Theorem 2.3, for $u \in D(\Delta_p^s)$, we have for any $\varepsilon > 0$,

$$\|Mx \cdot \nabla u\|_p \; \leq \; C \|\nabla u\|_p \; \leq \; C(\varepsilon \left\|\nabla^2 u\right\|_p + C(\varepsilon) \|u\|_p) \leq \; \varepsilon \|\Delta u\|_p + C(\varepsilon) \|u\|_p .$$

□

As in the case of an arbitrary bounded Lipschitz domain, our generation result for bounded Lipschitz domains satisfying a uniform outer ball condition is now a direct consequence of the proposition and results from the perturbation theory of generators of contractive and analytic C_0-semigroups.

Corollary 3.8. *For $1 < p \leq 2$, on bounded Lipschitz domains D satisfying a uniform outer ball condition, the operator A_D as defined in (3.15) generates an analytic quasi-contractive semigroup T with $\omega(T) \leq -\frac{\operatorname{tr} M}{p}$. For $\operatorname{Re}\lambda > -\frac{\operatorname{tr} M}{p}$, there exists a constant C_λ depending on λ such that we have the estimate*

$$\|u\|_{W^{2,p}(D)} + \|Mx \cdot \nabla u\|_{L^p(D)} \leq C_\lambda \|f\|_{L^p(D)} \tag{3.16}$$

for solutions of the resolvent problem $(\lambda - A_D)u = f$.

We now show that we also get so-called L^p-L^q-smoothing estimates for the generated semigroups. This kind of estimates is useful when studying semilinear equations. For example, in [15], L^p-L^q-smoothing estimates for the solution of the Stokes equation with linearly growing initial data are used to obtain mild solutions for the corresponding Navier-Stokes equations.

Lemma 3.9. *Let D be a bounded Lipschitz domain satisfying a uniform outer ball condition. Let $1 < p \le q \le 2$ and denote the semigroup generated by the operator A_D by e^{tA_D}. Then for any $\omega > -\min\left\{\frac{\operatorname{tr} M}{p}, \frac{\operatorname{tr} M}{q}\right\}$ and all $f \in L^p(D)$,*

$$\left\|e^{tA_D}f\right\|_q \le Ct^{-\frac{n}{2}(\frac{1}{p}-\frac{1}{q})}e^{\omega t}\left\|f\right\|_p, \; t>0, \tag{3.17}$$

and

$$\left\|\nabla e^{tA_D}f\right\|_q \le Ct^{-\frac{1}{2}-\frac{n}{2}(\frac{1}{p}-\frac{1}{q})}e^{\omega t}\left\|f\right\|_p, \; t>0. \tag{3.18}$$

Proof. The proof is standard and relies on analyticity of the semigroup and the Gagliardo-Nirenberg inequality. We include it here for completeness. We first note that the semigroups generated by A_D on the spaces $L^p(\Omega)$, $1 < p \le 2$, are consistent which can be shown similarly as in the proof of Proposition 2.17 for the Laplacian. By (3.16) we have

$$\left\|\nabla^2 u\right\|_s \le C\left\|(\lambda - A_D)u\right\|_s \le C(\left\|A_D u\right\|_s + \left\|u\right\|_s),$$

for sufficiently large λ, any $u \in D(A_D)$ and $1 < s \le 2$.

Now choose

$$q_1 \in [p,q] \quad \text{such that} \quad \frac{1}{q_1} \le \frac{1}{n} + \frac{1}{q} \quad \text{and}$$

$$\widetilde{\omega} \quad \text{such that} \quad -\min\left\{\frac{\operatorname{tr} M}{p}, \frac{\operatorname{tr} M}{q}\right\} < \widetilde{\omega} < \omega.$$

Then we have that $a := \frac{n}{2}(\frac{1}{q_1} - \frac{1}{q}) \le \frac{1}{2}$. Using the Gagliardo-Nirenberg inequality (cf. [22, Section 1.4.8]) with $j = 0$, $l = 2$, we then get

$$\begin{aligned}
\left\|e^{tA_D}f\right\|_q &\le C\left\|\nabla^2 e^{tA_D}f\right\|_{q_1}^a \left\|e^{tA_D}f\right\|_{q_1}^{1-a} + C\left\|e^{tA_D}f\right\|_{q_1} \\
&\le C\left\|A_D e^{tA_D}f\right\|_{q_1}^a \left\|e^{tA_D}f\right\|_{q_1}^{1-a} + C\left\|e^{tA_D}f\right\|_{q_1} \\
&\le C\left\|A_D e^{\frac{t}{2}A_D}e^{\frac{t}{2}A_D}f\right\|_{q_1}^a \left\|e^{\frac{t}{2}A_D}e^{\frac{t}{2}A_D}f\right\|_{q_1}^{1-a} + C\left\|e^{\frac{t}{2}A_D}e^{\frac{t}{2}A_D}f\right\|_{q_1} \\
&\le Ct^{-a}e^{\frac{a\widetilde{\omega}t}{2}}\left\|e^{\frac{t}{2}A_D}f\right\|_{q_1}^a e^{\frac{(1-a)\widetilde{\omega}t}{2}}\left\|e^{\frac{t}{2}A_D}f\right\|_{q_1}^{1-a} + Ce^{\frac{\widetilde{\omega}t}{2}}\left\|e^{\frac{t}{2}A_D}f\right\|_{q_1} \\
&\le (Ct^{-\frac{n}{2}(\frac{1}{q_1}-\frac{1}{q})}e^{\frac{\widetilde{\omega}t}{2}} + Ce^{\frac{\widetilde{\omega}t}{2}})\left\|e^{\frac{t}{2}A_D}f\right\|_{q_1} \\
&\le Ct^{-\frac{n}{2}(\frac{1}{q_1}-\frac{1}{q})}e^{\frac{\omega t}{2}}\left\|e^{\frac{t}{2}A_D}f\right\|_{q_1}
\end{aligned}$$

for $t > 0$ and $f \in L^p(\Omega)$. Now choose $q_2 \in [p, q_1]$ such that $\frac{1}{q_2} \le \frac{1}{n} + \frac{1}{q_1} \le \frac{2}{n} + \frac{1}{q}$. By a similar calculation we see that

$$\left\|e^{tA_D}f\right\|_q \le Ct^{-\frac{n}{2}(\frac{1}{q_2}-\frac{1}{q})}e^{\frac{3\omega t}{4}}\left\|e^{\frac{t}{4}A_D}f\right\|_{q_2}$$

Iterating this procedure we obtain q_m such that $\frac{1}{q_m} \le \frac{m}{n} + \frac{1}{q}$ and

$$\left\|e^{tA_D}f\right\|_q \le Ct^{-\frac{n}{2}(\frac{1}{q_m}-\frac{1}{q})}e^{\frac{(2^m-1)\omega t}{2^m}}\left\|e^{\frac{t}{2^m}A_D}f\right\|_{q_m}.$$

After finitely many steps, we reach p and then (3.17) obviously is satisfied.

For (3.18), we use the Gagliardo-Nirenberg inequality in the form

$$\|\nabla u\|_q \leq C(\|\nabla^2 u\|_{q_1} + \|u\|_{q_1})^a \|u\|_{q_1}^{1-a}$$

where we choose $q_1 \in [p,q]$ such that $\frac{1}{q_1} \leq \frac{1}{2n} + \frac{1}{q}$ and $a := \frac{n}{2}(\frac{1}{q_1} - \frac{1}{q}) + \frac{1}{2} \leq \frac{3}{4}$. Proceeding similarly to above then leads to the desired gradient estimate. □

3.5.2. Exterior domains. Let Ω be an exterior Lipschitz domain in $\mathbb{R}^n$ satisfying a uniform outer ball condition. By this we mean that the complement of Ω is a compact set and that Ω itself is a Lipschitz domain and satisfies a uniform outer ball condition. Let the operator A_Ω be defined by

$$\begin{cases} A_\Omega u(x) = \Delta u(x) + Mx \cdot \nabla u(x), \ x \in \Omega, \\ D(A_\Omega) = \{u \in W^{2,p}(\Omega) : Mx \cdot \nabla u \in L^p(\Omega)\} \cap W_0^{1,p}(\Omega). \end{cases} \tag{3.19}$$

Then, as by Corollary 2.11 the strong Dirichlet-Laplacian is dissipative in $L^p(\Omega)$ for $1 < p \leq 2$, Proposition 3.4 implies that $A_\Omega + \frac{\operatorname{tr} M}{p}$ is dissipative in $L^p(\Omega)$ for $1 < p \leq 2$. We now have to show that $(\lambda - A_\Omega)D(A_\Omega) = L^p(\Omega)$ is satisfied for some $\lambda > 0$ and $1 < p \leq 2$. Using the same notation and proceeding just as before, we note that now D is a bounded Lipschitz domain satisfying a uniform outer ball condition. Therefore, $u_* \in W^{2,p}(D) \cap W_0^{1,p}(D)$ and $u := \varphi u_0 + (1-\varphi)u_* \in W^{2,p}(\Omega) \cap W_0^{1,p}(\Omega)$. Instead of (3.14), we have the stronger estimate

$$\|u\|_{W^{2,p}(\Omega)} + \|Mx \cdot \nabla u\|_{L^p(\Omega)} \leq C_{\lambda,\varphi} \|f\|_{L^p(\Omega)} \, .$$

We summarise the generator result of this section in the following theorem.

Theorem 3.10. *Let Ω be an exterior Lipschitz domain satisfying a uniform outer ball condition. Then for $1 < p \leq 2$, the operator A_Ω generates a quasi-contractive C_0-semigroup T on $L^p(\Omega)$. The growth bound can be estimated by $\omega(T) \leq -\frac{\operatorname{tr} M}{p}$.*

Finally, we can extend the L^p-L^q-smoothing estimates for the generated semigroups to exterior domains. In order to do this, we use a lemma on iterated convolutions proved in [11]:

Lemma 3.11. *Let X, Y be Banach spaces and let $T : (0,\infty) \to \mathcal{L}(Y,X)$ and $S : (0,\infty) \to \mathcal{L}(Y)$ be strongly continuous functions. Assume that*

$$\|T(t)\|_{\mathcal{L}(Y,X)} \leq C_0 t^\alpha e^{\omega t}, \quad \|S(t)\|_{\mathcal{L}(Y)} \leq C_0 t^\beta e^{\omega t}, \quad t > 0,$$

for some $C_0, \omega > 0$ and $\alpha, \beta > -1$. For $f \in Y$, set $T_0(t)f := T(t)f$ and

$$T_n(t)f := \int_0^t T_{n-1}(t-s)S(s)f \mathrm{d}s, \quad n \in \mathbb{N}, \ t > 0.$$

Then there exist $C, \tilde{\omega} > 0$ such that $\sum_{n=0}^{\infty} \|T_n(t)f\|_X \leq C t^\alpha e^{\tilde{\omega} t} \|f\|_Y, \quad t > 0.$

Just as in the case of bounded domains, the statement of the L^p-L^q-smoothing properties in exterior domains reads as follows:

Theorem 3.12. *Let $1 < p \le q \le 2$ and let Ω be an exterior Lipschitz domain satisfying a uniform outer ball condition. Denote the semigroup generated by A_Ω by e^{tA_Ω}. Then there exist constants C, ω such that for all $f \in L^p(\Omega)$,*

$$\left\|e^{tA_\Omega} f\right\|_q \le Ct^{-\frac{n}{2}(\frac{1}{p}-\frac{1}{q})} e^{\omega t} \left\|f\right\|_p, \ t > 0, \tag{3.20}$$

and

$$\left\|\nabla e^{tA_\Omega} f\right\|_q \le Ct^{-\frac{1}{2}-\frac{n}{2}(\frac{1}{p}-\frac{1}{q})} e^{\omega t} \left\|f\right\|_p, \ t > 0. \tag{3.21}$$

Proof. The proof is the same as the proof in [12] for $C^{1,1}$-domains and is included here for completeness. Recall from the construction in Section 3.4 that for sufficiently large λ, the resolvent of $\lambda - A_\Omega$ has the form

$$(\lambda - A_\Omega)^{-1} f = \Theta_\lambda (I - T_\lambda)^{-1} f, \ f \in L^p(\Omega)$$

with Θ_λ and T_λ as in Section 3.4. Let f_0 denote the trivial extension of f to $\mathbb{R}^n$ and f_D denote the restriction of f to D. Then the Laplace transforms of the strongly continuous functions $U : [0,\infty) \to \mathcal{L}(L^p(\Omega))$ and $V : [0,\infty) \to \mathcal{L}(L^p(\Omega))$ given by

$$\begin{aligned} U(t)f &:= \varphi e^{tA_{\mathbb{R}^n}} f_0 + (1-\varphi) e^{tA_D} f_D, \\ V(t)f &:= -2\nabla\varphi \cdot \nabla(e^{tA_{\mathbb{R}^n}} f_0 - e^{tA_D} f_D) \\ &\quad -[\Delta\varphi + (Mx \cdot \nabla\varphi)](e^{tA_{\mathbb{R}^n}} f_0 - e^{tA_D} f_D) \end{aligned}$$

are given by Θ_λ and T_λ, respectively. By the estimates (3.4), (3.6), (3.17) and (3.18), there exist constants C, ω such that

$$\|U(t)\|_{\mathcal{L}(L^p(\Omega))} \le Ce^{\omega t}, \qquad \|V(t)\|_{\mathcal{L}(L^p(\Omega))} \le Ct^{-\frac{1}{2}} e^{\omega t}, \qquad t > 0.$$

For $f \in L^p(\Omega)$ we set $T_0(t)f := U(t)f$ and define

$$T_n(t)f := \int_0^t T_{n-1}(t-s)V(s)f \, ds, \quad n \in \mathbb{N}, \quad t > 0.$$

It then follows from Lemma 3.11 that, $T_\Omega(t)f := \sum_{n=0}^{\infty} T_n(t)f$ is well defined for all $t > 0$ and exponentially bounded. Thus, by Lebesgue's theorem

$$\int_0^\infty e^{-\lambda t} T_\Omega(t) dt = \sum_{n=0}^{\infty} \int_0^\infty e^{-\lambda t} \, T_n(t) dt = \sum_{n=0}^{\infty} \hat{U}(\lambda)\hat{V}(\lambda)^n = (\lambda - A_\Omega)^{-1}$$

for λ large enough and hence $T_\Omega(t) = e^{tA_\Omega}$ for $t \ge 0$.

Now, consider U as a mapping $U : [0,\infty) \to \mathcal{L}(L^p(\Omega), L^q(\Omega))$. Once more using the estimates (3.4), (3.6), (3.17) and (3.18), there exist constants C, ω such that

$$\|U(t)\|_{\mathcal{L}(L^p(\Omega), L^q(\Omega))} \le Ct^{-\frac{n}{2}(\frac{1}{p}-\frac{1}{q})} e^{\omega t}, \quad t > 0.$$

Then (3.20) follows as above by Lemma 3.11 if $\frac{n}{2}(\frac{1}{p} - \frac{1}{q}) < 1$. In the general case, choose q_1 such that $\frac{n}{2}(\frac{1}{p} - \frac{1}{q_1}) < 1$ and proceed as above. Then repeat this procedure until q is reached.

Estimate (3.21) follows in a similar way by defining $W(t)f := \nabla U(t)f$, replacing $U(t)$ in the above proof by $W(t)$ and using the estimates (3.5) and (3.18) on the gradient of the semigroup. □

Remark 3.13. Note that due to the representation of the semigroup using only the semigroups on the whole space $\mathbb{R}^n$ and on the bounded domain D which are consistent, the semigroups on exterior domains are also consistent.

References

[1] R.A. Adams: *Sobolev Spaces*, Academic Press, New York, 1975.

[2] W. Arendt, C.J.K. Batty, M. Hieber, F. Neubrander: *Vector-valued Laplace Transforms and Cauchy Problems.* Monographs in Mathematics **96**, Birkhäuser, Basel 2001.

[3] W. Arendt, P. Bénilan: Wiener regularity and heat semigroups on spaces of continuous functions. *Progress in Nonlinear Diff. Eq. and their applications* **35** (1999), 29–49.

[4] D. Bakry: L'hypercontractivité et son utilisation en théorie des semigroupes. In: *Lectures on probability theory (Saint-Flour, 1992).* Lecture Notes in Math., vol. 1581, (1994) Springer, Berlin, 1–114.

[5] J. Bergh, J. Löfström: *Interpolation Spaces.* Springer, Berlin 1976.

[6] P. Clément, H.J.A.M. Heijmans et al.: *One-Parameter Semigroups.* CWI Monographs **5**, North-Holland, Amsterdam (1987).

[7] G. Da Prato, A. Lunardi: On the Ornstein-Uhlenbeck operator in spaces of continuous functions. *J. Func. Anal.* **131** (1995), no. 1, 94–114.

[8] G. Dore: L^p-regularity for abstract differential equations. In: Functional Analysis and Related Topics, H. Komatsu (ed.), Lecture Notes in Math. 1540. *Springer*, 1993.

[9] K.-J.Engel, R. Nagel: *One-Parameter Semigroups for Linear Evolution Equations.* Springer, 2000.

[10] S.J. Fromm: Potential space estimates for Green potentials in convex domains. *Proc. Amer. Math. Soc.* **119**, 1 (1993), 225–233.

[11] M. Geissert, H. Heck, M. Hieber: L^p-theory of the Navier-Stokes flow in the exterior of a moving or rotating obstacle. *J. Reine Angew. Math.* **596**, (2006), 45–62.

[12] M. Geissert, H. Heck, M. Hieber, I. Wood: The Ornstein-Uhlenbeck semigroup in exterior domains. *Arch. Math.*, **85**, 6 (2005), 554–562.

[13] B. Goldys: On analyticity of Ornstein-Uhlenbeck semigroups. *Atti Accad. Naz. Lincei Cl. Sci. Fis. Mat. Natur. Rend. Lincei (9) Mat. Appl.* **10, 3** (1999), 131–140.

[14] P. Grisvard: *Elliptic Problems in Nonsmooth Domains.* Pitman, Boston, 1985.

[15] M. Hieber, O. Sawada: The Navier-Stokes equations in $\mathbb{R}^n$ with linearly growing initial data. *Arch. Rat. Mech. Anal.*, **175**, 2 (2005), 269–285.

[16] T. Hishida: An existence theorem for the Navier-Stokes flow in the exterior of a rotating obstacle. *Arch. Rational Mech. Anal.* **150** (1999), 307–348.

[17] T. Hishida: The Stokes operator with rotation effect in exterior domains. *Analysis* **19** (1999), 51–67.

[18] T. Hishida: L^2-theory for the operator $\Delta + (k \times x) \cdot \nabla$ in exterior domains. *Nihonkai Math. J.* **11** (2000), 103–135.

[19] D. Jerison, C.Kenig: The inhomogeneous Dirichlet problem in Lipschitz domains. *J. Func. Anal.* **130** (1995), 161–219.

[20] P.C. Kunstmann: Heat kernel estimates and L^p spectral independence of elliptic operators. *Bull. London Math. Soc.* **31** (1999), 345–353.

[21] A. Lunardi, V. Vespri: Generation of strongly continuous semigroups by elliptic operators with unbounded coefficients in $L^p(\mathbb{R}^n)$. *Rend. Instit. Mat. Univ. Trieste* **28** (1996), no. suppl., 251–279.

[22] V.G. Maz'ja: *Sobolev Spaces.* Springer, Berlin, 1985.

[23] G. Metafune: L^p-spectrum of Ornstein-Uhlenbeck operators. *Ann. Scuola Norm. Sup. Pisa Cl. Sci.* (4) **30** (2001), 97–124.

[24] G. Metafune, J. Prüss, A. Rhandi, R. Schnaubelt: The domain of the Ornstein-Uhlenbeck operator on an L^p-space with invariant measure. *Ann. Scuola Norm. Sup. Pisa Cl. Sci.* (5) **1** (2002), 471–485.

[25] G. Metafune, C. Spina: An integration by parts formula in Sobolev spaces. *Ulmer Seminare über Funktionalanalysis und Differentialgleichungen* (2004).

[26] A. Pazy: *Semigroups of Linear Operators and Applications to Partial Differential Equations.* Springer, 1983.

[27] J. Prüss, A. Rhandi, R. Schnaubelt: The domain of elliptic operators on $L^p(\mathbb{R}^d)$ with unbounded drift coefficients. *Preprint.*

[28] L. Weis: A new approach to maximal L^p-regularity. In: Evolution Equations and Appl. Physical Life Sciences, G. Lumer, L. Weis (eds.), Lect. Notes in Pure & Appl. Math., Vol. 215 *Marcel Dekker, New York* (2001), 195–214.

[29] I. Wood: Maximal L^p-regularity for the Laplacian on Lipschitz domains, *Math. Z.* (4) **255** (2007), 855–875.

Ian Wood
Institute of Mathematical and Physical Sciences
Aberystwyth University
Penglais
Aberystwyth SY23 3BZ, UK
e-mail: ian.wood@aber.ac.uk

Operator Theory:
Advances and Applications, Vol. 186, 437–443

List of Participants

Organizers

Jan Janas
Institute of Mathematics PAN
Branch in Cracow
Sw.Tomasz 30, 31-027, Cracow, Poland
najanas@cyf-kr.edu.pl

Pavel Kurasov
Dept of Mathematics, LTH
Sölvegatan 18, Box 118
221 00 Lund, Sweden
kurasov@maths.lth.se

Ari Laptev
Department of Mathematics
Royal Institute of Technology
100 44 Stockholm, Sweden
laptev@math.kth.se

Sergey Naboko
Department of Mathematical Physics
Institute of Physics
NIIF, St. Petersburg State University
ul.Ulianovskaya 1, St. Petergoff
198504 St. Petersburg, Russia
naboko@math.su.se

Gunter Stolz
Department of Mathematics, CH 542
University of Alabama at Birmingham
1300 University Blvd
Birmingham, AL 35294-1170, USA
stolz@math.uab.edu

Participants

Vadym Adamyan
Department of Theor. Physics
Odessa National I.I. Mechnikov University
vul. Dvoryanska 2, 65026 Odessa, Ukraine
vadamyan@paco.net

Sergei A. Avdonin
Department of Mathematics and Statistics
University of Alaska Fairbanks
Fairbanks, AK 99775-6660, USA
ffsaa@uaf.edu

Ingrid Beltiţă
Institute of Mathematics "Simion Stoilow"
of the Romanian Academy
Calea Grivitei 21, PO Box 1–764
RO 014700, Bucharest, Romania
Ingrid.Beltita@imar.ro

Jussi Behrndt
Institut für Mathematik
MA 6-4, TU Berlin
Strasse des 17. Juni 136
10623 Berlin, Germany
behrndt@math.tu-berlin.de

Christer Bennewitz
Department of Mathematics
Lund University, P.O. Box 118
221 00 Lund, Sweden
Christer.Bennewitz@math.lu.se

Tanja Bergkvist
Department of Mathematics
Stockholm University
106 91 Stockholm, Sweden
tanjab@math.su.se

Gregory Berkolaiko
Dept of Mathematics
Texas A&M University
College Station, TX 77843-3368, USA
Gregory.Berkolaiko@math.tamu.edu

Jan Boman
Department of Mathematics
Stockholm University
106 91 Stockholm, Sweden
jabo@math.su.se

Julius Borcea
Department of Mathematics
Stockholm University
106 91 Stockholm, Sweden
julius@math.su.se

Denis I. Borisov
Department of Theoretical Physics
Nuclear Physics Institute
250 68 Rez, Czech Republic
borisovdi@yandex.ru

Jonathan Breuer
Einstein Institute of Mathematics
Edmond J. Safra Campus, Givat Ram
The Hebrew University of Jerusalem
Jerusalem, 91904, Israel
jbreuer@math.huji.ac.il

Malcolm B. Brown
Computer Science, Cardiff University,
Queen's Buildings, 5 The Parade, Roath
Cardiff CF24 3AA, UK
Malcolm.Brown@cs.cardiff.ac.uk

Vsevolod L. Chernyshev
Fundamental Sciences
Moscow State Technical University
n.a. N.E. Bauman
2-nd Baumanskaya 5
105005 Moscow, Russia
vchern@bmstu.ru

Evgeney V. Cheremnikh
Institute of Mathematics
12 Bandera Str.
79013 Lviv, Ukraine
echeremn@polynet.lviv.ua

Jacob S. Christiansen
Caltech, Mathematics
253-37, Pasadena, CA 91125, USA
stordal@caltech.edu

Monique Combescure
IPNL, 4 rue Enrico Fermi
F-69622 Villeurbanne , France
mcombe@ipnl.in2p3.fr

Rafael R. del Rio
IIMAS-UNAM, Circuito Escolar
Ciudad Universitaria
04510 Mexico D.F., Mexico
delrio@leibniz.iimas.unam.mx

Nils Dencker
Center for Mathematical Sciences
Division of Mathematics
(Faculty of Science) Lund University
Box 118, 221 00 Lund, Sweden
dencker@maths.lth.se

Jan Dereziński
Dep. Math. Meth. in Physics, Warsaw University
Hoza 74, 00-682 Warszawa, Poland
jan.derezinski@fuw.edu.pl

Joanne M. Dombrowski
Department of Mathematics and Statistics
Wright State University
Dayton, Ohio 45434, USA
Joanne.dombrowski@wright.edu

Tomas Ekholm
Complexo Interdisciplinar
da Universidade de Lisboa
Av. Prof. Gama Pinto 2
1649-003 Lisboa, Portugal
tomase@math.kth.se

Mats Ehrnström
Centre for Mathematical Sciences
Lund University
P.O. Box 118, 221 00 Lund, Sweden
mats.ehrnstrom@math.lu.se

Domenico Finco
Institut fur Angewandte Mathematik
Abteilung Stochastik
Wegelerstr. 6, D-53115 Bonn, Germany
finco@wiener.iam.uni-bonn.de

Magnus Fontes
Department of Mathematics (LTH)
Lund University
P.O. Box 118, 221 00 Lund, Sweden
Magnus.Fontes@math.lth.se

Rupert Frank
KTH Stockholm
Department of Mathematics
100 44 Stockholm, Sweden
rupert@math.kth.se

David Färm
Mathematics LTH
Box 118, 221 00 Lund, Sweden
david@maths.lth.se

Clemens Förster
Institut für Analysis
Dynamik und Modellierung
Fakultät für Mathematik und Physik
Universität Stuttgart
Pfaffenwaldring 57
70569 Stuttgart, Germany
Clemens.Foerster
@mathematik.uni-stuttgart.de

Jeffrey S. Geronimo
School of Mathematics, Georgia Tech
Atlanta, Ga 30332, USA
geronimo@math.gatech.edu

Fritz Gesztesy
Department of Mathematics
University of Missouri
Columbia, MO-65211, USA
fritz@math.missouri.edu

Daphne J. Gilbert
Department of Pure
and Applied Mathematics
Dublin Institute of Technology
Kevin Street, Dublin 8, Ireland
daphne.gilbert@dit.ie

Daniel Grieser
Institut für Mathematik
Carl von Ossietzky Universität Oldenburg
D-26111 Oldenburg, Germany
grieser@mathematik.uni-oldenburg.de

Thomas Guhr
Mathematical Physics, LTH
Lund University
Box 118, 221 00 Lund, Sweden
thomas.guhr@matfys.lth.se

Anders Hansson
KTH Matematik
100 44 Stockholm, Sweden
anhan@math.kth.se

Anders Holst
Mathematics LTH, Lund University
P.O. Box 118, 221 00 Lund, Sweden
ah@maths.lth.se

Alexei Iantchenko
Malmö Högskola, Teknik och Samhälle
205 06 Malmö, Sweden
ai@ts.mah.se

Per-Anders Ivert
Lund University, Centre for Mathematical Sciences
P.O. Box 118, 221 00 Lund, Sweden
ivert@maths.lth.se

Yulia Karpeshina
Department of Mathematics, UAB
1300 University blvd
Birmingham, AL 35294-1170, USA
karpeshi@math.uab.edu

Steffen Klassert
Technische Universität Chemnitz
Fakultät für Mathematik
Reichenhainerstr. 41
09107 Chemnitz, Germany
klassert@mathematik.tu-chemnitz.de

Torbjörn Kolsrud
Department of Mathematics
Royal Institute of Technology
100 44 Stockholm, Sweden
kolsrud@math.kth.se

Sylwia M. Kondej
Department of Physics and Astronomy
Institute of Physics
ul. Szafrana 4a (room 428)
65-516 Zielona Gora, Poland
skondej@proton.if.uz.zgora.pl

Evgeny Korotyaev
Institut für Mathematik
Humboldt Universität zu Berlin
Rudower Chaussee 25
12489 Berlin, Germany
evgeny@math.hu-berlin.de

Hynek Kovařík
Institute for Analysis
Dynamics and Modelling
Pfaffenwaldring 57
D-70569 Stuttgart, Germany
kovarik@mathematik.uni-stuttgart.de

Stanislav Kupin
CMI, Université de Provence
39 rue Joliot-Curie
13453 Marseille Cedex 13, France
kupin@cmi.univ-mrs.fr

Yoram Last
Institute of Mathematics
The Hebrew University
Givat Ram, 91904 Jerusalem, Israel
ylast@math.huji.ac.il

Tatiana Levitina
Abteilung Partielle Differentialgleichungen
Institut Computational Mathematics
TU Braunschweig, Pockelsstr. 14 - Forum
D-38106 Braunschweig, Germany
t.levitina@tu-bs.de

Annemarie Luger
Technische Universität Berlin
Sekretariat MA 674
Strasse des 17. Juni 136
10623 Berlin, Germany
aluger@mail.zserv.tuwien.ac.at

Marco Marletta
Cardiff School of Mathematics
Senghennydd Road
Cardiff CF24 4AG
Wales, United Kingdom
MarlettaM@cardiff.ac.uk

Asya A. Metelkina
LAGA, Institut Galilée
Université Paris 13
99 avenue Jean-Baptiste Clément
93430 Villetaneuse, France
assiaaton@mail.ru

Victor S. Mikhaylov
University os Alaska
101 Chapman Hall
Department of Mathematics
Fairbanks, AK 99775, USA
ftvsm@uaf.edu

Stanislav A. Molchanov
Department of Mathematics and Statistics
University of North Carolina at Charlotte
Charlotte, NC 28223, USA
smolchan@uncc.edu

Marcin Moszyński
Wydzial Matematyki Informatyki
i Mechaniki UW
Instytut Matematyki Stosowanej
i Mechaniki
ul. Banacha 2
02-097 Warszawa, Poland
mmoszyns@mimuw.edu.pl

Wojciech Motyka
Institute of Mathematics
of the Polish Academy of Sciences
ul. Sw. Tomasza 30
31-027 Krakow, Poland
namotyka@cyf-kr.edu.pl

Hagen Neidhardt
WIAS-Berlin, Mohrenstr. 39
D-10117 Berlin, Germany
neidhard@wias-berlin.de

Johan Nilsson
Mathematics LTH
Box 118, 221 00 Lund, Sweden
johann@maths.lth.se

Marlena Nowaczyk
Mathematics LTH
Box 118, 221 00 Lund, Sweden
marlena@maths.lth.se

Andrey Osipov
6 Parkovaya St., 29 "A", flat 18
105043 Moscow, Russia
osipa68@yahoo.com

Jorgen S. Ostensson
Department of Mathematics
Katholieke Universiteit Leuven
Celestijnenlaan 200B
B3001 Leuven, Belgium
ostensson@wis.kuleuven.be

Elin Ottergren
Department of Mathematics
Roslagsv. 101
106 91 Stockholm, Sweden
ottergren@math.su.se

Kateřina Ožanová
Department of Mathematical
Sciences Chalmers
University of Technology Chalmers
Tvargata 3
412 96 Goteborg, Sweden
nemco@math.chalmers.se

Konstantin Pankrashkin
Institute of Mathematics
Humboldt-University of Berlin
Rudower Chaussee 25
12489 Berlin, Germany
const@mathematik.hu-berlin.de

Boris S. Pavlov
Department of Mathematics
University of Auckland
private box 92019
Auckland, New Zealand
pavlov@math.auckland.ac.nz

Tomas Persson
Mathematics LTH
Box 118, 221 00 Lund, Sweden
tomasp@maths.lth.se

Irina Peterburgsky
Department of Mathematics
and Computer Science
Suffolk University
41 Temple Street
Boston, MA 02114, USA
IrPeterbur@aol.com

Pelle Pettersson
Department of Mathematics (LTH)
Lund University
P.O. Box 118, 221 00 Lund, Sweden
Pelle@maths.lth.se

Victoria Rayskin
Department of Mathematics
PO Box 750561, 28725 Bremen, Germany
v.rayskin@iu-bremen.de

Roman V. Romanov
Laboratory of Quantum Networks
Institute for Physics
Saint Petersburg State University
198504 St. Petersburg, Russia
roma@rvr.stud.pu.ru

Aida M. Sahmurova
Istanbul University, Engineering Faculty
Department of Environmental Engineering
34320 Avcilar/Istanbul, Turkey
aida@istanbul.edu.tr

Jörg Schmeling
Mathematics LTH
Box 118, 221 00 Lund, Sweden
Jorg.Schmeling@math.lth.se

Achim Schroll
Centre for Mathematical Sciences
Box 118, 221 00 Lund, Sweden
achim.schroll@na.lu.se

Veli B. Shakhmurov
Istanbul University, Engineering Faculty
Department of Electrical
& Electronics Engineering
Avcilar, 34320 Istanbul, Turkey
sahmurov@istanbul.edu.tr

Luis O. Silva P.
Departamento de Métodos Matemáticos
y Numéricos IIMAS
Univ. Nacional Autónoma de México
Apdo. Postal 20-726
México, D.F. 01000, México
silva@leibniz.iimas.unam.mx

Sergei Silvestrov
Centre for Mathematical Sciences LTH
Lund University
Box 118, 221 00 Lund, Sweden
Sergei.Silvestrov@math.lth.se

Anna-Maria Simbotin
Centre for Mathematical Sciences
Box 118, 221 00 Lund, Sweden
simbotin@maths.lth.se

Barry Simon
Mathematics 253-37, Caltech
Pasadena, CA 91125, USA
bsimon@caltech.edu

Sergey A. Simonov
Department of Mathematical Physics
Institute of Physics
St. Petersburg University
Ulianovskaia 1
198504, St. Petergoff
St. Petersburg, Russia
sergey_simonov@mail.ru

Jan Stochel
Instytut Matematyki, ul. Reymonta 4
30-059 Krakow, Poland
stochel@im.uj.edu.pl

Jan E. Sträng
Abteilung Theoretische Physik
Universität Ulm
Albert-Einstein-Allee 11
89069 Ulm, Germany
eric.straeng@uni-ulm.de

Vladimir V. Sukhanov
Dep. of Mathematical Physics
St. Petersburg Univ.
Ulyanovskaya 1
198504 St. Petersburg, Russia
vvsukhanov@mail.ru

Alexander Teplyaev
Department of Mathematics
University of Connecticut
196 Auditorium Road U-3009
Storrs, CT 06269-3009, USA
teplyaev@math.uconn.edu

Gerald Teschl
Faculty of Mathematics
Nordbergsr. 15
A-1090 Vienna, Austria
gerald.teschl@univie.ac.at

Alexey S. Tikhonov
Dep. of Mathematics
4 Yaltinskaya
95007 Simferopol, Ukraine
tikhonov@club.cris.net

Françoise Truc
Department of Mathematics
Institut Fourier
100 rue des maths
38402 St Martin d'Hères, France
trucfr@ujf-grenoble.fr

Tatyana Turova
Department of Mathematical Statistics
Lund University
Box 118, 221 00 Lund, Sweden
tatyana@maths.lth.se

Thomas Vallier
Centre for Mathematical Sciences
Lund University
Box 118, 221 00 Lund, Sweden
vallier@maths.lth.se

Olivier Verdier
Mathematics LTH
Box 118, 221 00 Lund, Sweden
olivier@maths.lth.se

Ivan Veselic'
TU Chemnitz, Fakultät für Mathematik
Room 428 in Reichenheiner Str. 39
09107 Chemnitz, Germany
ivan.veselic
@mathematik.tu-chemnitz.de

Javier A. Villamizar V.
Department of Mathematics
Simon Bolivar University
Apdo 89000
Caracas 1080-A, Venezuela
jvilla@ma.usb.ve

Erik Wahlén
Lund University
Centre for Mathematical Sciences
P.O. Box 118
S-221 00 Lund, Sweden
erik.wahlen@math.lu.se

Kazuo Watanabe
Department of Mathematics
Gakushuin Univ.
1-5-1 Mejiro, Toshima-ku
Tokyo 171-8588, Japan
kazuo.watanabe@gakushuin.ac.jp

Ricardo Weder
IIMAS-UNAM
Apartado Postal 20-726
Mexico DF 01000, Mexico
weder@servidor.unam.mx

Rudi Weikard
Department of Mathematics
University of Alabama at Birmingham
Birmingham, AL 35294-1170, USA
rudi@math.uab.edu

Jens Wittsten
Centre for Mathematical Sciences
Lund University
P.O. Box 118, 221 00 Lund, Sweden
jens.wittsten@math.lu.se

Ian G. Wood
Cardiff School of Mathematics
Cardiff University
Senghennydd Road
Cardiff CF24 4AG UK
woodig@cardiff.ac.uk

Peter Yuditskii
Mathematical Department
of Bar Ilan University
52900 Ramat Gan, Israel
yuditski@macs.biu.ac.il

Alessandro Zampini
SiSSA, via Beirut 4
34014 Trieste, Italy
zampini@sissa.it